Lecture Notes in Computer Science 13027

More information about this subseries at http://www.springer.com/series/7407

Uli Fahrenberg · Mai Gehrke ·
Luigi Santocanale · Michael Winter (Eds.)

Relational and Algebraic Methods in Computer Science

19th International Conference, RAMiCS 2021
Marseille, France, November 2–5, 2021
Proceedings

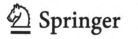

Springer

Editors
Uli Fahrenberg
École polytechnique
Palaiseau, France

Mai Gehrke
Université Côte d'Azur
Nice, France

Luigi Santocanale
Aix-Marseille University
Marseille, France

Michael Winter
Brock University
St Catharines, ON, Canada

ISSN 0302-9743 ISSN 1611-3349 (electronic)
Lecture Notes in Computer Science
ISBN 978-3-030-88700-1 ISBN 978-3-030-88701-8 (eBook)
https://doi.org/10.1007/978-3-030-88701-8

LNCS Sublibrary: SL1 – Theoretical Computer Science and General Issues

This Springer imprint is published by the registered company Springer Nature Switzerland AG
The registered company address is: Gewerbestrasse 11, 6330 Cham, Switzerland

Preface

This volume contains the proceedings of the 19th International Conference on Relational and Algebraic Methods in Computer Science (RAMiCS 2021), which was held at the Centre International de Rencontres Mathématiques in Marseille, France, during November 2–5, 2021.

The RAMiCS conferences aim to bring together a community of researchers to advance the development and dissemination of relation algebras, Kleene algebras, and similar algebraic formalisms. Topics covered range from mathematical foundations to applications as conceptual and methodological tools in computer science and beyond. More than 30 years after its foundation in 1991 in Warsaw, Poland, RAMiCS, initially named "Relational Methods in Computer Science," remains a main venue in this field. The series merged with the workshops on Applications of Kleene Algebra in 2003 and adopted its current name in 2009. Previous events were organized in Dagstuhl, Germany (1994), Paraty, Brazil (1995), Hammamet, Tunisia (1997), Warsaw, Poland (1998), Québec, Canada (2000), Oisterwijk, The Netherlands (2001), Malente, Germany (2003), St. Catharines, Canada (2005), Manchester, UK (2006), Frauenwörth, Germany (2008), Doha, Qatar (2009), Rotterdam, The Netherlands (2011), Cambridge, UK (2012), Marienstatt, Germany (2014), Braga, Portugal (2015), Lyon, France (2017), Groningen, The Netherlands (2018), and Palaiseau, France (2020, online).

RAMiCS 2021 attracted 35 submissions, of which 29 were selected for presentation by the Program Committee. Each submission was evaluated according to high academic standards by at least three independent reviewers and scrutinized further during two weeks of intense electronic discussion. The organizers are very grateful to all Program Committee members for this hard work, including the lively and constructive debates, and to the external reviewers for their generous help and expert judgments. Without this dedication, we could not have assembled such a high-quality program; we hope that all authors have benefitted from these efforts.

Apart from the submitted articles, this volume features the abstracts of the presentations of the three invited speakers: Marcelo Frias, Barbara König, and Dmitriy Zhuk. We are delighted that all three invited speakers accepted our invitation to present their work at the conference.

Last, but not least, we would like to thank the members of the RAMiCS Steering Committee for their support and advice. We gratefully acknowledge financial and administrative support by the Centre International de Rencontres Mathématiques and the Laboratoire d'Informatique et Systèmes and financial support by the Institut Archimède Mathématiques-Informatique, Aix-Marseille Université, the Métropole Aix-Marseille Provence, and the Conseil Départemental des Bouches-du-Rhône.

We also appreciate the excellent facilities offered by the EasyChair conference administration system and Anna Kramer's help in publishing this volume with Springer. Finally, we are indebted to all authors and participants for supporting this conference.

August 2021

Uli Fahrenberg
Mai Gehrke
Luigi Santocanale
Michael Winter

Organization

Program Committee

Bahareh Afshari	University of Gothenburg, Sweden
Christel Baier	TU Dresden, Germany
Manuel Bodirsky	TU Dresden, Germany
Uli Fahrenberg (Co-chair)	École polytechnique, France
Hitoshi Furusawa	Kagoshima University, Japan
Ignacio Fábregas	Universidad Complutense de Madrid, Spain
Mai Gehrke (Co-chair)	CNRS and Université Côte d'Azur, France
Silvio Ghilardi	Università degli Studi di Milano, Italy
Roland Glück	Deutsches Zentrum für Luft- und Raumfahrt, Augsburg, Germany
Walter Guttmann	University of Canterbury, New Zealand
Robin Hirsch	University College London, UK
Peter Höfner	Australian National University, Sydney, Australia
Ali Jaoua	Qatar University, Qatar
Peter Jipsen	Chapman University, USA
Sebastiaan Joosten	Dartmouth College, USA
Laura Kovacs	Vienna University of Technology, Austria
Tadeusz Litak	Friedrich-Alexander-Universität Erlangen-Nürnberg, Germany
Roger Maddux	Iowa State University, USA
Dale Miller	Inria and École polytechnique, France
Martin Mueller	University of Augsburg, Germany
Daniela Petrisan	Université de Paris, France
Damien Pous	CNRS and ENS Lyon, France
Luigi Santocanale (Co-chair)	Aix-Marseille Université, France
Georg Struth	The University of Sheffield, UK
Sam van Gool	Université de Paris, France
Michael Winter (Co-chair)	Brock University, Canada

Additional Reviewers

Backhouse, Roland
Borlido, Célia
Cockett, Robin
de Groot, Jim
Fujii, Soichiro
Gaubert, Stephane
Knäuer, Simon
Kuznetsov, Stepan
Lapenta, Serafina
Linkhorn, Deacon
McIver, Annabelle
Mimram, Samuel
Montoli, Andrea
Mühle, Henri

Nishizawa, Koki
Palmigiano, Alessandra
Reggio, Luca
Robinson-O'Brien, Nicolas
Rowe, Reuben
San Pietro, Pierluigi
Sarkar, Saptarshi
Schlicht, Leopold
Sedlar, Igor
Tsumagari, Norihiro
Valencia, Frank
van den Berg, Benno
Vucaj, Albert
Youssef, Youssef Mahmoud
Zaïdi, Medhi
Zetzsche, Georg

Abstracts of Invited Talks

Relational Tight Field Bounds for Distributed Analysis of Programs

Marcelo F. Frias

Instituto Tecnológico de Buenos Aires and CONICET, Argentina

Relational tight field bounds [1] are an abstraction of the semantics of data structures. In the presence of appropriate symmetry-breaking predicates, these bounds can be computed automatically and allow to dramatically speed up bug-finding using SAT-solving. In this lecture, after giving an introduction to tight field bounds and symmetry-breaking predicates, I will present a general technique for distributing program analyses. As examples, I will show how the technique allows one to distribute SAT-based bug-finding [2] as well as symbolic execution over complex data types.

References

1. Galeotti, J.P., Rosner, N., Pombo, C.G.L., Frias, M.F.: TACO: efficient SAT-based bounded verification using symmetry breaking and tight Bounds. IEEE Trans. Softw. Eng. **39**(9), 1283–1307 (2013)
2. Rosner, N., et al.: Parallel bounded analysis in code with rich invariants by refinement of field bounds. ISSTA 2013, 23–33 (2013)

Fixpoint Games

Barbara König

Universität Duisburg-Essen

Solving fixpoint equations is a recurring problem in several domains: the result of a dataflow analysis can be characterized as either a least or greatest fixpoint. It is well-known that bisimilarity - the largest bisimulation - admits a characterization as a greatest fixpoint and furthermore μ-calculus model-checking requires to solve systems of nested fixpoint equations.

Often, these fixpoint equations or equation systems are defined over powerset lattices, however in several applications - such as lattice-valued or real-valued μ-calculi - the lattice under consideration is not a powerset.

Hence we extend the notion of fixpoint games (or unfolding games, introduced by Venema) to games for equation systems over more general lattices. In particular continuous lattices admit a very elegant characterization of the solution.

We will also describe how to define progress measures which describe winning strategies for the existential players and explain how abstractions and up-to functions can be integrated into the framework.

(Joint work with Paolo Baldan, Tommaso Padoan, Christina Mika-Michalski).

Quantified Constraint Satisfaction Problem: Towards the Classification of Complexity

Dmitriy Zhuk

Lomonosov Moscow State University

The *Quantified Constraint Satisfaction Problem (QCSP)* is the generalization of the Constraint Satisfaction problem (CSP) where we are allowed to use both existential and universal quantifiers. Formally, the QCSP over a constraint language Γ is the problem to evaluate a sentence of the form

$$\forall x_1 \exists y_1 \forall x_2 \exists y_2 \dots \forall x_n \exists y_n \, (R_1(\dots) \wedge \dots \wedge R_s(\dots)),$$

where $R_1, \dots, R_s$ are relations from Γ. While CSP remains in NP for any Γ, QCSP(Γ) can be PSpace-hard, as witnessed by Quantified 3-Satisfiability or Quantified Graph 3-Colouring. For many years there was a hope that for any constraint language the QCSP is either in P, NP-complete, or PSpace-complete. Moreover, a very simple conjecture describing the complexity of the QCSP was suggested by Hubie Chen. However, in 2018 together with Mirek Olšák and Barnaby Martin we discovered constraint languages for which the QCSP is coNP-complete, DP-complete, and even Θ_2^P-complete, which refutes the Chen conjecture. Despite the fact that we described the complexity for each constraint language on a 3-element domain with constants, we did not hope to obtain a complete classification.

This year I obtained several results that make me believe that such a classification is closer than it seems. First, I obtained an elementary proof of the PGP reduction, which allows to reduce the QCSP to the CSP. Second, I showed that there is a gap between Π_2^P and PSpace, and found a criterion for the QCSP to be PSpace-hard. In the talk I will discuss the above and some other results.

Contents

Amalgamation Property for Varieties of BL-algebras Generated by One Chain with Finitely Many Components ... 1
Stefano Aguzzoli and Matteo Bianchi

Unary-Determined Distributive ℓ-magmas and Bunched Implication Algebras ... 19
Natanael Alpay, Peter Jipsen, and Melissa Sugimoto

Effect Algebras, Girard Quantales and Complementation in Separation Logic ... 37
Callum Bannister, Peter Höfner, and Georg Struth

Relational Computation of Sets of Relations ... 54
Rudolf Berghammer

Experimental Investigation of Sufficient Criteria for Relations to Have Kernels ... 72
Rudolf Berghammer and Mitja Kulczynski

ℓr-Multisemigroups, Modal Quantales and the Origin of Locality ... 90
Cameron Calk, Uli Fahrenberg, Christian Johansen, Georg Struth, and Krzysztof Ziemiański

Abstract Strategies and Coherence ... 108
Cameron Calk, Eric Goubault, and Philippe Malbos

Algorithmic Correspondence for Relevance Logics, Bunched Implication Logics, and Relation Algebras via an Implementation of the Algorithm PEARL ... 126
Willem Conradie, Valentin Goranko, and Peter Jipsen

The Class of Representable Semilattice-Ordered Monoids Is Not a Variety ... 144
Amina Doumane

Accretive Computation of Global Transformations ... 159
Alexandre Fernandez, Luidnel Maignan, and Antoine Spicher

Some Modal and Temporal Translations of Generalized Basic Logic ... 176
Wesley Fussner and William Zuluaga Botero

Isolated Sublattices and Their Application to Counting Closure Operators 192
Roland Glück

Second-Order Properties of Undirected Graphs 209
Walter Guttmann

Relation-Algebraic Verification of Borůvka's Minimum Spanning Tree Algorithm 225
Walter Guttmann and Nicolas Robinson-O'Brien

Deciding FO-definability of Regular Languages 241
Agi Kurucz, Vladislav Ryzhikov, Yury Savateev, and Michael Zakharyaschev

Relational Models for the Lambek Calculus with Intersection and Unit 258
Stepan L. Kuznetsov

Free Modal Riesz Spaces are Archimedean: A Syntactic Proof 275
Christophe Lucas and Matteo Mio

Polyadic Spaces and Profinite Monoids 292
Jérémie Marquès

Time Warps, from Algebra to Algorithms 309
Sam van Gool, Adrien Guatto, George Metcalfe, and Simon Santschi

On Algebra of Program Correctness and Incorrectness 325
Bernhard Möller, Peter O'Hearn, and Tony Hoare

Computing Least and Greatest Fixed Points in Absorptive Semirings 344
Matthias Naaf

A Variety Theorem for Relational Universal Algebra 362
Chad Nester

On Tools for Completeness of Kleene Algebra with Hypotheses 378
Damien Pous, Jurriaan Rot, and Jana Wagemaker

Skew Metrics Valued in Sugihara Semigroups 396
Luigi Santocanale

Computing Distributed Knowledge as the Greatest Lower Bound of Knowledge 413
Carlos Pinzón, Santiago Quintero, Sergio Ramírez, and Frank Valencia

Relational Sums and Splittings in Categories of L-fuzzy Relations 433
Michael Winter

Change of Base Using Arrow Categories 448
Michael Winter

Automated Reasoning for Probabilistic Sequential Programs with Theorem Proving 465
Kangfeng Ye, Simon Foster, and Jim Woodcock

Domain Range Semigroups and Finite Representations 483
Jaš Šemrl

Author Index 499

Amalgamation Property for Varieties of BL-algebras Generated by One Chain with Finitely Many Components

Stefano Aguzzoli[1] and Matteo Bianchi[2](✉)

[1] Department of Computer Science, Università degli Studi di Milano, Via Celoria 18, 20133 Milano, Italy
aguzzoli@di.unimi.it
[2] Independent researcher, Milano, Italy
matteob@gmail.com

Abstract. BL-algebras are the algebraic semantics for Hájek's Basic Logic BL, the logic of all continuous t-norms and their residua. Every BL-chain can be decomposed (up to isomorphism) as an ordinal sum of non-trivial Wajsberg hoops - called components - with the first bounded. In this paper we study the amalgamation property for the varieties of BL-algebras generated by one BL-chain with finitely many components.

Keywords: BL-algebras · Hoops · Amalgamation property · Ordinal sums · Lattices of varieties

1 Introduction

BL-algebras are the algebraic semantics for the Basic Logic BL, introduced in [Háj98]. The class of all BL-algebras forms a variety, which is called $\mathbb{BL}$. As shown in [CEGT00] BL is the logic of all continuous t-norms and their residua, whence it is one of the major mathematical fuzzy logic (see [CHN11]) which are useful to formally deal with vagueness and uncertainty.

In [AM03] there is a general result concerning the structure of the totally ordered BL-algebras, BL-chains. Every BL-chain is isomorphic to an ordinal sum of totally ordered Wajsberg hoops, with the first bounded. Moreover, the same paper provides a full description of the subdirectly irreducible members of every variety of BL-algebras generated by one BL-chain with finitely many components. Using these ingredients, in our recent work [AB21a] we classified the finite model property for all the BL-algebras generated by a finite set of BL-chains with finitely many components.

In this paper we focus on the amalgamation property (AP), for some varieties of BL-algebras. We recall that the AP for a variety of BL-algebras corresponds to the deductive interpolation property for the associated logic. This is a well established topic, since the amalgamation property for $\mathbb{BL}$ and some of its subvarieties was shown in [Mon06], and in [CMM11] the analysis was further

U. Fahrenberg et al. (Eds.): RAMiCS 2021, LNCS 13027, pp. 1–18, 2021.
https://doi.org/10.1007/978-3-030-88701-8_1

extended. In [MMT14] the case of GBL-algebras (a far-reaching generalization of BL-algebras) was also tackled, by providing a partial classification. Nevertheless, the study of the AP for varieties of BL-algebras is far from over, since the lattice of varieties of BL-algebras itself is uncountable and its structure is still poorly understood. The present paper is an additional contribution to this topic, since we will provide a full classification of the AP for all the varieties of BL-algebras generated by one BL-chain with finitely many components.

The paper is structured as follows. After introducing in Sect. 2 some preliminary results, in Sect. 3 we tackle the investigation of the AP for the varieties of BL-algebras generated by one finite chain with finitely many components. In Sect. 4 we discuss some open problems and future work.

2 Preliminaries

2.1 BL-algebras

Definition 1 ([Háj98]). *A BL-algebra is an algebra* $(A, *, \Rightarrow, \wedge, \vee, 0, 1)$ *such that:*

(i) $(A, \wedge, \vee, 0, 1)$ *is a bounded lattice with minimum* 0 *and maximum* 1.
(ii) $(A, *, 1)$ *is a commutative monoid.*
(iii) $(*, \Rightarrow)$ *forms a* residuated pair: $z * x \leq y$ *iff* $z \leq x \Rightarrow y$ *for all* $x, y, z \in A$.
(iv) The following identities hold, for all $x, y \in A$:

$$(x \Rightarrow y) \vee (y \Rightarrow x) = 1. \qquad \text{(Prelinearity)}$$

$$x \wedge y = x * (x \Rightarrow y). \qquad \text{(Divisibility)}$$

A totally ordered BL-algebra is called a BL-chain.

Every algebra $([0,1], *, \Rightarrow, \min, \max, 0, 1)$, where $*$ is a continuous t-norm, and $\Rightarrow$ is its residuum, is a BL-algebra ([CHN11]), called standard BL-algebra. Two well-known examples are the standard MV-algebra $[0,1]_{\text{Ł}}$ and the standard Gödel-algebra $[0,1]_{\text{G}}$. In $[0,1]_{\text{Ł}}$ we have $x * y = \max\{0, x + y - 1\}$, and $x \Rightarrow y = \min\{1, 1 - x + y\}$. In $[0,1]_{\text{G}}$ it holds that $x * y = \min\{x, y\}$, whilst $x \Rightarrow y = 1$ if $x \leq y$, and $x \Rightarrow y = y$ if $x > y$. We define $\neg x \stackrel{\text{def}}{=} x \Rightarrow 0$.

2.2 BL-algebras and Ordinal Sums

Every BL-chain can be decomposed as an *ordinal sum* of hoops. Before stating the result, we need some preparation.

Definition 2 ([EGHM03]). *A* hoop *is an algebra* $\mathcal{A} = (A, *, \Rightarrow, 1)$ *of type* $(2, 2, 0)$ *such that:*

(i) $(A, *, 1)$ *is a commutative monoid,*

(ii) $\Rightarrow$ *is a binary operation satisfying the following properties:*
- $x \Rightarrow x = 1$,
- $x * (x \Rightarrow y) = y * (y \Rightarrow x)$,
- $x \Rightarrow (y \Rightarrow z) = (x * y) \Rightarrow z$.

A bounded hoop is an algebra $\mathcal{A} = (A, *, \Rightarrow, 0, 1)$ such that $(A, *, \Rightarrow, 1)$ is a hoop, and $0 \leq x$ for all $x \in A$. The binary relation $\leq$ on $\mathcal{A}$ is defined as $x \leq y$ if and only if $x \Rightarrow y = 1$. It follows from the hoop axioms that this binary relation is indeed a partial order. An unbounded hoop is a hoop without minimum.

A *Wajsberg hoop* is a hoop $\mathcal{A}$ satisfying

$$(x \Rightarrow y) \Rightarrow y = (y \Rightarrow x) \Rightarrow x.$$

A *cancellative hoop* is a hoop satisfying

$$x \Rightarrow (x * y) = y.$$

It is well known that bounded Wajsberg hoops are term-equivalent to MV-algebras (see [AFM07], and [CDM99] for MV-algebras). We also recall that the variety of Wajsberg hoops $\mathbb{WH}$ contains all cancellative hoops. In particular, the class of totally ordered cancellative hoops coincides with the class of totally ordered unbounded Wajsberg hoops. The class of all cancellative hoops forms a variety, called $\mathbb{CH}$. Of course $\mathbb{CH} \subsetneq \mathbb{WH}$. BL-chains can be obtained by means of the *ordinal sum construction.*

Definition 3. *Let* $(I, \leq)$ *be a totally ordered set with minimum* 0. *For all* $i \in I$, *let* $\mathcal{A}_i = (A_i, *_i, \Rightarrow_i, 1)$ *be a hoop such that for* $i \neq j$, $A_i \cap A_j = \{1\}$. *Then* $\bigoplus_{i \in I} \mathcal{A}_i$ *is called the* ordinal sum *of the family* $\{\mathcal{A}_i\}_{i \in I}$, *whose universe is given by* $\bigcup_{i \in I} A_i$, *and whose operations* $\Rightarrow, *$ *are given by:*

$$x \Rightarrow y \stackrel{def}{=} \begin{cases} x \Rightarrow_i y & \text{if } x, y \in A_i, \\ y & \text{if } j < i,\ x \in A_i,\ y \in A_j, \\ 1 & \text{if } i < j,\ 1 \neq x \in A_i,\ y \in A_j. \end{cases}$$

$$x * y \stackrel{def}{=} \begin{cases} x *_i y & \text{if } x, y \in A_i, \\ y & \text{if } j < i,\ x \in A_i,\ 1 \neq y \in A_j, \\ x & \text{if } i < j,\ 1 \neq x \in A_i,\ y \in A_j. \end{cases}$$

The hoops $\mathcal{A}_i$ *are called* components. *When* I *is finite, for example* $I = \{0, \ldots, k\}$, *we sometimes use the notation* $\mathcal{A}_0 \oplus \cdots \oplus \mathcal{A}_k$, *in place of* $\bigoplus_{i \in I} \mathcal{A}_i$.

As shown in [AM03] every BL-chain is canonically representable as an ordinal sum of hoops.

Theorem 1 ([AM03]). *For every BL-chain* $\mathcal{A}$ *there are a unique (up to order-isomorphisms) totally ordered set* $(I, \leq)$ *with minimum* 0 *and a unique (up to isomorphisms) family* $\{\mathcal{A}_i \mid i \in I\}$ *of non-trivial totally ordered Wajsberg hoops where* $\mathcal{A}_0$ *is bounded, such that* $\mathcal{A} \cong \bigoplus_{i \in I} \mathcal{A}_i$.

Observe that the idempotent elements in any ordinal sum of Wajsberg hoops are exactly 1 and the bottoms of every component with minimum. Let $\mathcal{A}$ be a BL-chain. With $\#\mathcal{A}$ we denote the number of the components of $\mathcal{A}$, *i.e.*, $\#\mathcal{A} = |I|$, in the decomposition of $\mathcal{A}$ described in Theorem 1.

Remark 1. – By slight abuse of terminology we shall often consider ordinal sums $\bigoplus_{i\in I} \mathcal{A}_i$, where there are some $\mathcal{A}_i$ (with $i \neq \min I$) being MV-chains, with the obvious meaning that we are actually considering the 0-free reduct of each such $\mathcal{A}_i$.

- By slight abuse of notation we shall sometimes consider ordinal sums $\bigoplus_{i\in I} \mathcal{A}_i$ where two or more components have elements in common distinct from 1 (for example, $\mathcal{A}_i = \mathcal{A}_j$ for some $i \neq j \in I$). In such cases we tacitly mean to consider an ordinal sum $\bigoplus_{i\in I} \mathcal{B}_i$, with $\mathcal{B}_i \simeq \mathcal{A}_i$ for every $i \in I$ and $B_i \cap B_j = \{1\}$ for $i \neq j$.
- Unless stated otherwise, from now on we assume that all the ordinal sums of Wajsberg hoops that we consider have non-trivial components.

It is possible to capture the property that a BL-chain $\mathcal{A}$ has at most n components ($\#\mathcal{A} \leq n$), equationally.

Lemma 1 ([AM03, Lemma 4.2]). *Let $\mathcal{A}$ be a BL-chain. Then $\#\mathcal{A} \leq n$ if and only if it satisfies the following equation:*

$$\bigwedge_{i=0}^{n-1}((x_{i+1} \Rightarrow x_i) \Rightarrow x_i) \Rightarrow \left(\bigvee_{i=0}^{n} x_i\right) = 1. \qquad (\lambda_n)$$

Consider the set $C_\infty = \{x \in \mathbb{Z} : x \leq 0\}$. The hoop $\mathcal{C}_\infty = (C_\infty, *, \Rightarrow, 1)$ is defined as follows, for $x, y \in C_\infty$:

- $1^{\mathcal{C}_\infty} = 0$,
- $x *^{\mathcal{C}_\infty} y = x + y$,
- $x \Rightarrow^{\mathcal{C}_\infty} y = \begin{cases} 0 & \text{if } x \leq^{\mathbb{Z}} y, \\ y - x & \text{otherwise.} \end{cases}$

A direct inspection shows that $\mathcal{C}_\infty$ is a cancellative hoop, and it is known that $\mathbf{V}(\mathcal{C}_\infty) = \mathbb{CH}$. In general, $\mathbb{CH}$ is generated by each of its non-trivial chains. We assume that the reader is acquainted with some basic notions of universal algebra, and we refer to [BS81] for more details. If K is a class of BL-chains, $\mathbf{H}(K), \mathbf{S}(K), \mathbf{P}(K), \mathbf{I}(K), \mathbf{P}_u(K)$ denote, respectively, the classes of all the homomorphic images, subalgebras, direct products, isomorphic algebras and ultraproducts of members of K. If $\mathcal{A}$ is a BL-chain, with $\mathbf{V}(\mathcal{A})$ we denote the variety generated by $\mathcal{A}$, *i.e.*, $\mathbf{HSP}(\mathcal{A})$ [BS81]: similarly, if K is a class of BL-chains, then $\mathbf{V}(K)$ indicates the variety generated by them. For example $\mathbf{V}(\mathbf{2}) = \mathbb{B}$, where $\mathbf{2}$ is the two-element Boolean algebra, and $\mathbb{B}$ is the variety of Boolean algebras. Let $\mathbf{R}$, $\mathbf{Q}$, $\mathbf{Z}$ be the additive totally ordered abelian groups over, respectively, real, rational and integer numbers. For $k \geq 2$, let $\mathbf{Q}_k$ be the totally ordered abelian subgroup of $\mathbf{Q}$, with carrier $\{\frac{a}{k-1} : a \in \mathbb{Z}\}$. As it is customary, given two lattice ordered abelian groups $\mathcal{S}, \mathcal{T}$, with $\mathcal{S} \times_{lex} \mathcal{T}$ we denote the lattice ordered abelian group obtained as the lexicographic product of $\mathcal{S}$ and $\mathcal{T}$. With Γ we denote Mundici's gamma functor[1]: see [CDM99] for details. For $n \geq 2$ we define $\mathcal{L}_n \stackrel{\text{def}}{=} \Gamma(\mathbf{Q}_n, 1)$ and $\mathcal{K}_n \stackrel{\text{def}}{=} \Gamma(\mathbf{Z} \times_{\text{lex}} \mathbf{Z}, (n-1, 0))$. Finally, we define $[0,1]_{\text{Ł}} \stackrel{\text{def}}{=} \Gamma(\mathbf{R}, 1)$. The radical of a totally ordered Wajsberg hoop (MV-chain) $\mathcal{A}$, is the intersection of all the maximal filters of $\mathcal{A}$, and will be denoted by $Rad(\mathcal{A})$. Let $\mathcal{A}$ be an MV-chain. We say that $\mathcal{A}$ has a finite rank if $\mathcal{A}/Rad(\mathcal{A}) \simeq \mathcal{L}_k$, for some k (in this case $rank(\mathcal{A}) = k$), whilst $\mathcal{A}$ has infinite rank if $\mathcal{A}/Rad(\mathcal{A})$ is an infinite simple MV-chain. We can define *mutatis mutandis* the same notion for a totally ordered Wasjberg hoop[2]: the only difference is that $\mathcal{A}/Rad(\mathcal{A})$ would be the 0-free reduct of a simple MV-chain (finite or infinite). For $k \geq 1$ we define $\mathcal{P}_k$ as $\mathbf{2} \oplus \underbrace{\mathcal{C}_\infty \oplus \cdots \oplus \mathcal{C}_\infty}_{k \text{ times}}$. For $k \geq 1$, $\mathcal{P}_k$ generates a variety, called $\mathbb{P}_k$, where $\mathbb{P}_1$ is the variety of product algebras. We refer the reader to [AB19] for further details. For $k \geq 2$ we define $\mathcal{G}_k$ as $\underbrace{\mathbf{2} \oplus \cdots \oplus \mathbf{2}}_{k-1 \text{ times}}$. $\mathcal{G}_k$ is a Gödel-chain (G-chain, for short) with k elements: see [CHN11] for further details. Let $\mathcal{A}$ be a BL-chain or the 0-free reduct of an MV-chain (*i.e.*, a totally ordered Wajsberg hoop with minimum). We define $A^+ \stackrel{\text{def}}{=} \{x \in A : x > \neg x\}$ and $A^- \stackrel{\text{def}}{=} \{x \in A : x \leq \neg x\}$. If $\mathcal{A}$ is a totally ordered Wajsberg hoop with minimum m, here $\neg x$ stands for $x \Rightarrow m$. Let $\mathcal{A}$ be a BL-chain or a totally ordered Wajsberg hoop. With $Si(\mathcal{A})$ we denote the class of the subdirectly irreducible algebras in $\mathbf{V}(\mathcal{A})$. Finally, given a non-trivial variety $\mathbb{L}$ of BL-algebras (Wajsberg hoops), with $Ch(\mathbb{L})$ we denote the class of all the non-trivial chains in $\mathbb{L}$. Every non-trivial variety $\mathbb{L}$ of BL-algebras is generated by its chains, *i.e.*, $\mathbb{L} = \mathbf{V}(Ch(\mathbb{L}))$. To simplify the notation, if $\mathcal{A}$ is a BL-chain (a totally ordered Wajsberg hoop), we will write $Ch(\mathcal{A})$ instead of $Ch(\mathbf{V}(\mathcal{A}))$.

3 Amalgamation Property for Varieties Generated by One BL-chain with Finitely Many Components

We start with the definition of the amalgamation property. With $\hookrightarrow$ we denote an embedding between algebras.

Definition 4. *We say that a class K of BL-algebras has the* amalgamation property *(AP) if for every 5-tuple (called V-formation) $(\mathcal{A}, \mathcal{B}, \mathcal{C}, i, j)$, where $\mathcal{A}, \mathcal{B}, \mathcal{C} \in K$ and $\mathcal{A} \stackrel{i}{\hookrightarrow} \mathcal{B}$, $\mathcal{A} \stackrel{j}{\hookrightarrow} \mathcal{C}$, there is a triple (called amalgam) $(\mathcal{D}, h, k)$, with $\mathcal{D} \in K$, $\mathcal{B} \stackrel{h}{\hookrightarrow} \mathcal{D}$, $\mathcal{C} \stackrel{k}{\hookrightarrow} \mathcal{D}$, such that $h \circ i = k \circ j$.*

[1] Γ establishes a categorical equivalence between abelian l-groups with a strong order unit $(\mathcal{G}, u)$ and MV-algebras, by equipping the interval $[0, u]$ of $\mathcal{G}$ with MV-algebraic operations obtained by truncation of the group ones. On arrows Γ acts by restriction.

[2] Note that every non-trivial totally ordered cancellative hoop $\mathcal{A}$ does not have rank, since $\mathcal{A}/Rad(\mathcal{A})$ is an infinite cancellative hoop.

For the varieties of BL-algebras a sufficient condition for the AP is the following.

Theorem 2 ([Mon06, MMT14]). *Let $\mathbb{L}$ be a non-trivial variety of BL-algebras. If $Ch(\mathbb{L})$ enjoys the AP then the same holds for $\mathbb{L}$.*

In this section we will provide a full classification of the AP for the varieties of BL-algebras generated by one BL-chain with finitely many components. We start with the following result.

Lemma 2 ([AB21b]). *Let $\mathbb{L}$ be a variety of BL-algebras such that every chain has finitely many components. Then there are k, h such that:*

(i) Every chain in $\mathbb{L}$ has at most k components.
(ii) Every chain in $\mathbb{L}$ is such that the rank of a component, if finite, is at most h.

Lemma 3. *Let $\mathbb{L}$ be a variety of BL-algebras. Then the following are equivalent.*

(i) $\mathbb{L}$ contains neither $\mathcal{G}_4$ nor $\mathcal{P}_2$.
(ii) Every chain $\mathcal{A} = \bigoplus_{i\in I} \mathcal{A}_i$ in $\mathbb{L}$ is such that $|I| \leq 3$, there is at most one $i \in I \setminus \{0\}$ such that $\mathcal{A}_i$ is infinite, and there is at most one $j \in I \setminus \{0\}$ such that $\mathcal{A}_j$ is bounded.

Proof. Let $\mathbb{L}$ be a variety of BL-algebras.

(ii) $\Rightarrow$ *(i)* Immediate, as $\mathcal{G}_4$ and $\mathcal{P}_2$ do not satisfy condition *(ii)*.
(i) $\Rightarrow$ *(ii)* Assume that condition *(i)* holds true, and pick $\mathcal{A} = \bigoplus_{i\in I} \mathcal{A}_i \in Ch(\mathbb{L})$. If $|I| > 3$, then there are $i \neq j \in I \setminus \{0\}$ such that $\mathcal{A}_i, \mathcal{A}_j$ are either both bounded or both cancellative. But then $\mathcal{G}_4 \hookrightarrow \mathcal{A}$ (since the subalgebra of $\mathcal{A}$ generated by its idempotent elements is a G-chain with at least 4 elements) or $\mathcal{P}_2 \hookrightarrow \mathcal{A}$ (since $\mathcal{C}_\infty$ embeds into every infinite totally ordered cancellative hoop, see [AFM07]), in contrast with condition *(i)*. So we must have $|I| \leq 3$. Suppose now that there are $i, j \in I \setminus \{0\}$ such that $\mathcal{A}_i, \mathcal{A}_j$ are both infinite. Assume *w.l.o.g.* that $i < j$. By [AM03, Theorem 7.9], $\mathcal{P}_2 \in Si(\mathbf{2}\oplus\mathcal{A}_i\oplus\mathcal{A}_j)$, but then $\mathcal{P}_2 \in \mathbb{L}$, in contrast with condition *(i)*. So there is at most one $i \in I\setminus\{0\}$ such that $\mathcal{A}_i$ is infinite. Suppose now that there are $i, j \in I \setminus \{0\}$ such that $\mathcal{A}_i, \mathcal{A}_j$ are both bounded. This would imply that $\mathcal{G}_4 \hookrightarrow \mathcal{A}$, in contrast with condition *(i)*. Whence all the requirements of condition *(ii)* are satisfied, and the proof is settled.

□

Theorem 3. *Let $\mathbb{L}$ be a variety of BL-algebras such that every chain has finitely many components. If $\mathbb{L}$ contains $\mathcal{G}_4$ or $\mathcal{P}_2$, then $\mathbb{L}$ does not have the AP.*

Proof. Let $\mathbb{L}$ be a variety of BL-algebras such that every chain has finitely many components. By Lemma 2 there is $n \in \mathbb{N}$ which is the largest number of components of a chain in $\mathbb{L}$.

Suppose first that $\mathbb{L}$ contains $\mathcal{G}_4$. Since every chain has at most n components, then every G-chain in $\mathbb{L}$ has at most $n+1$ elements. Let $4 \leq k \leq n+1$ be the cardinality of the largest G-chain in $\mathbb{L}$.

Pick now the V-formation $(\mathcal{B}, \mathcal{C}, \mathcal{D}, i, j)$, such that $\mathcal{B} \simeq \mathcal{G}_{k-1}$, $\mathcal{C} \simeq \mathcal{G}_k$ and $\mathcal{D} \simeq \mathcal{G}_k$. Assume that the lattice reducts of $\mathcal{B}, \mathcal{C}, \mathcal{D}$ are, respectively, $b_1 < \cdots < b_{k-1}$, $c_1 < \cdots < c_k$, $d_1 < \cdots < d_k$. Let us define i, j as follows:

- $i(b_1) = c_1$, $i(b_{k-1}) = c_k$.
- For $2 \leq r \leq k-2$, $i(b_r) = c_r$.
- $j(b_1) = d_1$, $j(b_{k-1}) = d_k$.
- For $2 \leq r \leq k-2$, $j(b_r) = d_{r+1}$.

It is immediate to see that i, j are embeddings.

We now show that there is no amalgam in $\mathbb{L}$ for the V-formation $(\mathcal{B}, \mathcal{C}, \mathcal{D}, i, j)$.

Suppose by contradiction that there is an amalgam $(\mathcal{E}, l, m)$ for $(\mathcal{B}, \mathcal{C}, \mathcal{D}, i, j)$, with $\mathcal{E} \in \mathbb{L}$. Then we must have that $l(i(b)) = m(j(b))$, for every $b \in \mathcal{B}$, and hence an easy computation shows that $S = l(\mathcal{C}) \cup m(\mathcal{D})$ has $k+1$ elements, and its elements are ordered (in $\mathcal{E}$) as follows[3]: $0 = l(c_1) = m(d_1) < l(c_2) < l(c_3) = m(d_2) < \cdots < l(c_{k-1}) = m(d_{k-2}) < m(d_{k-1}) < 1 = l(c_k) = m(d_k)$. We now show that S is a subuniverse of $\mathcal{E}$. Note that every element of S is idempotent. Also, since $l(i(\mathcal{B})) = S \setminus \{m(d_{k-1})\}$ and $m(j(\mathcal{B})) = S \setminus \{l(c_2)\}$, then the operations $*$ and $\Rightarrow$ coincide with the ones of a G-chain, over $S \setminus \{m(d_{k-1})\}$ and $S \setminus \{l(c_2)\}$. Then, to show that S is a subuniverse of $\mathcal{E}$ we only need to check that $m(d_{k-1}) \Rightarrow l(c_2) \in S$. Clearly $m(d_{k-1}) \Rightarrow l(c_2) \geq l(c_2)$, and by monotonicity, $m(d_{k-1}) \Rightarrow l(c_2) \leq l(c_{k-1}) \Rightarrow l(c_2) = l(c_2)$. This proves $m(d_{k-1}) \Rightarrow l(c_2) = l(c_2) \in S$. Whence S is a subuniverse of $\mathcal{E}$.

Let $\mathcal{S}$ be the subalgebra of $\mathcal{E}$ with carrier S. It is immediate to see that $\mathcal{S} \simeq \mathcal{G}_{k+1}$, and clearly $\mathcal{S} \in \mathbb{L}$. However this is not possible, as the largest G-chain (up to isomorphisms) in $\mathbb{L}$ is $\mathcal{G}_k$.

Whence we conclude that $(\mathcal{B}, \mathcal{C}, \mathcal{D}, i, j)$ cannot have an amalgam in $\mathbb{L}$, and the AP fails for $\mathbb{L}$.

Suppose, finally, that $\mathbb{L}$ contains $\mathcal{P}_2$. The proof strategy is very similar to the $\mathcal{G}_4$ case, with the difference that instead of the idempotent elements we take in account the number of cancellative components. Remember that every chain in $\mathbb{L}$ has at most n components. Let h be the maximum number of cancellative components of a chain in $\mathbb{L}$: clearly $2 \leq h \leq n-1$. This means that $\mathbb{L}$ contains $\mathcal{P}_1, \ldots, \mathcal{P}_h$.

Pick now the V-formation $(\mathcal{B}, \mathcal{C}, \mathcal{D}, i, j)$, such that $\mathcal{B} \simeq \mathcal{P}_{h-1}$, $\mathcal{C} \simeq \mathcal{P}_h$ and $\mathcal{D} \simeq \mathcal{P}_h$.

More specifically we have that $\mathcal{B} = \mathbf{2} \oplus \mathcal{B}_1 \oplus \cdots \oplus \mathcal{B}_{h-1}, \mathcal{C} = \mathbf{2} \oplus \mathcal{C}_1 \oplus \cdots \oplus \mathcal{C}_h, \mathcal{D} = \mathbf{2} \oplus \mathcal{D}_1 \oplus \cdots \oplus \mathcal{D}_h$, and all the $\mathcal{B}_r$'s, $\mathcal{C}_s$'s, $\mathcal{D}_t$'s are isomorphic to $\mathcal{C}_\infty$. Let us define i, j as follows:

- $i(0) = 0$, $i(1) = 1$.
- For $1 \leq r \leq h-1$, i maps isomorphically $\mathcal{B}_r$ in $\mathcal{C}_r$.

[3] To use this proof strategy is essential that $k \geq 4$.

– $j(0) = 0$, $j(1) = 1$.
– For $1 \leq r \leq h-1$, j maps isomorphically $\mathcal{B}_r$ in $\mathcal{D}_{r+1}$.

It is immediate to see that i, j are embeddings.

We now show that there is no amalgam in $\mathbb{L}$ for the V-formation $(\mathcal{B}, \mathcal{C}, \mathcal{D}, i, j)$.

Suppose by contradiction that there is an amalgam $(\mathcal{E}, l, m)$ for $(\mathcal{B}, \mathcal{C}, \mathcal{D}, i, j)$, with $\mathcal{E} \in \mathbb{L}$. Then we must have that $l(i(b)) = m(j(b))$, for every $b \in \mathcal{B}$, and hence an easy check shows that the elements of $S = l(C) \cup m(D)$ are ordered (in $\mathcal{E}$) as follows[4]: $0 < l(C_1) \setminus \{1\} < l(C_2) \setminus \{1\} = m(D_1) \setminus \{1\} < \cdots < l(C_h) \setminus \{1\} = m(D_{h-1}) \setminus \{1\} < m(D_h) \setminus \{1\} < 1$. So, S contains the elements of $h+1$ totally ordered cancellative hoops, with the top element 1 in common. We now show that S is a subuniverse of $\mathcal{E}$. Since $l(i(B)) = S \setminus (m(D_h) \setminus \{1\})$ and $m(j(B)) = S \setminus (l(C_1) \setminus \{1\})$, then the operations $*$ and $\Rightarrow$ of $l(i(\mathcal{B}))$ and $m(j(\mathcal{B}))$ coincide with the ones of a chain isomorphic to $\mathcal{P}_h$, over $S \setminus (m(D_h) \setminus \{1\})$ and $S \setminus (l(C_1) \setminus \{1\})$, respectively. Then, to show that S is a subuniverse of $\mathcal{E}$ we only need to check that $x \Rightarrow y \in S$, for every $x \in m(D_h) \setminus \{1\}$, and every $y \in l(C_1) \setminus \{1\}$. Clearly $x \Rightarrow y \geq y$, and by monotonicity, $x \Rightarrow y \leq z \Rightarrow y = y$, for every $z \in l(C_h) \setminus \{1\}$. This proves $x \Rightarrow y = y \in S$. Whence S is a subuniverse of $\mathcal{E}$. Let $\mathcal{S}$ be the subalgebra of $\mathcal{E}$ with carrier S. It is immediate to see that $\mathcal{S} \simeq \mathcal{P}_{h+1}$, and clearly $\mathcal{S} \in \mathbb{L}$. However this is not possible, as no chain in $\mathbb{L}$ has more than h cancellative components. Whence we conclude that $(\mathcal{B}, \mathcal{C}, \mathcal{D}, i, j)$ cannot have an amalgam in $\mathbb{L}$, and the AP fails for $\mathbb{L}$. The proof is settled. □

Lemma 4. *Let $\mathcal{A}$ be a simple MV-chain, and let $\mathcal{B} = \bigoplus_{i \in I} \mathcal{B}_i$ be a BL-chain such that $\mathcal{B}_0 = \mathbf{2}$ and $|I| \geq 2$ (*i.e., *$\mathcal{B}$ is an SBL-chain with at least two components). Suppose that there is a BL-algebra $\mathcal{C}$ such that $\mathcal{A} \hookrightarrow \mathcal{C}$ and $\mathcal{B} \hookrightarrow \mathcal{C}$. Then $\mathcal{A} \oplus \bigoplus_{i \in I \setminus \{0\}} \mathcal{B}_i \hookrightarrow \mathcal{C}$.*

Proof. Let $\mathcal{A}$, $\mathcal{B}$ as above, and assume that there is a BL-algebra $\mathcal{C}$ such that $\mathcal{A} \overset{i}{\hookrightarrow} \mathcal{C}$ and $\mathcal{B} \overset{j}{\hookrightarrow} \mathcal{C}$. Let $\mathcal{D}$ be the subalgebra of $\mathcal{C}$ generated by $i(\mathcal{A}) \cup j(\mathcal{B})$.

We now show that $x *_D y = x$, for every $x \in i(A \setminus \{1\})$ and $y \in j(B \setminus \{0\})$.

Pick $x \in i(A \setminus \{1\})$ and $y \in j(B \setminus \{0\})$. It is easy to check that $\neg\neg x = x$ and $\neg y = 0$. Now, $\mathcal{D}$ is isomorphic to a subdirect product $\mathcal{E}$ of a family of subdirectly irreducible BL-chains $\{\mathcal{E}_r : r \in R\}$. Then there are two tuples $\langle x_r \rangle_{r \in R}, \langle y_r \rangle_{r \in R} \in \mathcal{E}$ which correspond - via the isomorphism between $\mathcal{D}$ and $\mathcal{E}$ - to x and y, respectively.

Since $\neg y = 0$, every y_r in $\langle y_r \rangle_{r \in R}$ is either 1 or it belongs to a component of $\mathcal{E}_r$ which is different from the first-one. Indeed, if not we would have $\neg y_s > 0$, for some $s \in R$, and this would imply $\neg y > 0$, a contradiction.

Since $\neg\neg x = x$, every x_r in $\langle x_r \rangle_{r \in R}$ belongs to the first component of $\mathcal{E}_r$. Indeed, if not we would have $\neg\neg x_s \neq x_s$, for some $s \in R$, and then $x \neq \neg\neg x$, a contradiction. Since $\mathcal{A}$ is simple, then x is nilpotent, *i.e.*, $x^n = 0$, for some n. This implies that $x_r < 1$ for every $r \in R$.

[4] To use this proof strategy it is essential that $h \geq 2$. For every $l(C_i)$ ($m(D_i)$, respectively), the elements are ordered as in the chain $l(\mathcal{C}_i)$ ($m(\mathcal{D}_i)$, respectively).

From the previous observations we have that $x_r < y_r$, for every x_r in $\langle x_r \rangle_{r \in R}$ and every y_r in $\langle y_r \rangle_{r \in R}$. Then an easy computation shows that $x_r * y_r = x_r$, for every $r \in R$, and hence $x * y = x$. Moreover we have that $y \Rightarrow x = x$, since $y_r \Rightarrow x_r = x_r$, for every x_r in $\langle x_r \rangle_{r \in R}$ and every y_r in $\langle y_r \rangle_{r \in R}$. This means that $i(A) \cup j(B)$ is closed under $*_D$ and $\Rightarrow_D$, and it contains 0 and 1. Moreover we have that $x \wedge_D y = x$, as $x *_D y \leq x \wedge_D y \leq x$, and since $i(\mathcal{A})$ and $j(\mathcal{B})$ are both chains we conclude that $i(A) \cup j(B)$ is a totally ordered subuniverse of $\mathcal{C}$. Then we have $\mathcal{D} \simeq \mathcal{A} \oplus_{i \in I \setminus \{0\}} \mathcal{B}_i$, and hence the theorem's claim is an immediate consequence. □

Remark 2. Note that Lemma 4 does not hold, in general, if we remove the assumption that $\mathcal{A}$ is simple. For example, consider Chang's MV-algebra $\mathcal{K}_2$ and $\mathbf{2} \oplus \mathcal{L}_3$, as well as their direct product $\mathcal{K}_2 \times \mathbf{2} \oplus \mathcal{L}_3$. It is very easy to check that $\mathcal{K}_2 \hookrightarrow \mathcal{K}_2 \times \mathbf{2} \oplus \mathcal{L}_3$ and $\mathbf{2} \oplus \mathcal{L}_3 \hookrightarrow \mathcal{K}_2 \times \mathbf{2} \oplus \mathcal{L}_3$. However $\mathcal{K}_2 \oplus \mathcal{L}_3 \not\hookrightarrow \mathcal{K}_2 \times \mathbf{2} \oplus \mathcal{L}_3$.

Proposition 1. *Let $\mathcal{A} = \bigoplus_{i \in I} \mathcal{A}_i$ be a BL-chain. Suppose that:*

- $|I| \geq 2$, i.e., *$\mathcal{A}$ has at least two components.*
- *$\mathcal{A}_0$ is an MV-chain with infinite rank such that $\mathcal{L}_k \not\hookrightarrow \mathcal{A}_0$, for some $k \geq 3$ or $\mathcal{A}_0$ is an infinite MV-chain with rank $k \geq 3$, and $\mathcal{L}_k \not\hookrightarrow \mathcal{A}_0$.*

Then $\mathbf{V}(\mathcal{A})$ does not have the AP.

Proof. Let $\mathcal{A}$ be a BL-chain as above. Since $\mathcal{A}_0$ is an infinite MV-chain with infinite rank or rank k, then $\mathcal{L}_k \in \mathbf{V}(\mathcal{A}_0) \subsetneq \mathbf{V}(\mathcal{A})$. Pick now the V-formation $(\mathbf{2}, \mathcal{A}, \mathcal{L}_k, i, j)$, where i, j are defined in the unique and obvious way. Suppose that there is an amalgam $(\mathcal{D}, l, m)$, with $\mathcal{D} \in \mathbf{V}(\mathcal{A})$. By Lemma 4, $\mathcal{D}$ contains $\mathcal{E} = \mathcal{L}_k \oplus \bigoplus_{i \in I \setminus \{0\}} \mathcal{A}_i$ as a subalgebra. Since $\mathcal{L}_k \not\hookrightarrow \mathcal{A}_0$, by [AM03, Lemma 4.6] the equation[5] $e \stackrel{\text{def}}{=} (((y \Rightarrow x) \Rightarrow x) * ((k-2)x \Leftrightarrow \neg x)) \Rightarrow (x \vee y) = 1$ is such that $\mathcal{A} \models e$, whilst $\mathcal{E} \not\models e$. Whence $\mathcal{E} \notin \mathbf{V}(\mathcal{A})$, and we conclude that $\mathbf{V}(\mathcal{A})$ does not have the AP. □

We recall the following construction, introduced in [Jen03].

Definition 5. *Let $\mathcal{A}$ be a totally ordered Wajsberg hoop. The* disconnected rotation *of $\mathcal{A}$ is an algebra denoted by $\mathcal{A}^*$ and defined as follows. Let $A' = \{(a, 0) : a \in A\}$. We define an order $\leq_{A'}$ on A' such that $(A', \leq_{A'})$ and $(A, \leq_A)$ are dually isomorphic. Let $A^* \stackrel{\text{def}}{=} A \cup A'$. We extend the orders $\leq_A$ and $\leq_{A'}$ to an order $\leq_{A^*}$ in A^*, by putting $a <_{A^*} b$ for every $a \in A', b \in A$. For every $a \in A^*$ we define $a' = (a, 0)$ if $a \in A$, and $a' = b$ if $a = (b, 0) \in A'$.*

[5] Here $x \Leftrightarrow y$ stands for $(x \Rightarrow y) * (y \Rightarrow x)$. Moreover $x \uplus y \stackrel{\text{def}}{=} (x \Rightarrow (x * y)) \Rightarrow y$, whilst nx is defined inductively by $0x = 0$ and $n(x) = (n-1)x \uplus x$.

Finally, we take the following operations in $\mathcal{A}^*$. $1_{\mathcal{A}^*} \stackrel{def}{=} 1_{\mathcal{A}}$, $0_{\mathcal{A}^*} \stackrel{def}{=} (1_{\mathcal{A}})'$, $\wedge_{\mathcal{A}^*}$ *is the minimum w.r.t.* $\leq_{\mathcal{A}^*}$, $\vee_{\mathcal{A}^*}$ *is the maximum w.r.t.* $\leq_{\mathcal{A}^*}$,

$$a *_{\mathcal{A}^*} b \stackrel{def}{=} \begin{cases} a *_{\mathcal{A}} b & \text{if } a, b \in A, \\ (a \Rightarrow_{\mathcal{A}} b')' & \text{if } a \in A, b \in A', \\ (b \Rightarrow_{\mathcal{A}} a')' & \text{if } a \in A', b \in A, \\ 0_{\mathcal{A}^*} & \text{if } a, b \in A'. \end{cases}$$

$$a \Rightarrow_{\mathcal{A}^*} b \stackrel{def}{=} \begin{cases} a \Rightarrow_{\mathcal{A}} b & \text{if } a, b \in A, \\ (a *_{\mathcal{A}} b')' & \text{if } a \in A, b \in A', \\ 1_{\mathcal{A}^*} & \text{if } a \in A', b \in A, \\ b' \Rightarrow_{\mathcal{A}} a' & \text{if } a, b \in A'. \end{cases}$$

Theorem 4 ([NEG05, Theorem 9]). *Let* $\mathcal{A}$ *be a totally ordered cancellative hoop. Then* $\mathcal{A}^*$ *is isomorphic to a perfect MV-chain. Conversely, every perfect MV-chain is isomorphic to the disconnected rotation of a totally ordered cancellative hoop.*

Let $\mathcal{A}$ be a totally ordered Wajsberg hoop. With $\mathcal{A}^r$ we denote the 0-free reduct of $\mathcal{A}^*$.

Lemma 5. *(i) Let* $\mathcal{A}$ *be a totally ordered cancellative hoop, and let* $\mathcal{B}$ *be a totally ordered Wajsberg hoop with minimum. If* $\mathcal{A} \hookrightarrow \mathcal{B}$*, then* $\mathcal{A}^r \hookrightarrow \mathcal{B}$.
(ii) Let $\mathcal{A}, \mathcal{B}$ *be two totally ordered cancellative hoops. If* $\mathcal{A} \hookrightarrow \mathcal{B}$*, then* $\mathcal{A}^r \hookrightarrow \mathcal{B}^r$.
(iii) Let $\mathcal{A}$ *be an infinite totally ordered Wajsberg hoop with minimum. Then, for every* $\mathcal{B} \in Ch(\mathbb{CH})$, $\mathcal{B}^r \in \mathbf{V}(\mathcal{A})$.

Proof. (i) Let $\mathcal{A}, \mathcal{B}$ as above, and suppose that $\mathcal{A} \stackrel{i}{\hookrightarrow} \mathcal{B}$. Let $\mathcal{C}$ be the subalgebra of $\mathcal{B}$ generated by $i(\mathcal{A}) \cup \{0\}$, where 0 is the minimum of $\mathcal{B}$. Of course $i(A) \cup \neg i(A) \subseteq C$, where $\neg i(A) = \{\neg x : x \in i(A)\}$. We show that also the other inclusion holds. Since $\mathcal{A}$ is a cancellative hoop, then $i(A) \subseteq Rad(\mathcal{B}) \subseteq B^+$: this means that $i(A)$ is closed under $*, \Rightarrow$, and $i(A) \cap \neg i(A) = \emptyset$. As an easy check shows, $\neg i(A)$ cannot contain a negation fixpoint, and hence $\neg i(A) \subseteq B^-$, which implies that $\neg i(A)$ is closed under $*$ (notice that $0 \in \neg i(A)$). Since $\mathcal{B}$ has an involutive negation, then $i(A) \cup \neg i(A)$ is closed under $*, \neg$. To show that $i(A) \cup \neg i(A)$ is a subuniverse of B, it remains to show that given $x, y \in \neg i(A)$, with $x > y > 0$, and $y \in \neg i(A)$, $x \Rightarrow y \in i(A) \cup \neg i(A)$. Suppose first $x \in \neg i(A)$. As it is well known, on MV-algebras it holds that $x \Rightarrow y = (\neg y) \Rightarrow (\neg x)$. since the negation is involutive $\neg y, \neg x \in i(A)$, and hence $x \Rightarrow y = (\neg y) \Rightarrow (\neg x) \in i(A)$. Suppose, finally, that $x \in i(A)$. Notice that $x \Rightarrow y = \neg x \oplus y = \neg(x * \neg y)$. Since the negation is involutive then $\neg y \in i(A)$, and since $i(A)$ is closed under $*$ we conclude that $x * \neg y \in i(A)$. Then $x \Rightarrow y = \neg(x * \neg y) \in \neg i(A)$.
This means that $i(A) \cup \neg i(A)$ is a subuniverse of $\mathcal{B}$, and hence $C = i(A) \cup \neg i(A)$. By [Jen03, Definition 5] an easy check shows that $\mathcal{C} \simeq \mathcal{A}^r$. Then we conclude that $\mathcal{A}^r \hookrightarrow \mathcal{B}$.

(ii) Immediate, by the definition of disconnected rotation.
(iii) Let $\mathcal{A}$ be an infinite totally ordered Wajsberg hoop with minimum. By [AP02, Corollary 2.3] we know that there is an MV-chain $\mathcal{A}'$ such that $\mathcal{A}$ is the 0-free reduct of $\mathcal{A}'$, and $Ch(\mathcal{A}) = Sh(Ch(\mathcal{A}'))$, where $Sh(Ch(\mathcal{A}'))$ denotes the class of 0-free subreducts of $Ch(\mathcal{A}')$. Since $\mathcal{A}'$ is infinite, then $\mathbb{C} \subseteq \mathbf{V}(\mathcal{A}')$. By Theorem 4 we conclude that for every $\mathcal{B} \in Ch(\mathbb{CH})$, $\mathcal{B}^r \in \mathbf{V}(\mathcal{A})$. □

Theorem 5. *(i) Let $\mathcal{A}$ be an MV-chain. Then $Ch(\mathcal{A})$ has the AP.*
(ii) Let $\mathcal{A}$ be a totally ordered Wajsberg hoop. Then $Ch(\mathcal{A})$ has the AP.

Proof. (i) Assume first that $\mathcal{A}$ is an MV-chain. If $\mathcal{A}$ has infinite rank, then $\mathbf{V}(\mathcal{A}) = \mathbb{MV}$. Then, as shown in [Mun88, Mon06] $Ch(\mathcal{A})$ has the AP.
If $\mathcal{A}$ has a finite rank, then (see [CDM99]) either $\mathbf{V}(\mathcal{A}) = \mathbf{V}(\mathcal{K}_n)$ or $\mathbf{V}(\mathcal{A}) = \mathbf{V}(\mathcal{L}_n)$, for some $n \geq 2$. The rest of the proof is very similar to the one [NL00, Proposition 4], with some modifications. Let $(\mathcal{B}, \mathcal{C}, \mathcal{D}, i, j)$ be a V-formation with $\mathcal{B}, \mathcal{C}, \mathcal{D} \in Ch(\mathcal{A})$. By [Mun88, Mon06] there exists an amalgam $(\mathcal{E}, h, k)$, where $\mathcal{E}$ is an MV-chain. By [NL00, Lemma 6, Proposition 7], $Ch(\mathcal{A}) \cap \mathbf{S}(\mathcal{E})$ has a largest element $\mathcal{E}_0$. Since $h(\mathcal{C}), k(\mathcal{D}) \in \mathbf{IS}(\mathcal{E})$, then both $\mathcal{C}$ and $\mathcal{D}$ embeds into $\mathcal{E}_0$. Consider now $(\mathcal{E}_0, h_1, k_1)$, where $h_1 : \mathcal{C} \to \mathcal{E}_0$ and $k_1 : \mathcal{D} \to \mathcal{E}_0$ are maps such that $h_1(x) = h(x)$, and $k_1(y) = k(y)$, for every $x \in C, y \in D$. An easy check shows that $(\mathcal{E}_0, h_1, k_1)$ is an amalgam for $(\mathcal{B}, \mathcal{C}, \mathcal{D}, i, j)$, and clearly $\mathcal{E}_0 \in Ch(\mathcal{A})$.
(ii) Suppose first that $\mathcal{A}$ is an unbounded Wajsberg hoop, *i.e.*, $\mathcal{A}$ is an infinite cancellative hoop. Then by [AFM07, Theorem 6.3], $\mathbf{V}(\mathcal{A})$ is the variety $\mathbb{CH}$ of cancellative hoops, and as shown in [Mon06] $Ch(\mathcal{A})$ has the AP.
Finally, assume that $\mathcal{A}$ is a totally ordered Wajsberg hoop with minimum. This means that $\mathcal{A}$ is the 0-free reduct of an MV-chain, say $\mathcal{A}'$. Then $\mathbf{V}(\mathcal{A})$ can contain 0-free reducts of MV-chains and, possibly, cancellative hoops. Let $(\mathcal{B}, \mathcal{C}, \mathcal{D}, i, j)$ be a V-formation with $\mathcal{B}, \mathcal{C}, \mathcal{D} \in Ch(\mathcal{A})$. Let us define $(\mathcal{B}', \mathcal{C}', \mathcal{D}', i_1, j_1)$ as follows. For $\mathcal{E} \in \{\mathcal{B}, \mathcal{C}, \mathcal{D}\}$, $\mathcal{E}' = \mathcal{E}^r$ if $\mathcal{E}$ is a cancellative hoop, and $\mathcal{E}' = \mathcal{E}$ if $\mathcal{E}$ is a Wajsberg hoop with minimum. The maps $i_1 : \mathcal{B}' \to \mathcal{C}'$, and $j_1 : \mathcal{B}' \to \mathcal{D}'$ are defined as follows.

$$i_1(x) = \begin{cases} i(x) & \text{if } \mathcal{B} = \mathcal{B}' \\ i(x) & \text{if } \mathcal{B}' = \mathcal{B}^r \text{ and } x > \neg_{\mathcal{B}'} x. \\ \neg_{\mathcal{B}'}(i(\neg_{\mathcal{B}'} x)) & \text{if } \mathcal{B}' = \mathcal{B}^r \text{ and } x \leq \neg_{\mathcal{B}'} x. \end{cases}$$

$$j_1(x) = \begin{cases} j(x) & \text{if } \mathcal{B} = \mathcal{B}' \\ j(x) & \text{if } \mathcal{B}' = \mathcal{B}^r \text{ and } x > \neg_{\mathcal{B}'} x. \\ \neg_{\mathcal{B}'}(j(\neg_{\mathcal{B}'} x)) & \text{if } \mathcal{B}' = \mathcal{B}^r \text{ and } x \leq \neg_{\mathcal{B}'} x. \end{cases}$$

The maps i_1, j_1 are well-defined, since by Lemma 5, $\mathcal{B} \hookrightarrow \mathcal{B}'$, $\mathcal{C} \hookrightarrow \mathcal{C}'$, and $\mathcal{D} \hookrightarrow \mathcal{D}'$. Since $(\mathcal{B}, \mathcal{C}, \mathcal{D}, i, j)$ is a V-formation, using Lemma 5 it can be easily checked that $(\mathcal{B}', \mathcal{C}', \mathcal{D}', i_1, j_1)$ is a V-formation as well. By Lemma 5 we also have that $\mathcal{B}', \mathcal{C}', \mathcal{D}' \in Ch(\mathcal{A})$. By the construction we have that

$\mathcal{B}', \mathcal{C}', \mathcal{D}'$ are the 0-free reducts of three MV-chains, say $\mathcal{B}'', \mathcal{C}'', \mathcal{D}''$. Then we can construct a V-formation $(\mathcal{B}'', \mathcal{C}'', \mathcal{D}'', i_2, j_2)$, where i_2 and j_2 maps the elements in the same way of i_1 and j_1, respectively. By 1) there is an amalgam $(\mathcal{E}, h, k)$ for $(\mathcal{B}'', \mathcal{C}'', \mathcal{D}'', i_2, j_2)$, where $\mathcal{E} \in Ch(\mathcal{A}')$. Then, by calling $\mathcal{E}'$ the 0-free reduct of $\mathcal{E}$, we can construct an amalgam $(\mathcal{E}', h_1, k_1)$ for the V-formation $(\mathcal{B}', \mathcal{C}', \mathcal{D}', i_1, j_1)$, where h_1 and j_1 maps the elements in the same way of h and k, respectively. Clearly $\mathcal{E}' \in Ch(\mathcal{A})$.
Consider now $(\mathcal{E}', h_2, j_2)$, where $h_2 : \mathcal{C} \to \mathcal{E}'$, and $k_2 : \mathcal{D} \to \mathcal{E}'$ are such that $h_2(x) = h_1(x)$, for every $x \in \mathcal{C}$, and $k_2(y) = k_1(y)$, for every $y \in \mathcal{D}$. The maps h_2, k_2 are well-defined, since by Lemma 5, $\mathcal{C} \hookrightarrow \mathcal{C}'$, and $\mathcal{D} \hookrightarrow \mathcal{D}'$. It is straightforward to check that $(\mathcal{E}', h_2, j_2)$ is an amalgam for $(\mathcal{B}, \mathcal{C}, \mathcal{D}, i, j)$.
The proof is settled. □

Proposition 2 ([AB21b]). *Let* $\mathcal{A} = \bigoplus_{i=0}^{k} \mathcal{A}_i$ *be a BL-chain. Define* $\mathcal{A}^s = \bigoplus_{i=0}^{k-1} \mathcal{A}_i \oplus \mathcal{B}_k$, *where:*

- $\mathcal{B}_k = \mathcal{A}_k$ *if* $\mathcal{A}_k$ *is finite.*
- $\mathcal{B}_k$ *is the* 0*-free reduct of* $[0,1]_{\mathcal{L}}$ *if* $\mathcal{A}_k$ *has infinite rank.*
- $\mathcal{B}_k$ *is the* 0*-free reduct of* $\mathcal{K}_n$ *if* $\mathcal{A}_k$ *is non-simple and with rank* n.
- $\mathcal{B}_k = \mathcal{C}_\infty$ *if* $\mathcal{A}_k$ *is an infinite cancellative hoop.*

Then $\mathcal{A}^s$ *is subdirectly irreducible, and* $\mathbf{V}(\mathcal{A}) = \mathbf{V}(\mathcal{A}^s)$.

Lemma 6 ([AM03]).

- *Let* $\bigoplus_{i\in I} \mathcal{A}_i$ *be a non-trivial BL-chain. Then* $\mathbf{ISP}_u(\bigoplus_{i\in I} \mathcal{A}_i) = \mathbf{I}(\bigoplus_{i\in I} \mathbf{SP}_u(\mathcal{A}_i))$, *where* $\bigoplus_{i\in I} \mathbf{SP}_u(\mathcal{A}_i) = \{\bigoplus_{i\in I} \mathcal{B}_i : \mathcal{B}_i \in \mathbf{SP}_u(\mathcal{A}_i)\}$.
- *If* $\mathcal{A}$ *is an infinite totally ordered cancellative hoop, then* $\mathbf{ISP}_u(\mathcal{A}) = Ch(\mathbb{CH})$.
- *If* $\mathcal{A}$ *is a totally ordered Wajsberg hoop with infinite rank, and for every* $n \geq 2$, $\mathcal{L}_n \hookrightarrow \mathcal{A}$, *then* $\mathbf{ISP}_u(\mathcal{A}) = Ch(\mathcal{A})$.
- *If* $\mathcal{A}$ *is a totally ordered Wajsberg hoop with* $rank(\mathcal{A}) = n$, *and* $\mathcal{L}_n \hookrightarrow \mathcal{A}$, *then* $\mathbf{ISP}_u(\mathcal{A}) = Ch(\mathcal{A})$. *If in addition* $\mathcal{A}$ *is also finite, then*[6] $\mathbf{ISP}_u(\mathcal{A}) = \mathbf{IS}(\mathcal{A}) = Ch(\mathcal{A})$.

We can finally state our main result.

Theorem 6. *Let* $\mathbb{L}$ *be a variety of BL-algebras generated by one chain with finitely many components. Then the following are equivalent:*

(i) $\mathbb{L}$ *has the AP.*
(ii) Every BL-chain $\mathcal{A} = \bigoplus_{i\in I} \mathcal{A}_i$ *such that* $\mathbf{V}(\mathcal{A}) = \mathbb{L}$ *satisfies the following conditions.*
  - $|I| \leq 3$.

[6] The assumption that $Ch(\mathcal{A})$ does not contain trivial chains is essential. Indeed, if $\mathcal{A}$ is non-trivial, then $\mathbf{ISP}_u(\mathcal{A})$ does not contain trivial algebras.

- *There is at most one $i \in I \setminus \{0\}$ such that $\mathcal{A}_i$ is infinite, and there is at most one $j \in I \setminus \{0\}$ such that $\mathcal{A}_j$ is bounded.*
- *If $|I| \geq 2$ then the following ones hold.*
  * *If $\mathcal{A}_0$ has infinite rank, then $\mathcal{L}_k \hookrightarrow \mathcal{A}_0$, for every $k \geq 2$.*
  * *If $\mathcal{A}_0$ is infinite and $rank(\mathcal{A}_0) = k$, then $\mathcal{L}_k \hookrightarrow \mathcal{A}_0$.*

Proof. Let $\mathbb{L}$ be a variety of BL-algebras generated by one chain with finitely many components.

$(i) \Rightarrow (ii)$ Suppose that condition 2 does not hold. The results of Lemma 3 plus an easy check show that the hypothesis of Theorem 3 or Proposition 1 are satisfied, and we conclude that $\mathbb{L}$ does not have the AP.
$(ii) \Rightarrow (i)$ Let $\mathcal{A} = \bigoplus_{i \in I} \mathcal{A}_i$ be a BL-chain such that $\mathbf{V}(\mathcal{A}) = \mathbb{L}$. By hypothesis $\mathcal{A}$ satisfies condition *(ii)*, which implies $\#\mathcal{A} \leq 3$. Our proof strategy is to prove the AP for $Ch(\mathcal{A})$. By Theorem 2 this implies that $\mathbb{L} = \mathbf{V}(\mathcal{A})$ has the AP as well. We distinguish the cases $\#\mathcal{A} = 1$, $\#\mathcal{A} = 2$, and $\#\mathcal{A} = 3$.
$\#\mathcal{A} = 1$. In this case $\mathcal{A}$ is an MV-chain, and by [NL00], $\mathbb{L}$ has the AP.
$\#\mathcal{A} = 2$. We have that $\mathcal{A} = \mathcal{A}_0 \oplus \mathcal{A}_1$. By Proposition 2, $\mathbf{V}(\mathcal{A}) = \mathbf{V}(\mathcal{A}^s)$, and $\mathcal{A}^s = \mathcal{A}_0 \oplus \mathcal{A}_1^s$. An easy check shows that $\mathcal{A}^s$ satisfies 2) as well as the hypothesis of [AM03, Theorems 7.4,7.6]: by inspecting the proof of [AM03, Theorem 7.6] we have $Ch(\mathcal{A}) = Ch(\mathcal{A}^s) = \mathbf{ISP}_u(\mathcal{A}^s)$. Let $(\mathcal{B}, \mathcal{C}, \mathcal{D}, i, j)$ be a V-formation in $Ch(\mathcal{A})$. Since, by Lemma 1, every chain in $\mathbf{V}(\mathcal{A})$ has at most two components, without loss of generality we can assume that $\mathcal{B} = \mathcal{B}_0 \oplus \mathcal{B}_1$, $\mathcal{C} = \mathcal{C}_0 \oplus \mathcal{C}_1$, $\mathcal{D} = \mathcal{D}_0 \oplus \mathcal{D}_1$, possibly with the trivial algebra as second component. Because of the operations of an ordinal sum, i and j can only map the first (second) component of $\mathcal{B}$ into the first (second) component of, respectively, $\mathcal{C}$ and $\mathcal{D}$. By Theorem 5 and Lemma 6, for $n \in \{0, 1\}$ we can find an amalgam $(\mathcal{E}_n, h_n, k_n)$ of $(\mathcal{B}_n, \mathcal{C}_n, \mathcal{D}_n, i_{\restriction \mathcal{B}_n}, j_{\restriction \mathcal{B}_n})$, with $\mathcal{E}_0 \in Ch(\mathcal{A}_0) = \mathbf{ISP}_u(\mathcal{A}_0)$, and $\mathcal{E}_1 \in Ch(\mathcal{A}_1^s) = \mathbf{ISP}_u(\mathcal{A}_1^s)$. Pick now $(\mathcal{E}_0 \oplus \mathcal{E}_1, h, k)$, where:
- $h(x) = h_0(x)$ if $x \in \mathcal{C}_0$, and $h(x) = h_1(x)$ otherwise.
- $k(x) = k_0(x)$ if $x \in \mathcal{D}_0$, and $k(x) = k_1(x)$ otherwise.

From this and Lemma 6, a direct inspection shows that $(\mathcal{E}_0 \oplus \mathcal{E}_1, h, k)$ is an amalgam of $(\mathcal{B}, \mathcal{C}, \mathcal{D}, i, j)$, and $\mathcal{E}_0 \oplus \mathcal{E}_1 \in \mathbf{ISP}_u(\mathcal{A}^s) = Ch(\mathcal{A}^s) = Ch(\mathcal{A})$.
$\#\mathcal{A} = 3$. We have that $\mathcal{A} = \mathcal{A}_0 \oplus \mathcal{A}_1 \oplus \mathcal{A}_2$. Since 2) holds true, the only possibility is that exactly one among $\mathcal{A}_1$ and $\mathcal{A}_2$ is finite, and the other one is an infinite cancellative hoop. Let us assume that $\mathcal{A}_1$ is finite, and $\mathcal{A}_2$ is cancellative: we will omit the other case since the proof remains basically identical, mutatis mutandis. An easy check shows that $\mathcal{A}$ satisfies both 2) and the hypothesis of [AM03, Theorems 7.4,7.6]. By inspecting the proof of [AM03, Theorem 7.6] we have $Ch(\mathcal{A}) = \mathbf{ISP}_u(\mathcal{A}) = \mathbf{I}(\{\mathcal{R} \oplus \mathcal{S} \oplus \mathcal{T} : \mathcal{R} \in \mathbf{SP}_u(\mathcal{A}_0), \mathcal{S} \in \mathbf{SP}_u(\mathcal{A}_1), \mathcal{T} \in \mathbf{SP}_u(\mathcal{A}_2)\}$, where the last equality is due to Lemma 6. By Lemma 6, if $\mathcal{H}, \mathcal{K}$ are, respectively, a cancellative totally ordered hoop, and a finite totally ordered Wajsberg hoop, then $\mathbf{ISP}_u(\mathcal{H}) = \{\mathcal{R} : \mathcal{R} \in Ch(\mathbb{CH})\}$, and $\mathbf{ISP}_u(\mathcal{K}) = \mathbf{IS}(\mathcal{K})$. This means that for every $\mathcal{C} \in Ch(\mathcal{A})$, if $\mathcal{C} = \mathcal{C}_0 \oplus \mathcal{C}_1$ (*i.e.*, $\#\mathcal{C} = 2$), then either $\mathcal{C}_1 \in \mathbf{ISP}_u(\mathcal{A}_2)$ and hence it is a cancellative hoop or $\mathcal{C}_1 \in \mathbf{ISP}_u(\mathcal{A}_1) =$

$\mathbf{IS}(\mathcal{A}_1)$, *i.e.*, $\mathcal{C}_1$ is a finite Wajsberg hoop embeddable into $\mathcal{A}_1$. Of course, if $\mathcal{C} \in Ch(\mathcal{A})$ has three components, *i.e.*, $\mathcal{C} = \mathcal{C}_0 \oplus \mathcal{C}_1 \oplus \mathcal{C}_2$, then $\mathcal{C}_1$ is a cancellative hoop, whilst $\mathcal{C}_2$ is a finite Wajsberg hoop embeddable into $\mathcal{A}_1$. Let $(\mathcal{B}, \mathcal{C}, \mathcal{D}, i, j)$ be a V-formation in $Ch(\mathcal{A})$. If both $\mathcal{C}$ and $\mathcal{D}$ have at most two components, then we can find an amalgam by using the same argument used for the case $\#\mathcal{A} = 2$. Suppose then that either $\mathcal{C}$ or $\mathcal{D}$ has three components, w.l.o.g. $\#\mathcal{C} = 3$. We have three subcases.

$*$ $\#\mathcal{D} = 1$. This implies that $\#\mathcal{B} = 1$. By our previous observations we know that $\mathcal{B}_0, \mathcal{C}_0, \mathcal{D}_0 \in Ch(\mathcal{A}_0) = \mathbf{ISP}_u(\mathcal{A}_0)$. Whence by Theorem 5 we can find an amalgam $(\mathcal{E}_0, h, k)$ for the V-formation $(\mathcal{B}_0, \mathcal{C}_0, \mathcal{D}_0, i{\restriction_{\mathcal{B}_0}}, j{\restriction_{\mathcal{B}_0}})$, where $\mathcal{E}_0 \in Ch(\mathcal{A}_0) = \mathbf{ISP}_u(\mathcal{A}_0)$. Consider now $(\mathcal{E}_0 \oplus \mathcal{C}_1 \oplus \mathcal{C}_2, r, s)$, where $r : \mathcal{C} \to \mathcal{E}_0 \oplus \mathcal{C}_1 \oplus \mathcal{C}_2$, and $s : \mathcal{D} \to \mathcal{E}_0 \oplus \mathcal{C}_1 \oplus \mathcal{C}_2$ are maps defined as follows.

- $r(i) = h(i)$, for $i \in C_0$, and $r(i) = i$ if $i \in C_1 \cup C_2$.
- $s(i) = k(i)$, for $i \in D_0$, and $s(i) = i$ if $i \in D_1 \cup D_2$.

A direct inspection shows that $(\mathcal{E}_0 \oplus \mathcal{C}_1 \oplus \mathcal{C}_2, r, s)$ is an amalgam for $(\mathcal{B}, \mathcal{C}, \mathcal{D}, i, j)$, and $\mathcal{E}_0 \oplus \mathcal{C}_1 \oplus \mathcal{C}_2 \in \mathbf{ISP}_u(\mathcal{A}) = Ch(\mathcal{A})$.

$*$ $\#\mathcal{D} = 2$. This implies that $1 \leq \#\mathcal{B} \leq 2$. We can assume that $\#\mathcal{B} = 2$, *i.e.*, $\mathcal{B} = \mathcal{B}_0 \oplus \mathcal{B}_1$, as the proof for the case $\#\mathcal{B} = 1$ can be obtained by replacing $\mathcal{B}_1$ with the trivial Wajsberg hoop. As in the $\#\mathcal{D} = 1$ case, we can find an amalgam $(\mathcal{E}_0, h, k)$ for the V-formation $(\mathcal{B}_0, \mathcal{C}_0, \mathcal{D}_0, i{\restriction_{\mathcal{B}_0}}, j{\restriction_{\mathcal{B}_0}})$, where $\mathcal{E}_0 \in Ch(\mathcal{A}_0) = \mathbf{ISP}_u(\mathcal{A}_0)$. By hypothesis $\mathcal{D} = \mathcal{D}_0 \oplus \mathcal{D}_1$, and from the previous parts of the proof we know that $\mathcal{D}_1$ is either an infinite cancellative hoop or a finite Wajsberg hoop such that $\mathcal{D}_1 \in \mathbf{IS}(\mathcal{A}_1)$. We now analyze these two subcases.

- $\mathcal{D}_1$ is cancellative. Then both $\mathcal{C}_2$ and $\mathcal{D}_1$ belongs to $\mathbf{ISP}_u(\mathcal{A}_2)$. Since we assumed $\#\mathcal{B} = 2$, then $\mathcal{B}_1 \in \mathbf{ISP}_u(\mathcal{A}_2)$ (*i.e.*, $\mathcal{B}_1$ is an infinite cancellative hoop). Whence $i(\mathcal{B}_1) \subseteq \mathcal{C}_2$, and $j(\mathcal{B}_1) \subseteq \mathcal{D}_1$. Consider the V-formation $(\mathcal{B}_1, \mathcal{C}_2, \mathcal{D}_1, i{\restriction_{\mathcal{B}_1}}, j{\restriction_{\mathcal{B}_1}})$. By Theorem 5 we can find an amalgam $(\mathcal{E}_1, v, w)$, where $\mathcal{E}_1 \in Ch(\mathbb{WH}) = \mathbf{ISP}_u(\mathcal{A}_2)$. Consider now $(\mathcal{E}_0 \oplus \mathcal{C}_1 \oplus \mathcal{E}_1, r, s)$, where $r : \mathcal{C} \to \mathcal{E}_0 \oplus \mathcal{C}_1 \oplus \mathcal{E}_1$, and $s : \mathcal{D} \to \mathcal{E}_0 \oplus \mathcal{C}_1 \oplus \mathcal{E}_1$ are maps defined as follows.

$$r(i) = \begin{cases} h(i) & \text{if } i \in C_0, \\ i & \text{if } i \in C_1 \\ v(i) & \text{if } i \in C_2 \end{cases} \qquad s(i) = \begin{cases} k(i) & \text{if } i \in D_0, \\ w(i) & \text{if } i \in D_1 \end{cases}$$

  A direct inspection shows that $(\mathcal{E}_0 \oplus \mathcal{C}_1 \oplus \mathcal{E}_1, r, s)$ is an amalgam for $(\mathcal{B}, \mathcal{C}, \mathcal{D}, i, j)$, and $\mathcal{E}_0 \oplus \mathcal{C}_1 \oplus \mathcal{E}_1 \in \mathbf{ISP}_u(\mathcal{A}) = Ch(\mathcal{A})$.
- $\mathcal{D}_1$ is finite. Then both $\mathcal{C}_1$ and $\mathcal{D}_1$ belongs to $\mathbf{ISP}_u(\mathcal{A}_1) = \mathbf{IS}(\mathcal{A}_1)$. Since we assumed $\#\mathcal{B} = 2$, then $\mathcal{B}_1 \in \mathbf{IS}(\mathcal{A}_1)$ (*i.e.*, $\mathcal{B}_1$ is a finite Wajsberg hoop). Whence $i(\mathcal{B}_1) \subseteq \mathcal{C}_1$, and $j(\mathcal{B}_1) \subseteq \mathcal{D}_1$. Consider the V-formation $(\mathcal{B}_1, \mathcal{C}_1, \mathcal{D}_1, i{\restriction_{\mathcal{B}_1}}, j{\restriction_{\mathcal{B}_1}})$. By Theorem 5 we can find an amalgam $(\mathcal{E}_1, v, w)$, where $\mathcal{E}_1 \in Ch(\mathcal{A}_1) = \mathbf{IS}(\mathcal{A}_1)$.

Consider now $(\mathcal{E}_0 \oplus \mathcal{E}_1 \oplus \mathcal{C}_2, r, s)$, where $r : \mathcal{C} \to \mathcal{E}_0 \oplus \mathcal{E}_1 \oplus \mathcal{C}_2$, and $s : \mathcal{D} \to \mathcal{E}_0 \oplus \mathcal{E}_1 \oplus \mathcal{C}_2$ are maps defined as follows.

$$r(i) = \begin{cases} h(i) & \text{if } i \in C_0, \\ v(i) & \text{if } i \in C_1 \\ i & \text{if } i \in C_2 \end{cases} \qquad s(i) = \begin{cases} k(i) & \text{if } i \in D_0, \\ w(i) & \text{if } i \in D_1 \end{cases}$$

it is easy to check that $(\mathcal{E}_0 \oplus \mathcal{E}_1 \oplus \mathcal{C}_2, r, s)$ is an amalgam for $(\mathcal{B}, \mathcal{C}, \mathcal{D}, i, j)$, and $\mathcal{E}_0 \oplus \mathcal{E}_1 \oplus \mathcal{C}_2 \in \mathbf{ISP}_u(\mathcal{A}) = Ch(\mathcal{A})$.

$*$ $\#\mathcal{D} = 3$. By Theorem 5 we can find an amalgam $(\mathcal{E}_0, h, k)$ for the V-formation $(\mathcal{B}_0, \mathcal{C}_0, \mathcal{D}_0, i_{\restriction \mathcal{B}_0}, j_{\restriction \mathcal{B}_0})$, where $\mathcal{E}_0 \in Ch(\mathcal{A}_0) = \mathbf{ISP}_u(\mathcal{A}_0)$. We distinguish the cases $\#\mathcal{B} = 2$, and $\#\mathcal{B} = 3$. The case $\#\mathcal{B} = 1$ can be omitted, since it is a subcase of $\#\mathcal{B} = 2$, when the second component of $\mathcal{B}$ is the trivial Wajsberg hoop.

· $\#\mathcal{B} = 2$. We have that $\mathcal{B}_1$ is either cancellative or finite.
If $\mathcal{B}_1$ is cancellative, then $\mathcal{B}_1, \mathcal{C}_2, \mathcal{D}_2 \in \mathbf{ISP}_u(\mathcal{A}_2) = Ch(\mathcal{A}_2)$, which means that they are all cancellative hoops. Since $\mathcal{C}_1$ and $\mathcal{D}_1$ are finite Wajsberg hoops, then $i(\mathcal{B}_1) \subseteq \mathcal{C}_2$ and $j(\mathcal{B}_1) \subseteq \mathcal{D}_2$.
By Theorem 5 we can find an amalgam $(\mathcal{E}_2, v, w)$, with $\mathcal{E}_2 \in \mathbf{ISP}_u(\mathcal{A}_2) = Ch(\mathcal{A}_2)$ for the V-formation $(\mathcal{B}_1, \mathcal{C}_2, \mathcal{D}_2, i_{\restriction \mathcal{B}_1}, j_{\restriction \mathcal{B}_1})$.
Moreover, by hypothesis $\mathcal{C}_1, \mathcal{D}_1 \in \mathbf{IS}(\mathcal{A}_1)$, which means that there are two embeddings l, m such that $\mathcal{C}_1 \overset{l}{\hookrightarrow} \mathcal{A}_1$ and $\mathcal{D}_1 \overset{m}{\hookrightarrow} \mathcal{A}_1$.
Consider now $(\mathcal{E}_0 \oplus \mathcal{A}_1 \oplus \mathcal{E}_2, r, s)$, where $r : \mathcal{C} \to \mathcal{E}_0 \oplus \mathcal{A}_1 \oplus \mathcal{E}_2$, and $s : \mathcal{D} \to \mathcal{E}_0 \oplus \mathcal{A}_1 \oplus \mathcal{E}_2$ are maps defined as follows.

$$r(i) = \begin{cases} h(i) & \text{if } i \in C_0, \\ l(i) & \text{if } i \in C_1 \\ v(i) & \text{if } i \in C_2 \end{cases} \qquad s(i) = \begin{cases} k(i) & \text{if } i \in D_0, \\ m(i) & \text{if } i \in D_1, \\ w(i) & \text{if } i \in C_2 \end{cases}$$

A direct inspection shows that $(\mathcal{E}_0 \oplus \mathcal{A}_1 \oplus \mathcal{E}_2, r, s)$ is an amalgam for $(\mathcal{B}, \mathcal{C}, \mathcal{D}, i, j)$, and $\mathcal{E}_0 \oplus \mathcal{A}_1 \oplus \mathcal{E}_2 \in \mathbf{ISP}_u(\mathcal{A}) = Ch(\mathcal{A})$.
If $\mathcal{B}_1$ is finite, then $\mathcal{B}_1, \mathcal{C}_1, \mathcal{D}_1 \in \mathbf{ISP}_u(\mathcal{A}_1) = Ch(\mathcal{A}_1)$, which means that they are all finite Wajsberg hoops. Since $\mathcal{C}_2$ and $\mathcal{D}_2$ are cancellative hoops, then $i(\mathcal{B}_1) \subseteq \mathcal{C}_1$ and $j(\mathcal{B}_1) \subseteq \mathcal{D}_1$.
By Theorem 5 we can find an amalgam $(\mathcal{E}_1, l, m)$, with $\mathcal{E}_1 \in \mathbf{IS}(\mathcal{A}_1) = Ch(\mathcal{A}_1)$ for the V-formation $(\mathcal{B}_1, \mathcal{C}_1, \mathcal{D}_1, i_{\restriction \mathcal{B}_1}, j_{\restriction \mathcal{B}_1})$.
By Theorem 5 we can find an amalgam $(\mathcal{E}_2, v, w)$, with $\mathcal{E}_2 \in \mathbf{ISP}_u(\mathcal{A}_2) = Ch(\mathcal{A}_2)$ for the V-formation $(\mathcal{T}_H, \mathcal{C}_2, \mathcal{D}_2, i_{\restriction \mathcal{T}_H}, j_{\restriction \mathcal{T}_H})$, where $\mathcal{T}_H$ is the trivial Wajsberg hoop.
Consider now $(\mathcal{E}_0 \oplus \mathcal{A}_1 \oplus \mathcal{E}_2, r, s)$, where $r : \mathcal{C} \to \mathcal{E}_0 \oplus \mathcal{E}_1 \oplus \mathcal{E}_2$, and $s : \mathcal{D} \to \mathcal{E}_0 \oplus \mathcal{A}_1 \oplus \mathcal{E}_2$ are maps defined as follows.

$$r(i) = \begin{cases} h(i) & \text{if } i \in C_0, \\ l(i) & \text{if } i \in C_1 \\ v(i) & \text{if } i \in C_2 \end{cases} \qquad s(i) = \begin{cases} k(i) & \text{if } i \in D_0, \\ m(i) & \text{if } i \in D_1, \\ w(i) & \text{if } i \in C_2 \end{cases}$$

A direct inspection shows that $(\mathcal{E}_0 \oplus \mathcal{E}_1 \oplus \mathcal{E}_2, r, s)$ is an amalgam for $(\mathcal{B}, \mathcal{C}, \mathcal{D}, i, j)$, and $\mathcal{E}_0 \oplus \mathcal{E}_1 \oplus \mathcal{E}_2 \in \mathbf{ISP}_u(\mathcal{A}) = Ch(\mathcal{A})$.

- $\#\mathcal{B} = 3$. In this case $\mathcal{B}_1, \mathcal{C}_1, \mathcal{D}_1 \in \mathbf{IS}(\mathcal{A}_1) = Ch(\mathcal{A}_1)$, and hence they are all finite Wajsberg hoops, whilst $\mathcal{B}_2, \mathcal{C}_2, \mathcal{D}_2 \in \mathbf{ISP}_u(\mathcal{A}_2) = Ch(\mathcal{A}_2)$, which means that they are all cancellative hoops. In particular $i(\mathcal{B}_1) \subseteq \mathcal{C}_1$, $i(\mathcal{B}_2) \subseteq \mathcal{C}_2$, $j(\mathcal{B}_1) \subseteq \mathcal{D}_1$, and $i(\mathcal{B}_2) \subseteq \mathcal{D}_2$; moreover $i(\mathcal{B}_0) \subseteq \mathcal{C}_0$, and $j(\mathcal{B}_0) \subseteq \mathcal{D}_0$. By Theorem 5 we can find an amalgam $(\mathcal{E}_1, l, m)$, with $\mathcal{E}_1 \in \mathbf{IS}(\mathcal{A}_1) = Ch(\mathcal{A}_1)$, for the V-formation $(\mathcal{B}_1, \mathcal{C}_1, \mathcal{D}_1, i_{\restriction \mathcal{B}_1}, j_{\restriction \mathcal{B}_1})$. Using again Theorem 5 we can also find an amalgam $(\mathcal{E}_2, v, w)$, with $\mathcal{E}_2 \in \mathbf{ISP}_u(\mathcal{A}_2) = Ch(\mathcal{A}_2)$ for the V-formation $(\mathcal{B}_2, \mathcal{C}_2, \mathcal{D}_2, i_{\restriction \mathcal{B}_2}, j_{\restriction \mathcal{B}_2})$.
  Consider now $(\mathcal{E}_0 \oplus \mathcal{E}_1 \oplus \mathcal{E}_2, r, s)$, where $r : \mathcal{C} \to \mathcal{E}_0 \oplus \mathcal{E}_1 \oplus \mathcal{E}_2$, and $s : \mathcal{D} \to \mathcal{E}_0 \oplus \mathcal{E}_1 \oplus \mathcal{E}_2$ are maps defined as follows.

$$r(i) = \begin{cases} h(i) & \text{if } i \in C_0, \\ l(i) & \text{if } i \in C_1 \\ v(i) & \text{if } i \in C_2 \end{cases} \qquad s(i) = \begin{cases} k(i) & \text{if } i \in D_0, \\ m(i) & \text{if } i \in D_1, \\ w(i) & \text{if } i \in C_2 \end{cases}$$

  A direct inspection shows that $(\mathcal{E}_0 \oplus \mathcal{E}_1 \oplus \mathcal{E}_2, r, s)$ is an amalgam for $(\mathcal{B}, \mathcal{C}, \mathcal{D}, i, j)$, and $\mathcal{E}_0 \oplus \mathcal{E}_1 \oplus \mathcal{E}_2 \in \mathbf{ISP}_u(\mathcal{A}) = Ch(\mathcal{A})$.

The proof is settled. □

4 Discussion and Open Problems

In this paper we studied the AP for the varieties of BL-algebras generated by one BL-chain with finitely many components. Future works will be devoted to generalize these results to a larger family of varieties. A first step could be the study of the AP for the varieties of BL-algebras generated by a finite set S of BL-chains with finitely many components. As shown in [AB21b] these varieties coincide with the small varieties of BL-algebras, *i.e.*, the varieties of BL-algebras whose lattice of subvarieties is finite. We have some partial results in this direction, but the main issue is that for many of these varieties the AP for the class of chains fails to holds, making the analysis of the AP for the whole variety harder. One may argue that the requirement on the finiteness of S is quite strong, but as the following result shows, one should be careful when removing this restriction.

Theorem 7. *Every variety of BL-algebras is generated by some set of BL-chains with finitely many components.*

Proof. As shown in [AB21b] every variety of BL-algebras $\mathbb{L}$ is equal to the join of a family of strictly join irreducible varieties, let us say $\mathbb{L} = \bigvee_{i \in I} \mathbb{L}_i$. Further it is shown in [AB21b] that every $\mathbb{L}_i$ is generated by one BL-chain with finitely many components, say $\mathcal{A}_i$. As a consequence $\mathbb{L} = \mathbf{V}(S)$, where $S = \{\mathcal{A}_i : i \in I\}$, and clearly S is a set of BL-chains with finitely many components. □

So, by removing the restriction on the finiteness of S we get the whole lattice of varieties of BL-algebras $\mathcal{L}(\mathbb{BL})$, which is uncountable and poorly known.

References

[AB19] Aguzzoli, S., Bianchi, M.: On linear varieties of MTL-algebras. Soft Comput. **23**(7), 2129–2146 (2018). https://doi.org/10.1007/s00500-018-3423-3

[AB21a] 2 Aguzzoli, S., Bianchi, M.: Finite model property and varieties of BL-algebras. In: Computational Intelligence and Mathematics for Tackling Complex Problems 3, Studies in Computational Intelligence, vol. 959, pp. 22–30. Springer, Heidelberg (2022). https://doi.org/10.1007/978-3-030-74970-5_4

[AB21b] Aguzzoli, S., Bianchi, M.: Strictly join irreducible varieties of BL-algebras: the missing pieces. Fuzzy Sets Syst. **418**, 84–100 (2021). https://doi.org/10.1016/j.fss.2020.12.008

[AFM07] Aglianò, P., Ferreirim, I., Montagna, F.: Basic hoops: an algebraic study of continuous *t*-norms. Studia Logica **87**, 73–98 (2007). https://doi.org/10.1007/s11225-007-9078-1

[AM03] Aglianò, P., Montagna, F.: Varieties of BL-algebras I: general properties. J. Pure Appl. Algebra **181**(2–3), 105–129 (2003). https://doi.org/10.1016/S0022-4049(02)00329-8

[AP02] Aglianò, P., Panti, G.: Geometrical methods in Wajsberg hoops. J. Algebra **256**(2), 352–374 (2002). https://doi.org/10.1016/S0021-8693(02)00085-6

[BS81] Burris, S., Sankappanavar, H.P.: A Course in Universal Algebra, vol. 78, Springer, Heidelberg (1981). http://tinyurl.com/zaxeopo

[CDM99] Cignoli, R., D'Ottaviano, I., Mundici, D.: Algebraic Foundations of Many-Valued Reasoning. Trends in Logic, vol. 7. Kluwer Academic Publishers, Dordrecht (1999)

[CEGT00] Cignoli, R., Esteva, F., Godo, L., Torrens, A.: Basic Fuzzy Logic is the logic of continuous t-norms and their residua. Soft Comput. **4**(2), 106–112 (2000). https://doi.org/10.1007/s005000000044

[CHN11] Cintula, P., Hájek, P., Noguera, C.: Handbook of Mathematical Fuzzy Logic, vol. 1 and 2. College Publications, London (2011)

[CMM11] Cortonesi, T., Marchioni, E., Montagna, F.: Quantifier elimination and other model-theoretic properties of BL-algebras. Notre Dame J. Formal Logic **52**, 339–379 (2011). https://doi.org/10.1215/00294527-1499336

[EGHM03] Esteva, F., Godo, L., Hájek, P., Montagna, F.: Hoops and fuzzy logic. J. Log. Comput. **13**(4), 532–555 (2003). https://doi.org/10.1093/logcom/13.4.532

[Háj98] Hájek, P.: Metamathematics of Fuzzy Logic. Trends in Logic, vol. 4. Kluwer Academic Publishers, Dordrecht (1998)

[Jen03] Jenei, S.: On the structure of rotation-invariant semigroups. Arch. Math. Log. **42**, 489–514 (2003). https://doi.org/10.1007/s00153-002-0165-8

[MMT14] Metcalfe, G., Montagna, F., Tsinakis, C.: Amalgamation and interpolation in ordered algebras. J. Alg. **402**, 21–82 (2014). https://doi.org/10.1016/j.jalgebra.2013.11.019

[Mon06] Montagna, F.: Interpolation and Beth's property in propositional many-valued logics: a semantic investigation. Ann. Pure. Appl. Log. **141**(1–2), 148–179 (2006). https://doi.org/10.1016/j.apal.2005.11.001

[Mun88] Mundici, D.: Free products in the category of abelian l-groups with strong unit. J. Algebra **113**(1), 89–109 (1988). https://doi.org/10.1016/0021-8693(88)90185-8

[NEG05] Noguera, C., Esteva, F., Gispert, J.: Perfect and bipartite IMTL-algebras and disconnected rotations of prelinear semihoops. Arch. Math. Logic **44**(7), 869–886 (2005). https://doi.org/10.1007/s00153-005-0276-0

[NL00] Di Nola, A., Lettieri, A.: One chain generated varieties of MV-Algebras. J. Alg. **225**(2), 667–697 (2000). https://doi.org/10.1006/jabr.1999.8136

Unary-Determined Distributive ℓ-magmas and Bunched Implication Algebras

Natanael Alpay, Peter Jipsen(✉), and Melissa Sugimoto

Chapman University, Orange, CA, USA
jipsen@chapman.edu

Abstract. A distributive lattice-ordered magma ($d\ell$-magma) $(A, \wedge, \vee, \cdot)$ is a distributive lattice with a binary operation $\cdot$ that preserves joins in both arguments, and when $\cdot$ is associative then $(A, \vee, \cdot)$ is an idempotent semiring. A $d\ell$-magma with a top $\top$ is *unary-determined* if $x{\cdot}y = (x{\cdot}\top \wedge y) \vee (x \wedge \top{\cdot}y)$. These algebras are term-equivalent to a subvariety of distributive lattices with $\top$ and two join-preserving unary operations p, q. We obtain simple conditions on p, q such that $x{\cdot}y = (px \wedge y) \vee (x \wedge qy)$ is associative, commutative, idempotent and/or has an identity element. This generalizes previous results on the structure of doubly idempotent semirings and, in the case when the distributive lattice is a Heyting algebra, it provides structural insight into unary-determined algebraic models of bunched implication logic. We also provide Kripke semantics for the algebras under consideration, which leads to more efficient algorithms for constructing finite models.

Keywords: Distributive lattice-ordered magmas · Bunched implication algebras · Idempotent semirings · Enumerating finite models

1 Introduction

Idempotent semirings $(A, \vee, \cdot)$ play an important role in several areas of computer science, such as network optimization, formal languages, Kleene algebras and program semantics. In this setting they are often assumed to have constants $0, 1$ that are the additive and multiplicative identity respectively, with 0 also being an absorbing element. However semirings are usually only assumed to have two binary operations $+, \cdot$ that are associative such that $+$ is also commutative and $\cdot$ distributes over $+$ from the left and right [9]. They are (additively) idempotent if $x + x = x$, hence $+$ is a (join) semilattice, and *doubly idempotent* if $x \cdot x = x$ as well. If $\cdot$ is also commutative then it defines a meet semilattice. The special case when these two semilattices coincide corresponds exactly to the variety of distributive lattices, which have a well understood structure theory.

In [1] a complete structural description was given for finite commutative doubly idempotent semirings where either the multiplicative semilattice is a chain, or the additive semilattice is a Boolean algebra. Here we show that the second

U. Fahrenberg et al. (Eds.): RAMiCS 2021, LNCS 13027, pp. 19–36, 2021.
https://doi.org/10.1007/978-3-030-88701-8_2

description can be significantly generalized to the setting where the additive semilattice is a distributive lattice, dropping the assumptions of finiteness, multiplicative commutativity and idempotence in favor of the algebraic condition $x{\cdot}y = (px \wedge y) \vee (x \wedge qy)$ for two unary join-preserving operations p, q. While this property is quite restrictive in general, it does hold in all idempotent Boolean magmas and expresses a binary operation in terms of two simpler unary operations. A full structural description of all (finite) idempotent semirings is unlikely, but in the setting of unary-determined idempotent semirings progress is possible.

In Sect. 2 we provide the needed background and prove a term-equivalence between a subvariety of top-bounded $d\ell$-magmas and a subvariety of top-bounded distributive lattices with two unary operators. This is then specialized to cases where $\cdot$ is associative, commutative, idempotent or has an identity element. In the next section we show that when the distributive lattice is a Brouwerian algebra or Heyting algebra, then $\cdot$ is residuated if and only if both p and q are residuated. This establishes a connection with bunched implication algebras (BI-algebras) that are the algebraic semantics of bunched implication logic [14], used in the setting of separation logic for program verification, including reasoning about pointers [16] and concurrent processes [13]. Section 4 contains Kripke semantics for $d\ell$-magmas, called Birkhoff frames, and for the two unary operators p, q. This establishes the connection to the previous results in [1] and leads to the main result (Theorem 15) that preorder forest P-frames capture a larger class of multiplicatively idempotent BI-algebras and doubly idempotent semirings. Although the heap models of BI-algebras used in applications are not (multiplicatively) idempotent, they contain idempotent subalgebras and homomorphic images, hence a characterization of unary-determined idempotent BI-algebras does provide insight into the general case. In Sect. 5, as an application, we count the number of such algebras up to isomorphism if their partial order is an antichain and also if it is a chain.

2 A Term-Equivalence Between Distributive Lattices with Operators

A *distributive lattice-ordered magma*, or *$d\ell$-magma*, is an algebra $(A, \wedge, \vee, \cdot)$ such that $(A, \wedge, \vee)$ is a distributive lattice and $\cdot$ distributes over $\vee$, i. e., $x(y \vee z) = xy \vee xz$ and $(x \vee y)z = xz \vee yz$ for all $x, y, z \in A$. If the distributive lattice has a top element $\top$ or a bottom element $\bot$ then it is called *$\top$-bounded* or *$\bot$-bounded*, or simply *bounded* if both exist. A $d\ell$-magma A is *normal* and $\cdot$ is a *normal* operation if A is $\bot$-bounded and satisfies $x{\cdot}\bot = \bot = \bot{\cdot}x$. Similarly, a unary operation f on A is an *operator* if it satisfies $f(x \vee y) = fx \vee fy$, and it is *normal* if $f\bot = \bot$. For brevity and to reduce the number of nested parentheses, we write function application as fx rather than $f(x)$, with the convention that it has priority over $\cdot$ hence, e.g., $fxy = (f(x)){\cdot}y$ (this convention ensures unique readability). Note that since operators distribute over $\vee$ in each argument, they are order-preserving in each argument. The operation f is said to be *inflationary* if $x \leq fx$ for all $x \in A$.

A binary operation $\cdot$ is said to be *idempotent* if $xx = x$ for all $x \in A$, *commutative* if $xy = yx$ and *associative* if $(xy)z = x(yz)$. A *semigroup* is a set with an associative operation, a *band* is a semigroup that is also idempotent, and a *semilattice* is a commutative band. As usual, a semilattice is partially ordered by $x \sqsubseteq y \iff xy = x$, and in this case xy is the meet operation with respect to $\sqsubseteq$. We also use this terminology with the prefix *dℓ*, in which case the magma operation satisfies the corresponding identities.

A *dℓ*-magma is called *unary-determined* if it is $\top$-bounded and satisfies the identity

$$x \cdot y = (x \cdot \top \wedge y) \vee (x \wedge \top \cdot y).$$

As examples, we mention that all doubly-idempotent semirings with a Boolean join-semilattice are unary-determined (see Lemma 3). Complete and atomic versions of such semirings are studied in [1], and the results from that paper are generalized here to unary-determined *dℓ*-magmas with point-free algebraic proofs. This is an improvement since the algebraic results apply to all members of the variety, while the previous results applied only to complete and atomic algebras.

A *dℓpq-algebra* is a $\top$-bounded distributive lattice with two unary operators p, q that satisfy

$$x \wedge p\top \leq qx, \quad x \wedge q\top \leq px.$$

These two equational axioms are needed for our first result which shows that unary-determined *dℓ*-magmas and *dℓpq*-algebras are term-equivalent. This means that although the two varieties are based on different sets of fundamental operations (called the signature of each class), each fundamental operation of an algebra in one variety is identical to a term-operation constructed from fundamental operations of an algebra in the other variety (and vice versa). From the point of view of category theory, term-equivalent varieties are model categories of the same Lawvere theory.

Although unary-determined *dℓ*-magmas and *dℓpq*-algebras seem rather special, they are simpler than general *dℓ*-magmas, yet include interesting idempotent semirings (as reducts).

Theorem 1. (1) *Let $(A, \wedge, \vee, \top, p, q)$ be a dℓpq-algebra and define $x \cdot y = (px \wedge y) \vee (x \wedge qy)$. Then $(A, \wedge, \vee, \top, \cdot)$ is a unary-determined dℓ-magma and p, q are given by $px = x \cdot \top$ and $qx = \top \cdot x$.*

(2) *Let $(A, \wedge, \vee, \top, \cdot)$ be a unary-determined dℓ-magma and define $px = x \cdot \top$, $qx = \top \cdot x$. Then $(A, \wedge, \vee, \top, p, q)$ is a dℓpq-algebra and $\cdot$ is definable from p, q via $x \cdot y = (px \wedge y) \vee (x \wedge qy)$.*

Proof. (1) Assume p, q are unary operators on a $\top$-bounded distributive lattice $(A, \wedge, \vee, \top)$, and $xy = (px \wedge y) \vee (x \wedge qy)$. Then

$$\begin{aligned} x(y \vee z) &= (px \wedge (y \vee z)) \vee (x \wedge q(y \vee z)) \\ &= (px \wedge y) \vee (px \wedge z) \vee (x \wedge qy) \vee (x \wedge qz) \\ &= (px \wedge y) \vee (x \wedge qy) \vee (px \wedge z) \vee (x \wedge qz) \\ &= xy \vee xz. \end{aligned}$$

A similar calculation shows that $(x \vee y)z = xz \vee yz$, hence $\cdot$ is an operator.

Since p, q satisfy $x \wedge q{\cdot}\top \leq px$, it follows that $x{\cdot}\top = (px \wedge \top) \vee (x \wedge q{\cdot}\top) = px \vee (x \wedge q\top) = px$, and similarly $\top{\cdot}x = qx$ is implied by $x \wedge p\top \leq qx$. Now the identity $xy = (x{\cdot}\top \wedge y) \vee (x \wedge \top{\cdot}y)$ holds by definition.

(2) Assume $(A, \wedge, \vee, \top, \cdot)$ is a unary-determined $d\ell$-magma, and define $px = x{\cdot}\top$, $qx = \top{\cdot}y$. Then p, q are unary operators and $px = x{\cdot}\top = (x{\cdot}\top \wedge \top) \vee (x \wedge \top\top) = px \vee (x \wedge q\top)$, hence $x \wedge q\top \leq px$. The inequation $x \wedge p\top \leq qx$ is proved similarly. The operation $\cdot$ can be recovered from p, q since $xy = (px \wedge y) \vee (x \wedge qy)$ follows from the identity we assumed. □

The preceding theorem shows that unary-determined $d\ell$-magmas and $d\ell pq$-algebras are "essentially the same", and we can choose to work with the signature that is preferred in a given situation. The unary operators of $d\ell pq$-algebras are simpler to handle, while the binary operator $\cdot$ is familiar in the semiring setting. Next we examine how standard properties of $\cdot$ are captured by identities in the language of $d\ell pq$-algebras.

Lemma 2. *Let $(A, \wedge, \vee, \top, p, q)$ be a $d\ell pq$-algebra and define $x{\cdot}y = (px \wedge y) \vee (x \wedge qy)$.*

(1) *The operator $\cdot$ is commutative if and only if $p = q$.*
(2) *If $p = q$ then $\cdot$ is associative if and only if $p((px \wedge y) \vee (x \wedge py)) = (px \wedge py) \vee (x \wedge ppy)$.*
(3) *The operator $\cdot$ is idempotent if and only if p and q are inflationary, if and only if $p\top = \top = q\top$.*
(4) *If $\cdot$ is idempotent then it is associative if and only if*

$$p((px \wedge y) \vee (x \wedge qy)) = (px \wedge py) \vee (x \wedge qy) \text{ and}$$
$$q((px \wedge y) \vee (x \wedge qy)) = (px \wedge y) \vee (qx \wedge qy).$$

(5) *The operator $\cdot$ has an identity 1 if and only if $p1{=}\top{=}q1$ and $(px \vee qx) \wedge 1 \leq x$.*
(6) *If $\cdot$ has an identity then $\cdot$ is idempotent.*

Proof. (1) Assuming $xy = yx$, we clearly have $x{\cdot}\top = \top{\cdot}x$, hence $px = qx$. The converse makes use of commutativity of $\wedge$ and $\vee$: $xy = (px \wedge y) \vee (x \wedge py) = (py \wedge x) \vee (y \wedge px) = yx$.

(2) Assume $p = q$. If $\cdot$ is associative then $(xy)\top = x(y\top)$, so by the previous theorem, $p(xy) = xpy$, which translates to

$$p((px \wedge y) \vee (x \wedge py)) = (px \wedge py) \vee (x \wedge ppy) \quad (*).$$

Conversely, suppose $(*)$ holds, and note that $p(xy) = p(yx)$ by (1), hence

$$p((px \wedge y) \vee (x \wedge py)) = (px \wedge py) \vee (ppx \wedge y) = (px \wedge py) \vee (x \wedge ppy) \vee (ppx \wedge y) \quad (**).$$

It suffices to prove $(xy)z \leq x(yz)$ since then $z(yx) \leq (zy)x$ follows by commutativity. Now

$$\begin{aligned}(xy)z &= [p((px \wedge y) \vee (x \wedge py)) \wedge z] \vee [((px \wedge y) \vee (x \wedge py)) \wedge pz] \\ &= [((px \wedge py) \vee (x \wedge ppy)) \wedge z] \vee [px \wedge y \wedge pz] \vee [x \wedge py \wedge pz] \text{ using } (*) \\ &= [px \wedge py \wedge z] \vee [x \wedge ppy \wedge z] \vee [px \wedge y \wedge pz] \vee [x \wedge py \wedge pz] \\ &\leq [px \wedge py \wedge z] \vee [px \wedge y \wedge pz] \vee [x \wedge py \wedge pz] \vee [x \wedge y \wedge ppz] \vee [x \wedge ppy \wedge z] \\ &= [px \wedge py \wedge z] \vee [px \wedge y \wedge pz] \vee [x \wedge ((py \wedge pz) \vee (y \wedge ppz) \vee (ppy \wedge z))] \\ &= [px \wedge ((py \wedge z) \vee (y \wedge pz))] \vee [x \wedge p((py \wedge z) \vee (y \wedge pz))] \text{ using } (**) \\ &= x(yz).\end{aligned}$$

(3) If $\cdot$ is idempotent, then $x = xx \leq x{\cdot}\top = px$ and $x \leq \top{\cdot}x = qx$. Conversely, if p, q are inflationary then $xx = (px \wedge x) \vee (x \wedge qy) = x \vee x = x$, hence $\cdot$ is idempotent. For the second equivalence, if $p\top = \top = q\top$ then p, q are inflationary since they satisfy $x \wedge p\top \leq qx, x \wedge q\top \leq px$. The reverse implication holds because $x \leq px, qx$ implies $\top \leq p\top, q\top$.

(4) Assume $\cdot$ is idempotent and associative. Then $(\top{\cdot}x)\top = \top(x{\cdot}\top)$, hence $qpx = pqx$. Furthermore, $pqx = \top{\cdot}x{\cdot}\top = \top xx\top = (qx)(px) = (pqx \wedge px) \vee (qx \wedge qpx)$. By (3) p, q are inflationary, so $px \leq pqx$ and $qx \leq qpx$. Therefore $pqx = px \vee qx$. Now we translate $(xy)\top = x(y\top)$ to obtain $p(xy) = x(py)$, hence

$$\begin{aligned}&p((px \wedge y) \vee (x \wedge qy)) = (px \wedge py) \vee (x \wedge qpy) = (px \wedge py) \vee (x \wedge (py \vee qy)) \\ &= (px \wedge py) \vee (x \wedge py) \vee (x \wedge qy) = (px \wedge py) \vee (x \wedge qy) \text{ since } x \leq px \text{ by (3).}\end{aligned}$$

The identity $q((px \wedge y) \vee (x \wedge qy)) = (px \wedge y) \vee (qx \wedge qy)$ has a similar proof. Conversely, assume the two identities hold. Then using distributivity

$$\begin{aligned}(xy)z &= [p((px \wedge y) \vee (x \wedge qy)) \wedge z] \vee [((px \wedge y) \vee (x \wedge qy)) \wedge qz] \\ &= [px \wedge py \wedge z] \vee [x \wedge qy \wedge z] \vee [px \wedge y \wedge qz] \vee [x \wedge qy \wedge qz] \\ &= [px \wedge py \wedge z] \vee [px \wedge y \wedge qz] \vee [x \wedge qy \wedge qz] \text{ since } x \wedge qy \wedge z \leq x \wedge qy \wedge qz \\ &= [px \wedge py \wedge z] \vee [px \wedge y \wedge qz] \vee [x \wedge py \wedge z] \vee [x \wedge qy \wedge qz] \\ &= [px \wedge ((py \wedge z) \vee (y \wedge qz))] \vee [x \wedge q((py \wedge z) \vee (y \wedge qz))] = x(yz).\end{aligned}$$

(5) Assume x has an identity 1. Then $p1 = 1\top = \top = \top 1 = q1$ and $x = x1 = (px \wedge 1) \vee (x \wedge q1) = (px \wedge 1) \vee x$, so $px \wedge 1 \leq x$ and similarly $qx \wedge 1 \leq x$. Therefore $(px \vee qx) \wedge 1 = (px \wedge 1) \vee (qx \wedge 1) \leq x$.
Conversely, suppose $p1 = \top = q1$ and $(px \vee qx) \wedge 1 \leq x$. Then $x1 = (px \wedge 1) \vee (x \wedge q1) = (px \wedge 1) \vee x = x$ since $px \wedge 1 \leq x$. Likewise $1x = x$.

(6) This follows from (3) since $x = x1 \leq x{\cdot}\top = px$ and $x = 1x \leq qx$. □

Note that if A also has a bottom bound $\bot$ then p, q are normal if and only if $\cdot$ is normal, hence the term-equivalence preserves normality.

Table 1. The number of algebras of cardinality n up to isomorphism.

	Cardinality $n =$	2	3	4	5	6	7	8
1	Normal $d\ell$-magmas	2	20	1116				
2	Normal $d\ell pq$-algebras	2	6	46	3435			
3	Normal comm. $d\ell$-semigroups	2	8	57	392	3212		
4	Normal assoc. $d\ell p$-algebras	2	4	13	35	109	315	998
5	Normal comm. idem. $d\ell$-semigroups	1	2	8	25	97	366	
6	Normal assoc. idem. $d\ell p$-algebras	1	2	7	18	57	163	521
7	Normal comm. idem. $d\ell$-monoids	1	2	6	15	44	115	326
8	Normal assoc. idem. $d\ell p1$-algebras	1	2	5	10	24	47	108
9	Distributive lattices	1	1	2	3	5	8	15

This term-equivalence is useful since distributive lattices with unary operators are considerably simpler than distributive lattices with binary operators. In particular, (2) and (4) show that associativity can be replaced by one or two 2-variable identities in this variety. This provides more efficient ways to construct associative operators from a (pair of) unary operator(s) on a distributive lattice. The variety of $\top$-bounded distributive lattices is obtained as a subvariety of $d\ell pq$-algebras that satisfy $px = x = qx$, or a subvariety of unary determined $d\ell$-magmas that satisfy $x \cdot y = x \wedge y$.

For small cardinalities, Table 1 shows the number of algebras that are unary-determined (shown in the even numbered rows) for several subvarieties of normal $d\ell$-magmas. As seen from rows 5–8, under the assumption of associativity, commutativity and idempotence of $\cdot$, the property of being unary-determined is a relatively mild restriction compared to the general case of normal $d\ell$-magmas.

A *Boolean magma* is a Boolean algebra with a binary operator. The next lemma shows that if the operator is idempotent, then it is always unary-determined, hence the results in the current paper generalize the theorems about idempotent Boolean nonassociative quantales in [1].

Lemma 3. *Every idempotent Boolean magma* $(A, \wedge, \vee, \neg, \bot, \top, \cdot)$ *is unary-determined, i.e., satisfies* $xy = (x{\cdot}\top \wedge y) \vee (x \wedge \top{\cdot}y)$.

Proof. Idempotence is equivalent to $x \wedge y \leq xy \leq x \vee y$ since $(x \wedge y)^2 \leq xy \leq (x \vee y)^2$ holds in all partially ordered algebras where $\cdot$ is an order-preserving binary operation. The following calculation

$$\begin{aligned} x{\cdot}\top \wedge y &= x(y \vee \neg y) \wedge y = (xy \wedge y) \vee (x(\neg y)) \wedge y) \\ &\leq xy \vee ((x \vee \neg y) \wedge y) = xy \vee (x \wedge y) \vee (\neg y \wedge y) = xy \end{aligned}$$

and a similar one for $x \wedge \top{\cdot}y \leq xy$ prove that $xy \geq (x{\cdot}\top \wedge y) \vee (x \wedge \top{\cdot}y)$.

Using Boolean negation, the opposite inequation is equivalent to

$$xy \wedge \neg(x{\cdot}\top \wedge y) \leq x \wedge \top{\cdot}y.$$

By De Morgan's law it suffices to show $(xy \wedge \neg(x{\cdot}\top)) \vee (xy \wedge \neg y) \leq x \wedge \top{\cdot}y$. Since $xy \leq x{\cdot}\top$, the first meet disappears. Next, by idempotence, $xy \wedge \neg y \leq (x \vee y) \wedge \neg y = (x \wedge \neg y) \vee (y \wedge \neg y) \leq x$ and finally $xy \wedge \neg y \leq xy \leq \top{\cdot}y$. □

3 BI-algebras from Heyting Algebras and Residuated Unary Operations

We now recall some basic definitions about residuated operations, adjoints and residuated lattices. For an overview and additional details we refer to [6]. A *Brouwerian algebra* $(A, \wedge, \vee, \to, \top)$ is a $\top$-bounded lattice such that $\to$ is the *residual* of $\wedge$, i.e.,

$$x \wedge y \leq z \quad \iff \quad y \leq x \to z.$$

Since $\to$ is the residual of $\wedge$, we have that $\wedge$ is join-preserving, so the lattice is distributive [6, Lemma 4.1]. The $\top$-bound is included as a constant since it always exists when a meet-operation has a residual: $x \wedge y \leq x$ always holds, hence $y \leq (x \to x) = \top$. A *Heyting algebra* is a bounded Brouwerian algebra with a constant $\bot$ denoting the bottom element.

A dual operator is an n-ary operation on a lattice that preserves meet in each argument. A *residual* or *upper adjoint* of a unary operation p on a poset A is a unary operation p^* such that

$$px \leq y \iff x \leq p^*y$$

for all $x, y \in A$. If A is a lattice, then the existence of a residual guarantees that p is an operator and p^* is a dual operator [6, Lemma 3.5]. Moreover, if A is bounded, then $p\bot = \bot$ and $p^*\top = \top$.

A binary operation $\cdot$ on a poset is *residuated* if there exist a *left residual* $\backslash$ and a *right residual* $/$ such that

$$x \cdot y \leq z \iff y \leq x\backslash z \iff x \leq z/y.$$

A *residuated ℓ-magma* $(A, \wedge, \vee, \cdot, \backslash, /)$ is a lattice with a residuated binary operation. In this case $\cdot$ is an operator and $\backslash, /$ are dual operators in the "numerator" argument. In the "denominator" $\backslash, /$ map joins to meets, hence they are order reversing. A *residuated Brouwerian-magma* is a residuated ℓ-magma expanded with $\to, \top$ such that $(A, \wedge, \vee, \to, \top)$ is a Brouwerian algebra.

A *residuated lattice* is a residuated ℓ-magma with $\cdot$ associative and a constant 1 that is an identity element, i.e., $(A, \cdot, 1)$ is a monoid. A *generalized bunched implication algebra*, or GBI-algebra, $(A, \wedge, \vee, \to, \top, \cdot, 1, \backslash, /)$ is a $\top$-bounded residuated lattice with a residual $\to$ for the meet operation, i.e., $(A, \wedge, \vee, \to, \top)$ is a Brouwerian algebra. A GBI-algebra is called a *bunched implication algebra* (BI-algebra) if $\cdot$ is commutative and A also has a bottom element, denoted by the constant $\bot$, hence a BI-algebra has a Heyting algebra reduct. These algebras are the algebraic semantics for bunched implication logic, which is the propositional part of separation logic, a Hoare logic used for reasoning about memory references in computer programs. In this setting the operation $\cdot$ is usually denoted by $*$, the left residual $\backslash$ is denoted $-\!\!*$, and $/$ can be omitted since $x/y = y -\!\!* x$.

Note that the property of being a residual can be expressed by inequalities (p^* is a residual of p if and only if $p(p^*x) \leq x \leq p^*(px)$ for all x, and p, p^* are order preserving), hence the classes of all Brouwerian algebras, Heyting algebras, residuated ℓ-magmas, residuated Brouwerian-magmas, residuated lattices, (G)BI-algebras, and pairs of residuated unary maps on a lattice are varieties (see e.g. [6, Theorem 2.7, Lemma 3.2.]). Recall also that a $\top$-bounded magma is unary-determined if it satisfies the identity $xy = (x{\cdot}\top \wedge y) \vee (x \wedge \top{\cdot}y)$.

We are now ready to prove a result that upgrades the term-equivalence of Theorem 1 to Brouwerian algebras with two pairs of residuated maps and unary-determined residuated Brouwerian-magmas.

Theorem 4. (1) *Let $(A, \wedge, \vee, \rightarrow, \top, p, p^*, q, q^*)$ be a Brouwerian algebra with unary operators p, q and their residuals p^*, q^* such that $x \wedge p\top \leq qx$, $x \wedge q\top \leq px$. If we define $x{\cdot}y = (px \wedge y) \vee (x \wedge qy)$,*

$$x\backslash y = (px \rightarrow y) \wedge q^*(x \rightarrow y) \quad \text{and} \quad x/y = p^*(y \rightarrow x) \wedge (qy \rightarrow x)$$

then $(A, \wedge, \vee, \top, \cdot, \backslash, /)$ is a unary-determined residuated Brouwerian-magma and the unary operations are recovered by $px = x{\cdot}\top$, $p^*x = x/\top$, $qx = \top{\cdot}x$ and $q^*x = \top\backslash x$.

(2) *Let $(A, \wedge, \vee, \rightarrow, \top, \cdot, \backslash, /)$ be a unary-determined residuated Brouwerian-magma and define $px = x{\cdot}\top$, $p^*x = x/\top$, $qx = \top{\cdot}x$ and $q^*x = \top\backslash x$. Then $(A, \wedge, \vee, \rightarrow, \top, p, p^*, q, q^*)$ is a Brouwerian algebra with a unary operators p, q and dual operators p^*, q^* that satisfies $x \wedge p\top \leq qx$, $x \wedge q\top \leq px$.*

Proof. (1) The following calculation shows that $\cdot$ is residuated.

$$\begin{aligned} x \cdot y \leq z &\iff (px \wedge y) \vee (x \wedge qy) \leq z && \iff px \wedge y \leq z \text{ and } x \wedge qy \leq z \\ &\iff y \leq px \rightarrow z \text{ and } y \leq q^*(x \rightarrow z) && \iff y \leq (px \rightarrow z) \wedge q^*(x \rightarrow z) \end{aligned}$$

hence $x\backslash z = (px \rightarrow z) \wedge q^*(x \rightarrow z)$ and similarly $z/y = p^*(y \rightarrow z) \wedge (qy \rightarrow z)$. By Theorem 1 it follows that $px = x{\cdot}\top, qx = \top{\cdot}x$ and $xy = (x{\cdot}\top \wedge y) \vee (x \wedge \top{\cdot}y)$. Since $x{\cdot}\top \leq y \iff x \leq y/\top$ we obtain $p^*(x) = x/\top$, and similarly $q^*(x) = \top\backslash x$.

(2) Since $\cdot$ is residuated it follows that p^* and q^* are the unary residuals of p, q respectively. The remaining parts hold by Theorem 1. □

Recall that a closure operator p is an order-preserving unary function on a poset such that $x \leq px = ppx$. A $d\ell p$-algebra where p is a closure operator is called a $d\ell p$-closure algebra. If $\cdot$ is idempotent and associative then $x{\cdot}\top = x(\top\top) = (x\top)\top$, so $px = x{\cdot}\top$ is a closure operator.

Lemma 5. *Assume **A** is a $d\ell p$-closure algebra and let $x{\cdot}y = (px \wedge y) \vee (x \wedge py)$. Then $\cdot$ is associative if and only if $px \wedge py \leq p((px \wedge y) \vee (x \vee py))$.*

Proof. By Lemma 2 $\cdot$ is associative if and only if the identity $p((px \wedge y) \vee (x \vee py)) = (px \wedge py) \vee (x \wedge py)$ holds. This is equivalent to $px \wedge py \leq p((px \wedge y) \vee (x \vee py))$ since $x \wedge py \leq px \wedge py$, $p(px \wedge y) \leq ppx \wedge py = px \wedge py$ and similarly $p(x \wedge py) \leq px \wedge py$. □

Hence the preceding theorems specialize to a term-equivalence for a subvariety of unary-determined BI-algebras as follows:

Corollary 6. *1. Let $(A, \wedge, \vee, \to, \top, \bot, p, p^*, 1)$ be a Heyting algebra with a closure operator p, residual p^* and constant 1 such that $px \wedge py \leq p((px \wedge y) \vee (x \wedge py))$, $p1 = \top$ and $px \wedge 1 \leq x$. If we define $x{*}y = (px \wedge y) \vee (x \wedge py)$ and $x{-\!*}y = (px \to y) \wedge p^*(x \wedge y)$ then $(A, \wedge, \vee, \top, \to, *, -\!*, 1)$ is a unary-determined BI-algebra and $x{*}\top \wedge y{*}\top \leq ((x{*}\top \wedge y) \vee (x \wedge y{*}\top)){*}\top$ holds.*

2. Let $(A, \wedge, \vee, \to, \top, \bot, *, -\!*, 1)$ be a unary-determined BI-algebra, and define $px = x{*}\top$ and $p^*x = \top{-\!*}x$. Then $(A, \wedge, \vee, \to, \top, \bot, p, p^*, 1)$ is a Heyting algebra with a closure operator p that has p^* as residual and satisfies $px \wedge py \leq p((px \wedge y) \vee (x \wedge py))$, $p1 = \top$ and $px \wedge 1 \leq x$.

By Lemma 2 (6) unary-determined BI-algebras satisfy $x{*}x = x$, which does not hold in BI-algebras that model applications (e.g., heap storage). However, as mentioned in the introduction, they are members of the variety of BI-algebras, and understanding their properties via this term-equivalence is useful for the general theory. E.g., structural results about algebraic object (such as rings) often start by investigating the idempotent algebras, followed by sets of idempotent elements in more general algebras. Line 8 in Table 1 also shows that finite unary-determined BI-algebras are not rare (normal join-preserving operators are automatically residuated in the finite case, hence the algebras counted in Line 8 are indeed term-equivalent to unary-determined BI-algebras).

4 Relational Semantics for $d\ell$-magmas

We now briefly recall relational semantics for bounded distributive lattices with operators and then apply correspondence theory to derive first-order conditions for the equational properties of the preceding sections.

An element in a lattice is *completely join-irreducible* if it is not the supremum of all the elements strictly below it. The set of all completely join-irreducible elements of a lattice A is denoted by $J(A)$, and it is partially ordered by restricting the order of A to $J(A)$. For example, if A is a Boolean lattice, then $J(A) = At(A)$ is the antichain of *atoms*, i.e., all elements immediately above the bottom element. The set $M(A)$ of completely meet-irreducible elements is defined dually. A lattice is *perfect* if it is complete (i.e., all joins and meets exist) and every element is a join of completely join-irreducibles and a meet of completely meet-irreducibles. For a Boolean algebra, the notion of perfect is equivalent to being *complete* (i.e., joins and meets of all subsets exist) and *atomic* (i.e., every non-bottom element has an atom below it).

Recall that for a poset $\mathbf{W} = (W, \leq)$, a *downset* is a subset X such that $y \leq x \in X$ implies $y \in X$. As in modal logic, W is considered a set of "worlds" or states. We let $D(\mathbf{W})$ be the set of all downsets of $\mathbf{W}$, and $(D(\mathbf{W}), \cap, \cup)$ the *lattice of downsets*. The collection $D(\mathbf{W})$ is a perfect distributive lattice with infinitary meet and join given by (arbitrary) intersections and unions. The following result, due to Birkhoff [2] for lattices of finite height, shows that up to isomorphism all perfect distributive lattices arise in this way. The poset $J(D(\mathbf{W}))$ contains exactly the principal downsets $\downarrow x = \{y \in W \mid y \leq x\}$.

Theorem 7 ([3, 10.29]). *For a lattice A the following are equivalent:*

1. *A is distributive and perfect.*
2. *A is isomorphic to the lattice of downsets of a partial order.*

Note that the set of upsets of a poset is also a perfect distributive lattice, and if it is ordered by reverse inclusion then this lattice is isomorphic to the downset lattice described above. It is also well known that the maps J and D are functors for a categorial duality between the category of posets with order-preserving maps and the category of perfect distributive lattices with complete lattice homomorphisms (i.e., maps that preserve arbitrary joins and meets).

A *complete* operator on a complete lattice is an operation that is either completely join-preserving, completely meet-preserving, maps all arbitrary meets to joins or all arbitrary joins to meets in each argument. A lattice-ordered algebra is called *perfect* if its lattice reduct is perfect and every fundamental operation on it is a complete operator. The duality between the category of perfect distributive lattices and posets extends to the category of perfect distributive lattices with (a fixed signature of) complete operators. The corresponding poset category has additional relations of arity $n + 1$ for each operator of arity n, and the relations have to be upward or downward closed in each argument. For example, a binary relation $Q \subseteq W^2$ is upward closed in the second argument if $xQy \leq z \implies xQz$. Here $xQy \leq z$ is an abbreviation for xQy and $y \leq z$.

Perfect distributive lattices with operators are algebraic models for many logics, including relevance logic, intuitionistic logic, Hajek's basic logic, Łukasiewicz logic and bunched implication logic [6,7]. In such an algebra $\mathbf{A}$, a join-preserving binary operation is determined by a ternary relation R on $J(\mathbf{A})$ given by

$$xRyz \iff x \leq yz.$$

The notation $xRyz$ is shorthand for $(x, y, z) \in R$. For $b, c \in A$ the product bc is recovered as $\bigvee\{x \in J(\mathbf{A}) \mid xRyz \text{ for some } y \leq b \text{ and } z \leq c\}$.

The relational structure $(J(\mathbf{A}), \leq, R)$ is an example of a Birkhoff frame. In general, a *Birkhoff frame* [5] is a triple $\mathbf{W} = (W, \leq, R)$ where $(W, \leq)$ is a poset, and $R \subseteq W^3$ satisfies the following three properties (downward closure in the 1st, and upward closure in the 2nd and 3rd argument):

$$\text{(R1) } u \leq xRyz \implies uRyz$$
$$\text{(R2) } xRyz \ \&\ y \leq v \implies xRvz$$
$$\text{(R3) } xRyz \ \&\ z \leq w \implies xRyw.$$

A Birkhoff frame $\mathbf{W}$ defines the downset algebra $\mathbf{D}(\mathbf{W}) = (D(\mathbf{W}), \bigcap, \bigcup, \cdot)$ by

$$Y \cdot Z = \{x \in W \mid xRyz \text{ for some } y \in Y \text{ and } z \in Z\}.$$

The property (R1) ensures that $Y \cdot Z \in D(\mathbf{W})$.

In relevance logic [4] similar ternary frames are known as Routley-Meyer frames. In that setting upsets are used to recover the distributive lattice-ordered relevance algebra, and this choice implies that $J(A)$ with the induced order from A is dually isomorphic to $(W, \leq)$. Another difference is that Routley-Meyer frames have a unary relation and axioms to ensure it is a left identity element of the $\cdot$ operation.

The duality between perfect $d\ell$-magmas and Birkhoff frames is recalled below. Here we assume that the binary operation on a complete $d\ell$-magma is a complete operator, i.e., distributes over arbitrary joins in each argument. Such algebras are also known as *nonassociative quantales* or *prequantales*.

Theorem 8 ([5]).

1. *If* $\mathbf{A}$ *is a perfect* $d\ell$*-magma and* $R \subseteq J(A)^3$ *is defined by* $xRyz \Leftrightarrow x \leq yz$ *then* $J(\mathbf{A}) = (J(A), \leq, R)$ *is a Birkhoff frame, and* $\mathbf{A} \cong \mathbf{D}(J(\mathbf{A}))$.
2. *If* $\mathbf{W}$ *is a Birkhoff frame then* $\mathbf{D}(\mathbf{W})$ *is a perfect* $d\ell$*-magma, and* $\mathbf{W} \cong (J(D(\mathbf{W})), \subseteq, R_\downarrow)$*, where* $(\downarrow x, \downarrow y, \downarrow z) \in R_\downarrow \Leftrightarrow xRyz$.

A ternary relation R is called *commutative* if $xRyz \implies xRzy$ for all x, y, z. The justification for this terminology is provided by the following result.

Lemma 9. *For any Birkhoff frame* $\mathbf{W}$, $\mathbf{D}(\mathbf{W})$ *is commutative if and only if* R *is commutative.*

Lemma 10. *Let* $\mathbf{W}$ *be a Birkhoff frame. Then* $\mathbf{D}(\mathbf{W})$ *is idempotent if and only if* $xRxx$ *and* $(xRyz \implies x \leq y$ *or* $x \leq z)$ *for all* $x, y, z \in W$.

Proof. Assume $\mathbf{D}(\mathbf{W})$ is idempotent, and let $x \in W$. Then $\downarrow x \cdot \downarrow x = \downarrow x$ since $\downarrow x \in D(\mathbf{W})$. From $x \in \downarrow x$ we deduce $x \in \downarrow x \cdot \downarrow x$, whence it follows that $xRyz$ for some $y \in \downarrow x, z \in \downarrow x$. Therefore $xRyz$ for $y \leq x, z \leq x$, which implies $xRxx$ by (R2) and (R3).

Next assume $xRyz$ holds. Then $x \in \downarrow\{y, z\} \cdot \downarrow\{y, z\} = \downarrow\{y, z\}$ by idempotence. Hence for some $w \in \{y, z\}$ we have $x \leq w$, and it follows that $x \leq y$ or $x \leq z$.

For the converse, assume $xRxx$ and $(xRyz \implies x \leq y$ or $x \leq z)$ for all $x, y, z \in W$ and let $X \in D(\mathbf{W})$. From $xRxx$ we obtain $X \subseteq X \cdot X$.

For the reverse inclusion, let $x \in X \cdot X$. Then $xRyz$ holds for some $y, z \in X$. By assumption $xRyz$ implies $x \leq y$ or $x \leq z$. Since X is a downset, $x \leq y \implies x \in X$ and $x \leq z \implies x \in X$. Hence $X \cdot X = X$. □

The previous two results are examples of correspondence theory, since they show that an equational property on a perfect $d\ell$-magma corresponds to a first-order condition on its Birkhoff frame.

The relational semantics of a perfect $d\ell pq$-magma is given by a *PQ-frame*, which is a partially-ordered relational structure $(W, \leq, P, Q)$ such that P, Q are binary relations on W, $u \leq xPy \leq v \implies uPv$ and $u \leq xQy \leq v \implies uQv$. Relations with this property are called *weakening relations* [5,11], and this is what ensures that if we define $p(Y) = \{x \mid \exists y(xPy \;\&\; y \in Y)\}$ for a downset Y, then p is a complete normal join-preserving operator that produces a downset, and P is uniquely determined by $xPy \Leftrightarrow x \in p({\downarrow}y)$. Similarly, a normal operator q is defined from Q, and uniquely determines Q. The residual p' of p is a completely meet-preserving operator, defined by $p'(Y) = \{x \mid \forall y(yPx \Rightarrow y \in Y)\}$, and likewise for q'. If $P = Q$ then we omit Q and refer to $(W, \leq, P)$ simply as a *P-frame.*

We now list some correspondence results for $d\ell pq$-magmas. We begin with a theorem that restates the term-equivalence of Theorem 1 as a definitional equivalence on frames. A direct proof of this result is straightforward, but it also follows from Theorem 1 by correspondence theory.

Theorem 11. (1) *Let $(W, \leq, P, Q)$ be a PQ-frame such that $x \leq y \;\&\; xPz \Rightarrow xQy$ and $x \leq y \;\&\; xQz \Rightarrow xPy$. If we define $xRyz \Leftrightarrow (xPy \;\&\; x \leq z)$ or $(x \leq y \;\&\; xQz)$ then $(W, \leq, R)$ is a Birkhoff frame, and P, Q are obtained from R via $xPy \Leftrightarrow \exists z(xRyz)$ and $xQy \Leftrightarrow \exists z(xRzy)$.*

(2) *Let $(W, \leq, R)$ be a Birkhoff frame that satisfies $xRyz \Leftrightarrow (\exists z(xRyz) \;\&\; x \leq z)$ or $(x \leq y \;\&\; \exists z(xRzy))$ and define $xPy \Leftrightarrow \exists z(xRyz)$, $xQy \Leftrightarrow \exists z(xRzy)$. Then $(W, \leq, P, Q)$ is a PQ-frame in which $x \leq y \;\&\; xPz \Rightarrow xQy$ and $x \leq y \;\&\; xQz \Rightarrow xPy$ hold.*

Note that the universal formula $x \leq y \;\&\; xPz \implies xQy$ corresponds to the $d\ell pq$-magma axiom $Y \wedge p\top \leq qY$.

A significant advantage of *PQ*-frames over Birkhoff frames is that binary relations have a graphical representation in the form of directed graphs (whereas ternary relations are 3-ary hypergraphs that are more complicated to draw). Equational properties from Lemma 2, Corollary 6 correspond to the following first-order properties on *PQ*-frames.

Lemma 12. *Assume* $\mathbf{A}$ *is a perfect $d\ell pq$-algebra and* $\mathbf{W} = (W, \leq, P, Q)$ *is its corresponding PQ-frame. The constant $1 \in A$ (when present) is assumed to correspond to a downset $E \subseteq W$. Then*

(1) *$a \leq pa$ holds in* $\mathbf{A}$ *if and only if P is reflexive,*
(2) *$ppa \leq pa$ holds in* $\mathbf{A}$ *if and only if P is transitive,*
(3) *$pa = qa$ holds in* $\mathbf{A}$ *if and only if $P = Q$,*
(4) *$p1 = \top$ holds in* $\mathbf{A}$ *if and only if $\forall x \exists y(y \in E \;\&\; xPy)$ holds in* $\mathbf{W}$,
(5) *$pa \wedge 1 \leq a$ holds in* $\mathbf{A}$ *if and only if $x \in E \;\&\; xPy \Rightarrow x \leq y$ holds in* $\mathbf{W}$,
(6) *$pa \wedge pb \leq p((pa \wedge b) \vee (a \wedge pb))$ holds in* $\mathbf{A}$ *if and only if*

$$wPx \;\&\; wPy \Rightarrow \exists v(wPv \;\&\; (vPx \;\&\; v \leq y \text{ or } v \leq x \;\&\; vPy)) \quad \text{holds in } \mathbf{W}.$$

Proof. (1)–(3) These correspondences are well known from modal logic.

(4) For $x \in J(A)$ and $E = {\downarrow}1$ we have $x \leq p1$ if and only if there exists $y \in J(A)$ such that $y \leq 1$ and $x \leq py$, or equivalently, $y \in E$ and xPy.

(5) In the forward direction, let $a = {\downarrow}y$. Then it follows that $x \in p({\downarrow}y) \cap E$ implies $x \in {\downarrow}y$, and consequently $x \in E$ & $xPy \implies x \leq y$.

In the other direction, let Y be a downset of W and assume $x \in pY \cap E$. Then $x \in E$ and xPy for some $y \in Y$. Hence $x \leq y$, or equivalently $x \in {\downarrow}y \subseteq Y$. Thus, $pY \cap E \subseteq Y$, so the algebra $\mathbf{A}$ satisfies $pa \wedge 1 \leq a$ for all $a \in A$.

(6) In the forward direction, let $a = {\downarrow}x$ and $b = {\downarrow}y$. Then it follows from the inequation that $w \in p{\downarrow}x \cap {\downarrow}y \implies w \in p((p{\downarrow}x \cap {\downarrow}y) \cup ({\downarrow}x \cap p{\downarrow}y))$ for all $w \in W$. This in turn implies wPx & $wPy \implies \exists v(wPv$ & $v \in (p{\downarrow}x \cap {\downarrow}y) \cup ({\downarrow}x \cap p{\downarrow}y))$, which translates to the given first-order condition.

In the reverse direction, let X, Y be downsets of W and assume $w \in pX \cap pY$. Then wPx and wPy for some $x \in X$ and $y \in Y$. It follows that there exists a $v \in W$ such that $(wPv$ & $(vPx$ & $v \leq y$ or $v \leq x$ & $vPy))$, hence $v \in (pX \cap Y) \cup (X \cap pY)$. Therefore $w \in p(pX \cap Y) \cup (X \cap pY)$. □

Recall that a ternary relation R is commutative if $xRyz \Leftrightarrow xRzy$ for all x, y. From Theorem 11 we also obtain the following result.

Corollary 13. *Let $(W, \leq, P, Q)$ be a PQ-frame and define R as in Theorem 11(1). Then R is commutative if and only if $xPy \Leftrightarrow xQy$ for all $x, y \in W$.*

This corollary shows that in the commutative setting a PQ-frame only needs one of the two binary relations. Hence we define $\mathbf{W} = (W, \leq, P)$ to be a *P-frame* if P is a weakening relation, i.e., $u \leq xPy \leq v \implies uPv$.

We now turn to the problem of ensuring that the binary operation of a $d\ell$-magma is associative. For Birkhoff frames the following characterization of associativity is well known from relation algebras [10] (in the Boolean case) and from the Routley-Meyer semantics for relevance logic [4] in general.

Lemma 14. *Let $\mathbf{W} = (W, \leq, R)$ be a Birkhoff frame. Then $\mathbf{D}(\mathbf{W})$ is an associative ℓ-magma if and only if $\forall wxyz(\exists u(uRxy \,\&\, wRuz) \Leftrightarrow \exists v(vRyz \,\&\, wRxv))$. If R is commutative then the equivalence can be replaced by the implication $\forall uwxyz(uRxy \,\&\, wRuz \Rightarrow \exists v(vRyz \,\&\, wRxv))$.*

This lemma is another correspondence result that follows from translating $w \in (XY)Z \Leftrightarrow w \in X(YZ)$ for $X, Y, Z \in D(\mathbf{W})$. In the commutative case $(XY)Z \subseteq X(YZ)$ implies the reverse inclusion, hence only one of the implications is needed. We now show that for a large class of P-frames the 5-variable universal-existential formula for associativity can be replaced by simpler universal formulas with only three variables.

A *preorder forest P-frame* is a P-frame such that P is a preorder (i.e. reflexive and transitive) and satisfies the formula

$$\text{(Pforest)} \quad xPy \text{ and } xPz \implies x \leq y \text{ or } x \leq z \text{ or } yPz \text{ or } zPy.$$

Note that since P is a weakening relation, reflexivity of P implies that $\leq \subseteq P$ because xPx and $x \leq y$ implies xPy.

It is interesting to visualize the properties that define preorder forest P-frames by implications between Hasse diagrams with $\leq$-edges (solid) and P-edges (dotted) as in Fig. 1. However, one needs to keep in mind that dotted lines could be horizontal (if xPy and yPx) and that any line could be a loop if two variables refer to the same element.

(Pforest) 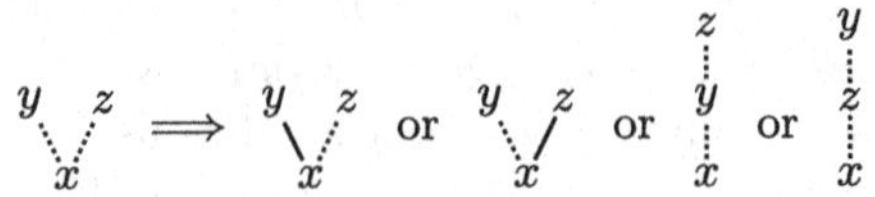

Fig. 1. The (Pforest) axiom. The partial order $\leq$ and the preorder P are denoted by solid lines and dotted lines respectively.

We are now ready to state the main result. We use the algebraic characterization of associativity in Lemma 2.

Theorem 15. *Let* $\mathbf{W} = (W, \leq, P)$ *be a preorder forest* P*-frame and* $\mathbf{D}(\mathbf{W})$ *its corresponding downset algebra. Then the operation* $x{\cdot}y = (px \wedge y) \vee (x \wedge py)$ *is associative in* $\mathbf{D}(\mathbf{W})$.

Proof. Let $\mathbf{W} = (W, \leq, P)$ be a preorder forest P-frame and $\mathbf{D}(\mathbf{W})$ its $d\ell p$-algebra of downsets with operator p. Since P is a preorder, $\mathbf{D}(\mathbf{W})$ is a $d\ell p$-closure algebra. By Lemma 5, a $d\ell p$-closure algebra is associative if and only if $p(x) \wedge p(y) \leq p(p(x) \wedge y) \vee (x \wedge p(y))$. By Lemma 12 this is equivalent to the frame property

$$(*) \qquad xPy \ \&\ xPz \Rightarrow \exists w(xPw \ \&\ (wPy \ \&\ w \leq z \text{ or } w \leq y \ \&\ wPz)).$$

We now show that this frame property holds in $\mathbf{W}$. We know that P is reflexive and (Pforest) holds.

Assume xPy and xPz. By (Pforest) there are four cases:

1. $x \leq y$: take $w = x$. Then xPx, $x \leq y$ and xPz, hence $(*)$ holds.
2. $x \leq z$: again take $w = x$. Then the other disjunct of $(*)$ holds.
3. yPz: take $w = y$. Then xPy, $y \leq y$ and yPz, hence $(*)$ holds.
4. zPy: take $w = z$. Then xPz, zPy and $y \leq y$, hence again $(*)$ holds. □

The universal class of preorder forest P-frames is strictly contained in the class of all P-frames in which $x{\cdot}y$ is associative. In fact the latter class is not closed under substructures, hence not a universal class: $W = \{0, 1, 2, 3\}$, $\leq = id_W \cup \{(0,1), (0,2), (0,3)\}$, $P = \leq \cup \{(1,0), (1,2), (1,3)\}$ is a P-frame with associative $\cdot$ (use e.g. Lemma 5), but restricting $\leq, P$ to the subset $\{1, 2, 3\}$ gives a P-frame where $\cdot$ fails to be associative, hence (Pforest) also fails.

A *$d\ell$-semilattice* is an associative commutative idempotent distributive ℓ-magma. The point of the previous result is that it allows the construction of perfect associative commutative idempotent $d\ell$-magmas and idempotent bunched

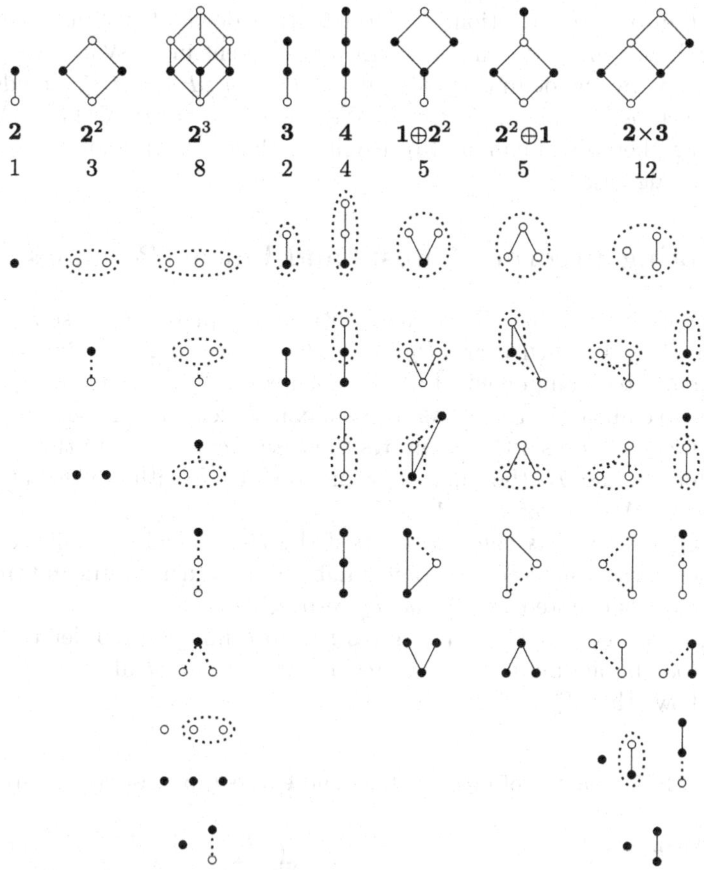

Fig. 2. All 40 preorder forest P-frames $(W, \leq, P)$ with up to 3 elements. Solid lines show $(W, \leq)$, dotted lines show the additional edges of P, and the identity (if it exists) is the set of black dots. The first row shows the lattice of downsets, and the Boolean quantales from [1] appear in the first three columns.

implication algebras from preorder forest P-frames. This is much simpler than constructing the ternary relation R of the Birkhoff frame of such algebras. For example the Hasse diagrams for all the preorder forest P-frames with up to 3 elements are shown in Fig. 2, with the preorder P given by dotted lines and ovals. The corresponding ternary relations can be calculated from P, but would have been hard to include in each diagram.

We now examine when a preorder forest P-frame will have an identity element. For any P-frame $\mathbf{W}$ we define $E = \{x \in W \mid \forall y(xPy \Rightarrow x \leq y)\}$.

Lemma 16. *Let* $\mathbf{W}$ *be a* P*-frame. Then* E *is an identity element for* $\cdot$ *in the downset algebra* $D(\mathbf{W})$ *if and only if* E *is a downset and* $pE = W$*.*

Proof. In the forward direction, E is certainly a downset and it follows from Lemma 2(5) that $pE = W$ since W is the top element in $D(\mathbf{W})$.

Conversely, by the definition of E, if $x \in E$ then $xPy \Rightarrow x \leq y$ holds for all $y \in W$. Hence by Lemma 12(5) for all $X \in D(\mathbf{W})$ we have $pX \cap E \subseteq X$. Since $pE = W$ together with Lemma 2(5), it follows that E is an identity element in the downset algebra. □

5 Counting Preorder Forests and Linear *P*-frames

In the case when the poset $(W, \leq)$ is an antichain, a preorder forest P is simply a preorder $P \subseteq W^2$ such that xPy and xPz implies yPz or zPy. A *preorder tree* is a connected component of a preorder forest. A *rooted* preorder forest is defined to have an equivalence class of P-maximal elements in each component. For finite preorder forests this is always the case. Let F_n denote the number of preorder forests and T_n the number of preorder trees with n elements (up to isomorphism). We also let $F_0 = 1$.

A preorder forest *has singleton roots* if the P-maximal equivalence class of each component is a singleton set. The number of preorder forests and trees with singleton roots is denoted by F_n^s and T_n^s respectively.

Note that every preorder forest gives rise to a unique preorder tree with a singleton root by adding one new element r such that for all $x \in W$ we have xPr. It follows that $T_n^s = F_{n-1}$ (Table 2).

Table 2. Number of preorder trees and forests (up to isomorphism)

Cardinality $n =$	1	2	3	4	5	6	7
Preorder trees $T_n =$	1	2	5	13	37	108	337
$c_n =$	1	5	16	57	186	668	
Preorder forests $F_n =$	1	3	8	24	71	224	
Preorder trees with singleton roots $T_n^s =$	1	1	3	8	24	71	224
$c_n^s =$	1	3	10	35	121	438	
Preorder forests with singleton roots $F_n^s =$	1	2	5	14	41	127	

Every preorder tree with a non-singleton root equivalence class and n elements is obtained from a preorder tree with $n-1$ elements by adding one more element to the root equivalence class. Hence for $n > 0$ we have $T_n = F_{n-1} + T_{n-1}$. The Euler transform of T_n is used to calculate the next value of F_n as follows:

$$c_n = \sum_{d|n} d \cdot T_n \qquad\qquad F_n = \frac{1}{n}\sum_{k=1}^{n} c_k \cdot F_{n-k}.$$

Since preorder forests with singleton roots are disjoint unions of preorder trees with singleton roots, F_n^s is calculated by an Euler transform from T_n^s.

Corollary 17. *The sequence F_n^s is the Euler transform of T_n^s.*

While it is difficult to count preorder forest P-frames in general, it is simple to count the linear ones. Let L_n be the number of linearly ordered preorder forest P-frames with n elements. Note that $(P3)$ is actually redundant for linearly ordered frames.

Theorem 18. *For linearly ordered forest P-frames $L_n = 2^{n-1}$. In the algebraic setting there are 2^{n-2} unary-determined commutative doubly idempotent linear semirings with n elements, and $n-1$ of them have an identity element.*

Proof. Let $\mathbf{W}$ be a linearly ordered P-frame with elements $W = \{1 < 2 < \cdots < n\}$ such that P is transitive and $(P0)$ holds. Then each possible relation P on W is determined by choosing a subset S of the edges $\{(2,1), (3,2), \ldots, (n, n-1)\}$ and defining P to be the transitive closure of $S \cup \leq$. Since there are $n-1$ such edges to choose from, the number of p-frames is 2^{n-1}.

Let $\mathbf{A}$ be a unary-determined commutative doubly idempotent linear semiring with n elements. Then the P-frame $\mathbf{W}$ associated with $\mathbf{A}$ has $n-1$ elements, is linearly ordered, and P is reflexive and transitive since $\cdot$ is idempotent and associative. Hence there are 2^{n-2} such algebras.

By Lemma 16, the subset $E = \{x \in W \mid \forall y(xPy \Rightarrow x \leq y)\}$ will be an identity of the downset algebra if and only if it is a downset of W and $p(E) = W$. This will only be the case if there exists an element $w \in W$ such that for all $y \in W$ we have $y \geq w$ if and only if wPy. Every choice of $w \in W$ determines one such P, hence there are $n-1$ algebras with an identity element. □

6 Conclusion

We showed that unary-determined $d\ell$-magmas have a simple algebraic structure given by two unary operators and that their relational frames are definitionally equivalent to frames with two binary relations. The complex algebras of these frames are complete distributive lattices with completely distributive operators, hence they have residuals and can be considered Kripke semantics for unary-determined bunched implication algebras and bunched implication logic. Associativity of the binary operator for idempotent unary-determined algebras can be checked by an identity with 2 rather than 3 variables, and for the frames by a 3-variable universal formula rather than a 6-variable universal-existential formula. All idempotent Boolean magmas are unary-determined, hence these results significantly extend the structural characterization of idempotent atomic Boolean quantales in [1] and relate them to bunched implication logic. As an application we counted the number of preorder forest P-frames with n elements for which the partial order is an antichain, as well as the number of linearly ordered preorder P-frames.

Acknowledgments. The investigations in this paper made use of Prover9/ Mace4 [12]. In particular, parts of Lemma 2 and Theorem 11 were developed with the help of

Prover9 (short proofs were extracted from the output) and the results in Table 1 were calculated with Mace4. The remaining results in Sections 2–4 were proved manually, and later also checked with Prover9.

References

1. Alpay, N., Jipsen, P.: Commutative doubly-idempotent semirings determined by chains and by preorder forests. In: Fahrenberg, U., Jipsen, P., Winter, M. (eds.) RAMiCS 2020. LNCS, vol. 12062, pp. 1–14. Springer, Cham (2020). https://doi.org/10.1007/978-3-030-43520-2_1
2. Birkhoff, G.: Lattice Theory, 3rd edn. American Mathematical Society, Providence (1967)
3. Davey, B., Priestley, H.: Introduction to Lattices and Order. Cambridge (2002)
4. Dunn, J., Restall, G.: Relevance logic. In: Gabbay, D., Guenthner, F. (eds.) Handbook of Philosophical Logic, 2nd ed., vol. 6, pp. 1–128. Springer, Heidelberg (2002)
5. Galatos, N., Jipsen, P.: The structure of generalized BI-algebras and weakening relation algebras. Algebra Universalis **81**(3), 1–35 (2020). https://doi.org/10.1007/s00012-020-00663-9
6. Galatos, N., Jipsen, P., Kowalski, T., Ono, H.: Residuated Lattices: An Algebraic Glimpse at Substructural Logics. Elsevier, Amsterdam (2007)
7. Gehrke, M., Nagahashi, H., Venema, Y.: A Sahlqvist theorem for distributive modal logic. Ann. Pure Appl. Logic **131**, 65–102 (2005)
8. Gil-Férez, J., Jipsen, P., Metcalfe, G.: Structure theorems for idempotent residuated lattices. Algebra Universalis **81**(2), 1–25 (2020). https://doi.org/10.1007/s00012-020-00659-5
9. Hebisch, U., Weinert, H.J.: Semirings: Algebraic Theory and Applications in Computer Science. World Scientific, Singapore (1998)
10. Maddux, R.: Some varieties containing relation algebras. Trans. Am. Math. Soc. **272**(2), 501–526 (1982)
11. Kurz, A., Velebil, J.: Relation lifting, a survey. J. Logical Algebraic Methods Program. **85**(4), 475–499 (2016)
12. McCune, W.: Prover9 and Mace4 (2005–2010). http://cs.unm.edu/~mccune/prover9
13. O'Hearn, P.W.: Resources, concurrency, and local reasoning. Theor. Comput. Sci. **375**, 271–307 (2007)
14. O'Hearn, P.W., Pym, D.J.: The logic of bunched implications. Bull. Symb. Logic **5**(2), 215–244 (1999)
15. Raney, G.N.: Completely distributive complete lattices. Proc. Am. Math. Soc. **3**, 677–680 (1952)
16. Reynolds, J.C.: Separation logic: a logic for shared mutable data structures. In: Proceedings 17th Annual IEEE Symposium on Logic in Computer Science, pp. 55–74 (2002)

Effect Algebras, Girard Quantales and Complementation in Separation Logic

Callum Bannister[1,3](✉), Peter Höfner[2,3], and Georg Struth[2,3]

[1] University of Queensland, Brisbane, Australia
c.bannister@uq.edu.au
[2] Australian National University, Canberra, Australia
[3] University of Sheffield, Sheffield, UK

Abstract. We study convolution and residual operations within convolution quantales of maps from partial abelian semigroups and effect algebras into value quantales, thus generalising separating conjunction and implication of separation logic to quantitative settings. We show that effect algebras lift to Girard convolution quantales, but not the standard partial abelian monoids used in separation logic. It follows that the standard assertion quantales of separation logic do not admit a linear negation relating convolution and its right adjoint. We consider alternative dualities for these operations on convolution quantales using boolean negations, some old, some new, relate them with properties of the underlying partial abelian semigroups and outline potential uses.

1 Introduction

Separation logic and linear logic reason about resources. Both, in fact, have powerset quantale semantics that lift certain monoids. The phase quantale semantics of linear logic is even a Girard quantale [26]: it admits a dualising element that relates the quantalic mutltiplication with its residuals in the way negation relates conjunction and implication in classical logic. For the standard statelet and heaplet models of separation logic [6], previous work [3,4] suggests that such a linear negation between separating conjunction and implication is impossible. But an algebraic account is missing.

We investigate the relationship between the standard models of separation logic and Girard quantales in the more general setting of convolution quantales formed by spaces of functions from partial monoids to quantales [7,12,14]. These yield quantale-valued semantics for linear and separation logic with applications in quantitative, for instance probabilistic program verification [16].

The classical heaplet models of separation logic are generalised effect algebras [17], but lack the greatest element present in effect algebras [15]. Effect algebras, in turn, are equipped with an orthosupplementation that seems suitable for extending previous lifting results from generalised effect algebras to convolution quantales to those from effect algebras and Girard quantales.

We prove that this extension works: effect algebras lift to commutative Girard quantales and in particular phase semantics for linear logic. We also show that

U. Fahrenberg et al. (Eds.): RAMiCS 2021, LNCS 13027, pp. 37–53, 2021.
https://doi.org/10.1007/978-3-030-88701-8_3

it is impossible to lift generalised effect algebras *without* a greatest element that way. This rules out a linear negation between separating conjunction and implication over the classical heaplet models. Further, we present a read-only heaplet model that forms an effect algebra and makes linear negation available to separation logic in some situations, and we outline its use.

We generalise these lifting and impossibility results to cover partial abelian monoids with several units, as in the statelet models of separation logic [6], and from powersets to convolution quantales, for quantitative applications.

Beyond these results, we show how separating conjunction and implication in convolution quantales relate to operations in value quantales and partial abelian monoids. In the absence of linear negation, we follow [5] in studying the effect of boolean negation on separating conjunction and implication. This leads to operations of septraction and coimplication [2,5] as well as some new ones. We also expose the symmetries and dualities between these operations in boolean convolution quantales. Boolean negation may not be the most natural duality for quantales, but the resulting operations are at least useful for program verification [2]. Finally, we constrast these results with a non-boolean assertion quantale for separation logic based on Alexandrov topologies for posets that captures the sub-heaplet and sub-statelet orderings more faithfully than the standard one.

Our main results have been checked with the Isabelle/HOL proof assistant.[1] Our Isabelle theories already contain more general lifting results for non-commutative partial monoids and Girard quantales appropriate for the non-commutative linear logics originally studied by Yetter [26]. These, however, are beyond the scope of this paper.

2 Partial Abelian Monoids and Effect Algebras

We recall the basics of partial abelian monoids. Most of the development has been formalised with Isabelle [11]. Most results are known in the special case of generalised effect algebras [17].

A *partial abelian semigroup* (PAS) is a structure $(S, \oplus, D)$ with *domain of definition* $D \subseteq S \times S$ for the partial *composition* $\oplus : S \times S \to S$ (or $\oplus : D \to S$) such that, for all $x, y, z \in S$, $D\,x\,y$ and $D\,(x \oplus y)\,z$ imply that $D\,y\,z$, $D\,x\,(y \oplus z)$ and $(x \oplus y) \oplus z = x \oplus (y \oplus z)$, and $D\,x\,y$ implies that $D\,y\,x$ and $x \oplus y = y \oplus x$.

We identify sets and predicates. The above associativity and commutativity axioms state that if one side of the equation is defined, then so is the other, and both are equal. This notion of equality is known as *Kleene equality.* We write $x \simeq y$ for it. Hence, more briefly, $(x \oplus y) \oplus z \simeq x \oplus (y \oplus z)$ and $x \oplus y \simeq y \oplus x$.

Units of a PAS S can be defined like for (object-free) categories: $e \in S$ is a *unit* in S if there exists an $x \in S$ such that $x \simeq e \oplus x$ and for all $x, y \in S$ if $y \simeq e \oplus x$ then $y = x$. A *partial abelian monoid* (PAM) is a PAS S in which every element has a unit: $\forall x \in S. \exists e \in E.\ D\,e\,x$, writing E for the set of units of S.

[1] Most results on partial abelian monoids, more generally relational monoids, and (convolution) quantales can be found in the Archive of Formal Proofs [11,24]. The complete formalisation can be found online http://hoefner-online.de/ramics21.

Every element of a PAM has precisely one unit, different units cannot be composed and total PAMs have precisely one unit [8].

PAMs and related partial algebras appear across mathematics. They are instances of relational semigroups and monoids or multisemigroups and multimonoids, see [7,13] for details. Relational monoids, in particular, are monoids in the category **Rel** equipped with the canonical monoidal structure.

In any PAM S, the *divisibility preorder* is defined, for all $x, y \in S$, by $x \preceq y$ iff $x \oplus z \simeq y$ for some $z \in S$. Hence $x \preceq y$ iff $x \oplus z \simeq y$ has a solution in z. This preorder is a precongruence: $x \preceq y$ and $D\, z\, x$ imply $z \oplus x \preceq z \oplus y$ (and $D\, y\, z$). A subtraction can now be defined.

A PAM S is *cancellative* if $x \oplus z \simeq y \oplus z$ imply $x = y$ for all $x, y, z \in S$.

Lemma 2.1. *In a cancellative* PAM, $x \preceq y$ *implies* $x \oplus z \simeq y$ *for exactly one* z.

One can thus write $y \ominus x$ for this solution.

Lemma 2.2. *In a cancellative* PAM,

1. $x \oplus z \simeq y \Leftrightarrow x \preceq y \wedge z = y \ominus x$,
2. $D\, x\, y \Rightarrow (x \oplus y) \ominus x = y$ *and* $x \preceq y \Rightarrow x \oplus (y \ominus x) = y$,
3. $D\, x\, y \Rightarrow x \preceq x \oplus y$ *and* $x \preceq y \Rightarrow y \ominus x \preceq y$.

By Lemma 2.2 (1) and (2), $x \oplus (\_)$ and $(\_) \ominus x$ are inverses up-to definedness.

Finally, a PAM is *positive* if $D\, x\, y$ and $x \oplus y \in E$ imply $x \in E$.

Lemma 2.3. *In any positive cancellative* PAM, $\preceq$ *is a partial order in which all units are* $\preceq$*-minimal.*

Cancellative positive PAMs with a single unit $E = \{0\}$ are known as *generalised effect algebras* (GEAs) [17] in the foundations of quantum mechanics. The resource monoids used in separation logic [6] are nothing but GEAs.

Example 2.4 (Heaplets). Partial maps $X \rightharpoonup Y$ form a GEA H with $D\, \eta_1\, \eta_2$ iff $dom\, \eta_1 \cap dom\, \eta_2 = \emptyset$, $\eta_1 \oplus \eta_2 = \eta_1 \cup \eta_2$ and $E = \{\varepsilon\}$, where $\varepsilon : X \rightharpoonup Y$ is the empty partial function. By definition, $dom\, \varepsilon = \emptyset$. These are the heaplets of separation logic. Alternatively, heaplets have been modelled as a GEA of *finite* partial maps $X \rightharpoonup_{\text{fin}} Y$. The latter captures the fact that programs use finitely many variables and heaps can always be extended. The former admits full heaps where no additional memory can be allocated. □

Example 2.5 (Generalised Heaplets). Heaplet models readily generalise to additions defined as union whenever heaplets coincide where they overlap: $D\, \eta_1\, \eta_2$ iff $\eta_1\, x = \eta_2\, x$ for all $x \in \eta_1 \cap dom\, \eta_2$. The resulting PAM is not cancellative. □

An *effect algebra* (EA) [15] is a PAM S with single unit 0 and orthosupplement $(\_)^{\perp} : S \to S$ such that for each $x \in S$, $x^{\perp}$ is the unique element satisfying $x \oplus x^{\perp} = 0^{\perp}$ and if $D\, x\, 0^{\perp}$, then $x = 0$. It is standard to write 1 for $0^{\perp}$. It follows that $x^{\perp\perp} = x$. The following fact is well known.

Proposition 2.6. *Every* EA *is a* GEA *with greatest element* $1 = 0^\perp$ *while every* GEA *with greatest element* 1 *is an* EA *with* $(\_)^\perp = 1 \ominus (\_)$.

Example 2.7. PAM H from Example 2.4 is not an EA: it is cancellative positive, but has no greatest element when $|Y| > 1$. Replacing any $m \mapsto n \in \eta$ by $m \mapsto n'$ with $n \neq n'$ in heaplet η yields an incomparable heaplet. □

Statelet models of separation logic [6] are based on the following coproduct.

Lemma 2.8. *Let* X *be a set and* $(S, \oplus, D, E)$ *a* PAM.

1. $(X \times S, \hat{\oplus}, \hat{D}, \hat{E})$ *forms a* PAM *with* $\hat{D}\,(x_1, y_1)\,(x_2, y_2)$ *iff* $x_1 = x_2$ *and* $D\, y_1\, y_2$, $(x_1, y_1) \hat{\oplus} (x_2, y_2) = (x_1, y_1 \oplus y_2)$ *and* $\hat{E}\,(x, e)$ *iff* $x \in X$ *and* $e \in E$.
2. *If* S *is cancellative or positive, then so is* $X \times S$.

Example 2.9 (*Statelets*). The PAM H from Example 2.4 is formed by (finite) partial functions $X \rightharpoonup Y$. Program stores can be modelled as a set Z (e.g. a function from variables to values). Lemma 2.8 then shows that $Z \times (X \rightharpoonup Y)$ forms a cancellative positive PAM with many units $E = \{(z, \varepsilon) \mid z \in Z\}$. □

3 Convolution Quantales over PAMs

We apply a lifting construction for functions from partial monoids, and even ternary relations with suitable algebraic properties, to quantales, so that a generalised quantale-weighted separating conjunction arises as a convolution and a quantale-weighted separating implication as its right adjoint [13,14]. A simple instance yields the assertion algebra of separation logic [12]—a convolution quantale of functions from the PAM of statelets into the quantale of booleans.

A *quantale* [23] is a structure $(Q, \leq, \cdot, 1)$ such that $(Q, \leq)$ is a complete lattice, $(Q, \cdot, 1)$ a monoid, and $\cdot$ preserves arbitrary sups in both arguments. We write $\bigvee X$ for the sup of $X \subseteq Q$, $\bigwedge X$ for its inf, $\vee$ for the binary sup and $\wedge$ for the binary inf. We write $\bot = \bigvee \emptyset$ for the least element of the lattice and $\top = \bigwedge \emptyset$ for its greatest element. It follows that $\bot$ is a zero of multiplication.

A quantale is *commutative* if its monoid is abelian, and *boolean* if its complete lattice is a boolean algebra. We write $\overline{x}$ for the boolean complement of x in Q.

As quantalic multiplication preserves sups in both arguments, it has two right adjoints, $x \backslash (\_)$ of $x \cdot (\_)$ and $(\_)/x$ of $(\_) \cdot x$, for all $x \in Q$, given, as usual, by

$$x \backslash z = \bigvee \{y \mid x \cdot y \leq z\} \qquad \text{and} \qquad z/x = \bigvee \{y \mid y \cdot x \leq z\},$$

and related by the Galois connection $y \leq x \backslash z \Leftrightarrow x \cdot y \leq z \Leftrightarrow x \leq z/y$. The residuals coincide in commutative quantales: $y/x = x \backslash y$. As right adjoints, $x \backslash (\_)$ and $(\_)/x$ preserve infs and therefore $x \cdot y = \bigwedge \{z \mid y \leq x \backslash z\} = \bigwedge \{z \mid x \leq z/y\}$.

Example 3.1

1. Every frame is a commutative quantale and hence every complete boolean algebra. In the latter, finite sups and infs are related by De Morgan duality; the residual is definable as $x \to y = \overline{x} \vee y$.

2. The booleans $\mathbb{B} = \{\mathsf{f}, \mathsf{t}\}$ form a two-element commutative quantale with $\cdot$ as $\wedge$/min, $\bigvee$ as max and $\backslash$ as boolean implication $\rightarrow$. Predicates over a PAM S are functions $S \rightarrow \mathbb{B}$; $\mathbb{B}^S$ is isomorphic to $\mathcal{P}S$. □

We now fix a PAM $(S, \oplus, D, E)$ and a commutative quantale $(Q, \leq, \cdot, 1)$. We equip the function space Q^S with quantalic operations following [14]. Sups, infs and the order extend pointwise from Q to Q^S. Thus $\bot = \lambda x.\ \bot$ and $\top = \lambda x.\ \top$ in Q^S. We define the *convolution* of $f, g : S \rightarrow Q$ and the unit $id_E : S \rightarrow Q$ as

$$(f * g)\, x = \bigvee_{x \simeq y \oplus z} f\, y \cdot g\, z \qquad \text{and} \qquad id_E\, x = \begin{cases} 1 & \text{if } x \in E, \\ \bot & \text{otherwise.} \end{cases}$$

The following lifting result characterises the *convolution algebra* on Q^S.

Theorem 3.2 ([14]). *If S is a* PAM *and Q a commutative quantale, then the convolution algebra $(Q^S, \leq, *, id_E)$ is a commutative quantale.*

In addition, properties, such as being boolean lift from Q to the convolution quantale Q^S. As an instance of Theorem 3.2, $Q = \mathbb{B}$ yields the commutative powerset quantale $(\mathcal{P}S, \subseteq, *, E)$ over the PAM S.

Cancellative PAMs give us an arguably more elegant variant of convolution.

Lemma 3.3. *If S is cancellative, then $(f * g)\, x = \bigvee_{y \preceq x} f\, y \cdot g\, (x \ominus y)$.*

Remark 3.4. Lemma 2.8 yields the following instance of Theorem 3.2: if X is a set, then $Q^{X \times S}$ is a quantale with $(f * g)\,(x, y) = \bigvee_{y \simeq y_1 \oplus y_2} f\,(x, y_1) \cdot g\,(x, y_2)$ and $id_E\,(x, y) = id_E\, y$, where, in the second identity, the left E is on $X \times S$ and the right one on S.

The right adjoint $f \mathrel{-\!\!*} (\_)$ of $f * (\_)$ in Q^S is $f \mathrel{-\!\!*} h = \bigvee\{g \mid f * g \leq h\}$. In quantalic notation, $f \mathrel{-\!\!*} g = f\backslash g$.

Theorem 3.5. *In every* PAM *S,*

1. $(f \mathrel{-\!\!*} g)\, x = \bigwedge_{z = x \oplus y} f\, y \backslash g\, z = \bigwedge_{D\, x\, y} f\, y \backslash g\, (x \oplus y)$,
2. $(f \mathrel{-\!\!*} g)\, x = \bigwedge_{x = z \ominus y} f\, y \backslash g\, z$, *if S is cancellative.*

Proof.

1. Suppose $D\, x\, y$. Then $f\, y \cdot (f \mathrel{-\!\!*} g)\, x \leq (f * (f \mathrel{-\!\!*} g))\,(x \oplus y) \leq g\,(x \oplus y)$, thus $\forall y.\ (f \mathrel{-\!\!*} g)\, x \leq f\, y \backslash g\,(x \oplus y)$ and finally $(f \mathrel{-\!\!*} g)\, x \leq \bigwedge\{f\, y \backslash g\,(x \oplus y), \mid D\, x\, y\}$ by the adjunction and properties of inf.
Conversely, suppose $D\, x\, z$ and let $\varphi\, x = \bigwedge\{f\, y \backslash g\,(x \oplus y) \mid D\, x\, y\}$. Then $\varphi\, x \leq f\, z \backslash g\,(x \oplus z)$, $f\, z \cdot \varphi\, x \leq g\,(x \oplus z)$ by the adjunction and $f * \varphi \leq g$ by definition of convolution. Finally, $\varphi\, x \leq (\bigvee\{h \mid f * h \leq g\})\, x = (f \mathrel{-\!\!*} g)\, x$.
2. Immediate from (1) using Lemma 2.2(1).

□

Example 3.6 (*Powerset Lifting*). Theorem 3.2 shows that the convolution algebra $(\mathcal{P}S, \subseteq, *, E)$, for $Q = \mathbb{B}$, is a commutative quantale of predicates over any PAM $(S, \oplus, E)$, in fact a boolean atomic one. For the PAM on $X \times S$ and in particular for statelets, convolution is separating conjunction and its residual separating implication (a.k.a. magic wand):

$$\begin{aligned} f * g &= \{(x, y_1) \oplus (x, y_2) \mid (x, y_1) \in f \wedge (x, y_2) \in g \wedge D\, y_1\, y_2\}, \\ f \mathrel{-\!\!*} g &= \{(x, y) \mid \forall y'.\, (x, y') \in f \wedge D\, y\, y' \rightarrow (x, y \oplus y') \in g\} \\ &= \{(x, y_1 \ominus y_2) \mid (x, y_2) \in f \wedge y_2 \preceq y_1 \rightarrow (x, y_1) \in g\}, \end{aligned}$$

where the second step requires cancellation. This powerset quantale is the standard assertion algebra of separation logic. These set-based operations are also described in [9]. □

4 PAMs and Girard Quantales

Additional operations have been defined on quantales. A linear negation is inspired by linear logic—a classical multiplicative negation that coincides with boolean negation if $\cdot$ is $\wedge$.

Formally, an element d of a quantale Q is *dualising* if $(d/x)\backslash d = x = d/(x\backslash d)$ for all $x \in Q$. An element $c \in Q$ is *cyclic* if $c/x = x\backslash c$ for all $x \in Q$. A *Girard quantale* [23,26] is a quantale with a cyclic dualising element d.

This definition is meant for non-commutative quantales; in the commutative case all elements are cyclic. A *linear negation* can be defined as $x^d = x\backslash d$ (which is then the same as d/x). It has many features of classical negation: it is involutive, reverses the order and all sups and infs, hence in particular 0 and $\top$; and it allows expressing residuation in terms of multiplication and vice versa:

$$x\backslash y = \left(y^d \cdot x\right)^d \qquad \text{and} \qquad x \cdot y = \left(y\backslash x^d\right)^d.$$

Moreover, $d^d = 1$ and therefore $1^d = d$, $(\bigvee X)^d = \bigwedge\{x^d \mid x \in X\}$ and $(\bigwedge X)^d = \bigvee\{x^d \mid x \in X\}$ [23]. Also $d = \top$ implies $\top = \bot$. In a boolean Girard quantale, where the underlying complete lattice is a boolean algebra, both negations commute: $\overline{x^d} = \overline{x}^d$.

First we show that any EA gives rise to a commutative Girard quantale. In any EA S we define $X^\perp = \{x^\perp \mid x \in X\}$ for $X \subseteq S$. Then $X^\perp = \{x \mid x^\perp \in X\}$, $X^{\perp\perp} = X$ and $\overline{X}^\perp = \overline{X^\perp}$ because $x^{\perp\perp} = x$. Also note that $0^\perp = 1$ [23].

Proposition 4.1. *Let* $(S, \oplus, 0, ^\perp)$ *be an* EA. *Then* $(\mathcal{P}S, \subseteq, *, \{0\})$ *is a commutative Girard quantale with dualising element* $\Delta = S - \{1\}$.

Proof. Theorem 3.2 implies that every PAM lifts to a powerset quantale. It thus remains to check that Δ is a dualising element, that is, $X^{\Delta\Delta} = X$ for any $X \subseteq S$.

First we compute X^{Δ}:

$$\begin{aligned} X^{\Delta} &= \{y \mid \forall x \in X.\, D\,x\,y \rightarrow x \oplus y \in \Delta\} \\ &= \{y \mid \neg\exists x \in X.\, x \oplus y \simeq 1\} \\ &= \{y \mid \neg\exists x \in X.\, x = y^{\perp}\} \\ &= \{y \mid y^{\perp} \in \overline{X}\} \\ &= \overline{X}^{\perp}, \end{aligned}$$

using the definition of $(\_)^{\perp}$ in the second step. Then $X^{\Delta\Delta} = X$ follows immediately from the equations preceding this theorem. □

As a sanity check, $\Delta^{\Delta} = \overline{\Delta}^{\perp} = \{1\}^{\perp} = \{1^{\perp}\} = \{0\}$. Next we show that commutative Girard quantales contain EAs.

Lemma 4.2. *Let $x \notin \Delta$ in a commutative powerset Girard quantale over set S with unit $\{0\}$. Then $\{x\}^{\Delta} = \overline{\{0\}}$ and $\{x\} = \overline{\Delta}$.*

Proof. We have $x \notin \Delta \Leftrightarrow \{x\} \subseteq \overline{\Delta} \Leftrightarrow \overline{\Delta^{\Delta}} \subseteq \overline{\{x\}^{\Delta}} \Leftrightarrow \overline{\{0\}} \subseteq \{x\}^{\Delta}$. It then follows that $\{x\}^{\Delta} = \overline{\{0\}}$ because if $S - \{0\} = \overline{\{0\}} \subset \{x\}^{\Delta}$, then $\{x\}^{\Delta} = S$ and therefore $\{x\}^{\Delta\Delta} = S^{\Delta} = \emptyset \neq \{x\}$, a contradiction. Finally, therefore, $\{x\} = \{x\}^{\Delta\Delta} = \overline{\{0\}}^{\Delta} = \overline{\{0\}^{\Delta}} = \overline{\Delta}$. □

It follows that $\overline{\Delta}$ is a singleton set. We call its element 1.

Proposition 4.3. *Let S be a positive* PAM *and $\mathcal{P}S$ a commutative Girard quantale with unit $\{0\}$ and dualising element Δ. Then S is an* EA.

Proof. For every convolution quantale Q^S, S forms a PAM [7]. It remains to check the two EA axioms. For any $x \in S$, we abbreviate $\{x\}^{\perp} = \overline{\{x\}}^{\Delta}$. By Lemma 4.2, $\overline{\Delta} = \{1\}$. Then $\{x\}^{\perp} = \overline{\{x\}^{\Delta}} = \{y \mid D\,x\,y \wedge x \oplus y \in \overline{\Delta}\} = \{y \mid x \oplus y \simeq 1\}$, for all $x \in S$. Also, $\{x\}^{\perp} \neq \emptyset$ because otherwise $\overline{\{x\}}^{\Delta\Delta} = S \neq \overline{\{x\}}$. For each $x \in S$ there thus is a $y \in S$ such that $x \oplus y \simeq 1$, that is, 1 is the greatest element of S. It also follows that $\{x\} * \{x\}^{\perp} = \{1\}$ and $\{0\}^{\perp} = \{1\}$ using Lemma 4.2.

Next we show that S is cancellative. Suppose $\{x\} * \{y\} = \{x\} * \{z\}$. Then, using $\{x\} * \{y\} = \{x \oplus y\}$, we have $\{x \oplus y\} * \{x \oplus y\}^{\perp} = \{x \oplus z\} * \{x \oplus y\}^{\perp} = \{1\}$ and therefore $\{y\} = (\{x\} * \{x \oplus y\}^{\perp})^{\perp} = \{z\}$.

Cancellativity implies that $x \oplus y \simeq 1$ for at most one y by Lemma 2.1. Thus $\{x\}^{\perp}$ is a singleton set, and we call its element $x^{\perp}$. It satisfies $\{x^{\perp}\} = \{x\}^{\perp}$ and therefore $\{x\} * \{x^{\perp}\} = \{1\}$, which verifies the first EA axiom.

Moreover, $\preceq$ is a partial order that extends to singleton sets. For the second EA axiom, now suppose $D\,x\,1$. Then $\{1\} \preceq \{x\} * \{1\}$ by Lemma 2.2(3) and therefore $\{1\} = \{x\} * \{1\}$. Yet $\{1\} = \{0\} * \{0^{\perp}\} = \{0\} * \{1\}$ and $x = 0$, once again by cancellativity. □

We leave the question whether positivity is derivable open.

Corollary 4.4. *Let S be a* GEA *without greatest element. Then the commutative quantale $\mathcal{P}S$ is not Girard.*

In particular, therefore, the standard heaplet model of separation logic, which is not an EA by Example 2.7, does not give rise to Girard quantales. Consequently, separating conjunction and implication over the heaplet models in Example 2.4 cannot be related by a quantalic linear negation. Next we give an alternative no-Girard proof for heaplets that extends to statelets.

Theorem 4.5. *The unital commutative quantale $(\mathcal{P}H, \subseteq, *, \{\varepsilon\})$ over the* PAM *H of heaplets is not Girard.*

Proof. Let Δ be a dualising element in $\mathcal{P}H$. By a remark above, we know $\Delta \neq H$. In fact, we can show that there are two (different) heaplets outside of Δ. Then, by Lemma 4.2, this yields a contradiction.

Claim: $\exists \upsilon_1, \upsilon_2$ with $\upsilon_1, \upsilon_2 \notin \Delta$ and $\upsilon_1 \neq \upsilon_2$.

Proof of Claim. There are exotic cases where this is not the case, e.g., when either X or Y of type $X \rightharpoonup Y$ have cardinality 1.

Example 2.4 shows two standard models for heaplets: arbitrary partial functions and finite mappings.

In the former we can characterise a full heap using heaplets η with $dom(\eta) = X$. When $|Y| > 1$—in most standard models it is $\mathbb{Z}$—there are at least two different full heaplets (Example 2.7). It suffices to show that any heaplet ζ with $dom(\zeta) = X$ is not part of Δ. We use the equality $\overline{x^d} = \overline{x}^d$ and the equivalence $\exists \eta'.\, D\,\eta\,\eta' \wedge \eta' \in X \wedge \eta \oplus \eta' \in \overline{\Delta} \Leftrightarrow \forall \eta'.\, D\,\eta\,\eta' \wedge \eta' \in \overline{X} \rightarrow \eta \oplus \eta' \in \Delta$, as in Proposition 4.1. Using ζ for η, $X = \{\varepsilon\}$ and the fact that the only heaplet that can be added to ζ is the empty heaplet ε ($D\,\zeta\,\eta \Leftrightarrow \eta = \varepsilon$) yields

$$\begin{aligned} \zeta \notin \Delta &\Leftrightarrow \forall \eta'.\, D\,\zeta\,\eta' \wedge \eta' \neq \varepsilon \rightarrow \zeta \oplus \eta' \in \Delta \\ &\Leftrightarrow \forall \eta'.\, \mathsf{f} \rightarrow \zeta \oplus \eta' \in \Delta \Leftrightarrow \mathsf{t} \end{aligned}$$

Now consider the model of finite mappings. We follow [22] and assume that the partial functions are of type $\mathbb{Z} \rightharpoonup_{\mathsf{fin}} Y$. We know there exists one heaplet υ with $\upsilon \notin \Delta$, for otherwise the algebra collapses. Next we assume that $\overline{\{\upsilon\}}$ is a dualising element and derive a contradiction. In the heaplet model we have

$$(\overline{\{\upsilon\}}/\{\upsilon\})\backslash\overline{\{\upsilon\}} = (\{\upsilon\} \mathrel{-\!\!*} \overline{\{\upsilon\}}) \mathrel{-\!\!*} \overline{\{\upsilon\}} = \{\eta \mid \forall \eta'.\, D\,\eta\,\eta' \wedge \eta \oplus \eta' = \upsilon \rightarrow \eta' = \varepsilon\}$$

If $\overline{\{\upsilon\}}$ is a dualising element, this set equates to $\{\upsilon\}$. However, every heaplet υ' that is strictly larger than υ, i.e. $\upsilon \prec \upsilon'$ is an element of this set as well, since $\upsilon' \oplus \eta' \neq \upsilon$, for all η', and therefore the antecedent inside the set evaluates to false. Since υ is a finite mapping and the set of locations is $\mathbb{Z}$, we can always find a larger heaplet υ'. □

Theorem 4.5 holds in the standard heaplet models (Examples 2.4 and 2.5) of separation logic and generalises easily to statelets. It shows in particular that separating conjunction and separating implications over PAMs of statelets cannot be related by a linear negation. Similar results are claimed in [3,4], but not in a quantalic setting.

5 Binative PAMs and Girard Convolution Quantales

We now generalise Proposition 4.1 to PAMs with multiple units and general convolution quantales. This yields the main lifting theorem in this paper.

EAs generalise to several units. An element x of a PAS S is *maximal* if $x \oplus y \simeq x$ for all y. A PAS S is *orthosupplemented* if $x \oplus x^\perp$ is defined and maximal for all $x \in S$, and if z is maximal, then $x \oplus y \simeq z$ iff $y = x^\perp$. Orthosupplemented PAMs are automatically PASs, and $e^\perp$ is maximal for each $e \in E$.

Example 5.1. (Read-only-heaplets). Heaplets become an orthosupplemented PAM H^{ro} with many units when switching to total maps $X \to Y \times \mathbb{B}$. We define $dom_{\mathsf{t}}\, \eta = \{x \in X \mid \exists y \in Y.\ \eta\, x = (y, \mathsf{t})\}$ and write π_1, π_2 for the cartesian projections. We define heaplet composition by $D\, \eta_1\, \eta_2$ iff $dom_{\mathsf{t}}\, \eta_1 \cap dom_{\mathsf{t}}\, \eta_1 = \emptyset$ and $\pi_1 \circ \eta_1 = \pi_1 \circ \eta_2$, and $(\eta_1 \oplus \eta_2)\, x = ((\pi_1 \circ \eta_1)\, x, (\pi_2 \circ \eta_1 \vee \pi_2 \circ \eta_2)\, x)$. We define the set of units as $E = \{\eta \in H^{ro} \mid dom_{\mathsf{t}}\, \eta = \emptyset\}$. Finally, we define orthosupplementation by $\eta^\perp\, x = (y, b) \Leftrightarrow \eta\, x = (y, \overline{b})$.

In this model, we denote ownership of locations by the boolean flag. Such heaplets are "read-only" in the sense that if the composition of heaplets η_1 and η_2 is defined, they must agree on the values at each location in memory, and updating one requires updating the other. Hence, for any $f : H^{ro} \to H^{ro}$, we have $D\, \eta_1\, \eta_2 \Rightarrow D\, (f\, \eta_1)\, \eta_2 \Rightarrow \pi_1 \circ (f\, \eta_1) = \pi_1 \circ \eta_2$. □

A generalised heaplet model mapping natural numbers to sets of integers and a second model reminiscent of a PAM with abelian group-like negative elements have been studied by Brotherston and Calcagno [3] among many other models relevant to separation logic. Both yield models of classical logic of bunched implication and thus probably Girard quantales. Both of these models have a single unit. They are thus quite different to the one in Example 5.1.

We generalise orthosupplementation further to cover more models. A PAS S is *binative* if it is equipped with a function $(\_)^\perp : S \to S$ such that $D\, x\, x^\perp$, for all $x \in S$, and, for all $x, y, z \in S$, $x \oplus x^\perp \simeq y \oplus z$ implies $y = z^\perp$. Thus $x = x^{\perp\perp}$ holds because $x \oplus x^\perp \simeq x \oplus x^\perp$. We call $(x, x^\perp)$ the *binates* of S.

Intuitively, binativity generalises positivity for PAMs from units to binates.

Lemma 5.2. *Every binative* PAS S *is a cancellative* PAM *with*

$$E = \{(x \oplus x^\perp)^\perp \mid x \in S\}.$$

Proof. For cancellation, suppose $x \oplus y \simeq x \oplus z$. Then $(x \oplus y) \oplus (x \oplus y)^\perp \simeq (x \oplus y) \oplus (x \oplus z)^\perp$, hence $(x \oplus y) \oplus (x \oplus y)^\perp \simeq y \oplus (x \oplus (x \oplus z)^\perp)$ and therefore $z = y = (x \oplus (x \oplus z)^\perp)^\perp$ by binativity.

For the units, $x \oplus x^\perp \oplus (x \oplus x^\perp)^\perp \simeq x^\perp \oplus x \oplus (x \oplus x^\perp)^\perp$ by commutativity. Then, by binativity, $x = x^{\perp\perp} = x \oplus (x \oplus x^\perp)^\perp$. □

Example 5.3 (Binative PAMs).

1. Orthosupplemented PASs are binative PASs where compositions of binates are maximal. Equivalently, positive binative PASs are orthosupplemented.

2. EAs are binative PASs with single unit 0 and greatest element 1.
3. Abelian Groupoids are binative semigroups with $(\_)^{\perp}$ a inverse and binates composing to units.
4. Partial deterministic CBI models [21] are precisely binative PASs with single unit 0 and the composition of any binate equals $0^{\perp}$.[2]

We now generalise Theorem 4.1.

Theorem 5.4. *Let S be a binative* PAM *and $(Q, \cdot, \leq, d)$ a commutative Girard quantale. Then $(Q^S, \leq, *, \Delta)$ is a commutative Girard quantale with*

$$\Delta\, x = \begin{cases} d & \text{if } x = y \oplus y^{\perp} \text{ for some } y \in S, \\ \top & \text{otherwise.} \end{cases}$$

Proof. Relative to Theorem 3.2 we need to check $f^{\Delta\Delta} = f$ for all $f : S \to Q$. Define $f^{\perp}\, x = f\,(x^{\perp})$ and $f^{d}\, x = (f\, x)^{d}$. Then $f^{\perp\perp} = f = f^{dd}$ and $f^{\perp d} = f^{d\perp}$. First we compute

$$f^{\Delta}\, x = \bigwedge_{D\, x\, y} f\, y \backslash \Delta\, (x \oplus y) = f^{d}\, x^{\perp} \wedge \bigwedge_{\substack{D\, x\, y \\ y \neq x^{\perp}}} f\, y \backslash \top = f^{\perp d}\, x \wedge \top = f^{\perp d}\, x.$$

Binativity is used in the second step. Hence $f^{\Delta\Delta} = f^{\perp d \perp d} = f^{\perp\perp dd} = f$. □

A natural question is whether Theorem 5.4 could be generalised by restricting $D\, x\, x^{\perp}$ while avoiding the collapse into a monoid. But if Q^S and Q are both unital and $1 \neq \perp$ in Q, then the underlying PAS must be unital, too, and thus a PAM [7, Proposition 4.1]. Girard quantales, in particular, are unital [23].

Theorem 5.4 generalises further to non-abelian binative semigroups and non-commutative Girard quantales, yet this is beyond the scope of this paper. A proof can be found in our Isabelle theories.

6 Using Linear Negation in Separation Logic

Statelets do not lift to a Girard quantale. It is therefore natural to ask how the lifting results in the previous section might be applied. We show that lifting assertions on ordinary heaps to those on read-only heaps makes it possible to use linear negation for reasoning about resources that lack binativity.

Separation logic allows enriching a Hoare triple with a frame

$$\forall R.\ \{P * R\}\, C\, \{Q * R\},$$

which states that the execution of C only modifies the resources whose ownership is asserted by P. If these are assertions over a standard heap, then the validity of adding a frame means the only variables that C touches are claimed

[2] CBI models are relational monoids, deterministic means that results of compositions are singletons, *partial* deterministic that they are singletons or empty.

by P. However, if they are assertions on a read-only heap, a triple can only be enriched with a frame if C does *not* mutate the heap. This restriction is somewhat artificial: if a frame R only specifies the values of the heap portion it owns, then C would be free to mutate.

Hence we lift R to $\langle R \rangle$, where $\langle \_ \rangle : \mathcal{P}\, H \to \mathcal{P}\, H^{ro}$ asserts R over the heap where (v,t) is kept and (v,f) is discarded. Note that $\langle \_ \rangle$ is an injective quantale morphism: $\langle P * Q \rangle = \langle P \rangle * \langle Q \rangle$, $\langle \bigvee S \rangle = (\bigvee x \in S. \langle x \rangle)$ and even$\langle \top \rangle = \top$, hence a quantale embedding. We can thus obtain triples

$$\forall R.\ \{P * \langle R \rangle\}\, C\, \{Q * \langle R \rangle\},$$

where C is free to mutate the resource described by P. What does linear negation mean in this setting? If we take $(p \to -)$ to be the assertion that *only* the address p is allocated, and $(p \hookrightarrow -)$ says that at *least* p is allocated, then for boolean negation we have that $\overline{(p \to -)}$ says that if p is allocated, then some other address is, and $\overline{(p \hookrightarrow -)}$ says that p is not allocated. For *linear* negation we have $(p \hookrightarrow -)^d = \overline{(p \hookrightarrow -)}^{\perp} = (p \hookrightarrow -)$, and $(p \to -)^d = \overline{(p \to -)}^{\perp}$, which says that if p is *not* allocated, then some other address is not.

With a PAM that cannot be lifted to a Girard quantale, and a binative PAM seemingly unsuitable for standard applications of separation logic, we have obtained an enriched assertion language taking the best parts of both. It might therefore be fruitful to find binative semigroups that can serve as targets for embeddings, rather than taken as resource models directly.

7 Other Residuals

A linear negation is not available in separation logic, but $\mathbin{-\!\!*}$ has been dualised, perhaps less naturally, with respect to boolean negation on the boolean assertion quantale. The resulting operation is known as septraction [5,25]. We study it in convolution quantales over a PAM without boolean complementation.

We define the *septraction* operation more generally as the convolution of $f, g : S \to Q$, where S is a PAM and Q a commutative quantale, as

$$(f \mathbin{-\!\!\circledast} g)\ x = \bigvee_{x \oplus y = z} f\ y \cdot g\ z.$$

The only difference to separating conjunction is that the supremum in y and z is now taken over $x \oplus y = z$ rather than $x = y \oplus z$. In the ternary relation $(\_) \oplus (\_) = (\_)$, septraction is thus separating conjunction up-to an exchange of variables. In such a general relational setting it has been shown that a convolution $\mathbin{-\!\!\circledast}$ is associative if and only if the dual ternary relation satisfies a relational associativity law [7]. For S, this clearly cannot be expected. Similarly, a unit exists in the convolution algebra if and only if the underlying PAM or relational structure has units [7]. It has also been shown that associativity of the ternary relation is not needed to make the convolution operation sup-preserving in both arguments [13]. These results specialise as follows.

Lemma 7.1

1. *If S is a* PAM *and Q a complete lattice equipped with sup-preserving operation $\cdot$, then $-\!\circledast$ preserves all sups on Q^S.*
2. *The operation $-\!\circledast$ need not be associative, commutative or have a unit, even for the* PAM*s of heaplets and statelets and for $Q = \mathbb{B}$.*

It follows that $-\!\circledast$ has two residuals: the right adjoints of $f -\!\circledast (\_)$ and $(\_) -\!\circledast f$. The first one has already been studied for the PAM H and $Q = \mathbb{B}$ as *(separating) coimplication* [2]; it is suitable for symbolic reasoning within separation logic. Here we define it abstractly on the convolution quantale Q^S as

$$f \rightsquigarrow\!\!* \, h = \bigvee\{g \mid f -\!\circledast g \leq h\}.$$

As a right adjoint, coimplication preserves infs, but is neither associative nor commutative. It does not have a unit either.

In Sect. 3 we have related separating conjunction and implication in Q^S to corresponding operations in S and Q. For $-\!\circledast$, a simple substitution yields

$$(f -\!\circledast g)\; x = \bigvee_{D\,x\,y} f\; y \cdot g\; (x \oplus y).$$

The name "septraction" is motivated by the following fact.

Lemma 7.2. *If S is a cancellative* PAM *and Q a quantale, then*

$$(f -\!\circledast g)\; x = \bigvee_{x \preceq z} f\; (z \ominus x) \cdot g\, z.$$

Similar results hold for $\rightsquigarrow\!\!*$.

Theorem 7.3. *If S is a* PAM *and Q a quantale, then*

1. $(f \rightsquigarrow\!\!* \, g)\, x = \bigwedge_{x = y \oplus z} f\; y \backslash g\; z$,
2. $(f \rightsquigarrow\!\!* \, g)\, x = \bigwedge_{y \preceq x} f\; y \backslash g\; (x \ominus y)$ *if S is cancellative.*

So far, we have considered septraction and coimplication in isolation. Even when they occur together with separating conjunction and magic wand in one single PAM, the target algebra Q could still be a double quantale with different monoidal multiplications for separating conjunction and septraction and different residuals for magic wand and coimplication—yet these two operations could also coincide, like in the following example.

Example 7.4. (*Powerset Lifting*) For the PAM on $X \times S$ and in particular for statelets,

$$\begin{aligned} f -\!\circledast g &= \{(x,y) \mid \exists y'.\; D\,y\,y' \wedge (x,y') \in f \wedge (x, y \oplus y') \in g\},\\ f \rightsquigarrow\!\!* \, g &= \{(x,y) \mid \forall y', y''.\; y \simeq y' \oplus y'' \wedge (x,y') \in f \rightarrow (x,y'') \in g\}\\ &= \{(x,y) \mid \forall y'.\; y' \preceq y \wedge (x,y') \in f \rightarrow (x, y \ominus y') \in g\}, \end{aligned}$$

where the second step for $\rightsquigarrow\!\!*$ requires cancellation. □

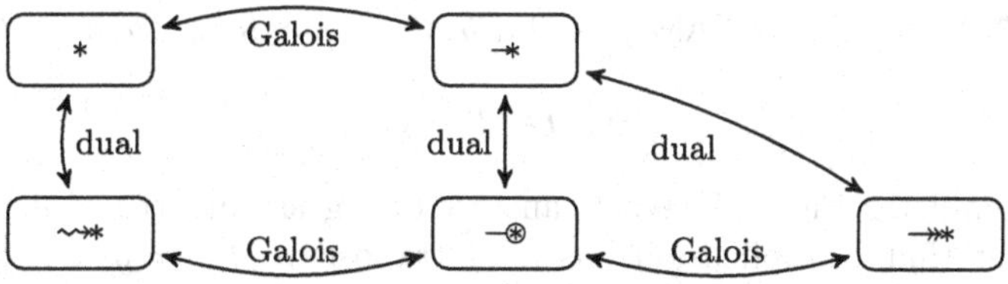

Fig. 1. Relationship between operators of separation logic

In boolean quantales, boolean complementation relates separating conjunction and coimplication on one hand, and septraction and magic wand on the other hand. In fact, this is how septraction and coimplication were originally defined for powerset quantales [2,5].

Theorem 7.5. *Let S be a* PAM *and Q a boolean quantale, Then, in Q^S,*

$$f \rightsquigarrow\!\!* \, g = \overline{\overline{f} * \overline{g}} \qquad \textit{and} \qquad f \mathrel{-\!\circledast} g = \overline{\overline{f} \mathrel{-\!*} \overline{g}}.$$

The relation between separating conjunction, implication, septraction and coimplication is also shown in Fig. 1.

Using the adjunction between $\mathrel{-\!\circledast}$ and $\rightsquigarrow\!\!*$, a complete method for generating strongest postconditions in separation logic is available [2]. It enables the transformation of any given Hoare triple—enriched with a frame—into a rule for forward reasoning. Symmetrically, the adjunction between $*$ and $\mathrel{-\!*}$ yield a method for backward reasoning, generating weakest preconditions. Ideas for this go back to the origins of separation logic [22].

To the best of our knowledge, the second right adjoint of septraction mentioned above has not been studied within the setting of separation logic. We define it abstractly on the convolution quantale Q^S as

$$g \twoheadrightarrow\!\!* \, h = \bigvee \{ f \mid f \mathrel{-\!\circledast} g \leq h \}.$$

The adjunction implies that preserves infs, but is neither associative nor commutative. It does not have a unit either.

Theorem 7.6. *If S is a* PAM *and Q a quantale, then*

$$(f \twoheadrightarrow\!\!* \, g)\, x = \bigwedge_{D\, x\, y} f\, (x \oplus y) \backslash g\, y.$$

Example 7.7 (*Powerset Lifting*). For the PAM on $X \times S$,

$$f \twoheadrightarrow\!\!* \, g = \{ (x, y) \mid \forall y'.\ D\, y\, y' \wedge (x, y \oplus y') \in f \rightarrow (x, y') \in g \}.$$

□

Boolean complementation relates this right adjoint back to magic wand.

Theorem 7.8. *Let S be a* PAM *and Q a boolean quantale, Then, in Q^S,*

$$f \mathrel{\twoheadrightarrow\!\!*} g = \overline{f} \mathrel{-\!\!*} \overline{g}.$$

This fact completes Fig. 1. I reveals an interesting asymmetry, which emerges from the fact that coimplication has a contrapositive $f \mathrel{\rightsquigarrow\!\!*} g = -g \mathrel{\rightsquigarrow\!\!*} -f$, whereas $\mathrel{-\!\!*}$ does not.

Theorems 7.5 and 7.8 suggest looking at the boolean dual $\overline{\overline{f} \mathrel{-\!\!*} g}$ as well. In the PAM on $X \times S$ this equates to $\{(x, y) \mid \exists y'.D\, y\, y' \wedge (x, y') \in \overline{f} \wedge y \oplus y' \in \overline{g}\}$. However, this is the same as $\overline{f} \mathrel{-\!\!\circledast} \overline{g}$ and and therefore the residual is $g \mathrel{\rightsquigarrow\!\!*} f$. Hence we stay in the setting of the well-known operators of separation logic.

Finally, we summarise these residuals in *boolean* Girard quantales.

Theorem 7.9. *Let S be a binative* PAS *and Q a boolean Girard quantale. Then in Q^S,*

1. $f \mathrel{\rightsquigarrow\!\!*} g = \overline{f * \overline{g}}$ *and* $f \mathrel{-\!\!\circledast} g = \overline{f \mathrel{-\!\!*} \overline{g}}$,
2. $f * g = (f \mathrel{-\!\!*} g^d)^d$ *and* $f \mathrel{-\!\!\circledast} g = (f \mathrel{\rightsquigarrow\!\!*} g^d)^d$,
3. $f \mathrel{\rightsquigarrow\!\!*} g = (f \mathrel{-\!\!*} g^\perp)^\perp$ *and* $f * g = (f \mathrel{-\!\!\circledast} g^\perp)^\perp$.

Item (1) has been copied from Theorem 7.5. The first part of (2) follows directly from linear negation of Girard quantales (see Sect. 4). The remaining identities follow Theorem 5.4. This combines the different strands of this paper, showing that in a boolean Girard quantale, such as the assertions over an effect algebra, there are three dualities—a boolean, a linear, and a binative one.

8 Another Assertion Quantale for Separation Logic

The standard assertion quantale of separation logic is also somewhat unnatural mathematically in that it does not reflect the order $\preceq$ on heaplets and statelets: it is not the case that $\{x\} \subseteq \{y\}$ iff $x \preceq y$. We present an alternative that supports such more fine-grained comparisons.

We fix a cancellative positive PAM S. Then $\preceq$ is a partial order for which the units are minimal by Lemma 2.3. For each $x \in S$, $x{\downarrow} = \{y \mid y \preceq x\}$; for each $X \subseteq S$, $X{\downarrow}$ is the image of X under $\downarrow$. We write $\mathcal{P}_{\downarrow} S$ for the set of downsets in S—the closed sets of the Alexandrov topology over $\preceq$.

We also need the following *Riesz decomposition property* [18] of S: for all $x, y_1, y_2 \in S$, $x \preceq y_1 \oplus y_2$ implies that there exist $x_1, x_2 \in S$ such that $x_1 \preceq y_1$, $x_2 \preceq y_2$ and $x_1 \oplus x_2 \preceq y_1 \oplus y_2$. It obviously holds in the heaplet and statelet models of separation logic.

Proposition 8.1. *Let S be a cancellative positive* PAM *that satisfies the Riesz decomposition property. Then $\mathcal{P}_{\downarrow} S$ forms a commutative quantale.*

Proof. Relative to Theorem 3.2 we need to check that $\{e\}$ is closed for each $e \in E$, which is the case due to positivity (Lemma 2.3), and that the quantalic multiplication and sups preserve downsets. First, using Riesz decomposition,

$$\begin{aligned}(X * Y){\downarrow} &= \{z \mid \exists x \in X, y \in Y.\ z \preceq x \oplus y \land D\,x\,y\}\\ &\subseteq \{x' \oplus y' \mid \exists x \in X, y \in Y.\ x' \preceq x \land y' \preceq y \land D\,x'\,y'\}\\ &= X{\downarrow} * Y{\downarrow}.\end{aligned}$$

Hence $(X{\downarrow} * Y{\downarrow}){\downarrow} = X{\downarrow} * Y{\downarrow}$, by extensivity and transitivity of ${\downarrow}$. Second, it is routine to check that $(\bigcup_{i \in I} X_i){\downarrow} = \bigcup_{i \in I}(X_i{\downarrow})$, for all I and therefore $(\bigcup_i X_i{\downarrow}){\downarrow} = \bigcup_i (X_i{\downarrow})$ by transitivity of ${\downarrow}$. (Similarly, $(\bigcap_i X_i{\downarrow}){\downarrow} = \bigcap_i (X_i{\downarrow})$, which is not strictly needed in the proof). □

Obviously, $(\overline{X{\downarrow}}){\downarrow}$ need not be equal to $\overline{X{\downarrow}}$: in the two-element poset defined by $p \prec q$, for instance, $\{q\} = \overline{\{p\}{\downarrow}}$ is not closed. The quantale $\mathcal{P}_{\downarrow} S$ is therefore not boolean in general. Many of the theorems in Sect. 7 fail. Whether this quantale is Girard is open as well. On one hand, the dualising set Δ used in Proposition 4.1 is closed. On the other hand, residuals are sups taken on the whole of $\mathcal{P}S$, so we should not expect that they preserve downsets. Similar models based on upclosed PAMs are well known as intuitionistic or affine assertion algebras of separation logic. See [19] for an overview and general approach.

9 Conclusion

In the context of convolution algebras of functions from partial abelian semigroups into commutative quantales, we have explored the standard operations of separation logic—separating conjunction and implication—and some less known ones (septraction, coimplication and a second right adjoint of septraction). Due to the generality of the approach, it can be used with weighted assertions. The Lawvere quantale makes them available in fuzzy settings, the well known isomorphic quantale on the unit interval to probabilistic reasoning.

As the combination of boolean complementation with the quantalic multiplication is somewhat unnatural, we have also investigated the link with the linear negation of Girard quantales. We have established a correspondence between effect algebras and commutative powerset Girard quantales, but shown that generalised effect algebras, where a greatest element is missing, cannot be lifted to such quantales. Our results imply that the classical heaplet and statelet models of separation logic do not admit a linear negation; separating conjunction and implication are therefore independent. Yet we have also shown how these models can be embedded into effect algebras and thus made linear negation available for separation logic in some cases.

We have generalised the lifting of effect algebras to binative partial semigroups and extended it from powerset quantales to arbitrary convolution Girard quantales. In this paper we only consider commutative algebras, but liftings for non-commutative algebras can be found in our Isabelle theories. We believe that these results are only stepping stones towards more general ones for binative relational monoids or multimonoids. In this setting one may consider the binary operations of separation logic as binary modalities and the underlying monoidal structures as ternary Kripke frames, as in the Jónsson-Tarski duality for boolean algebras with operators. The correspondence between effect algebras and commutative powerset Girard quantales is then a modal correspondence based on this duality. For convolution algebras we expect modal correspondence triangles between properties of relational monoids, value quantales and convolution quantales [7]. All this, and the relationship of other models that appear within the vast literature on separation logic (see for instance [1,20]), remains to be explored with a view on linear negation.

Other research questions relate to the generalisation of the adjunctions and dualities between the operations in Sect. 7 to non-commutative algebras, to their counterparts in convolution Girard quantales over effect algebras, where linear negation is present, to their status in the setting of non-boolean quantales, as the one introduced in Sect. 8 or those for affine and intuitionistic separation logic, and finally their generalisation to the setting of enriched categories [10].

Acknowledgments. This research was supported in part by Australian Research Council (ARC) Grant DP190102142. The third author acknowledges sponsorship by Labex DigiCosme for an invited professorship at Laboratoire d'informatique de l'École polytechnique. The authors are grateful for helpful comments by the anonymous reviewers.

References

1. Appel, A.W.: Program Logics - for Certified Compilers. Cambridge University Press, Cambridge (2014)
2. Bannister, C., Höfner, P., Klein, G.: Backwards and forwards with separation logic. In: Avigad, J., Mahboubi, A. (eds.) ITP 2018. LNCS, vol. 10895, pp. 68–87. Springer, Cham (2018). https://doi.org/10.1007/978-3-319-94821-8_5
3. Brotherston, J., Calcagno, C.: Classical BI: its semantics and proof theory. Logical Methods Comput. Sci. **6**(3), 1–42 (2010)
4. Brotherston, J., Villard, J.: Sub-classical boolean bunched logics and the meaning of par. In: CSL 2015, vol. 41 of LIPIcs, pp. 325–342. Schloss Dagstuhl - Leibniz-Zentrum für Informatik (2015)
5. Calcagno, C., Gardner, P., Zarfaty, U.: Context logic as modal logic: completeness and parametric inexpressivity. In: POPL 2007, pp. 123–134. ACM (2007)
6. Calcagno, C., O'Hearn, P.W., Yang, H.: Local action and abstract separation logic. In: LICS 2007, pp. 366–378. IEEE Computer Society (2007)
7. Cranch, J., Doherty, S., Struth, G.: Convolution and concurrency (2020). arXiv:2002.02321
8. Cranch, J., Doherty, S., Struth, G.: Relational semigroups and object-free categories (2020). arXiv:2001.11895
9. Dang, H.-H., Höfner, P., Möller, B.: Algebraic separation logic. J. Logic Algebraic Program. **80**(6), 221–247 (2011)

10. Day, B., Street, R.: Quantum categories, star autonomy, and quantum groupoids. In: Galois Theory, Hopf Algebras, and Semiabelian Categories, vol. 43 of Fields Institute Communications, pp. 193–231. American Mathematical Society (2004)
11. Dongol, B., Gomes, V.B.F., Hayes, I.J., Struth, G.: Partial semigroups and convolution algebras. Arch. Formal Proofs (2017)
12. Dongol, B., Gomes, V.B.F., Struth, G.: A program construction and verification tool for separation logic. In: Hinze, R., Voigtländer, J. (eds.) MPC 2015. LNCS, vol. 9129, pp. 137–158. Springer, Cham (2015). https://doi.org/10.1007/978-3-319-19797-5_7
13. Dongol, B., Hayes, I., Struth, G.: Convolution algebras: Relational convolution, generalised modalities and incidence algebras. Logical Methods Comput. Sci. **17**(1) (2021)
14. Dongol, B., Hayes, I.J., Struth, G.: Convolution as a unifying concept: applications in separation logic, interval calculi, and concurrency. ACM Trans. Comput. Logic **17**(3), 15:1-15:25 (2016)
15. Foulis, D.J., Bennett, M.K.: Effect algebras and unsharp quantum logics. Found. Phys. **24**, 1331–1352 (1994)
16. Haslbeck, M.P.L.: Verified Quantiative Analysis of Imperative Algorithms. PhD thesis, Fakuktät für Informatik, Technische Universität München (2021)
17. Hedlíková, J., Pulmannová, S.: Generalized difference posets and orthoalgebras. Acta Mathematica Universitatis Comenianae **LXV**, 247–279 (1996)
18. Jenča, G., Pulmannová, S.: Quotients of partial abelian monoids and the Riesz decomposition property. Algebra Universalis **47**, 443–447 (2002)
19. Jipsen, P., Litak, T.: An algebraic glimpse at bunched implications and separation logic (2017). CoRR, abs/1709.07063
20. Jung, R., Krebbers, R., Jourdan, J.-H., Bizjak, A., Birkedal, L., Dreyer, D.: Iris from the ground up: A modular foundation for higher-order concurrent separation logic. J. Funct. Program. **28**, e20 (2018)
21. Larchey-Wendling, D.:An alternative direct simulation of Minsky machines into classical bunched logics via group semantics. In: MFPS 2010, vol. 265 of ENTCS, pp. 369–387. Elsevier (2010)
22. O'Hearn, P., Reynolds, J., Yang, H.: Local reasoning about programs that alter data structures. In: Fribourg, L. (ed.) CSL 2001. LNCS, vol. 2142, pp. 1–19. Springer, Heidelberg (2001). https://doi.org/10.1007/3-540-44802-0_1
23. Rosenthal, K.L.: Quantales and their Applications. Longman Scientific & Technical, Harlow (1990)
24. Struth, G.: Quantales. Arch. Formal Proofs (2018)
25. Vafeiadis, V., Parkinson, M.: A marriage of rely/guarantee and separation logic. In: Caires, L., Vasconcelos, V.T. (eds.) CONCUR 2007. LNCS, vol. 4703, pp. 256–271. Springer, Heidelberg (2007). https://doi.org/10.1007/978-3-540-74407-8_18
26. Yetter, D.N.: Quantales and (noncommutative) linear logic. J. Symb. Logic **55**(1), 41–64 (1990)

Relational Computation of Sets of Relations

Rudolf Berghammer(✉)

Institut für Informatik, Universität Kiel, 24098 Kiel, Germany
rub@informatik.uni-kiel.de

Abstract. We present a technique for the relational computation of sets $\mathcal{R}$ of relations. It is based on a specification of a relation R to belong to $\mathcal{R}$ by means of an inclusion $\mathfrak{s} \subseteq \mathfrak{t}$, where $\mathfrak{s}$ and $\mathfrak{t}$ are relation-algebraic expressions constructed from a vector model of R in a specific way. To get the inclusion, we apply properties of a mapping that transforms relations into their vectors models and, if necessary, point-wise reasoning. The desired computation of $\mathcal{R}$ via a relation-algebraic expression $\mathfrak{r}$ is then immediately obtained from $\mathfrak{s} \subseteq \mathfrak{t}$ using a result of [3]. Compared with a direct development of $\mathfrak{r}$ from a logical specification of R to belong to $\mathcal{R}$, the proposed technique is much more simple. We demonstrate its use by some classes of specific relations and also show some applications.

1 Introduction

Reduced ordered binary decision diagrams (ROBDDs) are a very efficient data structure for the representation of sets and relations. This is also proved by numerous applications of RelView, a ROBDD-based tool for the manipulation and visualisation of relations and relational programming. The use of ROBDDs often leads to an amazing computational power of RelView, in particular, if the solution of a hard problem is based on the computation of a subset $\mathcal{R}$ of a powerset 2^X. See e.g., [2–4] for such applications. In certain situations X is a direct product, which means that RelView is used to compute a set $\mathcal{R}$ of relations. This is e.g., the case in [2], where $\mathcal{R}$ is the set of solutions of a timetabling problem and RelView is used to get solutions. Also in [5] a set $\mathcal{R}$ of relations is computed, viz. the set of up-closed multirelations on a given set. Experiments with RelView then lead to an appropriate definition of an approximation order for modeling computations which also may be infinite. A third example is [4]. Here $\mathcal{R}$ equals the set of pre-orders on a given set and RelView-results show a variant of the Kuratowski closure-complement-theorem, with closure-interior relatives instead of closure-complement relatives.

In this paper we present a general technique for the relational computation of sets $\mathcal{R}$ of relations. It is based on a specification of a relation R to belong to $\mathcal{R}$ by means of an inclusion $\mathfrak{s} \subseteq \mathfrak{t}$, where $\mathfrak{s}$ and $\mathfrak{t}$ are relation-algebraic expressions constructed from a vector model s of R in a specific way. Such expressions are introduced in [3] and can be seen as the syntactical counterparts of the vector

U. Fahrenberg et al. (Eds.): RAMiCS 2021, LNCS 13027, pp. 54–71, 2021.
https://doi.org/10.1007/978-3-030-88701-8_4

predicates of [7]. To get the specification $\mathfrak{s} \subseteq \mathfrak{t}$, we frequently apply properties of a mapping that transforms a relation to its vector model, but also argue point-wisely if this helps. The desired computation of $\mathcal{R}$ via a relation-algebraic expression $\mathfrak{r}$ is then obtained from $\mathfrak{s} \subseteq \mathfrak{t}$ in one step using a general result proved in [3]. Compared with the technique that directly develops $\mathfrak{r}$ from a logical specification of the relation R to belong to $\mathcal{R}$, the proposed technique is much more simple. We demonstrate its use by some classes of specific relations and also show some applications and how the expressions $\mathfrak{r}$ can be implemented as RelView-programs and then evaluated using the tool.

2 Relation-Algebraic Preliminaries

Given sets X and Y, we write $R : X \leftrightarrow Y$ if R is a (binary) relation with source X and target Y, i.e., a subset of the direct product $X \times Y$. If the sets X and Y of the *type* $X \leftrightarrow Y$ of R are finite, we may consider R as a Boolean matrix with $|X|$ rows and $|Y|$ columns. Since a matrix interpretation of relations is well suited for many purposes and also used by RelView as the main possibility to visualise relations, in the following we often use matrix terminology and notation. Especially, we speak about entries/components, rows and columns of a relation/matrix and write $R_{x,y}$ instead of $(x, y) \in R$ or $x\,R\,y$.

We assume the reader to be familiar with the five basic operations on relations, written as R^{T} (*transposition*), $\overline{R}$ (*complementation*), $R \cup S$ (*union*), $R \cap S$ (*intersection*) and $R;S$ (*composition*), the two basic predicates, written as $R \subseteq S$ (*inclusion*) and $R = S$ (*equality*), and the three special relations, written as O (*empty relation*), L (*universal relation*) and I (*identity relation*).

We denote the set of relations of type $X \leftrightarrow Y$ by $[X \leftrightarrow Y]$ instead of $2^{X \times Y}$. For each type $X \leftrightarrow Y$ then $[X \leftrightarrow Y]$ together with the Boolean operations $\cup$, $\cap$ and $^-$ and the constants $\mathsf{O} : X \leftrightarrow Y$ and $\mathsf{L} : X \leftrightarrow Y$ forms a complete Boolean lattice, where the lattice order is given by inclusion. Further well-known laws of relations are, for instance, $(R^{\mathsf{T}})^{\mathsf{T}} = R$, $(R;S)^{\mathsf{T}} = S^{\mathsf{T}};R^{\mathsf{T}}$, $Q;(R \cap S) \subseteq Q;R \cap Q;S$, $Q;(R \cup S) = Q;R \cup Q;S$ and $\overline{R}^{\mathsf{T}} = \overline{R^{\mathsf{T}}}$, for all relations Q, R and S (of fitting types). We assume that transposition and complementation bind stronger than composition and composition binds stronger than union and intersection.

The theoretical framework for these laws and many others to hold is that of a (heterogeneous) relation algebra, introduced in [9] as generalisation of a (homogeneous) relation algebra in the sense of A. Tarski and further developed in [10,11], for example. The five operations and three constants of this algebraic structure are denoted as those of the set-theoretic relations. As usual, in relation-algebraic expressions we overload the symbols O, L and I, i.e., avoid the binding of types to them. The axioms of a relation algebra are those of a complete Boolean lattice for $\cup$, $\cap$, $^-$, O and L (with lattice order $\subseteq$), that composition is associative and possesses identity relations as neutral elements, that the *Schröder equivalences* $Q^{\mathsf{T}};\overline{S} \subseteq \overline{R}$ iff $Q;R \subseteq S$ iff $\overline{S};R^{\mathsf{T}} \subseteq \overline{Q}$ hold, for all relations Q, R and S, and that $R \neq \mathsf{O}$ implies $\mathsf{L};R;\mathsf{L} = \mathsf{L}$, for all relations R. In later proofs we shall mention only the Schröder equivalences and "non-obvious" consequences of the axioms like $Q; R \cap S \subseteq (Q \cap S; R^{\mathsf{T}}); (R \cap Q^{\mathsf{T}}; S)$, for all relations Q, R and S, called *Dedekind rule*. Well-known laws as those presented above or in Sect. 2.1 to 2.3 of [10] remain unmentioned.

Relation algebra as just introduced can express exactly the formulae of first-order predicate logic with at most two free variables and all in all at most three variables. The expressive power of full first-order predicate logic is obtained if projection relations or equivalent notions are assumed to exist. Because of the intended applications in later sections, in this paper we consider set-theoretic relations only. But we treat them, as far as possible, with relation-algebraic means. So, for all direct products $X \times Y$ the two *projection relations* exist, which we denote as $\pi : X \times Y \leftrightarrow X$ and $\rho : X \times Y \leftrightarrow Y$. We always assume a pair u from a direct product to be of the form $u = (u_1, u_2)$. This allows to describe π and ρ point-wisely by $\pi_{u,x}$ iff $u_1 = x$ and $\rho_{u,y}$ iff $u_2 = y$, for all $u \in X \times Y$, $x \in X$ and $y \in Y$. In [10] it is shown that the formulae $\pi^{\mathsf{T}}; \pi = \mathsf{I}$, $\rho^{\mathsf{T}}; \rho = \mathsf{I}$, $\pi; \pi^{\mathsf{T}} \cap \rho; \rho^{\mathsf{T}} = \mathsf{I}$ and $\pi^{\mathsf{T}}; \rho = \mathsf{L}$ specify projection relations up to isomorphism and imply π and ρ to be surjective functions. Recall that R is a *function* (in the relational sense) if it is *univalent*, specified by $R^{\mathsf{T}}; R \subseteq \mathsf{I}$, and *total*, specified by $R; \mathsf{L} = \mathsf{L}$ or, equivalently, by $\mathsf{I} \subseteq R; R^{\mathsf{T}}$, and R is *surjective* iff R^{T} is total.

Based on the projection relations $\pi : X \times Y \leftrightarrow X$ and $\rho : X \times Y \leftrightarrow Y$ for two relations $R : X \leftrightarrow Z$ and $S : Y \leftrightarrow Z$ their *left pairing* is defined by $[\![R, S]\!] := \pi; R \cap \rho; S$, thereby being of type $X \times Y \leftrightarrow Z$. Using point-wise notation, from this we get that $[\![R, S]\!]_{u,z}$ iff $R_{u_1,z}$ and $S_{u_2,z}$, for all $u \in X \times Y$ and $z \in Z$. The counterpart to the left pairing, with now $Z \leftrightarrow X \times Y$ as type, is the *right pairing* $[R, S]\!]$ of two relations $R : Z \leftrightarrow X$ and $S : Z \leftrightarrow Y$. To get the desired property that $[R, S]\!]_{z,u}$ iff R_{z,u_1} and S_{z,u_2}, for all $z \in Z$ and $u \in X \times Y$, the notion is defined as $[R, S]\!] := R; \pi^{\mathsf{T}} \cap S; \rho^{\mathsf{T}}$. In Sect. 3 we will use that for all relations $R_1 : Z_1 \leftrightarrow X$, $R_2 : X \leftrightarrow Z_2$, $S_1 : Z_1 \leftrightarrow Y$ and $S_2 : Y \leftrightarrow Z_2$ from the univalence of R_1 and S_1 or the univalence of R_2 and S_2 the subsequent Eq. (1) follows:

$$[R_1, S_1]\!]; [\![R_2, S_2]\!] = R_1; R_2 \cap S_1; S_2 \tag{1}$$

A relation-algebraic proof of this fact can be found in [1]. For a point-wise proof of (1) it is not necessary that R_1 and S_1 are both univalent or R_2 and S_2 are both univalent. But, as shown by R. Maddux in [8], a proof of (1) that only is based on the axioms of a relation algebra, the above axiomatisation of projection relations and the definitions of left pairings and right pairings is impossible.

If the two arguments of a right pairing with reference to the projection relations $\pi : X \times Y \leftrightarrow X$ and $\rho : X \times Y \leftrightarrow Y$ are compositions $\alpha; R$ and $\beta; S$, respectively, with $\alpha : Z \times U \leftrightarrow Z$ and $\beta : Z \times U \leftrightarrow U$ as projection relations of a further direct product $Z \times U$ and relations $R : Z \leftrightarrow X$ and $S : U \leftrightarrow Y$, then we define $R \,\|\, S := [\alpha; R, \beta; S]\!]$ as the *parallel composition* of R and S, thereby being of type $Z \times U \leftrightarrow X \times Y$. Using a point-wise notation we have $(R \,\|\, S)_{u,v}$ iff R_{u_1,v_1} and S_{u_2,v_2}, for all $u \in Z \times U$ and $v \in X \times Y$.

In the next sections we also will use the relation-level equivalents of the set-theoretic symbol '$\in$' as basic relations. These are the *membership relations*

$\mathsf{M} : X \leftrightarrow 2^X$ and point-wisely described by $\mathsf{M}_{x,Y}$ iff $x \in Y$, for all $x \in X$ and $Y \in 2^X$. There exists a relation-algebraic axiomatisation of membership relations which specifies these up to isomorphism. See e.g., [11]. But for the applications of the present paper the above point-wise description suffices.

3 Vectors and the Modeling of Sets and Relations

In this paper we use (relational) vectors to model sets. A *vector* is a relation s such that $s = s; \mathsf{L}$. In the Boolean matrix interpretation this means that each row of s consists only of ones or only of zeros. As the targets of vectors are irrelevant, we only consider vectors of type $X \leftrightarrow \mathbf{1}$, with a specific singleton set $\mathbf{1} := \{\perp\}$ as common target. Such vectors correspond to Boolean column vectors and, therefore, as in linear algebra we write s_x instead of $s_{x,\perp}$.

By definition $s : X \leftrightarrow \mathbf{1}$ *models* (or *is a vector model of*) the subset Y of X if for all $x \in X$ we have $x \in Y$ iff s_x. This means that precisely those entries of s are 1 that correspond to an element of Y. Obviously, the mapping (in the usual mathematical sense) $set : [X \leftrightarrow \mathbf{1}] \to 2^X$, defined by $set(s) = \{x \in X \mid s_x\}$, for all $s : X \leftrightarrow \mathbf{1}$, is a Boolean lattice isomorphism from $([X \leftrightarrow \mathbf{1}], \cup, \cap, ^-, \mathsf{O}, \mathsf{L})$ to $(2^X, \cup, \cap, ^-, \emptyset, X)$, with the inverse mapping $set^{-1} : 2^X \to [X \leftrightarrow \mathbf{1}]$ given by $set^{-1}(Y) = Y \times \{\perp\}$, for all $Y \in 2^X$.

A general technique to compute for a subset Y of a given set X a vector model $s : X \leftrightarrow \mathbf{1}$ is to start with an arbitrary element x from X and a logical specification $\varphi(x)$ of x to belong to Y. Using the point-wise descriptions of relational constants and operations, e.g., of those introduced in Sect. 2, then the formula $\varphi(x)$ is transformed step-by-step into the form $\mathfrak{s}_x$, where now $\mathfrak{s}$ is a relation-algebraic expression of type $X \leftrightarrow \mathbf{1}$, i.e., a *vector expression*. From the equivalence of $\varphi(x)$ and the relationship $\mathfrak{s}_x$, for all $x \in X$, it follows $set(\mathfrak{s}) = Y$ and, hence, $\mathfrak{s}$ relation-algebraically specifies the vector s we are looking for.

If the just sketched technique is applied for the development of a vector that models a subset $\mathcal{R}$ of a powerset 2^X, then the starting point is a formula $\varphi(Y)$ that specifies the arbitrarily given set Y from 2^X to belong to $\mathcal{R}$. In such a case the development of the relationship $\mathfrak{s}_Y$ from $\varphi(Y)$ frequently becomes lengthy, cumbersome and error-prone if carried out by hand. To considerable simplify the development of $\mathfrak{s}$, in [3] instead of Y and $\varphi(Y)$ a vector model $s : X \leftrightarrow \mathbf{1}$ of Y and a relation-algebraic specification of $set(s)$ to belong to $\mathcal{R}$ are taken as starting point. E.g., if $R : X \leftrightarrow X$ is the adjacency relation of a directed graph $G = (X, R)$ and the goal is to get a vector model $\mathfrak{stable}(R) : 2^X \leftrightarrow \mathbf{1}$ of the set $\mathcal{R}$ of stable vertex sets of G, then instead of starting with $\forall x, y : R_{x,y} \land y \in Y \Rightarrow x \notin Y$ as formula $\varphi(Y)$ one starts with $R; s \subseteq \overline{s}$, with an arbitrary vector $s : X \leftrightarrow \mathbf{1}$, since $set(s) \in \mathcal{R}$ iff $R; s \subseteq \overline{s}$. Decisive for this approach to work are vector expressions of a specific syntactic form, which in [3] are introduced as follows.

Definition 3.1. *Given* $s : X \leftrightarrow \mathbf{1}$*, the set* $\mathfrak{V}(s)$ *of typed* column-wise extendible vector expressions *over* s *is inductively defined as follows:*

a) *We have* $s \in \mathfrak{V}(s)$ *and its type is* $X \leftrightarrow \mathbf{1}$.
b) *If* $v : Y \leftrightarrow \mathbf{1}$ *is different from* s, *then* $v \in \mathfrak{V}(s)$ *and its type is* $Y \leftrightarrow \mathbf{1}$.
c) *If* $\mathfrak{s} \in \mathfrak{V}(s)$ *is of type* $Y \leftrightarrow \mathbf{1}$, *then* $\overline{\mathfrak{s}} \in \mathfrak{V}(s)$ *and its type is* $Y \leftrightarrow \mathbf{1}$.
d) *If* $\mathfrak{s}, \mathfrak{t} \in \mathfrak{V}(s)$ *are of type* $Y \leftrightarrow \mathbf{1}$, *then* $\mathfrak{s} \cup \mathfrak{t} \in \mathfrak{V}(s)$ *and* $\mathfrak{s} \cap \mathfrak{t} \in \mathfrak{V}(s)$ *and their types are* $Y \leftrightarrow \mathbf{1}$.
e) *If* $\mathfrak{s} \in \mathfrak{V}(s)$ *is of type* $Y \leftrightarrow \mathbf{1}$ *and* $\mathfrak{R}$ *is a relation-algebraic expression of type* $Z \leftrightarrow Y$ *in which* s *does not occur, then* $\mathfrak{R};\mathfrak{s} \in \mathfrak{V}(s)$ *and its type is* $Z \leftrightarrow \mathbf{1}$.

So, the vector expressions from $\mathfrak{V}(s)$ are built from s using other vectors and as operations only complementation, union, intersection and left-composition with a relation different from s. They can be seen as syntactical counterparts of B. Kehden's vector predicates, which are introduced in [7] as mappings in the usual mathematical sense for the relational treatment of evolutionary algorithms.

In $\mathfrak{s} \in \mathfrak{V}(s)$ the vector s can be seen as a variable in the logical sense. Using this interpretation, we define next the replacement of s in $\mathfrak{s}$ by R as in [3].

Definition 3.2. *Given* $s : X \leftrightarrow \mathbf{1}$, $\mathfrak{s} \in \mathfrak{V}(s)$ *and* $R : X \leftrightarrow Z$, *we define* $\mathfrak{s}[R/s]$ *as follows, using induction on the structure of* $\mathfrak{s}$:

a) $s[R/s] = R$.
b) $v[R/s] = v;\mathsf{L}$, *with* $\mathsf{L} : \mathbf{1} \leftrightarrow Z$.
c) $\overline{\mathfrak{t}}[R/s] = \overline{\mathfrak{t}[R/s]}$.
d) $(\mathfrak{t} \cup \mathfrak{u})[R/s] = \mathfrak{t}[R/s] \cup \mathfrak{u}[R/s]$ *and* $(\mathfrak{t} \cap \mathfrak{u})[R/s] = \mathfrak{t}[R/s] \cap \mathfrak{u}[R/s]$.
e) $(\mathfrak{R};\mathfrak{t})[R/s] = \mathfrak{R};(\mathfrak{t}[R/s])$.

For all $s : X \leftrightarrow \mathbf{1}$, $\mathfrak{s} \in \mathfrak{V}(s)$ of type $Y \leftrightarrow \mathbf{1}$ and $R : X \leftrightarrow Z$ the type of $\mathfrak{s}[R/s]$ is $Y \leftrightarrow Z$; see [3]. So, for R as membership relation $\mathsf{M} : X \leftrightarrow 2^X$ we get $Y \leftrightarrow 2^X$ as type of $\mathfrak{s}[\mathsf{M}/s]$. As main result in [3] the following theorem is shown. It states a general procedure for developing a vector model of a subset $\mathcal{R}$ of 2^X.

Theorem 3.1. *Assume* $\mathcal{R}$ *to be a subset of the powerset* 2^X. *If it is specified as* $\mathcal{R} = \{set(s) \mid s : X \leftrightarrow \mathbf{1} \wedge \mathfrak{s} \subseteq \mathfrak{t}\}$, *with* $\mathfrak{s}, \mathfrak{t} \in \mathfrak{V}(s)$ *both of type* $Y \leftrightarrow \mathbf{1}$, *then*

$$\mathfrak{r} := \overline{\mathsf{L};(\mathfrak{s}[\mathsf{M}/s] \cap \overline{\mathfrak{t}[\mathsf{M}/s]})}^{\mathsf{T}} : 2^X \leftrightarrow \mathbf{1}$$

(where $\mathsf{L} : \mathbf{1} \leftrightarrow Y$ *and* $\mathsf{M} : X \leftrightarrow 2^X$*) is a vector model of* $\mathcal{R}$.

For $\mathcal{R}$ being the set of stable vertex sets of the directed graph $G = (X, R)$ we get $\mathcal{R} = \{set(s) \mid s : X \leftrightarrow \mathbf{1} \wedge R;s \subseteq \overline{s}\}$. As $R;s \in \mathfrak{V}(s)$ and $\overline{s} \in \mathfrak{V}(s)$, Theorem 3.1 yields $\mathfrak{stable}(R) = \overline{\mathsf{L};(R;\mathsf{M} \cap \mathsf{M})}^{\mathsf{T}}$ as vector model of the set $\mathcal{R}$.

In Sect. 4 we will apply Theorem 3.1 for the computation of sets of relations. This means that $\mathcal{R}$ is a subset of a set $[X \leftrightarrow Y]$, the vector s of the specification of $\mathcal{R}$ in Theorem 3.1 has type $X \times Y \leftrightarrow \mathbf{1}$ and $\mathfrak{r}$ has type $[X \leftrightarrow Y] \leftrightarrow \mathbf{1}$. Since in such a case s models a relation, instead of $set(s)$ the notation $rel(s)$ is used. This leads to s_u iff $rel(s)_{u_1,u_2}$, for all $u \in X \times Y$. In [10] a relation-algebraic specification of the mapping $rel : [X \times Y \leftrightarrow \mathbf{1}] \to [X \leftrightarrow Y]$ is given, viz.

$rel(s) = \pi^{\mathsf{T}};(s;\mathsf{L} \cap \rho)$, for all $s : X \times Y \leftrightarrow \mathbf{1}$, where $\pi : X \times Y \leftrightarrow X$ and $\rho : X \times Y \leftrightarrow Y$ are the projection relations of $X \times Y$ and $\mathsf{L} : \mathbf{1} \leftrightarrow Y$. In [10] also the inverse mapping, which we denote as $vec : [X \leftrightarrow Y] \to [X \times Y \leftrightarrow \mathbf{1}]$, is specified with relation-algebraic means. Translating the specification of [10] into a version with a left pairing, we get $vec(R) = [\![R, \mathsf{I}]\!];\mathsf{L}$, for all $R : X \leftrightarrow Y$, where $\mathsf{I} : Y \leftrightarrow Y$ and $\mathsf{L} : Y \leftrightarrow \mathbf{1}$. With relation-algebraic means the following theorem is shown in [10].

Theorem 3.2. *The mappings rel and vec are Boolean lattice isomorphisms from* $([X \times Y \leftrightarrow \mathbf{1}], \cup, \cap, ^{-}, \mathsf{O}, \mathsf{L})$ *to* $([X \leftrightarrow Y], \cup, \cap, ^{-}, \mathsf{O}, \mathsf{L})$ *and vice versa and mutually inverse.*

For typing reasons, the mapping *vec* neither can distribute over compositions nor commutate with transpositions. Instead we have the following two results, which also decisively will be used later. The first one, published in [7] deals with vector models of compositions.

Theorem 3.3. *Assume* Q, R *and* S *(of fitting types) to be given. Then we have* $vec(Q;R;S) = (Q \parallel S^{\mathsf{T}});vec(R)$.

The second result shows how $vec(R^{\mathsf{T}})$ can be reduced to $vec(R)$. We formulate and prove it only for relations for which source and target coincide, so-called *homogeneous relations*, since this suffices for our later applications.

Theorem 3.4. *Given* $R : X \leftrightarrow X$, *we have* $vec(R^{\mathsf{T}}) = [\rho, \pi]\!];vec(R)$, *where* $\pi : X^2 \leftrightarrow X$ *and* $\rho : X^2 \leftrightarrow X$ *are the projection relations of* X^2.

Proof. First, we apply the definition of $vec(R)$ in combination with (1) and get

$$[\rho, \pi]\!];vec(R) = [\rho, \pi]\!];[\![R, \mathsf{I}]\!];\mathsf{L} = (\rho;R \cap \pi;\mathsf{I});\mathsf{L} = (\rho;R \cap \pi);\mathsf{L}.$$

Next, the definitions of $vec(R^{\mathsf{T}})$ and $[\![R^{\mathsf{T}}, \mathsf{I}]\!]$ yield

$$vec(R^{\mathsf{T}}) = [\![R^{\mathsf{T}}, \mathsf{I}]\!];\mathsf{L} = (\pi;R^{\mathsf{T}} \cap \rho;\mathsf{I});\mathsf{L} = (\pi;R^{\mathsf{T}} \cap \rho);\mathsf{L}.$$

Now, the proof is concluded by the following calculation:

$$\begin{aligned} (\rho;R \cap \pi);\mathsf{L} &\subseteq (\rho \cap \pi;R^{\mathsf{T}});(R \cap \rho^{\mathsf{T}};\pi);\mathsf{L} && \text{Dedekind rule} \\ &\subseteq (\pi;R^{\mathsf{T}} \cap \rho);\mathsf{L} \\ &\subseteq (\pi \cap \rho;R);(R^{\mathsf{T}} \cap \pi^{\mathsf{T}};\rho);\mathsf{L} && \text{Dedekind rule} \\ &\subseteq (\rho;R \cap \pi);\mathsf{L} \end{aligned}$$

□

If a vector model of a subset $\mathcal{R}$ of $[X \leftrightarrow Y]$ is specified as vector expression $\mathfrak{r}$, it is simple to compute a relation of $\mathcal{R}$. We select an injective and surjective vector p, i.e., a (relational) *point* p, such that $p \subseteq \mathfrak{r}$. With the membership relation $\mathsf{M} : X \times Y \leftrightarrow [X \leftrightarrow Y]$ then $\mathsf{M};p : X \times Y \leftrightarrow \mathbf{1}$ is the vector model of a relation $R \in \mathcal{R}$ and, hence, R itself is obtained via $R = rel(\mathsf{M};p)$. In RelView for the selection of a point from a vector there exists a pre-defined operation `point`.

4 Computing Sets of Specific Relations

In Sect. 2 we have mentioned relation-algebraic specifications of a relation R to be univalent, total and surjective, respectively. R is *injective* iff R^{T} is univalent. Based on these specifications and using the Schröder equivalences, Theorem 3.2 and Theorem 3.3, in [2] the following equivalences are shown.

Theorem 4.1. *Assume* $s : X \times Y \leftrightarrow \mathbf{1}$ *to be given and let* $\pi : X \times Y \leftrightarrow X$ *and* $\rho : X \times Y \leftrightarrow Y$ *be the projection relations of* $X \times Y$*. Then we have:*

$$rel(s) \textit{ univalent} \iff (\mathsf{I} \,\|\, \overline{\mathsf{I}}); s \subseteq \overline{s} \qquad rel(s) \textit{ total} \iff \mathsf{L} \subseteq \pi^{\mathsf{T}}; s$$
$$rel(s) \textit{ injective} \iff (\overline{\mathsf{I}} \,\|\, \mathsf{I}); s \subseteq \overline{s} \qquad rel(s) \textit{ surjective} \iff \mathsf{L} \subseteq \rho^{\mathsf{T}}; s$$

Each side of the four inclusions of Theorem 4.1 is a column-wise extendible vector expression over s. Hence, Theorem 3.1 is applicable and immediately yields the following vector models of type $[X \leftrightarrow Y] \leftrightarrow \mathbf{1}$ of the sets of univalent, total, injective and surjective relations of type $X \leftrightarrow Y$:

$$\mathfrak{unival}(X,Y) := \overline{\mathsf{L}; ((\mathsf{I} \,\|\, \overline{\mathsf{I}}); \mathsf{M} \cap \mathsf{M})}^{\mathsf{T}} \qquad \mathfrak{total}(X,Y) := \overline{\mathsf{L}; \overline{\pi^{\mathsf{T}}; \mathsf{M}}}^{\mathsf{T}}$$
$$\mathfrak{injec}(X,Y) := \overline{\mathsf{L}; ((\overline{\mathsf{I}} \,\|\, \mathsf{I}); \mathsf{M} \cap \mathsf{M})}^{\mathsf{T}} \qquad \mathfrak{surjec}(X,Y) := \overline{\mathsf{L}; \overline{\rho^{\mathsf{T}}; \mathsf{M}}}^{\mathsf{T}}$$

The types of the basic relations of these four specifications easily can be derived from the typing rules of the relational operations. E.g., in case of $\mathfrak{unival}(X,Y)$ the universal relation has type $\mathbf{1} \leftrightarrow X \times Y$, the left identity relation of $\mathsf{I} \,\|\, \overline{\mathsf{I}}$ has type $X \leftrightarrow X$, the right one has type $Y \leftrightarrow Y$ and the type of the membership relation is $X \times Y \leftrightarrow [X \leftrightarrow Y]$. Therefore, in the remainder of the paper we make no mention of types of basic relations in such specifications.

Obviously, conjunction of relational properties corresponds to intersection of the corresponding vector models such that, for example, vector models of the sets of functions and bijective relations of type $X \leftrightarrow Y$ can be specified as follows:

$$\mathfrak{funct}(X,Y) := \mathfrak{unival}(X,Y) \cap \mathfrak{total}(X,Y)$$
$$\mathfrak{bijec}(X,Y) := \mathfrak{injec}(X,Y) \cap \mathfrak{surjec}(X,Y)$$

Now, we consider some important properties of homogeneous relations. Recall that $R : X \leftrightarrow X$ is *reflexive* iff $\mathsf{I} \subseteq R$, *irreflexive* iff $R \subseteq \overline{\mathsf{I}}$, *symmetric* iff $R \subseteq R^{\mathsf{T}}$, *antisymmetric* iff $R \cap R^{\mathsf{T}} \subseteq \mathsf{I}$, *asymmetric* iff $R \cap R^{\mathsf{T}} \subseteq \mathsf{O}$, *transitive* iff $R; R \subseteq R$ and *complete* iff $\overline{\mathsf{I}} \subseteq R \cup R^{\mathsf{T}}$. Except transitivity, from Theorem 3.2 and Theorem 3.4 we immediately get the following specifications of these properties of a relation R by means of the vector model $s := vec(R)$.

Theorem 4.2. *Assume* $s : X^2 \leftrightarrow \mathbf{1}$ *to be given and let* $\pi : X^2 \leftrightarrow X$ *and* $\rho : X^2 \leftrightarrow X$ *be the projection relations of* X^2*. Then we have:*

$$rel(s) \textit{ reflexive} \iff vec(\mathsf{I}) \subseteq s$$
$$rel(s) \textit{ irreflexive} \iff s \subseteq vec(\overline{\mathsf{I}})$$
$$rel(s) \textit{ symmetric} \iff s \subseteq [\rho, \pi]; s$$
$$rel(s) \textit{ antisymmetric} \iff s \cap [\rho, \pi]; s \subseteq vec(\mathsf{I})$$
$$rel(s) \textit{ asymmetric} \iff s \cap [\rho, \pi]; s \subseteq \mathsf{O}$$
$$rel(s) \textit{ complete} \iff vec(\overline{\mathsf{I}}) \subseteq s \cup [\rho, \pi]; s$$

Again all sides of the six inclusions of Theorem 4.2 are column-wise extendible vector expressions over s. So, a combination of Theorem 4.2 with Theorem 3.1 (and $vec(\overline{\mathsf{I}});\mathsf{L} = \overline{vec(\mathsf{I});\mathsf{L}}$ in the second case) immediately yields the following vector models of type $[X \leftrightarrow X] \leftrightarrow \mathbf{1}$ of the sets of reflexive, irreflexive, symmetric, antisymmetric, asymmetric and complete relations of type $X \leftrightarrow X$:

$$\mathfrak{refl}(X) := \overline{\mathsf{L};(vec(\mathsf{I});\mathsf{L} \cap \overline{\mathsf{M}})}^{\mathsf{T}}$$

$$begineqnarray* - 0.0mm]\mathfrak{irrefl}(X) := \overline{\mathsf{L};(\mathsf{M} \cap vec(\mathsf{I});\mathsf{L})}^{\mathsf{T}}$$

$$\begin{aligned}
\mathfrak{symm}(X) &:= \overline{\mathsf{L};(\mathsf{M} \cap \overline{[\![\rho,\pi]\!];\mathsf{M}})}^{\mathsf{T}} \\
\mathfrak{antisymm}(X) &:= \overline{\mathsf{L};(\mathsf{M} \cap [\![\rho,\pi]\!];\mathsf{M} \cap \overline{vec(\mathsf{I});\mathsf{L}})}^{\mathsf{T}} \\
\mathfrak{asymm}(X) &:= \overline{\mathsf{L};(\mathsf{M} \cap [\![\rho,\pi]\!];\mathsf{M})}^{\mathsf{T}} \\
\mathfrak{compl}(X) &:= \overline{\mathsf{L};(vec(\overline{\mathsf{I}});\mathsf{L} \cap \overline{\mathsf{M}} \cap \overline{[\![\rho,\pi]\!];\mathsf{M}})}^{\mathsf{T}}
\end{aligned}$$

The right-hand side of $\mathfrak{refl}(X)$ already can be found in [4], with a direct derivation from the point-wise description $\forall x,y : R_{x,y}$ of the reflexivity of R. The other vector models are not part of [4].

Given $s : X^2 \leftrightarrow \mathbf{1}$, in the next theorem we also specify the transitivity of $rel(s)$ by means of an inclusion between column-wise extendible vector expressions over s in such a way that Theorem 3.1 again can be used and directly yields a vector model $\mathfrak{trans}(X) : [X \leftrightarrow X] \leftrightarrow \mathbf{1}$ of the set of transitive relations of type $X \leftrightarrow X$. The two-fold occurrence of R within the left-hand side of $R;R \subseteq R$ prevents an application of Theorem 3.3. Instead we use the following lemma. It is a special case of an unpublished theorem of M. Winter (where R and S may be heterogeneous), told to the author as a private communication. The relation-algebraic proof of M. Winter is too complex and too long to be presented here.

Lemma 4.1. *Assume $R : X \leftrightarrow X$ and $S : X \leftrightarrow X$ to be given. Furthermore, let $\pi : X^2 \leftrightarrow X$ and $\rho : X^2 \leftrightarrow X$ be the projection relations of X^2 and $\alpha : X^2 \times X^2 \leftrightarrow X^2$ and $\beta : X^2 \times X^2 \leftrightarrow X^2$ be the projection relations of $X^2 \times X^2$. Then we have $vec(R;S) = C^{\mathsf{T}};[\![vec(R), vec(S)]\!]$, where*

$$C := (\mathsf{I} \cap \alpha;\rho;\pi^{\mathsf{T}};\beta^{\mathsf{T}});(\pi \parallel \rho) : X^2 \times X^2 \leftrightarrow X^2.$$

Proof. Let arbitrary pairs $(u,v) \in X^2 \times X^2$ and $(w,z) \in X^2 \times X^2$ be given. The definition of the projection relations and of the relational composition yields

$$(\alpha;\rho;\pi^{\mathsf{T}};\beta^{\mathsf{T}})_{(u,v),(w,z)} \iff u_2 = z_1.$$

To enhance readability, in the following we abbreviate in formulae conjunctions of equations as equational chain. Then the above equivalence implies

$$(\mathsf{I} \cap \alpha;\rho;\pi^{\mathsf{T}};\beta^{\mathsf{T}})_{(u,v),(w,z)} \iff u_1 = w_1 \wedge u_2 = w_2 = z_1 = v_1 \wedge v_2 = z_2.$$

Using this equivalence and the definition of C in the first step, for all pairs $(u, v) \in X^2 \times X^2$ and $(x, y) \in X^2$ we now calculate as follows:

$$\begin{aligned} C_{(u,v),(x,y)} &\iff \exists w, z : \\ &\quad u_1 = w_1 \wedge u_2 = w_2 = z_1 = v_1 \wedge v_2 = z_2 \wedge (\pi \,\|\, \rho)_{(w,z),(x,y)} \\ &\iff \exists w, z : \\ &\quad u_1 = w_1 \wedge u_2 = w_2 = z_1 = v_1 \wedge v_2 = z_2 \wedge w_1 = x \wedge z_2 = y \\ &\iff u_1 = x \wedge v_2 = y \wedge u_2 = v_1 \end{aligned}$$

As a consequence, we get for all pairs $(x, y) \in X^2$ the following equivaleces, which show the claim:

$$\begin{aligned} (C^{\mathsf{T}}; [\![vec(R), vec(S)]\!])_{(x,y)} &\iff \exists u, v : C_{(u,v),(x,y)} \wedge [\![vec(R), vec(S)]\!]_{(u,v)} \\ &\iff \exists u, v : u_1 = x \wedge v_2 = y \wedge \\ &\qquad u_2 = v_1 \wedge vec(R)_u \wedge vec(S)_v \\ &\iff \exists a : vec(R)_{(x,a)} \wedge vec(S)_{(a,y)} \\ &\iff \exists a : R_{x,a} \wedge S_{a,y} \\ &\iff (R; S)_{x,y} \\ &\iff vec(R; S)_{(x,y)} \end{aligned}$$

□

Here is the announced specification of transitivity,

Theorem 4.3. *Assume* $s : X^2 \leftrightarrow \mathbf{1}$ *to be given. With* $C : X^2 \times X^2 \leftrightarrow X^2$ *as defined in Lemma 4.1 we have:*

$$rel(s) \textit{ transitive} \iff C^{\mathsf{T}}; [\![s, s]\!] \subseteq s$$

Proof. The claim follows from the calculation

$$\begin{aligned} rel(s) \text{ transitive} &\iff rel(s); rel(s) \subseteq rel(s) \\ &\iff vec(rel(s); rel(s)) \subseteq vec(rel(s)) \\ &\iff C^{\mathsf{T}}; [\![vec(rel(s)), vec(rel(s))]\!] \subseteq vec(rel(s)) \\ &\iff C^{\mathsf{T}}; [\![s, s]\!] \subseteq s, \end{aligned}$$

where we use the definition of transitivity, then Theorem 3.2, then Lemma 4.1 and, finally, again Theorem 3.2. □

Left pairings of column-wise extendible vector expressions are again column-wise extendible vector expressions. As a consequence, both sides of the inclusion of Theorem 4.3 are column-wise extendible vector expressions over s. Hence, Theorem 3.1 is applicable and at once yields the following specification:

$$\mathfrak{trans}(X) := \overline{\mathsf{L}; (C^{\mathsf{T}}; [\![\mathsf{M}, \mathsf{M}]\!] \cap \overline{\mathsf{M}})}^{\mathsf{T}}$$

A lengthy direct derivation of a specification of $\mathfrak{trans}(X)$ from the point-wise description $\forall x, y, z : R_{x,y} \wedge R_{y,z} \Rightarrow R_{x,z}$ of the transitivity of R can be found in [4]. Experiments with the RelView-implementations of both versions have shown that the second one is less efficient than the above specification.

Intersections of the above specifications directly allow to get vector models of the set of pre-orders, partial orders, equivalence relations, tournaments and many other well-known classes of homogeneous relations. To give an example, if we want to model the set of *proper involutions* (that is, of self-inverse permutations without fixpoints) on X, we can do this as follows:

$$\mathfrak{propInvolut}(X) := \mathfrak{funct}(X,X) \cap \mathfrak{bijec}(X,X) \cap \mathfrak{symm}(X) \cap \mathfrak{irrefl}(X)$$

Next, we treat three classes of specific strict-orders which play a prominent role in preference modeling, viz. weak-orders, semi-orders and interval-orders. A relation $R : X \leftrightarrow X$ is a *weak-order* if it is asymmetric and *negatively transitive*, where the latter means $\overline{R}$ to be transitive. For defining semi-orders and interval-orders we need the notions of *semi-transitivity* of R, defined as $R; R; \overline{R}^{\mathsf{T}} \subseteq R$, and of R to be a *Ferrers relation*, defined as $R; \overline{R}^{\mathsf{T}}; R \subseteq R$. Then R is a *semi-order* if it is irreflexive, semi-transitive and a Ferrers relation. If it is only an irreflexive Ferrers relation, it is an *interval-order*. Usually, these classes of relations are not defined in such a way. For instance, interval-orders $<$ are defined by assigning intervals of the real line to the elements of the carrier sets and then $x < y$ holds iff the interval assigned to x is completely left of that assigned to y. The above relation-algebraic specifications can be found in [11].

We want to apply our technique also to specify vector models $\mathfrak{weakOrd}(X)$, $\mathfrak{semiOrd}(X)$ and $\mathfrak{intOrd}(X)$ of type $[X \leftrightarrow X] \leftrightarrow \mathbf{1}$ of the sets of weak-orders, semi-orders and interval-orders of type $X \leftrightarrow X$. Decisive for that is the following theorem. In it we treat the basic properties of the above definitions not been addressed until now.

Theorem 4.4. *Assume $s : X^2 \leftrightarrow \mathbf{1}$ to be given. With $C : X^2 \times X^2 \leftrightarrow X^2$ as defined in Lemma 4.1 and $\pi : X^2 \leftrightarrow X$ and $\rho : X^2 \leftrightarrow X$ as the projection relations of X^2 we have:*

$$\begin{aligned} rel(s) \textit{ negatively transitive} &\iff C^{\mathsf{T}}; [\![\overline{s}, \overline{s}]\!] \subseteq \overline{s} \\ rel(s) \textit{ semi-transitive} &\iff C^{\mathsf{T}}; [\![\overline{s}, \overline{s}]\!] \subseteq \overline{C^{\mathsf{T}}; [\![s, s]\!]} \\ rel(s) \textit{ Ferrers relation} &\iff C^{\mathsf{T}}; [\![[\rho,\pi]\!]; s, \overline{s}]\!] \subseteq \overline{C^{\mathsf{T}}; [\![[\rho,\pi]\!]; \overline{s}, s]\!]} \end{aligned}$$

Proof. To prove the first claim, we start as follows, where we use the definition of negative transitivity and Theorem 3.2:

$$\begin{aligned} rel(s) \text{ negatively transitive} &\iff \overline{rel(s)}; \overline{rel(s)} \subseteq \overline{rel(s)} \\ &\iff rel(\overline{s}); rel(\overline{s}) \subseteq rel(\overline{s}) \end{aligned}$$

The remaining steps are as in the proof of Theorem 4.3, with $\overline{s}$ instead of s.

In case of the second claim, we start with the definition of semi-transitivity, use then one of the Schröder equivalences and then Theorem 3.2, leading to the following calculation:

$$\begin{aligned} rel(s) \text{ semi-transitive } &\iff rel(s);rel(s);\overline{rel(s)}^{\mathsf{T}} \subseteq rel(s) \\ &\iff \overline{rel(s)};\overline{rel(s)} \subseteq \overline{rel(s);rel(s)} \\ &\iff vec(\overline{rel(s)};\overline{rel(s)}) \subseteq vec(\overline{rel(s);rel(s)}) \end{aligned}$$

Next, we transform the left-hand side of the last inclusion as follows:

$$\begin{aligned} vec(\overline{rel(s)};\overline{rel(s)}) &= C^{\mathsf{T}};\llbracket vec(\overline{rel(s)}), vec(\overline{rel(s)})\rrbracket && \text{Lemma 4.1} \\ &= C^{\mathsf{T}};\llbracket \overline{vec(rel(s))}, \overline{vec(rel(s))}\rrbracket && \text{Theorem 3.2} \\ &= C^{\mathsf{T}};\llbracket \overline{s}, \overline{s}\rrbracket && \text{Theorem 3.2} \end{aligned}$$

In a similar way $vec(\overline{rel(s);rel(s)}) = \overline{C^{\mathsf{T}};\llbracket s,s\rrbracket}$ can be shown and we are done.

Also in case of the third claim we start with the definition of a Ferrers relation, use then one of the Schröder equivalences and then Theorem 3.2. This yields:

$$\begin{aligned} rel(s) \text{ Ferrers relation } &\iff rel(s);\overline{rel(s)}^{\mathsf{T}};rel(s) \subseteq rel(s) \\ &\iff rel(s)^{\mathsf{T}};\overline{rel(s)} \subseteq \overline{\overline{rel(s)}^{\mathsf{T}};rel(s)} \\ &\iff vec(rel(s)^{\mathsf{T}};\overline{rel(s)}) \subseteq vec(\overline{\overline{rel(s)}^{\mathsf{T}};rel(s)}) \end{aligned}$$

The treatment of both sides of the last inclusion is rather similar to the calculation in the proof of the second claim. In case of the left-hand side we have:

$$\begin{aligned} vec(rel(s)^{\mathsf{T}};\overline{rel(s)}) &= C^{\mathsf{T}};\llbracket vec(rel(s)^{\mathsf{T}}), vec(\overline{rel(s)})\rrbracket && \text{Lemma 4.1} \\ &= C^{\mathsf{T}};\llbracket \llbracket\rho,\pi\rrbracket;vec(rel(s)), vec(\overline{rel(s)})\rrbracket && \text{Theorem 3.4} \\ &= C^{\mathsf{T}};\llbracket \llbracket\rho,\pi\rrbracket;vec(rel(s)), \overline{vec(rel(s))}\rrbracket && \text{Theorem 3.2} \\ &= C^{\mathsf{T}};\llbracket \llbracket\rho,\pi\rrbracket;s, \overline{s}\rrbracket && \text{Theorem 3.2} \end{aligned}$$

Eqution $vec(\overline{\overline{rel(s)}^{\mathsf{T}};rel(s)}) = \overline{C^{\mathsf{T}};\llbracket\llbracket\rho,\pi\rrbracket;\overline{s},s\rrbracket}$ can be shown in a similar way. □

All sides of the three inclusions of Theorem 4.4 are column-wise extendible vector expressions over s. Hence, Theorem 3.1 immediately yields the following vector models of type $[X \leftrightarrow X] \leftrightarrow \mathbf{1}$ of the sets of relations Theorem 4.4 deals with:

$$\begin{aligned} \mathfrak{negTrans}(X) &:= \overline{\mathsf{L};(C^{\mathsf{T}};\llbracket\overline{\mathsf{M}},\overline{\mathsf{M}}\rrbracket \cap \mathsf{M})}^{\mathsf{T}} \\ \mathfrak{semiTrans}(X) &:= \overline{\mathsf{L};(C^{\mathsf{T}};\llbracket\overline{\mathsf{M}},\overline{\mathsf{M}}\rrbracket \cap C^{\mathsf{T}};\llbracket\mathsf{M},\mathsf{M}\rrbracket}^{\mathsf{T}} \\ \mathfrak{ferrers}(X) &:= \overline{\mathsf{L};(C^{\mathsf{T}};\llbracket\llbracket\rho,\pi\rrbracket;\mathsf{M},\overline{\mathsf{M}}\rrbracket \cap C^{\mathsf{T}};\llbracket\llbracket\rho,\pi\rrbracket;\overline{\mathsf{M}},\mathsf{M}\rrbracket)}^{\mathsf{T}} \end{aligned}$$

From the definitions of weak-orders, semi-orders and Ferrers relations we now obtain the vector models $\mathfrak{weakOrd}(X)$, $\mathfrak{semiOrd}(X)$ and $\mathfrak{intOrd}(X)$ we are looking for by means of intersections as follows:

$$\begin{aligned} \mathfrak{weakOrd}(X) &:= \mathfrak{asymm}(X) \cap \mathfrak{negTrans}(X) \\ \mathfrak{semiOrd}(X) &:= \mathfrak{irrefl}(X) \cap \mathfrak{semiTrans}(X) \cap \mathfrak{ferrers}(X) \\ \mathfrak{intOrd}(X) &:= \mathfrak{irrfl}(X) \cap \mathfrak{ferrers}(X) \end{aligned}$$

There are situations where each relation of the set $\mathcal{R}$ of relations we want to compute is contained in a given relation R or contains R. This is e.g., the case if $R : X \leftrightarrow X$ is a partial order and $\mathcal{R}$ is the set of *linear extensions* of R, i.e., of the superrelations of R which are partial orders and complete. Having a vector model $\mathfrak{suprel}(R)$ of type $[X \leftrightarrow X] \leftrightarrow \mathbf{1}$ of the set of superrelations of R at hand, it immediately allows to specify a vector model of type $[X \leftrightarrow X] \leftrightarrow \mathbf{1}$ of the set of linear extensions of R as follows:

$$\mathfrak{linExt}(R) := \mathfrak{refl}(X) \cap \mathfrak{antisymm}(X) \cap \mathfrak{trans}(X) \cap \mathfrak{compl}(X) \cap \mathfrak{suprel}(R)$$

To get a specification of $\mathfrak{suprel}(R)$ and also of the vector model $\mathfrak{subrel}(R)$ of the set of subrelations of R, we use the following theorem, which is a direct consequence of Theorem 3.2.

Theorem 4.5. *Assume $s : X \times Y \leftrightarrow \mathbf{1}$ and $R : X \leftrightarrow Y$ to be given. Then we have:*

$$R \subseteq rel(s) \iff vec(R) \subseteq s \qquad rel(s) \subseteq R \iff s \subseteq vec(R)$$

The sides of the two inclusions of Theorem 4.5 are column-wise extendible vector expressions over s. Using Theorem 3.1 we, therefore, immediately get the following specifications of $\mathfrak{suprel}(R)$ and $\mathfrak{subrel}(R)$:

$$\mathfrak{suprel}(R) := \overline{\mathsf{L}; (vec(R); \mathsf{L} \cap \overline{\mathsf{M}})}^{\mathsf{T}} \qquad \mathfrak{subrel}(R) := \overline{\mathsf{L}; (\mathsf{M} \cap \overline{vec(R); \mathsf{L}}}^{\mathsf{T}}$$

We close this section with an application of $\mathfrak{subrel}(R)$, where $R : X \leftrightarrow X$ is again a partial order. Recall that a closure operator with respect to R is an extensive, idempotent and monotone mapping on X. If it is considered as a function $C : X \leftrightarrow X$ in the relational sense, then extensiveness is described by $C \subseteq R$, idempotency by $C; C \subseteq C$ and monotonicity by $R; C \subseteq C; R$. In the next theorem we show how the *homomorphism property* $R; F \subseteq F; S$ of F with respect to R and S can be specified in terms of the vector model of F.

Theorem 4.6. *Assume $s : X \times Y \leftrightarrow \mathbf{1}$ and $R : X \leftrightarrow X$ and $S : Y \leftrightarrow Y$ to be given. Then we have:*

$$R;rel(s) \subseteq rel(s);S \iff (R \,\|\, \mathsf{I});s \subseteq (\mathsf{I} \,\|\, S^{\mathsf{T}});s$$

Proof. The following calculation shows the claim, where we use Theorem 3.2, then Theorem 3.3 and finally again Theorem 3.2:

$$\begin{aligned} R; rel(s) \subseteq rel(s); S &\iff vec(R; rel(s); \mathsf{I}) \subseteq vec(\mathsf{I}; rel(s); S) \\ &\iff (R \,\|\, \mathsf{I}^{\mathsf{T}}); vec(rel(s)) \subseteq (\mathsf{I} \,\|\, S^{\mathsf{T}}); vec(rel(s)) \\ &\iff (R \,\|\, \mathsf{I}); s \subseteq (\mathsf{I} \,\|\, S^{\mathsf{T}}); s \end{aligned}$$

□

As both sides of the inclusion of Theorem 4.6 are column-wise extendible vector expressions over s, Theorem 3.1 yields the following vector model of type $[X \leftrightarrow Y] \leftrightarrow \mathbf{1}$ for the set of relations which satisfy the homomorphism property with respect to $R : X \leftrightarrow X$ and $S : Y \leftrightarrow Y$:

$$\mathfrak{homProp}(R, S) := \overline{\mathsf{L}; ((R \,\|\, \mathsf{I}); \mathsf{M} \cap \overline{(\mathsf{I} \,\|\, S^{\mathsf{T}}); \mathsf{M}})}^{\mathsf{T}}$$

And here is the specification of the vector model of type $[X \leftrightarrow X] \leftrightarrow \mathbf{1}$ for the set of closure operators with respect to the partial order $R : X \leftrightarrow X$:

$$\mathfrak{closOp}(R) := \mathfrak{funct}(X, X) \cap \mathfrak{suprel}(R) \cap \mathfrak{trans}(X) \cap \mathfrak{homProp}(R, R)$$

5 Applications and Implementation

Each specification of the vector models of Sect. 4 immediately can be implemented within the programming language of RelView. To demonstrate this, we consider the following RelView-program for $\mathfrak{injec}(X, Y)$, where (since RelView knows relations as the only data type) the two inputs X and Y are homogeneous relations the carrier sets of which determine the sets X and Y.

```
injec(X,Y)
  DECL XxY = PROD(X,Y);
       pi, M, L
  BEG pi = p-1(XxY); M = epsi(pi); L = Ln1(pi)^
      RETURN -(L*(parcomp(-I(X),I(Y))*M & M))^
  END.
```

In the program's declaration part a relational direct product XxY for $X \times Y$ and variables pi for the projection relation $\pi : X \times Y \leftrightarrow X$, M for the membership relation $\mathsf{M} : X \times Y \leftrightarrow [X \leftrightarrow Y]$ and L for the universal relation $\mathsf{L} : \mathbf{1} \leftrightarrow X \times Y$ are introduced. The three assignments of the body then compute these relations by means of three pre-defined RelView-operations and store them in pi, M and L. Finally, the return-clause – a direct translation of the specification of $\mathfrak{injec}(X, Y)$ into RelView-code – computes the result. In this RelView-expression a small RelView-program parcomp for computing parallel compositions is used.

If relation algebra is extended by projection relations and membership relations, the expressive power of full second-order predicate logic is obtained. This logic allows to specify for each set X the *size-comparison relation* $\mathsf{S} : 2^X \leftrightarrow 2^X$ such that $\mathsf{S}_{A,B}$ iff $|A| \leq |B|$, for all $A, B \in 2^X$. As usual in set theory, $|A| \leq |B|$ means that there exists an injective mapping $f : A \to B$. As first application of the results of Sect. 4 we present a relation-algebraic specification of S. To this end, we introduce the *right residual* $R \setminus S := \overline{R^{\mathsf{T}}; \overline{S}}$ and the *symmetric quotient* $syq(R, S) := (R \setminus S) \cap (\overline{R} \setminus \overline{S})$ of relations $R : X \leftrightarrow Y$ and $S : X \leftrightarrow Z$. Then both, $R \setminus S$ and $syq(R, S)$ have type $Y \leftrightarrow Z$ and for all $y \in Y$ and $z \in Z$ they point-wisely are described as follows:

$$(R \setminus S)_{y,z} \iff \forall x : R_{x,y} \Rightarrow S_{x,z} \qquad syq(R, S)_{y,z} \iff \forall x : R_{x,y} \Leftrightarrow S_{x,z} \quad (2)$$

Besides these two derived relational operations we need the projection relations $\pi : X^2 \leftrightarrow X$ and $\rho : X^2 \leftrightarrow X$ of X^2 and two membership relations. Since the latter have different types, we use different symbols. For that of type $X \leftrightarrow 2^X$ we use M as before, for that of type $X^2 \leftrightarrow [X \leftrightarrow X]$ we use $\mathbf{M}$. After these preparations we can prove the following result.

Theorem 5.1. *For all size-comparison relations* $\mathsf{S} : 2^X \leftrightarrow 2^X$ *we have:*

$$\mathsf{S} = (syq(\mathsf{M}, \pi^{\mathsf{T}};\mathbf{M}) \cap \mathsf{L};inject(X,X)^{\mathsf{T}});((\rho^{\mathsf{T}};\mathbf{M}) \setminus \mathsf{M})$$

Proof. Assume arbitrary sets $A, B \in 2^X$ to be given. Using the right description of (2), for all $R : X \leftrightarrow X$ we obtain

$$\begin{aligned} syq(\mathsf{M}, \pi^{\mathsf{T}};\mathbf{M})_{A,R} &\iff \forall x : \mathsf{M}_{x,A} \Leftrightarrow (\pi^{\mathsf{T}};\mathsf{M})_{x,R} \\ &\iff \forall x : x \in A \Leftrightarrow \exists u : \pi_{u,x} \wedge \mathbf{M}_{u,R} \\ &\iff \forall x : x \in A \Leftrightarrow \exists y : R_{x,y} \\ &\iff A = dom(R), \end{aligned}$$

where $dom(R)$ denotes the domain of R. Similarly, using the left description of (2), for all $R : X \leftrightarrow X$ we get $((\rho^{\mathsf{T}};\mathbf{M}) \setminus \mathsf{M})_{R,B}$ iff $ran(R) \subseteq B$, where $ran(R)$ denotes the range of R. As a consequence, the relationship

$$((syq(\mathsf{M}, \pi^{\mathsf{T}};\mathbf{M}) \cap \mathsf{L};injec(X,X)^{\mathsf{T}});((\rho^{\mathsf{T}};\mathbf{M}) \setminus \mathsf{M}))_{A,B} \tag{3}$$

holds iff there exists an injective relation $R : X \leftrightarrow X$ such that $A = dom(R)$ and $ran(R) \subseteq B$. Restricting the source of R to A and the target of R to B we get that (3) holds iff there exists an injective relation $S : A \leftrightarrow B$ such that $A = dom(S)$ and $ran(S) \subseteq B$. Hence, S is also total. The Axiom of Choice implies that S contains a function as a subrelation. So, an injective relation $S : A \leftrightarrow B$ exists iff an injective function $F : A \leftrightarrow B$ exists, i.e., iff $\mathsf{S}_{A,B}$. □

Translated into a RelView-program, the specification of the size-comparison relations of Theorem 5.1 looks as follows:

```
sizeComp(X)
  DECL XxX = PROD(X,X);
       pi, rho, M, MM
  BEG pi = p-1(XxX); rho = p-2(XxX); M = epsi(X); MM = epsi(pi)
      RETURN (syq(M,pi^*MM) & Ln1(M)*injec(X,X)^)*((rho^*MM)\M)
  END.
```

Systematic experiments are an accepted means for doing science and meanwhile they have also become important in mathematics and computer science. They are used, e.g., for gaining insight and intuition, for identifying properties and for testing conjectures. In the following we demonstrate how RelView and the results of Sect. 4 can be used in that regard.

For given $R : X \leftrightarrow X$ and $S : Y \leftrightarrow Y$ a relation $F : X \leftrightarrow Y$ is a *homomorphism* from R to S if it is a function and satisfies the homomorphism property $R;F \subseteq F;S$. If F is a bijective function and satisfies $R;F = F;S$, then it is an *isomorphism* from R to S. By an intersection of the vector models $\mathfrak{funct}(X, Y)$ and $\mathfrak{homProp}(R, S)$ of Sect. 4 we get the following vector model of type $[X \leftrightarrow Y] \leftrightarrow \mathbf{1}$ for the set of homomorphisms from R to S:

$$\mathfrak{hom}(R, S) := \mathfrak{funct}(X, Y) \cap \mathfrak{homProp}(R, S)$$

The equation $R; F = F; S$ is equivalent to $R; F \subseteq F; S$ and $F; S \subseteq R; F$. To obtain a vector model of the set of relations F which satisfy $F; S \subseteq R; F$, we can proceed as in the case of $\mathfrak{homProp}(R, S)$. Doing so, we finally get the following vector model of type $[X \leftrightarrow Y] \leftrightarrow \mathbf{1}$ for the set of isomorphisms from R to S:

$$\mathfrak{iso}(R, S) := \mathfrak{hom}(R, S) \cap \mathfrak{bijec}(X, Y) \cap \mathsf{L}; \overline{((\mathsf{I} \,||\, S^\mathsf{T}); \mathsf{M} \cap \overline{(R \,||\, \mathsf{I}); \mathsf{M}})}^\mathsf{T}$$

There exist relations R and S with a bijective homomorphism from R to S that is not an isomorphism – even if the types of R and S are equal. We have investigated the still stronger restriction $R = S$. To this end, we formulated the equation $\mathfrak{hom}(R, R) \cap \mathfrak{bijec}(X, X) = \mathfrak{iso}(R, R)$ as a RelView-program. Then we executed it for all relations $R : X \leftrightarrow X$ on $X = \{1, \ldots, n\}$, with $1 \leq n \leq 5$. Doing so, the relations R were computed by a loop through all points p contained in $\mathsf{L} : [X \leftrightarrow X] \leftrightarrow \mathbf{1}$ and using $R = rel(\mathsf{M}; p)$ as explained at the end of Sect. 3. In each case the result was 'true'. This and a generalisation to again two relations R and S (which was obtained by an analysis of a previous proof of the specific case $R = S$) led to Theorem 5.2 below. To our knowledge, it seems not been published until now. For the proof of Theorem 5.2 we need the following properties, for all relations Q, R and S, where, again as in set theory, $|A| = |B|$ means that there exists a bijective mapping $f : A \to B$:

$$|R^\mathsf{T}| = |R| \qquad Q \text{ univalent} \implies |R \cap Q^\mathsf{T}; S| \leq |Q; R \cap S| \tag{4}$$

The equation is obvious. The implication is shown in [6] by Y. Kawahara. Both properties are part of an axiomatisation of the cardinality of relations in [6].

Theorem 5.2. *Assume $R : X \leftrightarrow X$, $S : Y \leftrightarrow Y$ and $F : X \leftrightarrow Y$ to be given such that $|R| = |S|$ and F is a bijective function. Then $|R; F| = |F; S|$. Furthermore, if R is finite and $R; F \subseteq F; S$, then $R; F = F; S$.*

Proof. The following calculation shows $|F; S| = |S|$:

$$\begin{aligned}
|S| &= |S \cap F^\mathsf{T}; \mathsf{L}| && F \text{ surjective} \\
&\leq |F; S \cap \mathsf{L}| && F \text{ univalent, implication of (4)} \\
&= |\mathsf{L} \cap F^{\mathsf{T}^\mathsf{T}}; S| && \\
&\leq |F^\mathsf{T}; \mathsf{L} \cap S| && F \text{ injective, implication of (4)} \\
&= |S| && F \text{ surjective}
\end{aligned}$$

If we replace in this calculation S by R^T and F by F^T, we get $|F^\mathsf{T}; R^\mathsf{T}| = |R^\mathsf{T}|$ and the equation of (4) yields $|R; F| = |(R; F)^\mathsf{T}| = |F^\mathsf{T}; R^\mathsf{T}| = |R^\mathsf{T}| = |R|$. Altogether, we have $|R; F| = |R| = |S| = |F; S|$.

Now, let R be finite. Then $|R| = |R;F| = |F;S|$ shows that $F;S$ is finite. By the usual definition of finitenes in set theory there is no Q such that $Q \subset F;S$ and $|Q| = |F;S|$ or, equivalently, for all Q from $Q \subseteq F;S$ and $|Q| = |F;S|$ it follows $Q = F;S$. So, $R;F \subseteq F;S$ and $|R;F| = |F;S|$ imply $R;F = F;S$. □

In particular, already a bijective homomorphism on a finite relation (or a graph) is an automorphism. Notice that neither X nor Y nor F have to be finite. Finiteness of R and S, however, is necessary as the following example by M. Winter shows. Consider $R : \mathbb{Z} \leftrightarrow \mathbb{Z}$ and $F : \mathbb{Z} \leftrightarrow \mathbb{Z}$, defined by $R_{x,y}$ iff $x \geq 0$ and $x = y$ and $F_{x,y}$ iff $x + 1 = y$, for all $x, y \in \mathbb{Z}$. Then F is a bijective function. Simple calculations show $(R;F)_{x,y}$ iff $x \geq 0$ and $x + 1 = y$ and $(F;R)_{x,y}$ iff $x \geq -1$ and $x+1 = y$, for all $x, y \in \mathbb{Z}$. From these properties we get $R;F \subset F;R$.

In 1969 D. Scott presented a partially ordered set $(D, \leq)$ that is isomorphic to the partially ordered set $([D \rightarrow D], \sqsubseteq)$ of continuous mappings on D ordered by the function order induced by $\leq$, i.e., by $f \sqsubseteq g$ iff $f(x) \leq g(x)$, for all $x \in D$. The set D is the inverse limit of a retraction sequence starting with $(X_0, \leq_0)$, where $X_0 := \{\bot, \top\}$ and $\bot \leq_0 \top$, and continued by X_{n+1} as set of monotone mappings on X_n and $\leq_{n+1}$ as function order induced by $\leq_n$. In [12] the construction is described in detail, the partially ordered sets $(X_1, \leq_1)$ and $(X_2, \leq_2)$ are presented and it is noted that $|X_3| = 120\,549$. We have used RelView to verify this number and even have been able to compute $\leq_3$. Decisive for that is the following relation-algebraic specification of the function order, where $inj(v) : \mathcal{F} \leftrightarrow [X \leftrightarrow X]$ is the *injective embedding* induced by v, that is, the identity function.

Theorem 5.3. *Let $R : X \leftrightarrow X$ be a partial order, $v : [X \leftrightarrow X] \leftrightarrow \mathbf{1}$ be the vector model of a set $\mathcal{F}$ of functions and $\mathsf{F} : \mathcal{F} \leftrightarrow \mathcal{F}$ be the function order induced by R. Then we have $\mathsf{F} = \overline{inj(v); \mathsf{M}^\mathsf{T}; (R \,\|\, \overline{R}); \mathsf{M}; inj(v)^\mathsf{T}}$.*

Proof. For all $F, G \in \mathcal{F}$ we calculate as follows to show the claim:

$$\begin{aligned}
\mathsf{F}_{F,G} &\iff \forall u, v : F_{u_1,u_2} \wedge G_{v_1,v_2} \wedge R_{u_1,v_1} \Rightarrow R_{u_2,v_2} \\
&\iff \neg\exists u, v : F_{u_1,u_2} \wedge G_{v_1,v_2} \wedge R_{u_1,v_1} \wedge \overline{R}_{u_2,v_2} \\
&\iff \neg\exists u, v : F_{u_1,u_2} \wedge (R \,\|\, \overline{R})_{u,v} \wedge G_{v_1,v_2} \\
&\iff \neg\exists u : u \in F \wedge \exists v : (R \,\|\, \overline{R})_{u,v} \wedge v \in G \\
&\iff \neg\exists u : (\mathsf{M}; inj(v)^\mathsf{T})_{u,F} \wedge \exists v : (R \,\|\, \overline{R})_{u,v} \wedge (\mathsf{M}; inj(v)^\mathsf{T})_{v,G} \\
&\iff \neg\exists u : (inj(v); \mathsf{M}^\mathsf{T})_{F,u} \wedge ((R \,\|\, \overline{R}); \mathsf{M}; inj(v)^\mathsf{T})_{u,G} \\
&\iff \overline{inj(v); \mathsf{M}^\mathsf{T}; (R \,\|\, \overline{R}); \mathsf{M}; inj(v)^\mathsf{T}}_{F,G}
\end{aligned}$$

□

RelView contains a pre-defined operation `inj` for computing injective embeddings. If we take $\mathcal{F}$ as set of monotone functions and suppose `hom` to implement $\mathfrak{hom}(R, S)$, then the following RelView-program computes $\leq_{n+1}$ from $\leq_n$.

```
functOrder(R)
  DECL XxX = PROD(R,R);
       Inj, M
  BEG Inj = inj(hom(R,R)); M = epsi(p-1(XxX))*Inj^
      RETURN -(M^*parcomp(R,-R)*M)
  END.
```

The following three pictures show the partial orders $\leq_0$, $\leq_1$ and $\leq_2$ as Boolean RelView-matrices, where a black (white) square means a 1-entry (0-entry).

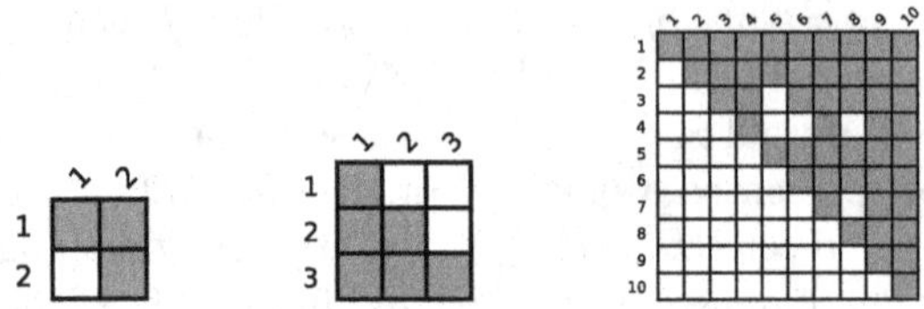

On a PC with 2 CPUs of type Intel® Xeon® E5-2698, each with 20 cores and 3.60 GHz base frequency, 512 GByte RAM and running Arch Linux 5.2.0, RelView needs 31.36 s to compute the partial order $\leq_3$ as a ROBDD with 2 500 126 nodes and to report that its carrier set consists of 120 549 elements and its Boolean matrix has 1 805 247 020 1-entries. It is noteworthy that only 0.02 s suffice for the evaluation of `hom(R,R)`, i.e., for the computation of the vector model of the set of mappings on X_2 which are $\leq_2$-monotone.

6 Concluding Remarks

We have applied our technique to some other classes of relations including rectangles, matchings, Aumann contact relations, difunctional relations, Noetherian relations and bipartitions respectively bichromatic partitions of relations. By it we also obtained the vector model of [5] for the set of up-closed multirelations in a simple way.

During the past 30 years RelView proved to be an excellent tool for supporting work with relations, especially the development of relational algorithms. But it only is able to treat set-theoretic relations on finite carrier sets. For this reason results that seem to confirm an abstract relation-algebraic property are to handle with some care. Namely, it may happen that the considered property holds for all set-theoretic relations on finite carrier sets but not in case of infinite carrier sets. It is known that there are even properties that hold for all set-theoretic relations (i.e., also with infinite carrier sets) but not in axiomatic relation algebras. Theorem 5.2 is an example for the first situation and Eq. (1) is an examples for the second one. But, fortunately, such situations are rare.

Acknowledgment. I want to thank W. Guttmann and M. Winter for the cooperation concerning the applications presented in Sect. 5 and the referees for their very helpful comments and suggestions.

References

1. Berghammer, R., Zierer, H.: Relational algebraic semantics of deterministic and nondeterministic programs. Theor. Comput. Sci. **43**, 123–147 (1986)
2. Berghammer, R., Kehden, B.: Relation-algebraic specification and solution of special university timetabling problems. J. Logic Algebraic Progr. **79**, 722–739 (2010)
3. Berghammer, R.: Column-wise extendible vector expressions and the relational computation of sets of sets. In: Hinze, R., Voigtländer, J. (eds.) MPC 2015. LNCS, vol. 9129, pp. 238–256. Springer, Cham (2015). https://doi.org/10.1007/978-3-319-19797-5_12
4. Berghammer, Rudolf: Tool-based relational investigation of closure-interior relatives for finite topological spaces. In: Höfner, Peter, Pous, Damien, Struth, Georg (eds.) RAMICS 2017. LNCS, vol. 10226, pp. 60–76. Springer, Cham (2017). https://doi.org/10.1007/978-3-319-57418-9_4
5. Guttmann, W.: Multirelations with infinite computations. J. Logic Algebraic Progr. **83**, 194–211 (2014)
6. Kawahara, Y.: On the cardinality of relations. In: Schmidt, R.A. (ed.) RelMiCS 2006. LNCS, vol. 4136, pp. 251–265. Springer, Heidelberg (2006). https://doi.org/10.1007/11828563_17
7. Kehden, B.: Evaluating sets of search points using relational algebra. In: Schmidt, R.A. (ed.) RelMiCS 2006. LNCS, vol. 4136, pp. 266–280. Springer, Heidelberg (2006). https://doi.org/10.1007/11828563_18
8. Maddux, R.: On the derivation of identities involving projection functions. In: Cirmaz, L., Gabbay, D., de Rijke, M. (eds.) Logic Colloquium 92, Studies in Logic, Language and Information, pp. 143–163. CSLI Publications (1995)
9. Schmidt, G.: Programs as partial graphs I: flow equivalence and correctness. Theor. Comput. Sci. **15**, 1–25 (1981)
10. Schmidt, G., Ströhlein, T.: Relations and Graphs. Springer, Heidelberg (1993). https://doi.org/10.1007/978-3-642-77968-8
11. Schmidt, G.: Relational Mathematics. Cambridge University Press, Cambridge (2010)
12. Stoy, J.E.: Denotational Semantics. The MIT Press, Cambridge (1977)

Experimental Investigation of Sufficient Criteria for Relations to Have Kernels

Rudolf Berghammer(✉) and Mitja Kulczynski

Department of Computer Science, Kiel University, Kiel, Germany
rub@informatik.uni-kiel.de

Abstract. We investigate four well-known criteria for the existence of kernels in directed graphs/relations which can be tested efficiently, viz. to be irreflexive and symmetric, to be progressively finite, to be bipartite and to satisfy Richardson's criterion. The numerical data, obtained by the evaluation of relation-algebraic problem specifications using RelView show that even the most general of them is very far away from a characterisation of the class of directed graphs/relations having kernels.

1 Introduction

When written as a logical formula, most mathematical theorems have the form

$$\forall x : \Phi(x) \Rightarrow \Psi(x),$$

where x is a list of variables, each variable ranges over a certain class of mathematical objects, $\Psi(x)$ describes the property one is actually interested in and $\Phi(x)$ describes a property that ensures $\Psi(x)$. Mostly, one tries to get $\Phi(x)$ as general as possible. Whenever $\Psi(x)$ is equivalent to $\Phi(x)$ it characterises the class of mathematical objects for which $\Psi(x)$ holds. An example is the fixpoint theorem of A. Tarski (see [12]). Here there is only one variable x that ranges over the class of lattices, $\Phi(x)$ describes that x is complete and $\Psi(x)$ describes that each monotonic function on x has a least fixpoint. That in this case $\Phi(x)$ and $\Psi(x)$ are equivalent is an immediate consequence of a theorem of A. Davis, published in [6]. Other prominent examples are characterisations of classes of mathematical objects by means of forbidden substructures, e.g., that a lattice is modular iff it does not contain a sublattice isomorphic to the pentagon-lattice N_5 (R. Dedekind, see [7]) and that a finite graph is planar iff it does not contain a subgraph that is a subdivision of the Kuratowski graph K_5 or the Kuratowski graph $K_{3,3}$ (K. Kuratowski, see [9]).

In this paper we investigate kernels within graphs. A kernel of a directed graph is a subset K of the set of vertices such that no pair of vertices of K is connected by an edge and from each vertex outside of K there is an edge to a vertex of K. This concept is introduced in [13] by J. von Neumann and O. Morgenstern as a generalisation of a solution of a cooperative game. In [5] V. Chvatal shows that determining whether a directed graph possesses a kernel is NP-complete.

U. Fahrenberg et al. (Eds.): RAMiCS 2021, LNCS 13027, pp. 72–89, 2021.
https://doi.org/10.1007/978-3-030-88701-8_5

Mapping kernels the aforementioned formula results in x ranging over the class of directed graphs, $\Psi(x)$ describes that x has a kernel and $\Phi(x)$ describes a sufficient criterion for this property. Hence, with $\Psi(x)$ as just introduced, it is very unlikely to get a $\Phi(x)$ such that $\Phi(x)$ and $\Psi(x)$ are equivalent and $\Phi(x)$ can be computed efficiently.

There exist a series of sufficient criteria for the existence of kernels which can be tested efficiently. An interesting question is how close these are to a characterisation of the class of directed graphs having kernels. To this end, in this paper we present for all vertex sets X up to 7 vertices the number of directed graphs $g = (X, R)$ having kernels. Then we consider the four most popular criteria for the existence of kernels and present for each criterion the number of directed graphs $g = (X, R)$ which satisfy it. These numerical data show that even in case of the most general of the four criteria, the absence of cycles of odd length (as shown by M. Richardson in [10]), only a very small portion of the directed graphs with kernels satisfy the criterion. We may conclude that the criteria are very far away from a characterisation of the class of directed graphs having kernels.

In case of 7 vertices there are $5.62 \cdot 10^{14}$ directed graphs and $1.88 \cdot 10^{14}$ of them have kernels. Only $1.62 \cdot 10^{10}$ of them satisfy Richardson's criterion. We have been able to compute the numerical data for such large numbers of directed graphs using only their adjacency relations R, relation-algebraic problem specifications and RelView for the evaluation of the latter. RelView is a tool for the manipulation and visualisation of relations and relational programming. It uses reduced ordered binary decision diagrams (ROBDDs) for implementing relations. See [3,4] for more details. Besides the excellent and manifold capabilities of relations and relation algebra in problem solving, this paper again demonstrates the amazing computational power of RelView.

2 Relational Preliminaries

If X and Y are given sets, a subset of the direct product $X \times Y$ is a relation with source X and target Y. We denote the set of all relations with source X and target Y (i.e., the powerset $2^{X \times Y}$) by $[X \leftrightarrow Y]$ and write $R : X \leftrightarrow Y$ instead of $R \in [X \leftrightarrow Y]$. In such a case $X \leftrightarrow Y$ is called the type of R. A (typed) relation corresponds to a Boolean matrix. This interpretation is well suited for many purposes and also used as one of the graphical representations of relations within RelView. Therefore, in this paper we also use matrix terminology and notation for relations. In particular, we write $R_{x,y}$ instead of $(x, y) \in R$ or $x\,R\,y$.

We will use the following five basic operations on relations: $\overline{R}$ (*complementation*), $R \cup S$ (*union*), $R \cap S$ (*intersection*), R^{T} (*transposition*) and $R; S$ (*composition*). We assume that transposition and complementation bind stronger than composition and composition binds stronger than union and intersection. As derived operation we will use the *right residual* of two relations with the same source, defined by $R \setminus S := \overline{R^{\mathsf{T}}; \overline{S}}$. If $R : X \leftrightarrow Y$ and $S : X \leftrightarrow Z$, from the typing rules and the point-wise definitions of complementation, transposition and composition we get $R \setminus S : Y \leftrightarrow Z$ and, given arbitrary $y \in Y$ and $z \in Z$, that $(R \setminus S)_{y,z}$ iff for all $x \in X$ from $R_{x,y}$ it follows $S_{x,z}$.

Besides the just mentioned operations, we will use the three special relations O (*empty relation*), L (*universal relation*) and I (*identity relation*). Here we overload the symbols, i.e., avoid the binding of types to them. Finally, if R is included in S we write $R \subseteq S$ and $R = S$ means their equality.

Relation algebra as just introduced can express exactly those formulae of first-order predicate logic which contain at most two free variables and all in all at most three variables. The expressive power of full first-order predicate logic is obtained by means of projection relations or equivalent notions. In this paper we always assume that a pair u from a direct product is of the form $u = (u_1, u_2)$. This allows to describe the meaning of the *projection relations* $\pi : X \times Y \leftrightarrow X$ and $\rho : X \times Y \leftrightarrow Y$ of a direct product $X \times Y$ by $\pi_{u,x}$ iff $u_1 = x$ and $\rho_{u,y}$ iff $u_2 = y$, for all $u \in X \times Y$, $x \in X$ and $y \in Y$. Based on the projection relations $\pi : X \times Y \leftrightarrow X$ and $\rho : X \times Y \leftrightarrow Y$ for $R : X \leftrightarrow Z$ and $S : Y \leftrightarrow Z$ their *left pairing* is defined by $[\![R, S] := \pi; R \cap \rho; S$, thereby being of type $X \times Y \leftrightarrow Z$. Using point-wise notation we have $[\![R, S]_{u,z}$ iff $R_{u_1,z}$ and $S_{u_2,z}$, for all $u \in X \times Y$ and $z \in Z$. The counterpart to the left pairing, with now $Z \leftrightarrow X \times Y$ as type, is the *right pairing* of $R : Z \leftrightarrow X$ and $S : Z \leftrightarrow Y$, defined as $[R, S]\!] := R; \pi^\mathsf{T} \cap S; \rho^\mathsf{T}$. Point-wisely we get $[R, S]\!]_{z,u}$ iff R_{z,u_1} and S_{z,u_2}, for all $u \in X \times Y$ and $z \in Z$. The *parallel composition* (or *product*) $R \,\|\, S : X \times X' \leftrightarrow Y \times Y'$ of $R : X \leftrightarrow Y$ and $S : X' \leftrightarrow Y'$, such that $(R \,\|\, S)_{u,v}$ iff R_{u_1,v_1} and S_{u_2,v_2}, for all $u \in X \times X'$ and $v \in Y \times Y'$, can be defined by means of the right pairing. We get the desired property if we define $R \,\|\, S := [\pi; R, \rho; S]\!]$, where $\pi : X \times X' \leftrightarrow X$ and $\rho : X \times X' \leftrightarrow X'$ are the projection relations of $X \times X'$ and the right pairing is formed with respect to the projection relations of $Y \times Y'$.

Assume the projection relations $\pi : X \times Y \leftrightarrow X$ and $\rho : X \times Y \leftrightarrow Y$ of $X \times Y$ and $R : X \times Y \leftrightarrow Z$ to be given. A property that we will use frequently in Sect. 4 is the equivalence of $([\rho, \pi]\!]; R)_{u,z}$ and $R_{(u_2,u_1),z}$, for all $u \in X \times Y$ and $z \in Z$.

The relation-level equivalents of the set-theoretic symbol "$\in$" are the *membership relations* $\mathsf{M} : X \leftrightarrow 2^X$, point-wisely defined by $\mathsf{M}_{x,Y}$ iff $x \in Y$, for all $x \in X$ and $Y \in 2^X$. By means of projection relations and membership relations the expressive power of full second-order predicate logic is obtained and this suffices for our later applications. If the source of a membership relation is a direct product and, hence, its target is a set of relations, we use the symbol $\mathbf{M}$ instead of M. An important property of such a membership relation $\mathbf{M} : X \times Y \leftrightarrow [X \leftrightarrow Y]$ is the equivalence of $\mathbf{M}_{u,R}$ and R_{u_1,u_2}, for all $u \in X \times Y$ and $R : X \leftrightarrow Y$, which we also will use frequently later in Sect. 4.

At the end of this section it should be mentioned that – except the parallel composition – all specific relations and all relational operations and tests we have introduced in this section are available in the programming language of RelView. Details will be presented in Sect. 5. See also the Web-site [14].

3 The Experiments and Their Results

In this section we present the numerical data we already have mentioned in the introduction. By means of RelView we have been able to count for a given set X having at most 7 elements the number of directed graphs $g = (X, R)$ possessing kernels. These numbers are presented in the third column in Table 1. In the second column the numbers of all directed graphs with vertex set X are given, i.e., the numbers $2^{|X|^2}$, where $1 \leq |X| \leq 7$. The percentages of the directed graphs having kernels with regard to the total number of directed graphs are given in the last column of the table. Notice that the last number of the second and the last number of the third column of this table are the exact values of the approximations $5.62 \cdot 10^{14}$ and $1.88 \cdot 10^{14}$ mentioned in the introduction.

Table 1. Occurrences of kernels within graphs having at most 7 vertices.

$\|X\|$	All relations	Rel. with kernel	Percentage
1	2	1	50.00 %
2	16	8	50.00 %
3	512	230	44.92 %
4	65 536	26 346	40.19 %
5	33 554 432	12 378 964	39.98 %
6	68 719 476 736	23 921 882 920	34.80 %
7	562 949 953 421 312	188 553 949 010 868	33.49 %

We investigate four sufficient criteria for the existence of kernels in a directed graph which can be tested efficiently. That each of them indeed ensures the existence of kernels is shown in [11] with relation-algebraic means.

The first criterion is that the adjacency relation R is irreflexive and symmetric, that is, $g = (X, R)$ is the directed version of an undirected graph, where each undirected edge is replaced by two parallel directed edges with opposite directions. The corresponding numbers are presented in the second column of Table 2 corresponding to $2^{\frac{|X|(|X|-1)}{2}}$, where $1 \leq |X| \leq 7$. The second criterion is that $R : X \leftrightarrow X$ is a progressively finite relation in the sense of [11] which means that there is no non-empty subset A of X such that for each $x \in A$ there exists a $y \in A$ with $R_{x,y}$. In other words, R is progressively finite iff R^{T} is Noetherian iff there is no infinite sequence $(x_n)_{n \in \mathbb{N}}$ in X such that $R_{x_n, x_{n+1}}$, for all $n \in \mathbb{N}$. This criterion generalises the criterion "to be cycle-free" of [13] since on a finite set X the relation R is progressively finite iff it is cycle-free; see [11]. The data for this criterion (i.e., the number of cycle-free directed graphs $g = (X, R)$ with $1 \leq |X| \leq 7$), are presented in the third column. The fourth column of the table shows the number of bipartite directed graphs $g = (X, R)$, with $1 \leq |X| \leq 7$, since "to be bipartite" is also a sufficient criterion for the existence of kernels. In the introduction we already have mentioned Richardson's criterion stating that a graph has no cycles of odd length. The data for this fourth criterion can be found in the last column of the table. Notice that the theorem of [10] on the existence of kernels in directed graphs without cycles of odd length assumes finite graphs. In case of an undirected graph the set of kernels equals the set of maximal stable sets such that kernels exist if the vertex set is finite. The other two criteria also hold for infinite graphs.

Each directed graph with a progressively finite relation is cycle-free and, as a consequence, also does not contain cycles of odd length. From a well-known theorem of D. König (see [8]) we immediately get that each bipartite directed graph has no cycles of odd length. Therefore, on finite directed graphs (which are important in practical applications) Richardson's criterion is more general than the criteria "to be cycle-free" and "to be bipartite", which also is demonstrated by the numerical data given in Table 2. The last number of the last column is the exact value of the approximation $1.62 \cdot 10^{10}$ mentioned in the introduction. A comparison of the second and the last column shows that Richardson's criterion is also much more general than the first criterion "R is irreflexive and symmetric". Notice that, however, neither the first criterion implies Richardson's criterion nor vice versa.

Table 2. Number of graphs for the four criteria having at most 7 vertices.

$\|X\|$	Irr., symm.	Progr. finite	Bipartite	Richardson
1	1	1	1	1
2	2	3	4	4
3	8	25	37	49
4	64	543	829	1 699
5	1 024	29 281	36 616	150 736
6	32 768	3 781 503	3 327 499	32 398 249
7	2 097 152	1 138 779 265	581 809 537	16 230 843 049

At the end of this section it should be mentioned that each of the above four criteria not only can be tested efficiently but also can be used to obtain efficient algorithms for computing a kernel of a graph that satisfies the criterion. In case of Richardson's criterion such an algorithm is presented in [2]. It is formulated as a relational while-program and formally derived by means of the assertion technique and reconstructing a proof of Richardson's theorem.

4 Computing Classes of Relations Having Kernels

In this section vectors play a central role. A *(relational) vector* as introduced in [11] is a relation $s : X \leftrightarrow Y$ such that $s = s; \mathsf{L}$, for $\mathsf{L} : Y \leftrightarrow Y$. In the Boolean matrix interpretation this means that each row of s consists only of ones or only of zeros. Consequently, the targets of vectors are irrelevant and we only consider vectors of type $X \leftrightarrow \mathbf{1}$, with a specific singleton set $\mathbf{1} := \{\perp\}$ as common target. Such vectors correspond to Boolean column vectors and, therefore, as in linear algebra we write s_x instead of $s_{x,\perp}$. For $R : \mathbf{1} \leftrightarrow X$ we retain the notation $R_{\perp,x}$.

Given a set X and a subset S of X, we call $s : X \leftrightarrow \mathbf{1}$ a *vector model* of S (for short *s models* S) if for all $x \in X$ it holds $x \in S$ iff s_x. If X is a direct product, say $Y \times Z$, then s models a relation $S : Y \leftrightarrow Z$ and we have S_{u_1,u_2} iff s_u, for all $u \in Y \times Z$. The computation of $s := vec(S)$ from S can be done relation-algebraically, as $vec(S) = [\![S, \mathsf{I}]\!]; \mathsf{L}$, where $\mathsf{I} : Z \leftrightarrow Z$ and $\mathsf{L} : Z \leftrightarrow \mathbf{1}$.

Convention 4.1. *For the following we fix a set X. Throughout this section then $\pi : X^2 \leftrightarrow X$ and $\rho : X^2 \leftrightarrow X$ denote the two projection relations of the direct product X^2 and $\mathsf{M} : X \leftrightarrow 2^X$ and $\mathbf{M} : X^2 \leftrightarrow [X \leftrightarrow X]$ are membership relations.*

Instead of working with directed graphs $g = (X, R)$ in the following we work with their adjacency relations $R : X \leftrightarrow X$ (in [11] called associated relation) and use the notions kernel, cycle, bipartite etc. for R in an obvious way. The computations we will present consist of relation-algebraic specifications of vector models of those sets of relations on X which satisfy the first, second, third respectively fourth of the four sufficient criteria for the existence of kernels we have mentioned in Sect. 3. In Sect. 5 we will demonstrate how these specifications rather straightforwardly can be implemented in the programming language of RelView and the executions of these RelView-programs led to the numerical data of Sect. 3.

Given $R : X \leftrightarrow X$, from the description of kernels in Sect. 1 we get that a subset K of X is a kernel of R iff the following two formulae hold, where the variables x and y range over X:

$$\neg\exists x, y : x \in K \wedge y \in K \wedge R_{x,y} \qquad\qquad \forall x : x \notin K \Rightarrow \exists y : y \in K \wedge R_{x,y}$$

The first formula defines K as *R-stable* and the second one as *R-absorbant*. As a consequence, kernels of R are precisely those subsets of X which are R-stable and R-absorbant at the same time. Based on two auxiliary specifications for R-stable and R-absorbant subsets, in the following theorem we specify relation-algebraically a vector model $\mathfrak{kernel} : [X \leftrightarrow X] \leftrightarrow \mathbf{1}$ for the set of relations on X having kernels. Besides the relations of Convention 4.1 the second projection relation $\beta : X \times 2^X \leftrightarrow 2^X$ of the direct product $X \times 2^X$ is used. Notice that the backslash-symbol used in the second auxiliary specification $\mathfrak{absorb}$ denotes the right residual operation.

Theorem 4.1. *We consider the following three relation-algebraic specifications:*

$$\mathfrak{stable} := \overline{[\![\mathsf{M}^\mathsf{T}, \mathsf{M}^\mathsf{T}]\!];\mathbf{M}} : 2^X \leftrightarrow [X \leftrightarrow X]$$
$$\mathfrak{absorb} := (\beta \cap vec(\overline{\mathsf{M}});\mathsf{L}) \setminus ((\mathsf{I} \,||\, \mathsf{M}^\mathsf{T});\mathbf{M}) : 2^X \leftrightarrow [X \leftrightarrow X]$$
$$\mathfrak{kernel} := (\mathsf{L};(\mathfrak{stable} \cap \mathfrak{absorb}))^\mathsf{T} : [X \leftrightarrow X] \leftrightarrow \mathbf{1}$$

For all $A \in 2^X$ and $R : X \leftrightarrow X$ then $\mathfrak{stable}_{A,R}$ *iff A is R-stable,* $\mathfrak{absorb}_{A,R}$ *iff A is R-absorbant and* $\mathfrak{kernel}_R$ *iff R has a kernel.*

Proof. Assume arbitrary $A \in 2^X$ and $R : X \leftrightarrow X$. Then the first claim is shown by the following calculation, where the variable u ranges over X^2:

$$\begin{aligned}
\mathfrak{stable}_{A,R} &\iff \overline{[\![\mathsf{M}^\mathsf{T}, \mathsf{M}^\mathsf{T}]\!]; \mathbf{M}}_{A,R} \\
&\iff \neg\exists u : [\![\mathsf{M}^\mathsf{T}, \mathsf{M}^\mathsf{T}]\!]_{A,u} \wedge \mathbf{M}_{u,R} \\
&\iff \neg\exists u : \mathsf{M}^\mathsf{T}_{A,u_1} \wedge \mathsf{M}^\mathsf{T}_{A,u_2} \wedge R_{u_1,u_2} \\
&\iff \neg\exists u : u_1 \in A \wedge u_2 \in A \wedge R_{u_1,u_2} \\
&\iff A \text{ is } R\text{-stable}
\end{aligned}$$

To prove the second claim, we calculate as follows, where the variables x and y range over X, the variable u ranges over X^2 and the variable B ranges over 2^X:

$$\begin{aligned}
\mathfrak{absorb}_{A,R} &\iff ((\beta \cap vec(\overline{\mathsf{M}}); \mathsf{L}) \setminus ((\mathsf{I} \,||\, \mathsf{M}^\mathsf{T}); \mathbf{M}))_{A,R} \\
&\iff \forall x, B : (\beta \cap vec(\overline{\mathsf{M}}); \mathsf{L})_{(x,B),A} \Rightarrow ((\mathsf{I} \,||\, \mathsf{M}^\mathsf{T}); \mathbf{M})_{(x,B),R} \\
&\iff \forall x, B : B = A \wedge \overline{\mathsf{M}}_{x,B} \Rightarrow \exists u : (\mathsf{I} \,||\, \mathsf{M}^\mathsf{T})_{(x,B),u} \wedge \mathbf{M}_{u,R} \\
&\iff \forall x, B : B = A \wedge x \notin B \Rightarrow \exists u : x = u_1 \wedge \mathsf{M}^\mathsf{T}_{B,u_2} \wedge R_{u_1,u_2} \\
&\iff \forall x : x \notin A \Rightarrow \exists y : y \in A \wedge R_{x,y} \\
&\iff A \text{ is } R\text{-absorbant}
\end{aligned}$$

Finally, we calculate as follows, where the variable A ranges over 2^X:

$$\begin{aligned}
\mathfrak{kernel}_R &\iff (\mathsf{L}; (\mathfrak{stable} \cap \mathfrak{absorb}))_{\perp,R} \\
&\iff \exists A : \mathsf{L}_{\perp,A} \wedge (\mathfrak{stable} \cap \mathfrak{absorb})_{A,R} \\
&\iff \exists A : \mathfrak{stable}_{A,R} \wedge \mathfrak{absorb}_{A,R}
\end{aligned}$$

Together with the first two claims this implies the third claim. □

We have used the prevalent mathematical theorem-proof-style to emphasise the result of this theorem and to enhance readability. However, in fact, we have obtained the relation-algebraic specifications by developing them formally from the corresponding logical specifications by replacing step-by-step logical constructions by equivalent relational ones. This remark also holds for the other theorems of this section.

The next theorem presents relation-algebraic specifications of vector models $\mathfrak{irrefl} : [X \leftrightarrow X] \leftrightarrow \mathbf{1}$ and $\mathfrak{sym} : [X \leftrightarrow X] \leftrightarrow \mathbf{1}$ for the set of irreflexive respectively symmetric relations on X such that the intersection $\mathfrak{irrefl} \cap \mathfrak{sym}$ models the set of relations on X which satisfy the first sufficient criterion for the existence of kernels we have mentioned in Sect. 3. Only the three relations $\pi : X^2 \leftrightarrow X$ and $\rho : X^2 \leftrightarrow X$ and $\mathbf{M} : X^2 \leftrightarrow [X \leftrightarrow X]$ of Convention 4.1 are used.

Theorem 4.2. *We consider the following relation-algebraic specifications:*

$$\mathfrak{irrefl} := \overline{\mathsf{L}; (\mathbf{M} \cap vec(\mathsf{I}); \mathsf{L})}^{\mathsf{T}} : [X \leftrightarrow X] \leftrightarrow \mathbf{1}$$

$$\mathfrak{sym} := \overline{\mathsf{L}; (\mathbf{M} \cap \overline{[\![\rho, \pi]\!]; \mathbf{M}})}^{\mathsf{T}} : [X \leftrightarrow X] \leftrightarrow \mathbf{1}$$

For all $R : X \leftrightarrow X$ *then* $\mathfrak{irrefl}_R$ *iff* R *is irreflexive and* $\mathfrak{sym}_R$ *iff* R *is symmetric.*

Proof. Assume an arbitrary $R : X \leftrightarrow X$. Then the following calculation shows the first claim, where the variable u ranges over X^2:

$$\begin{aligned}
\mathfrak{irrefl}_R &\iff \overline{\mathsf{L}; (\mathbf{M} \cap vec(\mathsf{I}); \mathsf{L})}_{\perp,R} \\
&\iff \neg\exists u : \mathsf{L}_{\perp,u} \wedge \mathbf{M}_{u,R} \wedge vec(\mathsf{I})_u \\
&\iff \neg\exists u : R_{u_1,u_2} \wedge \mathsf{I}_{u_1,u_2} \\
&\iff \neg\exists u : R_{u_1,u_2} \wedge u_1 = u_2 \\
&\iff R \text{ is irreflexive}
\end{aligned}$$

Also in the following calculation the variable u ranges over X^2:

$$\begin{aligned}
\mathfrak{sym}_R &\iff \overline{\mathsf{L}; (\mathbf{M} \cap \overline{[\![\rho, \pi]\!]; \mathbf{M}})}_{\perp,R} \\
&\iff \neg\exists u : \mathsf{L}_{\perp,u} \wedge \mathbf{M}_{u,R} \wedge \neg([\![\rho, \pi]\!]; \mathbf{M})_{u,R} \\
&\iff \forall u : \mathbf{M}_{u,R} \Rightarrow ([\![\rho, \pi]\!]; \mathbf{M})_{u,R} \\
&\iff \forall u : R_{u_1,u_2} \Rightarrow \mathbf{M}_{(u_2,u_1),R} \\
&\iff \forall u : R_{u_1,u_2} \Rightarrow R_{u_2,u_1} \\
&\iff R \text{ is symmetric}
\end{aligned}$$

With this verification of the second claim the proof is complete. □

The second sufficient criterion for the existence of kernels we have mentioned in Sect. 3 is "to be progressively finite". In the following we show how to specify relation-algebraically a vector model $\mathfrak{progFin} : [X \leftrightarrow X] \leftrightarrow \mathbf{1}$ of the set of progressively finite relations on X. As in the case of Theorem 4.1 besides the relations of Convention 4.1 we use the second projection relation $\beta : X \times 2^X \leftrightarrow 2^X$ of the direct product $X \times 2^X$.

Theorem 4.3. *We consider the following relation-algebraic specification:*

$$\mathfrak{progFin} := \overline{\mathsf{L}; \mathsf{M}; \overline{(\beta^{\mathsf{T}} \cap \mathsf{L}; vec(\mathsf{M})^{\mathsf{T}}); \overline{[\![\pi, \rho; \mathsf{M}]\!]^{\mathsf{T}}; \mathbf{M}}}}^{\mathsf{T}} : [X \leftrightarrow X] \leftrightarrow \mathbf{1}$$

For all $R : X \leftrightarrow X$ *then* $\mathfrak{progFin}_R$ *iff* R *is progressively finite.*

Proof. To structure the proof, we define the following auxiliary relation:

$$\mathfrak{R} := \overline{(\beta^{\mathsf{T}} \cap \mathsf{L}; vec(\mathsf{M})^{\mathsf{T}}); \overline{[\![\pi, \rho; \mathsf{M}]\!]^{\mathsf{T}}; \mathbf{M}}} : 2^X \leftrightarrow [X \leftrightarrow X]$$

Now, assume an arbitrary $R : X \leftrightarrow X$. For all $A \in 2^X$ we then calculate as follows, where the variables x and y range over X, the variable B ranges over 2^X and the variable u ranges over X^2:

$$
\begin{aligned}
\mathfrak{R}_{A,R} &\iff \overline{(\beta^\mathsf{T} \cap \mathsf{L}; vec(\mathsf{M})^\mathsf{T}); \overline{[\pi, \rho; \mathsf{M}]\!]^\mathsf{T}; \mathbf{M}}}_{A,R} \\
&\iff \neg\exists x, B : (\beta^\mathsf{T} \cap \mathsf{L}; vec(\mathsf{M})^\mathsf{T})_{A,(x,B)} \wedge \overline{[\pi, \rho; \mathsf{M}]\!]^\mathsf{T}; \mathbf{M}}_{(x,B),R} \\
&\iff \neg\exists x, B : A = B \wedge vec(\mathsf{M})_{(x,B)} \wedge \neg([\pi, \rho; \mathsf{M}]\!]^\mathsf{T}; \mathbf{M})_{(x,B),R} \\
&\iff \neg\exists x : vec(\mathsf{M})_{(x,A)} \wedge \neg\exists u : [\pi, \rho; \mathsf{M}]\!]^\mathsf{T}_{(x,A),u} \wedge \mathbf{M}_{u,R} \\
&\iff \neg\exists x : \mathsf{M}_{x,A} \wedge \neg\exists u : [\pi, \rho; \mathsf{M}]\!]_{u,(x,A)} \wedge R_{u_1,u_2} \\
&\iff \forall x : x \in A \Rightarrow \exists u : u_1 = x \wedge u_2 \in A \wedge R_{u_1,u_2} \\
&\iff \forall x : x \in A \Rightarrow \exists y : y \in A \wedge R_{x,y}
\end{aligned}
$$

Using this result, we now calculate as follows, where the variable A ranges over 2^X and, as above, the variables x and y range over X:

$$
\begin{aligned}
\mathfrak{progFin}_R &\iff \overline{\mathsf{L}; \mathsf{M}; \mathfrak{R}}_{\perp,R} \\
&\iff \neg\exists A : (\mathsf{L}; \mathsf{M})_{\perp,A} \wedge \mathfrak{R}_{A,R} \\
&\iff \neg\exists A : (\mathsf{M}^\mathsf{T}; \mathsf{L})_A \wedge \forall x : x \in A \Rightarrow \exists y : y \in A \wedge R_{x,y} \\
&\iff \neg\exists A : A \neq \emptyset \wedge \forall x : x \in A \Rightarrow \exists y : y \in A \wedge R_{x,y}
\end{aligned}
$$

The last formula is the logical specification of R being progressively finite; see the definition given in Sect. 3. □

We continue with the third sufficient criterion for the existence of kernels we have mentioned in Sect. 3, viz. "to be bipartite". A corresponding relation-algebraic specification of a vector model $\mathfrak{bipartite} : [X \leftrightarrow X] \leftrightarrow \mathbf{1}$ of the set of bipartite relations on X is given in the theorem below. In this theorem only the two membership relations of Convention 4.1 are used.

Theorem 4.4. *We consider the following relation-algebraic specification:*

$$\mathfrak{bipartite} := (\mathbf{M} \setminus ([\mathsf{M}, \overline{\mathsf{M}}]\!] \cup [\overline{\mathsf{M}}, \mathsf{M}]\!])); \mathsf{L} : [X \leftrightarrow X] \leftrightarrow \mathbf{1}$$

For all $R : X \leftrightarrow X$ *then* $\mathfrak{bipartite}_R$ *iff* R *is bipartite.*

Proof. Assume an arbitrary $R : X \leftrightarrow X$. Then we calculate as follows, where the variable A ranges over 2^X and the variable u ranges over X^2:

$$
\begin{aligned}
\mathfrak{bipartite}_R &\iff ((\mathbf{M} \setminus ([\mathsf{M}, \overline{\mathsf{M}}]\!] \cup [\overline{\mathsf{M}}, \mathsf{M}]\!])); \mathsf{L})_R \\
&\iff \exists A : (\mathbf{M} \setminus ([\mathsf{M}, \overline{\mathsf{M}}]\!] \cup [\overline{\mathsf{M}}, \mathsf{M}]\!]))_{R,A} \wedge \mathsf{L}_A \\
&\iff \exists A : \forall u : \mathbf{M}_{u,R} \Rightarrow [\mathsf{M}, \overline{\mathsf{M}}]\!]_{u,A} \vee [\overline{\mathsf{M}}, \mathsf{M}]\!]_{u,A} \\
&\iff \exists A : \forall u : R_{u_1,u_2} \Rightarrow (\mathsf{M}_{u_1,A} \wedge \overline{\mathsf{M}}_{u_2,A}) \vee (\overline{\mathsf{M}}_{u_1,A} \wedge \mathsf{M}_{u_2,A}) \\
&\iff \exists A : \forall u : R_{u_1,u_2} \Rightarrow (u_1 \in A \wedge u_2 \notin A) \vee (u_1 \notin A \wedge u_2 \in A) \\
&\iff R \text{ is bipartite}
\end{aligned}
$$

This completes the proof. □

Concerning Richardson's criterion, we have not been able to specify a vector model of the set of relations on X without cycles of odd length with purely relation-algebraic means. Experiments with the RELVIEW tool have shown that the RELVIEW-implementation of the vector model $\mathfrak{kernel}$ of Theorem 4.1 seems to be successfully executable up to $|X| = 7$ only. For $|X| = 8$ we cancelled the computation after about 20 h. Based on this fact, we have decided to consider one after the other the lengths 1, 3, 5 and 7 of cycles. If $|X| \leq 2$, then a relation on X has no cycle of odd length iff it is irreflexive. As a consequence, the first two numbers of the last column of the second table of Sect. 3, i.e., the numbers for $|X| = 1$ and $|X| = 2$, are $2^{|X|(|X|-1)}$, since this expression specifies the number of irreflexive relations on X. The next three theorems present relation-algebraic specifications of three vector models with the following meanings:

a) $\mathfrak{cyc3} : [X \leftrightarrow X] \leftrightarrow \mathbf{1}$ models the set of relations on X which have a cycle of length 3.
b) $\mathfrak{cyc5} : [X \leftrightarrow X] \leftrightarrow \mathbf{1}$ models the set of relations on X which have a cycle of length 5.
c) $\mathfrak{cyc7} : [X \leftrightarrow X] \leftrightarrow \mathbf{1}$ models the set of relations on X which have a cycle of length 7.

Since the complement $\overline{\mathfrak{cyc3}}$ models the set of relations on X without cycles of length 3 and for the complements $\overline{\mathfrak{cyc5}}$ and $\overline{\mathfrak{cyc7}}$ the same applies for length 5 and 7, respectively, the vector

$$\mathfrak{irrefl} \cap \overline{\mathfrak{cyc3}} \cap \overline{\mathfrak{cyc5}} \cap \overline{\mathfrak{cyc7}} : [X \leftrightarrow X] \leftrightarrow \mathbf{1}$$

models the set of relations on X which have no cycles of length 1, 3, 5 and 7. Consequently, we get for $|X| \leq 8$ that it models the set of relations on X without cycles of odd length. This way we have obtained the numbers of the last column of Table 2 for $3 \leq |X| \leq 7$.

The following relation-algebraic specification of the vector model $\mathfrak{cyc3}$ uses the relations of Convention 4.1 except the membership relation $\mathsf{M} : X \leftrightarrow 2^X$. Furthermore, it uses the two projection relations of the direct product $X^2 \times X^2$, which we denote as $\gamma : X^2 \times X^2 \leftrightarrow X^2$ and $\delta : X^2 \times X^2 \leftrightarrow X^2$.

Theorem 4.5. *We consider the following relation-algebraic specification:*

$$\mathfrak{cyc3} := (\mathsf{L}; ([\delta, \gamma]; (\rho \,\|\, \pi); \mathbf{M} \cap [\![\mathbf{M}, \mathbf{M}]\!] \cap vec(\rho; \pi^{\mathsf{T}}); \mathsf{L}))^{\mathsf{T}} : [X \leftrightarrow X] \leftrightarrow \mathbf{1}$$

For all $R : X \leftrightarrow X$ then $\mathfrak{cyc3}_R$ iff R has a cycle of length 3.

Proof. Assume an arbitrary $R : X \leftrightarrow X$. Furthermore, let $u, v \in X^2$. Then we have

$$([\delta, \gamma]; (\rho \,\|\, \pi); \mathbf{M})_{(u,v),R} \iff ((\rho \,\|\, \pi); \mathbf{M})_{(v,u),R} \iff \mathbf{M}_{(v_2,u_1),R} \iff R_{v_2,u_1}$$

and

$$[\![\mathbf{M}, \mathbf{M}]\!]_{(u,v),R} \iff \mathbf{M}_{u,R} \wedge \mathbf{M}_{v,R} \iff R_{u_1,u_2} \wedge R_{v_1,v_2}$$

and

$$(vec(\rho;\pi^{\mathsf{T}});\mathsf{L})_{(u,v),R} \iff vec(\rho;\pi^{\mathsf{T}})_{(u,v)} \iff (\rho;\pi^{\mathsf{T}})_{u,v} \iff u_2 = v_1.$$

From these equivalences we get

$$\begin{aligned}
\mathfrak{cyc3}_R &\iff (\mathsf{L};([\delta,\gamma]];(\rho\,\|\,\pi);\mathbf{M} \cap [\![\mathbf{M},\mathbf{M}]\!] \cap vec(\rho;\pi^{\mathsf{T}});\mathsf{L}))_{\perp,R}\\
&\iff \exists u,v : \mathsf{L}_{\perp,(u,v)} \wedge R_{v_2,u_1} \wedge R_{u_1,u_2} \wedge R_{v_1,v_2} \wedge u_2 = v_1\\
&\iff \exists x,y,z : R_{x,y} \wedge R_{y,z} \wedge R_{z,x},
\end{aligned}$$

where the variables u and v range over X^2 and the variables x, y and z range over X. The last formula of this calculation is the logical description of R having a cycle of length 3. □

In the next theorem we present a relation-algebraic specification of the vector model $\mathfrak{cyc5}$. Precisely, it is based on the same projection relations and membership relations as Theorem 4.5 and uses two auxiliary specifications for the construction of paths.

Theorem 4.6. *We consider the following relation-algebraic specifications:*

$$\begin{aligned}
\mathfrak{R} &:= (\pi;\pi^{\mathsf{T}} \,\|\, \rho;\rho^{\mathsf{T}});([\![\mathbf{M},\mathbf{M}]\!] \cap vec(\rho;\pi^{\mathsf{T}});\mathsf{L}) : X^2 \times X^2 \leftrightarrow [X \leftrightarrow X]\\
\mathfrak{S} &:= [\![[\rho,\pi]\!];\mathbf{M}, [\rho,\pi]\!];\mathbf{M}] \cap [\delta,\gamma]\!];(\pi\,\|\,\rho);\mathbf{M} : X^2 \times X^2 \leftrightarrow [X \leftrightarrow X]\\
\mathfrak{cyc5} &:= (\mathsf{L};(\mathfrak{R} \cap \mathfrak{S}))^{\mathsf{T}} : [X \leftrightarrow X] \leftrightarrow \mathbf{1}
\end{aligned}$$

For all $R : X \leftrightarrow X$ then $\mathfrak{cyc5}_R$ iff R has a cycle of length 5.

Proof. Assume an arbitrary $R : X \leftrightarrow X$. Furthermore, let $u, v \in X^2$. First, we treat $\mathfrak{R}$ and calculate as given below, where the variables a and b range over X^2 and the variable x ranges over X:

$$\begin{aligned}
\mathfrak{R}_{(u,v),R} &\iff \exists a,b : (\pi;\pi^{\mathsf{T}} \,\|\, \rho;\rho^{\mathsf{T}})_{(u,v),(a,b)} \wedge ([\![\mathbf{M},\mathbf{M}]\!] \cap vec(\rho;\pi^{\mathsf{T}});\mathsf{L})_{(a,b),R}\\
&\iff \exists a,b : (\pi;\pi^{\mathsf{T}})_{u,a} \wedge (\rho;\rho^{\mathsf{T}})_{v,b} \wedge [\![\mathbf{M},\mathbf{M}]\!]_{(a,b),R} \wedge vec(\rho;\pi^{\mathsf{T}})_{(a,b)}\\
&\iff \exists a,b : u_1 = a_1 \wedge v_2 = b_2 \wedge \mathbf{M}_{a,R} \wedge \mathbf{M}_{b,R} \wedge a_2 = b_1\\
&\iff \exists x : \mathbf{M}_{(u_1,x),R} \wedge \mathbf{M}_{(x,v_2),R}\\
&\iff \exists x : R_{u_1,x} \wedge R_{x,v_2}
\end{aligned}$$

Hence, $\mathfrak{R}_{(u,v),R}$ specifies that there exists a path (u_1, x, v_2) in R. With regard to $\mathfrak{S}$ we calculate as follows:

$$\begin{aligned}
\mathfrak{S}_{(u,v),R} &\iff [\![[\rho,\pi]\!];\mathbf{M}, [\rho,\pi]\!];\mathbf{M}]_{(u,v),R} \wedge ([\delta,\gamma]\!];(\pi\,\|\,\rho);\mathbf{M})_{(u,v),R}\\
&\iff ([\rho,\pi]\!];\mathbf{M})_{u,R} \wedge ([\rho,\pi]\!];\mathbf{M})_{v,R} \wedge ((\pi\,\|\,\rho);\mathbf{M})_{(v,u),R}\\
&\iff \mathbf{M}_{(u_2,u_1),R} \wedge \mathbf{M}_{(v_2,v_1),R} \wedge \mathbf{M}_{(v_1,u_2),R}\\
&\iff R_{v_2,v_1} \wedge R_{v_1,u_2} \wedge R_{u_2,u_1}
\end{aligned}$$

So, $\mathfrak{S}_{(u,v),R}$ specifies that (v_2, v_1, u_2, u_1) is a path in R. After these preparations we now prove the claim. We start with the following calculation, where the variables u and v range over X^2 and the variable x ranges over X:

$$\begin{aligned}\mathfrak{cyc5}_R &\iff (\mathsf{L};(\mathfrak{R}\cap\mathfrak{S}))_{\perp,R}\\ &\iff \exists u,v : \mathsf{L}_{\perp,(u,v)} \wedge \mathfrak{R}_{(u,v),R} \wedge \mathfrak{S}_{(u,v),R}\\ &\iff \exists u,v : \mathfrak{R}_{(u,v),R} \wedge \mathfrak{S}_{(u,v),R}\\ &\iff \exists u,v : (\exists x : R_{u_1,x} \wedge R_{x,v_2}) \wedge (v_2,v_1,u_2,u_1) \text{ is a path in } R\end{aligned}$$

It remains to verify that the last formula holds iff R has a cycle of length 5. For the direction "$\Longrightarrow$", let the formula be true. Then $(u_1, x, v_2, v_1, u_2, u_1)$ is a cycle of length 5 in R. For the converse, suppose that R possesses a cycle $(c_1, c_2, c_3, c_4, c_5, c_1)$ of length 5. We define $u := (c_1, c_5)$, $v := (c_4, c_3)$ and $x := c_2$. Then (u_1, x, v_2) and (v_2, v_1, u_2, u_1) are paths in R and the formula holds. □

The relation-algebraic specification of the vector model $\mathfrak{cyc7}$ is given in the next theorem. We follow the ideas of Theorem 4.6, but the realisation is far more complex. We use a further projection relation, viz. the second projection relation of the direct product $(X^2 \times X^2)^2$, which we denote as $\mu : (X^2 \times X^2)^2 \leftrightarrow X^2 \times X^2$. Furthermore, we use the auxiliary specification $\mathfrak{R}$ of Theorem 4.6 and three further auxiliary specifications.

Theorem 4.7. *With $\mathfrak{R} : X^2 \times X^2 \leftrightarrow [X \leftrightarrow X]$ as defined in Theorem 4.6 we consider the following relation-algebraic specifications:*

$$\begin{aligned}\mathfrak{S} &:= [\![[\rho,\pi]\!];\mathbf{M}, [\rho,\pi]\!];\mathbf{M}] : X^2 \times X^2 \leftrightarrow [X \leftrightarrow X]\\ \mathfrak{T} &:= [\![[\![\rho;\rho^{\mathsf{T}}, \rho;\pi^{\mathsf{T}}]\!];\mathbf{M}, [\![\pi;\pi^{\mathsf{T}}, \pi;\rho^{\mathsf{T}}]\!];\mathbf{M}] : (X^2 \times X^2)^2 \leftrightarrow [X \leftrightarrow X]\\ \mathfrak{U} &:= (\gamma^{\mathsf{T}} \parallel \gamma^{\mathsf{T}});(\mathfrak{T} \cap \mu;\delta;\mathbf{M} \cap vec(\delta;\delta^{\mathsf{T}});\mathsf{L}) : X^2 \times X^2 \leftrightarrow [X \leftrightarrow X]\\ \mathfrak{cyc7} &:= (\mathsf{L};(\mathfrak{R}\cap\mathfrak{S}\cap\mathfrak{U}))^{\mathsf{T}} : [X \leftrightarrow X] \leftrightarrow \mathbf{1}\end{aligned}$$

For all $R : X \leftrightarrow X$ then $\mathfrak{cyc7}_R$ iff R has a cycle of length 7.

Proof. Assume an arbitrary $R : X \leftrightarrow X$. Furthermore, let $u, v \in X^2$. From the proof of Theorem 4.6 we already know the following facts, where the variable x of the left equivalence ranges over X:

$$\mathfrak{R}_{(u,v),R} \iff \exists x : R_{u_1,x} \wedge R_{x,v_2} \qquad\qquad \mathfrak{S}_{(u,v),R} \iff R_{u_2,u_1} \wedge R_{v_2,v_1}$$

So, we have $(\mathfrak{R}\cap\mathfrak{S})_{(u,v),R}$ iff there exists a path (u_2, u_1, x, v_2, v_1) in R. In the remainder of the proof we show that $\mathfrak{U}_{(u,v),R}$ iff there is a path (v_1, w_1, w_2, u_2) in R, from which then the claim follows similarly to the last step of the proof of Theorem 4.6.

First, we concentrate on $\mathfrak{T}$. Guided by its source we assume arbitrary pairs $(a, b) \in X^2 \times X^2$ and $(c, d) \in X^2 \times X^2$ to be given. For all $e \in X^2$ we then have

$$[\![\rho;\rho^{\mathsf{T}}, \rho;\pi^{\mathsf{T}}]\!]_{(a,b),e} \iff (\rho;\rho^{\mathsf{T}})_{a,e} \wedge (\rho;\pi^{\mathsf{T}})_{b,e} \iff a_2 = e_2 \wedge b_2 = e_1$$

and this implies

$$([\![\rho;\rho^{\mathsf{T}},\rho;\pi^{\mathsf{T}}]\!];\mathbf{M})_{(a,b),R} \iff \mathbf{M}_{(b_2,a_2),R} \iff R_{b_2,a_2}.$$

In the same way we show that

$$([\![\pi;\pi^{\mathsf{T}},\pi;\rho^{\mathsf{T}}]\!];\mathbf{M})_{(c,d),R} \iff \mathbf{M}_{(c_1,d_1),R} \iff R_{c_1,d_1}.$$

So, altogether, we get:

$$\begin{aligned}
\mathfrak{T}_{((a,b),(c,d)),R} &\iff [\![[\![\rho;\rho^{\mathsf{T}},\rho;\pi^{\mathsf{T}}]\!];\mathbf{M},[\![\pi;\pi^{\mathsf{T}},\pi;\rho^{\mathsf{T}}]\!];\mathbf{M}]\!]_{((a,b),(c,d)),R}\\
&\iff ([\![\rho;\rho^{\mathsf{T}},\rho;\pi^{\mathsf{T}}]\!];\mathbf{M})_{(a,b),R} \wedge ([\![\pi;\pi^{\mathsf{T}},\pi;\rho^{\mathsf{T}}]\!];\mathbf{M})_{(c,d),R}\\
&\iff R_{b_2,a_2} \wedge R_{c_1,d_1}
\end{aligned}$$

Second, we concentrate on $\mathfrak{U}$ and calculate as follows, where the variables a, b, c, d and w range over X^2:

$$\begin{aligned}
\mathfrak{U}_{(u,v),R} &\iff ((\gamma^{\mathsf{T}} \parallel \gamma^{\mathsf{T}});(\mathfrak{T} \cap \mu;\delta;\mathbf{M} \cap vec(\delta;\delta^{\mathsf{T}});\mathsf{L}))_{(u,v),R}\\
&\iff \exists a,b,c,d : (\gamma^{\mathsf{T}} \parallel \gamma^{\mathsf{T}})_{(u,v),((a,b),(c,d))}\\
&\qquad\qquad \wedge (\mathfrak{T} \cap \mu;\delta;\mathbf{M} \cap vec(\delta;\delta^{\mathsf{T}});\mathsf{L})_{((a,b),(c,d)),R}\\
&\iff \exists a,b,c,d : (\gamma^{\mathsf{T}} \parallel \gamma^{\mathsf{T}})_{(u,v),((a,b),(c,d))} \wedge \mathfrak{T}_{((a,b),(c,d)),R}\\
&\qquad\qquad \wedge (\mu;\delta;\mathbf{M})_{((a,b),(c,d)),R} \wedge vec(\delta;\delta^{\mathsf{T}})_{((a,b),(c,d))}\\
&\iff \exists a,b,c,d : \gamma_{(a,b),u} \wedge \gamma_{(c,d),v} \wedge R_{b_2,a_2} \wedge R_{c_1,d_1}\\
&\qquad\qquad \wedge (\delta;\mathbf{M})_{(c,d),R} \wedge (\delta;\delta^{\mathsf{T}})_{(a,b),(c,d)}\\
&\iff \exists a,b,c,d : a = u \wedge c = v \wedge R_{b_2,a_2} \wedge R_{c_1,d_1} \wedge \mathbf{M}_{d,R} \wedge b = d\\
&\iff \exists w : R_{w_2,u_2} \wedge R_{v_1,w_1} \wedge \mathbf{M}_{w,R}\\
&\iff \exists w : R_{v_1,w_1} \wedge R_{w_1,w_2} \wedge R_{w_2,u_2}
\end{aligned}$$

Hence, we have $\mathfrak{U}_{(u,v),R}$ iff there exists a path (v_1, w_1, w_2, u_2) in R as required to conclude the proof. □

5 Implementation in RelView

RelView is a *specific purpose computer algebra system* for the manipulation and visualisation of relations, relational prototyping and relational programming. Computational tasks can be described by short and concise programs, which frequently consist of only a few lines that present the relation-algebraic expressions or formulae of the notions in question. At the beginning of Sect. 4 we have mentioned that all relation-algebraic specifications of the section rather straightforwardly can be implemented in the programming language of RelView. In the following we will demonstrate this by means of the specifications $\mathfrak{kernel}$, $\mathfrak{progFin}$ and $\mathfrak{bipartite}$.

Projection relations play a decisive role. Therefore, we start with the following two RelView-programs pr1 and pr2, which implement the two projection relations $\pi : X \times Y \leftrightarrow X$ and $\rho : X \times Y \leftrightarrow Y$ of the direct product $X \times Y$.

```
pr1(X,Y)
  DECL XY = PROD(X,Y)
  BEG  RETURN p-1(XY) END.
```

```
pr2(X,Y)
  DECL XY = PROD(X,Y)
  BEG  RETURN p-2(XY) END.
```

RelView knows relations as the only data type. In the above programs the parameters X and Y stand for homogeneous relations and X is assumed as carrier set of X and Y as carrier set of Y. The declaration XY = PROD(X,Y) introduces XY as name for the relational direct product $(X \times Y, \pi, \rho)$ in the sense of [11]. In pr1 the return-clause yields the first projection relation $\pi : X \times Y \leftrightarrow X$ by means of the pre-defined RelView-operation p-1 and in pr2 the second projection relation $\rho : X \times Y \leftrightarrow X$ is obtained via the pre-defined RelView-operation p-2.

The following RelView-program par implements the parallel composition of relations. It immediately is obtained from the definition $R \parallel S := [\pi; R, \rho; S]\!]$ using the above RelView-programs pr1 and pr2. A comparison with the definition of the parallel composition shows that * is the RelView-notation for composition, ^ that for transposition and [· , · |] that for right pairing.

```
par(R,S)
  DECL pi, rho
  BEG  pi = pr1(R*R^,S*S^);
       rho = pr2(R*R^,S*S^)
       RETURN [pi*R,rho*S|]
  END.
```

Also the following RelView-function vec immediately follows from the definition $vec(R) = [\![R, \mathsf{I}]; \mathsf{L}$, where the pre-defined RelView-operation I computes the identity relation of the same type as its argument, the pre-defined RelView-operation dom computes the composition of its argument with an universal vector of appropriate type (i.e., a vector that models the domain of the argument) and [| · , ·] is the RelView-notation for left pairing.

```
vec(R) = dom([|R,I(R^*R)]).
```

We now implement *kernel* as follows, where the parameter X of the RelView-program kernel stands for a homogeneous relation and the set X of Convention 4.1 is defined as the carrier set of X.

```
kernel(X)
  DECL M, MM, beta, stable, absorb
  BEG  M = epsi(X);
       MM = epsi(pr1(X,X));
       beta = pr2(X,M^*M);
       stable = -([M^,M^|]*MM);
       absorb = (beta & vec(-M)*L1n(M)) \ par(I(X),M^)*MM
       RETURN (L1n(M)*(stable & absorb))^
  END.
```

By means of the pre-defined RELVIEW-operation **epsi** and the first two assignments the two membership relations $\mathsf{M} : X \leftrightarrow 2^X$ and $\mathbf{M} : X^2 \leftrightarrow [X \leftrightarrow X]$ are computed and stored in the variables M and MM. The third assignment computes the second projection relation $\beta : X \times 2^X \leftrightarrow 2^X$ of the direct product $X \times 2^X$ and stores it in the variable beta. The right-hand sides of the following two assignments are the RELVIEW-versions of the relation-algebraic specifications of $\mathfrak{stable}$ and $\mathfrak{absorb}$ of Theorem 4.1. Finally, the expression of the return-clause is the RELVIEW-version of the relation-algebraic specification of $\mathfrak{kernel}$ of Theorem 4.1, where & means intersection, - means complementation and the pre-defined RELVIEW-operation L1n computes a transposed universal vector $\mathsf{L} : \mathbf{1} \leftrightarrow Y$ with the target Y equal to that of the argument.

In the same way the relation-algebraic specifications of $\mathfrak{progFin}$ and $\mathfrak{bipartite}$ of Theorem 4.3 and Theorem 4.4 immediately lead to the following two RELVIEW-programs for their computation. In progFin the variable R corresponds to the auxiliary relation $\mathfrak{R}$ of the proof of Theorem 4.3 and in bipartite the symbol | denotes union of relations.

```
progFin(X)
  DECL pi, rho, M, MM, beta, R
  BEG  pi = pr1(X,X);
       rho = pr2(X,X);
       M = epsi(X);
       MM = epsi(pi);
       beta = pr2(X,M^*M);
       R = -((beta^ & L1n(M)^*vec(M)^)*-([pi,rho*M|]^*MM))
       RETURN -(L1n(X)*M*R)^
  END.
```

```
bipartite(X)
  DECL M, MM
  BEG  M = epsi(X);
       MM = epsi(pr1(X,X))
       RETURN dom(MM \ ([|M,-M] | [|-M,M]))
  END.
```

When RELVIEW computes a relation and displays it in the relation window, it shows in the window's frame the number of rows, of columns and of 1-entries.

Using this feature, we have obtained the numerical data of Sect. 3. The running times (in seconds) of the computations are given in Table 3. A computation of the vector model of the set of relations on X means the computation of the universal vector $\mathsf{L} : [X \leftrightarrow X] \leftrightarrow \mathbf{1}$. In RelView this is possible via the expression `L1n(epsi(pr1(X,X)))^` and practically needs no time (see last column of the table). For the computation of the numerical data we have used a PC with 2 CPUs of type Intel® Xeon® E5-2698, each with 20 cores and 3.60 GHz base frequency, 512 GByte RAM and running Arch Linux 5.2.0, and version 8.2 of RelView. This newest version of the tool is described at the Web-site [14] and the source code is available from Github via [15] and from Zenodo via [16]. The virtual machine of [16] was built to ease running RelView not only using Linux but also Microsoft Windows and Mac OSX.

Table 3. Running times within RelView.

$\lvert X\rvert$	Irr., symm.	Progr. fin.	Bipartite	Richardson	With kernel	All rel.
1	0.0010	0.0012	0.0009	0.0012	0.0015	0.0006
2	0.0026	0.0032	0.0018	0.0067	0.0057	0.0007
3	0.0069	0.0082	0.0053	0.0117	0.0117	0.0007
4	0.0081	0.0172	0.0194	0.0150	0.0171	0.0008
5	0.0169	0.0262	0.0199	0.1807	0.0213	0.0010
6	0.0181	0.1211	0.0833	10.4710	0.3141	0.0011
7	0.0476	1.8771	2.3501	32220.5500	138.6700	0.0011

The amazing computational power obtained by the use of ROBDDs and RelView becomes clear if we compare the running times of Table 3 with the times needed in case of a "classical" brute-force approach. If we assume that some algorithm could generate every relation on a given finite set X and test the existence of a kernel in, say, 10^{-6} seconds, it would take $5.62 \cdot 10^{14} \cdot 10^{-6}$ seconds, i.e., more than 17 years, for this task in the case of $|X| = 7$.

6 Concluding Remarks

There exist some extensions of Richardson's theorem which allow the existence of cycles of odd length but demand certain properties for them to hold. In [1] C. Berge and P. Duchet prove that a finite directed graph $g = (X, R)$ has a kernel if every cycle of odd length *has all its arcs belonging to pairs of parallel arcs*, meaning for each cycle $(c_1, c_2, \ldots, c_n, c_1)$ of odd length of g also the reversed list $(c_1, c_n, \ldots, c_2, c_1)$ is a cycle of g, that is, all cycles of odd length are *symmetric*.

Although not explicitly mentioned, this criterion of Berge and Duchet includes g to be irreflexive. This becomes clear if one studies the proof of Proposition 1.1 of [1] in detail. Roughly the idea is as follows. Suppose $X = \{x_1, \ldots, x_n\}$.

From g then construct a graph g' by removing all edges (x_i, x_j) for which $i > j$ and also (x_j, x_i) is an edge of g. Since cycles of length 1 do not occur and all cycles of odd length are assumed as symmetric, this way each cycle of odd length is split into non-cyclic paths. Hence, the graph g' has no cycles of odd length. Richardson's theorem implies that it has a kernel K and K is also a kernel of g.

In contrast to the four criteria we have mentioned in Sect. 3, testing the criterion of Berge and Duchet seems to be rather expensive since it requires to check all cycles of odd length in view of symmetry. The same holds for all other extensions of Richardson's theorem mentioned in [1]. We also have been concerned with the question whether such weaker criteria are satisfied by much more graphs/relations with kernels than Richardson's criterion.

To get at least a feeling for their behaviour, we have applied our approach to the criterion of Berge and Duchet and computed, again for small sets X, the set of all irreflexive relations on X such that all cycles of odd length are symmetric. In case of $1 \leq |X| \leq 2$ the criterion of Berge and Duchet is equivalent to that of Richardson and, hence, is satisfied by 1 respectively 4 relations on X. For $|X| = 3$ the number of relations on X which satisfy the criterion of Berge and Duchet is 50; this are 2.04% more than the 49 relations on X which satisfy Richardson's criterion. For $4 \leq |X| \leq 6$ the numbers of relations on X which satisfy the criterion of Berge and Duchet are 1 778 (or 4.64% more than those which satisfy Richardson's criterion), 161 254 (or 6.97% more than those which satisfy Richardson's criterion) and 35 280 286 (or 8.89% more than those which satisfy Richardson's criterion). Hence, the criterion of Berge and Duchet seems to be only slightly more general than Richardson's criterion.

In [1] it is also mentioned that the existence of kernels already follows from the fact that (besides irreflexivety) every cycle of odd length *has at least two arcs belonging to pairs of parallel arcs.* This criterion is ascribed to P. Duchet. We also have checked it and RelView computed for $1 \leq |X| \leq 6$ the following numbers of relations on X which satisfy it: 1, 4, 56, 2 534, 348 064 and 138 636 886. Compared with Richardson's criterion we get for $1 \leq |X| \leq 6$ that Duchet's criterion is satisfied by 0%, 0%, 14.28%, 49.14%, 130.90% respectively 327.91% more relations on X than Richardson's criterion. Despite these better percentages it still seems to be very far away from a characterisation of the class of directed graphs having kernels. E.g., in case $|X| = 6$ it is satisfied by only 0.9% of the graphs of this class.

Acknowledgment. We thank the referees for their very helpful comments and suggestions.

References

1. Berge, C., Duchet, P.: Recent problems and results aboutkernels in directed graphs. Disc. Math. **86**, 27–31 (1990)
2. Berghammer, R., Hoffmann, T.: Deriving relational programs for computing kernels by reconstructing a proof of Richardson's theorem. Sci. Comput. Program. **38**, 1–25 (2000)

3. Berghammer, R., Leoniuk, B., Milanese, U.: Implementation of relational algebra using binary decision diagrams. In: de Swart, H.C.M. (ed.) RelMiCS 2001. LNCS, vol. 2561, pp. 241–257. Springer, Heidelberg (2002). https://doi.org/10.1007/3-540-36280-0_17
4. Berghammer, R., Neumann, F.: RelView – an OBDD-based computer algebra system for relations. In: Ganzha, V.G., Mayr, E.W., Vorozhtsov, E.V. (eds.) CASC 2005. LNCS, vol. 3718, pp. 40–51. Springer, Heidelberg (2005). https://doi.org/10.1007/11555964_4
5. Chvatal, V.: On the computational complexity of finding a kernel. Technical report CRM-300, Centre de Recherche Mathématiques, Université de Montréal (1973)
6. Davis, A.C.: A characterization of complete lattices. Pac. J. Math. **5**(2), 311–319 (1955)
7. Dedekind, R.: Über die von drei Moduln erzeugte Dualgruppe. Mathematische Annalen **53**(1), 371–403 (1900)
8. König, D.: Über Graphen und ihre Anwendung auf Determinantentheorie und Mengenlehre. Mathematische Annalen **77**, 453–465 (1916)
9. Kuratowski, K.: Sur le probleme des courbes gauches en topologie. Fundamenta Mathematicae **15**(1), 271–283 (1930)
10. Richardson, M.: Solutions of irreflexive relations. Ann. Math. **58**(3), 573–590 (1953)
11. Schmidt, G., Ströhlein, T.: Relations and Graphs. Springer, Heidelberg (1993). https://doi.org/10.1007/978-3-642-77968-8
12. Tarski, A.: A lattice-theoretical fixpoint theorem and its applications. Pac. J. Math. **5**(2), 285–309 (1955)
13. von Neumann, J., Morgenstern, O.: Theory of Games and Economic Bevaviour. Princeton University Press, Princeton (1944)
14. https://www.rpe.informatik.uni-kiel.de/en/research/relview (homepage of Rel View)
15. https://github.com/relview (source code of Rel View)
16. https://zenodo.org/record/4708085#.YICAmS0RppR (virtual machine for Rel View)

ℓr-Multisemigroups, Modal Quantales and the Origin of Locality

Cameron Calk[1], Uli Fahrenberg[1], Christian Johansen[2], Georg Struth[3(✉)], and Krzysztof Ziemiański[4]

[1] École Polytechnique, Palaiseau, France
[2] Norwegian University of Science and Technology, Trondheim, Norway
[3] University of Sheffield, Sheffield, UK
g.struth@sheffield.ac.uk
[4] University of Warsaw, Warszawa, Poland

Abstract. We introduce ℓr-multisemigroups as duals of modal quantales and study modal correspondences between equations in these multisemigroups and the domain and codomain axioms of modal quantales. Our results yield new insights on the origin of locality in modal semirings and quantales. They also yield construction principles for modal powerset quantales that cover a wide range of models and applications.

1 Introduction

This work adds to a series on convolution semirings and quantales built over relational monoids and multimonoids [3,8,12]. It explains the structure of modal semirings and quantales [7,11], not generally for convolution algebras [12], but specifically for modal powerset quantales—the standard setting for computational models in this context. We consider such quantales as boolean algebras with operators [19]. The quantalic composition is then a binary modality; the domain and codomain operations needed for defining modal operators are unary ones. We ask about the dual relational structure in the sense of Jónsson and Tarski [19] and its equational properties corresponding to the modal quantale axioms for domain and codomain [7,11] in the sense of modal correspondence theory. For plain quantales, this is well known: the dual monoidal structure is a ternary relation equipped with a relational monoid structure and many units [3,8]—a monoid in the category **Rel** with the standard tensor. Yet which relational structure corresponds to domain and codomain?

The standard models of modal semirings and quantales give us a hint: modal quantales of binary relations, for instance, are powerset liftings of pair groupoids; modal quantales of paths lift from path categories. We might therefore try to lift (object-free) categories [23, Chap. XII.5] to modal quantales so that their

U. Fahrenberg—Supported by the *Chaire ISC : Engineering Complex Systems* – École polytechnique – Thales – FX – DGA – Dassault Aviation – Naval Group – ENSTA ParisTech – Télécom ParisTech.

U. Fahrenberg et al. (Eds.): RAMiCS 2021, LNCS 13027, pp. 90–107, 2021.
https://doi.org/10.1007/978-3-030-88701-8_6

source and target maps match the domain and codomain operations of modal quantales. Categories, however, are partial monoids, whereas relational monoids are isomorphic to multimonoids, whose composition maps pairs of elements to sets, like the shuffle of words. Other examples, such as the lifting of partial abelian monoids of heaplets to assertion quantales of separation logic, do not fall into this lifting scheme with categories either. A generalisation is desirable.

We introduce ℓr-multisemigroups as relational structures in disguise and the appropriate dual structures to modal powerset quantales. Categories then arise as partial ℓr-semigroups (where the image of the multioperation is suitably restricted) that satisfy a locality property capturing the categorical composition pattern: two arrows are composable precisely if the target of the first equals the source of the second. Thus, ℓr-multisemigroups generalise object-free categories and related structures such as function systems [28], ordered semigroupoids [21] and modal semigroups [5] from (partial) operations to multioperations.

Our second main contribution lies in modal correspondences between identities in families of modal quantales with axioms of varying strength and those of families of ℓr-multisemigroups. The most intriguing one holds between the well studied locality axioms for domain and codomain in modal semirings and quantales and similar identities in ℓr-multisemigroups, which in turn are equivalent to the composition pattern for categories mentioned. This explains the origin of locality of domain and codomain in modal semirings and quantales in terms of this fundamental pattern. It also makes local ℓr-multisemigroups the algebras of choice for constructing modal quantales axiom by axiom.

Our results thus provide a generic construction recipe for modal quantales from simpler structures: every ℓr-multisemigroup gives us a modal powerset quantale for free—and even modal convolution quantales capturing weighted variants of the models presented in this text. This generalisation is briefly outlined at the end of this article, see [12] for details.

All results about ℓr-multisemigroups and the lifting to modal powerset quantales have been formalised with Isabelle/HOL[1]. The proofs for ℓr-multisemigroups are straightforward equational calculations that do not need to be shown on paper. The proof of the powerset lifting has been added because it yields an intuition for the more complex construction of modal convolution quantales. Additional proofs, definitions and explanations can be found in [12], including a glossary of the algebraic structures featured in this text.

2 ℓr-Multisemigroups and Object-Free Categories

As mentioned in the introduction, the dual of the binary composition of a quantale is a ternary relation. For powerset quantales it is defined on their atom structure of singleton sets. But instead of a ternary relation $R \subseteq X \times X \times X$ on a set X, say, we work with the isomorpic multioperation $\odot : X \times X \to \mathcal{P}X$ and the resulting multisemigroups. See [22] for an overview. Henceforth we are using

[1] https://github.com/gstruth/lr-multisemigroups.

"set" naively, so that we can speak, for instance, about the set of all posets and include large categories as examples.

We extend the multioperation $\odot$ to $\mathcal{P}X \times \mathcal{P}X \to \mathcal{P}X$ by

$$A \odot B = \bigcup\{x \odot y \mid x \in A \text{ and } y \in B\} \qquad \text{for all } A, B \subseteq X.$$

We write $x \odot B$ instead of $\{x\} \odot B$, $A \odot x$ instead of $A \odot \{x\}$, $f(A)$ for the image of A under f and drop $\odot$ when convenient. Finally, $\odot$ is a *partial operation* if $|x \odot y| \leq 1$ and a *(total) operation* if $|x \odot y| = 1$, for all $x, y \in X$.

A *multimagma* $(X, \odot)$ is a set X with a multioperation $\odot$ on X. A *multisemigroup* X is an associative multimagma, it satisfies $x \odot (y \odot z) = (x \odot y) \odot z$ for all $x, y, z \in X$. Partial semigroups and semigroups are defined by restricting the image of $\odot$ as just explained.

Object-free categories are obtained either by defining source and target maps on partial semigroups or by equipping partial semigroups with many units [23]. We explore both ways more generally for multisemigroups.

An *ℓr-multimagma* is a multimagma X with operations $\ell, r : X \to X$ that satisfy, for all $x, y \in X$,

$$x \odot y \neq \emptyset \Rightarrow r(x) = \ell(y), \qquad \ell(x) \odot x = \{x\}, \qquad x \odot r(x) = \{x\}.$$

An *ℓr-multisemigroup* is an associative ℓr-multimagma. We call ℓ the source operation and r the target operation of X. The letters indicate "left" and "right".

Alternatively, a multimagma X is *unital* if there exists a set $E \subseteq X$ such that $E \odot x = \{x\} = x \odot E$ for all $x \in X$. A *multimonoid* is then a unital multisemigroup. See [12] for a more detailed discussion.

We briefly summarise the relationship between the two structures. First, in unital multimagmas, every $e \in E$ satisfies $e \odot e = \{e\}$ and, if $e, e' \in E$, then $e \odot e' \neq \emptyset \Leftrightarrow e = e'$. Units are thus "orthogonal" idempotents. In multimonoids, every element has therefore precisely one left and one right unit, and this allows defining source and target maps. Second, the set $\ell(X)$ of all source elements in any ℓr-multisemigroup X equals the set $r(X)$ of all target elements and the elements of those sets satisfy the unit axioms for multimonoids (see also Sect. 4). Third, ℓr-multisemigroups and multimonoids form categories with morphisms satisfying $f(x \odot_1 y) \subseteq f(x) \odot_2 f(y)$ for multisemigroups $(X_i, \odot_i)$ with $i \in \{1, 2\}$. For ℓr-multisemigroups, morphisms need to preserve ℓ and r as well; for multimonoids they need to preserve units. It is then easy to see that the categories of ℓr-multisemigroups and multimonoids are isomorphic [12].

Partial ℓr-semigroups are not yet (object-free) categories—see Examples 7 and 8 below. We need to impose the typical composition pattern of categories: two morphisms can be composed if the target of the first equals the source of the second. So we call an ℓr-multimagma *ℓr-local* if

$$r(x) = \ell(y) \Rightarrow x \odot y \neq \emptyset \qquad \text{for all } x, y \in X.$$

We relate this property with notions of locality known from modal semigroups and semirings in Sect. 4. Example 6 below shows a local ℓr-multisemigroup with a proper multioperation that does not form an object-free category.

An ℓr-multisemigroup X is ℓr-local if and only if

$$u \in x \odot y \wedge y \odot z \neq \emptyset \Rightarrow u \odot z \neq \emptyset \qquad \text{for all } u, x, y, z \in X.$$

This implication is expressible in any multimagma. The connection to the two equivalent formalisations of (object-free) categories in Mac Lane's book [23] is thus as follows.

Proposition 1 ([4]). *The categories of object-free categories [23, Chap. I.1] and those of local partial monoids are isomorphic.*

Proposition 2. *The categories of object-free categories [23, Chap. XII.5] and those of ℓr-local partial ℓr-semigroups are isomorphic.*

The morphisms used are those outlined above. Hence local partial ℓr-semigroups *are* categories (when these structures are defined over classes).

3 Examples of ℓr-Multisemigroups

We start with concrete instances of categories.

Example 3 (Monoids). Monoids are one-object categories. The monoid $1 \stackrel{a}{\to} 1$, for instance, corresponds to a partial monoid $X = \{1, a\}$ with composition defined by $11 = \{1\}$ and $1a = a1 = aa = \{a\}$. Obviously, $\ell(a) = 1 = r(a)$ and locality follows from totality of composition. □

Multimonoids must have precisely one unit if the multioperation is total (in the sense that images of compositions cannot be empty).

Example 4 (Pair Groupoids). The pair groupoid $(X \times X, \odot, Id_X)$ on set X (or the universal relation on X) is a local partial ℓr-semigroup with

$$(w, x) \odot (y, z) = \begin{cases} \{(w, z)\} & \text{if } x = y, \\ \emptyset & \text{otherwise,} \end{cases}$$

identity relation Id_X on X, $\ell((x, y)) = (x, x)$ and $r((x, y)) = (y, y)$. □

Pair groupoids lift to quantales of binary relations.

Example 5 (Matrix Theories). Elgot's matricial theories [9] consist of sets $\mathbb{M}S = \bigcup_{n,m \geq 0} S^{n \times m}$ of matrices over a semiring S with matrix multiplication as partial composition. These form a category with natural numbers as objects and $n \times m$-matrices as morphisms. Defining ℓ and r to map any $M \in S^{n \times m}$ to the identity matrices $\ell(M) = I_n$ and $r(M) = I_m$ of the appropriate dimensions, $\mathbb{M}S$ forms a local partial ℓr-semigroup. Matrix theories become categories of finite relations if S is the semiring of booleans. □

The next example presents a local proper ℓr-multisemigroup.

Example 6 (Shuffle Algebras). The shuffle multimonoid $(\Sigma^*, \|, \{\varepsilon\})$ over the free monoid Σ^* has the empty word ε as its unit, and the proper multioperation $\| : \Sigma^* \times \Sigma^* \to \mathcal{P}\Sigma^*$ models the standard interleaving of words that respects the orders of their letters. The shuffle multimonoid is local because $\|$ is total (defined everywhere) and $\ell(w) = \varepsilon = r(w)$. □

Finally, here are two non-local partial semigroups.

Example 7 (Broken Monoid). The monoid in Example 3 becomes a non-local partial ℓr-semigroup when composition is broken by imposing $aa = \emptyset$. □

Example 8 (Heaplets). The partial abelian monoid of heaplets $(H, \odot, \varepsilon)$ from separation logic is formed by the set of partial functions $X \rightharpoonup Y$. Its partial operation $f \odot g$ equals $f \cup g$ if $dom(f) \cap dom(g)$ is empty and $\emptyset$ otherwise. The unit is the empty partial function ε with empty domain. Locality fails because $\ell(f) = \varepsilon = r(g)$ always holds while $f \odot g = \emptyset$ if domains of f and g overlap. □

4 ℓr-Multisemigroups in Context

We have already seen that local partial ℓr-semigroups are categories. Here we relate them with Schweizer and Sklar's function systems [28] and modal semigroups [5]. The following property gives us half of our results for free.

Duality (by opposition) for ℓr-multimagmas arises by interchanging ℓ and r as well as the arguments of $\odot$. The classes of ℓr-multimagmas and ℓr-multisemigroups are closed under this transformation. Locality and partiality are self-dual. Hence the dual of any property that holds in any of these classes holds as well.

Lemma 9. *In any ℓr-multimagma, the following laws hold:*

1. $\ell \circ r = r$, $r \circ \ell = \ell$ (compatibility),
2. $\ell \circ \ell = \ell$, $r \circ r = r$ (retraction),
3. $\ell(x)\ell(x) = \{\ell(x)\}$ (idempotency),
4. $r(x)\ell(y) = \ell(y)r(x)$ (commutativity),
5. $\ell(\ell(x)y) = \ell(x)\ell(y)$ *and* $r(xr(y)) = r(x)r(y)$ (export),
6. $\ell(xy)x \subseteq x\ell(y)$ *and* $xr(yx) \subseteq r(y)x$ (weak twisted).

All proofs have been checked with Isabelle. All laws in Lemma 9 correspond to axioms for Schweizer and Sklar's function systems [28] (see [12] for a detailed comparison), yet generalised to multioperations.

The compatibility laws imply that $\ell(x) = x \Leftrightarrow r(x) = x$ and further that

$$X_\ell = \{x \mid \ell(x) = x\} = \{x \mid r(x) = x\} = X_r.$$

Moreover, by the retraction laws, $X_\ell = \ell(X)$ and $X_r = r(X)$.

Lemma 9 also implies that $\ell(x)\ell(y) = \ell(y)\ell(x)$, $r(x)r(y) = r(y)r(x)$ and $r(x)r(x) = \{r(x)\}$. Further, the orthogonality law $\ell(x)\ell(y) \neq \emptyset \Leftrightarrow \ell(x) = \ell(y)$ and its dual hold. As ℓr-Multimagmas are unital, we may write E for X_ℓ or X_r.

Lemma 10. *In any ℓr-multisemigroup, the following laws hold:*

1. $\ell(xy) \subseteq \ell(x\ell(y))$ *and* $r(xy) \subseteq r(r(x)y)$ (weak locality),
2. $xy \neq \emptyset \Rightarrow \ell(xy) = \ell(x\ell(y))$ *and* $xy \neq \emptyset \Rightarrow r(xy) = r(r(x)y)$ (cond. locality),
3. $\ell(xy) \subseteq \{\ell(x)\}$ *and* $r(xy) \subseteq \{r(y)\}$,
4. $xy \neq \emptyset \Rightarrow \ell(xy) = \{\ell(x)\}$ *and* $xy \neq \emptyset \Rightarrow r(xy) = \{r(y)\}$,
5. $xy \neq \emptyset \Rightarrow \ell(xy)x = x\ell(y)$ *and* $xy \neq \emptyset \Rightarrow yr(xy) = r(x)y$ (cond. twisted).

Proofs have again been checked with Isabelle. The locality and twisted laws generalise the remaining axioms of function systems. Function systems without the twisted laws correspond to modal semigroups [5] and therefore semigroups of binary relations. The twisted laws are specific to semigroups of functions. ℓr-Multisemigroups thus generalise function systems and modal semigroups beyond totality. See [5] for a discussion of related structures studied in semigroups theory and applications.

5 ℓr-Locality in Context

Next we return to locality, the specific difference between object-free categories and partial ℓr-semigroups according to Sect. 2.

Lemma 11. *In any local ℓr-multisemigroup, the following laws hold:*

1. $\ell(xy) = \ell(x\ell(y))$ *and* $r(xy) = r(r(x)y)$ *(equational locality),*
2. $\ell(xy)x = x\ell(y)$ *and* $yr(xy) = r(x)y$ *(twisted).*

Once again, all proofs have been done with Isabelle. In fact, ℓr-locality, the composition pattern of categories, is an equational property. We henceforth refer to equational locality simply as *locality.*

Proposition 12. *An ℓr-multisemigroup is ℓr-local if and only if*

$$\ell(x\ell(y)) \subseteq \ell(xy) \qquad \text{and} \qquad r(r(x)y) \subseteq r(xy).$$

Proof. Isabelle confirms that the equational locality laws imply ℓr-locality in any ℓr-multimagma. Equality in ℓr-multisemigroups follows from Lemma 11. □

Locality and weak locality are known from (pre)domain and (pre)codomain operations for modal semirings [6]. Predomain and precodomain operations are weakly local, domain and codomain are local. Relative to ℓr-multisemigroups, these laws are at powerset level. Modal semirings are meant to model semirings of binary relations. These in turn are based on pair groupoids, as we shall see. Equational locality and the equivalent variant

$$xy \neq \emptyset \Leftrightarrow r(x) = \ell(y)$$

of ℓr-locality thus describe the origin of locality in categories and more generally ℓr-multisemigroups. The precise relationship to modal semirings and quantales is explained in the following sections.

Our final lemma on ℓr-multisemigroups yields a more fine-grained view on definedness conditions and ℓr-locality.

Lemma 13

1. *In any ℓr-multimagma,*

$$r(x) = \ell(y) \Leftrightarrow r(x)\ell(y) \neq \emptyset \qquad \text{and} \qquad r(x)\ell(y) = \emptyset \Rightarrow xy = \emptyset.$$

2. *In any local ℓr-multisemigroup, $xy = \emptyset \Leftrightarrow r(x)\ell(y) = \emptyset$.*

A property analogous to Lemma 13(2) is well known from modal semirings [6]. An analogue to ℓr-locality fails already in the one-element modal semiring.

6 Modal Quantales

We have already extended $\odot : X \times X \to \mathcal{P}X$ to $\mathcal{P}X \times \mathcal{P}X \to \mathcal{P}X$ and the functions $\ell, r : X \to X$ to $\mathcal{P}X \to \mathcal{P}X$ by taking images. We wish to explore the algebraic structure of such powerset liftings over ℓr-multimagmas and related structures. Powerset liftings of relational monoids, and therefore those of ℓr-multisemigroups, yield unital quantales [8,26]. But the precise lifting of source and target operations remains to be explored. This requires some preparation.

A *quantale* [25] $(Q, \leq, \cdot, 1)$ is a complete lattice $(Q, \leq)$ with a monoidal composition $\cdot$ with unit 1 that preserves all sups in both arguments. A quantale is *boolean* if its lattice reduct is a complete boolean algebra—a complete lattice and a boolean algebra. Some applications require weaker notions. A *prequantale* is a quantale where the associativity law is absent [25].

We write $\bigvee$ for the sup and $\bigwedge$ for the inf operator, and $\vee$, $\wedge$ for their binary variants. We also write $\bot = \bigwedge Q = \bigvee \emptyset$ for the least and $\top = \bigvee Q = \bigwedge \emptyset$ for the greatest element of Q, and $-$ for boolean complementation (both unary and binary) if Q is boolean. We write $Q_1 = \{\alpha \in Q \mid \alpha \leq 1\}$ for the set of *subidentities* of Q. In a boolean quantale, Q_1 is a complete boolean subalgebra with complementation $\lambda x.\ 1 - x$ and composition coinciding with meet [11].

We lift the source and target operations of ℓr-multisemigroups to domain and codomain operations at powerset level. Modal quantales of relations, which are formally lifted from pair groupoids below, provide some intuition:

$$dom(R) = \{(a, a) \mid \exists b.\ (a, b) \in R\}, \qquad cod(R) = \{(b, b) \mid \exists a.\ (a, b) \in R\}$$

and hence $dom(R) = \ell(R)$ and $cod(R) = r(R)$.

More generally, a *domain quantale* [11] is a quantale $(Q, \leq, \cdot, 1)$ equipped with a domain operation $dom : Q \to Q$ that satisfies, for all $\alpha, \beta \in Q$,

$$\alpha \leq dom(\alpha) \cdot \alpha, \qquad dom(\alpha \cdot dom(\beta)) = dom(\alpha \cdot \beta), \qquad dom(\alpha) \leq 1,$$
$$dom(\bot) = \bot, \qquad dom(\alpha \vee \beta) = dom(\alpha) \vee dom(\beta).$$

We call these equations the *absorption*, *locality*, *subidentity*, *strictness* and *(binary) sup-preservation* axiom, respectively. Absorption can be strengthened to $dom(\alpha)\alpha = \alpha$. These domain axioms are precisely those of domain semirings [7]. Domain quantales are thus quantales that are also domain semirings.

Properties of domain semirings therefore translate [11,12]. Interestingly, domain axioms for $\bigvee$ are not needed in domain quantales [11] because *dom* preserves arbitrary sups. The interaction of *dom* with $\bigwedge$ is weaker and not our concern.

Much of the structure of the domain algebra induced by *dom* is inherited from domain semirings as well. In particular, $Q_{dom} = \{x \mid dom(x) = x\} = dom(Q)$, and it follows that the *domain algebra* $(Q_{dom}, \leq, \cdot, 1)$ is a subquantale of Q that forms a bounded distributive lattice with $\cdot$ as binary inf [7]. The elements of Q_{dom} are called *domain elements* of Q. Yet, by contrast to modal semirings, the lattice Q_{dom} is complete [11], and if Q is boolean, then $Q_{dom} = Q_1$ is a complete boolean algebra. For powerset quantales, this complete boolean algebra is atomic.

Quantales are closed under opposition: interchanging the order of composition in Q yields the quantale Q^{op}; properties translate under this duality. The dual of *dom* on a domain quantale is of course a codomain operation *cod*.

A *codomain quantale* (Q, cod) is thus simply a domain quantale (Q^{op}, dom). It satisfies the dual domain axioms. A *modal quantale* is a domain and codomain quantale $(Q, \leq, \cdot, 1, dom, cod)$ that satisfies the *compatibility* axioms

$$dom \circ cod = cod \qquad \text{and} \qquad cod \circ dom = dom.$$

These force $Q_{dom} = Q_{cod}$.

Some ℓr-structures of interest fail to yield associativity or locality laws when lifted. This requires more general notions.

- A *modal prequantale* is a prequantale in which the locality axioms for *dom* and *cod* are replaced by the export axiom $dom(dom(\alpha)\beta) = dom(\alpha)dom(\beta)$ and its dual for *cod*. Then $Q_{dom} = dom(Q) = cod(Q) = Q_{cod}$ is still a complete distributive lattice, but locality laws for *dom* and *cod* are not even derivable as inequalities.
- A *weakly local modal quantale* is a modal quantale that satisfies the previous axioms for *dom* and *cod*. The *weak locality* law $dom(\alpha\beta) \leq dom(\alpha dom(\beta))$ and its dual for *cod* are now derivable, but not the equational laws.

7 Constructing Modal Powerset Quantales

We now construct modal powerset quantales from ℓr-multisemigroups in the spirit of modal correspondence theory for boolean algebras with operators. First we recall the quantalic part.

Proposition 14. *Let* $(X, \odot, \ell, r)$ *be an* ℓr*-multisemigroup. Then* $(\mathcal{P}X, \subseteq, \odot, E)$ *forms a boolean quantale whose underlying lattice is boolean atomic.*

Proof. If $(X, \odot, \ell, r)$ is an ℓr-multisemigroup, then it is isomorphic to a multimonoid and further to a relational monoid, and its powerset algebra forms a quantale [8,26]. The complete lattice on $\mathcal{P}X$ is trivially boolean atomic. □

Similarly, ℓr-multimagmas lift to prequantales.

Example 15 (Powerset Quantales over ℓr-Semigroups). The powerset lifting of any category yields a powerset quantale. It is boolean and has the arrows of the category as atoms. The pair groupoid on set X lifts to the quantale of binary relations over X. Its elements are possibly infinite-dimensional boolean-valued square matrices in which the quantalic composition is matrix multiplication. □

The fact that groupoids can be lifted to algebras of binary relations with an additional operation of converse was known to Jónsson and Tarski [20].

Proposition 14 combines source and target elements into the unit E of the powerset quantale. The lifting to modal quantales is more refined. In the following theorems, we identify $dom(A)$ with $\ell(A)$ and $cod(A)$ with $r(A)$ for $A \subseteq X$. We develop our main theorem step-by-step from ℓr-multimagmas.

Lemma 16. *Let X be an ℓr-multimagma. For $A, B \subseteq X$ and $\mathcal{A} \subseteq \mathcal{P}X$,*

1. *$\ell(r(A)) = r(A)$ and $r(\ell(A)) = \ell(A)$ (compatibility),*
2. *$\ell(A) \cdot A = A$ and $A \cdot r(A) = A$ (absorption),*
3. *$\ell\left(\bigcup \mathcal{A}\right) = \bigcup_{A \in \mathcal{A}} \ell(A)$ and $r\left(\bigcup \mathcal{A}\right) = \bigcup_{A \in \mathcal{A}} r(A)$ (sup-preservation),*
4. *$f(A)g(B) = g(B)f(A)$ hold for $f, g \in \{\ell, r\}$ (commutativity),*
5. *$\ell(A) \subseteq X_\ell$ and $r(A) \subseteq X_r$ (subidentity),*
6. *$\ell(\ell(A) \cdot B) = \ell(A)\ell(B)$ and $r(A \cdot r(B)) = r(A)r(B)$ (export).*

Proof. We show proofs up-to duality.

1. $\ell(r(A)) = \{\ell(r(x)) \mid x \in A\} = \{r(x) \mid x \in A\} = r(A)$.
2. $$\begin{aligned} \ell(A)A &= \bigcup \{\ell(x)y \mid x, y \in A \text{ and } \ell(x)y \neq \emptyset\} \\ &= \bigcup \{\ell(x)y \mid x, y \in A,\ \ell(x)y \neq \emptyset \text{ and } r(\ell(x)) = \ell(y)\} \\ &= \bigcup \{\ell(x)y \mid x, y \in A,\ \ell(x)y \neq \emptyset \text{ and } \ell(x) = \ell(y)\} \\ &= \bigcup \{\ell(y)y \mid y \in A\} \\ &= \bigcup \{\{y\} \mid y \in A\} = A. \end{aligned}$$
3. $\ell\left(\bigcup \mathcal{A}\right) = \{\ell(x) \mid x \in \bigcup \mathcal{A}\} = \bigcup \{\ell(A) \mid A \in \mathcal{A}\}$.
4. We only prove the identity for $\ell(A)r(B)$. The others then follow from (1).
$$\begin{aligned} \ell(A)r(B) &= \bigcup \{\ell(x)r(y) \mid x \in A \text{ and } y \in B\} \\ &= \bigcup \{r(y)\ell(x) \mid x \in A \text{ and } y \in B\} \\ &= r(B)\ell(A). \end{aligned}$$
5. $\ell(A) = \{\ell(x) \mid x \in A\} \subseteq \{\ell(x) \mid x \in X\} = \{x \mid \ell(x) = x\} = E$.

6.
$$\begin{aligned}\ell(\ell(A)B) &= \bigcup\{\ell(\ell(x)y) \mid x \in A,\, y \in B \text{ and } \ell(x)y \neq \emptyset\}\\ &= \bigcup\{\ell(x)\ell(y) \mid x \in A,\, y \in B,\, \ell(x)y \neq \emptyset \text{ and } r(\ell(x)) = \ell(y)\}\\ &= \bigcup\{\ell(x)\ell(y) \mid x \in A,\, y \in B,\, \ell(x)y \neq \emptyset \text{ and } \ell(x) = \ell(y)\}\\ &= \bigcup\{\ell(x)\ell(y) \mid x \in A,\, y \in B \text{ and } \ell(y)y \neq \emptyset\}\\ &= \bigcup\{\ell(x)\ell(y) \mid x \in A \text{ and } y \in B\}\\ &= \ell(A)\ell(B).\end{aligned}$$
□

The proof has also been checked with Isabelle. And now for locality.

Lemma 17. *Let X be an ℓr-multisemigroup and $A, B \subseteq X$. Then*

$$\ell(AB) \subseteq \ell(A\ell(B)) \qquad \text{and} \qquad r(AB) \subseteq r(r(A)B).$$

The converse inclusions of these weak locality laws hold if X is local.

Proof. The inclusions hold in any quantale that satisfies the laws of Lemma 16. For the opposite direction, suppose that X is a local ℓr-multisemigroup. Then, writing $r(x) = \ell(y)$ in place of $x \odot y \neq \emptyset$ owing to locality,

$$\begin{aligned}\ell(A\ell(B)) &= \bigcup\{\ell(x\ell(y)) \mid x \in A,\, y \in B \text{ and } r(x) = \ell(\ell(y))\}\\ &= \bigcup\{\ell(xy) \mid x \in A,\, y \in B \text{ and } r(x) = \ell(y)\} = \ell(AB)\end{aligned}$$

and the opposite result for r is obvious. □

The proofs have again been checked with Isabelle. We can now summarise.

Theorem 18. *Let X be an ℓr-multimagma.*

1. *Then $(\mathcal{P}X, \subseteq, \odot, E, dom, cod)$ is a boolean modal prequantale, and the complete boolean algebra is atomic.*
2. *It is a weakly local modal quantale if X is an ℓr-multisemigroup.*
3. *It is a modal quantale if X is a local ℓr-multisemigroup.*

This result highlights the role of weak locality and locality in the three stages of lifting. Its construction follows one direction of Jónsson-Tarski duality between relational structures and boolean algebras with operators [13,19], which it refines. Like in modal logic, it leads to correspondences between identities in relational structures and boolean algebras with operators. Those lifted in Lemma 16 and 17 are one direction of these. Their converses are explored in Sect. 8.

Example 19. (Modal Powerset Quantales over ℓr-Semigroups)

1. Any category as a local partial ℓr-semigroup can be lifted to a modal powerset quantale. The domain algebra is the entire boolean subalgebra of subidentities of the quantale, the set of all objects of the category (or its identity arrows). A modal quantale can thus be obtained from any category.
2. An instance is the modal powerset quantale of binary relations over X lifted from the pair groupoid on X. Domain and codomain elements are precisely the subidentity relations below Id_X. In the associated matrix algebras, these correspond to (boolean-valued) sub-diagonal matrices (which may have zeros along the diagonal) and further to predicates.
3. Recall that the partial ℓr-semigroup of the broken monoid is only weakly local. The powerset quantale is only weakly local, too. To check this, we simply replay the non-locality proof for the partial ℓr-semigroup with $A = \{a\}$: $dom(AA) = dom(\emptyset) = \emptyset \subset \{1\} = dom(A\{1\}) = dom(Adom(A))$. Locality of codomain is ruled out by duality. □

Most models of domain and modal semirings considered in the literature are powerset structures lifted from categories. Theorem 18 yields a uniform construction recipe for all of them. The final example of this section shows that the twisted laws for function systems do not lift to powersets.

Example 20. The category $1 \xrightarrow{a} 2$ is a partial local ℓr-semigroup with $X = \{1, a, 2\}$, ℓ and r defined by $\ell(1) = r(1) = 1 = \ell(a)$ and $\ell(2) = r(2) = 2 = r(a)$ and composition $11 = 1$, $1a = a = a2$ and $22 = 2$. Then, for $A = \{1, a\}$ and $B = \{2\}$, $A \cdot dom(B) = A \cdot B = \{a\} \subset A = \{1\} \cdot A = dom(A \cdot B) \cdot A$ refutes the twisted law in $\mathcal{P}X$. The opposite law for cod is refuted by a dual example. □

8 Recovering ℓr-Multisemigroups

We know from Jónsson-Tarski duality that one can find an ℓr-multisemigroup within each modal powerset quantale, using its atom structure. Here we prove correspondence results in this direction. These strengthen the relationship between locality in modal quantales and ℓr-multisemigroups further. Parts of these results are special cases of more general theorems for convolution algebras [3,12].

Proposition 21

1. *If $\mathcal{P}X$ is a prequantale in which $\emptyset \neq E$, then X is an ℓr-multimagma.*
2. *If $\mathcal{P}X$ is a quantale in which $\emptyset \neq E$, then X is an ℓr-multisemigroup.*

Proof. The results are known for unital relational magmas and relational monoids [3, Proposition 4.1 and Corollary 4.7]. They thus hold for ℓr-multimagmas and multisemigroups via the isomorphisms. □

The ℓr-semigroup X is thus completely determined by the subidentites below E in $\mathcal{P}X$. We calculate the absorption law for ℓ explicitly as an example:

$$\ell(x) \odot x = \{\ell(x)\} \odot \{x\} = dom(\{x\}) \odot \{x\} = \{x\},$$

where the second step uses domain absorption in modal quantales. The fact that *dom* appears in the calculation does not go beyond Proposition 21 because $dom(\{x\}) = \{\ell(x)\} \subseteq E$ in $\mathcal{P}X$.

The next statement adds locality to the picture.

Theorem 22. *If $\mathcal{P}X$ is a modal quantale in which $\emptyset \neq E$, then X is a local ℓr-multisemigroup.*

Proof. Relative to Proposition 21 it remains to consider locality:

$$\ell(x \odot \ell(y)) = dom(\{x\}) \odot dom(\{y\}) = dom(\{x\} \odot \{y\}) = \ell(x \odot y).$$

Locality of r follows by duality. □

In light of Jónsson-Tarski duality, these results extend to atomic boolean quantales. With the lifting results from Sect. 7 they yield in particular a correspondence between locality in ℓr-multisemigroups and modal powerset quantales. To construct such a quantale one should therefore look for the underlying ℓr-multisemigroup, and often, more specifically, the underlying category.

9 Further Examples

We apply our construction to further examples of ℓr-multisemigroups and modal convolution quantales. We start with those based on categories.

Path Quantales. A quiver (or digraph) K is formed by a set V_K of vertices, a set E_K of edges and source/target maps $s, t : E_K \to V_K$. The path category of K has vertices as objects and sequences $\pi = (v_1, e_1, v_2, \dots, v_{n-1}, e_{n-1}, v_n) : v_1 \to v_n$ as arrows in which vertices and edges alternate. Composition $\pi_1 \cdot \pi_2$ of $\pi_1 : v_3 \to v_4$ and $\pi_2 : v_1 \to v_2$ is defined if $v_2 = v_3$. It concatenates the two paths while gluing the common end $v_2 = v_3$. Sequences (v) of length 0 are identities. Path categories are local partial ℓr-semigroups, with $\ell(\pi) = (v_1)$ and $r(\pi) = (v_n)$ for π as above. Theorem 18 shows that the powerset algebra over the path category of any quiver is a modal quantale—a modal quantale of path languages.

The path category generated by the one-point quiver with n arrows represents the free monoid with n generators. The ℓr-structure and hence the modal structure is then trivial. Lifting along Theorem 18 yields the quantale of formal languages. Path categories are relevant to computing: they capture execution sequences of programs, automata or transition systems.

Interval Quantales. Pair groupoids over X become poset categories in which pairs represent (closed) segments or intervals when the universal relations used for pair groupoids are generalised to partially or totally ordered relations. Segments or intervals can be composed like the elements of the pair groupoid; the units are the singleton intervals. Modal powerset quantales over such categories yield algebraic semantics for interval logics [17] and interval temporal logics [24] via the isomorphism between sets and predicates [8]. The modalities lifted from source and target maps express properties of endpoints of segments and intervals.

Pomset Quantales. Finite posets form partial ℓr-multisemigroups with respect to serial composition, which is the disjoint union of posets with all elements of the first poset preceding that of the second one in the order of the composition. The only unit is the empty poset, the algebra is therefore non-local and the modal structure of the powerset quantale trivial.

Partial words [14] or pomsets are isomorphism classes of finite node-labelled posets. The serial composition becomes total on pomsets, which yields a monoid and establishes locality. Pomsets and pomset languages, obtained by powerset lifting, form a standard model of concurrency.

Pomsets can be equipped with interfaces [29]. The source interface of a pomset consists of its minimal elements (with their labels); its target interface of its maximal elements (again with their labels). Pomsets with interfaces form partial ℓr-semigroups with ℓ and r mapping pomsets to their source and target interfaces, and composition defined by gluing pomsets on their interfaces whenever they match and extending the order as in serial composition. The partial ℓr-semigroups are local and therefore categories. The modal structure at powerset level is no longer trivial.

Path Quantales in Topology. A *path* in a topological space X is a continuous map $f : [0,1] \to X$. The source of path f is $\ell(f) = f(0)$, its target $r(f) = f(1)$. Paths f and g in X can be composed whenever $r(f) = \ell(g)$, and then

$$(f \cdot g)(x) = \begin{cases} f(2x) & \text{if } 0 \leq x \leq \frac{1}{2}, \\ g(2x-1) & \text{if } \frac{1}{2} \leq x \leq 1. \end{cases}$$

The parameterisation destroys associativity; $(X^{[0,1]}, \cdot, \ell, r)$ is therefore only a local partial ℓr-magma. The powerset lifting to $\mathcal{P}(X^{[0,1]})$ satisfies the properties of Lemma 16, but even weak locality fails due to the absence of associativity in $X^{[0,1]}$ and, accordingly, $\mathcal{P}(X^{[0,1]})$. This leads to modal prequantales.

Yet path composition is associative up-to homotopy. The associated local partial ℓr-semigroup then lifts to a modal quantale like any other category. Alternatively, categories of paths can be defined on intervals of arbitrary length [1].

Higher Path Algebras. *A 2-polygraph* is a quiver $\Sigma = (\overline{s}_0, \overline{t}_0 : \Sigma_1 \to \Sigma_0)$, whose edges (or 1-*cells*) are equipped with a *cellular extension.* This consists of a quiver $(\overline{s}_1, \overline{t}_1 : \Gamma \to \Sigma^*)$, where Σ^* is the free category generated by Σ and Γ is a set of globular 2-*cells* relating parallel 1-cells. A 2-polygraph generates a 2-category pictured in the following diagram:

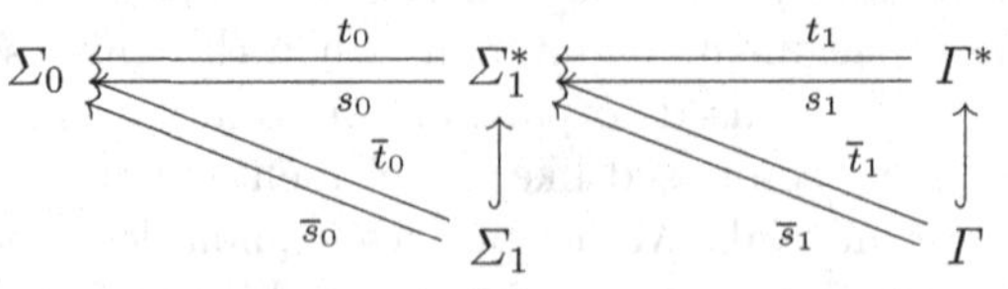

Here, s_i, t_i are the source and target maps induced by the free category construction, and the globular equations $s_0 t_1 = s_0 s_1$ and $t_0 t_1 = t_0 s_1$ hold, see [16] for details. In the example of abstract rewriting systems, Σ_0 is a carrier set, Σ_1 a set of generating rewrite rules, Γ a set of relations between these rules.

For $i \in \{0,1\}$, the set Γ^* of freely generated 2-cells forms a local partial ℓr-semigroup $(\Gamma^*, \odot_i, \ell_i, r_i)$, where $\odot_i$ is forward i-composition of 2-cells and $\ell_i = s_i$, $r_i = t_i$. By Theorem 18, $(\mathcal{P}\Gamma^*, \subseteq, \odot_i, E_i, \mathit{dom}_i, \mathit{cod}_i)$ is a modal quantale with $E_0 = \{1_{1_x} \mid x \in \Sigma_0\}$ and $E_1 = \{1_u \mid u \in \Sigma_1\}$. Beyond Theorem 18, we get a *globular 2-quantale* [2] when combining the two structures. For all $A, A', B, B' \in \mathcal{P}\Gamma^*$, a lax interchange law $(A \odot_1 B) \odot_0 (A' \odot_1 B') \subseteq (A \odot_0 A') \odot_1 (B \odot_0 B')$ holds, and also $E_1 \odot_0 E_1 = E_1$. The absorption laws $\mathit{dom}_1 \circ \mathit{dom}_0 = \mathit{dom}_0$ and $\mathit{cod}_1 \circ \mathit{cod}_0 = \mathit{cod}_0$ hold as well. Finally, we recover the globularity conditions that $\mathit{dom}_0 \circ \mathit{cod}_1 = \mathit{dom}_0$, that $\mathit{cod}_0 \circ \mathit{dom}_1 = \mathit{cod}_0$ and that dom_1 as well as cod_1 are morphisms for $\odot_0$. This construction generalises to n-polygraphs and *globular n-quantales* [2]. Applications include higher dimensional algebraic rewriting [16].

Δ-sets. *A presimplicial set* [27] K is a sequence of sets $(K_n)_{n\geq 0}$, called *simplices*, equipped with face maps $d_{i,n} : K_n \to K_{n-1}$, $i \in \{0,\dots,n\}$, satisfying the simplicial identities $d_{i,n-1} \circ d_{j,n} = d_{j-1,n-1} \circ d_{i,n}$ for all $i < j \leq n$ (we omit the extra indices n from now). The set $K = \bigsqcup_{n\geq 0} K_n$ forms an ℓr-multisemigroup $(K, \odot, \ell, r)$ with

$$x \in y \odot z \Leftrightarrow \exists i.\ y = s_i(x) \wedge z = t_{n-i}(x)$$

and $\ell(x) = s_0(x)$, $r(x) = t_0(x)$, where $s_i(x) = (d_{i+1} \circ d_{i+2} \circ \cdots \circ d_n)(x)$ and $t_i(x) = (d_0 \circ d_1 \circ \cdots \circ d_{n-i-1})(x)$ stand for the initial and the final i–face of $x \in K_n$, respectively. In general, $(K, \odot, \ell, r)$ is neither local nor partial. Locality and partiality hold if K is the nerve of a category (we omit degeneracies).

Also, the set of triples $(s_i(x), x, t_j(x))$ $(x \in K_n, 0 \leq i, j \leq n)$, called *simplices with interfaces*, forms an ℓr-multisemigroup $\mathrm{Int}(K)$ with

$$\begin{aligned}&(s_p(x), x, t_q(x)) \in (s_i(y), y, t_j(y)) \odot (s_k(z), z, t_l(z))\\ &\Leftrightarrow s_p(x) = s_i(y) \wedge t_j(y) = s_k(z) \wedge t_q(x) = t_l(z) \wedge y = s_u(x) \wedge z = t_{n-u+j}(x),\end{aligned}$$

for $x \in K_n$, $y \in K_u$, $z \in K_{n-u+j}$. There is an obvious embedding $K \ni x \mapsto (s_0(x), x, t_0(x)) \in \mathrm{Int}(K)$ of ℓr-multisemigroups. Hence $\mathrm{Int}(K)$ is again neither partial nor local in general.

Precubical Sets. *A precubical set* X [15] is a sequence of sets $(X_n)_{n\geq 0}$ equipped with face maps $d_i^\varepsilon : X_n \to X_{n-1}$, $1 \leq i \leq n$, $\varepsilon \in \{0,1\}$, satisfying the identities $d_i^\varepsilon \circ d_j^\eta = d_{j-1}^\eta \circ d_i^\varepsilon$ for $i < j$ and $\varepsilon, \eta \in \{0,1\}$. Denote $d_A^\varepsilon = d_{a_1}^\varepsilon \circ \cdots \circ d_{a_k}^\varepsilon$ for $A = \{a_1 < \cdots < a_k\} \subseteq [n]$ and $\varepsilon \in \{0,1\}$. The precubical set X forms an ℓr-semigroup $(X, \odot, \ell, r)$ with

$$x \in y \odot z \Leftrightarrow \exists A \subseteq [n].\ y = d_A^0(x) \wedge z = d_{[n]\setminus A}^1(x),$$

$\ell(x) = d_{[n]}^0(x) \in X_0$, $r(x) = d_{[n]}^1(x) \in X_0$ for all $x \in X_n$. Like the previous example, the ℓr-multisemigroup X is neither partial nor local.

A special case of this example is the shuffle multimonoid (Example 6). Let Σ be a finite alphabet, X_n the set of all words of length n, and $d_i^\varepsilon : X_n \to X_{n-1}$ the map that removes the i–th letter. Then $X = (X_n, d_i^\varepsilon)$ is a precubical set and the associated ℓr-multisemigroup $(X, \odot, \ell, r)$ is the shuffle multimonoid on Σ. The domain/codomain structure of the quantale of shuffle languages is trivial, as there is no element between $\emptyset$ and the set containing the empty word.

Assertions Quantales of Separation Logic. The non-local partial ℓr-semigroups of heaplets lift to weakly local modal powerset quantales, but once again with trivial domain/codomain structure. The set $\{\varepsilon\}$ containing the empty heaplet is the only unit. These form the assertion quantales of separation logic. The modal structure is again trivial as there is no element between $\emptyset$ and $\{\varepsilon\}$.

10 Discussion

We summarise some additional results and generalisations in this section. See [12] for details.

Extension to Convolution Algebras. The powerset lifting in Sect. 7 can be seen as a lifting to function spaces $X \to 2$ and generalised to $X \to Q$ for an arbitrary (modal) quantale Q. The composition $\odot : 2^X \times 2^X \to 2^X$ then generalises to a convolution $* : Q^X \to Q^X \to Q^X$ with $\bigvee$ and $\cdot$ taken in Q:

$$(f * g)(x) = \bigvee_{x \in y \odot z} f(y) \cdot g(z).$$

Theorem 18 also generalises: if X is a local ℓr-multisemigroup and Q a modal quantale, then Q^X is a modal quantale with

$$Dom(f) = \bigvee_{x \in X} dom(f(x)) \cdot \delta_{\ell(x)},$$

where $\delta_x(y)$ is 1 if $x = y$ and $\bot$ otherwise, and Cod given by duality. The monoidal identity in Q^X, $id_E(x)$ is 1 if $x \in E$ and $\bot$ otherwise. Beyond lifting, there is now a triangle of correspondences between identities in X, Q and Q^X. The results in this text thus generalise to modal quantales of weighted languages or weighted relations, and towards group, incidence or category algebras.

Finite Decomposability. Some ℓr-multisemigroups in our examples are *finitely decomposable*: for every x the fiber $\odot^{-1}(x) = \{(y, z) \mid x \in y \odot z\}$ is finite. Examples are shuffle quantales, where each word can only be decomposed into finitely many pairs of words, or quantales of $n \times m$-matrices, where multiplications sum over finitely many indices. The sups in convolutions can then be replaced by sums and quantales by semirings. In modal settings, one can then use modal semirings [7] and, if X is a finitely decomposable local ℓr-multisemigroup and S a modal semiring, then S^X forms again a modal semiring.

Modal Concurrent Quantales. Concurrent semirings and quantales [18] can be constructed as convolution algebras from concurrent relational semigroups [3], hence from concurrent local ℓr-multisemigroups equipped with two multioperations that satisfy a weak interchange law. In combination with the corresponding results for modal structures we can construct modal concurrent semirings and quantales as convolution algebras. Target models are categories of pomsets with interfaces, with applications in concurrency theory [10, 29], and n-polygraphs [2].

Algebras of Modalities. The domain and codomain operations in convolution algebras support definitions of modal box and diamond operators along the lines of modal semirings [7] as $|f\rangle\pi = Dom(f * \pi)$, where $f \in Q^X$ and $\pi \in (Q^X)_{Dom}$, and dually $\langle f|\pi = Cod(\pi * f)$. In modal quantales, diamonds preserve arbitrary sups and box operators exist as right adjoints, even if $(Q^X)_{Dom}$ is not a boolean algebra. For box and diamond modalities, locality in ℓr-multisemigroups is crucial. Without it, the action laws $|f * g\rangle = |f\rangle \circ |g\rangle$, $\langle f * g| = \langle g| \circ \langle f|$ and their analogues for boxes would not exist. Our results thus lead to uniform construction principles for dynamic algebras and predicate transformer algebras based on more general semantics than Kripke frames, including arbitrary categories, and weighted variants.

11 Conclusion

We have introduced ℓr-multisemigroups, related them with categories, and shown how their source and target operations give rise to the domain and codomain operations studied previously in the contexts of function systems, modal semigroups, modal semirings and modal quantales. In particular, we have explained how the typical composition pattern of categories corresponds to the well-known locality axioms that appear in such modal algebras. This analysis is based on a generic lifting construction from ℓr-multisemigroups to modal quantales and the modal correspondences to which it leads. It captures most known models of computational interest of modal semirings and quantales, and explains how additional models for them could be built, including modal-concurrent ones. For every local ℓr-multisemigroup we find, we get a dual modal quantale for free. The approach extends to modal convolution algebras that seem relevant to quantitative verification, but this requires concepts and proofs beyond these pages [12].

Acknowledgments. The third and fourth author would like to thank the *Laboratoire d'informatique de l'École polytechnique*, where part of this work has been conducted, for their hospitality and financial support.

References

1. Brown, R.: Topology and Groupoids (2006). www.groupoids.org

2. Calk, C., Goubault, E., Malbos, P., Struth, G.: Algebraic coherent confluence and globular Kleene algebras. CoRR abs/2006.16129 (2020)
3. Cranch, J., Doherty, S., Struth, G.: Convolution and concurrency. CoRR abs/2002.02321 (2020)
4. Cranch, J., Doherty, S., Struth, G.: Relational semigroups and object-free categories. CoRR abs/2001.11895 (2020)
5. Desharnais, J., Jipsen, P., Struth, G.: Domain and antidomain semigroups. In: Berghammer, R., Jaoua, A.M., Möller, B. (eds.) RelMiCS 2009. LNCS, vol. 5827, pp. 73–87. Springer, Heidelberg (2009). https://doi.org/10.1007/978-3-642-04639-1_6
6. Desharnais, J., Möller, B., Struth, G.: Kleene algebra with domain. ACM TOCL **7**(4), 798–833 (2006)
7. Desharnais, J., Struth, G.: Internal axioms for domain semirings. Sci. Comput. Program. **76**(3), 181–203 (2011)
8. Dongol, B., Hayes, I.J., Struth, G.: Convolution algebras: relational convolution, generalised modalities and incidence algebras. Log. Meth. Comput. Sci. **17**(1), 13:1–13:34 (2021)
9. Elgot, C.C.: Matricial theories. J. Algebra **42**(2), 391–421 (1976)
10. Fahrenberg, U., Johansen, C., Struth, G., Bahadur Thapa, R.: Generating posets beyond N. In: Fahrenberg, U., Jipsen, P., Winter, M. (eds.) RAMiCS 2020. LNCS, vol. 12062, pp. 82–99. Springer, Cham (2020). https://doi.org/10.1007/978-3-030-43520-2_6
11. Fahrenberg, U., Johansen, C., Struth, G., Ziemiański, K.: Domain semirings united. CoRR abs/2011.04704 (2020)
12. Fahrenberg, U., Johansen, C., Struth, G., Ziemiański, K.: lr-Multisemigroups and modal convolution algebras. CoRR abs/2105.00188 (2021)
13. Goldblatt, R.: Varieties of complex algebras. Ann. Pure Appl. Log. **44**, 173–242 (1989)
14. Grabowski, J.: On partial languages. Fund. Inform. **4**(2), 427 (1981)
15. Grandis, M.: Directed Algebraic Topology: Models of Non-reversible Worlds. New Mathematical Monographs. Cambridge University Press (2009)
16. Guiraud, Y., Malbos, P.: Polygraphs of finite derivation type. Math. Struct. Comput. Sci. **28**(2), 155–201 (2016)
17. Halpern, J.Y., Shoham, Y.: A propositional modal logic of time intervals. J. ACM **38**(4), 935–962 (1991)
18. Hoare, T., Möller, B., Struth, G., Wehrman, I.: Concurrent Kleene algebra and its foundations. J. Log. Algebraic Program. **80**(6), 266–296 (2011)
19. Jónsson, B., Tarski, A.: Boolean algebras with operators. Part I. Am. J. Math. **73**(4), 891–939 (1951)
20. Jónsson, B., Tarski, A.: Boolean algebras with operators. Part II. Am. J. Math. **74**(1), 127–162 (1952)
21. Kahl, W.: Relational semigroupoids: abstract relation-algebraic interfaces for finite relations between infinite types. J. Log. Algeb. Meth. Program. **76**(1), 60–89 (2008)
22. Kudryavtseva, G., Mazorchuk, V.: On multisemigroups. Portugaliae. Mathematica **71**(1), 47–80 (2015)
23. Mac Lane, S.: Categories for the Working Mathematician. Springer, Heidelberg (1998)
24. Moszkowski, B.C.: A complete axiom system for propositional interval temporal logic with infinite time. Log. Meth. Comput. Sci. **8**(3), 1–56 (2012)
25. Rosenthal, K.L.: Quantales and Their Applications. Longman (1990)

26. Rosenthal, K.L.: Relational monoids, multirelations, and quantalic recognizers. Cahiers Topologie Géom. Différentielle Catég. **38**(2), 161–171 (1997)
27. Rourke, C.P., Sanderson, B.J.: Δ-sets I: Homotopy theory. Q. J. Math. **22**(3), 321–338 (1971)
28. Schweizer, B., Sklar, A.: Function systems. Mathem. Annalen **172**, 1–16 (1967)
29. Winkowski, Józef.: Algebras of partial sequences—a tool to deal with concurrency. In: Karpiński, Marek (ed.) FCT 1977. LNCS, vol. 56, pp. 187–196. Springer, Heidelberg (1977). https://doi.org/10.1007/3-540-08442-8_85

Abstract Strategies and Coherence

Cameron Calk[1(✉)], Eric Goubault[1], and Philippe Malbos[2]

[1] Laboratoire d'Informatique de l'École Polytechnique, École Polytechnique, Palaiseau, France
cameron.calk@lix.polytechnique.fr

[2] Institut Camille Jordan, Université Claude Bernard Lyon 1, Villeurbanne, France

Abstract. Normalisation strategies give a categorical interpretation of the notion of contracting homotopy via confluent and terminating rewriting. This approach relates standardisation to coherence results in the context of higher-dimensional rewriting systems. On the other hand, globular 2-Kleene algebras provide a formal setting for reasoning about coherence proofs in abstract rewriting systems. In this setting, we formalise the notion of normalisation strategy and we prove a formal coherence theorem for convergent abstract rewriting systems.

Keywords: Normalisation strategies · Kleene algebras · Formalisation · Coherence · Higher-dimensional rewriting

1 Introduction

As pointed out in [5,29] a central difficulty in formal mathematics is in balancing readability of specifications and proficient automated proof search. Capturing intuitions while remaining formally rigorous constitutes a first stumbling block, which ideally should result in a setting that provides correct, automated proofs which are readable and even illuminating. A powerful formalisation of abstract rewriting theory may be found in the theory of Kleene algebras. Algebraic abstraction allows for simple proofs in which deduction is replaced by calculation [29]. Proofs in this setting reconstruct intuitive proofs by *diagrammatic reasoning*, making Kleene algebras a formal setting well suited to capture abstract rewriting results. Modal Kleene algebras (MKAs) formalise abstract rewriting systems (ARS), abstractions of graphs of (1-dimensional) transitions, especially with respect to termination and normalisation properties [5,29]. This setting does not suffice to formalise more subtle properties of normalisation strategies [24], such as standardisation properties, nor for dealing with inherently higher-dimensional transition systems. Indeed, these need a formalisation of equivalences between paths. This line of work started in [12,20], culminating in the introduction of a specific axiomatics on a 2-dimensional refinement of ARSs.

In this work, we are going one step further by giving a formalisation of a *coherent* extension of diagrammatic reasoning in the algebraic style of MKAs,

U. Fahrenberg et al. (Eds.): RAMiCS 2021, LNCS 13027, pp. 108–125, 2021.
https://doi.org/10.1007/978-3-030-88701-8_7

inspired by coherent presentations in categorical algebra [23], or in algebra [10], and using a rewriting approach in the line of [27]. In a higher categorical structure, certain algebraic properties, *e.g.* associativity of composition, may only hold up to the existence of higher-dimensional morphisms. Given a collection of such higher morphisms, *coherence* is the requirement that the whole structure is contractible, *i.e.* all parallel morphisms are linked by higher morphisms. A *coherence theorem* states that, given a (generating) collection of such morphisms, coherence is satisfied. The objective is to obtain a minimal collection of generating higher morphisms. Graph-theoretical methods on string rewriting systems (SRS) were initiated by Squier in [27] to study coherence problems for monoids, a two dimensional word problem. The main point is to compute extensions of a SRS by *homotopy generators* which take the relations amongst the rewriting paths into account. That is, every pair of zig-zag sequences of rewriting paths with same source and same target can be paved by compositions of these generators. In Squier's approach, when the SRS is convergent, the homotopy generators are defined by the confluence diagrams of the critical branchings of the SRS. This rewriting method for coherence was applied to solve coherence problems in algebra [4,10,17], and for monoidal categories [14]. Thereby, the homotopy generators constitute the bottom part of a cofibrant replacement of the monoid presented by the SRS [10,15]. Squier's constructions were formulated in the categorical language of polygraphs in [16] for monoids and in [13] for higher categories.

In this work, we consider the case of ARS. The extension to the case of SRS will be done in a subsequent work because requires a formalisation of algebraic contexts and of the critical branching lemma, which constitutes a further development of the theory presented here. An ARS is represented by a quiver Φ, aka a 1-*polygraph*, see Sect. 2. Parallel zig-zag sequences of rewriting paths are pairs of 1-cells in the free groupoid $\Phi^\top$ on Φ with same source and same target. Homotopical generators for the ARS consist of such pairs and form a *cellular extension* X of $\Phi^\top$, see Sect. 2. The *coherence theorem for* (Φ, X) states that all parallel 1-cells in $\Phi^\top$ are equal modulo X. When Φ is convergent and X is the set of confluence diagrams of (critical) branchings, Squier's method gives a proof of the coherence theorem for Φ. It is exactly this proof that we formalise in this article.

This work uses the algebraic setting of a 2-dimensional (globular) version of MKAs, which model relation algebras and relations among relations, introduced in [3]. Interestingly enough, these 2-dimensional MKAs are close to Concurrent Kleene Algebras (CKAs), which introduce an extra algebraic operation modelling parallel composition, hence equivalences between (1-dimensional) paths.

Structure of the Article, and Main Results. This article is about formalising normalisation strategies and coherence properties in view of automating proofs. In Sect. 2, we present the categorical formulation of relations among relations in terms of cellular tilings, and based on Squier's coherence result. We then recap the MKA approach for ARS in Sect. 3. Coherent rewriting in globular modal 2-Kleene algebra, which we introduced in [3], is recalled in Sects. 4 and 5.

Sections 6 and 7 form the core of our new results, where we first model normalisation strategies in 2-MKAs, and prove abstract coherence properties therein. Our first result, Theorem 1, gives a formalisation of a coherent normalising Newman's lemma. We thereby deduce our main result, Theorem 2, which formalises a proof of contractibility via normalisation strategies.

2 Squier's Theorem for ARS

We consider an ARS as a *quiver*, *i.e.* a directed graph with parallel and looping transitions, which we call a 1-*polygraph* from the terminology of higher-dimensional rewriting [2,28]. Denote by $\Phi := (\Phi_0, \Phi_1)$ a 1-polygraph with source and target maps $s_0, t_0 : \Phi_1 \to \Phi_0$. We model the *reflexive, transitive closure* of Φ by the *free* 1-*category* Φ^* generated by Φ, the underlying graph of which consists of the directed paths in Φ. Similarly, the *symmetric, reflexive, transitive closure of* Φ is modelled by the *free* 1-*groupoid* $\Phi^\top$ generated by Φ, its underlying graph consisting of undirected paths. In both cases the source and target maps are obtained by naturally extending those of Φ. The vertices (resp. edges) of such structures will henceforth be referred to as 0-cells (resp. 1-cells), and the set of i-cells of Φ^* (resp. $\Phi^\top$) will be denoted by Φ_i^* (resp. $\Phi_i^\top$). The 0-*composition* of 1-cells x, y is defined when $t_0(x) = s_0(y)$ and is denoted by $x \star_0 y$. The identity 1-cell on $a \in \Phi_0$ is denoted by 1_a and the inverse of a 1-cell x is denoted by x^-. Two 1-cells are *parallel* when they have the same 0-source and 0-target. Directed paths correspond to compositions $x_1 \star_0 \cdots \star_0 x_k$, with $x_i \in \Phi_1$. Similarly, undirected paths correspond to finite compositions of elements of Φ_1 and their formal inverses, quotiented by the relations $x \star_0 x^- \sim 1_{s_0(x)}$, for $x \in \Phi_1$.

A *cellular extension* X of $\Phi^\top$ is a quiver on the edges of $\Phi^\top$, *i.e.* a pair $(\Phi_1^\top, X)$ with source (resp. target) map s_1 (resp. t_1), such that the *globular relations* $t_0 \circ s_1 = t_0 \circ t_1$ and $s_0 \circ s_1 = s_0 \circ t_1$ are satisfied. The elements of X are called *generating* 2-*cells* and may be thought of as (directed) tiles filling the space between parallel 1-cells. The pair (Φ, X) is called a $(2,0)$-*polygraph*.

Recall that the 2-cells in a 2-category may be composed in two different ways. The 0-composition of $\gamma : x \Rightarrow y$ and $\delta : x' \Rightarrow y'$, where $x, y : a \to b$ and $x', y' : b \to c$ are pairs of parallel 1-cells, is a 2-cell $\gamma \star_0 \delta : x \star_0 x' \Rightarrow y \star_0 y'$. The 1-composition of 2-cells $\alpha : x \Rightarrow y$ and $\beta : y \Rightarrow z$, where x, y, z are parallel 1-cells, is a 2-cell $\alpha \star_1 \beta : x \Rightarrow z$. A 2-*groupoid* is a 2-category in which all 1- and 2-cells are invertible for 0- and 1-composition, respectively. Given a $(2,0)$-polygraph (Φ, X), we consider the *free* 2-*groupoid generated by* (Φ, X), denoted by $X^\top$, which has $\Phi^\top$ as its underlying 1-groupoid and containing all finite 0- and 1-compositions of the generating 2-cells in X and their inverses, as well as 0-compositions with 1-cells of $\Phi^\top$.

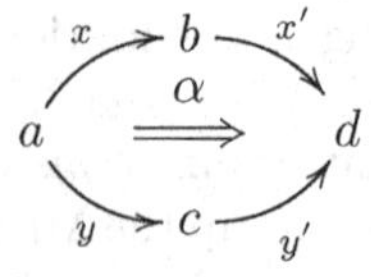

The confluence properties of an ARS Φ can be stated with respect to a cellular extension X of $\Phi^\top$. This approach first appeared in [20] under the terminology of *commuting diagrams*. A local branching (x, y) of Φ is X-*confluent* if there exist 1-cells x', y' in Φ_1^*, and a 2-cell α in the free 2-groupoid $X^\top$ as in the adjacent diagram. The ARS Φ is *locally* X-*confluent* when every local branching of Φ is X-confluent. We say that parallel 1-cells f and g of $\Phi^\top$ are X*-congruent* if there exists a 2-cell $\alpha : f \Rightarrow g$ in $X^\top$, and that (Φ, X) is *acyclic* if all parallel 1-cells of $\Phi^\top$ are X-congruent. The ARS Φ *terminates* if it contains no infinite directed paths.

Let us recall the proof that a terminating, locally X-confluent $(2,0)$-polygraph (Φ, X) is acyclic. Firstly note that if an ARS Φ is locally X-confluent then it is locally confluent so, under the hypothesis of termination, is confluent by Newman's lemma. In this case, from every 0-cell a, a *normal form*, *i.e.* a 0-cell irreducible by Φ, may be reached in a finite number of steps. Since Φ is confluent, the normal form of a is unique; we denote it by $\hat{a}$.

By local X-confluence and termination, we may therefore choose, for every 0-cell a of Φ, a 1-cell $\sigma_a : a \to \hat{a}$ in Φ_1^*. A *normalisation strategy* σ is a function $\Phi_0 \to \Phi_1^\top$ which assigns such a σ_a to every 0-cell a, under the condition that $\sigma_b = 1_b$ for any normal form b. Just as normal forms provide a representative 0-cell for connected components in $\Phi^\top$, a normalisation strategy is the given of a representative 1-cell in $\Phi^\top$ among parallel reductions to normal forms.

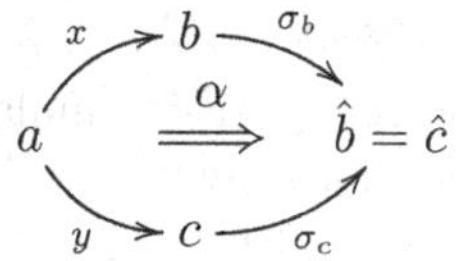

Now that we are equipped with a normalisation strategy σ, we prove by Noetherian induction on the distance from a normal form that for any branching (x, y) of Φ^*, there exists a 2-cell α as in the adjacent diagram. When $s_0(x) = s_0(y)$ is a normal form, we can simply use identity 1- and 2-cells to obtain the desired diagram. For the induction step, we observe that we can write x as $x = x_1 \star_0 x_r$, where x_1 is a 1-cell of Φ and x_r is one step closer to a normal form, and similarly for y. By the hypothesis of local confluence and the Noetherian induction hypothesis, we obtain the result by composing the 2-cells in the diagram on the left below:

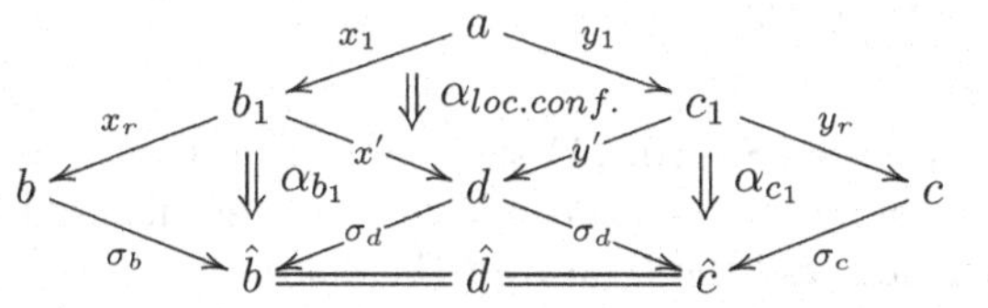

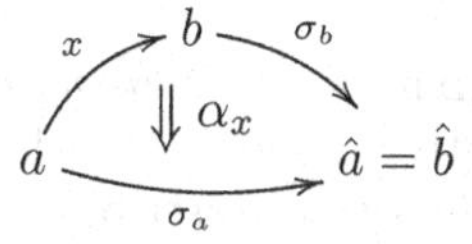

Let $x : a \to b$ be a 1-cell of Φ^*, consider the branching $(x \star_0 \sigma_b, \sigma_a)$ of Φ^*. Since we cannot reduce any further than normal forms, by the above result, as well as a rotation of the 2-cell by properties of 2-groupoids, we obtain a 2-cell α_x as pictured above on the right. A similar 2-cell for all inverses of 1-cells may be found, again using properties of 2-groupoids which we will not develop here. Note that every 1-cell $f : a \to b$ of $\Phi^\top$ can be factorised as $f = x_1 \star_0 y_1^- \star_0 \cdots \star_0 x_p \star_0 y_p^-$, where the x_i and y_i are 1-cells of Φ^*. Denote by α_f the composite 2-cell of $X^\top$:

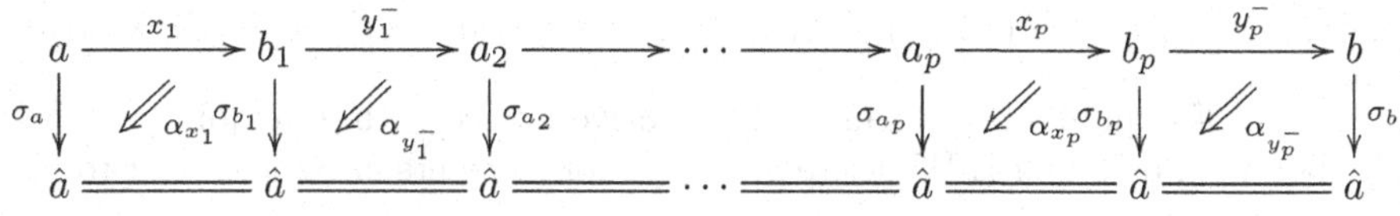

Compiling all of the above, we obtain the *coherence theorem for ARS*:

Theorem A. *Let Φ be a terminating ARS and X be a cellular extension of $\Phi^\top$. If Φ is locally X-confluent, then for every 1-cell $f : a \to b$ of $\Phi^\top$, there exists a 2-cell $\alpha_f : f \star_0 \sigma_b \Rightarrow \sigma_a$ in the free 2-groupoid generated by (Φ, X).*

Squier's theorem [27] is deduced from the above result. Indeed, we prove that for all parallel 1-cells $f, g : a \to b$ of $\Phi^\top$, the composite 2-cell

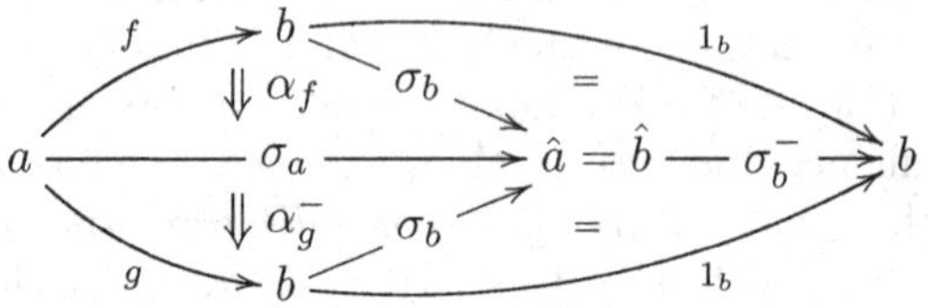

in $X^\top$ has source f and target g. This proves that the pair (Φ, X) is acyclic.

Theorem B. *Let Φ be a terminating ARS and X be a cellular extension of $\Phi^\top$. If Φ is locally X-confluent, then (Φ, X) is acyclic.*

This is Squier's formulation of the coherence theorem for ARSs, and is an immediate consequence of Theorem A, relying solely on the definitions of acyclicity and of 2-groupoids.

3 Modal 1-Kleene Algebras

In order to fix notation, we recall the definitions of Boolean modal Kleene algebras from [5,6] and of converse from [1]. We adapt one of the converse axioms in order to establish a natural relationship between domain and conversion akin to that of inverse semigroups, see e.g. [22].

Semirings. A *semiring* is a structure $(S, +, 0, \cdot, 1)$ such that $(S, +, 0)$ is a commutative monoid, $(S, \cdot, 1)$ is a monoid whose *multiplication* $\cdot$ (often denoted by juxtaposition) distributes on the left and the right over the *addition* $+$, and 0 is a left and right annihilator for $\cdot$. A *dioid* is a semiring in which addition is idempotent. In this case, the relation defined by $x \leq y \Leftrightarrow x + y = y$, for all $x, y \in S$, is a partial order on S, with respect to which addition and multiplication are monotone, and for which 0 is the minimum.

(Boolean) Domain Semirings. A *domain semiring* is a dioid S equipped with a *domain operation* $d : S \to S$ satisfying the following five axioms for all $x, y \in S$:

$$x \leq d(x)x, \quad d(xy) = d(xd(y)), \quad d(x) \leq 1, \quad d(0) = 0, \quad d(x+y) = d(x)+d(y).$$

The set S_d of fixpoints of d forms a distributive lattice with $+$ as join and $\cdot$ as meet, bounded by 0 and 1. We write $p, q, r, \ldots$ for elements of S_d and refer to S_d

as the *domain algebra* of S. A *Boolean domain semiring* is a dioid S equipped with an *antidomain operation* $ad : S \to S$ satisfying the following three axioms:

$$ad(x)x = 0, \qquad ad(xy) \leq ad(x\, ad^2(y)), \qquad ad^2(x) + ad(x) = 1,$$

for all $x, y \in S$. Setting $d = ad^2$, we recover a domain semiring. In the presence of an antidomain, $S_d = ad(S)$ and ad acts as Boolean complementation on S_d. We denote the restriction of ad to S_d by $\neg$.

Modal Semirings. We denote by S^{op} the *opposite* of a dioid S, in which the order of multiplication has been reversed. A *codomain* (resp. *Boolean codomain*) *semiring* is a dioid equipped with a map $r : S \to S$ (resp. $ar : S \to S$) such that (S^{op}, r) (resp. (S^{op}, ar)) is a domain (resp. Boolean domain) semiring. A *modal semiring* is a dioid S which is both a domain and codomain semiring, and satisfies for every $x \in S$, $d(r(x)) = r(x)$ and $r(d(x)) = d(x)$.

Modal Kleene Algebras. A *Kleene algebra* is a dioid K equipped with an operation $(-)^* : K \to K$ called the *Kleene star*, satisfying the following axioms:

i) $1 + xx^* \leq x^*$ and $1 + x^*x \leq x^*$ (*unfold axioms*),
ii) $z + xy \leq y \Rightarrow x^*z \leq y$ and $z + yx \leq y \Rightarrow zx^* \leq y$ (*induction axioms*),

for all $x, y, z \in K$. The *Kleene plus* is defined by $x^+ = xx^*$. (Anti-)domain and (anti-)codomain operations extend to Kleene algebras without additional axioms. We thus define a *(Boolean) modal Kleene algebra*, or (Boolean) MKA for short, as a Kleene algebra that is also a (Boolean) modal semiring.

Converse. A *Kleene algebra with converse* [1] is a Kleene algebra K equipped with an involution $\overline{(-)} : K \to K$ that satisfies, for all $x, y \in K$,

$$\overline{(x+y)} = \overline{x} + \overline{y}, \qquad \overline{(x \cdot y)} = \overline{y} \cdot \overline{x}, \qquad \overline{(x^*)} = (\overline{x})^*, \qquad \overline{(\overline{x})} = x, \qquad (1)$$

and the inequality $x \leq x\overline{x}x$. In this work, we alter the final axiom in order to relate conversion to the domain operation. We consider an involution $\overline{(-)} : K \to K$ satisfying axioms (1) and

$$x\overline{x} \geq d(x), \qquad (2)$$

a similar axiom to that found in inverse semigroups [22]. We observe that such a converse operation exchanges domain and codomain, *i.e.* $d(\overline{x}) = r(x)$ and $r(\overline{x}) = d(x)$, and that for $p \in K_d$, $\overline{p} = p$. A (Boolean) MKA with converse is a (Boolean) MKA equipped with such a converse operation.

Modalities in Dimension One. Let K be a MKA. For $x \in K$ and $p \in K_d$, we define *modal forward and backward diamond operators*:

$$|x\rangle p = d(xp), \qquad \langle x| p = r(px). \qquad (3)$$

When a statement holds for both forward and backward diamonds, we will write $\langle x \rangle$. Note that by monotonicity of domain, the assignment $x \mapsto \langle x \rangle$ is monotone for the point-wise order on operators. When K is a Boolean MKA, we additionally define *modal box operators*:

$$|x]p = \neg|x\rangle(\neg p), \qquad [x|p = \neg\langle x|(\neg p).$$

These are modal operators in the sense of Boolean algebras with operators [21]. For K with converse, we have $|\overline{x}\rangle = \langle x|$ and $\langle \overline{x}| = |x\rangle$, and similarly for boxes. Boxes and diamonds form a Galois connection, *i.e.*

$$|x\rangle p \le q \Leftrightarrow p \le [x|q \qquad \text{and} \qquad \langle x|p \le q \Leftrightarrow p \le |x]q. \tag{4}$$

We have $|xy\rangle = |x\rangle \circ |y\rangle$, $\langle xy| = \langle y| \circ \langle x|$, $|xy] = |x] \circ |y]$ and $[xy| = [y| \circ [x|$ for all $x, y \in K$; in what follows we will denote functional composition of modal operators simply by juxtaposition. Finally, star unfold and induction axioms lift to modalities:

$$|1\rangle + |x\rangle|x^*\rangle = |x^*\rangle, \qquad |1\rangle + |x\rangle|x^*\rangle = |x^*\rangle, \tag{5}$$

$$|y\rangle + |x\rangle|z\rangle \le |z\rangle \Rightarrow |x^*\rangle|y\rangle \le |z\rangle, \tag{6}$$

where the addition is the point-wise lifting of that in K_d.

Rewriting and Modal Kleene Algebras. We recall from [5] formalised properties of ARS expressed in MKA. An element $x \in K$ *terminates*, or is *Noetherian*, provided that for all $p \in K_d$ the implication $p \le |x\rangle p \Rightarrow p = 0$ holds. The set of Noetherian elements of K is denoted by $\mathcal{N}(K)$. The Galois connections (4) yield the following equivalent characterisation of termination:

$$\forall p \in K_d, \quad |x]p \le p \Rightarrow p = 1.$$

The *exhaustion* of an element $x \in K$, denoted by $exh(x)$, is defined by

$$exh(x) := x^* \cdot \neg d(x). \tag{7}$$

The *normal forms element* of $x \in K$, denoted by nf_x, is defined by

$$\mathrm{nf}_x := r(exh(x)) \in K_d. \tag{8}$$

Confluence properties are captured in MKA by semi-commutation. Given $x, y \in K$, we say that the ordered pair (x, y) *semi-commutes locally* if $xy \le y^*x^*$, *semi-commutes* if $x^*y^* \le y^*x^*$, and has the *Church-Rosser property* if $(x + y)^* \le y^*x^*$. An element $x \in K$ is (*locally*) *confluent* (resp. *Church-Rosser*) if the pair $(\overline{x}, x)$ semi-commutes (resp. has the Church-Rosser property). We say that x is *convergent* if it is both terminating and confluent. These properties are related to exhaustion as follows:

Lemma 1 ([5]). *Let K be a Boolean modal Kleene algebra and $x \in K$. If x terminates, then $d(exh(x)) = 1$. If x is confluent, then $exh(x)$ is deterministic,* i.e. $\langle exh(x)||exh(x)\rangle \le \langle 1\rangle$.

4 Globular 2-Kleene Algebras

In [3], the notion of p-Boolean globular n-Kleene algebra was introduced as a higher-dimensional extension of MKAs. Here we briefly recall the case of $p = 0$ and $n = 2$, and append the notion of converse.

A *modal 2-Kleene algebra* is a structure $(K, +, 0, \odot_i, 1_i, d_i, r_i, (-)^{*_i})_{i=0,1}$, such that for each $i \in \{0, 1\}$, K is a MKA with respect to i-operations, and in which the following additional axioms hold:

i) (*2-dioid axioms*) The *lax interchange law*: for all $A, A', B, B' \in K$,
$$(A \odot_1 A') \odot_0 (B \odot_1 B') \leq (A \odot_0 B) \odot_1 (A' \odot_0 B'),$$
and the 1-unit is an idempotent for 0-multiplication, *i.e.* $1_1 \odot_0 1_1 = 1_1$. Note that these correspond to the standard concurrent semiring axioms [18], except that the equality $1_0 = 1_1$ is normally assumed in this case.

ii) (*Domain 2-semiring axioms*) The (co-)domain operations satisfy absorption axioms $d_1 \circ d_0 = d_0$ and $r_1 \circ r_0 = r_0$. The set K_{d_i} is called the *i-dimensional domain algebra*, and is denoted by K_i.

iii) (*Kleene star axioms*) The 1-star $(-)^{*_1}$ is a *lax morphism* with respect to 0-multiplication of 1-dimensional elements on the right (resp. left), *i.e.* for all $A \in K$ and $\phi \in K_1$,
$$\phi \odot_0 A^{*_1} \leq (\phi \odot_0 A)^{*_1}, \qquad (\text{resp. } A^{*_1} \odot_0 \phi \leq (A \odot_0 \phi)^{*_1}).$$

For more details, see [3]. In order to distinguish elements of distinct dimensions, we denote elements of K_0 by $p, q, r, \ldots$, elements of K_1 by $\phi, \psi, \xi, \ldots$, and general elements of K by $A, B, C, \ldots$.

As additional conditions, we may ask that a modal 2-Kleene algebra be globular, Boolean or equipped with converses. These notions are recalled below.

Globular Axioms. A modal 2-Kleene algebra K is *globular* if the following *globular relations* hold for all $A, B \in K$:
$$d_0 \circ d_1 = d_0 \text{ and } d_0 \circ r_1 = d_0, \qquad d_1(A \odot_0 B) = d_1(A) \odot_0 d_1(B),$$
$$r_0 \circ d_1 = r_0, \text{ and } r_0 \circ r_1 = r_0, \qquad r_1(A \odot_0 B) = r_1(A) \odot_0 r_1(B).$$

As a consequence of the rightmost axioms, K_1 is a MKA with respect to 0-operations. An element A of K will be represented graphically by the adjacent diagram with respect to its 0- and 1-domains and codomains.

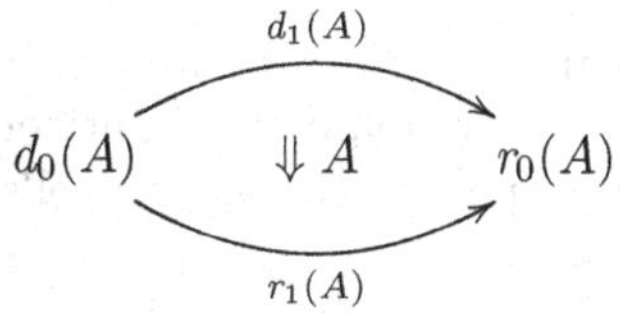

Boolean Axioms. A modal 2-Kleene algebra is *Boolean* if it is augmented with maps $ad_0 : K \to K$ and $ar_0 : K \to K$, such that $(K, +, 0, \odot_0, 1_0, ad_0, ar_0)$ is a Boolean MKA, *i.e.* ad_0 (resp. ar_0) satisfies the antidomain (resp. anticodomain) axioms and $d_0 = ad_0^2$ (resp. $r_0 = ar_0^2$). The domain algebra K_0 is thus a Boolean algebra whose complementation, denoted by $\neg$, is given by the restriction of ad_0 (and ar_0) to K_0.

Converses. We will consider modal 2-Kleene algebras with *0-converses, i.e.* equipped with an operation $\overline{(-)} : K_1 \to K_1$ such that $(K_1, +, 0, \odot_0, 1_0, (-)^{*_0}, \overline{(-)})$ is a MKA with converse. For a more general notion of converse in higher-dimensional Kleene algebra, we refer the reader to [3].

Modalities in 2-semirings. Recall from [3], that the *i-diamond operators* of a modal 2-Kleene algebra K are defined via the (co-)domain operators in each dimension. For $i \in \{0, 1\}$, $A \in K$ and $\phi \in K_i$,

$$|A\rangle_i(\phi) = d_i(A \odot_i \phi), \quad \text{and} \quad \langle A|_i(\phi) = r_i(\phi \odot_i A).$$

These modal operators have all of the properties recalled in Sect. 3 with respect to i-operations and elements of K_i. Since we are considering Boolean modal 2-Kleene algebras we may additionally define 0-boxes.

Polygraphic Model. Let (Φ, X) be a $(2, 0)$-polygraph. We define $K(\Phi, X)$, the *full 2-path algebra over* (Φ, X) as follows. Let $X_2^\top$ denote the set of 2-cells in $X^\top$. The carrier set of $K(\Phi, X)$ is the power set $\mathcal{P}(X_2^\top)$, whose elements, denoted by $A, B, C \ldots$ are sets of 2-cells, which in turn are denoted by $\alpha, \beta, \gamma \ldots$ Recall that for each 1-cell x of $X^\top$, there exists a unique 2-cell 1_x, its identity 2-cell, and similarly, for each 0-cell a there exists a unique 2-cell 1_{1_a}, the identity 2-cell on its identity 1-cell. For $i \in \{0, 1\}$, the i-composition, i-source and i-target maps are thereby defined for cells of any dimension.

For $i \in \{0, 1\}$, the multiplication $\odot_i$ on $K(\Phi, X)$ is the lifting of the composition operations of $X^\top$ to the power-set, *i.e.* for any $A, B \in K(\Phi, X)$,

$$A \odot_i B := \{\alpha \star_i \beta \mid \alpha \in A \wedge \beta \in B \wedge t_i(\alpha) = s_i(\beta)\}.$$

The units are the sets $\mathbb{1}_0 = \{1_{1_a} \mid a \in \Phi_0\}$, and $\mathbb{1}_1 = \{1_x \mid x \in \Phi_1^\top\}$. The addition in $K(\Phi, X)$ is given by set union; the ordering is therefore given by set inclusion. The domain and codomain maps are defined by

$$d_0(A) := \{1_{1_{s_0(\alpha)}} \mid \alpha \in A\}, \qquad r_0(A) := \{1_{1_{t_0(\alpha)}} \mid \alpha \in A\},$$
$$d_1(A) := \{1_{s_1(\alpha)} \mid \alpha \in A\}, \quad \text{and} \quad r_1(A) := \{1_{t_1(\alpha)} \mid \alpha \in A\},$$

and are thus given by lifting the source and target maps of $X^\top$ to the power set. The i-antidomain and i-anticodomain maps are then given by complementation with respect to the set of i-cells. The i-star is given by $A^{*_i} = \bigcup_{k \in \mathbb{N}} A^{k_i}$, where in the above, $A^{0_i} := \mathbb{1}_i$ and $A^{k_i} := A \odot_i A^{(k-1)_i}$. For $\psi \in K(\Phi, X)_1$, the converse is given by $\overline{\psi} := \{1_{x^-} \mid 1_x \in \psi\}$.

Proposition 1 ([3]). *Let (Φ, X) be a $(2, 0)$-polygraph. Then, $K(\Phi, X)$ is a globular Boolean modal 2-Kleene algebra.*

5 Coherent Rewriting and Modal 2-Kleene Algebras

We fix K a globular 2-Kleene algebra. Given $A \in K$ and $\phi, \phi' \in K_1$, $|A\rangle_1(\phi) \geq \phi'$ is equivalent to $d_1(A \odot_1 \phi) \geq \phi'$ by definition. In terms of quantification over collections of cells, this means that *for every* u *in* ϕ', *there exist* v in ϕ and α in A such that the 1-source (resp. 1-target) of α is u (resp. v). This observation motivates the following definitions from [3]. For ϕ, ψ in K_1, an element A in K is a *local confluence filler* for (ϕ, ψ) if $|A\rangle_1(\psi^{*_0} \odot_0 \phi^{*_0}) \geq \phi \odot_0 \psi$, is a *confluence filler* for (ϕ, ψ) if $|A\rangle_1(\psi^{*_0} \odot_0 \phi^{*_0}) \geq \phi^{*_0} \odot_0 \psi^{*_0}$, and is a *Church-Rosser filler* for (ϕ, ψ) if $|A\rangle_1(\psi^{*_0} \odot_0 \phi^{*_0}) \geq (\psi + \phi)^{*_0}$.

The *right* (resp. *left*) *whiskering* of an element $A \in K$ by $\phi \in K_1$ is the element $A \odot_0 \phi$ (resp. $\phi \odot_0 A$). Recall from [3] that whiskering commutes with 1-diamonds, that is, for all $A \in K$ and $\phi, \psi, \phi', \psi', \gamma \in K_1$ such that $\phi' \leq \phi$, $\psi' \leq \psi$, and $d_1(A) \leq \gamma$, we have:

$$\phi' \odot_0 |A\rangle_1(\gamma) \odot_0 \psi' = |\phi' \odot_0 A \odot_0 \psi'\rangle_1(\phi \odot_0 \gamma \odot_0 \psi). \tag{9}$$

Fix a (local) confluence filler A of a pair (ϕ, ψ) of elements in K_1. The *total whiskering of* A, denoted by $\hat{A}$, is the following element of K:

$$\hat{A} := (\phi + \psi)^{*_0} \odot_0 A \odot_0 (\phi + \psi)^{*_0}. \tag{10}$$

The 1-star of $\hat{A}$ is called the *completion* of A. Note that this element *absorbs whiskers*, that is, for every $\xi \leq (\phi + \psi)^{*_0}$,

$$\xi \odot_0 \hat{A}^{*_1} \leq \hat{A}^{*_1} \quad \text{and} \quad \hat{A}^{*_1} \odot_0 \xi \leq \hat{A}^{*_1}. \tag{11}$$

6 Formalisation of Normalisation Strategies

In this section, we formalise the notion of normalisation strategy, introduced in [15]. We first define notions of section, skeleton and strategy in one-dimensional Kleene algebras and show properties thereof. In what follows, we consider a Boolean MKA K with converse and an element $x \in K$.

- **i)** The *equivalence* generated by x is the element $x^\top := (x + \overline{x})^*$. For $p \in K_d$, the *x-saturation* of p is the element $|x^\top\rangle(p) \in K_d$.
- **ii)** A *covering set* for x is an element $q \in K_d$ such that $|x^\top\rangle(q) \geq 1$, *i.e.* whose x-saturation is total. A *section* of x is a minimal covering set.
- **iii)** A *wide sub* of x is an element $w \leq x$ such that $|w\rangle = |x\rangle$ and $\langle w| = \langle x|$. A *skeleton* of x is a minimal wide sub.
- **iv)** Given a section s_0 of x, a *strategy for x relative to* s_0 is a skeleton σ of $x^\top s_0$ such that $s_0\sigma \leq s_0$.

Note that when (Φ, X) is a $(2,0)$-polygraph, we describe Φ in $K(\Phi, X)$ as the element $\phi := \{1_x \,|\, x \in \Phi_1\} \cup \{1_{1_a} \,|\, a \in \Phi_0\}$. In $K(\Phi, X)_1$, which we recall is a Boolean MKA for 0-operations, the equivalence generated by ϕ corresponds to the 1-groupoid $\Phi^\top$, and a section corresponds to a choice of a representative 0-cell for each connected component in $\Phi^{\top}$. A wide sub of ϕ is a subset ψ such that for any 1-cell $x : a \to b \in \Phi_1$, there exists some parallel 1-cell $x' : a \to b \in \Phi_1$ such that $1_{x'} \in \psi$. A skeleton of ϕ therefore corresponds to the choice of a single 1-cell amongst the sets of parallel 1-cells in Φ; it is thus not unique and does not coincide with ϕ in general. When Φ is convergent and $\{\sigma_a\}_{a\in\phi_0}$ is a strategy in the sense of Sect. 2, then $\sigma = \{1_{\sigma_a} | a \in \phi_0\}$ is a strategy for ϕ in $K(\phi, X)$ with respect to nf_{ϕ}. This result is proved for any convergent element of a MKA in Proposition 2.

By definition, a strategy σ satisfies $d(\sigma) = d(x^{\top} s_0) = 1$, and $r(\sigma) = r(x^{\top} s_0) = s_0$. The following lemma states that a strategy contains the associated section:

Lemma 2. *Given a section s_0 of x and a strategy σ for x relative to s_0, we have $s_0\sigma = s_0$ and $s_0 \leq \sigma$.*

Proof. By hypothesis we have $s_0\sigma \leq s_0$. Showing that $s_0\sigma$ is a covering set allows us to deduce by minimality of s_0 that $s_0 \leq s_0\sigma \leq \sigma$, which gives both desired conclusions. Since σ is a strategy relative to x, we know that $\langle x^{\top} s_0| = \langle \sigma|$. We calculate the saturation of $s_0\sigma$

$$\langle x^{\top}|(s_0\sigma) = r(s_0\sigma x^{\top}) = \langle x^{\top}|\langle \sigma|(s_0) = \langle x^{\top}|\langle x^{\top} s_0|(s_0) \geq \langle x^{\top}|(s_0) \geq 1,$$

where we used properties of modalities for the first two steps, then the hypothesis that σ is a strategy. To conclude, we used that $\langle x^{\top} s_0|(s_0) \geq \langle s_0|(s_0) = s_0$ and that s_0 is a covering set. □

By conversion, we also get $\overline{\sigma} s_0 = s_0$ and $s_0 \leq \overline{\sigma}$. This immediately gives the following properties of a strategy σ relative to a section s_0:

$$\sigma \cdot \sigma = \sigma, \quad \overline{\sigma} \cdot \overline{\sigma} = \overline{\sigma}, \quad \sigma \leq \sigma \cdot \overline{\sigma} \quad \text{and} \quad \overline{\sigma} \leq \sigma \cdot \overline{\sigma}. \tag{12}$$

Indeed, $\sigma\sigma = \sigma s_0 \sigma = \sigma s_0 = \sigma$ by the fact that $r(\sigma) = s_0$ and Lemma 2, the case of $\overline{\sigma}$ follows by conversion. Additionally, $s_0 \leq \overline{\sigma}$ so $\sigma = \sigma s_0 \leq \sigma\overline{\sigma}$ and symmetrically for $\overline{\sigma}$.

Next, we will show that the normal forms and exhaustive iteration of a convergent element give us a section and a strategy, respectively. First, we show:

Lemma 3. *Let K a Boolean MKA. For a convergent element $x \in K$, we have $|x^{\top}\rangle = |exh(x)\rangle\langle exh(x)|$.*

Proof. One direction holds since $exh(x)\overline{exh(x)} \leq x^{*}\overline{x}^{*} \leq x^{\top}$ so by monotonicity of taking diamonds and reversal of diamonds by conversion, we get $|x^{\top}\rangle \geq |exh(x)\rangle\langle exh(x)|$. The other inequality is obtained via the star induction law for modalities (6). Indeed, it suffices to prove that

$$|1\rangle + |x + \overline{x}\rangle |exh(x)\overline{exh(x)}\rangle \leq |exh(x)\overline{exh(x)}\rangle.$$

We prove the inequality for each of the summands. We treat the case of $|1\rangle$ first: by definition,

$$|exh(x)\overline{exh(x)}\rangle(p) = d(x^{*}\neg d(x) r(px^{*})) = d(x^{*} r(px^{*})\neg d(x)),$$

where we used the so-called *import-export law* [5] $r(yp) = r(y)p$ for codomains and that multiplication is commutative in K_d. Since $p \leq 1$ we have

$$px^* r(px^*) \neg d(x) \leq x^* r(px^*) \neg d(x),$$

and since $(px^*)r(px^*) = px^*$, applying domain on both sides yields

$$|exh(x)\overline{exh(x)}\rangle(p) \geq d(px^* \neg d(x)) = pd(exh(x)) = p,$$

where we used the import-export law for domains $d(py) = pd(y)$ and Lemma 1. Thus $|exh(x)\overline{exh(x)}\rangle \geq |1\rangle$. The case of $|x\rangle$ follows by the star unfold axiom:

$$|x\rangle|x^* \neg d(x)\overline{x}^*\rangle = |xx^* \neg d(x)\overline{x}^*\rangle \leq |x^* \neg d(x)\overline{x}^*\rangle.$$

The final case follows by the hypothesis of confluence:

$$\begin{aligned} |\overline{x}\rangle|x^* \neg d(x)\overline{x}^*\rangle = \langle x||x^*\rangle\langle exh(x)| &\leq \langle x^*||x^*\rangle\langle exh(x)| \\ &\leq |x^*\rangle\langle x^*|\langle exh(x)| \\ &\leq |x^*\rangle\langle exh(x)x^*| = |x^* \neg d(x)\overline{x}^*\rangle, \end{aligned}$$

where we also used $exh(x)x^* = exh(x)$. Applying the star induction axiom for modalities, we obtain the result. □

Now we are ready to relate exhaustion and normal forms to strategies and sections, respectively:

Proposition 2. *If x is convergent, then nf_x is a section of x. Furthermore, any skeleton σ of $exh(x)$ is a strategy for x with respect to nf_x, and we have*

$$\sigma \leq nf_x + x^+, \qquad \overline{\sigma} \leq nf_x + \overline{x}^+ \qquad \text{and} \quad \overline{\sigma}\sigma = nf_x$$

Proof. First we show that nf_x is a section. It is a covering set since

$$|x^\top\rangle(\mathrm{nf}_x) \geq |exh(x)\rangle(\mathrm{nf}_x) = d(exh(x)) = 1$$

where the last step is by Lemma 1. Suppose now there is some $s \in K_d$ such that $s \leq \mathrm{nf}_x$ and s is a covering set. Since $s \leq \mathrm{nf}_x \leq \neg d(x)$, the star unfold and antidomain axioms give $s \cdot exh(x) = s$, so $\langle exh(x)|(s) = s$.

Therefore $1 = |x^\top\rangle(s) = |exh(x)\rangle\langle exh(x)|(s) = |exh(x)\rangle(s)$, where we used Lemma 3. This means that

$$s \geq \langle exh(x)||exh(x)\rangle(s) = \langle exh(x)|(1) = r(exh(a)) = \mathrm{nf}_x,$$

where the first inequality is by Lemma 1, so we may conclude $\mathrm{nf}_x = s$, *i.e.* nf_x is minimal.

Now we show that a skeleton σ of $exh(x)$ is a strategy for x relative to nf_x. Note that $|x^\top \mathrm{nf}_x\rangle = |x^\top\rangle\langle \mathrm{nf}_x\rangle$ and $\langle x^\top \mathrm{nf}_x| = \langle \mathrm{nf}_x\rangle\langle x^\top|$. By Lemma 3,

$$|x^\top \mathrm{nf}_x\rangle = |exh(x)\rangle\langle exh(x)|\langle \mathrm{nf}_x\rangle = |exh(x)\rangle\langle \mathrm{nf}_x\rangle = |exh(x)\rangle,$$

since $\mathrm{nf}_x exh(x) = \mathrm{nf}_x$, and $exh(x)\mathrm{nf}_x = exh(x)$. A symmetric proof gives $\langle x^\top \mathrm{nf}_x| = \langle exh(x)|$. Since σ is a skeleton of $exh(x)$, its diamonds coincide with those of $exh(x)$ and so, by what precedes, also with those of $x^\top \mathrm{nf}_x$. Since $exh(x) \leq x^\top \mathrm{nf}_x$, σ is a wide sub of $x^\top \mathrm{nf}_x$. Minimality of σ as a wide sub follows from that same inequality plus the hypothesis that it is a skeleton of $exh(x)$. To conclude, note that $\mathrm{nf}_x \sigma \leq \mathrm{nf}_x exh(x) = \mathrm{nf}_x$. The first inequality follows from

$$\sigma \leq exh(x) = x^* \mathrm{nf}_x = (1 + xx^*)\mathrm{nf}_x \leq \mathrm{nf}_x + xx^* = \mathrm{nf}_x + x^+,$$

where we used the definition of $exh(x)$, the left star unfold axiom, $\mathrm{nf}_x \leq 1$ and the definition of the Kleene plus. The inequality for $\overline{\sigma}$ is then obtained by conversion. Finally, since $\sigma \leq exh(x)$ and x is confluent, we get

$$\overline{\sigma}\sigma \leq \overline{exh(x)} exh(x) = \mathrm{nf}_x \overline{x}^* x^* \mathrm{nf}_x \leq \mathrm{nf}_x x^* \overline{x}^* \mathrm{nf}_x = \mathrm{nf}_x,$$

where we also used that $\mathrm{nf}_x \leq \neg d(x) = \neg r(\overline{x})$. □

7 Abstract Coherence in 2-MKA

Here we state and prove a formalisation of Theorem A in the context of globular modal 2-Kleene algebras. First we prove the main result of this paper:

Theorem 1 (Coherent normalising Newman's lemma). *Let K be a Boolean globular 2-Kleene algebra such that*

i) $(K_0, +, 0, \odot_0, 1_0, \neg_0)$ *is a complete Boolean algebra,*
ii) K_1 *is continuous with respect to 0-restriction, that is for all $\psi, \psi' \in K_1$ and $(p_\alpha)_\alpha \subseteq K_0$ we have $\psi \odot_0 \sup p_\alpha \odot_0 \psi' = \sup(\psi \odot_0 p_\alpha \odot_0 \psi')$.*

Let $\phi \in K_1$ be convergent and σ be a skeleton of $exh(\phi)$. If A is a local confluence filler for $(\overline{\phi}, \phi)$, then $|\hat{A}^{*_1}\rangle_1(\sigma \odot_0 \overline{\sigma}) \geq \overline{\phi}^{*_0} \odot_0 \phi^{*_0}$.

Proof. We denote 0-multiplication by juxtaposition. First, we define a predicate RNP expressing restricted normalised paving. Given $p \in K_0$, let

$$RNP(p) \quad \Leftrightarrow \quad |\hat{A}^{*_1}\rangle_1(\sigma\overline{\sigma}) \geq \overline{\phi}^{*_0} p \phi^{*_0}.$$

By completeness of K_0, we set $r := sup\{p \,|\, RNP(p)\}$ and by continuity of restriction we may infer $RNP(r)$. Furthermore, by downward closure of RNP, we have $RNP(p)$ if, and only if, $p \leq r$. We thereby deduce:

$$\begin{aligned} \forall p.\ (RNP(\langle\phi|_0 p) \Rightarrow RNP(p)) &\Leftrightarrow \forall p.\ (\langle\phi|_0 p \leq r \Rightarrow p \leq r) \\ &\Leftrightarrow \forall p.\ (p \leq |\phi]_0 r \Rightarrow p \leq r) \\ &\Leftrightarrow |\phi]_0 r \leq r \end{aligned}$$

where we used the Galois connection (4). Thus, it suffices to show that

$$\forall p.\ (RNP(\langle\phi|_0 p) \Rightarrow RNP(p))$$

in order to conclude that $r = 1_0$, by Noethericity of ϕ. This method constitutes formalised Noetherian induction for Boolean MKA.
Given $p \in K_0$, we denote by p_ϕ the element $\langle\phi|_0(p) = |\overline{\phi}\rangle_0(p)$. We have

$$p\phi = p\phi r_0(p\phi) = p\phi\langle\phi|_0(p) \le \phi p_\phi,$$

and similarly $\overline{\phi}p \le p_\phi\overline{\phi}$. Using the star unfold axioms, we thereby deduce that

$$\overline{\phi}^{*_0} p\phi^{*_0} \le \overline{\phi}^{*_0} p + \overline{\phi}^{*_0}\overline{\phi} p\phi\phi^{*_0} + p\phi^{*_0} \le \overline{\phi}^{*_0} p + \overline{\phi}^{*_0} p_\phi\overline{\phi}\phi p_\phi\phi^{*_0} + p\phi^{*_0}.$$

We first examine the middle summand:

$$\begin{aligned}
&\overline{\phi}^{*_0} p_\phi\overline{\phi}\phi p_\phi\phi^{*_0}\\
&\quad\le \overline{\phi}^{*_0} p_\phi |A\rangle_1(\phi^{*_0}\overline{\phi}^{*_0}) p_\phi\phi^{*_0}\\
&\quad\le |\overline{\phi}^{*_0} p_\phi A p_\phi\phi^{*_0}\rangle_1(\overline{\phi}^{*_0} p_\phi\phi^{*_0}\overline{\phi}^{*_0} p_\phi\phi^{*_0})\\
&\quad\le |\hat{A}\rangle_1(\overline{\phi}^{*_0} p_\phi\phi^{*_0}\overline{\phi}^{*_0} p_\phi\phi^{*_0})\\
&\quad\le |\hat{A}\rangle_1(|\hat{A}^{*_1}\rangle_1(\sigma\overline{\sigma})\overline{\phi}^{*_0} p_\phi\phi^{*_0})\\
&\quad\le |\hat{A}\rangle_1(|\hat{A}^{*_1}\rangle_1(\sigma\overline{\sigma}\overline{\phi}^{*_0} p_\phi\phi^{*_0}))\\
&\quad\le |\hat{A}\rangle_1(|\hat{A}^{*_1}\rangle_1(|\hat{A}^{*_1}\rangle_1(\sigma\overline{\sigma}\sigma\overline{\sigma})))\\
&\quad\le |\hat{A}\odot_1\hat{A}^{*_1}\odot_1\hat{A}^{*_1}\rangle_1(\sigma\overline{\sigma}\sigma\overline{\sigma}) \le |\hat{A}^{*_1}\rangle_1(\sigma\overline{\sigma}).
\end{aligned}$$

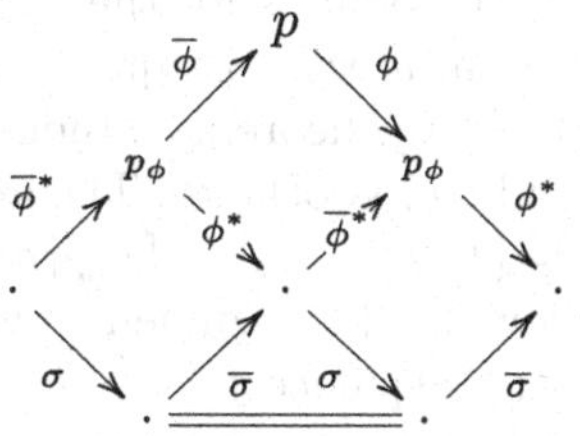

where we used that A is a local confluence filler for the first step, then commutation of modalities with whiskering (9) and the definition of $\hat{A}$ (10) for the second and third steps. We then use the induction hypothesis $RPN(p_\phi)$ on the left instance of $\overline{\phi}^{*_0} p_\phi\phi^{*_0}$, followed by commutation of modalities with whiskering and whisker absorption (11), and then repeat for the instance on the right. Finally, we used that $\hat{A}\odot_1\hat{A}^{*_1} \le \hat{A}^{*_1}\odot_1\hat{A}^{*_1} \le \hat{A}^{*_1}$, monotonicity of taking diamonds and $\overline{\sigma}\sigma = \mathrm{nf}_\phi = r(\sigma)$, a consequence of Proposition 2.
It remains to show that $\overline{\phi}^{*_0} p, p\phi^{*_0} \le |\hat{A}^{*_1}\rangle_1(\sigma\overline{\sigma})$. First, observe that we have

$$\begin{aligned}
\overline{\sigma}p\phi^{*_0} &= \overline{\sigma}p + \overline{\sigma}p\phi^{+_0}\\
&\le \overline{\sigma} + (\mathrm{nf}_\phi + \overline{\phi}^{+_0})p\phi^{+_0}\\
&= \overline{\sigma} + \overline{\phi}^{+_0} p\phi^{+_0} \le \sigma\overline{\sigma} + \overline{\phi}^{+_0} p\phi^{+_0} \le |\hat{A}^{*_1}\rangle_1(\sigma\overline{\sigma}).
\end{aligned}$$

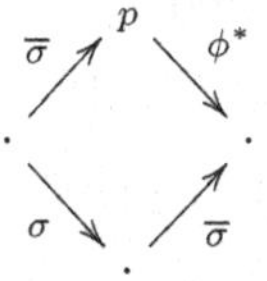

The first step is by the unfold axiom, the second uses Proposition 2 to bound $\overline{\sigma}$. The third step uses the fact that nf_ϕ is a left annihilator for ϕ^{+_0} since by definition we have $\mathrm{nf}_\phi \le \neg d_0(\phi)$. Finally we use the fact that $\overline{\sigma} \le \sigma\overline{\sigma}$ (12) coupled with $id_{K_1} = |1_1\rangle_1 \le |\hat{A}^{*_1}\rangle_1$, *i.e.* reflexivity of $\hat{A}^{*_1}$, as well as the bound established by the previous calculation.

For convergent ϕ, we have $d_0(exh(\phi)) = d_0(\phi^{*_0}\neg d_0(\phi)) = 1_0$ by Lemma 1. Since σ is a skeleton of $exh(\phi)$, we have $d_0(\sigma) = 1_0$. By the converse axiom (2), this means that $\sigma\overline{\sigma} \ge 1_0$. Therefore,

$$
\begin{aligned}
p\phi^{*_0} &\leq \sigma\overline{\sigma}p\phi^{*_0} \\
&\leq \sigma|\hat{A}^{*_1}\rangle_1(\sigma\overline{\sigma}) \\
&\leq |\hat{A}^{*_1}\rangle_1(\sigma\sigma\overline{\sigma}) = |\hat{A}^{*_1}\rangle_1(\sigma\overline{\sigma}),
\end{aligned}
$$

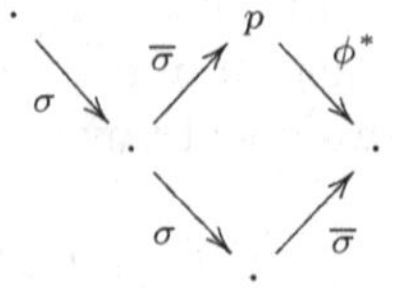

where we used commutation of whisker with modalities and whisker absorption, as well as $\sigma\sigma = \sigma$ (12). A symmetric argument yields $\overline{\phi}^{*_0}p \leq |\hat{A}^{*_1}\rangle_1(\sigma\overline{\sigma})$, concluding the proof. □

The use of formalised Noetherian induction, as well as the calculation establishing the upper bound for the middle summand, are similar to those in the proof of Newman's lemma in [5]. Due to the fact that our result involves confluences in σ, the bounds for the outer summands require a different approach.

As a direct consequence of Theorem 1, we obtain the following result, which formalises Theorem A. Indeed, if (Φ, X) is a $(2,0)$-polygraph satisfying the corresponding hypotheses, Theorem 2 lifts the result to the power set when applied to $\phi := \{1_x \mid x \in \Phi_1\} \cup \{1_{1_a} \mid a \in \Phi_0\}$ and $A = X$, viewed as elements of $K(\Phi, X)$. Following the argument given in Sect. 5, the conclusion asserts that *for every* zig-zag sequence $f : a \to b \in \Phi_1^\top$, *there exists* a 2-cell $\alpha_f : f \Rightarrow \sigma_a \star_0 \sigma_b^-$ obtained by whiskering and composing elements of X. In a 2-groupoid, this is equivalent to the existence of a 2-cell $f \star_0 \sigma_b \Rightarrow \sigma_a$.

Theorem 2 (Abstract coherence theorem). *Let K be a Boolean globular 2-Kleene algebra satisfying the additional hypotheses in Theorem 1 and $\phi \in K_1$ convergent. Given a normalisation strategy σ and a local confluence filler A for $(\overline{\phi}, \phi)$, we have*

$$|\hat{A}^{*_1}\rangle_1(\sigma \odot_0 \overline{\sigma}) \geq \phi^{\top_0} = (\phi + \overline{\phi})^{*_0}.$$

Proof. We denote 0-multiplication by juxtaposition. As a result of Theorem 1 we have $|\hat{A}^{*_1}\rangle_1(\sigma\overline{\sigma}) \geq \overline{\phi}^{*_0}\phi^{*_0}$. By the star induction axiom, it suffices to show:

$$1_0 + (\phi + \overline{\phi})|\hat{A}^{*_1}\rangle_1(\sigma\overline{\sigma}) \leq |\hat{A}^{*_1}\rangle_1(\sigma\overline{\sigma}).$$

By (2) and Proposition 2, we have $\sigma\overline{\sigma} \geq d_0(\sigma) = 1_0$, so by reflexivity of $\hat{A}^{*_1}$, *i.e.* $1_1 \leq \hat{A}^{*_1}$, we have $1_0 \leq |\hat{A}^{*_1}\rangle_1(\sigma\overline{\sigma})$. Furthermore, since $\phi \leq \overline{\phi}^{*_0}\phi^{*_0}$ we have:

$$\phi|\hat{A}^{*_1}\rangle_1(\sigma\overline{\sigma}) \leq \overline{\phi}^{*_0}\phi^{*_0}|\hat{A}^{*_1}\rangle_1(\sigma\overline{\sigma}) \leq |\hat{A}^{*_1}\rangle_1(\sigma\overline{\sigma})|\hat{A}^{*_1}\rangle_1(\sigma\overline{\sigma}) \leq |\hat{A}^{*_1}\rangle_1(\sigma\overline{\sigma}).$$

The case of $\overline{\phi}$ is identical. We conclude via the star induction axiom. □

8 Outlook

In this article, we have introduced a formalisation of the notion of strategy for convergent ARS and thereby obtained an abstract coherence theorem. This constitutes an initial result formalising cofibrant replacements of algebraic structures by rewriting, such as polygraphic resolutions from convergent SRS, [15].

In this perspective, the first step is to formalise the critical branching lemma, a coherent confluence result for SRS. Kleene algebra axioms only allow iteration on the left or right of expressions, but not in context. We expect a formalisation of coherent confluence for SRS using the structure of higher-dimensional quantales [26], similar to higher-dimensional semirings [3] but in which multiplication distributes over arbitrary sums. The second step consists in extending our formalisation of normalisation strategies to higher dimensions, necessary for constructing cofibrant replacements, for example polygraphic resolutions via convergent rewriting systems [15].

Another direction is found in the domain of concurrency theory. Concurrent Kleene algebras (CKA) [19] are a convenient extension of Kleene algebras. While similar to 2-MKAs, these are used to give semantics to concurrent languages and their corresponding proof systems. CKAs enrich classical Kleene algebras with an extra parallel composition operation alongside the classical sequential composition. In particular, CKAs have applications for validation of concurrent programs by formalising Hoare-like proof systems for parallel computations, similarly to MKAs which have applications to verification of hybrid systems [30] and program correctness [11]. We expect that our approach to abstract coherence proofs in 2-Kleene algebras can also find applications to formalisation of proof systems for verifying general concurrent systems, for example based on higher-dimensional trace semantics of Higher-Dimensional Automata [9,25] (a form of higher-dimensional rewriting system), see e.g. [7,8].

References

1. Bloom, S.L., Ésik, Z., Stefanescu, G.: Notes on equational theories of relations. Algebra Universalis **33**(1), 98–126 (1995)
2. Burroni, A.: Higher-dimensional word problems with applications to equational logic. Theoret. Comput. Sci. **115**(1), 43–62 (1993). 4th Summer Conference on Category Theory and Computer Science (Paris, 1991)
3. Calk, C., Goubault, E., Malbos, P., Struth, G.: Algebraic coherent confluence and higher-dimensional globular Kleene algebras (2020, preprint). arXiv:2006.16129
4. Curien, P.-L., Duric, A., Guiraud, Y.: Coherent presentations of a class of monoids admitting a Garside family (2021). arXiv:2107.00498
5. Desharnais, J., Möller, B., Struth, G.: Termination in modal Kleene algebra. In: Exploring New Frontiers of Theoretical Informatics. IFIP 18th World Computer Congress, TC1 3rd International Conference on Theoretical Computer Science (TCS2004), pp. 647–660. Kluwer Academic Publishers, Boston, MA (2004)
6. Desharnais, J., Struth, G.: Internal axioms for domain semirings. Sci. Comput. Program. **76**(3), 181–203 (2011)
7. Fahrenberg, U., Johansen, C., Struth, G., Ziemianski, K.: Languages of higher-dimensional automata. CoRR, abs/2103.07557 (2021)
8. Fahrenberg, U., Johansen, C., Struth, G., Ziemianski, K.: lr-multisemigroups and modal convolution algebras. CoRR abs/2105.00188 (2021)
9. Fajstrup, L., Goubault, E., Haucourt, E., Mimram, S., Raussen, M.: Directed Algebraic Topology and Concurrency. Springer, Cham (2016). https://doi.org/10.1007/978-3-319-15398-8

10. Gaussent, S., Guiraud, Y., Malbos, P.: Coherent presentations of Artin monoids. Compos. Math. **151**(5), 957–998 (2015)
11. Gomes, V.B.F., Struth, G.: Modal Kleene algebra applied to program correctness. In: Fitzgerald, J., Heitmeyer, C., Gnesi, S., Philippou, A. (eds.) FM 2016. LNCS, vol. 9995, pp. 310–325. Springer, Cham (2016). https://doi.org/10.1007/978-3-319-48989-6_19
12. Gonthier, G., Lévy, J., Melliès, P.: An abstract standardisation theorem. In: Proceedings of the 7th Annual Symposium on Logic in Computer Science, LICS 1992, pp. 72–81. IEEE Computer Society (1992)
13. Guiraud, Y., Malbos, P.: Higher-dimensional categories with finite derivation type. Theor. Appl. Categ. **22**(18), 420–478 (2009)
14. Guiraud, Y., Malbos, P.: Coherence in monoidal track categories. Math. Struct. Comput. Sci. **22**(6), 931–969 (2012)
15. Guiraud, Y., Malbos, P.: Higher-dimensional normalisation strategies for acyclicity. Adv. Math. **231**(3–4), 2294–2351 (2012)
16. Guiraud, Y., Malbos, P.: Polygraphs of finite derivation type. Math. Struct. Comput. Sci. **28**(2), 155–201 (2018)
17. Hage, N., Malbos, P.: Knuth's coherent presentations of plactic monoids of type A. Algebr. Represent. Theor. **20**(5), 1259–1288 (2017)
18. Hoare, T., Möller, B., Struth, G., Wehrman, I.: Concurrent Kleene algebra and its foundations. J. Log. Algebr. Program. **80**(6), 266–296 (2011)
19. Hoare, T., Möller, B., Struth, G., Wehrman, I.: Concurrent Kleene algebra and its foundations. J. Log. Algebraic Meth. Program. **80**(6), 266–296 (2011)
20. Huet, G.: Initiation à la théorie des catégories. INRIA. Notes de cours du DEA "Fonctionalité, Structures de Calcul et Programmation" donné à l'Université Paris VII en 1983–84 et 1984–1985 (1987)
21. Jónsson, B., Tarski, A.: Boolean algebras with operators. I. Am. J. Math. **73**, 891–939 (1951)
22. Lawson, M.: Inverse Semigroups. The Theory of Partial Symmetries. World Scientific Publishing Company (1998)
23. MacLane, S.: Natural associativity and commutativity. Rice Univ. Stud. **49**(4), 28–46 (1963)
24. Melliès, P.-A.: Axiomatic rewriting theory i: a diagrammatic standardization theorem. In: Middeldorp, A., van Oostrom, V., van Raamsdonk, F., de Vrijer, R. (eds.) Processes, Terms and Cycles: Steps on the Road to Infinity. LNCS, vol. 3838, pp. 554–638. Springer, Heidelberg (2005). https://doi.org/10.1007/11601548_23
25. Pratt, V.R.: Modeling concurrency with geometry. In: Wise, D.S. (eds.) Conference Record of the 18th Annual ACM Symposium on Principles of Programming Languages 1991, pp. 311–322. ACM Press (1991)
26. Rosenthal, K.I.: Quantales and their applications, volume 234 of Pitman Research Notes in Mathematics Series. Longman Scientific & Technical, Harlow; copublished in the United States with Wiley, New York (1990)
27. Squier, C.C., Otto, F., Kobayashi, Y.: A finiteness condition for rewriting systems. Theoret. Comput. Sci. **131**(2), 271–294 (1994)
28. Street, R.: Limits indexed by category-valued 2-functors. J. Pure Appl. Algebra **8**(2), 149–181 (1976)

29. Struth, G.: Calculating Church-Rosser proofs in Kleene Algebra. In: de Swart, H.C.M. (ed.) RelMiCS 2001. LNCS, vol. 2561, pp. 276–290. Springer, Heidelberg (2002). https://doi.org/10.1007/3-540-36280-0_19
30. Huerta y Munive, J.J., Struth, G.: Verifying hybrid systems with modal Kleene algebra. In: Desharnais, J., Guttmann, W., Joosten, S. (eds.) RAMiCS 2018. LNCS, vol. 11194, pp. 225–243. Springer, Cham (2018). https://doi.org/10.1007/978-3-030-02149-8_14

Algorithmic Correspondence for Relevance Logics, Bunched Implication Logics, and Relation Algebras via an Implementation of the Algorithm PEARL

Willem Conradie[1(✉)], Valentin Goranko[2,1], and Peter Jipsen[3]

[1] School of Mathematics, University of the Witwatersrand, Johannesburg, South Africa
willem.conradie@wits.ac.za

[2] Department of Philosophy, Stockholm University, Stockholm, Sweden
valentin.goranko@philosophy.su.se

[3] Department of Mathematics, Chapman University, Orange, USA
jipsen@chapman.edu

Abstract. The non-deterministic algorithmic procedure PEARL (acronym for 'Propositional variables Elimination Algorithm for Relevance Logic') has been recently developed for computing first-order equivalents of formulas of the language of relevance logics $\mathcal{L}_R$ in terms of the standard Routley-Meyer relational semantics. It succeeds on a large class of axioms of relevance logics, including all so called inductive formulas. In the present work we re-interpret PEARL from an algebraic perspective, with its rewrite rules seen as manipulating quasi-inequalities interpreted over Urquhart's relevant algebras, and report on its recent Python implementation. We also show that all formulae on which PEARL succeeds are canonical, i.e., preserved under canonical extensions of relevant algebras. This generalizes the "canonicity via correspondence" result in [37]. We also indicate that with minor modifications PEARL can also be applied to bunched implication algebras and relation algebras.

1 Introduction

This work relates two important areas of development in non-classical logics, viz. *relevance logics* and *algorithmic correspondence theory*, by applying the latter to the possible worlds semantics for relevance logic based on Routley-Meyer frames [31], by means of an implementation of the recently developed in [7] algorithm PEARL. That semantics is, in turn, duality-theoretically related to the algebraic semantics for relevance logic based on Urquhart's relevant algebras [37]. Routley-Meyer frames also capture the semantics of (positive) relation algebras [14,24], and of bunched implication algebras [30], hence the algorithm PEARL implemented here is also applicable to arrow logic [3,15] and bunched implication logics [30].

[1]Visiting professorship affiliation of the 2nd author.

U. Fahrenberg et al. (Eds.): RAMiCS 2021, LNCS 13027, pp. 126–143, 2021.
https://doi.org/10.1007/978-3-030-88701-8_8

Modal Correspondence Theory. The Sahlqvist-van Benthem theorem [2,32], proved in the mid 1970s, is a fundamental result in the model theory of modal logic. It gives a syntactic characterization of a class of modal formulas which define first-order conditions on Kripke frames and which are canonical, hence, when added to the basic normal modal logic K, they axiomatize logics which are strongly complete with respect to elementary frame classes. The Sahlqvist-van Benthem theorem sets the stage for the emergence and development of the so called *correspondence theory in modal logic*, cf. [4]. The literature on the topic contains many analogues of the Sahlqvist-van Benthem theorem for a wide range of non-classical logics. Various illuminating alternative proofs have appeared, including Jonsson's purely algebraic proof of the canonicity part [25], and the 'canonicity-via-correspondence' approach pioneered by Sambin and Vaccaro [33].

The Sahlqvist-van Benthem class of formulas has been significantly extended to the class of so called *inductive formulas* [21–23] which cover frame classes not definable by a Sahlqvist-van Benthem formula while enjoying the same properties of elementarity and canonicity. At about the same time, a new line of research known as *algorithmic correspondence theory* emerged. It involves the use of algorithms like SCAN and DLS to try and compute first-order frame correspondence for modal formulas by eliminating the second-order quantifiers from their standard second-order frame correspondents. In particular, the algorithm SQEMA [9] was developed for algorithmic correspondence in modal logic. It manipulates formulas in an extended hybrid language to eliminate propositional variables and thereby produces pure hybrid formulas which translate into first-order logic via the standard translation, and simultaneously proves their canonicity via an argument in the style of Sambin and Vaccaro. This approach was extended to logics algebraically captured by normal (distributive) lattice expansions [10,11] in a line of research known as *unified correspondence* [6].

Correspondence Theory for Relevance Logic. Much work has been done over the years on computing first-order equivalents and proving completeness of a range of specific axioms for relevance logics with respect to the *Routley-Meyer relational semantics* (cf. [31]). Routley-Meyer frames involve not a binary, but a ternary relation, with several conditions imposed on it, needed to ensure upward closedness of the valuations of all formulas. That makes the possible worlds semantics for relevance logic based on such frames technically more complex and proving correspondence results for it "by hand" can be significantly more elaborate than those for modal logics with their standard Kripke semantics, which calls for a systematic development of respective correspondence theory for relevance logics. Until recently, that problem remained little explored, with just a few works, incl. those of Seki [34] and Badia [1], defining some classes of Sahlqvist-van Benthem type formulas for relevance logics and proving correspondence results for them. Likewise, Suzuki [35,36], has established correspondence for the full Lambek calculus with respect to the so-called bi-approximation semantics, obtained via canonical extensions in the style of [16]. For closely related distributive substructural logics, such as bunched implication logics, an elegant categorical approach to canonicity and correspondence is based on duality theory and coalgebras [12].

A general algorithmic correspondence theory of relevance logics has recently been developed in [7], on which the presently reported work is based.

The Algorithm PEARL and Its Implementation. A non-deterministic algorithmic procedure PEARL (acronym for Propositional variables Elimination Algorithm for Relevance Logic) for computing first-order equivalents in terms of frame validity of formulas of the language $\mathcal{L}_R$ for relevance logics is developed in [7]. PEARL is an adaptation of the above mentioned procedures SQEMA [9] (for normal modal logics) and ALBA [10,11] (for distributive and non-distributive modal logics). Furthermore, a large syntactically defined class of *inductive relevance formulas* in $\mathcal{L}_R$ is defined in [7], based on specific order-theoretic properties of the algebraic interpretations of the connectives, following the general methodology of [11]. It is shown in [7] that PEARL succeeds for all such formulas and correctly computes for them equivalent with respect to frame validity first-order definable conditions on Routley-Meyer frames. This gives a general basis for comparing inductive and Sahlqvist formulas across different logics and for different relational semantics for the same logic. Thus, [11, Example 3.14] has shown that Suzuki's Sahlqvist class is properly included in the respective class of inductive formulas. Likewise, for the case of $\mathcal{L}_R$, it is shown in [7] that the class of inductive formulas properly extends the classes of Sahlqvist formulas of Seki [34] and Badia [1].

In the present work we re-interpret the algorithm PEARL from an algebraic perspective with its rewrite rules seen as manipulating quasi-inequalities interpreted over Urquhart's relevant algebras [37]. This enables us to complete the part of the Sahlqvist-van Benthem theorem still outstanding from the previous work, namely the fact that all inductive $\mathcal{L}_R$-formulas are canonical, i.e., are preserved under canonical extensions of relevant algebras. Via the discrete duality between perfect relevant algebras and Routley-Meyer frames, this establishes the fact that all inductive $\mathcal{L}_R$-formulas axiomatise logics which are complete with respect to first-order definable classes of Routley-Meyer frames. This generalizes the "canonicity via correspondence" result in [37] for (what we can now recognise as) a certain special subclass of Sahlqvist-van Benthem formulas in the "groupoid" sublanguage of $\mathcal{L}_R$ where fusion is the only connective. We then present an optimised and deterministic version of PEARL, which we have recently implemented in Python and applied to verify the first-order equivalents of a number of important axioms for relevance logics known from the literature, as well as on several new types of formulas. In this paper we report on the implementation and on some testing results.

Relevance Logics and Relation Algebras. Even though developed with different motivations, these two areas are technically closely related, as noted and explored in several papers besides [37], incl. [5,15,24,26,28]. We note that, by extending $\mathcal{L}_R$ with a Heyting implication (which is a residual of the meet operation), removing relevant negation, and adding commutativity and associativity as axioms of fusion, our results can also be applied to bunched implication algebras. Alternatively one can extend $\mathcal{L}_R$ with classical implication and apply the same algorithm to relation algebras. In this case the Routley-Meyer frames have the order of an antichain and are the same as atom structures of relation algebras. Further details are discussed at the end of Sect. 7.

Structure of the Paper. In Sect. 2 we provide the necessary background on the syntax, algebraic and relational semantics of relevance logic, define relevant algebras and then extend their language by adding adjoints and residuals of the standard operators of relevance logic. Then, in Sect. 3 we establish duality between perfect relevant algebras and complex algebras of Routley-Meyer frames. Section 4 presents the rules of the calculus on which PEARL is based, and Sect. 5 contains a concise description of the main phases of the algorithm itself. In Sect. 6 we give a brief description of the implementation of PEARL, and in Sect. 7 we state some results. We then conclude with Sect. 8. After the references we have included a short appendix containing some additional technicalities and some examples of the output of PEARL.

2 Preliminaries

In this section we provide background on the syntax and algebraic and relational semantics of relevance logic. For further details we refer the reader to [17,31] and (for relevance logics) to [37] and [7].

2.1 Relevance Logic and Its Algebraic Semantics

The language of propositional relevance logic $\mathcal{L}_R$ over a fixed set of propositional variables VAR is given by

$$A = p \mid \bot \mid \top \mid \mathbf{t} \mid {\sim}A \mid (A \wedge A) \mid (A \vee A) \mid (A \circ A) \mid (A \rightarrow A)$$

for $p \in \mathsf{VAR}$. The relevant connectives $\circ$, $\sim$ and $\rightarrow$ are called **fusion**, **(relevant) negation** and **(relevant) implication**, respectively. The constant $\mathbf{t}$ is referred to as **(relevant) truth**. We also add the constants $\top$ and $\bot$ for convenience. Equations and inequalities of $\mathcal{L}_R$-formulas can be algebraically interpreted in relevant algebras as defined by Urquhart in [37].

Definition 1 ([37]). *A structure* $\mathbb{A} = \langle A, \wedge, \vee, \circ, \rightarrow, \sim, \mathbf{t}, \top, \bot\rangle$ *is called a* ***relevant algebra*** *if it satisfies the following conditions:*

1. $\langle A, \wedge, \vee, \top, \bot\rangle$ *is a bounded distributive lattice,*
2. $a \circ (b \vee c) = (a \circ b) \vee (a \circ c)$,
3. $(b \vee c) \circ a = (b \circ a) \vee (c \circ a)$,
4. ${\sim}(a \vee b) = {\sim}a \wedge {\sim}b$,
5. ${\sim}(a \wedge b) = {\sim}a \vee {\sim}b$,
6. ${\sim}\top = \bot$ *and* ${\sim}\bot = \top$,
7. $a \circ \bot = \bot \circ a = \bot$,
8. $\mathbf{t} \circ a = a$, *and*
9. $a \circ b \leq c$ *iff* $a \leq b \rightarrow c$.

An $\mathcal{L}_R$-formula ϕ is **valid on a relevant algebra** $\mathbb{A}$ if the inequality $\mathbf{t} \leq \phi$ (implicitly universally quantified over all propositional variables) is valid on $\mathbb{A}$ and **valid on a class of relevant algebras** if it is valid on each member of that class. We also refer the reader to [37] for axiomatizations of the logic of the class of all relevant algebras.

2.2 Relational Semantics

Relevance logic can be given relational semantics based on structures called 'Routley-Meyer frames', which we will now define. A **relevance frame** is a tuple $\mathcal{F} = \langle W, O, R, ^* \rangle$, where:

- W is a non-empty set of states (possible worlds);
- $O \subseteq W$ is the subset of **normal** states;
- $R \subseteq W^3$ is a **relevant accessibility relation**;
- $^* : W \to W$ is a function, called the **Routley star**.

The binary relation $\preceq$ is defined in every relevance frame by specifying that $u \preceq v$ iff $\exists o(o \in O \wedge Rouv)$. A **Routley-Meyer frame** [1] (for short, **RM-frame**) is a relevance frame satisfying the following conditions for all $u, v, w, x, y, z \in W$:

1. $x \preceq x$
2. If $x \preceq y$ and $Ryuv$ then $Rxuv$.
3. If $x \preceq y$ and $Ruyv$ then $Ruxv$.
4. If $x \preceq y$ and $Ruvx$ then $Ruvy$.
5. If $x \preceq y$ then $y^* \preceq x^*$.
6. O is upward closed w.r.t. $\preceq$, i.e. if $o \in O$ and $o \preceq o'$ then $o' \in O$.

These properties ensure that $\preceq$ is reflexive and transitive, hence a preorder, and that the semantics of the logical connectives has the upward monotonicity property stated below.

A **Routley-Meyer model (RM-model)** is a tuple $\mathcal{M} = \langle W, O, R, ^*, V \rangle$, where $\langle W, O, R, ^* \rangle$ is a Routley-Meyer frame and $V : \mathsf{VAR} \to \mathcal{P}(W)$ is a mapping, called a **relevant valuation**, assigning to every atomic proposition $p \in \mathsf{VAR}$ a set $V(p)$ of states which is *upward closed* w.r.t. $\preceq$.

Truth of a formula A **in an RM-model** $\mathcal{M} = \langle W, O, R, ^*, V \rangle$ at a state $u \in W$, denoted $\mathcal{M}, u \Vdash A$, is defined as follows:

- $\mathcal{M}, u \Vdash p$ iff $u \in V(p)$;
- $\mathcal{M}, u \Vdash \mathbf{t}$ iff $u \in O$;
- $\mathcal{M}, u \Vdash {\sim}A$ iff $\mathcal{M}, u^* \nVdash A$;
- $\mathcal{M}, u \Vdash A \wedge B$ iff $\mathcal{M}, u \Vdash A$ and $\mathcal{M}, u \Vdash B$;
- $\mathcal{M}, u \Vdash A \vee B$ iff $\mathcal{M}, u \Vdash A$ or $\mathcal{M}, u \Vdash B$;
- $\mathcal{M}, u \Vdash A \to B$ iff for every v, w, if $Ruvw$ and $\mathcal{M}, v \Vdash A$ then $\mathcal{M}, w \Vdash B$.
- $\mathcal{M}, u \Vdash A \circ B$ iff there exist v, w such that $Rvwu$, $\mathcal{M}, v \Vdash A$ and $\mathcal{M}, w \Vdash B$.

Thus, the Routley-Meyer semantics follows a standard pattern for relational semantics of modal operators. In particular, the fusion is a binary 'diamond',

[1] The definition of Routley-Meyer frames takes the relation R and subset O as primary and defines the pre-order $\preceq$ in terms of them. This does not restrict the pre-orders that can occur within Routley-Meyer frames. Indeed, given an upward closed subset $O \subseteq W$ and a pre-order $\preceq$ on W one can define a respective ternary relation $R \subseteq W^3$ by specifying that, for all triples (x, y, z), $Rxyz$ iff $x \preceq o$ for some $o \in O$ and $x \preceq y$.

interpreted with a ternary relation, and negation is both a unary box and diamond, interpreted via a functional binary relation. One can show, by a routine structural induction on formulas, (cf. e.g. [31]) that this semantics satisfies **upward monotonicity**: for every RM-model $\mathcal{M}$ and a formula A of $\mathcal{L}_R$, the set $[\![A]\!]_{\mathcal{M}} = \{u \mid \mathcal{M}, u \Vdash A\}$ is upward closed.

A formula A is declared **true in an RM-model** $\mathcal{M}$, denoted by $\mathcal{M} \Vdash A$, if $\mathcal{M}, o \Vdash A$ for every $o \in O$. It is **valid in an RM-frame** $\mathcal{F}$, denoted by $\mathcal{F} \Vdash A$, iff it is true in every RM-model over that frame, and A is **RM-valid**, denoted by $\Vdash A$, iff it is true in every RM-model.

All semantic notions of truth and validity defined above can be translated to FOL, resp. universal monadic second order, by means of a **standard translation**, analogous to the one applied to modal logic (cf. [4]). See the details in the full paper [8].

2.3 Perfect Relevant Algebras and the Extended Language $\mathcal{L}_R^+$

Given a Routley-Meyer frame $\mathcal{F} = \langle W, R, ^*, O\rangle$, its **complex algebra** is the structure

$$\mathcal{F}^+ = \langle \mathcal{P}^\uparrow(W), \cap, \cup, \to, \circ, \sim, O, W, \varnothing\rangle$$

where $\mathcal{P}^\uparrow(W)$ is the set of all upwards closed subsets (hereafter called **up-sets**) of W, $\cap$ and $\cup$ are set-theoretic intersection and union, and for all $Y, Z \in \mathcal{P}^\uparrow(W)$ the following hold:

$Y \to Z = \{x \in W \mid \text{ for all } y, z \in W, \text{ if } Rxyz \text{ and } y \in Y, \text{ then } z \in Z\}$,

$Y \circ Z = \{x \in W \mid \text{ there exist } y \in Y \text{ and } z \in Z \text{ such that } Ryzx\}$,

${\sim}Y = \{x \in W \mid x^* \notin Y\}$.

It is easy to check that $\mathcal{F}^+$ is a relevant algebra.

An element a of a lattice $\mathbb{L}$ is **completely join-irreducible** (resp., **completely join-prime**) if whenever $a = \bigvee S$ ($a \leq \bigvee S$) for some $S \subseteq L$, then $a = s$ ($a \leq s$) for some $s \in S$. The notions of **meet-irreducibility** and **primality** are defined order-dually. Complete join/meet primality implies complete join/meet irreducibility and for complete distributive lattices the notions coincide.

A relevant algebra $\mathbb{A} = \langle A, \wedge, \vee, \circ, \to, \sim, \mathbf{t}, \top\bot\rangle$ is **perfect** if $\langle A, \wedge, \vee, \top\bot\rangle$ is a complete, completely distributive lattice that is join-generated (resp., meet-generated) by the set of its completely join-irreducible elements $J^\infty(\mathbb{A})$ (resp., the set of its completely meet-irreducible elements $M^\infty(\mathbb{A})$), while $\bigvee S \circ a = \bigvee_{s\in S}(s \circ a)$, $a \circ \bigvee S = \bigvee_{s\in S}(a \circ s)$, $\bigvee S \to a = \bigwedge_{s\in S}(s \to a)$, $a \to \bigwedge S = \bigwedge_{s\in S}(a \to s)$, ${\sim}\bigvee S = \bigwedge_{s\in S} {\sim}s$ and ${\sim}\bigwedge S = \bigvee_{s\in S} {\sim}s$ for all $S \subseteq A$ and $a \in A$. Now, in fact, every $\mathcal{F}^+$ is a *perfect* relevant algebra. Further, every relevant algebra $\mathbb{A}$ can be compactly and densely embedded in a unique perfect relevant algebra, namely in its *canonical extension* (cf. e.g. [16]) which we will denote $\mathbb{A}^\delta$.

For any perfect distributive lattice $\mathbb{A}$, the map $\kappa : J^\infty(\mathbb{A}) \to M^\infty(\mathbb{A})$ defined by $j \mapsto \bigvee\{u \in \mathbb{A} \mid j \not\leq u\}$ is an order isomorphism (cf. [19, Sec. 2.3]) when considering $J^\infty(\mathbb{A})$ and $M^\infty(\mathbb{A})$ as subposets of $\mathbb{A}$. The inverse of κ is $\lambda : M^\infty(\mathbb{A}) \to J^\infty(\mathbb{A})$, given by the assignment $m \mapsto \bigwedge\{u \in \mathbb{A} \mid u \not\leq m\}$. From these definitions, we immediately have that, for every $u \in \mathbb{A}$, every $j \in J^{\infty}(\mathbb{A})$ and every $m \in M^{\infty}(\mathbb{A})$,

$$j \not\leq u \text{ iff } u \leq \kappa(j), \tag{1}$$

$$u \not\leq m \text{ iff } \lambda(m) \leq u. \tag{2}$$

Since in perfect relevant algebras each of $\sim$, $\vee$, $\wedge$, $\circ$ and $\rightarrow$ preserves or reverses arbitrary meets and/or joins in each coordinate, they are residuated in each coordinate (see e.g. [18]). The algebra therefore supports the interpretation of an extended language with connectives for the residuals of these operations. In particular, we extend the language $\mathcal{L}_R$ to $\mathcal{L}_R^+$ by adding the **left adjoint** $\sim^\flat$ and the **right adjoint** $\sim^\sharp$ of $\sim$, the **intuitionistic (Heyting) implication** $\Rightarrow$ (as right residual of $\wedge$), the **coimplication** $-\!\!<$ as the left residual of $\vee$, and the operation $\hookrightarrow$ as the residual of $\circ$ in the second coordinate and of $\rightarrow$ in the first coordinate. Thus, in any perfect relevant algebra $\mathbb{A}$ we have that:

1. $\sim a \leq b$ iff $\sim^\flat b \leq a$
2. $a \leq \sim b$ iff $b \leq \sim^\sharp a$
3. $a \leq b \vee c$ iff $a -\!\!< b \leq c$
4. $a \wedge b \leq c$ iff $a \leq b \Rightarrow c$
5. $a \circ b \leq c$ iff $a \leq b \rightarrow c$
6. $a \circ b \leq c$ iff $b \leq a \hookrightarrow c$

We also include in $\mathcal{L}_R^+$ two countably infinite sets of special variables, $\mathsf{NOM} = \{\mathbf{j}_0, \mathbf{j}_1, \mathbf{j}_2, \ldots\}$ and $\mathsf{CNOM} = \{\mathbf{m}_0, \mathbf{m}_1, \mathbf{m}_2, \ldots\}$. These are respectively called **nominals** and **co-nominals** and will be interpreted as ranging respectively over completely join-irreducibles and completely meet-irreducibles. Informally, we will denote nominals by $\mathbf{i}, \mathbf{j}, \mathbf{k}$, possibly with indices, while co-nominals will be denoted by $\mathbf{m}, \mathbf{n}$, possibly with indices. To distinguish visually from $\mathcal{L}_R$, the formulas of the extended language $\mathcal{L}_R^+$ will be denoted by lowercase greek letters, typically $\alpha, \beta, \gamma, \phi, \psi, \xi$, etc. and are defined by the following grammar:

$$\phi = p \mid \mathbf{i} \mid \mathbf{m} \mid \top \mid \bot \mid \mathbf{t} \mid \sim\phi \mid (\phi \wedge \phi) \mid (\phi \vee \phi) \mid (\phi \circ \phi) \mid (\phi \rightarrow \phi) \mid$$
$$\sim^\flat\phi \mid \sim^\sharp\phi \mid (\phi -\!\!< \phi) \mid (\phi \Rightarrow \phi) \mid (\phi \hookrightarrow \phi)$$

where $p \in \mathsf{VAR}$, $\mathbf{i} \in \mathsf{NOM}$ and $\mathbf{m} \in \mathsf{CNOM}$. We denote $\mathsf{ATOMS} := \mathsf{VAR} \cup \mathsf{NOM} \cup \mathsf{CNOM}$. The elements of ATOMS will be called **atoms**. An $\mathcal{L}_R^+$-formula is called **pure** if it contains no propositional variables but only, possibly, nominals, co-nominals and constants. To each connective we assign a **polarity type**[2] indicating whether each coordinate of its interpretation in (perfect) relevant algebras is order-preserving or order-reversing, as follows:

1. $\epsilon_\sim = \epsilon_{\sim^\flat} = \epsilon_{\sim^\sharp} = (-)$
2. $\epsilon_\wedge = \epsilon_\vee = \epsilon_\circ = (+,+)$
3. $\epsilon_\rightarrow = \epsilon_\Rightarrow = \epsilon_\hookrightarrow = (-,+)$
4. $\epsilon_{-\!<} = (+,-)$

We write $\epsilon_h(i)$ for the i-th coordinate of ϵ_h. We now define the notions of **positive** and **negative occurrences** of atoms in $\mathcal{L}_R^+$-formulas recursively: an occurrence of an atom a is positive in a; an occurrence of a which is positive (negative) in ϕ is positive (negative) in $h(\psi_1, \ldots, \psi_{i-1}, \phi, \psi_{i+1}, \ldots \psi_n)$

[2] Also called an **order type** (e.g. [19]) or **monotonicity type** (e.g. [20]).

if $\epsilon_h(i) = +$ and negative (positive) in $h(\psi_1, \ldots, \psi_{i-1}, \phi, \psi_{i+1}, \ldots, \psi_n)$ if $\epsilon_h(i) = -$. We then say that a **formula** $\phi \in \mathcal{L}_R^+$ **is positive (negative) in an atom** a iff all occurrences of a in ϕ are positive (negative). An **inequality** $\phi \leq \psi$ **is positive (negative) in an atom** a if ϕ is negative (positive) in a while ψ is positive (negative) in a.

3 Duality Between Perfect Relevant Algebras and Complex Algebras of Routley-Meyer Frames

As already mentioned, the complex algebra $\mathcal{F}^+$ of any Routley-Meyer frame $\mathcal{F} = \langle W, R, ^*, O\rangle$ is a perfect relevant algebra. Moreover, $J^\infty(\mathcal{F}^+) = \{\uparrow x \mid x \in W\}$ the set of all **principal up-sets** $\uparrow x = \{y \in W \mid y \succeq x\}$ and $M^\infty(\mathcal{F}^+) = \{(\downarrow x)^c \mid x \in W\}$ the set of all set-theoretic complements of principal downwards closed subsets (hereafter called **co-downsets**) $\downarrow x = \{y \in W \mid x \succeq y\}$. Conversely, we will show that every perfect relevant algebra is isomorphic to the complex algebra of a Routley-Meyer frame.

Lemma 2. *In a perfect relevant algebra* $\mathbb{A}$, *it is the case that* $\sim^\sharp$ *maps* $J^\infty(\mathbb{A})$ *into* $M^\infty(\mathbb{A})$ *and* $\sim^\flat$ *maps* $M^\infty(\mathbb{A})$ *into* $J^\infty(\mathbb{A})$.

Proof. See proof in the full paper [8].

The following definition adapts a well-known method (see [16]) for obtaining dual relational structures from perfect algebras:

Definition 3. *The* ***prime structure*** *of a perfect relevant algebra* $\mathbb{A} = \langle A, \wedge, \vee, \circ, \rightarrow, \sim, \mathbf{t}, \top\bot\rangle$ *is the structure* $\mathbb{A}_\bullet = \langle J^\infty(\mathbb{A}), O_\mathbf{t}, R_\circ, ^{*\sim}\rangle$ *where:*
1. $R_\circ abc$ *iff* $c \leq a \circ b$ *2.* $O_\mathbf{t} = \{j \in J^\infty(\mathbb{A}) | j \leq \mathbf{t}\}$ *and 3.* $a^{*\sim} = \lambda(\sim^\sharp a)$

Lemma 4. $\mathbb{A}_\bullet$ *is a Routley-Meyer frame. Moreover the order* $\preceq$ *on* $\mathbb{A}_\bullet$ *coincides with the dual lattice order* $\geq$ *restricted to* $J^\infty(\mathbb{A})$.

Proof. We begin by noting that $b \preceq c$ iff there exists $j_0 \in O_\mathbf{t} = \{j \in J^\infty(\mathbb{A}) \mid j \leq \mathbf{t}\}$ such that $R_\circ j_0 bc$. By definition, the latter is equivalent to $c \leq j_0 \circ b$ for some completely join-irreducible $j_0 \leq \mathbf{t}$. By the monotonicity of $\circ$, this implies that $c \leq \mathbf{t} \circ b$ which is equivalent to $c \leq b$ by the clause 8 of Definition 1. Conversely, if $c \leq b$, then, by the same clause, we have $c \leq \mathbf{t} \circ b = \bigvee\{j \in J^\infty(\mathbb{A}) \mid j \leq \mathbf{t}\} \circ b = \bigvee\{j \circ b \in J^\infty(\mathbb{A}) \mid j \leq \mathbf{t}\}$. Since $c \in J^\infty(\mathbb{A})$, this means there is some $j_0 \in J^\infty(\mathbb{A})$ such that $j \leq \mathbf{t}$ and $c \leq j \circ b$, which implies $b \preceq c$. It is clear from the construction that $\mathbb{A}_\bullet$ is a relevance frame. In particular, the fact that $^{*\sim}$ maps elements of $J^\infty(\mathbb{A})$ into $J^\infty(\mathbb{A})$ follows from the definition of λ and Lemma 2. We verify the six defining properties of Routley-Meyer frames in [8].

Proposition 5. *For any perfect relevant algebra* $\mathbb{A}$ *it is the case that* $\mathbb{A} \simeq (\mathbb{A}_\bullet)^+$.

Proof. We show that the map $\theta : \mathbb{A} \rightarrow (\mathbb{A}_\bullet)^+$ given by $\theta(a) \mapsto \{j \in J^\infty(\mathbb{A}) \mid j \leq a\}$ is a relevant algebra isomorphism. See details in the full paper [8].

4 The Calculus of the Algorithm PEARL

In this section we present a calculus of rewrite rules[3], in the style of the algorithms SQEMA [9] and ALBA [10,11], which is sound and complete for deriving first-order frame correspondents and simultaneously proving canonicity for a large class of formulas of $\mathcal{L}_R$, viz. the class of *inductive (relevance) formulas* (see [7]). The algorithm PEARL and its implementation, described in the next section, are based on this calculus. The algorithm accepts (inequalities of) $\mathcal{L}_R^+$ formulas as input and, if it succeeds, it produces first-order formulas in the language of RM-frames that is valid in an RM-frame if and only if the original formulas are valid in the complex algebra of this RM-frame.

The rules manipulate **quasi-inequalities**[4] of $\mathcal{L}_R^+$ formulas, i.e., expressions of the form $\phi_1 \leq \psi_1, \ldots, \phi_n \leq \psi_n \implies \phi \leq \psi$ with $\phi, \psi, \phi_i, \psi_i \in \mathcal{L}_R^+$. In the setting of relevant algebras, quasi-inequalities are considered universally quantified over all propositional variables. Any formula $\phi \in \mathcal{L}_R^+$ can be treated as the inequality $\mathbf{t} \leq \phi$, which is a quasi-inequality with no assumptions. The inequalities not affected by the application of the rule are regarded as a **context**, which will be denoted by Γ. Given a set of inequalities Γ, we say that Γ is positive (negative) in an atom a whenever each member of Γ is positive (negative) in a. We will write $\Gamma(\alpha/p)$ for the set of inequalities obtained by uniformly substituting α for atom p in each member of Γ.

All rules that are indicated below by a double line are invertible, although the algorithm PEARL only applies the approximation rules in the downward direction.

Monotone Variable Elimination Rules

$$\frac{\Gamma(p) \implies \gamma(p) \leq \beta(p)}{\Gamma(\top/p) \implies \gamma(\bot/p) \leq \beta(\bot/p)}\ (\bot) \qquad \frac{\Delta(p) \implies \beta(p) \leq \gamma(p)}{\Delta(\bot/p) \implies \beta(\top/p) \leq \gamma(\top/p)}\ (\top)$$

where $\beta(p)$ and Γ are positive in p, while $\gamma(p)$ and $\Delta(p)$ are negative in p.

[3] These rules can be seen as instantiations of the rules of the general-purpose algorithm ALBA [11] in the context of perfect relevant algebras. However, the fact that the latter are distributive lattice expansions allows us to present simpler formulations of these rules closer to those in [10] and, to some extent, [9]. The approximation rules presented in [11] allow for the extraction of subformulas deep from within the consequents of quasi-inequalities, subject to certain conditions, rather than the connective-by-connective style of our presentation. Although the former style of rule is also sound in the present setting, we opted for the latter as we believe it is simpler to present since the formulation requires significantly fewer auxiliary notions.

[4] In [7] these are treated set-theoretically and are called 'quasi-inclusions'.

First Approximation Rule

$$\frac{\Gamma \;\Longrightarrow\; \phi \leq \psi}{\mathbf{j} \leq \phi,\ \psi \leq \mathbf{m},\ \Gamma \;\Longrightarrow\; \mathbf{j} \leq \mathbf{m}}$$

where $\mathbf{j}$ is a nominal and $\mathbf{m}$ is a co-nominal not occurring in the premise.

Approximation Rules

$$\frac{\chi \rightarrow \phi \leq \mathbf{m},\ \Gamma \;\Longrightarrow\; \alpha \leq \beta}{\mathbf{j} \leq \chi,\ \mathbf{j} \rightarrow \phi \leq \mathbf{m},\ \Gamma \;\Longrightarrow\; \alpha \leq \beta}\ (\rightarrow\text{Appr-L}) \qquad \frac{\chi \rightarrow \phi \leq \mathbf{m},\ \Gamma \;\Longrightarrow\; \alpha \leq \beta}{\phi \leq \mathbf{n},\ \chi \rightarrow \mathbf{n} \leq \mathbf{m},\ \Gamma \;\Longrightarrow\; \alpha \leq \beta}\ (\rightarrow\text{Appr-R})$$

$$\frac{\mathbf{i} \leq \chi \circ \phi,\ \Gamma \;\Longrightarrow\; \alpha \leq \beta}{\mathbf{j} \leq \chi,\ \mathbf{i} \leq \mathbf{j} \circ \phi,\ \Gamma \;\Longrightarrow\; \alpha \leq \beta}\ (\circ\text{Appr-L}) \qquad \frac{\mathbf{i} \leq \chi \circ \phi,\ \Gamma \;\Longrightarrow\; \alpha \leq \beta}{\mathbf{j} \leq \phi,\ \mathbf{i} \leq \chi \circ \mathbf{j},\ \Gamma \;\Longrightarrow\; \alpha \leq \beta}\ (\circ\text{Appr-R})$$

$$\frac{{\sim}\phi \leq \mathbf{m},\ \Gamma \;\Longrightarrow\; \alpha \leq \beta}{\phi \leq \mathbf{n},\ {\sim}\mathbf{n} \leq \mathbf{m},\ \Gamma \;\Longrightarrow\; \alpha \leq \beta}\ ({\sim}\text{Appr-L}) \qquad \frac{\mathbf{i} \leq {\sim}\phi,\ \Gamma \;\Longrightarrow\; \alpha \leq \beta}{\mathbf{j} \leq \phi,\ \mathbf{i} \leq {\sim}\mathbf{j},\ \Gamma \;\Longrightarrow\; \alpha \leq \beta}\ ({\sim}\text{Appr-R})$$

where $\mathbf{j}$ a nominal and $\mathbf{n}$ is a co-nominal not appearing in the premises.

Residuation Rules

$$\frac{\phi \leq \chi \vee \psi,\ \Gamma \;\Longrightarrow\; \alpha \leq \beta}{\phi \prec \chi \leq \psi,\ \Gamma \;\Longrightarrow\; \alpha \leq \beta}\ (\vee\text{Res}) \qquad \frac{\phi \wedge \chi \leq \psi,\ \Gamma \;\Longrightarrow\; \alpha \leq \beta}{\phi \leq \chi \Rightarrow \psi,\ \Gamma \;\Longrightarrow\; \alpha \leq \beta}\ (\wedge\text{Res})$$

$$\frac{\phi \leq \chi \rightarrow \psi,\ \Gamma \;\Longrightarrow\; \alpha \leq \beta}{\phi \circ \chi \leq \psi,\ \Gamma \;\Longrightarrow\; \alpha \leq \beta}\ (\rightarrow\text{Res}) \qquad \frac{\psi \leq \phi \hookrightarrow \chi,\ \Gamma \;\Longrightarrow\; \alpha \leq \beta}{\phi \circ \psi \leq \chi,\ \Gamma \;\Longrightarrow\; \alpha \leq \beta}\ (\hookrightarrow\text{Res})$$

Adjunction Rules

$$\frac{\phi \vee \chi \leq \psi,\ \Gamma \;\Longrightarrow\; \alpha \leq \beta}{\phi \leq \psi,\ \chi \leq \psi,\ \Gamma \;\Longrightarrow\; \alpha \leq \beta}\ (\vee\text{Adj}) \qquad \frac{\psi \leq \phi \wedge \chi,\ \Gamma \;\Longrightarrow\; \alpha \leq \beta}{\psi \leq \phi,\ \psi \leq \chi,\ \Gamma \;\Longrightarrow\; \alpha \leq \beta}\ (\wedge\text{Adj})$$

$$\frac{{\sim}\phi \leq \psi,\ \Gamma \;\Longrightarrow\; \alpha \leq \beta}{{\sim}^{\flat}\psi \leq \phi,\ \Gamma \;\Longrightarrow\; \alpha \leq \beta}\ ({\sim}\text{Adj-L}) \qquad \frac{\phi \leq {\sim}\psi,\ \Gamma \;\Longrightarrow\; \alpha \leq \beta}{\psi \leq {\sim}^{\sharp}\phi,\ \Gamma \;\Longrightarrow\; \alpha \leq \beta}\ ({\sim}\text{Adj-R})$$

Not to clutter the procedure with extra rules, we allow commuting the arguments of $\wedge$ and $\vee$ whenever needed before applying the rules ($\wedge$Adj) and ($\vee$Adj) above. These rules are applied exhaustively in the downward direction, and produce the same results regardless of how an expression is parenthesized.

Ackermann-Rules: The Right Ackermann-rule (RAR) and Left Ackermann-rule (LAR) are subject to the following conditions:

– p does not occur in α,
– β is positive in p,
– γ is negative in p,

– Γ is negative in p,
– Δ is positive in p,

$$\frac{\alpha \leq p,\ \Delta(p) \implies \gamma(p) \leq \beta(p)}{\Delta(\alpha/p) \implies \gamma(\alpha/p) \leq \beta(\alpha/p)}\ (\text{RAR}) \qquad \frac{p \leq \alpha,\ \Gamma(p) \implies \beta(p) \leq \gamma(p)}{\Gamma(\alpha/p) \implies \beta(\alpha/p) \leq \gamma(\alpha/p)}\ (\text{LAR})$$

Note that the rules ($\bot$) and ($\top$) are, in fact, special cases of the Ackermann-rules (RAR) and (LAR), respectively.

Simplification Rules: In the rules below Γ is a possibly empty list of inequalities.

$$\frac{\Gamma,\ \mathbf{i} \leq \phi \implies \mathbf{i} \leq \psi}{\Gamma \implies \phi \leq \psi}\ (\text{Simpl-Left}) \qquad \frac{\Gamma,\ \psi \leq \mathbf{m} \implies \phi \leq \mathbf{m}}{\Gamma \implies \phi \leq \psi}\ (\text{Simpl-Right})$$

In the rule (Simpl-Left) the nominal $\mathbf{i}$ must not occur in ϕ, or ψ, or any inequality in Γ. Likewise, in the rule (Simpl-Right) the co-nominal $\mathbf{m}$ must not occur in ϕ, or ψ, or any inequality in Γ. These rules are usually applied in the post-processing, to eliminate nominals and co-nominals introduced by the approximation rules.

Example 6. We illustrate an application of PEARL on the following formula (known as axiom B2 in [31]): $(p \rightarrow q) \wedge (q \rightarrow r) \rightarrow (p \rightarrow r)$. In the full paper [8] we show that the elimination phase of PEARL succeeds and produces the following pure quasi-inequality:

$$\mathbf{i} \circ (\mathbf{i} \circ \mathbf{j}_1) \leq \mathbf{n}_1,\ \mathbf{j}_1 \rightarrow \mathbf{n}_1 \leq \mathbf{m} \quad \implies \quad \mathbf{i} \leq \mathbf{m}.$$

5 Algorithmic Description of PEARL

5.1 Pre-processing and Main Phase of PEARL

Here we will present a deterministic algorithmic version of the procedure PEARL, which is used for the implementation.

1. Receive a formula ϕ in input.
2. If ϕ is an implication $\psi \rightarrow \theta$ set $X := \{\psi \leq \theta\}$, otherwise form the initial inequality $\mathbf{t} \leq \phi$ and set $X := \{\mathbf{t} \leq \phi\}$.
3. Now preprocess the set X by iterating steps 3a, 3b until a pass is reached in which none of the steps are applicable.

(a) For any $(\theta \leq \chi) \in X$, find the first positive occurrence of $\vee$ or negative occurrence of $\wedge$ in θ which is not in the scope of any positive occurrence of $\rightarrow$ or a negative occurrence of $\circ$. Letting $\theta(\alpha \diamond \beta)$ denote θ with the occurrence of the found subterm, where $\diamond \in \{\vee, \wedge\}$, replace $\theta \leq \chi$ in X by $\theta(\alpha) \leq \chi, \theta(\beta) \leq \chi$.
(b) For any $(\theta \leq \chi) \in X$, find the first positive occurrence of $\wedge$ or negative occurrence of $\vee$ in χ which is not in the scope of any negative occurrence of $\rightarrow$ or a positive occurrence of $\circ$. Again letting $\chi(\alpha \diamond \beta)$ denote χ with the found subterm, replace $\theta \leq \chi$ in X by $\theta \leq \chi(\alpha), \theta \leq \chi(\beta)$.
The preceding two "splitting" steps are justified by the distributivity of the operations $\circ, \rightarrow, \sim$ and the adjunction rules ($\vee$Adj) and ($\wedge$Adj).
(c) Apply the monotone variable elimination rules to all inequalities in X where they apply, replacing the involved inequalities in X with the results.

4. Proceed separately in each inequality $\phi_i \leq \psi_i$ in X. Apply the first-approximation rule to $\phi_i \leq \psi_i$ to produce the quasi-inequality $\mathbf{i} \leq \phi_i, \psi_i \leq \mathbf{m} \vdash \mathbf{i} \leq \mathbf{m}$.
5. As long as one of χ, ϕ in the approximation rules is matched by a subformula that is neither a nominal or conominal, apply these rules exhaustively to this quasi-inequality, interleaved with the splitting steps 3a–3b, where X is the set of premises. The resulting quasi-inequality has premises that are irreducible with respect to the approximation steps and splittings. This step terminates since approximation rules are only applied downwards and splittings eliminate a $\wedge$ or $\vee$-symbol.
6. For each variable p in the quasi-inequality, and for each choice of polarity, $+p$ or $-p$, check if the right Ackermann-rule (for $+p$) or the left Ackermann-rule (for $-p$) can be applied to eliminate p from the premises of the quasi-inequality. This is done by applying the residuation and $\sim$-adjunction rules exhaustively to all premises that contain exactly one occurrence of $+p$ (or $-p$) to solve the inequality for p (if possible) and checking that p only occurs (if at all) with the opposite sign in all other premises. If possible, apply the right or left Ackermann-rule. Otherwise, p cannot be eliminated, in which case the next variable is tried. Backtracking is used to attempt to eliminate all variables in all possible orders and with either positive or negative polarity. If a variable cannot be eliminated in some particular quasi-inequality, then the algorithm stops and reports this failure.
7. If the elimination phase has succeeded on all quasi-inequalities, the algorithm proceeds to post-processing, including simplification and translation phases.

5.2 Post-processing and Translation to First-Order Logic

This phase[5] applies if/when the algorithm succeeds to eliminate all variables, thus ending with **pure quasi-inequalities**, containing only nominals and co-nominals, but no variables. The purpose of the post-processing is to produce a first-order condition equivalent to the pure quasi-inequality produced as a

[5] This is an optimised version of the post-processing procedure outlined in [7].

result of the main phase described in Sect. 5.1, and hence to the input formula. Each pure quasi-inequality produced in the elimination phase is post-processed separately to produce a corresponding first-order condition, and all these are then taken conjunctively to produce the corresponding first-order condition of the input formula. So, we focus on the case of a single pure quasi-inequality. Computing a first-order equivalent of any pure quasi-inequality can be done by straightforward application of the standard translation, but the result would usually be unnecessarily long and complicated. This can be compensated by additional post-translation equivalent simplifications in first-order logic, also taking into account the monotonicity conditions in Routley-Meyer frames. Instead, we have chosen to first apply some pre-translation simplifications of the pure quasi-inequality, using again some of the PEARL rules, and then to modify the standard translation by applying it to pure inequalities, rather than to formulas, and by extending it with a number of additional clauses dependent on the type (main connective) of the formulas on both sides of these inequalities, thus applying simplifications on the fly. For lack of space we have omitted the list of these additional post-processing rules, which can be found in the full paper [8].

The resulting modified translation Tr is not restricted to pure quasi-inequalities and can be applied to arbitrary pure formulas.

The post-processing of the pure quasi-inequality produced in Example 6 using the translation Tr is illustrated in the full paper [8]. The resulting first-order formula is $\forall x_{\mathbf{i}}, x_{\mathbf{j}}, x_{\mathbf{j}_1}, y_{\mathbf{n}_1} (R x_{\mathbf{i}} x_{\mathbf{j}_1} y_{\mathbf{n}_1} \rightarrow \exists x_{\mathbf{j}} (R x_{\mathbf{i}} x_{\mathbf{j}_1} x_{\mathbf{j}} \wedge R x_{\mathbf{i}} x_{\mathbf{j}} y_{\mathbf{n}_1}))$ which is equivalent to the first-order condition known from [31] for the axiom B2, and to the one computed by the implementation of PEARL reported here.

6 Implementation of PEARL

Here we give a brief description of an implementation of PEARL in Python, based on the description given in Sect. 5. The input is a LaTeX string using the standard syntax of relevance logic expressions. Intuitionistic implication $\Rightarrow$, coimplication $-\!\!<$, the right residual $\hookrightarrow$ of $\circ$, and the adjoints $\sim^\sharp$ and $\sim^\flat$ can also appear in an input formula. The expression is parsed with a simple top-down Pratt parser [29] using standard rules of precedence. For well-formed formulas, an abstract syntax tree (AST) based on Python dictionaries and lists of arguments is created for each formula.

Five short recursive Python functions are used to transform the AST representation step-by-step according to the specific groups of PEARL transformation rules. The function `preprocess(st)` takes a LaTeX string `st` as input and parses it to an AST which we refer to as `A`. If the formula `A` is not well-formed, an error-string is returned. If it has a top-level $\rightarrow$ symbol, it is replaced with a $\leq$ to turn the formula into an inequality, and otherwise the equivalent inequality $\mathbf{t} \leq \mathtt{A}$ is constructed. Subsequently the splitting rules and monotonicity rules from Sect. 4 are applied and the resulting list of inequalities is returned.

For example, with `r"p\to q\land\mathbf t"` as input, the formula is parsed, rewritten as $p \leq q \wedge \mathbf{t}$, then the splitting rules produce the list $[p \leq q, p \leq \mathbf{t}]$ and monotonicity returns $[\top \leq \bot, \top \leq \mathbf{t}]$.

The function `approximate(As)` takes this list as input, and applies the first approximation rule to each formula, followed by all possible left and right approximations interleaved with further applications of the splitting rule. The result is a list of quasi-equations that always have conclusion $\mathbf{i} \leq \mathbf{m}$ and premises that are irreducible with respect to the approximation and splitting rules.

The function `eliminate(As)` then attempts to apply the Ackermann-rules to each quasi-equations by selecting each variable, first with positive polarity and, if that does not succeed, then with negative polarity. Backtracking is used to ensure that all variables are tried in all possible orders. If for some quasi-equations none of the variable orders allow all variables to be eliminated, then the function reports this result. On the other hand, if for each quasi-equations some variable order succeeds to eliminate all formula variables then the resulting list of pure quasi-equations (i.e., containing no formula variables, but only nominals or co-nominals) is returned.

Since these pure quasi-equations contain redundant premises, the function `simplify(As)` is used to eliminate them, and to also apply the left and right simplification rules. Finally the variant of the standard translation described in Sect. 5.2 is applied to the pure quasi-equations and produces a first-order formula on the Routley-Meyer frames.

The Python code can be used in any Jupyter notebook, with the output displayed in standard mathematical notation. No special installation is needed to use the program in a personal Jupyter notebook or in a public cloud-based notebook such as Colab.google.com, and the output can be pasted into standard LaTeX documents. Moreover the program can be easily extended to handle the syntax of other suitable logics and lattice-ordered algebras. The resulting formula can also be translated to TPTP, Prover9 or SPASS syntax. The Python code is available at github.com/jipsen/PEARL in the form of a Jupyter notebook. It can also be copied and used directly in a browser at https://colab.research.google.com/drive/1p0PTkmyq7vTWgYDxCTFHVRwjaLeT45uX?usp=sharing. In the full paper [8] we provide some examples of output from the PEARL implementation.

7 Canonicity and Applications to BI-Logic and Relation Algebras

Here we report on some new theoretical and practical results related to the theory and implementation of PEARL. We begin with a theoretical result, which, for lack of space, we only sketch here.

Theorem 7. *The validity of all $\mathcal{L}_R^+$-formulas on which PEARL succeeds is preserved under canonical extensions of relevant algebras.*

Proof. Let $\phi \leq \psi$ be an $\mathcal{L}_R$-inequality on which PEARL succeeds and let $\mathbb{A}$ be a relevant algebra. Let $\mathsf{PEARL}(\phi \leq \psi)$ denote the purified quasi-inequality produced from input $\phi \leq \psi$. For any $\mathcal{L}_R^+$ quasi-inequality $\Gamma \implies \alpha \leq \beta$, we write $\mathbb{A}^\delta \models_{\mathbb{A}} \Gamma \implies \alpha \leq \beta$ to indicate that $\Gamma \implies \alpha \leq \beta$ is true in

$\mathbb{A}^\delta$ under all assignments that send propositional variables to elements of the original algebra $\mathbb{A}$ (and nominals to $J^\infty(\mathbb{A})$ and co-nominals to $M^\infty(\mathbb{A})$) while, as usual, $\mathbb{A}^\delta \models \Gamma \implies \alpha \leq \beta$ indicates truth under *all* assignments. The following chain of equivalences establishes the canonicity of $\phi \leq \psi$:

$$\begin{array}{ccc}
\mathbb{A} \models \phi \leq \psi & & \mathbb{A}^\delta \models \phi \leq \psi \\
\Updownarrow & & \\
\mathbb{A}^\delta \models_{\mathbb{A}} \phi \leq \psi & & \Updownarrow \\
\Updownarrow & & \\
\mathbb{A}^\delta \models_{\mathbb{A}} \mathsf{PEARL}(\phi \leq \psi) & \Leftrightarrow & \mathbb{A}^\delta \models \mathsf{PEARL}(\phi \leq \psi)
\end{array}$$

The uppermost bi-implication on the left is immediate by the way we defined $\models_{\mathbb{A}}$ and the fact that $\mathbb{A}$ is a subalgebra of $\mathbb{A}^\delta$. The lower bi-implication on the left follows by that fact that, if a quasi-inequality $\Delta' \implies \gamma' \leq \chi'$ is obtained from another, $\Delta \implies \gamma \leq \chi$, through the application of PEARL rules, then $\mathbb{A}^\delta \models_{\mathbb{A}} \Delta \implies \gamma \leq \chi$ iff $\mathbb{A}^\delta \models_{\mathbb{A}} \Delta' \implies \gamma' \leq \chi'$. This is straightforward to check for all rules except the Ackermann-rules. We refer the reader to [10] and/or [11] for the details of the latter. The horizontal bi-implication follows from the facts that, by assumption, $\mathsf{PEARL}(\phi \leq \psi)$ is pure, and that restricting assignments of propositional variables to elements of $\mathbb{A}$ is vacuous for pure formulas, as they contain no propositional variables. The bi-implication on the right follows by the soundness of all PEARL rules on perfect algebras, which is routine to verity.

Via the discrete duality between perfect relevant algebras and Routley-Meyer frames established in Sect. 3, it follows that all $\mathcal{L}_R^+$-formulas on which PEARL succeeds axiomatise logics which are complete with respect to their respective first-order definable classes of Routley-Meyer frames.

As mentioned in the introduction, a large syntactically defined class of *inductive relevance formulas* in $\mathcal{L}_R$ is defined in [7], where it is shown that PEARL succeeds for all such formulas and correctly computes their equivalent with respect to frame validity first-order definable conditions on Routley-Meyer frames. Therefore, all inductive $\mathcal{L}_R^+$-formulas are canonical. This result generalizes the "canonicity via correspondence" result in [37], applied there to the fragment of $\mathcal{L}_R$ involving of all specific relevance logic connectives only the fusion.

We can now state the results above applied to the specific implementation of PEARL reported here. However, the proof of the correctness of the implementation is beyond the scope of this paper. Still, we can report that the implementation has succeeded on all axioms A1-A9, B1-B30, and D1-D8 listed in the appendix of [7], copied there from [31], and has computed first-order conditions equivalent to those known from the literature.

Bunched implication logic [30] is closely related to a negation-free relevance logic. The algebraic semantics of bunched implication logic is given by bunched implication algebras, or BI-algebras. They are defined by axioms *1–3* and *7–9* of Definition 1 together with a new binary operation symbol $\Rightarrow$ such that

10. $a \wedge b \leq c$ iff $a \leq b \Rightarrow c$ (hence $\Rightarrow$ is a Heyting algebra implication)
11. $(a \circ b) \circ c = a \circ (b \circ c)$,
12. $a \circ b = b \circ a$.

The steps of the PEARL algorithm are not affected by these addition axioms (although additional rules for the associativity and commutativity of $\circ$ could be added), and the relational semantic structures of BI-logic and BI-algebras are precisely Routley-Meyer frames. However in BI-logic the notation differs slightly, since $\rightarrow, \circ, \Rightarrow$ are replaced by $-\!\!*, *, \rightarrow$, and this alternative notation is user-selectable in the implementation.

Lastly, we note that the algorithm PEARL can also be applied to relation algebras, as they form a subvariety of relevant algebras extended with a Heyting implication $\Rightarrow$. An axiomatization of relation algebras in this setting consists of axioms of relevant algebras (*1–9* from Definition 1), *10, 11* above and[6]

13. $(x \Rightarrow \bot) \Rightarrow \bot = x$ (hence $\Rightarrow$ is a classical implication and $x \Rightarrow \bot$ is denoted $\neg x$),
14. $x \rightarrow y = {\sim}({\sim}y \circ x)$,
15. $x^{\smile} = {\sim}(x \Rightarrow \bot)$,
16. $(x \circ y)^{\smile} = y^{\smile} \circ x^{\smile}$.

Axiom *13* ensures that the lattice structure is a Boolean algebra, hence the partial order in the Routley-Meyer frames of a relation algebra is an antichain. In the theory of relation algebras these frames are known as 'atom structures', defined in [27, Def. 2.1]. For the application of PEARL to relation algebras, it suffices to replace the converse operation by the term ${\sim}(x \Rightarrow \bot)$ and to interpret any $\preceq$ symbol in the resulting first-order formula as an equality symbol. Note that relevant negation ${\sim}x$ can, in turn, also be defined via the relation algebra term $(\neg x)^{\smile}$. While there is a long history of Sahlqvist formulas and correspondence theory for Boolean algebras with operators [13,25], it is interesting to note that the PEARL algorithm and its implementation can be adapted to relation algebras and covers the more general class of inductive formulas.

8 Concluding Remarks

In this paper we have re-interpreted the algorithm PEARL from [7] as an algorithm which manipulates quasi-inequalities interpreted over perfect relevant algebras. Implementing the algorithm in a way that produces reasonably optimal (in size) versions of first-order correspondents required detailed specifications and strategic choices in the pre-processing, main, and post-processing phases (Sects. 5.1 and 5.2) and in the specialized post-processing and translation procedure, refining the normal standard translation, developed in Sect. 5.2. It is easy to see that the complexity of the problem solved by PEARL is in NP-time because, once the correct ordering or elimination of the variables is selected, PEARL completes its work in polynomial time. However, theoretically, it may

[6] While this equational basis for relation algebras appears to be quite long, it can be shown that axioms *3–7* are redundant. Hence, it is comparable in length to the original axiomatization of relation algebras.

take trying an exponential number of such orderings until success. Whether this is possible is not yet known, so the optimal complexity of the problem is still under investigation.

References

1. Badia, G.: On Sahlqvist formulas in relevant logic. J. Philos. Log. **47**(4), 673–691 (2018)
2. van Benthem, J.: Modal correspondence theory. Ph.D. thesis, Mathematisch Instituut & Instituut voor Grondslagenonderzoek, University of Amsterdam (1976)
3. van Benthem, J.: A note on dynamic arrow logic. Technical report LP-92-11, ILLC, University of Amsterdam (1992)
4. van Benthem, J.: Correspondence theory. In: Gabbay, D., Guenthner, F. (eds.) Handbook of Philosophical Logic, vol. 3, 2nd edn., pp. 325–408. Springer, Dordrecht (2001). https://doi.org/10.1007/978-94-017-0454-0_4
5. Bimbó, K., Dunn, J.M., Maddux, R.D.: Relevance logics and relation algebras. Rev. Symb. Log. **2**(1), 102–131 (2009)
6. Conradie, W., Ghilardi, S., Palmigiano, A.: Unified correspondence. In: Baltag, A., Smets, S. (eds.) Johan van Benthem on Logic and Information Dynamics. OCL, vol. 5, pp. 933–975. Springer, Cham (2014). https://doi.org/10.1007/978-3-319-06025-5_36
7. Conradie, W., Goranko, V.: Algorithmic correspondence for relevance logics I. The algorithm PEARL. In: Düntsch, I., Mares, E. (eds.) Alasdair Urquhart on Non-classical and Algebraic Logic and Complexity of Proofs, pp. 163–209. Springer, Heidelberg (2021). https://www2.philosophy.su.se/goranko/papers/PEARL.pdf
8. Conradie, W., Goranko, V., Jipsen, P.: Algorithmic correspondence for relevance logics, bunched implication logics, and relation algebras: the algorithm PEARL and its implementation (2021). Technical report https://arxiv.org/abs/2108.06603
9. Conradie, W., Goranko, V., Vakarelov, D.: Algorithmic correspondence and completeness in modal logic, I. The core algorithm SQEMA. Log. Methods Comput. Sci. **2**(1:5), 1–26 (2006)
10. Conradie, W., Palmigiano, A.: Algorithmic correspondence and canonicity for distributive modal logic. Ann. Pure Appl. Logic **163**(3), 338–376 (2012)
11. Conradie, W., Palmigiano, A.: Algorithmic correspondence and canonicity for non-distributive logics. Ann. Pure Appl. Logic **170**(9), 923–974 (2019). https://doi.org/10.1016/j.apal.2019.04.003
12. Dahlqvist, F., Pym, D.: Coalgebraic completeness-via-canonicity for distributive substructural logics. J. Log. Algebr. Methods Program. **93**, 1–22 (2017). https://doi.org/10.1016/j.jlamp.2017.07.002
13. de Rijke, M., Venema, Y.: Sahlqvist's theorem for Boolean algebras with operators with an application to cylindric algebras. Stud. Log. **54**(1), 61–78 (1995). https://doi.org/10.1007/BF01058532
14. Doumane, A., Pous, D.: Non axiomatisability of positive relation algebras with constants, via graph homomorphisms. In: Konnov, I., Kovács, L. (eds.) Proceedings of CONCUR 2020. LIPIcs, vol. 171, pp. 29:1–29:16. Schloss Dagstuhl (2020)
15. Dunn, J.M.: Arrows pointing at arrows: arrow logic, relevance logic, and relation algebras. In: Baltag, A., Smets, S. (eds.) Johan van Benthem on Logic and Information Dynamics. OCL, vol. 5, pp. 881–894. Springer, Cham (2014). https://doi.org/10.1007/978-3-319-06025-5_34

16. Dunn, J.M., Gehrke, M., Palmigiano, A.: Canonical extensions and relational completeness of some substructural logics. J. Symb. Logic **70**, 713–740 (2005)
17. Dunn, J., Restall, G.: Relevance logic. In: Gabbay, D., Guenthner, F. (eds.) Handbook of Philosophical Logic. HALO, vol. 6, 2nd edn., pp. 1–128. Springer, Dordrecht (2002). https://doi.org/10.1007/978-94-017-0460-1_1
18. Galatos, N., Jipsen, P., Kowalski, T., Ono, H.: Residuated Lattices: An Algebraic Glimpse at Substructural Logics. Elsevier, Amsterdam (2007)
19. Gehrke, M., Nagahashi, H., Venema, Y.: A Sahlqvist theorem for distributive modal logic. Ann. Pure Appl. Logic **131**, 65–102 (2005)
20. Gehrke, M., Jónsson, B.: Bounded distributive lattice expansions. Mathematica Scandinavica **94**(1), 13–45 (2004)
21. Goranko, V., Vakarelov, D.: Sahlqvist formulae in hybrid polyadic modal languages. J. Log. Comput. **11**(5), 737–754 (2001)
22. Goranko, V., Vakarelov, D.: Sahlqvist formulas unleashed in polyadic modal languages. In: Wolter, F., Wansing, H., de Rijke, M., Zakharyaschev, M. (eds.) Advances in Modal Logic, vol. 3, pp. 221–240. World Scientific, Singapore (2002)
23. Goranko, V., Vakarelov, D.: Elementary canonical formulae: extending Sahlqvist's theorem. Ann. Pure Appl. Logic **141**(1–2), 180–217 (2006)
24. Hirsch, R., Mikulás, S.: Positive fragments of relevance logic and algebras of binary relations. Rev. Symb. Log. **4**(1), 81–105 (2011)
25. Jónsson, B.: On the canonicity of Sahlqvist identities. Studis Logica **53**(4), 473–491 (1994). https://doi.org/10.1007/BF01057646
26. Kowalski, T.: Relevant logic and relation algebras. In: Galatos, N., Kurz, A., Tsinakis, C. (eds.) TACL 2013. Sixth International Conference on Topology, Algebra and Categories in Logic. EPiC Series in Computing, vol. 25, pp. 125–128 (2014)
27. Maddux, R.: Some varieties containing relation algebras. Trans. Am. Math. Soc. **272**, 501–526 (1982)
28. Maddux, R.D.: Relevance logic and the calculus of relations. Rev. Symb. Log. **3**(1), 41–70 (2010). https://doi.org/10.1017/S1755020309990293
29. Pratt, V.R.: Top down operator precedence. In: Fischer, P.C., Ullman, J.D. (eds.) Conference Record of the ACM Symposium on Principles of Programming Languages, Boston, Massachusetts, USA, October 1973, pp. 41–51. ACM Press (1973)
30. Pym, D.: The Semantics and Proof Theory of the Logic of Bunched Implications. APLS, vol. 26. Springer, Dordrecht (2002). https://doi.org/10.1007/978-94-017-0091-7
31. Routley, R., Meyer, R., Plumwood, V., Brady, R.: Relevant Logics and its Rivals (Volume I). Ridgeview, CA (1982)
32. Sahlqvist, H.: Correspondence and completeness in the first and second-order semantics for modal logic. In: Kanger, S. (ed.) Proceedings of the 3rd Scandinavian Logic Symposium, Uppsala 1973, pp. 110–143. Springer, Amsterdam (1975). https://doi.org/10.1016/S0049-237X(08)70728-6
33. Sambin, G., Vaccaro, V.: A new proof of Sahlqvist's theorem on modal definability and completeness. J. Symb. Log. **54**(3), 992–999 (1989)
34. Seki, T.: A Sahlqvist theorem for relevant modal logics. Stud. Logica. **73**(3), 383–411 (2003)
35. Suzuki, T.: Canonicity results of substructural and lattice-based logics. Rev. Symb. Log. **4**(1), 1–42 (2011). https://doi.org/10.1017/S1755020310000201
36. Suzuki, T.: A Sahlqvist theorem for substructural logic. Rev. Symb. Log. **6**(2), 229–253 (2013). https://doi.org/10.1017/S1755020313000026
37. Urquhart, A.: Duality for algebras of relevant logics. Studia Logica **56**(1/2), 263–276 (1996). https://doi.org/10.1007/BF00370149

The Class of Representable Semilattice-Ordered Monoids Is Not a Variety

Amina Doumane(✉)

CNRS - ENS Lyon, Lyon, France
amina.doumane@ens-lyon.fr

Abstract. We show a necessary and a sufficient condition for a quasivariety to be a variety. Using this, we show that the quasivariety of representable relation algebras over the signature $(\cdot, \cap, 1)$ is not avariety.

Keywords: Relations · Homomorphisms · Equational theories

1 Introduction

Relations can be equipped with several natural operations: union $\cup$, intersection $\cap$, complementation $^{-}$, composition $\cdot$, converse ^{c}, the empty relation 0, the full relation $\top$ and the identity relation 1. A set of relations closed under these operations forms a *proper relation algebra.* We call *representable relation algebras* (RRA) those algebras which are isomorphic to a proper relation algebra.

Representable relation algebras received a lot of attention since the seminal work of Tarski [11]. However, many of these results are actually negative results. For instance, RRA is not finitely axiomatizable [10] and its equational theory is undecidable [11, p88].

This motivated the investigation of the *subreducts* of RRA, that is, restrictions of RRA to smaller signatures, hoping that these negative results would turn out to be positive. The subreduct that we focus on in this work is $RRA(\cdot, \cap, 1)$, the restriction of RRA to the operations of composition, intersection and the identity relation, also known as the class of *representable semi-lattice ordered monoids.* It was deeply studied in [2] and [8]. For example, its equational theory is decidable [1] but not finitely axiomatizable [5].

Despite all the negative results about RRA, it enjoys an important positive result: it is a *variety* [12]. That is to say, membership in RRA can be characterized by (a possibly infinite) set of equations. It was then natural to ask whether this result holds also for its subreducts.

The positive subreducts[1] of RRA are known to be *quasi-varieties*, i. e., membership can be characterized by (a possibly infinite) set of Horn clauses [2]. However, some of them are *not* varieties [1, Thm.6].

[1] Those subreducts that do not use negation.

U. Fahrenberg et al. (Eds.): RAMiCS 2021, LNCS 13027, pp. 144–158, 2021.
https://doi.org/10.1007/978-3-030-88701-8_9

Despite an attempt in [8], the status of $RRA(\cdot, \cap, 1)$ is not known[2]. In this paper, we show that $RRA(\cdot, \cap, 1)$ is also not a variety.

We use a technique similar to [1]. Actually, we abstract their argument in a more general setting, then we apply it to our particular case. More precisely, we proceed in two steps.

First, we give a necessary and sufficient condition for a quasivariety to be a variety. In words, this condition says that a quasivariety $\mathcal{C}$ is a variety if and only if, for every Horn clause $(\mathcal{H} \Rightarrow t = u)$[3] which is valid in $\mathcal{C}$, its conclusion $(t = u)$ can be "deduced" from its hypothesis $\mathcal{H}$.

This gives us a strategy to show that $RRA(\cdot, \cap, 1)$ is not a variety. It is "enough" to find a Horn clause which is valid in $RRA(\cdot, \cap, 1)$ but whose conclusion cannot be deduced from its hypothesis. Of course, the difficulty here is to guess this Horn clause and to show that indeed its conclusion is not provable from its hypothesis. For this purpose, we rely on graph theoretical tools and intuitions coming from a well established *graph characterization* of the equational theory of $RRA(\cdot, \cap, 1)$ [1].

Outline. In Sect. 2, we define varieties, quasi-varieties and their equational and Horn theories. Then we introduce the quasivariety $RRA(\cdot, \cap, 1)$ and the graph characterization of its equational and Horn theories. We show in Sect. 3 the necessary and sufficient condition for a quasivariety to be a variety. Building on this, we show in Sect. 4 that $RRA(\cdot, \cap, 1)$ is not a variety.

2 Preliminaries

2.1 Algebras, Varieties and Quasi-varieties

Algebras. A *signature* is a pair $\mathcal{S} = (O, ar)$ where O is a set of *operations*, and $ar : O \to \omega$ is a function assigning to each operation an integer called its *arity*. An *algebra over* $\mathcal{S}$ consists of a set D called its *domain*, and for each operation o of $\mathcal{S}$ with arity n, a function $f_o : D^n \to D$.

Equations and Horn Clauses. We fix in the rest of the paper a set X of *variables*. *Terms* over a signature $\mathcal{S} = (O, ar)$ are generated by the following syntax:

$$t := x \mid o(t_1, \ldots, t_n) \qquad x \in X,\ o \in O \text{ and } n = ar(o).$$

We denote the set of terms by $\mathcal{T}_X(\mathcal{S})$, but if the signature and the set of variables are clear from the context, we denote it simply $\mathcal{T}$. An *equation* is a pair of terms

[2] The proof that $RRA(\cdot, \cap, 1)$ is not a variety in [8] relies on the claim that the equational theory of $RRA(\cdot, \cap, 1)$ is finitely axiomatizable [2], which turns out to be wrong, see [9] and [5].

[3] Here, $\mathcal{H}$ is a conjunction of equations called the *hypothesis*, the equation $(t = u)$ is the *conclusion*.

that we usually write $(t = u)$. A *Horn clause* consists of a finite set of equations $\mathcal{H}$ called its *hypothesis* and an equation called its *conclusion*, we usually write it like this $(\mathcal{H} \Rightarrow t = u)$. An equation can be seen as a Horn clause with an empty set of hypothesis.

Truth. Let $\mathcal{A}$ be an algebra over a signature $\mathcal{S}$, and let D be the domain of $\mathcal{A}$. An *interpretation* is a function $\sigma : X \to D$ mapping variables into elements of D. We can extend σ to all terms $\sigma : \mathcal{T} \to D$, by interpreting the operations of $\mathcal{S}$ as the corresponding functions of $\mathcal{A}$.

Let σ be an interpretation as above. An equation $(t = u)$ is *true in $\mathcal{A}$ under σ*, noted

$$\mathcal{A}, \sigma \models (t = u)$$

if $\sigma(t) = \sigma(u)$. A set of equations $\mathcal{H}$ *are true in $\mathcal{A}$ under σ*, noted

$$\mathcal{A}, \sigma \models \mathcal{H}$$

if this is the case for every equation in $\mathcal{H}$. A Horn clause

$$\varphi := (\mathcal{H} \Rightarrow t = u)$$

is *true in $\mathcal{A}$ under σ*, noted

$$\mathcal{A}, \sigma \models \varphi$$

if either $\mathcal{A}, \sigma \not\models \mathcal{H}$ or $\mathcal{A}, \sigma \models (t = u)$. We say that φ is *true in $\mathcal{A}$*, noted $\mathcal{A} \models \varphi$, if φ is true in $\mathcal{A}$ under all possible interpretations.

(Quasi-)Varieties. We have introduced individual algebras, now we focus on classes of algebras. Let $\mathcal{C}$ be a class of algebras over a signature $\mathcal{S}$. We say that an equation or a Horn clause φ is *valid in $\mathcal{C}$*, and write $\mathcal{C} \models \varphi$ if φ is true in every algebra of $\mathcal{C}$. The *equational theory* (resp. *Horn theory*) of $\mathcal{C}$ denoted $\mathcal{E}q(\mathcal{C})$ (resp. $\mathcal{H}orn(\mathcal{C})$) is the set of equations (resp. Horn clauses) which are valid in $\mathcal{C}$.

Let $\mathcal{C}$ be a class of algebras over $\mathcal{S}$ and let E be a set of equations or Horn clauses. We say that E *axiomatizes* $\mathcal{C}$ if for every algebra $\mathcal{A}$ over $\mathcal{S}$:

$$\mathcal{A} \in \mathcal{C} \qquad \text{iff} \qquad \mathcal{A} \models E$$

We say that a class of algebras is a *variety* (resp. *quasivariety*) if it can be axiomatized by a set of equations (resp. Horn clauses).

Remark 1. Note that if $\mathcal{C}$ is a variety (resp. quasivariety), then $\mathcal{C}$ is axiomatized by its equational (resp. Horn) theory.

2.2 Representable Relation Algebras

In this paper, we focus on the signature whose set of operations is $\{\cdot, \cap, 1\}$, the operations $\cdot$ and $\cap$ being of arity 2 and the operation 1 being of arity 0. We will write this signature $(\cdot, \cap, 1)$. To lighten notations of terms over this signature, we often write tu for $t \cdot u$, and assign priorities to operations so that $ab \cap c$ parses as $(a \cdot b) \cap c$.

A *proper relation algebra* is an algebra over $(\cdot, \cap, 1)$ whose universe U is a set of relations, that is $U \subseteq \mathcal{P}(B \times B)$, where B is a base set, the operations $\cdot$ and $\cap$ are respectively the composition and intersection of relations, and 1 is the identity relation over B. A *representable relation algebra* is an algebra over $(\cdot, \cap, 1)$ which is isomorphic to a proper relation algebra; we denote their set by $RRA(\cdot, \cap, 1)$.

The class of algebras $RRA(\cdot, \cap, 1)$ forms a quasivariety [2, p. 2]. The goal of this paper is to show that we cannot say more: $RRA(\cdot, \cap, 1)$ is not a variety.

In $RRA(\cdot, \cap, 1)$, it will be convenient to work with *inequations* instead of equations. An *inequation* is a pair of terms written as $(t \geq u)$, which is a shortcut for the equation $(t \cap u = u)$. Similarly, we will work with Horn clauses which use inequations in their hypothesis and conclusions instead of equations. By definition, every inequation is an equation, and conversely every equation $(t = u)$ is equivalent to the two equations $(t \geq u)$ and $(u \geq t)$. Similarly, every Horn clause is equivalent to a set of Horn clauses using inequations. In the sequel, when dealing with relation algebras, we will mostly work with inequations and Horn clauses using inequations. We call the *inequational (resp. Horn) theory of relations* the set of inequations (resp. Horn clauses using inequations) which are valid in $RRA(\cdot, \cap, 1)$.

2.3 Characterization of the Inequational and Horn Theory of $RRA(\cdot, \cap, 1)$

Graphs. A *2-pointed labeled graph* is a tuple (V, E, ι, o) where V is a set of vertices, $E \subseteq V \times A \times V$ is a set of edges, A is a set of labels and ι and o are two distinguished vertices called the input and output. We simply call them *graphs* in the sequel; we depict them as expected, with unlabeled ingoing and outgoing arrows to denote the input and the output, respectively. We denote by $\mathcal{G}$ the set of all graphs.

We define the operations $\cdot$ and $\cap$ on graphs as follows:

$G \cap H =$ 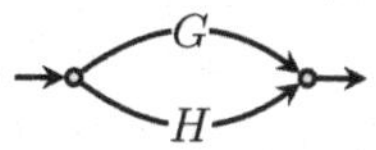

$G \cdot H =$ 

We associate to every term $t \in \mathcal{T}$ a graph $\mathcal{G}(t)$ called the *graph of* t, by letting:

$\mathcal{G}(1) =$ 

$\mathcal{G}(x) =$ 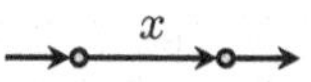

$(x \in X)$

and by interpreting the operations $\cdot$ and $\cap$ on graphs as above.

Example 1. The graphs of the terms $xy \cap xz$ and $xy \cap 1$ are respectively the following:

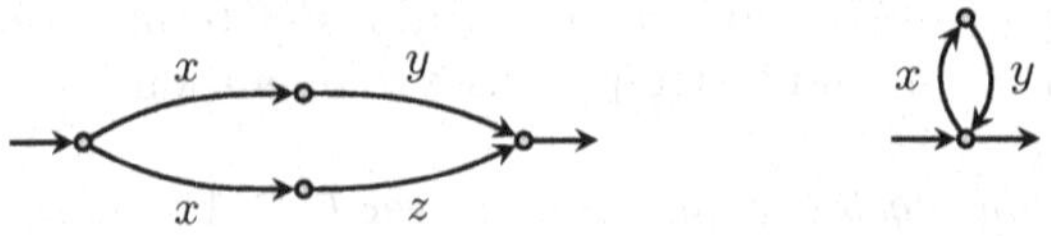

Graph Terms. A *graph term* is a graph which is the image of some term, and we denote by $\mathcal{G}_t$ the set of graph terms. Not every graph is a graph term. For instance, graph terms do not contain the graph K_4, the complete graph with four vertices (see Fig. 1(a)), as a minor[4] [3]. Graph terms also do not contain *back patterns* [5], which we recall below.

Definition 1 (Back pattern). *A* back pattern *in a graph is a pair of distinct nodes m, n together with three directed paths: π from the input to m, κ from n to m, and ρ from n to the output, such that π and κ intersect exactly on m and κ and ρ intersect exactly on n.*

Such a back pattern can be depicted as follows: $\iota \xrightarrow{\pi} m \xleftarrow{\kappa} n \xrightarrow{\rho} o$.

Proposition 1 ([3, Cor. 27], [5, Prop. 12]). *Graph terms do not contain back patterns, nor K_4 as a minor.*

Example 2. The graph of Fig. 1(b) is not a graph term, no matter how we label or orient the edges, because it contains K_4 as a minor. Indeed, if we remove the green edge and contract one of the two blue edges, we obtain K_4. The graph of Fig. 1(c) is not a graph term, no matter how we label its edges, because it contains a back pattern, colored in red.

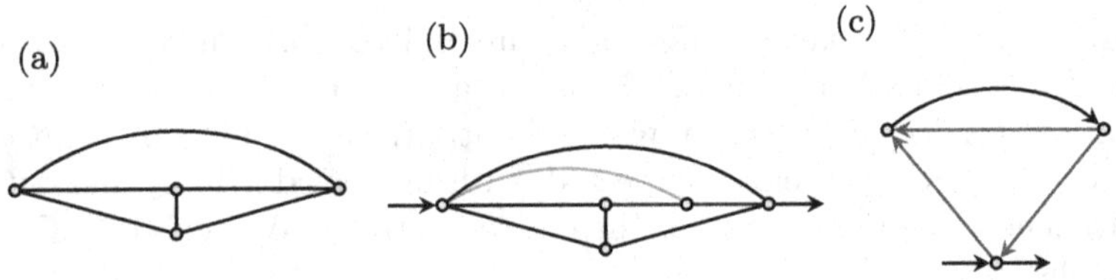

Fig. 1. (a) The graph K_4. (b) A graph containing K_4 as a minor. (c) A graph with a back pattern.

Homomorphisms. Given two graphs $G = \langle V, E, \iota, o\rangle$ and $G' = \langle V', E', \iota', o'\rangle$, a *homomorphism* $h : G \to G'$ is a mapping from $V \to V'$ that preserves labeled edges, i.e. if $(x, a, y) \in E$ then $(h(x), a, h(y)) \in E'$, and preserves input and output, i.e. $h(\iota) = \iota'$ and $h(o) = o'$. We say that the homomorphism h *identifies* the vertices x and y if $h(x) = h(y)$. We write $G \rhd G'$ if there exists a graph homomorphism from G to G'.

[4] A graph G is a minor of a graph H if G can be obtained from H by deleting some edges and vertices and contracting some edges.

Characterizing the Inequational Theory of Relations. The inequational theory of $RRA(\cdot, \cap, 1)$ can be characterized using graphs and homomorphisms as follows.

Theorem 1 (**[1, Thm. 1], [6, p. 208]**). *For all terms t, u,*

$$RRA(\cdot, \cap, 1) \models (t \geq u) \qquad \textit{iff} \qquad \mathcal{G}(t) \rhd \mathcal{G}(u)$$

Example 3. The validity of the inequation $xy \cap xz \geq x(y \cap z)$ is witnessed by the following homomorphism:

Characterizing the Horn Theory of Relations. A *graph context* is a graph with a distinguished edge labeled by a special letter $\bullet$, called its *hole*. If G is a graph and C a graph context, then $C[G]$ is the graph obtained by "plugging G in the hole" of C, that is, $C[G]$ is the graph obtained as the disjoint union of G and C, where we identify the input (resp. output) of G with the input (resp. output) of the edge labeled by $\bullet$ in C, and we remove the edge of C labeled $\bullet$. Here is an example:

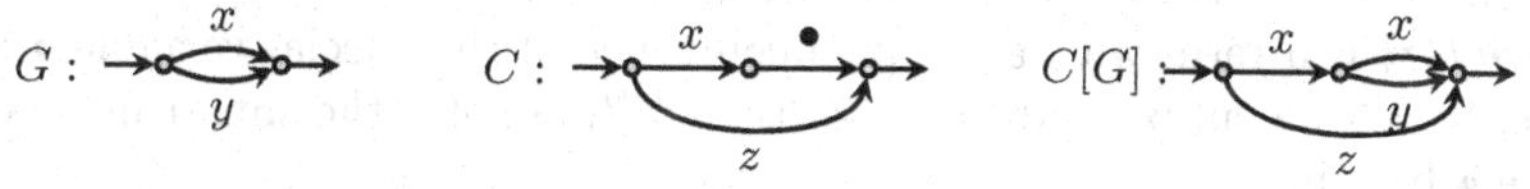

Definition 2 (The relation $\rhd_{\mathcal{H}}$). *Let $\mathcal{H}$ be a set of inequations. We define the relation $>_{\mathcal{H}}$ on graphs as follows. We set $G >_{\mathcal{H}} H$ if there exists a graph context C and an inequation $(t \geq u) \in \mathcal{H}$ such that:*

$$G = C[\mathcal{G}(t)] \quad \textit{and} \quad H = C[\mathcal{G}(u)]$$

We define $\rhd_{\mathcal{H}}$ as the transitive closure of $\rhd \cup >_{\mathcal{H}}$.

In the definition above, the graphs G, H and C are not necessarily graph terms.

Theorem 2 (**[4, Thm. 12]**). *For all terms t, u and set of inequations $\mathcal{H}$, we have:*

$$RRA(\cdot, \cap, 1) \models (\mathcal{H} \Rightarrow t \geq u) \qquad \textit{iff} \qquad \mathcal{G}(t) \rhd_{\mathcal{H}} \mathcal{G}(u)$$

Hence, in order to show that a Horn clause ($\mathcal{H} \Rightarrow t \geq u$) is valid, we need to find a sequence of graphs $G_0, \ldots, G_n$ such that $G_0 = \mathcal{G}(t)$, $G_n = \mathcal{G}(u)$ and for every $i \in [0, n-1]$ the graphs G_i and G_{i+1} are either related by homomorphism or by the relation $>_{\mathcal{H}}$. We say that this sequence *witnesses* the validity of this Horn sentence.

Example 4. The validity of the following Horn clause

$$xy \geq x \ \Rightarrow \ xyy \cap xz \geq x(y \cap z)$$

is witnessed by the following sequence:

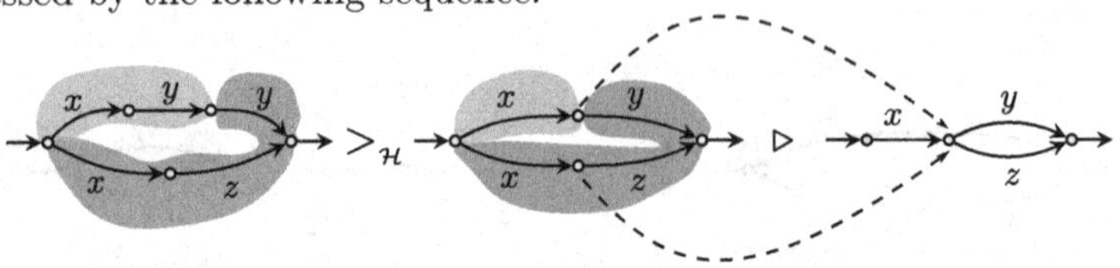

The picture should be read from left to right. First, we identified in the graph of $xyy \cap xz$ a context (in red) and the graph of xy (in green). As allowed by the hypothesis $xy \geq x$, we replaced the graph of xy by that of x, this is a $>_{\mathcal{H}}$ step. Finally we apply a homomorphism to get the graph of $x(y \cap z)$.

3 When Is a Quasivariety a Variety?

In the rest of this section we fix a signature $\mathcal{S}$ and a set of variables X. Algebras will be over $\mathcal{S}$ and terms over $\mathcal{S}$ and X. We omit the mention of the signature as it is clear from the context. In the following we give a necessary and sufficient condition for a quasivariety to be a variety.

A *context* is a term with a unique occurrence of the special variable $\bullet$ called its hole. If C is a context and t a term, then $C[t]$ denotes the substitution of the variable $\bullet$ by t in C.

Definition 3. *Let $\mathcal{C}$ be a class of algebras and $\mathcal{H}$ be a set of equations. We define the relation $\underset{\mathcal{H}}{=}$ on terms as follows. We set $(t \underset{\mathcal{H}}{=} u)$ if there exists a term context C and an equation $(t' = u') \in \mathcal{H}$ such that:*

$$t = C[t'] \quad \textit{and} \quad u = C[u']$$

We define the relation $\underset{\mathcal{H},\mathcal{C}}{=}$ as the transitive closure of the relation $(\mathcal{E}q(\mathcal{C}) \cup \underset{\mathcal{H}}{=})$ i. e., the union of the equational theory of $\mathcal{C}$ and the relation $\underset{\mathcal{H}}{=}$.

If $(t \underset{\mathcal{H},\mathcal{C}}{=} u)$ we say that the equation $(t = u)$ is a consequence of the hypothesis $\mathcal{H}$ in the algebras $\mathcal{C}$.

In words, we have $(t \underset{\mathcal{H},\mathcal{C}}{=} u)$ if there exists a sequence of terms $t_0, \ldots, t_n$ such that $t_0 = t$, $t_n = u$ and for every $i \in [0, n-1]$ the equation $t_i = t_{i+1}$ is either valid in $\mathcal{C}$ or is obtained as the application of a context to an equation of $\mathcal{H}$. We call such sequence a *witness* of $(t \underset{\mathcal{H},\mathcal{C}}{=} u)$.

Remark 2. An alternative definition for $\underset{\mathcal{H},\mathcal{C}}{=}$ would have been to use a proof system which uses the equations of $\mathcal{H}$ and $\mathcal{E}q(\mathcal{C})$ as axioms.

A *congruence on terms* is an equivalence relation on terms that is stable under contexts.

Lemma 1. *The relation* $\underset{\mathcal{H},\mathcal{C}}{=}$ *is a congruence on terms.*

Proof. The relation $\underset{\mathcal{H},\mathcal{C}}{=}$ is clearly an equivalence relation. Let us show that it is stable under contexts. We proceed by induction on the length of a witness sequence for $(t \underset{\mathcal{H},\mathcal{C}}{=} u)$. If $(t, u) \in \mathcal{E}q(\mathcal{C})$ then we have also $(C[t], C[u]) \in \mathcal{E}q(\mathcal{C})$ because $\mathcal{E}q(\mathcal{C})$ is a congruence. If $(t \underset{\mathcal{H}}{=} u)$ then $\left(C[t] \underset{\mathcal{H}}{=} C[u]\right)$, because the composition of two contexts is also a context. The inductive step is immediate.

The following theorem says that a quasivariety $\mathcal{C}$ is a variety if and only if the conclusion of every valid Horn clause is a consequence of its hypothesis in $\mathcal{C}$.

Theorem 3. *A quasivariety* $\mathcal{C}$ *is a variety if and only if the following holds*

$$(\mathcal{H} \Rightarrow t = u) \in \mathcal{H}orn(\mathcal{C}) \qquad \Rightarrow \qquad t \underset{\mathcal{H},\mathcal{C}}{=} u \qquad\qquad (\dagger)$$

Remark 3. Note that the reverse implication of (†) is true, regardless of whether $\mathcal{C}$ is a variety or not.

Proof. ($\Rightarrow$) Suppose that $\mathcal{C}$ is a variety and let $(\mathcal{H} \Rightarrow t = u)$ be a Horn clause valid in $\mathcal{C}$. Let $\mathcal{T}_{\mathcal{H},\mathcal{C}}$ be the algebra of terms quotiented by the congruence $\underset{\mathcal{H},\mathcal{C}}{=}$. The algebra $\mathcal{T}_{\mathcal{H},\mathcal{C}}$ satisfies all the equations in $\mathcal{E}q(\mathcal{C})$ because $\underset{\mathcal{H},\mathcal{C}}{=}$ contains $\mathcal{E}q(\mathcal{C})$. Since $\mathcal{C}$ is a variety, we have that $\mathcal{T}_{\mathcal{H},\mathcal{C}} \in \mathcal{C}$.

Consider the interpretation σ which assigns to every letter its equivalence class w.r.t. the relation $\underset{\mathcal{H},\mathcal{C}}{=}$. We have that

$$\mathcal{T}_{\mathcal{H},\mathcal{C}}, \sigma \models \mathcal{H}$$

because the relation $\underset{\mathcal{H},\mathcal{C}}{=}$ contains $\mathcal{H}$. Since $(\mathcal{H} \Rightarrow t = u) \in \mathcal{H}orn(\mathcal{C})$, we have that

$$\mathcal{T}_{\mathcal{H},\mathcal{C}}, \sigma \models (t = u)$$

which is the same thing as $(t \underset{\mathcal{H},\mathcal{C}}{=} u)$.

($\Leftarrow$) Suppose that (†) holds and let us show that $\mathcal{C}$ is a variety. Let $\mathcal{A}$ be an algebra satisfying all the equations in $\mathcal{E}q(\mathcal{C})$, we show that $\mathcal{A}$ is an algebra in $\mathcal{C}$. As $\mathcal{C}$ is a quasivariety, it is enough to show that $\mathcal{A}$ satisfies all Horn clauses valid in $\mathcal{C}$. Let $(\mathcal{H} \Rightarrow t = u)$ be such Horn clause and let σ be an interpretation such that $\mathcal{A}, \sigma \models \mathcal{H}$. By (†), we have that $(t \underset{\mathcal{H},\mathcal{C}}{=} u)$. We can show by a simple induction on the length of a sequence justifying $(t \underset{\mathcal{H},\mathcal{C}}{=} u)$ that we have $\mathcal{A}, \sigma \models (t = u)$. This concludes the proof.

4 The Quasivariety $RRA(\cdot, \cap, 1)$ Is Not a Variety

Let us first specify Theorem 3 for the quasivariety $RRA(\cdot, \cap, 1)$. For that, we define below the relation $\blacktriangleright_{\mathcal{H}}$ on graph terms. Recall that graph terms are those graphs coming from terms, and that we denote their set by $\mathcal{G}_t$.

Definition 4 (The relation $\blacktriangleright_{\mathcal{H}}$). *Let $\mathcal{H}$ be a set of inequations. We define the relation $\succ_{\mathcal{H}}$ on graph terms as follows. For $G, H \in \mathcal{G}_t$, we set $G \succ_{\mathcal{H}} H$ if there exists a context $C \in \mathcal{G}_t$ and an inequation $(t \geq u) \in \mathcal{H}$ such that:*

$$G = C[\mathcal{G}(t)] \quad \text{and} \quad H = C[\mathcal{G}(u)]$$

We define the relation $\blacktriangleright_{\mathcal{H}} \subseteq \mathcal{G}_t \times \mathcal{G}_t$ as the transitive closure of the relation $\rhd \cup \succ_{\mathcal{H}}$.

In words, $G \blacktriangleright_{\mathcal{H}} H$ if there exists a sequence of graph terms $G_0, \ldots, G_n$ such that $G_0 = G$, $G_n = H$ and for every $i \in [0, n-1]$ the graphs G_i and G_{i+1} are either related by homomorphism or by the relation $\succ_{\mathcal{H}}$. We say that this sequence is a witness for $G \blacktriangleright_{\mathcal{H}} H$. The relation $\blacktriangleright_{\mathcal{H}}$ should not be confused with the relation $\rhd_{\mathcal{H}}$. Indeed, for the latter, the graphs of a witness sequence may not be graph terms.

Using Theorem 1, we can adapt Theorem 3 to get the following version for relation algebras.

Theorem 4. *If $RRA(\cdot, \cap, 1)$ is a variety then the following holds*

$$RRA(\cdot, \cap, 1) \models (\mathcal{H} \Rightarrow t \geq u) \qquad \Rightarrow \qquad \mathcal{G}(t) \blacktriangleright_{\mathcal{H}} \mathcal{G}(u)$$

Proof. Suppose that $RRA(\cdot, \cap, 1)$ is a variety and let $(\mathcal{H} \Rightarrow t \geq u)$ be a Horn clause valid in $RRA(\cdot, \cap, 1)$. To simplify notations suppose that $\mathcal{H} = \{v \geq w\}$. The general case can be treated similarly. Note that the Horn clause above is a shortcut for

$$(v \cap w = w) \Rightarrow (t \cap u = u)$$

By Theorem 3, there is a sequence of terms $t_0, \ldots, t_n$ such that $t_0 = (t \cap u), t_n = u$ and for every $i \in [0, n-1]$ the equation $(t_i = t_{i+1})$ is either valid in $RRA(\cdot, \cap, 1)$ or is obtained as an application of the hypothesis $(v \cap w = w)$ under some context.

Let us show that $\mathcal{G}(t) \blacktriangleright_{\mathcal{H}} \mathcal{G}(u)$. We have that $\mathcal{G}(t) \rhd \mathcal{G}(t \cap u)$, so we only need to show that $\mathcal{G}(t \cap u) \blacktriangleright_{\mathcal{H}} \mathcal{G}(u)$. For that, we exploit the sequence above. Note that if $(t_i = t_{i+1})$ is valid in $RRA(\cdot, \cap, 1)$, then $\mathcal{G}(t_i) \rhd \mathcal{G}(t_{i+1})$ by Theorem 1. If $(t_i = t_{i+1})$ is obtained as an application of the hypothesis $(v \cap w = w)$ under some context C, we can distinguish two cases.

- Either $t_i = C[w]$ and $t_{i+1} = C[v \cap w]$. In this case we have $\mathcal{G}(t_i) \rhd \mathcal{G}(t_{i+1})$ because $\mathcal{G}(w) \rhd \mathcal{G}(v \cap w)$.
- Or $t_i = C[v \cap w]$ and $t_{i+1} = C[w]$. In this case we show that $\mathcal{G}(t_i) \blacktriangleright_{\mathcal{H}} \mathcal{G}(t_{i+1})$. Indeed, Let C' be the graph context $\mathcal{G}(C[\bullet \cap w])$. We have that

$$\mathcal{G}(t_i) = C'[\mathcal{G}(v)] >_{\mathcal{H}} C'[\mathcal{G}(w)] = \mathcal{G}(C[w \cap w]) \rhd \mathcal{G}(C[w]) = \mathcal{G}(t_{i+1})$$

This ends the proof.

This gives us a strategy to show that $RRA(\cdot, \cap, 1)$ is not a variety: it is enough to find a Horn clause $(\mathcal{H} \Rightarrow t \geq u)$ which is valid for relations, but for which $\mathcal{G}(t) \not\blacktriangleright_{\mathcal{H}} \mathcal{G}(u)$. We explicit such counter-example below.

Definition 5 (The counter-example). *Let* $\mathbb{X} = \{x, y, z, a, b, c\}$ *be a set of variables. We define the terms* η *and* ρ *and the set of equations* $\mathcal{K}$ *over* $\mathbb{X}$ *as follows:*

$$\eta := xy(z \cap a(bya \cap 1)bc(dxc \cap 1)d) \cap 1 \qquad \rho := xyz \cap 1 \qquad \mathcal{K} := \{ab \geq zx,\ cd \geq yz\}$$

We show that $(\mathcal{K} \Rightarrow \eta \geq \rho)$ is indeed a counter-example. This is Proposition 2 below.

Proposition 2. *We have that:*

$$RRA(\cdot, \cap, 1) \models (\mathcal{K} \Rightarrow \eta \geq \rho) \qquad \textit{but} \qquad \mathcal{G}(\eta) \not\blacktriangleright_{\mathcal{K}} \mathcal{G}(\rho).$$

Corollary 1. *The quasivariety* $RRA(\cdot, \cap, 1)$ *is not a variety.*

Proof (of the first part of Proposition 2). To prove that $(\mathcal{K} \Rightarrow \eta \geq \rho)$ is a valid Horn clause, we can either proceed by a direct but boring proof, or use Theorem 2. We choose the second option, and show that $\mathcal{G}(\eta) \rhd_{\mathcal{K}} \mathcal{G}(\rho)$. In Fig. 2, we explicit a sequence witnessing that $\mathcal{G}(\eta) \rhd_{\mathcal{K}} \mathcal{G}(\rho)$. It should be read from top left, then

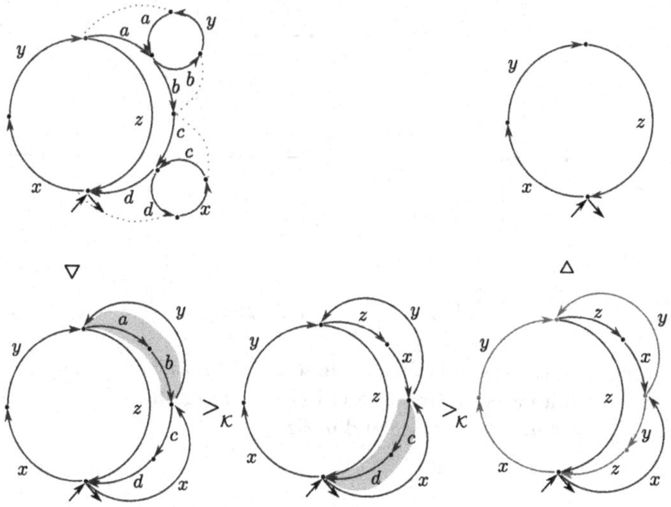

Fig. 2. A sequence witnessing that $\mathcal{G}(\eta) \rhd_{\mathcal{K}} \mathcal{G}(\rho)$.

down, then top right: we start by applying a homomorphism to the graph $\mathcal{G}(\eta)$; the nodes which are identified by the homomorphism are linked by the dotted lines. Then we apply the relation $>_{\mathcal{K}}$: the graph of ab is colored in green, we replace it by zx as allowed by the set $\mathcal{K}$. We apply again $>_{\mathcal{K}}$, this time by replacing the graph of cd by the graph of yz. We finally apply a homomorphism to get $\mathcal{G}(\rho)$.

Note that the intermediary graphs of the witness sequence of Fig. 2 are not graph terms. For example, the second graph from the right contains a back pattern (Definition 1), colored in red.

Let us now explicit our strategy to prove that $\mathcal{G}(\eta) \not\blacktriangleright_{\mathcal{K}} \mathcal{G}(\rho)$. First, let us remark that $\mathcal{G}(\eta) \not\triangleright \mathcal{G}(\rho)$. Indeed, no homomorphism can preserve the edges labeled a, b, c and d. We will show that under some constraints on G and H, we have the following result:

$$\text{If} \quad G \blacktriangleright_{\mathcal{K}} H \quad \text{and} \quad \mathcal{G}(\eta) \triangleright G \qquad \text{then} \qquad \mathcal{G}(\eta) \triangleright H \qquad \text{(Lem. 4)}$$

By taking G and H to be respectively $\mathcal{G}(\eta)$ and $\mathcal{G}(\rho)$ in this result, and using the remark above, we can show by contradiction that $\mathcal{G}(\eta) \not\blacktriangleright_{\mathcal{K}} \mathcal{G}(\rho)$.

As said before, Lemma 4 is true under some constraints on the graphs G and H. More precisely, it is true when they do not contain some specific graphs called *persistent patterns* as sub-graphs. In the following we define these persistent patterns and show Lemma 4.

Definition 6 (Persistent patterns). *Persistent patterns are the graphs of Fig. 3.*

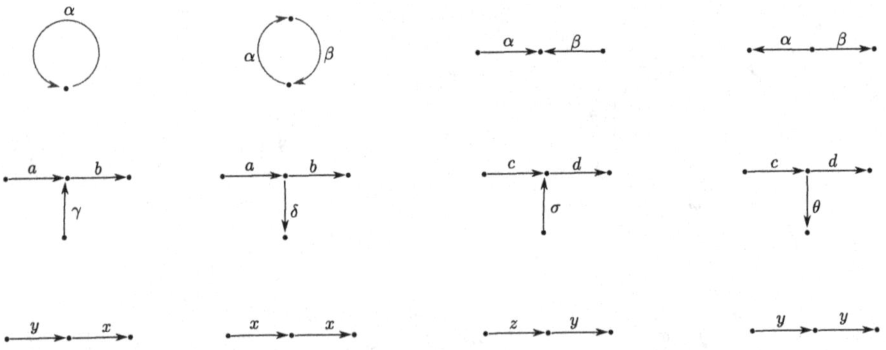

Fig. 3. Persistent patterns. The vertices of these graphs may not be distinct. All labels belong to $\mathbb{X}$, the variables used in the counter example (Definition 5) with the following constraints: $\alpha \neq \beta$, $\gamma \neq a$, $\delta \neq b$, $\sigma \neq c$ and $\theta \neq d$.

Persistent patterns are called so because they satisfy the following property, whose proof is a simple case distinction.

Lemma 2. *If $G \blacktriangleright_{\mathcal{K}} H$ and G contains a persistent pattern, then H also contains a persistent pattern.*

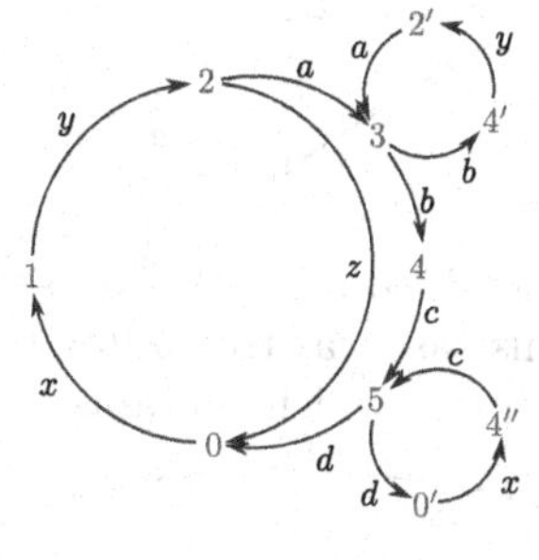

Fig. 4. Labeling the vertices of $\mathcal{G}(\eta)$.

For convenience, we label the nodes of $\mathcal{G}(\eta)$ as in Fig. 4. We say that a vertex of $\mathcal{G}(\eta)$ is *tagged by an integer i* if its label contains i. For instance, the vertex labeled by $2'$ is tagged by the integer 2. To prove Lemma 4, we need the following result, which says that a homomorphism from $\mathcal{G}(\eta)$ to any graph term not containing a persistent pattern cannot identify vertices tagged by distinct integers.

Lemma 3. *Let G be a graph term not containing a persistent pattern, and let $h : \mathcal{G}(\eta) \to G$ be a homomorphism from $\mathcal{G}(\eta)$ to G. The homomorphism h cannot identify two vertices tagged by distinct integers.*

Note however that h may identify two vertices tagged by the same integer, for instance the vertices labeled by 2 and $2'$.

Proof (of Lemma 3). We show that if h identifies two nodes tagged by distinct integers, then G contains necessarily a persistent pattern, a back pattern or K_4 as a minor. Figure 5 shows the persistent patterns that appear if we identify nodes with distinct tags. The gray cells are symmetric and the white cells correspond to vertices tagged by the same integer. To complete the proof, we need to prove that the vertex tagged by 1 cannot be identified by a vertex tagged by 4, these cases correspond to the cells $(\star)$, $(\dagger)$ and $(\ddagger)$. We show that if we do such identification, we create a back pattern or K_4 as a minor.

$(\star)$ If we identify the vertices labeled 1 and 4 we create the following back pattern, where the vertex labeled $(1, 4)$ is the image of 1 and 4 by h. It is a back pattern because its vertices are pairwise distinct. Indeed, if its vertices were not pairwise distinct, then we would create one of the patterns treated in Fig. 5.

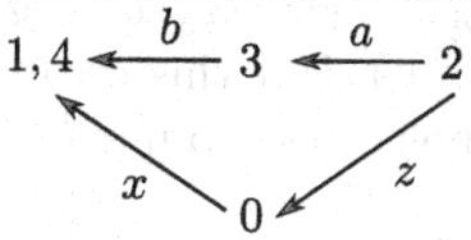

(†) If we identify the vertices labeled 1 and $4'$ we create the same back pattern.

(‡) If we identify the vertices labeled 1 and $4''$ we distinguish two cases. Either the vertices 0 and $0'$ are not identified by h, and in this case we create this back pattern:

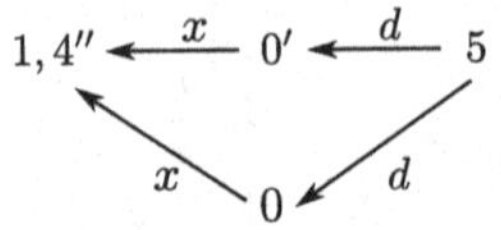

Otherwise, 0 and $0'$ are identified by h. We can suppose that 1 and 4 are not identified by h (otherwise, this case was treated by $(\star)$). In this case, we obtain the following graph as a subgraph, which contains the graph K_4 as a minor:

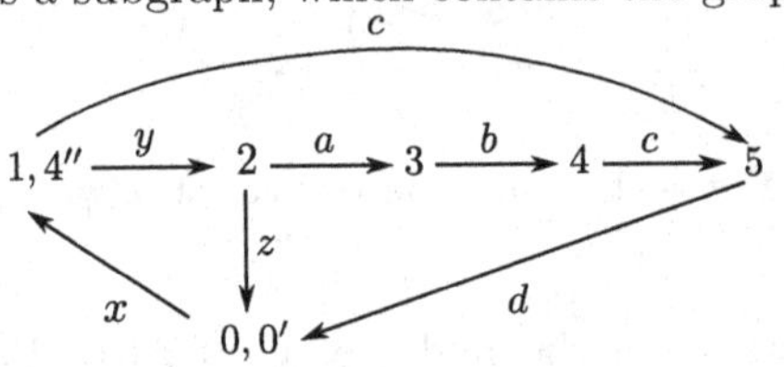

Lemma 4. *Let G, H be two graphs and suppose that H does not contain a persistent pattern. We have that:*

$$\text{If} \quad G \blacktriangleright_{\mathcal{K}} H \quad \text{and} \quad \mathcal{G}(\eta) \rhd G \qquad \text{then} \qquad \mathcal{G}(\eta) \rhd H$$

Proof. We proceed by induction on length of the sequence witnessing that $G \blacktriangleright_{\mathcal{K}} H$. Note that since H does not contain a persistent pattern, all the graphs of the witness sequence do not contain a persistent pattern as well thanks to Lemma 2.

The inductive step is easy, the interesting part are the base cases. The first one is when $G \rhd H$, and we have clearly that $\mathcal{G}(\eta) \rhd G$ implies that $\mathcal{G}(\eta) \rhd H$ because we can compose these two homomorphisms. Now suppose that $G \succ_{\mathcal{K}} H$. There are two cases to consider: we have either used the inequation $ab \geq zx$ or $cd \geq yz$ to justify $G \succ_{\mathcal{K}} H$. Suppose that we are in the first case, i. e., there exists a context $C \in \mathcal{G}_t$ such that:

$$G = C[\mathcal{G}(ab)] \qquad \text{and} \qquad H = C[\mathcal{G}(zx)]$$

Let h be a homomorphism from $\mathcal{G}(\eta)$ to G. Our goal is to show that the image of h lies in C. If we do so, we can easily prove that $\mathcal{G}(\eta) \rhd H$ because h can also be used to map $\mathcal{G}(\eta)$ to H.

Let m be the inner vertex of the graph $\mathcal{G}(ab)$, that is the vertex distinct from the input and output. We show that no vertex of $\mathcal{G}(\eta)$ can be mapped by h to m. Suppose for contradiction that there exits a vertex of $\mathcal{G}(\eta)$ whose image is m. This vertex is necessarily the vertex 3 (see Fig. 4), and we have necessarily that $h(2) = h(2')$ and $h(4) = h(4')$. But this creates the following back-pattern in G, where as usual a node labeled by two integers is the common image of the corresponding vertices of $\mathcal{G}(\eta)$:

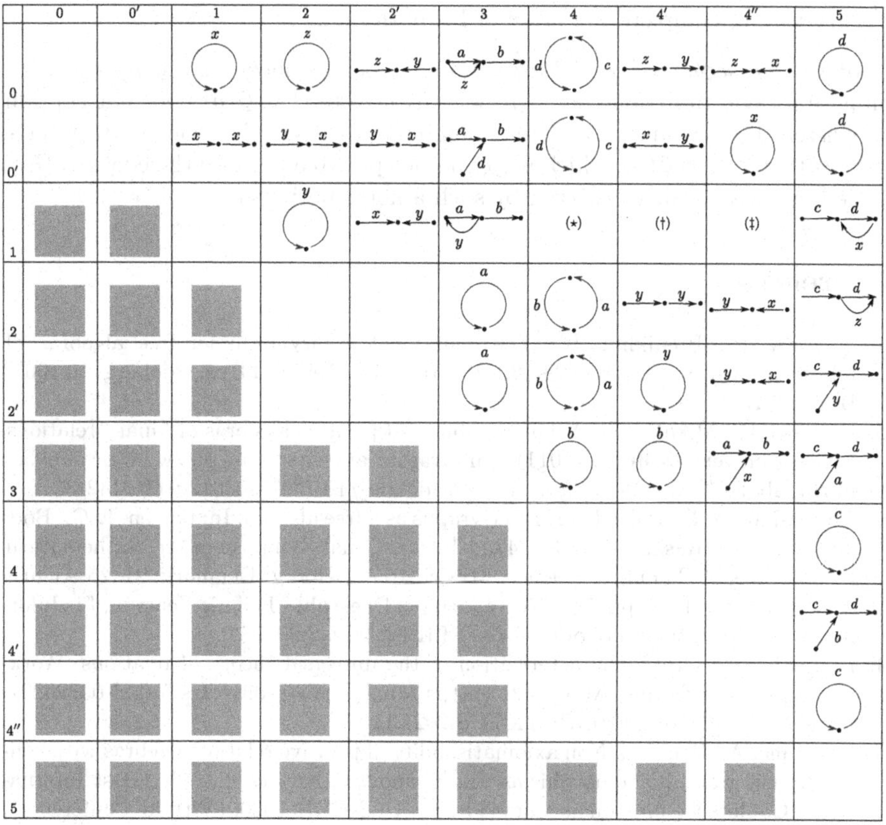

Fig. 5. Identifying two vertices tagged by distinct integers creates persistent patterns.

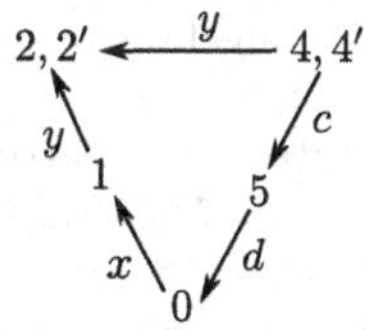

This is indeed a back pattern because its vertices are pairwise distinct thanks to Lemma 3.

We treat the case where the hypothesis $cd \geq yz$ was used to derive $G \succ_{\mathcal{K}} H$ in a similar way. In this case, the following back pattern appears:

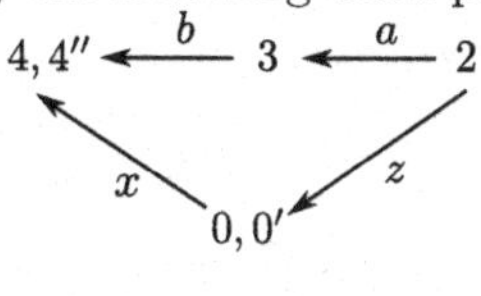

This concludes the proof of this lemma.

Now we can complete the proof of Proposition 2.

Proof (of the second part of Proposition 2). Let us show that $\mathcal{G}(\eta) \not\blacktriangleright_{\mathcal{K}} \mathcal{G}(\rho)$. Suppose for contradiction that $\mathcal{G}(\eta) \blacktriangleright_{\mathcal{K}} \mathcal{G}(\rho)$. Note that $\mathcal{G}(\eta)$ does not contain a persistent pattern and that $\mathcal{G}(\eta)$ is homomorphic to itself, hence by Lemma 4, we have that $\mathcal{G}(\eta) \triangleright \mathcal{G}(\rho)$, which is clearly not possible because the edges of $\mathcal{G}(\eta)$ labeled by a cannot be preserved by such a homomorphism.

References

1. Andréka, H., Bredikhin, D.: The equational theory of union-free algebras of relations. Algebra Universalis **33**(4), 516–532 (1995). https://doi.org/10.1007/BF01225472
2. Andréka, H., Mikulás, S.: Axiomatizability of positive algebras of binary relations. Algebra universalis **66**(7) (2011). An erratum appears at http://www.dcs.bbk.ac.uk/~szabolcs/AM-AU-err6.pdf. https://doi.org/10.1007/s00012-011-0142-3
3. Cosme-Llópez, E., Pous, D.: K4-free graphs as a free algebra. In: Larsen, K.G., Bodlaender, H.L., Raskin, J. (eds.) 42nd International Symposium on Mathematical Foundations of Computer Science, MFCS 2017, Aalborg, Denmark, 21–25 August 2017. LIPIcs, vol. 83, pp. 76:1–76:14. Schloss Dagstuhl - Leibniz-Zentrum für Informatik (2017). https://doi.org/10.4230/LIPIcs.MFCS.2017.76
4. Doumane, A.: Graph characterization of the universal theory of relations. Working paper or preprint, May 2021. https://hal.archives-ouvertes.fr/hal-03226221, https://doi.org/10.4230/LIPIcs.MFCS.2021.11
5. Doumane, A., Pous, D.: Non axiomatisability of positive relation algebras with constants, via graph homomorphisms. In: Konnov, I., Kovács, L. (eds.) 31st International Conference on Concurrency Theory, CONCUR 2020, (Virtual Conference), Vienna, Austria, 1–4 September 2020. LIPIcs, vol. 171, pp. 29:1–29:16. Schloss Dagstuhl - Leibniz-Zentrum für Informatik (2020). https://doi.org/10.4230/LIPIcs.CONCUR.2020.29
6. Freyd, P., Scedrov, A.: Categories, Allegories (1990)
7. Hirsch, R., Hodkinson, I.: Relation algebras by games. Bull. Symb. Log. **9**(4), 515–520 (2003)
8. Hirsch, R., Mikulas, S.: Representable semilattice-ordered monoids. Algebra universalis (2007). https://doi.org/10.1007/s00012-007-2055-8
9. Mikulás, S.: http://www.dcs.bbk.ac.uk/~szabolcs/AM-AU-err6.pdf
10. Monk, D.: On representable relation algebras. Michigan Math. J. **31**(3), 207–210 (1964). https://doi.org/10.1307/mmj/1028999131
11. Tarski, A.: On the calculus of relations. J. Symb. Log. **6**(3), 73–89 (1941)
12. Tarski, A.: Contributions to the theory of models. III (1954)

Accretive Computation of Global Transformations

Alexandre Fernandez(✉), Luidnel Maignan(✉), and Antoine Spicher(✉)

Univ Paris Est Creteil, LACL, 94000 Creteil, France
{alexandre.fernandez,luidnel.maignan,antoine.spicher}@u-pec.fr

Abstract. Global transformations form a categorical framework adapting graph transformations to describe fully synchronous rule systems on a given data structure. In this work we focus on data structures that can be captured as presheaves and study the computational aspects of such synchronous rule systems. To obtain an online algorithm, a complete study of the sub-steps within each synchronous step is done at the semantic level. This leads to the definition of accretive rule systems and a local criterion to characterize these systems. Finally an online computation algorithm for theses systems is given.

Keywords: Global transformation · Synchronous rule application · Rewriting system · Online algorithm · Category theory

1 Introduction

Classically, a graph rewriting system consists of a set of rewriting rules $l \Rightarrow r$ expressing that l should be replaced by r somewhere in an input graph. Usually rules are applied one after the other in a non-deterministic way [3,4,6]. Allowing multiple rules to be applied simultaneously has been the subject of multiple studies, leading to the concepts of parallel rule applications, concurrent rule applications [7], and amalgamation of rules [2]. For instance, amalgamation of rules is considered when two rules $l \Rightarrow r$ and $l' \Rightarrow r'$ are applicable but l and l' overlap. Basically, the behavior on the overlap is given by a third rule specifying how r and r' should consequently overlap. But some systems do not only require the amalgamation of a few, finite, number of rule applications, but the amalgamation of an unbounded number, the whole input being transformed. A simple example is triangular mesh refinement where the triangles of a mesh are all subdivided into many smaller triangles simultaneously, with a coherent behavior on the overlap between triangles [14]. In this extreme case, the notion of replacement is not appropriate, no part of the initial mesh is really kept identical.

Rethinking rewriting for those particular systems where the transformation is globally coherent leads to a generic and economical mathematical structure captured easily with categorical concepts, the so-called *global transformations* [14]. This point of view has been applied mathematically to examples like mesh refinements on abstract cell complexes [14], but also deterministic Lindenmayer systems acting on formal words [8], and cellular automata acting on labeled Caley

U. Fahrenberg et al. (Eds.): RAMiCS 2021, LNCS 13027, pp. 159–175, 2021.
https://doi.org/10.1007/978-3-030-88701-8_10

graphs [10]. In the present work, we tackle global transformations in an algorithmic perspective and show how they can be computed in an online fashion when transforming graphs, but also any generalization of graphs suitably captured by categories of presheaves (labeled graphs, higher-dimensional graphs, etc.). This online strategy saves memory during the computation, more memory being also saved through a condition allowing the modifications to happen *in place*: accretiveness.

The article is organized as follows. After adapting in Sect. 2 the definition of global transformations to presheaves, Sect. 3 unfolds all implications of the online and accretive perspective at the semantic level, and gathers all necessary formal results. This leads to the presentation of the algorithm in Sect. 4, followed by a discussion in Sect. 5. In the present version, facts are only stated. An extended version with all the proofs can be found at [9].

2 Background on Global Transformations

In the section, we adapt the definitions of global transformations given in [8,14] to fit with the context of presheaves and monomorphisms between them. The reader is assumed to be familiar with the definitions of categories, functors, monomorphisms, comma categories, diagrams, cocones, colimits and categories of presheaves. Refer to [13] for details. These constructions are also pedagogically introduced in the context of global transformation in [8].

In the following, we consider an arbitrary category $\mathbf{C}$ and write $\hat{\mathbf{C}}$ for the category $\mathbf{Set}^{\mathbf{C}^{\mathrm{op}}}$ of all presheaves on $\mathbf{C}$, $\hat{\mathbf{C}}_{\mathcal{M}}$ for the subcategory restricting morphisms to monomorphisms, and $\mathrm{U} : \hat{\mathbf{C}}_{\mathcal{M}} \to \hat{\mathbf{C}}$ for the obvious forgetful functor. Morphisms of $\hat{\mathbf{C}}_{\mathcal{M}}$ and monomorphisms of $\hat{\mathbf{C}}$ are written $p \hookrightarrow p'$. We write $\mathbf{y} : \mathbf{C} \to \hat{\mathbf{C}}$ for the Yoneda embedding, and call *representable presheaves* the image $\mathbf{y}c$ of any $c \in \mathbf{C}$. The category $\hat{\mathbf{C}}$ is cocomplete and for any diagram $D : \mathbf{I} \to \hat{\mathbf{C}}$, the colimit C of D is directly written $\mathrm{Colim}(D)$; C also abusively designates the apex and $C_i : D(i) \to C$ the cocone components for any $i \in \mathbf{I}$.

The examples are spelled out with $\mathbf{C}$ set to the category with two objects $\mathtt{v}$ and $\mathtt{e}$, and two morphisms $\mathtt{s}, \mathtt{t} : \mathtt{v} \to \mathtt{e}$. A presheaf $p \in \hat{\mathbf{C}}$ is then a directed multigraph with self-loops: $p(\mathtt{v})$ and $p(\mathtt{e})$ are respectively the sets of vertices and edges composing the graph, and $p(\mathtt{s})$ (resp. $p(\mathtt{t})$) is a function mapping each edge to its source (resp. target). The representable presheaves are the graph $\mathbf{y}\mathtt{v}$ with a single vertex and the graph $\mathbf{y}\mathtt{e}$ with two vertices and a single edge. We will make use of the following particular graphs: d_k the discrete graph with k vertices and no edge, p_k the path of length k, and c_k the cycle of length k, $k > 0$.

Given two categories A and B, a functor $F : A \to B$, and an object b in B, the comma category F/b sees its objects described as pairs $\langle a \in A, f : F(a) \to b \rangle$ and its morphisms from $\langle a, f \rangle$ to $\langle a', f' \rangle$ as pairs $\langle e : a \to a', f' \rangle$ such that $f = f' \circ F(e)$. The composition of $\langle e', f'' \rangle \circ \langle e, f' \rangle$ is therefore $\langle e' \circ e, f'' \rangle$.

Specification of Global Transformations. In this paper, we restrict ourselves to global transformations acting on $\hat{\mathbf{C}}_{\mathcal{M}}$. At a basic level, they are rewriting systems transforming presheaves into presheaves. As such, their specification is based on

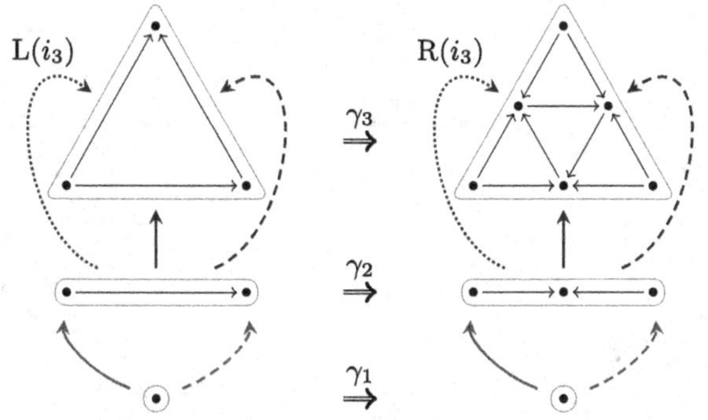

Fig. 1. Sierpinski rule system: one rule to divide the relevant triangles, the two others and their inclusions to specify the connections in the output based on the connections in the input. (Color figure online)

a set of rules. Each rule γ is a pair written $l \Rightarrow r$ with $l, r \in \hat{\mathbf{C}}_{\mathcal{M}}$. Given an input presheaf p, it expresses that any occurrence of the left hand side (l.h.s.) l in p produces the corresponding right hand side (r.h.s.) r in the associated output. The main feature of global transformations is to endow this set of rules with a structure of *category* where morphisms describe *inclusions of rules.* A rule inclusion $i : \gamma_1 \to \gamma_2$ from a *sub-rule* $\gamma_1 = l_1 \Rightarrow r_1$, to a *super-rule* $\gamma_2 = l_2 \Rightarrow r_2$ expresses how an occurrence of l_1 in l_2 is locally transformed into an occurrence of r_1 in r_2. So a rule inclusion i is a pair $\langle i_l : l_1 \to l_2, i_r : r_1 \to r_2 \rangle$. Formally, such a presentation is captured by a category and two functors.

Definition 1. *A* rule system T on $\hat{\mathbf{C}}_{\mathcal{M}}$ *is a tuple* $\langle \mathbf{\Gamma}_T, \mathrm{L}_T, \mathrm{R}_T \rangle$ *where* $\mathbf{\Gamma}_T$ *is a category whose objects are called* rules *and morphisms are called* rule inclusions, $\mathrm{L}_T : \mathbf{\Gamma}_T \to \hat{\mathbf{C}}_{\mathcal{M}}$ *is a full embedding functor called the* l.h.s. functor, *and* $\mathrm{R}_T : \mathbf{\Gamma}_T \to \hat{\mathbf{C}}_{\mathcal{M}}$ *is a functor called the* r.h.s. functor. *The subscript* T *is omitted when this does not lead to any confusion.*

Figure 1 illustrates a global transformation specification for generating a Sierpinski gasket. The rule system is composed of 3 rules transforming locally vertices (γ_1), edges (γ_2) and acyclic triangles (γ_3). These rules are related by 5 main rule inclusions: $i_1 : \gamma_1 \to \gamma_2$ (plain red), $i_2 : \gamma_1 \to \gamma_2$ (dashed red), $i_3 : \gamma_2 \to \gamma_3$ (dotted blue), $i_4 : \gamma_2 \to \gamma_3$ (plain blue), $i_5 : \gamma_2 \to \gamma_3$ (dashed blue). For instance, consider the inclusion i_3 which expresses that the left edge of triangle $\mathrm{L}(\gamma_3)$ is transformed into the left double-edge of $\mathrm{R}(\gamma_3)$. Formally, this is specified via the inclusion i_3 whose both components $\mathrm{L}(i_3)$ and $\mathrm{R}(i_3)$ are depicted in dotted blue arrows. The reader is invited to pay attention that even if Fig. 1 does not show them, the category $\mathbf{\Gamma}$ also contains compositions of the 5 main rule inclusions (*e.g.* $i_3 \circ i_1$), identities and symmetries, that, as functors, L and R do respect.

Computing with Global Transformations. Given a rule system T, its application on an arbitrary presheaf p is a three-step process. An illustration is given Fig. 2

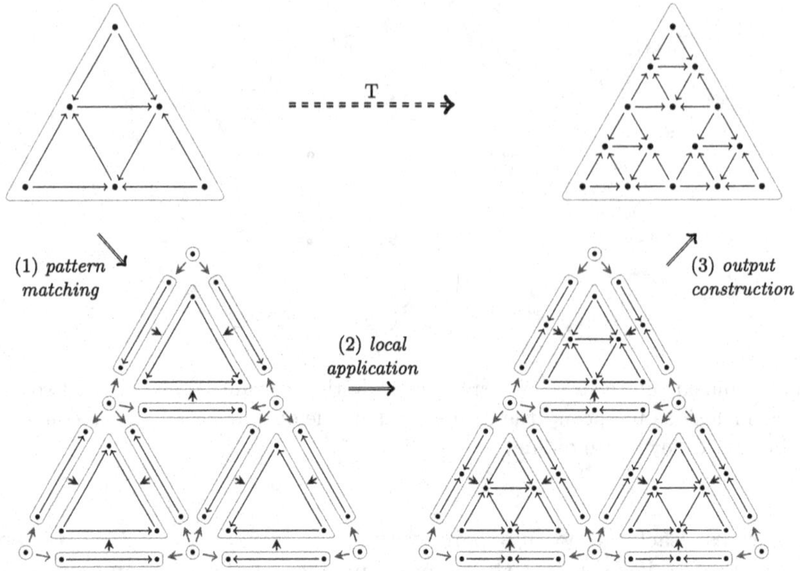

Fig. 2. Step of computation of the Sierpinski gasket using the rules of Fig. 1. (Color figure online)

based on the rule system of Fig. 1. The input is depicted at the top left and the output at the top right.

1. *Pattern matching* which consists in decomposing the input presheaf by mean of the rule l.h.s. It results a collection of l.h.s. instances, also called matches, structured by rule inclusions. This step is achieved by considering the comma category L_T/p: objects in that category are indeed all the morphisms from some l.h.s. to p; morphisms are the instantiations of the rule inclusions between those matches. See arrow (1) in Fig. 2 for an illustration. Formally, the figure at bottom left is a representation of $\mathrm{L}_T \circ \mathrm{Proj}[\mathrm{L}_T/p]$ where Proj designates the first projection of the comma category mapping each instance $\langle \gamma \in \mathbf{\Gamma}_T, f : \mathrm{L}_T(\gamma) \hookrightarrow p\rangle$ to the used rule γ. Notice the role of the rule inclusions (in red and blue) which are reminiscent of the input structure.
2. *Local application of rules* which consists in locally transforming each found l.h.s. into its corresponding r.h.s., the structure being conserved thanks to rule inclusions. This step is achieved by applying the r.h.s. functor R_T on each rule instance: $\mathrm{R}_T \circ \mathrm{Proj}[\mathrm{L}_T/p]$, as illustrated in Fig. 2.
3. *Output construction* which consists in assembling the output from the structured collection of r.h.s. The inclusions take here their full meaning as they are used to align the r.h.s. and drive the merge. See arrow (3) in Fig. 2 for an illustration. The resulting presheaf is formally the apex of a cocone from the diagram defined in the previous step which we used to obtain by colimit [8,14]. Since colimits are only guaranteed in $\hat{\mathbf{C}}$, we consider the following

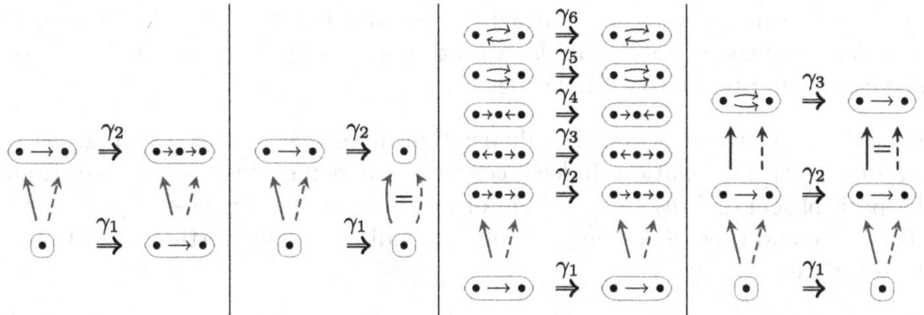

(a) Dualization of vertices and edges (b) Contraction of components (c) Removal of isolated vertices (d) Simplification of multi-edges

Fig. 3. Some rule-systems. Note that all of them remove self-loops. (Color figure online)

functor $\overline{T} : \hat{\mathbf{C}}_{\mathcal{M}} \to \hat{\mathbf{C}}$:

$$\overline{T}(-) = \mathrm{Colim}(\mathrm{D}_T(-)) \qquad \text{with} \qquad \mathrm{D}_T(-) = \mathrm{U} \circ \mathrm{R}_T \circ \mathrm{Proj}[\mathrm{L}_T/-] \qquad (1)$$

using the forgetful functor U, $\overline{T}(p)$ being the result of the application.

Remark 1. Notice that $\overline{T}$ is a complete functor also acting on morphisms. Consider a monomorphism $h : p \hookrightarrow p'$. By definition of colimits, $\overline{T}(p)$ is the universal cocone with components $\overline{T}(p)_{\langle\gamma,f\rangle} : \mathrm{D}_T(p)(\langle\gamma, f\rangle) \to \overline{T}(p)$ for each instance $\langle\gamma, f\rangle \in \mathrm{L}_T/p$. We have a similar construction for $\overline{T}(p')$ which gives rise to a cocone C as the restriction of $\overline{T}(p')$ on the diagram of $\overline{T}(p)$. Formally, C is defined with apex $C = \overline{T}(p')$ and components $C_{\langle\gamma,f\rangle} = \overline{T}(p')_{\langle\gamma,h\circ f\rangle}$. The image $\overline{T}(h)$ is the mediating morphism from colimit $\overline{T}(p)$ to C.

We focus on those rule systems where the results stay inside $\hat{\mathbf{C}}_{\mathcal{M}}$, *i.e.*, such that all previous mediating morphisms are monomorphisms. This leads to the following definition of global transformation for $\hat{\mathbf{C}}_{\mathcal{M}}$.

Definition 2. *A* global transformation T *is a rule system such that* $\overline{T}$ *factors through the forgetful functor* $\mathrm{U} : \hat{\mathbf{C}}_{\mathcal{M}} \to \hat{\mathbf{C}}$. *In this case, we denote* $T : \hat{\mathbf{C}}_{\mathcal{M}} \to \hat{\mathbf{C}}_{\mathcal{M}}$ *the functor such that* $\mathrm{U} \circ T = \overline{T}$.

The Sierpenski rule system of Fig. 1 is a global transformation. Its behavior is to split all edges and add some edges for acylic triangles. Thus, adding vertices and edges to an input only adds vertices and edges to the output. The induced functor maps monomorphisms to monomorphisms, so factors through U.

Figure 3 introduces four additional examples of rule systems that illustrate the various properties that we consider exhaustively. Let us see which of them are global transformations as a preparation for later considerations.

Example 1. The removal of isolated vertices (Fig. 3c) is a global transformation since removal is definitely a functorial behavior from $\hat{\mathbf{C}}_{\mathcal{M}}$ to $\hat{\mathbf{C}}_{\mathcal{M}}$.

Example 2. Simplification of multi-edges (Fig. 3d) is also a global transformation. For any two vertices a and b with at least an edge from a to b, it merges all edges from a to b into a single edge.

Example 3. On the contrary, the dualization of vertices and edges (Fig. 3a) is not a global transformation. Indeed, consider a monomorphism $i : p_2 \hookrightarrow c_3$ from the path of length 2 p_2 to the cycle of length 3 c_3. In this case $\overline{T}(p_2) = p_3$, $\overline{T}(c_3) = c_3$, and there is no monomorphism sending the four vertices of p_3 to the three vertices of c_3.

Example 4. Contraction of components (Fig. 3b) is not a global transformation. Consider the monomorphism $i : d_2 \hookrightarrow p_1$ from the graph d_2 with only vertices to the path p_1 of length 1. In this case, $\overline{T}(d_2) = d_2$ and $\overline{T}(p_1) = d_1$ and there is no monomorphism sending the two vertices of d_2 to the single vertex of d_1.

3 Accretion and Incrementality

We are interested in having an online algorithm computing Eq. (1) where the output $\overline{T}(p)$, for any p, is built while the comma category L_T/p is discovered by pattern matching. Informally, starting from a seed corresponding to the r.h.s. of some initial instance, L_T/p is visited from neighbor to neighbor, each instance providing a new piece of material to accumulate to the current intermediate result. We first give the formal features to be able to speak about intermediate results, leading to the notion of *accretive rule system*. Then we give a criterion, called *incrementality*, for a rule system to be an accretive global transformation.

In this algorithmic perspective, we restrict our discussion to finite presheaves and finite rule systems so that L_T/p is finite as well. Moreover we assume that L_T/p is connected; disconnected components can be processed independently. Finally, we fix a given finite rule system $T = \langle \mathbf{\Gamma}, \mathrm{L}, \mathrm{R} \rangle$ and a finite presheaf p. All omitted proofs are available in [9].

3.1 Accretive Rule Systems and Global Transformations

Informally an intermediate result consists in the application of $\overline{T}$ on an incomplete knowledge of L/p, *i.e.*, on a partial decomposition of the input. Our study of partial decompositions starts with some remarks. To begin, the comma category L/p is a preordered set.

Proposition 1. *For any category* $\mathbf{I}$*, any full embedding* $F : \mathbf{I} \to \hat{\mathbf{C}}_{\mathcal{M}}$*, and any presheaf* $p \in \hat{\mathbf{C}}_{\mathcal{M}}$*, the comma category* F/p *is* thin*, i.e., there is at most one morphism between any two objects.*

Therefore thinking in terms of maximal, non-maximal and minimal instances, sub- and super-instances actually makes sense for any comma category L/p.

Moreover, L/p being a preorder makes the colimit $\overline{T}(p)$ of the diagram $\mathrm{D}_T(p) : \mathrm{L}/p \to \hat{\mathbf{C}}$ special. Informally, only maximal instances matter, sub-instances being used to specify how to amalgamate the r.h.s. of maximal instances. Formally, whenever we have a morphism $\langle e, f' \rangle : \langle \gamma, f' \circ \mathrm{L}(e) \rangle \to \langle \gamma', f' \rangle \in \mathrm{L}/p$, the r.h.s. of γ' contains the r.h.s. of γ through $e : \gamma \to \gamma'$. With $f = f' \circ \mathrm{L}(e)$, this means that $\langle \gamma, f \rangle$ does not contribute more data to the output. The role of the non-maximal $\langle \gamma, f \rangle$ is to specify how the r.h.s. of some γ'' should be aligned with the r.h.s. of γ' in the resulting presheaf when there is a second morphism $\langle \gamma, f \rangle \to \langle \gamma'', f'' \rangle$ to another maximal instance $\langle \gamma'', f'' \rangle$.

As an example, consider the diagram depicted on bottom right of Fig. 2 and its colimit on the top right. All the elements of the colimit are given in the r.h.s. of the maximal instances (the 3 triangles). The minimal instances (the 6 one-vertex graphs) are used to specify how the 9 vertices of the 3 refined triangles should be merged to get the colimit.

Computing $\overline{T}$ Online. Given a presheaf $p \in \hat{\mathbf{C}}$, we call a *partial decomposition of p with respect to* L a subset M of maximal instances of L/p such that the restriction $\widetilde{M}$ of L/p to M and morphisms into M remains connected. We write $\widetilde{\mathrm{L}/p}$ for the category of partial decompositions of p with set inclusions as morphisms. $\widetilde{\mathrm{L}/p}$ represents the different ways L/p can be visited from maximal instance to maximal instance by the use of non-maximal instances to guide the merge. We extend the action of $\overline{T}$ on p into a function $\widetilde{T}_p : \widetilde{\mathrm{L}/p} \to \hat{\mathbf{C}}$ as follows:

$$\widetilde{T}_p(M) = \mathrm{Colim}(\mathrm{D}_T(p) \restriction \widetilde{M}). \tag{2}$$

The definition $\widetilde{T}_p$ is in fact a complete functor also acting on any morphism $M \subseteq M'$ of $\widetilde{\mathrm{L}/p}$ using the exact same construction as given in Remark 1 for $\overline{T}$.

In the case of Fig. 2, the maximal instances being the three triangles, the partial decompositions consist of subsets having 0, 1, 2 or 3 of these triangles. As an example, choose arbitrarily two of these triangles, say t_1 and t_2. The intermediate result $\widetilde{T}_p(\{t_1, t_2\})$ is the graph consisting of the two refinements of t_1 and t_2 glued by their common vertex.

The online computation of $\overline{T}(p)$ consists in iterating a simple step that builds $\widetilde{T}_p(M \cup \{m\})$ from $\widetilde{T}_p(M)$ as soon as a new maximal instance m has been discovered. This step is the local amalgamation of $\mathrm{D}_T(p)(m)$ with $\widetilde{T}_p(M)$ considering all the non-maximal sub-instances, say $\{n_1, \ldots, n_k\}$, shared by the elements of M and m. Indeed each such sub-instance n_i gives rise to a span $\widetilde{T}_p(M) \leftarrow \mathrm{D}_T(p)(n_i) \to \mathrm{D}_T(p)(m)$. Gathering all these spans leads to the *suture* diagram $S_{M,m}$ defined as follows:

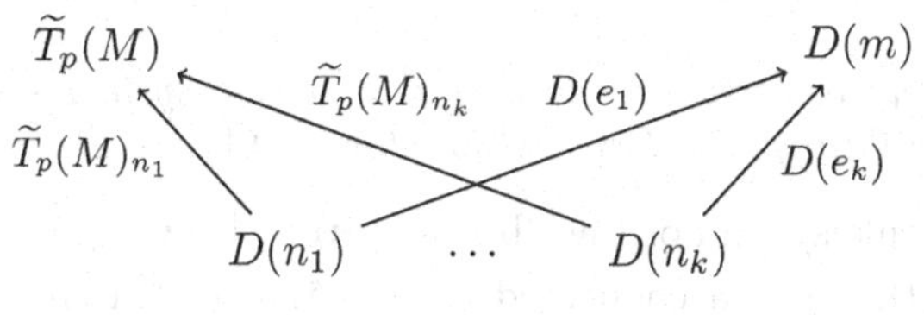

(3)

where D stands for $\mathrm{D}_T(p)$ for simplicity, notation that we will use from now on. The colimit of a single span being called a pushout, we call the colimit of this collection of spans a *generalized pushout*. The fact that this generalized pushout

$\mathrm{Colim}(S_{M,m})$ and the desired colimit $\widetilde{T}_p(M \cup \{m\})$ as given in Eq. (2) do coincide is formalized by the following proposition.

Proposition 2. *Let $M' = M \cup \{m\} \in \widetilde{\mathrm{L}/p}$. As a cocone, $\widetilde{T}_p(M')$ has the same apex as* $\mathrm{Colim}(S_{M,m})$ *and has components*

$$\widetilde{T}_p(M')_n = \begin{cases} \mathrm{Colim}(S_{M,m})_{\widetilde{T}_p(M)} \circ \widetilde{T}_p(M)_n & \textit{if } n \in \widetilde{M}, \\ \mathrm{Colim}(S_{M,m})_{D(m)} \circ D(e) & \textit{for any } e : n \hookrightarrow m, \end{cases}$$

where $S_{M,m}$ is diagram (3).

As an illustration, for computing $\widetilde{T}_p(\{t_1, t_2, t_3\})$ (which is in fact the output in Fig. 2), it is enough to amalgamate $\widetilde{T}_p(\{t_1, t_2\})$ (already considered) with the refinement $D(t_3)$ of the last triangle t_3, using as suture the two vertices of t_3 shared with t_1 and t_2 playing the role of n_1 and n_k in diagram (3).

Remark 2. To summarize, computing $\overline{T}(p)$ online is a matter of collecting the finite set of all maximal instances $\{m_1, m_2, \ldots, m_k\}$ of L/p in any order satisfying that m_{i+1} is connected to $\widetilde{M_i}$ where $M_i = \{m_1, \ldots, m_i\}$. This allows to replace the single colimit computation of the whole diagram, as in Eq (1), by a sequence of smaller colimit computations using the induction relation:

$$\widetilde{T}_p(M_i) = \begin{cases} D(m_1) & \text{if } i = 1 \\ \mathrm{Colim}(S_{M_{i-1}, m_i}) & \text{otherwise.} \end{cases} \tag{4}$$

The base case is obtained from Eq. 2 for a singleton set of maximal instance, and the inductive one is established by Proposition 2, S_{M_{i-1}, m_i} being a generalized pushout diagram linking $\widetilde{T}_p(M_{i-1})$ and $D(m_i)$. The final value $\widetilde{T}_p(M_k)$ of this sequence is the colimit of $\widetilde{M_k}$. $\widetilde{M_k}$ is exactly the diagram D without the arrows between the non-maximal instances. But the colimit $\widetilde{T}_p(M_k)$ of $\widetilde{M_k}$ is necessarily the same as the colimit $\overline{T}(p)$ of D by the following proposition.

Proposition 3. *The subcategory $\widetilde{M_k}$ of* L/p *given by all instances but only morphisms to maximal instances is final in* L/p, *in the sense of final functor.*

Accretive Rule Systems. We are interested in those rule systems where the intermediate results stay inside $\hat{\mathbf{C}}_{\mathcal{M}}$, *i.e.*, such that $\widetilde{T}_p(M \subseteq M')$ are monomorphisms for any p and any $M \subseteq M' \in \widetilde{\mathrm{L}/p}$. This leads to the following definition of accretive rule systems.

Definition 3. *An* accretive rule system *T is a rule system such that for any $p \in \hat{\mathbf{C}}$, $\widetilde{T}_p$ factors through the forgetful functor* $\mathrm{U} : \hat{\mathbf{C}}_{\mathcal{M}} \to \hat{\mathbf{C}}$.

Example 5. The rule system of Fig. 3b is accretive. Focusing on connected L/p, its l.h.s. implies that p is a connected graph. Any $M \in \widetilde{\mathrm{L}/p}$ corresponds to a connected sub-graph of p and is sent to the single vertex graph if it is not empty, or to the empty graph otherwise. So for any relation $M \subseteq M'$, $\widetilde{T}_p(M \subseteq M')$ is the empty morphism or the identity morphism, and both are monomorphisms.

Example 6. The rule system of Fig. 3d is also accretive. Again, p is a connected graph and any $M \in \widetilde{\mathrm{L}/p}$ corresponds to a connected sub-graph of p. Here M is sent to the same graph with parallel edges simplified into single edge. So for any relation $M \subseteq M'$, M' is sent either to the same thing as M when M' only adds more parallel edges, or to a strictly greater graph otherwise. In both case $\widetilde{T}_p(M \subseteq M')$ is a monomorphism.

Example 7. On the contrary, the rule system of Fig. 3a is not accretive. Consider the cycle c_3 of length 3, and the associated L/c_3. The latter contains 3 instances of rule γ_1 and 3 maximal instances of the rule γ_2. Consider the relation $M \subseteq M'$ where M' contains all three maximal instances and M only two of them. We have $\widetilde{T}_p(M) = p_3$ and $\widetilde{T}_p(M') = c_3$, but there is no monomorphisms between these two graphs.

Example 8. By the exact same reasoning, the rule system of Fig. 3c is also not accretive, disproving that being a global transformation implies being accretive.

3.2 Incremental Rule Systems and Global Transformations

We are interested in giving sufficient conditions for rule systems to be global transformations. These conditions also imply accretiveness.

Our strategy consists in preventing any super-rule to merge by itself the r.h.s. of its sub-rules. In other words, the rule only adds new elements to the r.h.s. of its sub-rules in an *incremental* way. A positive expression of this constraint is as follows: if the r.h.s. of two rules overlap in the r.h.s. of a common super-rule, this overlap must have been required by some common sub-rules.

Definition 4. *Given a rule system* $T = \langle \mathbf{\Gamma}, \mathrm{L}, \mathrm{R} \rangle$*, we say that a rule* $\gamma \in \mathbf{\Gamma}$ *is incremental if for any two sub-rules* $\gamma_1 \xrightarrow{i_1} \gamma \xleftarrow{i_2} \gamma_2$ *in* $\mathbf{\Gamma}$*, any representable presheaf* $\mathbf{yc}$*, and any* $\mathrm{R}(\gamma_1) \xleftarrow{x_1} \mathbf{yc} \xrightarrow{x_2} \mathrm{R}(\gamma_2)$ *such that* $\mathrm{R}(i_1) \circ x_1 = \mathrm{R}(i_2) \circ x_2$*, there are some* $\gamma_1 \xleftarrow{\pi_1} \gamma' \xrightarrow{\pi_2} \gamma_2$ *and* $x : \mathbf{yc} \to \mathrm{R}(\gamma')$ *such that the following diagrams commute.*

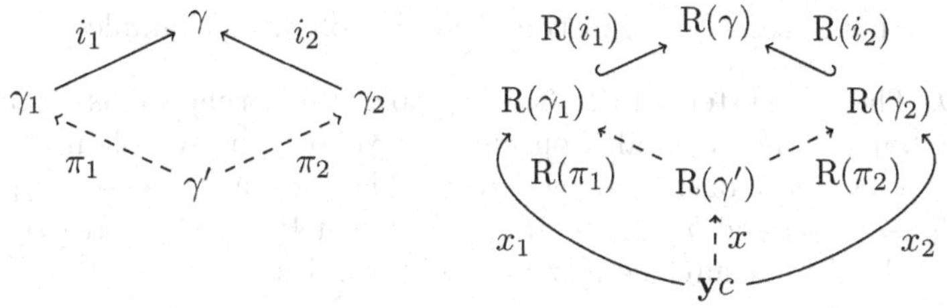

A rule system T *is said* incremental *if every* $\gamma \in \mathbf{\Gamma}$ *is incremental.*

The Sierpinski gasket rule system (Fig. 1) is incremental. The only non-trivial case is when the sub-rules γ_1 and γ_2 of Definition 4 are set to the edge rule of Fig. 1 and γ to be the triangle rule, such that the r.h.s. of γ_1 and γ_2 overlap on a common vertex in $\mathrm{R}(\gamma)$ (morphisms x_1 and x_2 of Definition 4). This vertex is nothing but the image of the vertex of $\mathrm{L}(\gamma)$ common to the inclusions of $\mathrm{L}(\gamma_1)$ and $\mathrm{L}(\gamma_2)$ in $\mathrm{L}(\gamma)$. This invites us to set γ' to the vertex rule of Fig. 1 and complete the requirements of Definition 4 to get incrementality.

The main constraint enforced by the incrementality criterion is that any merge is always required by sub-rules as stated by the following lemma.

Lemma 1. *Consider an incremental rule system $T = \langle \mathbf{\Gamma}, \mathrm{L}, \mathrm{R}\rangle$, an integer $k > 0$, a sequence of k rules $\langle \gamma_1, \ldots, \gamma_k\rangle$ in $\mathbf{\Gamma}$, a sequence of $k-1$ rule inclusions $\langle i_1, \ldots, i_{k-1}\rangle$ in $\mathbf{\Gamma}$ with $i_j : \gamma_j \to \gamma_{j+1}$ or $i_j : \gamma_{j+1} \to \gamma_j$ for $1 \leq j < k$, a representable presheaf $\mathbf{y}c$ and a cone $\langle x_j : \mathbf{y}c \to \mathrm{R}(\gamma_j)\rangle_{1\leq j \leq k}$ that commutes with each $\mathrm{R}(i_j)$. There are a rule γ' in $\mathbf{\Gamma}$, two rule inclusions $\pi_1 : \gamma' \to \gamma_1$ and $\pi_2 : \gamma' \to \gamma_k$, and a morphism $x : \mathbf{y}c \to \mathrm{R}(\gamma')$ such that the following diagram on the right commutes.*

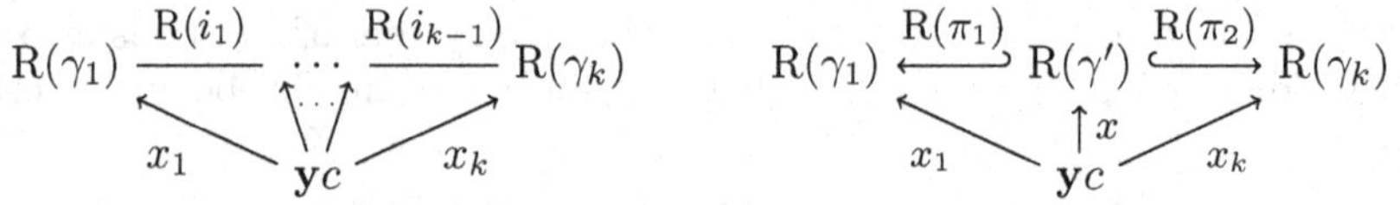

Consider any monomorphism $h : p \hookrightarrow p'$ of presheaves such that some merge is required by the computation of $\overline{T}(p')$ between some elements of r.h.s. instances also involved by $\overline{T}(p)$. Lemma 1 ensures that it is required by a sub-rule which must also be instantiated by $\overline{T}(p)$ so that the merge is also required by the computation of $\overline{T}(p)$. In other words, $\overline{T}(h)$ is a monomorphism as established by the following theorem.

Theorem 1. *Any incremental rule system is a global transformation.*

The previous remark also applies for intermediate results leading to the following theorem concerning accretiveness.

Theorem 2. *Any incremental rule system is accretive.*

However, the converses of these theorems do not hold so incrementality is sufficient but not necessary as illustrated by the following examples.

Example 9. The rule system of Fig. 3c is a global transformation as explained in Example 1, but not incremental. Consider $e_1 : \gamma_1 \to \gamma_2$ be the plain arrow into γ_2 and $e_2 : \gamma_1 \to \gamma_2$ the dashed arrow into γ_2. The cospan $\gamma_1 \hookrightarrow \gamma_2 \hookleftarrow \gamma_1$ is such that $h_1 : d_1 \hookrightarrow \mathrm{R}(\gamma_1)$ and $h_2 : d_1 \hookrightarrow \mathrm{R}(\gamma_1)$ such that $\mathrm{R}(e_1) \circ h_1 = \mathrm{R}(e_2) \circ h_2$ but there is no rule γ' to ensure the incrementality condition.

Example 10. Similarly, the rule system of Fig. 3d is a global transformation (Example 2) but is not incremental. Consider $e_1 : \gamma_2 \to \gamma_3$ be the plain arrow into γ_3 and $e_2 : \gamma_2 \to \gamma_3$ the dashed arrow into γ_3. The cospan $\gamma_2 \hookrightarrow \gamma_3 \hookleftarrow \gamma_2$ is such that $h_1 : l_1 \hookrightarrow \mathrm{R}(\gamma_2)$ and $h_2 : l_1 \hookrightarrow \mathrm{R}(\gamma_2)$ such that $\mathrm{R}(e_1) \circ h_1 = \mathrm{R}(e_2) \circ h_2$ but there is no rule γ' to ensure the incrementality condition.

Example 11. The rule system of Fig. 3b is accretive (Example 5) but non-incremental. Consider $e_1 : \gamma_1 \to \gamma_2$ the plain arrow into γ_3 and $e_2 : \gamma_1 \to \gamma_2$ the dashed arrow into γ_3. Observe that for the cospan $\gamma_1 \hookrightarrow \gamma_2 \hookleftarrow \gamma_1$ we have $h_1 : d_1 \hookrightarrow \mathrm{R}(\gamma_2)$ and $h_2 : d_1 \hookrightarrow \mathrm{R}(\gamma_2)$ such that $\mathrm{R}(e_1) \circ h_1 = \mathrm{R}(e_2) \circ h_2$ but there is no rule γ' to ensure the incremental condition.

Example 12. Similarly, the rule system of Fig. 3d is accretive (Example 6) but non-incremental.

Summarizing the properties collected with the four examples of Fig. 3 and with the one of Fig. 1 in a table, we can see that being a global transformation and being accretive are orthogonal properties, but incrementality forces the two.

	Non-incr.		Incr.	
	Non-G.T.	G.T.	Non-G.T.	G.T.
Non-accretive	ex. Fig. 3a	ex. Fig. 3c	None, Theorem 1/2	None, Theorem 2
Accretive	ex. Fig. 3b	ex. Fig. 3d	None, Theorem 1	Sierpenski

4 Computing Accretive Global Transformations

This section is devoted to the description of an effective implementation of the online procedure considered in Sect. 3.1. In this context, we focus on incremental global transformations. We first explain how the categorical concepts of Sect. 3 are represented computationally (Sect. 4.1). This is followed by a detailed presentation of the algorithm (Sect. 4.2).

4.1 Categorical Constructions Computationally

Up to now, we exposed everything formally using categorical concepts: the category of presheaves $\hat{\mathbf{C}}_{\mathcal{M}}$, finite incremental rule systems $T = \langle \mathbf{\Gamma}, \mathrm{L}, \mathrm{R} \rangle$, and the comma category L/p for some finite presheaf p. We now describe their computational counterparts. First, let us introduce some notations used in the algorithm:

- X^* stands for the set of finite words on the alphabet X; the empty word is denoted by ε and the concatenation by $u \cdot v$ for any two words $u, v \in X^*$.
- $\coprod_{a \in A} B(a)$ is the set of pairs (a, b) where $a \in A$ and $b \in B(a)$.
- $\prod_{a \in A} B(a)$ is the set of functions $f : A \to \bigcup_{a \in A} B(a)$ such that for any $a \in A$, $f(a) \in B(a)$. Such functions are also manipulated as sets of pairs. Those pairs are written $a \mapsto f(a)$.

The Category of Presheaves with Monomorphisms. The category $\hat{\mathbf{C}}_{\mathcal{M}}$ is the formal abstraction for a library providing a data structure suitably captured by presheaves (like sets, graphs, Petri nets, etc.) and how an instance of that data structure (presheaves) is part of another one (monomorphisms). Two functions $- \circ -$ and $- = -$ need to be provided to compute composition and equality test of sub-parts. The library also needs to come with a pattern matching procedure taking as input two finite presheaves p and p' and returning the set $\mathrm{Hom}_{\hat{\mathbf{C}}_{\mathcal{M}}}(p, p')$ of occurrences of p in p'. Finally, the library is assumed to provide a particular construction operation called `generalizedPushout`(p_1, p_2, S) computing the generalized pushout, *i.e.*, the colimit of the collection of spans S, each span being represented as a triplet $(p \in \hat{\mathbf{C}}_{\mathcal{M}}, f_1 : p \hookrightarrow p_1, f_2 : p \hookrightarrow p_2)$. The resulting colimit is returned as a triplet $(c \in \hat{\mathbf{C}}_{\mathcal{M}}, g_1 : p_1 \hookrightarrow c, g_2 : p_2 \hookrightarrow c)$ where c is the apex and g_1, g_2 the corresponding component morphisms.

Finite Incremental Rule System. A finite rule system is described as a finite graph whose vertices are rules $l \Rightarrow r$ as pairs of presheaves and edges are pairs of monomorphisms $\langle i_l : l_1 \hookrightarrow l_2, i_r : r_1 \hookrightarrow r_2 \rangle$. Functors L and R return the first and second components of these pairs respectively. At the semantic level, $\mathbf{\Gamma}$ is the category generated from this graph. Finally, incrementality as presented in Definition 4 is clearly decidable on finite rule systems, giving rise to an accretive global transformation by Theorems 1 and 2.

The Category of Instances. By Proposition 1, L/p is a preordered set, but in our implementation, any time an instance is matched, all of its isomorphic instances are taken care of at the same time. This corresponds informally to taking the poset of equivalence classes of the preordered set. Also, by Proposition 3, morphisms between non-maximal instances can be ignored. All in all, L/p is adequately thought of as an abstract undirected bipartite graph that we call *the network*.

Finally, the L/p is never entirely represented in memory (neither is the cocone associated to the resulting colimit). A first instance is constructed, and the others are built from neighbor to neighbor through the operation $\coprod_n \mathrm{Hom}_{\mathrm{L}/p}(n, m)$ and $\coprod_m \mathrm{Hom}_{\mathrm{L}/p}(n, m)$. The former lists the sub-instances of m and the latter lists the super-instances of n. For the "incoming neighbors" or sub-instances $\coprod_n \mathrm{Hom}_{\mathrm{L}/p}(n, \langle \gamma', f' \rangle)$, they are specified as

$$\{(\langle \gamma, f' \circ \mathrm{L}(e) \rangle, \langle e, f' \rangle) \mid e : \gamma \to \gamma'\}.$$

This corresponds simply to the composition $- \circ -$ in $\hat{\mathbf{C}}_{\mathcal{M}}$ discussed earlier. On the contrary, "outgoing neighbors" or super-instances $\coprod_m \mathrm{Hom}_{\mathrm{L}/p}(\langle \gamma', f' \rangle, m)$ correspond to extensions and are obtained by pattern matching.

$$\{\langle e', f'' \rangle : \langle \gamma', f' \rangle \to \langle \gamma'', f'' \rangle \mid \\ e' : \gamma' \to \gamma'', f'' \in \mathrm{Hom}_{\hat{\mathbf{C}}_{\mathcal{M}}}(\mathrm{L}(\gamma''), p) \text{ s.t. } f' = f'' \circ \mathrm{L}(e')\}.$$

Notice that these two specifications are obtained by simply unfolding the definition of the morphisms of the comma category. Also, the set of incoming morphisms $e : \gamma \to \gamma'$ and outgoing morphisms $e' : \gamma' \to \gamma''$ in $\mathbf{\Gamma}$ are directly available in the graph representation of the rule system T as said earlier. These operations are used to implement a breadth-first algorithm, earlier instances being dropped away as soon as their maximal super-instances have been found.

Algorithm 1:

Input: T : rule system on $\hat{\mathbf{C}}_{\mathcal{M}}$
Input: $p : \hat{\mathbf{C}}_{\mathcal{M}}$
Variable: $P : \hat{\mathbf{C}}_{\mathcal{M}}$
Variable: $N : (\mathrm{L}/p)^*$
Variable: $E \subseteq \coprod_{n \in N} \coprod_m \mathrm{Hom}_{\mathrm{L}/p}(n, m)$
Variable: $C : \prod_{n \in N} \mathrm{Hom}_{\hat{\mathbf{C}}_{\mathcal{M}}}(D(n), P)$

1 **let** $n =$ `findAnyMinimal`(T, g), *i.e.*, any minimal element in L/p
2 **let** $E = \emptyset$, $C = \{n \mapsto id_{D(n)}\}$, $N = n$, $P = D(n)$
3 **while** $N \neq \varepsilon$ **do**
4   **let** $n =$ `head`(N), *i.e.*, the first instance in the queue without modifying N
5   **let** $M' = \coprod_m \mathrm{Hom}_{\mathrm{L}/p}(n, m)$
6   **for** $(m, e) \in M'$ *s.t.* $(n, m, e) \notin E$ *and* m *is maximal* **do**
7   **let** $E' = \coprod_{n' \neq m} \mathrm{Hom}_{\mathrm{L}/p}(n', m)$
8   **let** $S = \{(n', C(n'), D(e')) \mid (n', e') \in E', n' \in N\}$
9   **let** $(P', t, c) =$ `generalizedPushout`$(P, D(m), S)$
10         $E := E \cup \{(n', m, e') \mid (n', e') \in E'\}$
11         $C := C \cup \{n' \mapsto c \circ D(e') \mid (n', e') \in E', n' \notin N\}$
12         $N := N \cdot \langle n' \mid (n', e') \in E', n' \notin N \rangle$
13         $P := P'$
14     $E := \{(n', m, e') \in E \mid n' \neq n\}$
15     $C := \{n' \mapsto C(n') \in C \mid n' \neq n\}$
16     $N :=$ `tail`(N), *i.e.*, removes the first instance n from the queue
17 **return** P

4.2 The Global Transformation Algorithm

Algorithm 1 gives a complete description of a procedure to compute $T(p)$ online. The algorithm manages four variables P, N, E and C. Variable P contains intermediate results and finally the output presheaf. The part of the network that is kept in memory is represented by variables N and E: N is a queue containing, in order of discovery, the non-maximal instances that might still have a role to play. E associates each instance in N to the set of its maximal super-instances that have already been processed. For simplicity, E is not represented as a function from N to sets but as a relation. The r.h.s. $D(n)$ of each instance $n \in N$ is already in the current result P through the morphism kept as $C(n)$.

Figure 4 illustrates the first steps of Algorithm 1 representing maximal instances as black dots, and non-maximal instances as white squares. The initialization step is to find a first instance (line 1). For that, we try each minimal pattern, and start with the first founded minimal instance, say n_1. At this point, the first intermediate result P_0 is simply the r.h.s. $D(n_1)$; we memorize the (identity) relationship between $D(n_1)$ and P_0, call it $C(n_1) : D(n_1) \to P_0$, and enqueue n_1 (line 2). Enqueued non-maximals are treated one after the other (lines 3, 4, 16). For each, we consider all maximal super-instances of n_1 (lines

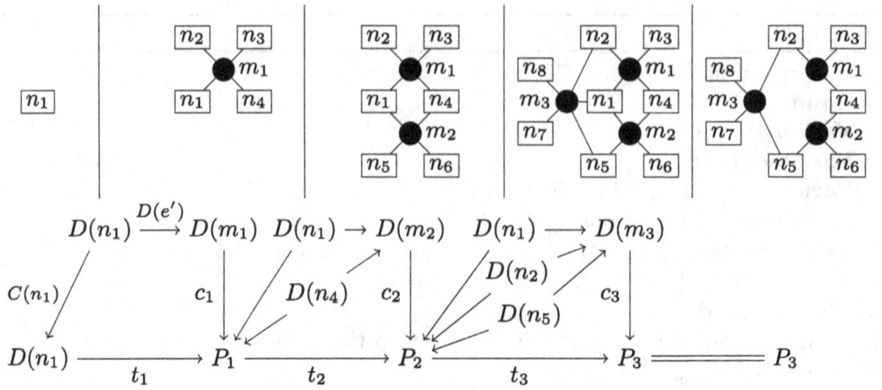

Fig. 4. Evolution of the data during the four firsts steps of the algorithm. From left to right: we start with a non-maximal instance, process its associated maximal instances successively, and finally drop the non-maximal. At each stage, the output is updated by generalized pushout.

5, 6). In Fig. 4, we assume three such super-instances m_1, m_2 and m_3. They are processed one after the other (line 6).

The first iteration processes m_1 by taking all its sub-instances $n_1, \ldots, n_4$ (line 7). The suture S is computed (line 8) by considering all already computed non-maximals (*i.e.*, in N) among these sub-instances. Here, only n_1 is already known and serves to define a one-span suture with morphisms $C(n_1) : D(n_1) \to P_0$ and $D(e') : D(n_1) \to D(m_1)$, where e' is the morphism from n_1 to m_1. The generalized pushout of P_0 and the r.h.s. $D(m_1)$ is therefore computed and gives the new intermediate result P_1 (lines 9, 13). Since P_1 includes the r.h.s. of all discovered non-maximals $N = \{n_1, \ldots, n_4\}$, we memorize as $C(n) : D(n) \to P_1$ for $n \in N$ the locations of these r.h.s. in P_1 (line 11). The newly discovered non-maximal instances n_2, n_3 and n_4 are enqueued (line 12).

The second iteration processes m_2 similarly and all its sub-instances n_1, n_4, n_5, and n_6 are computed. This time, n_1 and n_4 are used for computing the new intermediate result P_2 by generalized pushout using the two spans $\langle n_1, C(n_1) : D(n_1) \to P_1, D(j) : D(n_1) \to D(m_2)\rangle$ and $\langle n_4, C(n_4) : D(n_4) \to P_1, D(k) : D(n_4) \to D(m_2)\rangle$ as suture. The set N of discovered non-maximals is updated by adding n_5 and n_6 as well as the locations C of their r.h.s. in P_2.

The processing of m_3 is similar and shows no novelty. At this point non-maximal n_1 does not have any further role to play: the r.h.s. of all its associated maximals are already amalgamated to the current intermediate result. n_1 is dropped together with all data associated to it (lines 14–16), as shown in the last step of Fig. 4. Non-maximal instances being processed in the order of first discovery, the next one is n_2 in the example.

During these processings, other non-maximal instances see some of their associated maximals being processed. We have to keep track of this to avoid double processing of maximals which would cause infinite loops (condition at line 6).

This is the role of E to maintain this information. Clearly E contains only the useful part of the network: edges from maximals to their sub-instances are registered when discovered (line 10) but cleared up as soon as a non-maximal is dropped (line 14). Considering that non-maximal instances are treated in order of appearance, the algorithm will process the maximals at distance 1 from n_1 first, then those at distance 2, and so on, until the complete connected component of the network is processed. In memory, there are never stored more than four "radius" of instances d, $d+1$, $d+2$ and $d+3$ from n_1.

Theorem 3. *Algorithm 1 is correct, i.e., the final value of P is $\overline{T}(p)$.*

Proof. We ignore the case when L/p is composed of a single instance, since the algorithm behaves trivially in that case.

Ignoring lines 8, 9, 11, 13, and 15, variables P and C, and looking only at non-maximal instances (variable n), the algorithm behaves like a usual breadth-first search. Indeed, the search begins by enqueueing a first non-maximal instance at line 2. Each iteration of the while loop (line 3) processes the next non-maximal instance n in the queue (line 4), lists all its "neighbors via a maximal instance" (lines 5–7) and enqueues those that have not yet been visited (lines 10, 12) before popping n out of the queue (line 16). Variables E and N serve as the set of visited non-maximal instances. The reason line 14 can remove all occurrences of n in the set E without creating an infinite loop is that E memorizes the maximal instances m' from which each enqueued non-maximal instance n' has been reached (line 10). The constraint $(n, m, e) \notin E$ of the for loop (line 6) prevents this path to be taken in the other direction.

Since all non-maximal instances are assigned to n at line 4, and each maximal instance is a super-instance of some non-maximal instance, we have that m goes through all maximal instances as well (line 6). Let us call $m_1, \ldots, m_k$, the successive values taken by m and define the sequence of set of maximal instances $M_i = \{m_1, \ldots, m_i\}$ for $i \in \{1, \ldots, k\}$. The breath-first traversal ensures that each newly considered m_{i+1} is connected to some maximal instance in M_i by some non-maximal sub-instances. Let us show now that the successive values taken by P at line 13, numbered $P_1, P_2, \ldots, P_k$, are such that $P_i = \widetilde{T}_p(M_i)$. Using Remark 1, it is enough to show that $P_1 = D(m_1)$ and $P_{i+1} = \mathrm{Colim}(S_{M_i, m_{i+1}})$.

For P_1, consider the first steps of the algorithm before the first execution of line 13. Call n_1 the value of n at line 1 and note that P is assigned to $D(n_1)$ at line 2 and $C(n_1)$ to $id_{D(n_1)} : D(n_1) \to D(n_1)$. At lines 4, 5 and 6, n is assigned to n_1, m to m_1 and e to the corresponding morphism from n_1 to m_1. Lines 7 and 8 lead S to be $\{(n_1, id_{D(n_1)} : D(n_1) \to D(n_1), D(e) : D(n_1) \to D(m_1))\}$. So the first execution of line 9 computes this simple pushout and sets (P', t, c) to $(D(m_1), D(e), id_{D(m_1)})$, so $P_1 = D(m_1) = \widetilde{T}_p(M_1)$ at line 13.

To establish that $P_{i+1} = \mathrm{Colim}(S_{M_i, m_{i+1}})$ in the $(i+1)$-th execution of line 13, we need to show that the parameters $(P, D(m), S)$ provided in $(i+1)$-th execution of line 9 correspond to the diagram $S_{M_i, m_{i+1}}$ given in (3). Firstly, by induction hypothesis, we have that $P = P_i = \widetilde{T}_p(M_i)$ and $m = m_{i+1}$. The collection of spans S computed at line 8 is correct because E' is the set of sub-instances of m (line 7), and N contains all sub-instances of the maximal instances in M_i that could have a morphism to m_{i+1}. Indeed, a non-maximal instance is discarded from N, E and C (lines 14–16) only after that all of its maximal super-instances have been processed (for loop at lines 5 to 13). Line 11 and 15 ensure that C always contain the correct morphism $D(n) \to P$ for all non-maximal instances n contained in N.

Finally, line 11 modifies C without updating the cocone compounds already stored in C, resulting in mixing morphisms with codomain P and P'. It is correct considering the following fact. For accretive global transformations, t (line 9) is always a monomorphism and can be designed for t to be a trivial inclusion. In that case, any morphism to P is also a morphism to P', the latter materially including P. In other words, everything is implemented to ensure that the modification on intermediate results are realized *in place*. □

5 Conclusion

In this paper, we have presented an online algorithm for computing the application of global transformations on presheaves. Note that this work was originally restricted to global transformations of graphs but the extension to any category of presheaves appears to be straightforward. It is natural to expect the extensions to other well-known classes of categories, in particular for the class of ($\mathcal{M}$-)adhesive categories [5,12].

At the algorithmic level, there remain many interesting considerations that need to be settled. One of them is that the way this algorithm goes from maximal instances to maximal instances using common sub-instances reminds of the strategy of the famous Knuth-Morris-Pratt algorithm [11]: in both cases the content of one match is used to guide following subsequent pattern matching. This link is reinforced by the work of [15] that extend the Knuth-Morris-Pratt algorithm to sheaves. In Algorithm 1, we used pattern matching as a black-box but opening it should allow to mix the outer maximal-to-maximal strategy with the Knuth-Morris-Pratt considerations inside the pattern matching algorithm of [15]. Another important aspect is the complexity of this online approach and its natural extensions. Indeed, we described how a full input is decomposed in an online fashion, and the parts also treated online. The full picture includes the input itself being received by part, or even treated in a distributed way. Each of these versions deserve a careful study of their online complexity, *i.e.*, the complexity of the computation happening between each outputted data. We are also interested in the detailed study of the problem consisting of deciding, given a rule system, if it is a global transformation or not. Incremental rule systems form a particularly easy sub-class for this problem but we are talking here about the complete class of all rule systems.

The incremental criterion can be studied for itself. An alternative equivalent expression of Definition 4 is stated as follows: given a super-rule, its r.h.s. contains the r.h.s. of its sub-rules as if they were considered independently. Intuitively, this prevents from non-local behavior like collapsing non-empty graphs to a single vertex since the empty graph remains empty for example as in Fig. 3b. From that point of view, incremental global transformations follow the research direction of causal graph dynamics [1]. In this work any produced element in the output is attached to an element of the input graph and a particular attention is put on preventing two rule instances to produce a common element.

References

1. Arrighi, P., Martiel, S., Nesme, V.: Cellular automata over generalized Cayley graphs. Math. Struct. Comput. Sci. **18**, 340–383 (2018)
2. Boehm, P., Fonio, H.R., Habel, A.: Amalgamation of graph transformations: a synchronization mechanism. J. Comput. Syst. Sci. **34**(2–3), 377–408 (1987)
3. Corradini, A., Heindel, T., Hermann, F., König, B.: Sesqui-pushout rewriting. In: Corradini, A., Ehrig, H., Montanari, U., Ribeiro, L., Rozenberg, G. (eds.) ICGT 2006. LNCS, vol. 4178, pp. 30–45. Springer, Heidelberg (2006). https://doi.org/10.1007/11841883_4
4. Ehrig, H., Ehrig, K., Golas, U., Taentzer, G.: Fundamentals of Algebraic Graph Transformation, vol. XIV. Springer, Heidelberg (2006). https://doi.org/10.1007/3-540-31188-2
5. Ehrig, H., Golas, U., Hermann, F., et al.: Categorical frameworks for graph transformation and HLR systems based on the DPO approach. Bull. EATCS **102**, 111–121 (2010)
6. Ehrig, H., Pfender, M., Schneider, H.J.: Graph-grammars: an algebraic approach. In: 14th Annual Symposium on Switching and Automata Theory (SWAT 1973), pp. 167–180. IEEE (1973)
7. Ehrig, H., Rosen, B.K.: Parallelism and concurrency of graph manipulations. Theor. Comput. Sci. **11**(3), 247–275 (1980)
8. Fernandez, A., Maignan, L., Spicher, A.: Lindenmayer systems and global transformations. In: McQuillan, I., Seki, S. (eds.) UCNC 2019. LNCS, vol. 11493, pp. 65–78. Springer, Cham (2019). https://doi.org/10.1007/978-3-030-19311-9_7
9. Fernandez, A., Maignan, L., Spicher, A.: Accretive computation of global transformations - extended version. arXiv preprint arXiv:2103.09636 (2021)
10. Fernandez, A., Maignan, L., Spicher, A.: Cellular automata and Kan extensions. In: 27th IFIP WG 1.5 International Workshop on Cellular Automata and Discrete Complex Systems, AUTOMATA 2021, Aix-Marseille University, France. OASIcs, 12–14 July 2021, vol. 90, pp. 7:1–7:12. Schloss Dagstuhl - Leibniz-Zentrum für Informatik (2021). https://doi.org/10.4230/OASIcs.AUTOMATA.2021.7
11. Knuth, D.E., Morris, J.H., Jr., Pratt, V.R.: Fast pattern matching in strings. SIAM J. Comput. **6**(2), 323–350 (1977)
12. Lack, S., Sobociński, P.: Adhesive and quasiadhesive categories. RAIRO Theor. Inform. Appl. **39**(3), 511–545 (2005)
13. Mac Lane, S.: Categories for the Working Mathematician, vol. 5. Springer, New York (2013)
14. Maignan, L., Spicher, A.: Global graph transformations. In: GCM@ ICGT, pp. 34–49 (2015)
15. Srinivas, Y.V.: A sheaf-theoretic approach to pattern matching and related problems. Theor. Comput. Sci. **112**(1), 53–97 (1993)

Some Modal and Temporal Translations of Generalized Basic Logic

Wesley Fussner[1(✉)] and William Zuluaga Botero[1,2]

[1] Laboratoire J.A. Dieudonné, CNRS, and Université Côte d'Azur, Nice, France
wfussner@unice.fr, william.zuluaga@univ-cotedazur.fr
[2] Departamento de Matemática, Facultad de Ciencias Exactas, Universidad Nacional del Centro, Tandil, Argentina

Abstract. We introduce a family of modal expansions of Łukasiewicz logic that are designed to accommodate modal translations of generalized basic logic (as formulated with exchange, weakening, and falsum). We further exhibit algebraic semantics for each logic in this family, in particular showing that all of them are algebraizable in the sense of Blok and Pigozzi. Using this algebraization result and an analysis of congruences in the pertinent varieties, we establish that each of the introduced modal Łukasiewicz logics has a local deduction-detachment theorem. By applying Jipsen and Montagna's poset product construction, we give two translations of generalized basic logic with exchange, weakening, and falsum in the style of the celebrated Gödel-McKinsey-Tarski translation. The first of these interprets generalized basic logic in a modal Łukasiewicz logic in the spirit of the classical modal logic **S4**, whereas the second interprets generalized basic logic in a temporal variant of the latter.

Keywords: GBL-algebras · Modal logic · Modal translations

1 Introduction

Substructural logics make up a widely studied family of nonclassical logics originating in proof theory, and have found an array of applications in theoretical computer science (e.g., in the management of computational resources [22]). Generalized basic logic (see, e.g., [5,18]) is a prominent substructural logic that has been highly influential in the development of residuated lattices [14], which provide the algebraic semantics of substructural logics. Generalized basic logic is a common fragment of intuitionistic propositional logic and Hájek's basic fuzzy logic [16], and its algebraic models (viz. *GBL-algebras*) provide a natural common generalization of lattice-ordered groups, Heyting algebras, and continuous

This project received funding from the European Research Council (ERC) under the European Union's Horizon 2020 research and innovation program (grant agreement No. 670624).

U. Fahrenberg et al. (Eds.): RAMiCS 2021, LNCS 13027, pp. 176–191, 2021.
https://doi.org/10.1007/978-3-030-88701-8_11

t-norm based logic algebras (see [13] for a survey). This logic has also been proposed as a model of flexible resources [5], in keeping with resource-driven interpretations of substructural logics generally.

When extended by exchange, weakening, and falsum (as we do throughout the sequel), generalized basic logic may be regarded as an 'intuitionistic' variant of Hájek's basic fuzzy logic. In this formulation, generalized basic logic is related to Łukasiewicz logic [8] in much the same way that intuitionistic logic is related to classical logic. For instance, generalized basic logic admits a Kripke-style relational semantics [12] in which worlds are valued in MV-algebra chains, mirroring the well known Kripke semantics for intuitionistic logic (in which worlds are valued in the 2-element MV-algebra/Boolean algebra). It is evident from [12] that generalized basic logic may be viewed as a fragment of a modal Łukasiewicz logic, but the details of this modal connection are therein left implicit. On the other hand, [12] generalizes the temporal flow semantics for basic logic [1], which is deployed in [2] to obtain a modal translation of Gödel-Dummett logic into an extension of Prior's tense logic [23].

Inspired by this work, the present study makes the modal connection from [12] explicit and offers modal and temporal translations of generalized basic logic into certain expanded Łukasiewicz logics. The motivations for this study are threefold. First, due to the astounding diversity of substructural logics, understanding relationships among various substructural deductive systems is crucial to their general theory. The translation results of the present study deepen our understanding of the connection between generalized basic logic and Łukasiewicz logic, two of the most salient substructural logics. Second, the translations articulated in the sequel directly generalize the well known Gödel-McKinsey-Tarski translation of intuitionistic logic into the classical modal logic **S4**, adding to a long line of studies that generalize themes from intuitionistic logic to the substructural setting. In addition to clarifying the role of modality in generalized basic logic, this connects to the broader theory of modal companions of superintuitionistic logics. Third, because our translations target modal Łukasiewicz logics, this work adds to the emerging literature on fuzzy modal logics. Moreover, we expect that the results of this paper open up the application of tools from fuzzy modal logic (such as filtration [10]) to the analysis of generalized basic logic and its extensions.

Our contributions are as follows. First, we introduce in Sect. 2 a family of modal Łukasiewicz logics that serve as targets for our translations. This family includes both monounary modal systems, analogous to classical **S4**, as well as multimodal systems of temporal Łukasiewicz logic. This investigation is rooted in algebraic logic, and in Sect. 3 we provide pertinent information on algebras related to this study. In Sect. 3.2, we demonstrate that all of the logics introduced in Sect. 2 are algebraizable in the sense of Blok and Pigozzi (see [3]) and that the algebras introduced in Sect. 3.1 provide their equivalent algebraic semantics. Equipped with this algebra-logic bridge, Sect. 4 puts our algebraization theorem to work and establishes a local deduction detachment-theorem for our modal Łukasiewicz logics. The work of Sect. 4 is based on an analysis of congruences in the varieties of algebras introduced in Sect. 3.1, and in particular establishes the congruence extension property for each of these varieties. Finally, in Sect. 5 we introduce two translations of generalized basic logic, one into a Łukasiewicz version of **S4** and the other into a temporal Łukasiewicz logic. These translations both rely on the poset product construction of Jipsen and Montagna (see, e.g., [20]).

2 Generalized Basic Logic and Fuzzy Modal Logics

This section introduces the logical systems of our inquiry. The logics discussed in this paper are all defined over supersets of the propositional language $\mathcal{L}$ consisting of the binary connectives $\wedge, \vee, \cdot, \to$ and the constants $0, 1$. To the basic language $\mathcal{L}$ we will adjoin a set of box-like unary modal connectives. More specifically, given a set I of unary connective symbols with $I \cap \mathcal{L} = \emptyset$, we define a language $\mathcal{L}(I) = \mathcal{L} \cup I$. We further fix a countably-infinite set Var of propositional variables, and denote by $Fm_{\mathcal{L}(I)}$ the set of $\mathcal{L}(I)$-formulas over Var.[1] An $\mathcal{L}(I)$-*equation* is an ordered pair $(\varphi, \psi) \in Fm_{\mathcal{L}(I)}$, and we usually denote the equation (φ, ψ) by $\varphi \approx \psi$. The set of all $\mathcal{L}(I)$-equations is denoted by $Eq_{\mathcal{L}(I)}$. All of the logics we consider may be defined by Hilbert-style calculi using various selections from the axiom schemes and deduction rules depicted in Fig. 1. Observe that in Fig. 1 each of ($\mathrm{K}_\Box$), ($\mathrm{P}_\Box$), ($\mathrm{M}_\Box$), ($1_\Box$), ($0_\Box$), ($\mathrm{T}_\Box$), ($\mathbf{4}_\Box$), (GP), (HF), and ($\Box$-Nec) gives a family of axiom schemes/rules parameterized by the unary connectives $\Box$, G, H. Note that we write $\varphi \leftrightarrow \psi$ for $(\varphi \to \psi) \wedge (\psi \to \varphi)$ and $\neg\varphi$ for $\varphi \to 0$ as usual.

From [5], *generalized basic logic with exchange, weakening, and falsum* is the logic defined over $\mathcal{L}$ by the calculus with (A1)–(A13) and the modus ponens rule (MP). We denote this logic by **GBL**. Additionally including the prelinearity axiom (A14) yields Hájek's basic fuzzy logic [16], which we denote by **BL**. It follows from [9] that including both (A14) and (A15) gives an axiomatization of the infinite-valued Łukasiewicz logic **Ł** (see, e.g., [8]).

We will consider a number of different modal expansions of **Ł** in this study. For an arbitrary set I of unary connective symbols disjoint from $\mathcal{L}$, we denote by $\mathbf{Ł}(I)$ the logic with language $\mathcal{L}(I)$, axiom schemes (A1)–(A15), ($\mathrm{K}_\Box$), ($\mathrm{P}_\Box$), ($\mathrm{M}_\Box$), ($1_\Box$), and ($0_\Box$) (where $\Box$ ranges over I in all of the preceding axiom schemes), and rules (MP) and ($\Box$-Nec) (where again $\Box$ ranges over I). We denote by $\mathbf{S4Ł}(I)$ the logic resulting from adding to $\mathbf{Ł}(I)$ the axiom schemes ($\mathrm{T}_\Box$) and ($\mathbf{4}_\Box$) for all $\Box \in I$. If $I = \{\Box\}$ is a singleton, we write **S4Ł** for $\mathbf{S4Ł}(I)$. If $I = \{G, H\}$, then the logic defined by adding to $\mathbf{S4Ł}(I)$ the axioms (GP) and (HF) will be denoted by $\mathbf{S4_tŁ}$.

The logic **S4Ł** is a fuzzy analogue of the classical modal logic **S4**, whereas $\mathbf{S4_tŁ}$ is a temporal variant of **S4Ł** inspired by Prior's tense logic [23]. The names of the axioms (GP) and (HF) derive from the fact that—as usual in tense

[1] Recall that formulas are constructed recursively by stipulating that p is a formula for each $p \in \mathsf{Var}$, and further that if ω is an n-ary connective symbol and $\varphi_1, \dots, \varphi_n$ are formulas, then so is $\omega(\varphi_1, \dots, \varphi_n)$. As usual, we write binary connectives using infix notation.

Axiom schemes

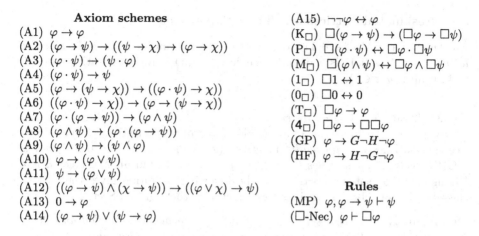

(A1) $\varphi \to \varphi$
(A2) $(\varphi \to \psi) \to ((\psi \to \chi) \to (\varphi \to \chi))$
(A3) $(\varphi \cdot \psi) \to (\psi \cdot \varphi)$
(A4) $(\varphi \cdot \psi) \to \psi$
(A5) $(\varphi \to (\psi \to \chi)) \to ((\varphi \cdot \psi) \to \chi))$
(A6) $((\varphi \cdot \psi) \to \chi)) \to (\varphi \to (\psi \to \chi))$
(A7) $(\varphi \cdot (\varphi \to \psi)) \to (\varphi \wedge \psi)$
(A8) $(\varphi \wedge \psi) \to (\varphi \cdot (\varphi \to \psi))$
(A9) $(\varphi \wedge \psi) \to (\psi \wedge \varphi)$
(A10) $\varphi \to (\varphi \vee \psi)$
(A11) $\psi \to (\varphi \vee \psi)$
(A12) $((\varphi \to \psi) \wedge (\chi \to \psi)) \to ((\varphi \vee \chi) \to \psi)$
(A13) $0 \to \varphi$
(A14) $(\varphi \to \psi) \vee (\psi \to \varphi)$
(A15) $\neg\neg\varphi \leftrightarrow \varphi$
($\mathrm{K}_\Box$) $\Box(\varphi \to \psi) \to (\Box\varphi \to \Box\psi)$
($\mathrm{P}_\Box$) $\Box(\varphi \cdot \psi) \leftrightarrow \Box\varphi \cdot \Box\psi$
($\mathrm{M}_\Box$) $\Box(\varphi \wedge \psi) \leftrightarrow \Box\varphi \wedge \Box\psi$
($1_\Box$) $\Box 1 \leftrightarrow 1$
($0_\Box$) $\Box 0 \leftrightarrow 0$
($\mathrm{T}_\Box$) $\Box\varphi \to \varphi$
($4_\Box$) $\Box\varphi \to \Box\Box\varphi$
(GP) $\varphi \to G\neg H\neg\varphi$
(HF) $\varphi \to H\neg G\neg\varphi$

Rules

(MP) $\varphi, \varphi \to \psi \vdash \psi$
($\Box$-Nec) $\varphi \vdash \Box\varphi$

Fig. 1. Axiom schemes and rules for the logics considered.

logic—we define modal diamond connectives P and F as abbreviations for $\neg H \neg$ and $\neg G \neg$, respectively. The typical intended interpretations of the modals G, P, H, F are:

- $G\varphi$: "It is always **g**oing to be the case that φ."
- $P\varphi$: "It was true at one point in the **p**ast that φ."
- $H\varphi$: "It always **h**as been the case that φ."
- $F\varphi$: "It will be true at some point in the **f**uture that φ."

In Sect. 5, we will exhibit translations of **GBL** into each of **S4Ł** and **S4$_\mathbf{t}$Ł**. These translations closely mirror the Gödel-McKinsey-Tarski translation of propositional intuitionistic logic into **S4**. Intuitively, **S4Ł** is a modal companion of **GBL** (see [7]). On the other hand, our translation into **S4$_\mathbf{t}$Ł** generalizes the translation presented in [2] of Gödel-Dummett logic into Prior's classical tense logic.

Given a logic **L**, we denote by $\vdash_{\mathbf{L}}$ the consequence relation corresponding to **L** (see [11] for background on consequence relations). As one may anticipate from the presence of the axioms ($\mathrm{K}_\Box$) and $\Box$-necessitation rules, the logics we have introduced above turn out to be algebraizable in the sense of Blok and Pigozzi [3] (see Theorem 1).

3 Algebraic Semantics for Ł(I) and Its Extensions

We now turn to providing algebraic semantics for the logics introduced in Sect. 2. In Sect. 3.1 we describe the pertinent algebraic structures, and then in Sect. 3.2 we give the algebraization results for the logics we have introduced. We assume familiarity with the basics of universal algebra [6], residuated lattices [14], and abstract algebraic logic [11], but where possible we provide specific references to some key background results that we invoke without full discussion.

3.1 Residuated Lattices and Their Expansions

An algebra $(A, \wedge, \vee, \cdot, \rightarrow, 0, 1)$ is called a *bounded commutative integral residuated lattice* if $(A, \wedge, \vee, 0, 1)$ is a bounded lattice, $(A, \cdot, 1)$ is a commutative monoid, and for all $x, y, z \in A$,

$$x \cdot y \leq z \iff x \leq y \rightarrow z.$$

We usually abbreviate $x \cdot y$ by xy.

By a *GBL-algebra* we mean a bounded integral commutative residuated lattice that satisfies the *divisibility* identity $x(x \rightarrow y) \approx x \wedge y$.[2] A *BL-algebra* is a GBL-algebra that satisfies $(x \rightarrow y) \vee (y \rightarrow x) \approx 1$, and an *MV-algebra* is a BL-algebra that satisfies $\neg\neg x \approx x$. The following definition gives the various classes of MV-algebra expansions that algebraize the logics of Sect. 2.

Definition 1. *Let I be a set of unary function symbols. We say that an algebra $\boldsymbol{A} = (A, \wedge, \vee, \cdot, \rightarrow, 0, 1, \{\Box\}_{\Box \in I})$ is an* MV(I)-algebra *provided that:*

1. *$(A, \wedge, \vee, \cdot, \rightarrow, 0, 1)$ is an MV-algebra.*
2. *For every $\Box \in I$, $\Box$ is a $\{\wedge, \cdot, 0, 1\}$-endomorphism of $(A, \wedge, \vee, \cdot, \rightarrow, 0, 1)$.*

If additionally $\Box$ is an interior operator for every $\Box \in I$, then we say that $\mathbf{A}$ is an S4MV(I)-algebra. *An* S4MV-algebra *is an S4MV(I)-algebra where $I = \{\Box\}$ is a singleton. An S4MV(I)-algebra for $I = \{G, H\}$ is called an S4$_t$MV*-algebra *if the map P defined by $P(x) = \neg H(\neg x)$ is the lower residual of G, i.e., for every $x, y \in A$,*

$$x \leq G(y) \iff P(x) \leq y.$$

In each S4$_t$MV-algebra, we also abbreviate $\neg G(\neg x)$ by $F(x)$.

The following summarizes some technical facts regarding S4$_t$MV-algebras. Its proof is straightforward and we omit it.

Lemma 1. *Let $\mathbf{A}$ be an S4$_t$MV-algebra and let $x, y \in A$. Then:*

1. *$P(x \vee y) = P(x) \vee P(y)$.*
2. *$P(0) = 0$ and $P(1) = 1$.*
3. *$x \leq H(y)$ if and only if $F(x) \leq y$.*
4. *$F(x \vee y) = F(x) \vee F(y)$.*
5. *$F(1) = 1$ and $F(0) = 0$.*
6. *$x \rightarrow GP(x) = 1$ and $PG(x) \rightarrow x = 1$.*
7. *$x \rightarrow HF(x) = 1$ and $FH(x) \rightarrow x = 1$.*
8. *P and F are closure operators.*

It is well known that bounded commutative integral residuated lattices form a variety, and hence so do the classes of GBL-algebras, BL-algebras, and MV-algebras. We denote these varieties by GBL, BL, and MV, respectively. The proof of the following lemma is straightforward from the definitions.

[2] Most studies refer to these algebras as *bounded commutative GBL-algebras* or *GBL$_{ewf}$-algebras*. Because we always assume boundedness and commutativity, we call them GBL-algebras in order to simplify terminology.

Lemma 2. *Let I be a set of unary function symbols with $\mathcal{L} \cap I = \emptyset$. The class of MV(I)-algebras forms a variety, and the class of S4MV(I)-algebras is a subvariety of the latter. Moreover, the class of $S4_t$MV-algebras forms a subvariety of the variety of S4MV(G,H)-algebras.*

We denote the varieties of MV(I)-algebras, S4MV(I)-algebras, S4MV-algebras, and $S4_t$MV-algebras by $\mathsf{MV}(I)$, $\mathsf{S4MV}(I)$, $\mathsf{S4MV}$, and $\mathsf{S4_tMV}$, respectively.

3.2 Algebraization

We now discuss algebraization of the logics of Sect. 2. Each of the logics **GBL**, **BL**, and **Ł** is algebraizable with the sole defining equation $\varphi \approx 1$ and sole equivalence formula $\varphi \leftrightarrow \psi$ (see, e.g., [14]). The equivalent variety semantics for **GBL**, **BL**, and **Ł** are, respectively, the varieties GBL, BL, and MV. Extending the algebraizability of **Ł** to its modal expansions boils down to showing that the consequence relation is compatible with the new connectives in the sense summarized in the following key lemma. We omit its straightforward proof.

Lemma 3. *Let I be a set of unary connectives with $I \cap \mathcal{L} = \emptyset$, and let* **L** *be an extension of* **Ł**(I). *Then $\varphi \leftrightarrow \psi \vdash_{\mathbf{L}} \Box\varphi \leftrightarrow \Box\psi$ for each $\Box \in I$.*

The following gives our main result on algebraization.

Theorem 1. *Let I be a set of unary connectives with $\mathcal{L} \cap I = \emptyset$. Then:*

1. **Ł**(I) *is algebraizable with the sole defining equation $\varphi \approx 1$ and sole equivalence formula $\varphi \leftrightarrow \psi$, and consequently so are* **S4Ł**(I), **S4Ł**, *and* $\mathbf{S4_tŁ}$.
2. *The equivalent variety semantics for* **Ł**(I), **S4Ł**(I), **S4Ł**, *and* $\mathbf{S4_tŁ}$ *are, respectively,* $\mathsf{MV}(I)$, $\mathsf{S4MV}(I)$, $\mathsf{S4MV}$, *and* $\mathsf{S4_tMV}$.

Proof. 1. It follows from [3, Theorem 4.7] that a logic **L** expanding **Ł** by a set of connectives Ω is algebraizable if for every n-ary $\omega \in \Omega$ we have

$$\varphi_0 \leftrightarrow \psi_0, \ldots, \varphi_{n-1} \leftrightarrow \psi_{n-1} \vdash_{\mathbf{L}} \omega(\varphi_0, \ldots, \varphi_{n-1}) \leftrightarrow \omega(\psi_0, \ldots, \psi_{n-1}).$$

Moreover, in this case **L** is algebraizable with sole defining equation $\varphi \approx 1$ and sole equivalence formula $\varphi \leftrightarrow \psi$. The result for **Ł**$(I)$ is thus immediate from Lemma 3. The claim for **S4Ł**(I), **S4Ł**, and $\mathbf{S4_tŁ}$ follows promptly because each of the latter logics is an axiomatic extension of **Ł**(I) for some I.

2. By [3, Theorem 2.17], the quasivariety K algebraizing **Ł**(I) is axiomatized by the following quasiequations: $\varphi \approx 1$ for all instances φ of the axiom schemes given in the calculus for **Ł**(I); $x \leftrightarrow x \approx 1$; $\varphi, \varphi \to \psi$ implies ψ; φ implies $\Box\varphi$; and $x \leftrightarrow y \approx 1$ implies $x \approx y$. It is easy to see from Definition 1 and the fact that MV algebraizes **Ł** that all of these quasiequations are valid in $\mathsf{MV}(I)$. Thus $\mathsf{MV}(I) \subseteq \mathsf{K}$. For the reverse inclusion, it suffices to show that all the defining equations of $\mathsf{MV}(I)$ follow from this list of quasiequations. Let $\mathbf{A} \in \mathsf{K}$. That the $\{\wedge, \vee, \cdot, \to, 0, 1\}$-reduct of $\mathbf{A}$ is an MV-algebra is immediate

from the fact that MV algebraizes **Ł**. On the other hand, for each $\Box \in I$ the equations $\Box(x \cdot y) \leftrightarrow \Box x \cdot \Box y \approx 1$, $\Box(x \wedge y) \leftrightarrow \Box x \cdot \Box y \approx 1$, $\Box 1 \leftrightarrow 1 \approx 1$, and $\Box 0 \leftrightarrow 0 \approx 0$ appear in the list of quasiequations, and together these imply that $\Box$ is a $\{\wedge, \cdot, 0, 1\}$-homomorphism of $\mathbf{A}$ for each $\Box \in I$. Thus $\mathsf{K} \subseteq \mathsf{MV}(\mathsf{I})$, giving equality. The result for the axiomatic extensions $\mathbf{S4Ł}(I)$, $\mathbf{S4Ł}$, and $\mathbf{S4_tŁ}$ follows by applying the formula-to-equation translation $\varphi \mapsto \varphi \approx 1$ to each formula φ axiomatizing the given logic relative to $\mathbf{Ł}(I)$. ∎

Recall that if K is a class of similar algebras and $\Theta \cup \{\epsilon \approx \delta\}$ is a set of equations in the type of K, then $\Theta \models_{\mathsf{K}} \epsilon \approx \delta$ means that for every $\mathbf{A} \in \mathsf{K}$ and every assignment h of variables into $\mathbf{A}$, if $h(\alpha) = h(\beta)$ for every $\alpha \approx \beta \in \Theta$, then $h(\epsilon) = h(\delta)$. Thanks to the finitarity of $\mathbf{Ł}(I)$, the following is a direct consequence of Theorem 1 (see [11, Corollary 3.40]).

Corollary 1. *Let I be a set of unary connectives with $\mathcal{L} \cap I = \emptyset$. There is a dual lattice isomorphism between the lattice of finitary extensions of* $\mathbf{Ł}(I)$ *and the lattice of subquasivarieties of* $\mathsf{MV}(I)$, *which restricts to a dual lattice isomorphism between the lattice of axiomatic extensions of* $\mathbf{Ł}(I)$ *and the lattice of subvarieties of* $\mathsf{MV}(I)$. *Moreover, suppose that* $\mathbf{L}$ *is a finitary extension of* $\mathbf{Ł}(I)$, *and let* K *be the equivalent algebraic semantics of* $\mathbf{L}$. *Then for any set* $\Gamma \cup \{\varphi\} \subseteq Fm_{\mathcal{L}(I)}$ *and any set* $\Theta \cup \{\epsilon \approx \delta\} \subseteq Eq_{\mathcal{L}(I)}$:

1. $\Gamma \vdash_{\mathbf{L}} \varphi \iff \{\gamma \approx 1 : \gamma \in \Gamma\} \models_{\mathsf{K}} \varphi \approx 1$.
2. $\Theta \models_{\mathsf{K}} \epsilon \approx \delta \iff \{\alpha \leftrightarrow \beta : \alpha \approx \beta \in \Theta\} \vdash_{\mathbf{L}} \epsilon \leftrightarrow \delta$.

In particular, this holds if $\mathbf{L} \in \{\mathbf{Ł}(I), \mathbf{S4Ł}(I), \mathbf{S4Ł}, \mathbf{S4_tŁ}\}$.

4 Characterizing Filters and a Deduction Theorem

If $\mathbf{L}$ is an algebraizable logic, there is a well known connection between the theories of $\mathbf{L}$, the deductive filters of algebraic models of $\mathbf{L}$, and the congruence relations of the equivalent algebraic semantics of $\mathbf{L}$ (see, e.g., [11,15]). Armed with the algebraizability results of Sect. 2, we now provide an analysis of congruences in the algebraic semantics given in Sect. 3 in terms of certain filters (see Definition 2 below). We use this description to establish local deduction-detachment theorems for the modal Łukasiewicz logics we have introduced.

Definition 2. *Let* $\mathbf{A}$ *be an MV(I)-algebra. We say that a non-empty subset* $\mathfrak{f}$ *of* A *is an* I-filter *provided that* $\mathfrak{f}$ *is an up-set,* $\mathfrak{f}$ *is closed under* $\cdot$, *and* $\mathfrak{f}$ *is closed under each* $\Box \in I$.

Let $\mathbf{A} \in \mathsf{MV}(I)$. We define a term operation $*$ by $x * y = (x \to y)(y \to x)$. We also write $\mathsf{Fi}(\mathbf{A})$ for the poset of I-filters of $\mathbf{A}$ ordered by inclusion and $\mathsf{Con}(\mathbf{A})$ for the congruence lattice of $\mathbf{A}$.

Lemma 4. *Let* $\mathbf{A}$ *be an MV(I)-algebra,* $\mathfrak{f} \in \mathsf{Fi}(\mathbf{A})$, *and* $\theta \in \mathsf{Con}(\mathbf{A})$. *Then the following hold:*

1. *$\mathfrak{f}_\theta = 1/\theta$ is an I-filter of* $\mathbf{A}$.
2. *The set $\theta_\mathfrak{f} = \{(x,y) \in A^2 \colon x * y \in \mathfrak{f}\} = \{(x,y) \in A^2 \colon x \leftrightarrow y \in \mathfrak{f}\}$ is a congruence on* $\mathbf{A}$.
3. *The maps $\mathfrak{f} \mapsto \theta_\mathfrak{f}$, $\theta \mapsto \mathfrak{f}_\theta$ define mutually-inverse poset isomorphisms between* $\mathsf{Con}(\mathbf{A})$ *and* $\mathsf{Fi}(\mathbf{A})$. *Consequently,* $\mathsf{Fi}(\mathbf{A})$ *is a lattice and these poset isomorphisms are lattice isomorphisms.*

Proof. 1. Note that $\mathfrak{f}_\theta$ is a deductive filter of the MV-algebra reduct of $\mathbf{A}$ (see, e.g., [14, Section 3.6]), so it suffices to show that $\mathfrak{f}_\theta$ is closed under $\Box$ for every $\Box \in I$. Observe that if $(1,x) \in \theta$ then since θ is a congruence we have $(\Box 1, \Box x) \in \theta$. But since $\Box 1 = 1$, it follows that $\Box x \in \mathfrak{f}_\theta$ as desired.

2. Observe first that $x * y \in \mathfrak{f}$ if and only if $x \leftrightarrow y \in \mathfrak{f}$, so the two sets displayed are equal. Now since $\mathfrak{f}$ is in particular a deductive filter of the MV-algebra reduct $\mathbf{A}$, it is immediate that $\theta_\mathfrak{f}$ respects all of the operations except for possibly those belonging to I. To show that $\theta_\mathfrak{f}$ respects these as well, it suffices to show the result for every $\Box \in I$. Suppose that $(x,y) \in \theta_\mathfrak{f}$, i.e., $x * y \in \mathfrak{f}$. Since $\mathfrak{f}$ is closed under $\Box$, and every $\Box \in I$ preserves $\cdot$, we have $\Box(x \to y) \cdot \Box(y \to x) \in \mathfrak{f}$. Residuation and the fact that $\Box$ preserves $\cdot$ gives $\Box(x \to y) \leq \Box x \to \Box y$ and $\Box(y \to x) \leq \Box y \to \Box x$, so

$$\Box(x * y) = \Box(x \to y)\Box(y \to x) \leq (\Box x \to \Box y)(\Box y \to \Box x).$$

Since $\mathfrak{f}$ is an up-set, we get $\Box x * \Box y \in \mathfrak{f}$. Hence $(\Box x, \Box y) \in \theta_\mathfrak{f}$ as required.

3. Direct computation shows $\mathfrak{f} = \mathfrak{f}_{\theta_\mathfrak{f}}$ and $\theta_{\mathfrak{f}_\theta} = \theta$ for every I-filter $\mathfrak{f}$ and congruence θ. Moreover, the given maps are clearly monotone. It follows that $\mathsf{Con}(\mathbf{A})$ and $\mathsf{Fi}(\mathbf{A})$ are isomorphic as posets. Because $\mathsf{Fi}(\mathbf{A})$ is isomorphic to the lattice $\mathsf{Con}(\mathbf{A})$, we get that $\mathsf{Fi}(\mathbf{A})$ is a lattice that is isomorphic to $\mathsf{Con}(\mathbf{A})$. ∎

The following gives a description of congruence generation in $\mathsf{MV}(I)$. Recall that if $(P, \leq)$ is a partially ordered set and $X \subseteq P$, then the smallest up-set containing X is the set $\uparrow X = \{y \in P : x \leq y \text{ for some } x \in X\}$.

Definition 3. *Let A be an MV(I)-algebra and let $X \subseteq A$.*

1. *An* I-block *is a nonempty word in the alphabet I. We denote the set of I-blocks by $\mathcal{B}_I$.*
2. $\mathsf{Fg}^\mathbf{A}(X) = \uparrow\{M_1(x_1) \cdot \dots \cdot M_n(x_n) \colon x_1, \dots, x_n \in X$ *and* $M_1, \dots, M_n \in \mathcal{B}_I\}$.

Lemma 5. *The set $\mathsf{Fg}^\mathbf{A}(X)$ is the least I-filter of $\mathbf{A}$ containing X.*

Proof. It is clear that $\mathsf{Fg}^\mathbf{A}(X)$ is an up-set. Note that if $y, y' \in \mathsf{Fg}^\mathbf{A}(X)$ then there exist $M_1, \dots, M_n, M_1', \dots, M_k' \in \mathcal{B}_I$ and $x_1, \dots, x_n, x_1', \dots, x_k' \in X$ with $M_1(x_1) \cdot \dots \cdot M_n(x_n) \leq y$ and $M_1'(x_1') \cdot \dots \cdot M_k'(x_k') \leq y'$, whence we obtain $M_1(x_1) \cdot \dots \cdot M_n(x_n) \cdot M_1'(x_1') \cdot \dots \cdot M_k'(x_k') \leq y \cdot y'$ since $\cdot$ preserve the order in each coordinate. It follows that $y \cdot y' \in \mathsf{Fg}^\mathbf{A}(X)$. To see that $\mathsf{Fg}^\mathbf{A}(X)$ is closed under every $\Box \in I$, observe that if $M_1(x_1) \cdot \dots \cdot M_n(x_n) \leq y$ then by the isotonicity of $\Box$ we have $\Box M_1(x_1) \cdot \dots \cdot \Box M_n(x_n) \leq \Box y$. As each $\Box M_i$ is an I-block, it follows that $\Box(y) \in \mathsf{Fg}^\mathbf{A}(X)$.

It remains to check that $\mathsf{Fg}^{\mathbf{A}}(X)$ is the least among the I-filters containing X. Suppose that $\mathfrak{f}$ is an I-filter and that $X \subseteq \mathfrak{f}$. If $y \in \mathsf{Fg}^{\mathbf{A}}(X)$, then there exist $M_1, \ldots, M_n \in \mathcal{B}_I$ and $x_1, \ldots, x_n \in X$ such that $M_1(x_1) \cdot \ldots \cdot M_n(x_n) \leq y$. Note that $x_1, \ldots, x_n \in \mathfrak{f}$, and since $\mathfrak{f}$ is closed under $\Box$ for every $\Box \in I$, we have that $M(x) \in \mathfrak{f}$ for every $M \in \mathcal{B}_I$ and every $x \in \mathfrak{f}$. In particular, this implies that $M_1(x_1), \ldots, M_n(x_n) \in \mathfrak{f}$. Thus $y \in \mathfrak{f}$ since $\uparrow\mathfrak{f} = \mathfrak{f}$, so $\mathsf{Fg}^{\mathbf{A}}(X) \subseteq \mathfrak{f}$ as claimed. ■

We abbreviate $\mathsf{Fg}^{\mathbf{A}}(\{x_1, \ldots, x_n\})$ by $\mathsf{Fg}^{\mathbf{A}}(x_1, \ldots, x_n)$. Also, for an algebra $\mathbf{A}$ and $x, y \in A$, we denote by $\mathsf{Cg}^{\mathbf{A}}(x, y)$ the congruence of $\mathbf{A}$ generated by (x, y).

Lemma 6. *Let* $\mathbf{A} \in \mathsf{MV}(I)$, *let* $x, y \in A$, *let* $Y \subseteq A$, *and consider the set* $X = \{(1, y) : y \in Y\}$. *Then:*

1. $\mathfrak{f}_{\mathsf{Cg}^{\mathbf{A}}(x,y)} = \mathsf{Fg}^{\mathbf{A}}(x * y) = \mathsf{Fg}^{\mathbf{A}}(x \leftrightarrow y)$.
2. $\mathfrak{f}_{\mathsf{Cg}^{\mathbf{A}}(X)} = \mathsf{Fg}^{\mathbf{A}}(Y)$.

Proof. 1. Note that $\mathsf{Cg}^{\mathbf{A}}(x, y) = \bigcap\{\theta \in \mathsf{Con}(\mathbf{A}) \colon (x, y) \in \theta\}$, and observe that for each $\theta \in \mathsf{Con}(\mathbf{A})$ we have $(x, y) \in \theta$ if and only if $x * y \in \mathfrak{f}_\theta$. Hence from the isomorphism given by Lemma 4(3) we obtain:

$$\mathfrak{f}_{\mathsf{Cg}^{\mathbf{A}}(x,y)} = \bigcap\{\mathfrak{f} \in \mathsf{Fi}(\mathbf{A}) \colon x * y \in \mathfrak{f}\} = \mathsf{Fg}^{\mathbf{A}}(x * y) = \mathsf{Fg}^{\mathbf{A}}(x \leftrightarrow y).$$

This proves 1.

2. Since $\mathsf{Cg}^{\mathbf{A}}(X) = \bigvee_{y \in Y} \mathsf{Cg}^{\mathbf{A}}(1, y)$, Lemma 4(3) and item 1 imply

$$\mathfrak{f}_{\mathsf{Cg}^{\mathbf{A}}(X)} = \bigvee_{y \in Y} \mathfrak{f}_{\mathsf{Cg}^{\mathbf{A}}(1,y)} = \bigvee_{y \in Y} \mathsf{Fg}^{\mathbf{A}}(y) = \mathsf{Fg}^{\mathbf{A}}(\bigcup_{y \in Y} \{y\}) = \mathsf{Fg}^{\mathbf{A}}(Y).$$

This proves 2. ■

Recall that an algebra $\mathbf{B}$ has the *congruence extension property* (or *CEP*) if for every subalgebra $\mathbf{A}$ of $\mathbf{B}$ and for any $\theta \in \mathsf{Con}(\mathbf{A})$, there exists $\xi \in \mathsf{Con}(\mathbf{B})$ such that $\xi \cap A^2 = \theta$. A variety V is said to have the congruence extension property if each $\mathbf{B} \in \mathsf{V}$ does.

Theorem 2. $\mathsf{MV}(I)$ *has the congruence extension property.*

Proof. Let $\mathbf{A}, \mathbf{B}$ be MV(I)-algebras, and assume that $\mathbf{A}$ is a subalgebra of $\mathbf{B}$. From Lemma 4, it follows that proving the congruence extension property for $\mathsf{MV}(I)$ is equivalent to showing that every I-filter of $\mathbf{A}$ can be extended by an I-filter of $\mathbf{B}$. For this, let $\mathfrak{f} \in \mathsf{Fi}(\mathbf{A})$ and set $\mathfrak{g} = \mathsf{Fg}^{\mathbf{B}}(\mathfrak{f})$. In order to prove $\mathfrak{f} = \mathfrak{g} \cap A$, let $y \in \mathfrak{g} \cap A$. Then since $y \in \mathfrak{g}$ there exist $M_1, \ldots, M_n \in \mathcal{B}_I$ and $x_1, \ldots, x_n \in \mathfrak{f}$ such that $M_1(x_1) \cdot \ldots \cdot M_n(x_n) \leq y$. Since $\mathfrak{f}$ is an I-filter of $\mathbf{A}$, we have $M_j(x_j) \in \mathfrak{f}$ for every $1 \leq j \leq n$. As $y \in A$, it follows that $y \in \mathfrak{f}$ and $\mathfrak{g} \cap A \subseteq \mathfrak{f}$. The reverse inclusion is obvious, and the result follows. ■

Of course, the CEP persists in subvarieties of a variety with the CEP. Thus:

Corollary 2. *Each of* $\mathsf{S4MV}(I)$, $\mathsf{S4MV}$, *and* $\mathsf{S4_tMV}$ *has the CEP.*

The CEP has far-reaching logical consequences. Recall that a logic $\mathbf{L}$ has the *local deduction-detachment theorem* (or *LDDT*) if there exists a family $\{d_j(p,q)\colon j \in J\}$ of sets $d_j(p,q)$ of formulas in at most two variables such that for every set $\Gamma \cup \{\varphi, \psi\}$ of formulas in the language of $\mathbf{L}$:

$$\Gamma, \varphi \vdash_{\mathbf{L}} \psi \Longleftrightarrow \Gamma \vdash_{\mathbf{L}} d_j(\varphi, \psi) \text{ for some } j \in J.$$

As a consequence of [4, Corollary 5.3], if $\mathbf{L}$ is an algebraizable logic with equivalent variety semantics V, then $\mathbf{L}$ has the LDDT if and only if V has the CEP. Therefore from Theorem 1, Theorem 2, and Corollary 2 we obtain:

Corollary 3. *Each of* $\mathbf{Ł}(I)$, $\mathbf{S4Ł}(I)$, $\mathbf{S4Ł}$, *and* $\mathbf{S4_tŁ}$ *has the LDDT.*

From our analysis of congruences in $\mathsf{MV}(I)$, we may give a more explicit rendering of this result. If V is a variety, we denote by $\mathbf{F}_{\mathsf{V}}(X)$ the V-free algebra over X. Further, if φ is a formula, denote by $\bar{\varphi}$ the image of φ under the natural map $\mathbf{Fm}(X) \to \mathbf{F}_{\mathsf{V}}(X)$ from the term algebra $\mathbf{F}m(X)$ over X onto $\mathbf{F}_{\mathsf{V}}(X)$. If Γ is a set of formulas, also denote by $\bar{\Gamma} = \{\bar{\varphi} : \varphi \in \Gamma\}$. The following lemma restates [21, Lemma 2].

Lemma 7. *Let* $\Theta \cup \{\varphi \approx \psi\}$ *be a set of equations in the language of* V, *and take* X *to be the set of variables appearing in* $\Theta \cup \{\varphi \approx \psi\}$. *Then the following are equivalent:*

1. $\Theta \models_{\mathsf{V}} \varphi \approx \psi$.
2. $(\bar{\varphi}, \bar{\psi}) \in \bigvee_{\epsilon \approx \delta \in \Theta} \mathsf{Cg}^{\mathbf{F}_{\mathsf{V}}(X)}(\bar{\epsilon}, \bar{\delta})$.

Theorem 3. *Let* I *be a set of unary connectives with* $I \cap \mathcal{L} = \emptyset$, *and suppose that* $\mathbf{L}$ *is an axiomatic extension of* $\mathbf{Ł}(I)$ *that is algebraized by the subvariety* V *of* $\mathsf{MV}(I)$. *Further, let* $\Gamma \cup \Delta \cup \{\psi\} \subseteq Fm_{\mathcal{L}(I)}$. *Then* $\Gamma, \Delta \vdash_{\mathbf{L}} \psi$ *if and only if for some* $n \geq 0$ *there exist I-blocks* $M_1, \ldots, M_n$ *and* $\psi_1, \ldots, \psi_n \in \Delta$ *such that* $\Gamma \vdash_{\mathbf{L}} \prod_{j=1}^{n} M_j(\psi_j) \to \psi$.

Proof. We give the proof of the left-to-right direction; the proof of the converse is similar. From Corollary 1(1) and Lemmas 7 and 4 we obtain:

$$\begin{aligned} \Gamma, \Delta \vdash_{\mathbf{L}} \psi &\Longrightarrow \{\alpha \approx 1 : \alpha \in \Gamma \cup \Delta\} \models_{\mathsf{V}} \psi \approx 1 \\ &\Longrightarrow (\bar{\psi}, 1) \in \textstyle\bigvee_{\alpha \in \Gamma \cup \Delta} \mathsf{Cg}^{\mathbf{F}_{\mathsf{V}}(X)}(\bar{\alpha}, 1) \\ &\Longrightarrow \bar{\psi} \in \mathsf{Fg}^{\mathbf{F}_{\mathsf{V}}(X)}(\bar{\Gamma} \cup \bar{\Delta}), \end{aligned}$$

where X is the set of variables appearing in $\Gamma \cup \Delta \cup \{\psi\}$. By Lemma 5 there exist $l \geq 0$, I-blocks $M_1, \ldots, M_l$, and $\bar{\chi}_1, \ldots, \bar{\chi}_l \in \bar{\Gamma} \cup \bar{\Delta}$ with $M_1(\bar{\chi}_1) \cdot \ldots \cdot M_l(\bar{\chi}_l) \leq \bar{\psi}$. Let $D = \{j \in \{1, \ldots, l\} : \bar{\chi}_j \in \bar{\Delta}\}$, and set $C = D \setminus \{1, \ldots, l\}$. Then by the commutativity of $\cdot$ we have

$$\prod_{j \in C} M_j(\bar{\chi}_j) \cdot \prod_{k \in D} M_k(\bar{\chi}_k) = M_1(\bar{\chi}_1) \cdot \ldots \cdot M_l(\bar{\chi}_l) \leq \bar{\psi},$$

whence by residuation $\prod_{j\in C} M_j(\bar{\chi}_j) \leq \prod_{k\in D} M_k(\bar{\chi}_k) \to \bar{\psi}$. Applying Lemma 5 again gives $\prod_{k\in D} M_k(\bar{\chi}_k) \to \bar{\psi} \in \mathsf{Fg}^{\mathbf{F}_\mathsf{V}(X)}(\bar{\Gamma})$. Hence by Lemmas 4 and 7 and Corollary 1(1) we obtain $\Gamma \vdash_{\mathbf{L}} \prod_{k\in D} M_k(\chi_k) \to \psi$. ■

Notice that the form of the local deduction-detachment theorem announced in Corollary 3 may be recovered from Theorem 3 by taking $\Delta = \{\varphi\}$ and taking $d_M(p,q) = Mp \to q$ for $M \in \mathcal{B}_I$.

In the monomodal logic **S4Ł**, I-blocks take an especially simple form. Because $I = \{\Box\}$ in this setting, each I-block M is a finite, nonempty string of occurrences of $\Box$. Since $\Box$ is idempotent in $\mathsf{S4MV}$, for each $\{\Box\}$-block M we have that $Mx \approx \Box x$ is satisfied in $\mathsf{S4MV}$. Due to this consideration and the fact that $\Box$ preserves $\cdot$, we may read off the following simplified form the LDDT for **S4Ł**:

Corollary 4. *Let $\Gamma \cup \Delta \cup \{\psi\} \subseteq Fm_{\mathcal{L}(\Box)}$. Then $\Gamma, \Delta \vdash_{\mathbf{S4Ł}} \psi$ if and only if for some $n \geq 0$ there exist $\psi_1, \ldots, \psi_n \in \Delta$ such that $\Gamma \vdash_{\mathbf{S4Ł}} \Box(\prod_{j=1}^n \psi_j) \to \psi$.*

If $I = \{\Box_1, \ldots, \Box_n\}$ is finite, then particular forms of the LDDT can be achieved for **S4Ł**(I) and its extensions by defining an operator $\lambda(x) = \prod_{i=1}^n \Box_i(x)$. Powers of λ are defined recursively by $\lambda^0(x) = x$ and $\lambda^{m+1}(x) = \lambda(\lambda^m(x))$ for $m > 0$. I-filters of S4MV(I)-algebras may be characterized with powers of λ instead of I-blocks. A full discussion of this alternative approach will appear in future work.

5 Two Translations

We now arrive at our main translation results. After discussing some necessary technical background regarding the Jipsen-Montagna poset product construction, we exhibit two translations. The first of these embeds **GBL** into **S4Ł**, and is conceptually in the spirit of the classical Gödel-McKinsey-Tarski translation of intuitionistic logic into **S4**. The second translation embeds **GBL** in **S4$_\mathbf{t}$Ł**.

5.1 Poset Products

The translation results of this paper rely heavily on the poset product construction of Jipsen and Montagna (see [19,20]), which we now sketch. Our discussion of poset products is drawn mainly from [12], to which we refer the reader for a more detailed summary.

Let $\mathbf{A}$ be a bounded commutative integral residuated lattice. A *conucleus* on $\mathbf{A}$ is an interior operator γ on the lattice reduct of $\mathbf{A}$ such that $\gamma(x)\gamma(y) \leq \gamma(xy)$ and $\gamma(x)\gamma(1) = \gamma(1)\gamma(x) = \gamma(x)$ for all $x, y \in A$. Given a conucleus γ on $\mathbf{A}$, the γ-image $\mathbf{A}_\gamma = (A_\gamma, \wedge_\gamma, \vee, \cdot, \to_\gamma, 0, \gamma(1))$ is a bounded commutative integral residuated lattice, where $A_\gamma = \gamma[A]$ and $x \star_\gamma y = \gamma(x \star y)$ for $\star \in \{\wedge, \to\}$.

Now let $(X, \leq)$ be a poset, let $\{\mathbf{A}_x : x \in X\}$ be an indexed collection of bounded commutative integral residuated lattices with a common least element

0 and a common greatest element 1, and let $\mathbf{B} = \prod_{x \in X} \mathbf{A}_x$. From [20, Lemma 9.4], one may define a conucleus σ on $\mathbf{B}$ by

$$\sigma(f)(x) = \begin{cases} f(x) & \text{if } f(y) = 1 \text{ for all } y > x \\ 0 & \text{if there exists } y > x \text{ with } f(y) \neq 1. \end{cases}$$

The algebra $\mathbf{B}_\sigma$ is called the *poset product* of $\{\mathbf{A}_x : x \in X\}$, and is denoted $\prod_{(X,\leq)} \mathbf{A}_x$. An element $f \in B_\sigma$ is called an *antichain labeling* or *ac-labeling*, and satisfies the condition that if $x, y \in X$ with $x < y$ then $f(x) = 0$ or $f(y) = 1$. The following is a direct consequence of [20, Corollary 5.4(i)] and its proof.

Lemma 8. *Let $\boldsymbol{A}$ be a GBL-algebra. Then there exists a poset $(X, \leq)$ and an indexed family $\{\boldsymbol{A}_x : x \in X\}$ of totally ordered MV-algebras such that $\boldsymbol{A}$ embeds in the poset product $\boldsymbol{B}_\sigma$, where $\boldsymbol{B} = \prod_{x \in X} \boldsymbol{A}_x$.*

Following [17], for a poset $(X, \leq)$ and indexed family $\{\mathbf{A}_x : x \in X\}$ we introduce a map δ on $\mathbf{B} = \prod_{x \in X} \mathbf{A}_x$ by

$$\delta(f)(x) = \begin{cases} f(x) & \text{if } f(y) = 0 \text{ for all } y < x \\ 1 & \text{if there exists } y < x \text{ with } f(y) \neq 0. \end{cases}$$

The following lemma illustrates that σ and δ provide algebraic interpretations of modals, which is crucial in obtaining our translation results.

Lemma 9. *Let $(X, \leq)$ be poset, let $\{\boldsymbol{A}_x : x \in X\}$ be an indexed family of bounded commutative integral residuated lattices, and set $\boldsymbol{B} = \prod_{x \in X} \boldsymbol{A}_x$ as above. Then:*

1. *σ and $\neg\delta\neg$ are $\{\wedge, \cdot, 0, 1\}$-endomorphisms of $\boldsymbol{B}$.*
2. *For all $f, g \in B$, $f \leq \sigma(g)$ if and only if $\delta(f) \leq g$.*
3. *$\neg\delta\neg$ is an interior operator.*
4. *If additionally $\boldsymbol{A}_x$ is an MV-algebra for all $x \in X$, then $(\boldsymbol{B}, \sigma)$ is an S4MV-algebra and $(\boldsymbol{B}, \sigma, \neg\delta\neg)$ is S4$_t$MV-algebra.*

Proof. 1. It is obvious that $\sigma(0) = 0$ and $\sigma(1) = 1$. Let $\star \in \{\wedge, \cdot\}$, $x \in X$, and $f, g \in B$, and observe that if $y > x$ then $(f \star g)(y) = 1$ if and only if $f(y) = g(y) = 1$. It follows that if $(f \star g)(y) = 1$ for all $y > x$, then $\sigma(f \star g)(x) = (f \star g)(x) = f(x) \star g(x) = \sigma(f)(x) \star \sigma(g)(x)$, and if otherwise then $\sigma(f \star g)(x) = 0 = \sigma(f)(x) \star \sigma(g)(x)$. Thus $\sigma(f \star g) = \sigma(f) \star \sigma(g)$.

To prove that $\neg\delta\neg$ is a $\{\wedge, \cdot, 0, 1\}$-endomorphism, again let $\star \in \{\wedge, \cdot\}$, $x \in X$, and $f, g \in B$. Note that for all $y < x$ we have $\neg(f \star g)(y) = 0$ if and only if $(f \star g)(y) = 1$, and as before this occurs if and only if $f(y) = g(y) = 1$. Thus we have $\neg(f \star g)(y) = 0$ for all $y < x$ if and only if $\neg f(y) = 0$ for all $y < x$ and $\neg g(y) = 0$ for all $y < x$. Hence if $\neg(f \star g)(y) = 0$ for all $y < x$, then we have $\neg\delta\neg(f \star g)(x) = \neg\neg(f \star g)(x) = f(x) \star g(x) = \neg\neg f(x) \star \neg\neg g(x) = \neg\delta\neg f(x) \star \neg\delta\neg g(x)$. On the other hand, if there exists $y < x$ with $\neg(f \star g)(y) \neq 0$, then $\neg\delta\neg(f \star g)(x) = \neg 1 = 0$, and $\neg\delta\neg f(x) \star \neg\delta\neg g(x) = 0$ since one of $\delta\neg f(y)$

or $\delta\neg g(y)$ must be 1. Since $\neg\delta\neg 0 = 0$ and $\neg\delta\neg 1 = 1$ by direct calculation, item 1 follows.

2. Suppose $f \leq \sigma(g)$ and let $x \in X$. Since σ is an interior operator, $f(x) \leq \sigma(g)(x) \leq g(x)$. If $\delta(f)(x) = f(x)$, then $\delta(f)(x) \leq g(x)$ is immediate. On the other hand, if $\delta(f)(x) \neq f(x)$ then there exists $y < x$ such that $f(y) \neq 0$. From $f \leq \sigma(g)$ we infer that $\sigma(g)(y) \neq 0$, so $\sigma(g)(x) = 1$ since $\sigma(g)$ is an ac-labeling. Thus $\delta(f)(x) \leq 1 = \sigma(g)(x) = g(x)$. It follows that $\delta(f) \leq g$. The proof that $\delta(f) \leq g$ implies $f \leq \sigma(g)$ is similar.

3. It is easy to see that δ is a closure operator. From this and the fact that $\neg$ is an antitone involution, it is a straightforward calculation to show that $\neg\delta\neg$ is an interior operator.

4. Under the hypothesis, $\mathbf{B}$ is a product of MV-algebras and is hence an MV-algebra. That $(\mathbf{B}, \sigma)$ is an S4MV-algebra follows promptly from item 1 and the fact that σ is an interior operator. That $(\mathbf{B}, \sigma, \neg\delta\neg)$ is a S4$_t$MV-algebra follows from items 1, 2, and 3. ■

Lemma 10. *Suppose that* $(\boldsymbol{A}, \Box)$ *is an S4MV-algebra and* $(\boldsymbol{B}, G, H)$ *is a S4$_t$MV-algebra. Then both* $\Box$ *and* G *are conuclei, and each of* $\boldsymbol{A}_\Box$ *and* $\boldsymbol{B}_G$ *is a GBL-algebra.*

Proof. Each of $\Box$ and G is a conucleus by definition. For each claim, it suffices to show that if $\mathbf{M}$ is an MV-algebra and γ is a conucleus on $\mathbf{M}$ preserving $\cdot$ and $\wedge$, then $\mathbf{M}_\gamma$ is a GBL-algebra. For this, it is enough to show that $\mathbf{M}_\gamma$ satisfies the divisibility identity $x \cdot (x \to y) \approx x \wedge y$. Let $x, y \in M_\gamma$. Since $\mathbf{M}$ is an MV-algebra, we have that $x \cdot^{\mathbf{M}} (x \to^{\mathbf{M}} y) = x \wedge^{\mathbf{M}} y$. Using the fact that γ preserves $\cdot$ and $\wedge$, and that x, y are γ-fixed, we have:

$$\begin{aligned} x \cdot^{\mathbf{M}_\gamma} (x \to^{\mathbf{M}_\gamma} y) &= x \cdot^{\mathbf{M}} \gamma(x \to^{\mathbf{M}} y) \\ &= \gamma(x) \cdot^{\mathbf{M}} \gamma(x \to^{\mathbf{M}} y) \\ &= \gamma(x \cdot^{\mathbf{M}} (x \to^{\mathbf{M}} y)) \\ &= \gamma(x \wedge^{\mathbf{M}} y) \\ &= \gamma(x) \wedge^{\mathbf{M}_\gamma} \gamma(y) \\ &= x \wedge^{\mathbf{M}_\gamma} y. \end{aligned}$$

This proves the claim. ■

5.2 The Translations

We define a pair of translations M and T from the language $\mathcal{L}$ into the languages $\mathcal{L}(\Box)$ and $\mathcal{L}(G, H)$, respectively. We set $M(p) = \Box p$ for each $p \in \mathsf{Var}$, $M(0) = 0$, $M(1) = 1$, and we extend M recursively by

$$M(\varphi \star \psi) = M(\varphi) \star M(\psi), \text{ for } \star \in \{\wedge, \vee, \cdot\}, \text{ and}$$

$$M(\varphi \to \psi) = \Box(M(\varphi) \to M(\psi)).$$

Further, if Γ is a set of formulas of $\mathcal{L}$ then we define

$$M(\Gamma) = \{M(\varphi) : \varphi \in \Gamma\}.$$

The translation T differs from M only by replacing $\Box$ by G and considering its codomain to be formulas of $\mathcal{L}(G, H)$ rather than those of $\mathcal{L}(\Box)$.

Lemma 11. *Let $(\boldsymbol{A}, \Box)$ be an S4MV-algebra, and let $(\boldsymbol{B}, G, H)$ be a $S4_t$MV-algebra.*

1. *Suppose that $h\colon \mathsf{Var} \to (\boldsymbol{A}, \Box)$ is an assignment, and define $\bar{h}\colon \mathsf{Var} \to \boldsymbol{A}_\Box$ by $\bar{h}(p) = \Box(h(p))$. If $\varphi \in Fm_{\mathcal{L}}$, then $\bar{h}(\varphi) = h(M(\varphi))$.*
2. *If $\varphi \in Fm_{\mathcal{L}}$, then $\varphi \approx 1$ is valid $\boldsymbol{A}_\Box$ if and only if $M(\varphi) \approx 1$ is valid in $\boldsymbol{A}$.*
3. *Suppose that $h\colon \mathsf{Var} \to (\boldsymbol{B}, G, H)$ is an assignment, and define $\bar{h}\colon \mathsf{Var} \to \boldsymbol{B}_G$ by $\bar{h}(p) = G(h(p))$. If $\varphi \in Fm_{\mathcal{L}}$, then $\bar{h}(\varphi) = h(T(\varphi))$.*
4. *If $\varphi \in Fm_{\mathcal{L}}$, then $\varphi \approx 1$ is valid $\boldsymbol{B}_G$ if and only if $T(\varphi) \approx 1$ is valid in $\boldsymbol{B}$.*

Proof. We will prove items 1 and 2. Item 3 follows by a proof identical to that of item 1 by replacing $\Box$ by G, M by T, and $(\mathbf{A}, \Box)$ by $(\mathbf{B}, G, H)$. Similarly, item 4 follows from the same proof given for item 2.

1. We argue by induction on the height of φ. If φ is a constant or $\varphi \in \mathsf{Var}$, then the statement is true by assumption. Now suppose that for all formulas φ' of height strictly less than the height of φ we have that $\bar{h}(\varphi') = h(M(\varphi'))$. If $\varphi = \psi \star \chi$ for $\star \in \{\cdot, \wedge, \vee\}$, then by definition $h(M(\varphi)) = h(M(\psi \star \chi)) = h(M(\psi) \star M(\chi)) = h(M(\psi)) \star h(M(\chi))$. By the inductive hypothesis, the latter is precisely $\bar{h}(\psi) \star \bar{h}(\chi) = \bar{h}(\psi \star \chi) = \bar{h}(\varphi)$ as desired. On the other hand, if $\varphi = \psi \to \chi$ then we have that $h(M(\varphi)) = h(M(\psi \to \chi)) = h(\Box(M(\psi) \to M(\chi))) = \Box(h(M(\psi)) \to h(M(\chi)))$. By the inductive hypothesis, this term is equal to $\Box(\bar{h}(\psi) \to \bar{h}(\chi)) = \bar{h}(\psi) \to^{\mathbf{A}_\Box} \bar{h}(\chi) = \bar{h}(\psi \to \chi) = \bar{h}(\varphi)$. The result follows by induction.

2. Suppose first that $\varphi \approx 1$ is valid in $\mathbf{A}_\Box$, and let $h\colon \mathsf{Var} \to (\mathbf{A}, \Box)$ be an assignment. By item 1, $\bar{h}\colon \mathsf{Var} \to \mathbf{A}_\Box$ is an assignment in $\mathbf{A}_\Box$ and $\bar{h}(\psi) = h(M(\psi))$ for all $\psi \in Fm_{\mathcal{L}}$. In particular, this shows that $h(M(\varphi)) = \bar{h}(\varphi) = 1$ since $\varphi \approx 1$ is valid in $\mathbf{A}_\Box$, so as h is arbitrary we have $M(\varphi) \approx 1$ is valid in $\mathbf{A}$.

For the converse, suppose that $M(\varphi) \approx 1$ is valid in $\mathbf{A}$ and let $h\colon \mathsf{Var} \to \mathbf{A}_\Box$ be an assignment. Because $A_\Box \subseteq A$, we may define a new assignment $k\colon \mathsf{Var} \to (\mathbf{A}, \Box)$ by $k(p) = h(p)$ for all $p \in \mathsf{Var}$. Since $M(\varphi) \approx 1$ is valid in $\mathbf{A}$, we have $k(M(\varphi)) = 1$. By item 1, we have that $k(M(\varphi)) = \hat{k}(\varphi)$, where $\hat{k}\colon \mathsf{Var} \to \mathbf{A}_\Box$ is defined by $\hat{k}(p) = \Box(k(p))$. Notice that since k has its image among the $\Box$-fixed elements of A, we have for all $p \in \mathsf{Var}$ that $\hat{k}(p) = \Box(k(p)) = k(p) = h(p)$, and thus $\hat{k} = h$. Thus $h(\varphi) = \hat{k}(\varphi) = k(M(\varphi)) = 1$, so $\varphi \approx 1$ is valid in $\mathbf{A}_\Box$. ∎

The following gives the main translation results of this paper.

Theorem 4. *Let $\Gamma \cup \{\varphi\} \subseteq Fm_{\mathcal{L}}$. Then:*

1. $\{\psi \approx 1 : \psi \in \Gamma\} \vDash_{\mathsf{GBL}} \varphi \approx 1 \iff \{M(\psi) \approx 1 : \psi \in \Gamma\} \vDash_{\mathsf{S4MV}} M(\varphi) \approx 1$.
2. $\Gamma \vdash_{\mathbf{GBL}} \varphi \iff M(\Gamma) \vdash_{\mathbf{S4Ł}} M(\varphi)$.

3. $\{\psi \approx 1 : \psi \in \Gamma\} \vDash_{\mathsf{GBL}} \varphi \approx 1 \iff \{T(\psi) \approx 1 : \psi \in \Gamma\} \vDash_{\mathsf{S4_tMV}} T(\varphi) \approx 1$.
4. $\Gamma \vdash_{\mathbf{GBL}} \varphi \iff T(\Gamma) \vdash_{\mathbf{S4_t Ł}} T(\varphi)$.

Proof. We first prove item 1. Suppose that $\{\psi \approx 1 : \psi \in \Gamma\} \vDash_{\mathbf{GBL}} \varphi \approx 1$, let $(\mathbf{A}, \Box)$ be an S4MV-algebra, and let $h\colon \mathsf{Var} \to (\mathbf{A}, \Box)$ be an assignment. We aim to show $\{M(\psi) \approx 1 : \psi \in \Gamma\} \vDash_{\mathsf{S4MV}} M(\varphi) \approx 1$, so suppose that for all $\psi \in \Gamma$ we have $h(M(\psi)) = 1$. By Lemma 11(2) we have that $1 = h(M(\psi)) = \bar{h}(\psi)$. Since $\bar{h}$ is an assignment in $\mathbf{A}_\Box$ (which is a GBL-algebra by Lemma 10), by hypothesis we have $\bar{h}(\varphi) = 1$. Applying Lemma 11(2) again yields $h(M(\varphi)) = 1$, showing that $\{M(\psi) \approx 1 : \psi \in \Gamma\} \vDash_{\mathsf{S4MV}} M(\varphi) \approx 1$.

For the converse, suppose that $\{M(\psi) \approx 1 : \psi \in \Gamma\} \vDash_{\mathsf{S4MV}} M(\varphi) \approx 1$. Let $\mathbf{A}$ be a GBL-algebra, let $h\colon \mathsf{Var} \to \mathbf{A}$ be an assignment, and suppose that $h(\psi) = 1$ for all $\psi \in \Gamma$. It is enough to show that $h(\varphi) = 1$. By Lemmas 8 and 9, there exists an S4MV-algebra $(\mathbf{B}, \Box)$ such that $\mathbf{A}$ embeds in $\mathbf{B}_\Box$, and without loss of generality we may assume that this embedding is an inclusion. Using the fact that $A \subseteq B_\Box \subseteq B$, we define a new assignment $k\colon \mathsf{Var} \to \mathbf{B}$ by $k(p) = h(p)$ for all $p \in \mathsf{Var}$. Notice that for all $p \in \mathsf{Var}$ we have $\bar{k}(p) = \Box k(p) = \Box h(p) = h(p)$ since the image of h consists of $\Box$-fixed elements, so by Lemma 11(2) we have $h(\chi) = k(M(\chi))$ for all χ. In particular, $k(M(\psi)) = 1$ for all $\psi \in \Gamma$, and by the hypothesis we have $k(M(\varphi)) = 1$. But this implies $h(\varphi) = 1$, proving the result.

Note that item 2 follows from Corollary 1 since we have:

$$\begin{aligned} \Gamma \vdash_{\mathbf{GBL}} \varphi &\iff \{\psi \approx 1 : \psi \in \Gamma\} \vDash_{\mathsf{GBL}} \varphi \approx 1 \\ &\iff \{M(\psi) \approx 1 : \psi \in \Gamma\} \vDash_{\mathsf{S4MV}} M(\varphi) \approx 1 \\ &\iff M(\Gamma) \vdash_{\mathbf{S4Ł}} M(\varphi). \end{aligned}$$

Items 3 and 4 follow analogously to items 1 and 2, respectively. ■

As a final remark, we note that the temporal translation of Theorem 4(3, 4) generalizes the translation offered in [2]. *Gödel-Dummett logic* is the extension of propositional intuitionistic logic by the axiom scheme $(\varphi \to \psi) \vee (\psi \to \varphi)$, and is algebraized by the variety of Gödel algebras (i.e., BL-algebras satisfying $x^2 \approx x$). In [2], the authors deploy the temporal flow semantics (see [1]) based on so-called bit sequences to exhibit a translation of Gödel-Dummett logic into an axiomatic extension of Prior's classical tense logic. This study was inspired by [12], which offers a relational semantics based on poset products that, among other things, generalizes the temporal flow semantics (see [12, Section 4.2]). Our development of the translations above can hence be thought of as extending the work of [2] along the generalization offered in [12]. Poset products give a powerful, unifying framework for inquiries of this kind, and we anticipate that they will find far-reaching application to translations. A thorough investigation of modal translations and modal companions for **GBL** remains to be conducted, but we expect the work in this paper to be an important preliminary step.

References

1. Aguzzoli, S., Bianchi, M., Marra, V.: A temporal semantics for basic logic. Studia Logica **92**, 147–162 (2009)

2. Aguzzoli, S., Gerla, B., Marra, V.: Embedding Gödel propositional logic into Prior's tense logic. In: Magdalena, L., Ojeda Aciego, M., Verdegay, J. (eds.) Proceedings of 12th International Conference Information Processing and Management of Uncertainty for Knowledge-Based Systems, pp. 992–999 (2008)
3. Blok, W., Pigozzi, D.: Algebraizable Logics, vol. 77. Memoirs of the American Mathematical Society, New York (1989)
4. Blok, W., Pigozzi, D.: Local deduction theorems in algebraic logic. In: Andréka, H., Monk, J., Németi, I. (eds.) Algebraic Logic, Colloquia Mathematica Societatis János Bolyai, vol. 54, pp. 75–109. North-Holland, Amsterdam (1991)
5. Bova, S., Montagna, F.: The consequence relation in the logic of commutative GBL-algebras is PSPACE-complete. Theor. Comput. Sci. **410**, 1143–1158 (2009)
6. Burris, S., Sankappanavar, H.: A Course in Universal Algebra. Springer, New York (1981)
7. Chagrov, A., Zakharyaschev, M.: Modal companions of intermediate propositional logics. Studia Logica **51**, 49–82 (1992)
8. Cignoli, R., D'Ottaviano, I., Mundici, D.: Algebraic Foundations of Many-Valued Reasoning. Trends in Logic-Studia Logica Library, Kluwer Academic Publishers, Dordrecht (2000)
9. Cignoli, R., Torrens, A.: Hájek's basic fuzzy logic and Łukasiewicz infinite-valued logic. Arch. Math. Logic **42**, 361–370 (2003)
10. Esteva, F., Godo, L., Rodríguez, R.: On the relation between modal and multimodal logics over Łukasiewicz logic. In: Proceedings of 2017 IEEE International Conference on Fuzzy Systems (FUZZ-IEEE), Naples, Italy, pp. 1–6 (2017)
11. Font, J.: Abstract Algebraic Logic: An Introductory Textbook. College Publications, London (2016)
12. Fussner, W.: Poset products as relational models. Studia Logica (2021). https://doi.org/10.1007/s11225-021-09956-z
13. Galatos, N., Jipsen, P.: A survey of generalized basic logic algebras. In: Cintula, P., Hanikova, Z., Svejdar, V. (eds.) Witnessed Years: Essays in Honour of Petr Hájek, pp. 305–331. College Publications, London (2009)
14. Galatos, N., Jipsen, P., Kowalski, T., Ono, H.: Residuated Lattices: An Algebraic Glimpse at Substructural Logics. Elsevier, Amsterdam (2007)
15. Galatos, N., Ono, H.: Algebraization, parametrized local deduction theorem and interpolation for substructural logics over FL. Studia Logica **83**, 279–308 (2006)
16. Hájek, P.: Metamathematics of Fuzzy Logic. Trends in Logic-Studia Logica Library, Kluwer, Dordrecht (1998)
17. Jipsen, P.: Generalizations of Boolean products for lattice-ordered algebras. Ann. Pure Appl. Logic **161**, 228–234 (2009)
18. Jipsen, P., Montagna, F.: On the structure of generalized BL-algebras. Algebra Universalis **55**, 226–237 (2006)
19. Jipsen, P., Montagna, F.: The Blok-Ferreirim theorem for normal GBL-algebras and its applications. Algebra Universalis **60**, 381–404 (2009)
20. Jipsen, P., Montagna, F.: Embedding theorems for classes of GBL-algebras. J. Pure Appl. Algebra **214**, 1559–1575 (2010)
21. Metcalfe, G., Montagna, F., Tsinakis, C.: Amalgamation and interpolation in ordered algebras. J. Algebra **402**, 21–82 (2014)
22. O'Hearn, P., Pym, D.: The logic of bunched implications. Bull. Symb. Logic **5**, 215–244 (1999)
23. Prior, A.: Time and Modality. Clarendon Press, Oxford (1957)

Isolated Sublattices and Their Application to Counting Closure Operators

Roland Glück(✉)

Deutsches Zentrum für Luft- und Raumfahrt, 86159 Augsburg, Germany
roland.glueck@dlr.de

Abstract. This paper investigates the interplay between isolated sublattices and closure operators. Isolated sublattices are a special kind of sublattices which can serve to diminish the number of elements of a lattice by means of a quotient. At the same time, there are simple formulae for the relationship between the number of closure operators in the original lattice and the quotient lattice induced by isolated sublattices. This connection can be used to derive an algorithm for counting closure operators, provided the lattice contains suitable isolated sublattices.

1 Introduction

Closure or hull operators, i.e., idempotent, isotone and extensive endofunctions on some carrier set, are a common and widespread concept in mathematics and computer science. Some of the most prominent examples are the (reflexive) transitive closure of a relation or a graph, the Kleene closure in language theory or the topological closure in traditional analysis. More sophisticated applications concern automated reasoning as in [11], database theory as in [10] or the algebraic description of connected components as in [12]. However, most of the work uses closures as a tool for specific purposes rather than investigating their general properties.

The biggest part of the work concerning closure properties deals with closure functions operating on the power set of some carrier set, here also under the term Moore family (see e.g. [7] for a survey and [6] for recent results). Other work deals mostly with properties of closure functions on lattices albeit the definition of a closure operator requires only a simple ordered set.

Counting structures of interest has become a rising theme of investigation in lattice theory and related areas of research. For example, [1] counts various kinds of doubly idempotent semirings, [16] deals with join-endomorphisms in lattices, the subject of [3] are topological spaces and [5] generates and counts a certain kind of bisemilattices. However, there is no work concerning the number of closure operators in general lattices or orders. The only results we are aware of concern the power set lattice $(\mathcal{P}(S), \subseteq)$. There, the exact number of closure operators is known only up to $|S| = 7$ (for curiosity, there are 14.087.648.235.707.352.472 of them, as shown in [8] only in 2010).

U. Fahrenberg et al. (Eds.): RAMiCS 2021, LNCS 13027, pp. 192–208, 2021.
https://doi.org/10.1007/978-3-030-88701-8_12

In this paper, we introduce a heuristic method for structuring lattices in a way that eases under certain circumstances the computation of the number of closure operators. It is based on so called isolated sublattices which are intuitively speaking sublattices which have contact with the rest of the lattice only via their top and bottom element. By means of quotient lattices we can reduce the number of elements of the lattice under consideration and maybe reach a lattice with a certain structure for which there are closed formulae for the number of closure operators.

The remainder of the paper is organized as follows: Sect. 2 gives an overview over the notation used in this paper and recalls some facts from lattice theory important for the further course. Section 3 introduces isolated sublattices and investigates their properties and relationships with closure operators. In Sect. 4 we first introduce closed formulas for the number of closure operators on particular lattices and then develop an algorithm for computing the number of closure operators using the previous results. As usual, we give an outlook to future work in the finishing Sect. 5.

2 Notations and Basic Properties

We assume that the reader is familiar with the basics of order and lattice theory and refer e.g. to [9,14,17] for more basic and to [4,15] for more advanced topics. However, we recapitulate some less used concepts and their properties and introduce the notation we will use in the sequel.

To denote orders, we use $\leq$ and indexed variants where appropriate. The relation $\nleq$ is defined by $x \nleq y \Leftrightarrow_{def} \neg(x \leq y)$. The signs $<$, $\geq$ and $>$ denote the associated strict order, reverse order and strict reverse order, resp. In a lattice, we use $\sqcap$ and $\sqcup$ for the binary infimum and supremum, resp., and index them if necessary. As usual, $\top$ and $\bot$ and variants thereof denote the greatest and least element, resp., if they exist. Given an ordered set $(S, \leq)$ we may refer slightly inaccurately also by S to the structure $(S, \leq)$. We say that x *majorizes* y if $x \geq y$ holds and use the notation $\mathsf{maj}(x, S') =_{def} \{y \in S' \mid y \geq x\}$. We call two elements x and y *comparable*, in signs $x \lesseqgtr y$, if $x \leq y$ or $y \leq x$ holds. x and y are called *incomparable*, denoted by $x \not\lesseqgtr y$, if they are not comparable. A *chain* is a set S such that every pair of elements of S is comparable. Given an ordered set $(S, \leq)$ we call a subset $S' \subseteq S$ *convex* if for all $x, y \in S'$ and all $z \in S$ the implication $x \leq z \leq y \Rightarrow z \in S'$ holds. We use the familiar notations $[a, b] =_{def} \{x \mid a \leq x \wedge x \leq b\}$ and $]a, b] =_{def} [a, b] \backslash \{a\}$ for intervals.

The equivalence class of x under an equivalence relation $\sim$ will be denoted by $[x]_\sim$. For a set of sets C we use the abbreviation $\bigcup C$ instead of $\bigcup\limits_{c \in C} c$. Conversely, $C^{\{\}}$ denotes the system of singleton sets $\{\{c\} \mid c \in C\}$.

An equivalence relation $\sim$ on a lattice S is called a *congruence* if for all $x_0, x_1, y_0, y_1 \in S$ the following implications hold:

1. $x_0 \sim y_0 \wedge x_1 \sim y_1 \Rightarrow x_0 \sqcap x_1 \sim y_0 \sqcap y_1$
2. $x_0 \sim y_0 \wedge x_1 \sim y_1 \Rightarrow x_0 \sqcup x_1 \sim y_0 \sqcup y_1$

In this case, the quotient lattice $S/\sim$ is a homomorphic image of S. In particular, if $|[x]_\sim| = |[y]_\sim| = |[z]_\sim| = 1$, then $x \sqcap y = z$ is equivalent to $\{x\} \sqcap_\sim \{y\} = \{z\}$ (where $\sqcap_\sim$ denotes the infimum operation on $S/\sim$; the respective orders, strict orders and their reverses are notated analogously). The analogous property holds for binary suprema. If $[x]_\sim$ and $[y]_\sim$ are disjoint the equivalence $x \leq y \Leftrightarrow [x]_\sim \leq_\sim [y]_\sim$ holds.

As already mentioned in Sect. 1, closures are widely known and used. There are two ways of characterizing closures. The first one gives a functional characterization:

Definition 2.1. *Given an ordered set S an endofunction c on S is called a* closure operator *if it fulfills the following properties for all $x, y \in S$:*

1. $x \leq c(x)$ *(extensitivity)*
2. $x \leq y \Rightarrow c(x) \leq c(y)$ *(isotony)*
3. $c(c(x)) = c(x)$ *(idempotence)*

The second one uses subsets of the carrier set of a lattice:

Definition 2.2. *Given a lattice $(S, \leq)$ a subset $S' \subseteq S$ is called a* closure system *if it fulfills the following properties:*

1. $x, y \in S' \Rightarrow x \sqcap y \in S'$
2. *for every $s \in S$ there is a smallest $x \in S'$ such that $s \leq x$ holds.*

The set of all closure systems of S is denoted by $\mathsf{C}(S)$.

In a finite lattice, the second condition of Definition 2.2 is equivalent to $\top \in S'$. On a lattice, closure operators and closure systems are cryptomorphic structures since there is a one-to-one correspondence between them: the set of fixpoints of a closure operator is always a closure system. Conversely, each closure system C determines a unique closure operator c with $\mathsf{fix}(c) = C$ (where $\mathsf{fix}(c)$ denotes the set of fixpoints of c). This correspondence makes only sense in the context of lattices; on general orders, we lack an infimum operation as used in Definition 2.2.

Remark: Definition 2.2 is taken from [14]. However, as one reviewer pointed out, this definition is redundant:

Lemma 2.3. *In Definition 2.2, the second condition implies the first one.*

Proof: Denote by $c(x)$ the function which maps x to the smallest element from S' majorizing x (this is well-defined due to the uniqueness of smallest elements) and consider arbitrary $y, z \in S'$. Clearly, c is extensive, so we have $y \sqcap z \leq c(y \sqcap z)$. Due to $y \in S'$ and $y \sqcap z \leq y$ we have $c(y \sqcap z) \leq y$ and symmetrically $c(y \sqcap z) \leq z$ from where $c(y \sqcap z) \leq y \sqcap z$ follows. Alltogether, we have $c(y \sqcap z) = y \sqcap z$ and hence $y \sqcap z \in S'$ because of $c(y \sqcap z) \in S'$. ∎

So we will use only the second condition of Definition 2.2 when reasoning about general, possibly infinite lattices. However, in the context of finite lattices, we will rather use the characterization given immediately after Definition 2.2.□

3 Closure Systems and Isolated Sublattices

This section introduces and investigates isolated sublattices. It contains a lot of results of rather technical nature so we give a short overview as a guideline to the reader.

Isolated sublattices induce congruence relations so they can be used to construct a quotient lattice. We will define and consider two different kinds of isolated sublattices: those including the top element of a lattice and isolated sublattices with bottleneck. In the algorithm we will introduce in Sect. 4 we use possibly series of quotient constructions. A crucial point to make the algorithm work is that all the isolated sublattices inducing the quotients in such a series are disjoint. This is ensured by Lemmata 3.11 and 3.12; the other results up to this point serve for the preparation of these lemmata.

The second part of this section, starting at Subsection 3.2, investigates the interplay between isolated sublattices and closures. Given a lattice S and an isolated sublattice S' of S, we establish connections between closures of S containing possibly an element of S' and closures on the quotient induced by S' containing S'. On the other hand, we show a relation between closures on S without an element in S' and closures on the quotient not containing S'. The main results of this part are the Theorems 3.20 and 3.21.

3.1 Isolated Sublattices

Definition 3.1. *Let $(S, \leq)$ be a lattice. A subset $S' \subseteq S$ is called an* isolated sublattice *if it fulfills the following properties:*

1. *S' is a sublattice with greatest element $\top_{S'}$ and least element $\bot_{S'}$.*
2. *$\forall x \notin S' \forall y' \in S' : y' \leq x \Rightarrow \top_{S'} \leq x$*
3. *$\forall x \notin S' \forall y' \in S' : x \leq y' \Rightarrow x \leq \bot_{S'}$*

Intuitively, this means that S' can be entered from below only via $\bot_{S'}$ and exited upwards only via $\top_{S'}$. We call an isolated sublattice S' a *summit isolated sublattice* if $\top_{S'} = \top_S$ holds. An isolated sublattice S' is called *nontrivial* if $S' \neq S$, and *useful* if $|S| > 1$ holds.

The next definition captures situations where intuitively speaking an order does not "branch upwards" at an element x:

Definition 3.2. *Given an ordered set $(S, \leq)$ we call an element $b \in S$ a* bottleneck *of an element $x \in S$ if the following conditions are fulfilled:*

1. *$b > x$,*
2. *$[x, b]$ is a chain, and*
3. *$y > x \Rightarrow (y \in [x, b] \vee y > b)$ holds for all $y \in S$.*

Remark: This definition os equivalent to the stipulation that x is meet-irreducible. However, we keep this definition because later proofs rely heavily on the properties from Definition 3.2. □

It is straightforward to see that if b is a bottleneck of x then every element in $]x, b]$ is also a bottleneck of x. An isolated sublattice S' is called an *isolated sublattice with bottleneck* b if b is a bottleneck of $\top_{S'}$. The purpose of a bottleneck of an isolated sublattice will be explained in the remark after Lemma 3.17. Figure 1 gives examples for different kinds of isolated sublattices.

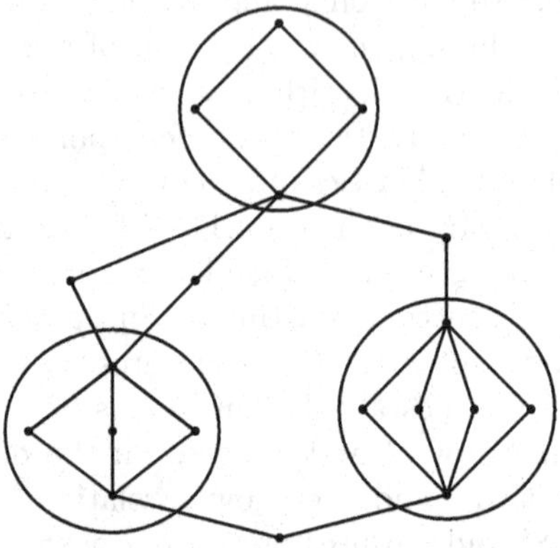

Fig. 1. An isolated sublattice (without bottleneck, left), an isolated sublattice with bottleneck (right) and a summit isolated sublattice (top)

Given a lattice $(S, \leq)$ and an isolated sublattice S' we define a relation $\sim_{S'}$ by $x \sim_{S'} y \Leftrightarrow_{def} x = y \vee (x \in S' \wedge y \in S')$. Clearly, $\sim_{S'}$ is an equivalence relation with S' as an equivalence class whereas all other equivalence classes consist of exactly one element in $S \backslash S'$. Moreover, $\sim_{S'}$ is even a congruence:

Lemma 3.3. *Let S' be an isolated sublattice of a lattice $(S, \leq)$. Then $\sim_{S'}$ is a congruence relation on $(S, \leq)$.*

Proof: Let us pick arbitrary $x_0, x_1, y_0, y_1 \in S$ and assume that both $x_0 \sim_{S'} x_1$ and $y_0 \sim_{S'} y_1$ hold. We distinguish the following cases:

1. $x_0 \notin S' \wedge y_0 \notin S'$: Here we have $x_0 = x_1$ and $y_0 = y_1$ by definition of $\sim_{S'}$ and hence $x_0 \sqcup y_0 = x_1 \sqcup y_1$.
2. $x_0 \in S' \wedge y_0 \in S'$: Then we have also $x_1 \in S'$ and $y_1 \in S'$ by definition of $\sim_{S'}$. Because S' is a sublattice this implies both $x_0 \sqcup y_0 \in S'$ and $x_1 \sqcup y_1 \in S'$.
3. $x_0 \in S' \wedge y_0 \notin S'$: Here $x_1 \in S'$ and $y_0 = y_1$ hold and we distinguish three cases:
   a) $y_0 \leq x_0$: By definition of an isolated sublattice this implies $y_0 \leq \perp_{S'}$. Consequently, we have also $y_0 \leq x_1$ due to $x_0 \in S'$ and $x_0 \sim_{S'} x_1$. So we have $x_0 \sqcup y_0 = x_0$ and $x_1 \sqcup y_1 = x_1 \sqcup y_0 = x_1$, hence $x_0 \sqcup y_0 \in S'$ and $x_1 \sqcup y_1 \in S'$.
   b) $x_0 \leq y_0$: By definition of an isolated sublattice this implies $\top_{S'} \leq y_0$ and hence also $x_1 \leq y_0$. From this in turn follows $x_0 \sqcup y_0 = y_0 = x_1 \sqcup y_0 = x_1 \sqcup y_1$.

c) $x_0 \not\leq y_0$: We claim that here $x_0 \sqcup y_0 = \top_{S'} \sqcup y_0$ holds. Due to $x_0 \in S'$ and hence $x_0 \leq \top_{S'}$, $\top_{S'} \sqcup y_0$ is an upper bound of both x_0 and y_0 so assume there is a $z < \top_{S'} \sqcup y_0$ with $x_0 \leq z$ and $y_0 \leq z$. Then z can not be an element of S': in this case we would have $y_0 \leq \bot_{S'}$ due to $y_0 \leq z$, $y_0 \notin S'$ and definition of an isolated sublattice. Now $x_0 \leq z$ and $z \notin S'$ imply $\top_{S'} \leq z$, hence z is an upper bound of both y_0 and $\top_{S'}$ strictly less than $\top_{S'} \sqcup y_0$ which is a clear contradiction. An analogous argumentation shows $x_1 \sqcup y_0 = y_0 \sqcup \top_{S'}$ and hence (due to $y_0 = y_1$) $x_0 \sqcup y_0 = x_1 \sqcup y_1$.

In all three cases we have $x_0 \sqcup y_0 \sim_{S'} x_1 \sqcup y_1$. This holds also by commutativity of the supremum operator for the case $x_0 \notin S' \wedge y_0 \in S'$, and a symmetrical argumentation shows the analogous claim for the infimum operator. Hence, $\sim_{S'}$ is indeed a congruence relation. ■

Remark: Note that not every equivalence class of a congruence is an isolated sublattice: consider for example the congruence on $(\mathbb{Z}, \leq)$, defined by $x\ y \Leftrightarrow_{def} (x \geq 0 \wedge y \geq 0) \vee (x < 0 \wedge y < 0)$. Both equivalence classes lack either a greatest or a least element, hence they do not fulfill the first point of Definition 3.1. However, under certain circumstances, congruences induce isolated sublattices:

Lemma 3.4. *Let be a congruence such that is at most one x with $|[x]_\sim| > 1$ and assume that $\top_{[x]_\sim}$ and $\bot_{[x]_\sim}$ exist. Then $[x]_\sim$ is an isolated sublattice.*

Proof: Lemma 10 in [14] shows that $[x]_\sim$ is a sublattice so we pick arbitrary $x' \in [x]_\sim$ and $y \notin [x]_\sim$ with $x \leq y$. By homomorphism we have $[x]_\sim \leq_\sim [y]_\sim$ and hence $\top_{[x]_\sim} \leq y$ because $[x]_\sim$ and $[y]$ are disjoint and due to $[y] = \{y\}$. The case $x \geq y$ can be treated symmetrically. ■

In particular, the existence of $\top[x]_\sim$ and $\bot[x]_\sim$ is ensured if the lattice under consideration is finite. □

Note that this means in particular that an isolated sublattice S' is convex because it is an equivalence class of a congruence (see e.g. [14]). This means that S' equals the interval $[\bot_{S'}, \top_{S'}]$. However, not every interval is an isolated sublattice; consider e.g. the interval $[\emptyset, \{1\}]$ in the lattice $(\mathcal{P}(\{1, 2\}), \subseteq)$ as a counterexample (here, $\emptyset \subseteq \{2\}$ holds but not $\{1\} \subseteq \{2\}$, contradicting point 2. of Definition 3.1).

Lemma 3.5. *Let S_1 and S_2 be two isolated sublattices with $S_1 \cap S_2 \neq \emptyset$. Then $\{\bot_{S_1}, \bot_{S_2}, \top_{S_1}, \top_{S_2}\}$ is a chain.*

Proof: Consider an arbitrary $s_{12} \in S_1 \cap S_2$. If $\top_{S_1} \in S_2$ holds then $\top_{S_1} \leq \top_{S_2}$ is obvious. In the case $\top_{S_1} \notin S_2$ we have $\top_{S_2} \leq \top_{S_1}$ due to $s_{12} \in S_2$, $s_{12} \leq \top_{S_1}$ and the definition of an isolated sublattice. A symmetric argumentation shows $\bot_{S_1} \lesseqgtr \bot_{S_2}$. The rest follows from $\bot_{S_1}, \bot_{S_2} \leq s_{12} \leq \top_{S_1}, \top_{S_2}$. ■

Lemma 3.6. *Let S_1 and S_2 be two isolated sublattices with $S_1 \cap S_2 \neq \emptyset$. Then $S_{12} =_{def} S_1 \cup S_2$ is an isolated sublattice, too.*

Proof: By Lemma 3.5 we know that $\{\bot_{S_1}, \bot_{S_2}, \top_{S_1}, \top_{S_2}\}$ is a chain so let us assume w.l.o.g. that $\bot_{S_1} \leq \bot_{S_2}$ holds. In the case $\top_{S_2} \leq \top_{S_1}$ we have $[\bot_{S_2}, \top_{S_2}] \subseteq [\bot_{S_1}, \top_{S_1}]$ and the claim follows immediately since isolated sublattices are intervals. The case $\top_{S_1} < \bot_{S_2}$ is ruled out by $S_1 \cap S_2 \neq \emptyset$ so the remaining case is $\bot_{S_1} \leq \bot_{S_2} \leq \top_{S_1} \leq \top_{S_2}$.

To show that in this case S_{12} is a lattice we pick two arbitrary $x, y \in S_{12}$. If $x, y \in S_1$ or $x, y \in S_2$ holds we are done due to the properties of a sublattice so the crucial case is w.l.o.g. $x \in S_2 \backslash S_1 \wedge y \in S_1 \backslash S_2$. By $y \in S_1$ we get $y \leq \top_{S_1}$ and by $\top_{S_1} \in S_2$ (this follows from $\bot_{S_2} \leq \top_{S_1} \leq \top_{S_2}$ and convexity of S_2) we obtain $y \in S_2 \vee y \leq \bot_{S_2}$. However, the case $y \in S_2$ is impossible due to $y \in S_1 \backslash S_2$ so $y \leq \bot_{S_2}$ has to hold which implies $y \leq x$. Hence we have $x \sqcap y = y \in S_{12}$. For the supremum the argumentation is dual. Clearly, $\bot_{S_{12}} = \bot_{S_1}$ and $\top_{S_{12}} = \top_{S_2}$ hold.

Consider now arbitrary $s_{12} \in S_{12}$ and $x \notin S_{12}$ with $s_{12} \leq x$. If $s_{12} \in S_2$ holds we obtain $x \geq \top_{S_2} = \top_{S_{12}}$ because S_2 is an isolated sublattice so we assume now that s_{12} is an element of S_1. Then we have $x \geq \top_{S_1}$ due to properties of S_1. But now we conclude $x \geq \top_{S_2}$ from $\top_{S_1} \in S_2$ and $x \notin S_2$. Again, the case $x \leq s_{12}$ can be treated symmetrically. ■

After this it is easy to see that $S_1 \cup S_2$ is an isolated sublattice with bottleneck if S_1 and S_2 are isolated sublattices with bottlenecks, and that $S_1 \cup S_2$ is a nontrivial summit isolated sublattice if S_1 and S_2 are nontrivial summit isolated sublattices. Since summit isolated sublattices share $\top$ as a common element we obtain the following theorem with the help of Lemma 3.6:

Theorem 3.7. *Let $(S, \leq)$ be a lattice.*

1. *Two different inclusion-maximal sublattices with bottleneck of $(S, \leq)$ are disjoint.*
2. *$(S, \leq)$ has at most one nontrivial inclusion-maximal summit isolated sublattice.*

The intuitive statement of the next lemma is that isolated sublattices in a quotient lattice induced by an isolated sublattice correspond to isolated sublattices in the original lattice:

Lemma 3.8. *Let S' be an isolated sublattice of $(S, \leq)$ and let $S_{S'}$ be an isolated sublattice of $S/{\sim_{S'}}$. Then $S'' =_{def} \bigcup S_{S'}$ is an isolated sublattice of S.*

Proof: First we show that S'' is indeed a sublattice of S, and therefore we pick two arbitrary $x, y \in S''$. If $x, y \in S'$ holds then we have $x \sqcap y \in S'$ because S' is a sublattice of $(S, \leq)$ and hence $x \sqcap y \in S''$. In the case $x, y \notin S'$ we have $\{x\} \sqcap_{\sim_{S'}} \{y\} \in S_{S'}$ (note that in this case both $[x]_{\sim_{S'}}$ and $[y]_{\sim_{S'}}$ are singleton sets) and hence $x \sqcap y \in S''$. W.l.o.g. the remaining case is $x \in S' \wedge y \notin S'$. If here $S' \sqcap_{\sim_{S'}} \{y\} = S'$ holds we conclude $x \leq y$ and hence $x \sqcap y = x \in S''$. Otherwise we have $S' \sqcap_{\sim_{S'}} \{y\} = \{z\}$ for some $\{z\} \in S_{S'}$ with $\{z\} \neq S'$ and claim that $x \sqcap y = z$ holds. By homomorphism, z is a lower bound of x and y so assume that there is a lower bound z' of x and y with $z \leq z'$. Again by homomorphism,

$[z']_{\sim_{S'}}$ is a lower bound of $\{x\}$ and $\{y\}$, and due to infimum properties we have $[z']_{\sim_{S'}} \leq_{S'} \{z\}$. By assumption we have $[z']_{\sim_{S'}} \neq S'$ which implies $[z']_{\sim_{S'}} = \{z'\}$ and by infimum properties also $\{z'\} \leq_{\sim_{S'}} \{z\}$. This in turn means $z' \leq z$ and hence $z = z'$. A symmetric argumentation can be carried out for the supremum. Obviously, we have $\bot_{S''} = \bot_{S'}$ if $\bot_{S_{S'}} = S'$ and $\bot_{S''} = x$ if $\bot_{S_{S'}} = \{x\}$ and a dual relationship for $\top_{S''}$.

Now we choose arbitrary $s \in S''$ and $x \notin S''$ with $s \leq x$. By construction of S'' we have $[s]_{\sim_{S'}} \in S_{S'}$ and $[x]_{\sim_{S'}} \notin S_{S'}$, and by homomorphism we get $[s]_{\sim_{S'}} \leq_{\sim_{S'}} [x]_{\sim_{S'}}$ from where we conclude $\top_{S_{S'}} \leq_{\sim_{S'}} [x]_{\sim_{S'}}$ because $S_{S'}$ is an isolated sublattice. Next we observe that $[x]_{\sim_{S'}}$ and $\top_{S_{S'}}$ are disjoint and consider first the case that $\top_{S_{S'}}$ is a singleton set. Then we have $\top_{S_{S'}} = \{\top_{S''}\}$ and hence $\top_{S''} \leq x$. If $\top_{S_{S'}}$ contains more than one element we have $\top_{S_{S'}} = S'$ and hence $\top_{S''} = \top_{S'}$. Also in this case $\top_{S''} \leq x$ holds by homomorphism and disjointness of $[x]_{\sim_{S'}}$ and $\top_{S_{S'}}$. The case $s \geq x$ can be treated symmetrically, so S'' satisfies indeed Definition 3.1. ■

This claim holds even for isolated sublattices with bottleneck:

Lemma 3.9. *Let S' be an isolated sublattice of $(S, \leq)$ and let $S_{S'}$ be an isolated sublattice with bottleneck of $S/{\sim_{S'}}$. Then $S'' =_{def} \bigcup S_{S'}$ is an isolated sublattice with bottleneck of S.*

Proof: From Lemma 3.8 we know already that S'' is an isolated sublattice so we pick an arbitrary bottleneck $B \in S/{\sim_{S'}}$ of $S_{S'}$. We now distinguish several cases:

1. $\top_{S_{S'}} = S'$: Here we have $\top_{S''} = \top_{S'}$. Moreover, $B = \{b\}$ holds for some $b \in S$, so b fulfills requirements of Definition 3.2 due to homomorphism properties (note that all elements of $]S', B]$ are singleton sets).
2. $B = S'$: Here we have $\top_{S_{S'}} = \{\top_{S''}\}$ and we claim that $\bot_{S'}$ is a bottleneck of S''. Because $[\{\top_{S''}\}, S']$ is a chain in $S/{\sim_{S'}}$, $[\top_{S''}, \bot_{S'}]$ is a chain in S (note that $S/{\sim_{S'}}$ consists only of singleton sets except possibly S'). The remaining properties of Definition 3.2 are now easy to check.
3. $\top_{S_{S'}} \neq S' \wedge B \neq S'$: If $S' \in [\top_{S_{S'}}, B]$ holds then B is also a bottleneck of $S_{S'}$ and we can proceed as in the previous case. Otherwise, all elements from $S/{\sim_{S'}}$ under consideration are singleton sets, and Definition 3.2 is easily verified by means of homomorphism. ■

Also for summit isolated sublattices we have an analogous lemma:

Lemma 3.10. *Let S' be an isolated sublattice of $(S, \leq)$ and let $S_{S'}$ be a summit isolated sublattice of $S/{\sim_{S'}}$. Then $S'' =_{def} \bigcup S_{S'}$ is a summit isolated sublattice of S.*

Proof: This is obvious due to $\top_{\sim_{S'}} \in S_{S'}$, Lemma 3.8, homomorphism properties and construction of S''. ■

In our algorithm we will be confronted with the iterative construction of quotients, so we investigate some properties thereof. We call a (finite or infinite) sequence $S_0, S_1, S_2, \ldots$ of lattices a *quotient sequence* if for all i the relationship $S_{i+1} = S_i/{\sim_{S'_i}}$ holds for some isolated sublattice S'_i of S_i.

The next lemma states intuitively that we can factor out an inclusion-maximal useful summit isolated sublattice at most once:

Lemma 3.11. *Let $S_0, S_1, S_2, \ldots$ be a quotient sequence such that $S_1 = S_0/{\sim_{S'_0}}$ holds for an inclusion-maximal useful summit isolated sublattice S'_0. Then no S_i with $i > 0$ contains a useful summit isolated lattice.*

Proof: Assume that some S_i with $i > 0$ contains a useful summit isolated lattice S'_i. Then we could construct an inclusion-maximal summit isolated sublattice $S''_0 \supsetneq$ of S_0 from S'_i backwards along the lines of Lemmata 3.8 and 3.10, contradicting the inclusion-maximality of S'_0 ■

The statement of the next lemma is that - disregarding the operation $\cdot^{\{\}}$ - an element can be part of an inclusion-maximal isolated sublattice at most once in a quotient sequence (the notation $\bigcup^n C$ is defined inductively by $\bigcup^0 C =_{def} C$ and $\bigcup^{n+1} C =_{def} \bigcup(\bigcup^n C)$; intuitively speaking, $\bigcup^n C$ strips n set braces from the elements of C):

Lemma 3.12. *Let $S_0, S_1, S_2, \ldots$ be a quotient sequence such that $S_{i+1} = S_i/{\sim_{S'_i}}$ holds for an inclusion-maximal useful isolated sublattice with bottleneck S'_i for all $i \geq 0$. Then S'_i and $\bigcup^{j-i} S'_j$ are disjoint for all i, j with $j > i$.*

Proof: Clearly, S'_i and $\bigcup S'_{i+1}$ are disjoint due to the first part of Theorem 3.7 and Lemma 3.9. The rest is simple induction. ■

3.2 Isolated Sublattices and Closure Systems

Now we will examine the interplay between isolated sublattices and closure systems. As a first observation, we note that by Lemma 3.3 $S/{\sim_{S'}}$ is a homomorphic image of $(S, \leq)$ if S' is an isolated sublattice of $(S, \leq)$. Hence it is easy to see that the two conditions from Definition 2.2 can be transferred from a closure system of S to a corresponding system in $S/{\sim_{S'}}$. This shows essentially the following lemma:

Lemma 3.13. *Let $(S, \leq)$ be a lattice, S' an isolated sublattice of $(S, \leq)$ and consider a closure system C of $(S, \leq)$.*

1. *If $C \cap S' = \emptyset$ then $C^{\{\}}$ is a closure system of $S/{\sim_{S'}}$.*
2. *If $C \cap S' \neq \emptyset$ then $(C \backslash S')^{\{\}} \cup \{S'\}$ is a closure system of $S/{\sim_{S'}}$.*

The reverse direction, i.e., reasoning about closure systems in S starting from closure systems in $S/{\sim_{S'}}$, is a little bit more elaborate. We start with the following definition:

Definition 3.14. *Let $(S, \leq)$ be a lattice with greatest element $\top$. A subset $C \subseteq S$ is called a* preclosure system *of $(S, \leq)$ if $C \cup \{\top\}$ is a closure system of $(S, \leq)$. The set of all preclosure systems of $(S, \leq)$ is denoted by $\mathcal{PC}(S)$.*

Note that every closure system is also a preclosure system and that the empty set is a closure system, too. Moreover, if $C(S)$ is finite then $|\mathcal{PC}(S)| = 2 \cdot |C(S)|$ holds. Clearly, a preclosure system is closed under $\sqcap$. Moreover, in a nonempty preclosure system we find a least element majorizing $\perp$ (which is also a least element of the preclosure system):

Lemma 3.15. *Let C be a nonempty preclosure system of a lattice $(S, \leq)$ with least element $\perp_S$ and greatest element $\top_S$. Then there is a least element $c \in C$ majorizing $\perp_S$.*

Proof: If $\top_S \in C$ holds then C is even a closure system and the claim follows immediately from Definition 2.2. Otherwise, $C' =_{def} C \dot{\cup} \{\top_S\}$ is a closure system hence there is a least $c' \in C'$ with $\perp_S \leq c'$. However, c' can not equal $\top_S$ because C' contains at least one other element except $\top_S$; hence we get $c' \in C$. ■

By means of preclosure systems we can characterize the structure of the intersection of a closure system and an isolated sublattice:

Lemma 3.16. *Let C be a closure system on a lattice S with greatest element $\top$, and let S' be an isolated sublattice of S with greatest element $\top_{S'}$ and least element $\perp_{S'}$. Then $C' =_{def} C \cap S'$ is a preclosure system of S'. Moreover, if S' is a summit isolated sublattice then C' is a closure system of S'.*

Proof: First we show that $C'_{\top_{S'}} =_{def} C' \dot{\cup} \{\top_{S'}\}$ is closed under binary infimum. Therefore we pick two arbitrary $x', y' \in C_{\top_{S'}}$. If one of them equals $\top_{S'}$ we have $x' \sqcap y' \in \{x', y'\} \subseteq C'_{\top_{S'}}$ and we are done; otherwise we have $x', y' \in C'$. By definition of a sublattice, we have $\perp_{S'} \leq x', y' \leq \top_{S'}$ and hence $\perp_{S'} \leq x' \sqcap y' \leq \top_{S'}$. By convexity of S' this implies $x' \sqcap y' \in S'$.

For the second criterion of Definition 2.2 we consider an arbitrary $s' \in S'$. Then there is a smallest $x \in C$ with $s' \leq x$. If x is an element of S' then it is by definition also an element of $C'_{\top_{S'}}$ so assume that $x \notin S'$ holds. Then, by definition of an isolated sublattice, we have $\top_{S'} \leq x$ so it is easy to see that $\top_{S'}$ is a smallest element of $C'_{\top_{S'}}$ majorizing s'. ■

Now we can formulate the "verse" lemma of Lemma 3.13 in the case of isolated sublattices with bottleneck:

Lemma 3.17. *Let $(S, \leq)$ be a lattice and S' an isolated sublattice of S such that $\top_{S'}$ has a least bottleneck b. Assume that $C_{S'}$ is a preclosure system of S' and let C' be a closure system of $S/{\sim_{S'}}$ with $S' \in C'$. Then $C =_{def} \bigcup(C' \backslash \{S'\}) \cup C_{S'}$ is a closure system of $(S, \leq)$.*

Proof: According to Lemma 2.3 it suffices to show that for every $x \in S$ there is a least $c \in C$ with $x \leq c$. Hence we distinguish the following cases:

1. $x \notin S'$: Then there is a least $c' \in C'$ with $\{x\} \leq_{\sim_{S'}} c'$. Again, there are two cases:
   a) $c' \neq S'$: Then c' has the form $\{c''\}$ for some $c'' \in C$. Clearly, $x \leq c''$ holds by homomorphism properties and let us assume there is a $c''' \in C$ with $c'' \not\leq c'''$ and $x \leq c'''$. If $c''' \notin C_{S'}$ then we have $\{x\} \leq_{\sim_{S'}} \{c'''\}$, contradicting the fact that c' is the least element majorizing $\{x\}$. In the case $c''' \in S'$ we have $\{x\} \leq_{S'}$ contradicting the fact $c' \neq S'$ because $\{c''\}$ was supposed to be the least element majorizing $\{x\}$.
   b) $c' = S'$: Here we have again two cases:
      i. $C_{S'} \neq \emptyset$: Because $S_{S'}$ is nonempty there is by Lemma 3.15 a least element $c'' \in C_{S'}$ majorizing $\bot_{S'}$. Clearly, $x \leq c''$ holds ($\{x\} \leq_{\sim_{S'}} S'$ implies $x \leq \bot_{S'}$) so let us assume there is a c''' with $c'' \not\leq c'''$ and $x \leq c'''$. c''' can not be an element of $C_{S'}$ (then it would be an element different from c'' majorizing $\bot_{S'}$) and hence no element from S' so $[c''']_{\sim_{S'}} = \{c'''\}$ has to hold. But then $\{c'''\}$ would majorize $\{x\}$ with $S' \not\leq_{\sim_{S'}} \{c'''\}$, contradicting the choice of c'.
      ii. $C_{S'} = \emptyset$: By homomorphism, $\{b\}$ is the least bottleneck of S' in $S/{\sim_{S'}}$. In particular, this means $\{y\} >_{\sim_{S'}} S' \Leftrightarrow \{y\} \geq_{\sim_{S'}} \{b\}$. Hence, $\mathsf{maj}(\{x\}, C')$ can be partitioned as $\mathsf{maj}(\{x\}, C') = \{S'\} \dot\cup \mathsf{maj}(\{b\}, C')$. Because C' is a closure system there is a least element $\{c''\}$ of $\mathsf{maj}(\{b\}, C')$. On the other hand, we have also $\mathsf{maj}(x, C) = \mathsf{maj}(b, C)$ by an analogous argument as above (note that $C \cap S' = \emptyset$ holds). By homomorphism, c'' is the least element of $\mathsf{maj}(b, C)$ and hence the least element of C majorizing x.
2. $x \in S'$: Because $C_{S'}$ is a preclosure on S' there is a least element c of $\mathsf{maj}(x, S' \cup \top_{S'})$. Again, we have two cases:
   a) $c \in C_{S'}$: By definition of on isolated sublattice, all elements in $S \backslash S'$ are strictly less than $\bot_{S'}$ (and can hence not majorize x) or strictly greater than $\top_{S'}$ (and can hence not be a minimal element majorizing x). So c is the least element of $\mathsf{maj}(x, C)$.
   b) $c \notin C_{S'}$: In this case, we have $c = \top_{S'}$. An argumentation analogous to case 1.a)ii. shows that c' is the least element of C majorizing x where $\{c'\}$ is the least element of $\mathsf{maj}(\{b\}, C')$. ■

Remark: The requirement that $\top_{S'}$ has a bottleneck element is necessary for the correctness of Lemma 3.17. To see this, take a look at Fig. 2. At the left, a preclosure system, indicated by encircled elements, on an isolated sublattice without bottleneck of its top element, indicated by an ellipse, is shown. The middle picture shows a closure system on the associated quotient lattice, indicated by circles. However, the construction from Lemma 3.17 leads to the set of encircled elements in the right picture which is not closed under binary infimum and hence is no closure system. This shows one effect of a bottleneck: it prevents the necessity that the top element of an isolated sublattice is the infimum of two elements above it. Moreover, the requirement that $\top_{S'}$ has even a least bottleneck is necessary: consider the lattice $S = ([0, 1], \leq)$ (where $\leq$ denotes the usual order on the reals) and the sublattice $S' = (\{0\}, \leq)$. We choose the empty set as preclosure $C_{S'}$ of S' and $[0,1]^{\{\}}$ as closure system C'. Then the construction yields for C the set $]0,1]$ which is no closure system since it contains no least element majorizing 0. □

In a summit isolated sublattice S', the top element $\top_{S'}$ obviously can not have a bottleneck, so Lemma 3.17 is not applicable in this situation. However, there is a slight variant of it:

Lemma 3.18. *Let $(S, \leq)$ be a lattice and S' a summit isolated sublattice of S. Assume that $C_{S'}$ is a closure system of S' on $(S, \leq)$ and let C' be a closure system of $S/{\sim_{S'}}$. Then $C =_{def} \bigcup C' \backslash S' \cup C_{S'}$ is a closure system of $(S, \leq)$.*

The proof is very similar to the one of Lemma 3.17 so we omit it here. Note that for every closure system $C_{S'}$ of S' we have $\top \in C_{S'}$ and hence also $S' \in C'$.

Finally, we consider the case of a closure system C' on $S/{\sim_{S'}}$ with $S' \notin C'$:

Lemma 3.19. *Let S be a lattice and S' an isolated sublattice with bottleneck of S and assume that C' is a closure system on $S/{\sim_{S'}}$ with $S' \notin C'$. Then $\bigcup C'$ is a closure system on S.*

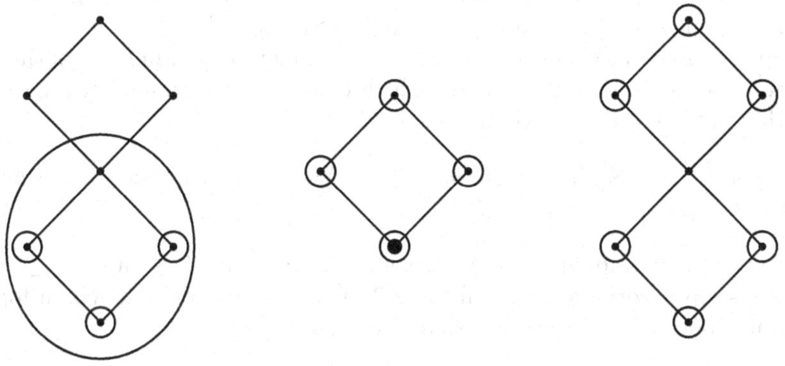

Fig. 2. A preclosure system on an isolated sublattice (left), a closure system on a quotient (middle) and no closure system on the original lattice (right)

The proof of this lemma is also very similar to the one of Lemma 3.17 so we leave it to the reader.

Using Lemmata 3.13, 3.16, 3.17 and 3.19 we obtain the following theorem:

Theorem 3.20. *Let S' be an isolated sublattice with bottleneck of a lattice $(S, \leq)$, and consider a set $C \subseteq S$.*

1. *Assume that $C' =_{def} C \cap S' \neq \emptyset$ holds. Then C is a closure system of S iff C' is a nonempty preclosure system of S' and $(C \backslash S')^{\{\}} \cup \{S'\}$ is a closure system of $S/{\sim_{S'}}$.*
2. *Assume that $C \cap S' = \emptyset$ holds. Then C is a closure system of S iff $C^{\{\}}$ is a closure system of $S/{\sim_{S'}}$.*

Analogously, Lemmata 3.13, 3.16 and 3.18 imply the following theorem:

Theorem 3.21. *Let S' be a summit isolated sublattice of a lattice $(S, \leq)$, and consider a set $C \subseteq S$. Then C is a closure system of S iff $C \cap S'$ is a closure system of S' and $(C \backslash S')^{\{\}} \cup \{S'\}$ is a closure system of $S/{\sim_{S'}}$.*

4 Counting Closure Operators

4.1 Closed Formulae for Special Cases

Since we are interested in counting the number of closure operators we assume from now on that every lattice under consideration is finite.

In the algorithm we will introduce in subsection 4.2 the recursive calls will have to compute closure system containing already processed elements. So for a subset $T \subseteq S$ of a lattice S we introduce the notations $\mathcal{C}(S)_T =_{def} \{C \in \mathcal{C}(S) \,|\, T \subseteq C\}$ and $\mathcal{C}(S)_{-x,T} =_{df} \{C \in \mathcal{C}(S)_T \,|\, x \notin C\}$ for $T \subseteq S$ and $x \in S$. Trivially, $\mathcal{C}(S) = \mathcal{C}(S)_\emptyset$ (the empty set imposes no constraints) and $\mathcal{C}(S)_T = \mathcal{C}(S)_{T \backslash \{\top_S\}} = \mathcal{C}(S)_{T \cup \{\top_S\}}$ (every closure system has to contain $\top_S$) hold. Moreover, $\mathcal{C}(S)_T$ is the disjoint union of $\mathcal{C}(S)_{T \cup \{x\}}$ and $\mathcal{C}(S)_{-x,T}$ so we have the equality $|\mathcal{C}(S)_T| = |\mathcal{C}(S)_{T \cup \{x\}}| + |\mathcal{C}(S)_{-x,T}|$.

Before we introduce a general divide-and-conquer algorithm for counting closures we examine some special cases which can serve as terminal cases for this algorithm. The first one concerns chains:

Lemma 4.1. *Let $(S, \leq)$ be a chain with n elements and consider an arbitrary $T \subseteq S$. Then we have $|\mathcal{C}(S)_T| = 2^{n-1-|T \backslash \{\top_S\}|}$.*

Proof: It is straightforward to see that for a finite chain $(S, \leq)$ a set $C \subseteq S$ is a closure system according to Definition 2.2 iff it contains $\top_S$. The claim follows now from the formula for the cardinality of power sets. ■

Next, we consider lattices with only one layer of elements between the bottom and top element:

Definition 4.2. *A lattice $(S, \leq)$ is called a* diamond lattice of width n *if its carrier set $S = \{\bot_S, \top_S, b_1, \ldots, b_n\}$ consists of $n+2$ pairwise different elements and $b_i \not\lesseqgtr b_j$ holds for all $i \neq j$. The elements $(b_i)_{1 \leq i \leq n}$ are called the* belt elements *of $(S, \leq)$.*

Lemma 4.3. *Let S be a diamond lattice of width n and let B be the set of its belt elements. Then the following holds:*

1. $\bot_S \in T \Rightarrow |\mathcal{C}(S)_T| = 2^{n-|T \backslash \{\top_S\}|+1}$
2. $\bot_S \notin T \wedge |T \cap B| > 1 \Rightarrow |\mathcal{C}(S)_T| = 2^{n-|T \backslash \{\top\}|}$
3. $\bot_S \notin T \wedge T \cap B = \{b_i\} \Rightarrow |\mathcal{C}(S)_T| = 2^{n-1} + 1$
4. $\bot_S \notin T \wedge T \cap B = \emptyset \Rightarrow |\mathcal{C}(S)_T| = 2^n + n + 1$

Proof: 1. In this case, the elements of $\mathcal{C}(S)_T$ have the form $\{\bot_S, \top_S\} \cup T \cup B$. The elements of T occupy already $|T \backslash \{\top_S\}| - 1$ places in B so the claim follows again from the cardinality formula for power sets.

2. Since closures systems are closed under infimum the condition $|T \cap B| > 1$ implies $\bot \in C$ for every $C \in \mathcal{C}(S)_T$ which reduces this case to the previous one.

3. Consider a closures system $C \in \mathcal{C}(S)_T$. If $\bot \notin C$ holds then $b_j \notin C$ has to hold for all $b_i \neq b_j \in B$ because C is closed under infimum so the only possibility in this case is $C = \{\bot_S, b_i, \top_S\}$. The case $\bot \in C$ can be treated analogously to the first case and the result follows from summing up.

4. We have 2^n closure systems of the form $\{\bot_S, \top_S\} \cup B'$ with $B' \subseteq B$, n of the form $\{b_i, \top_S\}$ and the trivial closure system $\{\top_S\}$. ∎

4.2 Simplifying Counting Closures Using Isolated Sublattices

Let us consider an isolated sublattice with bottleneck S' of a lattice $(S, \leq)$ and a subset $T \subseteq S$ with $T \cap S' = \emptyset$. The set $\mathcal{C}(S)_T$ can be partitioned into two disjoint sets $\mathcal{C}(S)_T^{S'}$ and $\mathcal{C}(S)_T^{\neg S'}$, the first one of them consisting of all elements from $\mathcal{C}(S)_T$ containing an element from S' and the second one consisting of all elements from $\mathcal{C}(S)_T$ containing no element from S'. By the first part of Theorem 3.20 we obtain the equation (the term -1 serves for discarding the empty preclosure of S')

$$|\mathcal{C}(S)_T^{S'}| = |\mathcal{C}(S/{\sim_{S'}})_{T\{\}\cup\{\{S'\}\}}| \cdot (|\mathcal{PC}(S')| - 1). \tag{1}$$

Analogously, the second part of Theorem 3.20 gives rise to the equation

$$|\mathcal{C}(S)_T^{\neg S'}| = |\mathcal{C}(S/{\sim_{S'}})_{-\{S'\},T\{\}}|. \tag{2}$$

Now the relationships $|\mathcal{C}(S/{\sim_{S'}})_{T\{\}}| = |\mathcal{C}(S/{\sim_{S'}})_{T\{\}\cup\{S'\}}| + |\mathcal{C}(S/{\sim_{S'}})_{-\{S'\},T\{\}}|$ and $|\mathcal{PC}(S')| = 2 \cdot |\mathcal{C}(S')|$, together with $|\mathcal{C}(S)_T| = |\mathcal{C}(S)_T^{S'}| + |\mathcal{C}(S)_T^{\neg S'}|$ lead to

$$|\mathcal{C}(S)_T| = |\mathcal{C}(S/{\sim_{S'}})_{T\{\}\cup\{\{S'\}\}}| \cdot 2(|\mathcal{C}(S')| - 1) + |\mathcal{C}(S/{\sim_{S'}})_{T\{\}}| \tag{3}$$

If we do analogous considerations for a summit isolated sublattice S' using Theorem 3.21 we obtain the much simpler formula

$$|\mathcal{C}(S)_T| = |\mathcal{C}(S/{\sim_{S'}})_{T\{\}}| \cdot |\mathcal{C}(S')| \tag{4}$$

Clearly, $|\mathcal{C}(S)| = |\mathcal{C}(S)_\emptyset|$ holds, so the relationships given in Eqs. (3) and (4) can be used for a recursive algorithm if the lattice under consideration contains a useful nontrivial summit isolated sublattice or a useful isolated sublattice with bottleneck. However, this is only feasible if the isolated sublattice S' and the set T are disjoint. Luckily, we can ensure this due to Lemmata 3.11 and 3.12 by choosing in the first step - if possible - an inclusion-maximal nontrivial sublattice followed by the choice of inclusion maximal isolated sublattices with bottleneck.

The details are given in Algorithm 1. Of course we have to resort to other methods if the lattice does not contain suitable isolated sublattices or is not of a special structure for which a closed formula is known.

Algorithm 1. Counting Closure Operators

function #CLOSURES(lattice S, set T)
  if Lemma 4.1 or Lemma 4.3 are applicable **then**
    return the respective number
  end if
  if S has a nontrivial useful summit isolated sublattice **then**
    $S' \leftarrow$ the inclusion maximal summit sublattice
    return #CLOSURES($S/{\sim_{S'}}$,$T^{\{\}}$)·#CLOSURES(S', $\emptyset$)
  end if
  if S has a useful isolated sublattice with bottleneck **then**
    $S' \leftarrow$ an inclusion maximal useful isolated sublattice with bottleneck
    return #CLOSURES($S/{\sim_{S'}}$, $T^{\{\}} \cup \{S'^{\{\}}\}$)·2(#CLOSURES($S'$, $\emptyset$)-1)+#CLOSURES($S/{\sim_{S'}}$,$T^{\{\}}$)
  end if
  compute and return $|C(S)_T|$ by some brute force algorithm
end function

Let us now briefly analyze this algorithm. In every recursive call of #CLOSURES the cardinality of the lattices in the first arguments is strictly smaller than the cardinality of the passed lattice: first, every isolated lattice with bottleneck or every nontrivial summit isolated lattice S' is a strict subset of S. Second, because we consider only useful isolated sublattices S', $S/{\sim_{S'}}$ contains strictly less element than S. This ensures termination in the sense that either a lattice is obtained for which a closed formula for the number of closures is known, or some other brute force algorithm is called.

For a short analysis of a possible speed-up we show first that isolated sublattices can be found in polynomial time Therefore we assume that the lattice $(S, \leq)$ is represented by its Hasse diagram. In a precomputation step, we determine for each $s \in S$ the sets $s \leq$ and $s \geq$, i.e., the elements greater (smaller) than or equal to s. This can be done by BFS or DFS in polynomial time. To compute the isolated sublattices of S, we loop over all tuples $(x, y) \in S \times S$ and determine the interval $[x, y]$ by intersecting $x \leq$ and $y \geq$. By Lemma 3.3, every sublattice is of the form $[x, y]$ so we test for each interval whether it is an isolated sublattice. To this end, test all elements $z \in [x, y] \backslash \{x, y\}$ if there is a direct predecessor or successor z' of z with $z' \notin [x, y]$. If such an z' exists, we discard $[x, y]$ as an isolated sublattice, otherwise we have found an isolated sublattice. Since there are quadratic many elements in $S \times S$, and we test at most $|S|$ elements as z' this can be executed in polynomial time.

Let us now assume that a brute force algorithm takes $c^{|S|}$ time for some $c > 1$. Furthermore, we consider a family of lattices which have a nontrivial useful summit isolated sublattice of cardinality $\frac{|S|}{2}$. Then Algorithm 1 makes two recursive calls with instances of size $\frac{|S|}{2}$ and $\frac{|S|}{2}+1$. In the worst case, these two instances have to be handled using a brute force algorithm. Then, using p as a polynomial caused by the computation of the isolated sublattice, the overall running time is $p(|S|) + c^{\frac{|S|}{2}} + c^{\frac{|S|}{2}+1} \in O(c^{\frac{|S|}{2}+1})$ which is clearly dominated by $c^{|S|}$, resulting from the immediate application of a brute force algorithm.

5 Conclusion and Further Work

We have shown that isolated sublattices can be used to simplify the computation of the number of closure operators. However, there is a lot of future work left. First, it seems realistic that more general structures than isolated sublattices can be used in a similar manner by means of quotient lattices. A possible class of candidates may be autobisimulations whose equivalence classes were already used for reducing the number of nodes by means of a quotient in many algorithmic contexts as e.g. model checking as in [2] or model refinement as in [13]. Second, the presented algorithm should be implemented and evaluated; also, a thorough analysis of its running time should be subject of further investigations. Third, special lattices other than chains and diamond lattices should be investigated in order to obtain analogous results as in Lemmata 4.1 and 4.3. Moreover, similar results could be obtained for general orders instead of lattices. Maybe similar ideas could be used for counting monads on categories since monads on categories are a generalization of closure operators on lattices.

Acknowledgments. The author is grateful to every anonymous (especially the first) reviewer and to Bernhard Möller for valuable hints and remarks.

References

1. Alpay, N., Jipsen, P.: Commutative doubly-idempotent semirings determined by chains and by preorder forests. In: Fahrenberg, U., Jipsen, P., Winter, M. (eds.) RAMiCS 2020. LNCS, vol. 12062, pp. 1–14. Springer, Cham (2020). https://doi.org/10.1007/978-3-030-43520-2_1
2. Baier, C., Katoen, J.-P.: Principles of Model Checking. MIT Press, Cambridge (2008)
3. Berghammer, R., Börm, S., Winter, M.: Algorithmic counting of zero-dimensional finite topological spaces with respect to the covering dimension. Appl. Math. Comput. **389**, 125523 (2021)
4. Birkhoff, G.: Lattice Theory, 3rd edn. American Mathematical Society, New York (1967)
5. Bonzio, S., Baldi, M.P., Valota, D.: Counting finite linearly ordered involutive bisemilattices. In: Desharnais, J., Guttmann, W., Joosten, S. (eds.) RAMiCS 2018. LNCS, vol. 11194, pp. 166–183. Springer, Cham (2018). https://doi.org/10.1007/978-3-030-02149-8_11
6. Brinkmann, G., Deklerck, R.: Generation of union-closed sets and Moore families. J. Integer Seq. **21**(1), 18.1.7 (2018)

7. Caspard, N., Monjardet, B.: The lattices of closure systems, closure operators, and implicational systems on a finite set: a survey. Discrete Appl. Math. **127**(2), 241–269 (2003)
8. Colomb, P., Irlande, A., Raynaud, O.: Counting of Moore families for n=7. In: Kwuida, L., Sertkaya, B. (eds.) ICFCA 2010. LNCS (LNAI), vol. 5986, pp. 72–87. Springer, Heidelberg (2010). https://doi.org/10.1007/978-3-642-11928-6_6
9. Davey, B.A., Priestley, H.A.: Introduction to Lattices and Order, 2nd edn. Cambridge University Press, Cambridge (2002)
10. Demetrovics, J., Hencsey, G., Libkin, L., Muchnik, I.B.: On the interaction between closure operations and choice functions with applications to relational database. Acta Cybern. **10**(3), 129–139 (1992)
11. Elloumi, S., Boulifa, B., Jaoua, A., Saleh, M., Al Otaibi, J., Frias, M.F.: Inference engine based on closure and join operators over truth table binary relations. J. Log. Algebraic Methods Program. **83**(2), 180–193 (2014)
12. Glück, R.: Algebraic investigation of connected components. In: Höfner, P., Pous, D., Struth, G. (eds.) RAMICS 2017. LNCS, vol. 10226, pp. 109–126. Springer, Cham (2017). https://doi.org/10.1007/978-3-319-57418-9_7
13. Glück, R., Möller, B., Sintzoff, M.: Model refinement using bisimulation quotients. In: Johnson, M., Pavlovic, D. (eds.) AMAST 2010. LNCS, vol. 6486, pp. 76–91. Springer, Heidelberg (2011). https://doi.org/10.1007/978-3-642-17796-5_5
14. Grätzer, G.: Lattice Theory: Foundation. Springer, Basel (2011). https://doi.org/10.1007/978-3-0348-0018-1
15. Jipsen, P., Rose, H.: Varieties of Lattices, 1st edn. Springer, Heidelberg (1992). https://doi.org/10.1007/BFb0090224
16. Quintero, S., Ramirez, S., Rueda, C., Valencia, F.: Counting and computing join-endomorphisms in lattices. In: Fahrenberg, U., Jipsen, P., Winter, M. (eds.) RAMiCS 2020. LNCS, vol. 12062, pp. 253–269. Springer, Cham (2020). https://doi.org/10.1007/978-3-030-43520-2_16
17. Roman, S.: Lattices and Ordered Sets, 1st edn. Springer, New York (2008). https://doi.org/10.1007/978-0-387-78901-9

Second-Order Properties of Undirected Graphs

Walter Guttmann(✉)

Department of Computer Science and Software Engineering,
University of Canterbury, Christchurch, New Zealand
walter.guttmann@canterbury.ac.nz

Abstract. We study second-order formalisations of graph properties expressed as first-order formulas in relation algebras extended with a Kleene star. The formulas quantify over relations while still avoiding quantification over elements of the base set. We formalise the property of undirected graphs being acyclic this way. This involves a study of various kinds of orientation of graphs. We also verify basic algorithms to constructively prove several second-order properties.

1 Introduction

Binary relations and relational operations provide convenient abstractions for expressing various kinds of logical specification in concise ways as the following examples demonstrate:

- Relation R is transitive if $RR \subseteq R$ (using relational composition), which is logically equivalent to $\forall x : \forall y : \forall z : (x, y) \in R \wedge (y, z) \in R \Rightarrow (x, z) \in R$.
- Point Q is reachable from point P in graph R if $P \subseteq R^*Q$ (using reflexive-transitive closure *), which is equivalent to: there is a number n and a sequence of vertices $x_0, \ldots, x_n$ with $\forall i : 0 \leq i < n \Rightarrow (x_i, x_{i+1}) \in R$, where x_0 and x_n correspond to P and Q, respectively. See Sect. 2 for the relational specification of points.
- Directed graph R is acyclic if $R^+ \subseteq \bar{\mathsf{I}}$ (using transitive closure $^+$ and the complement of the identity relation $\bar{\mathsf{I}}$), which is equivalent to: there is no number n and sequence of vertices $x_0, \ldots, x_n$ with $\forall i : 0 \leq i < n \Rightarrow (x_i, x_{i+1}) \in R$ and $(x_n, x_0) \in R$.

In these examples, conciseness is gained by eliminating quantifiers from logical specifications. The resulting expressions facilitate equational reasoning about entire relations rather than point-wise arguments involving elements of the base set.

The above logical formulas quantify over elements of the base set of the relation. Sometimes quantification over relations is used:

- A relation algebra is pair-dense if

$$\forall R : \mathsf{O} \neq R \subseteq \mathsf{I} \Rightarrow \exists Q : \mathsf{O} \neq Q \subseteq R \wedge Q\bar{\mathsf{I}}Q\bar{\mathsf{I}}Q \subseteq \mathsf{I}$$

(using the empty relation O) [19].

U. Fahrenberg et al. (Eds.): RAMiCS 2021, LNCS 13027, pp. 209–224, 2021.
https://doi.org/10.1007/978-3-030-88701-8_13

- The intermediate point theorem states

$$P \subseteq RSQ \Leftrightarrow \exists X : X \text{ is a point} \wedge P \subseteq RX \wedge X \subseteq SQ$$

for any relations R and S and any points P and Q [30].
- Two characterisations of difunctional relations are

$$R = RR^\mathsf{T}R \Leftrightarrow \exists P : \exists Q : P^\mathsf{T}P \subseteq \mathsf{I} \subseteq PP^\mathsf{T} \wedge Q^\mathsf{T}Q \subseteq \mathsf{I} \subseteq QQ^\mathsf{T} \wedge R = PQ^\mathsf{T}$$

(using relational converse $^\mathsf{T}$). The formula specifies that P and Q are mappings, that is, univalent and total relations; see Sect. 2. The above equivalence is from [28] which also characterises various types of orders by quantifying over relations.
- A form of the axiom of choice can be expressed as

$$\forall R : \mathsf{I} \subseteq RR^\mathsf{T} \Rightarrow \exists Q : Q \subseteq R \wedge Q^\mathsf{T}Q \subseteq \mathsf{I} \subseteq QQ^\mathsf{T}$$

This considers the set of R-image sets of each element of the base set, and selects one element from each according to choice function Q. The formula specifies that R is total and Q is a mapping.

Of course, already the axioms of relation algebras universally quantify over relations, but the above properties also use existential quantification. We call properties that quantify over relations 'second-order' to distinguish them from logical formulas that quantify over elements of the base set. We express these properties in the language of relation algebras extended with a Kleene star, which abstracts from elements of the base set. Hence, in this language, we can use first-order formulas with variables ranging over the elements of a relation algebra.

In this paper we study second-order properties that are useful for the application area of graphs. One of the motivations for this work is that while $R^+ \subseteq \overline{\mathsf{I}}$ concisely states that directed graph R is acyclic, no similarly compact formalisation of acyclicity is known for *undirected* graphs represented by symmetric relations. This complicates the formalisation of graph algorithms and their verification [13,14]. The focus of this paper is to present a number of second-order properties and study their relationships; future work will use these properties to simplify relational correctness proofs of graph algorithms.

Relation algebras are frequently associated with the aim of eliminating logical quantifiers and thereby enabling point-free equational reasoning at a higher abstraction level. The present work does not contradict this aim by reintroducing quantifiers. The quantifiers in our formulas are second-order, that is, they quantify over relations not elements of the base set. For comparison, consider the map-fusion law for lists in functional programming [4]. Its point-wise form uses functions f and g and a list xs:

$$\text{map } f \text{ (map } g \text{ } xs) = \text{map } (f \circ g) \text{ } xs$$

Its point-free form eliminates the list argument xs:

$$\text{map } f \circ \text{map } g = \text{map } (f \circ g)$$

It still involves implicit quantification over functions f and g, but the law can now be understood as talking about functions rather than lists. The variables f and g are 'higher-order points' and not usually eliminated from this law, though they could be removed in formalisms like combinatory logic [7,31].

The contributions of this paper are:

- We study and compare various notions of orientability of undirected graphs in Sect. 3. They serve as a basis for formalising more specific properties.
- We introduce several second-order formalisations of the property that an undirected graph is acyclic in Sect. 4. We prove a number of relationships between these formulas and give counterexamples in cases where formulas are not equivalent in relation algebras extended with a Kleene star.
- We give several equivalent formalisations of general and specific transitively orientable graphs in Sect. 5. We also formalise the property that an undirected graph contains only simple paths.
- We verify the correctness of several basic algorithms in Sect. 6 to constructively prove a number of axioms used throughout this paper.

Moreover, all results in this paper except the counterexamples have been formally verified in Isabelle/HOL [25]. The corresponding proofs are omitted and can be found in the Archive of Formal Proofs [15].

2 Relational and Algebraic Basics

This section recalls algebras we will use for reasoning about properties of directed and undirected graphs in the remainder of the paper. In particular we discuss Boolean algebras, relation algebras and Kleene relation algebras. We also recall basic relational definitions and give a number of general results.

A *Boolean algebra* [9] is a structure $(S, \sqcup, \sqcap, \overline{}, \bot, \top)$ such that

$$\begin{aligned} x \sqcup (y \sqcup z) &= (x \sqcup y) \sqcup z & \qquad x \sqcap (y \sqcap z) &= (x \sqcap y) \sqcap z \\ x \sqcup y &= y \sqcup x & x \sqcap y &= y \sqcap x \\ x \sqcup x &= x & x \sqcap x &= x \\ x \sqcup \bot &= x & x \sqcap \top &= x \\ x \sqcup \top &= \top & x \sqcap \bot &= \bot \\ x \sqcup (x \sqcap y) &= x & x \sqcap (x \sqcup y) &= x \\ x \sqcup (y \sqcap z) &= (x \sqcup y) \sqcap (x \sqcup z) & x \sqcap (y \sqcup z) &= (x \sqcap y) \sqcup (x \sqcap z) \\ x \sqcup \overline{x} &= \top & x \sqcap \overline{x} &= \bot \end{aligned}$$

for each $x, y, z \in S$. The axioms specify that the operations $\sqcup$ and $\sqcap$ are associative, commutative and idempotent, have units $\bot$ and $\top$, have zeros $\top$ and $\bot$, absorb each other, distribute over each other and are complementary.

The lattice order is obtained by $x \sqsubseteq y \Leftrightarrow x \sqcup y = y$ or the equivalent $x \sqsubseteq y \Leftrightarrow x \sqcap y = x$. The join $x \sqcup y$ is the $\sqsubseteq$-least upper bound of x and y; their meet or $\sqsubseteq$-greatest lower bound is $x \sqcap y$. The $\sqsubseteq$-least element is $\bot$; the $\sqsubseteq$-greatest element is $\top$. The element $\overline{x}$ is the complement of x.

A *relation algebra* [33] is a structure $(S, \sqcup, \sqcap, \cdot, ^{-}, ^{\mathsf{T}}, \bot, \top, 1)$ such that the reduct $(S, \sqcup, \sqcap, ^{-}, \bot, \top)$ is a Boolean algebra and

$$x \cdot (y \cdot z) = (x \cdot y) \cdot z \qquad (x \cdot y)^{\mathsf{T}} = y^{\mathsf{T}} \cdot x^{\mathsf{T}}$$
$$x \cdot 1 = x \qquad (x^{\mathsf{T}})^{\mathsf{T}} = x$$
$$(x \sqcup y) \cdot z = (x \cdot z) \sqcup (y \cdot z) \qquad (x \sqcup y)^{\mathsf{T}} = x^{\mathsf{T}} \sqcup y^{\mathsf{T}}$$
$$x^{\mathsf{T}} \cdot \overline{x \cdot y} \sqsubseteq \overline{y}$$

for each $x, y, z \in S$. It follows that composition $\cdot$ is a monoid with identity 1 and distributes over join, transpose ^{T} is involutive, antidistributes over composition and distributes over join and meet, and De Morgan's Theorem K holds. We abbreviate $x \cdot y$ by xy.

An element x of a relation algebra is reflexive if $1 \sqsubseteq x$, irreflexive if $x \sqsubseteq \overline{1}$, symmetric if $x^{\mathsf{T}} = x$, asymmetric if $x \sqcap x^{\mathsf{T}} = \bot$, antisymmetric if $x \sqcap x^{\mathsf{T}} \sqsubseteq 1$, transitive if $xx \sqsubseteq x$, a partial order if x is reflexive and antisymmetric and transitive, a strict order if x is irreflexive and transitive, a total order if $x \sqcup x^{\mathsf{T}} = \top$ and x is a partial order, a strict total order if $x \sqcup x^{\mathsf{T}} = \overline{1}$ and x is a strict order, univalent if $x^{\mathsf{T}} x \sqsubseteq 1$, injective if $xx^{\mathsf{T}} \sqsubseteq 1$, total if $1 \sqsubseteq xx^{\mathsf{T}}$, surjective if $1 \sqsubseteq x^{\mathsf{T}} x$, bijective if x is injective and surjective, a vector if $x\top = x$, a point if x is a bijective vector, and an arc if $x\top$ and $x^{\mathsf{T}}\top$ are points. The symmetric closure of x is $x \sqcup x^{\mathsf{T}}$. See [30] for further details about these properties.

A *Kleene relation algebra* is a structure $(S, \sqcup, \sqcap, \cdot, ^{-}, ^{\mathsf{T}}, ^{*}, \bot, \top, 1)$ such that the reduct $(S, \sqcup, \sqcap, \cdot, ^{-}, ^{\mathsf{T}}, \bot, \top, 1)$ is a relation algebra and

$$1 \sqcup xx^* \sqsubseteq x^* \qquad xy \sqsubseteq y \Rightarrow x^* y \sqsubseteq y$$
$$1 \sqcup x^* x \sqsubseteq x^* \qquad yx \sqsubseteq y \Rightarrow yx^* \sqsubseteq y$$

for each $x, y \in S$. It follows that $x^* y$ is the $\sqsubseteq$-least fixpoint of $\lambda z.xz \sqcup y$ and yx^* is the $\sqsubseteq$-least fixpoint of $\lambda z.zx \sqcup y$. The above unfold and induction axioms for the Kleene star * are from [17]. The transitive closure of x is $x^+ = xx^*$ and x^* models the reflexive-transitive closure of relations. Relation algebras with transitive closure have been studied in [22].

An element x of a Kleene relation algebra is acyclic if x^+ is irreflexive, and a forest if x is injective and acyclic.

The following theorem states a number of general results in (Kleene) relation algebras. Theorems 1.2 and 1.3 appear in [28,30].

Theorem 1. *Let S be a Kleene relation algebra and let $x, y \in S$. Then*

1. *Every acyclic element is asymmetric.*
2. *Every asymmetric element is irreflexive.*
3. *Acyclic, asymmetric and irreflexive are equivalent for transitive elements.*
4. *x is asymmetric if and only if xx is irreflexive.*
5. *x is a strict order if and only if x is transitive and acyclic.*

6. *x is a strict total order if and only if x is transitive and* $x \sqcup x^\mathsf{T} = \overline{1}$.
7. *x is acyclic if and only if x is irreflexive and x^* is antisymmetric.*
8. *x is acyclic if and only if x^+ is asymmetric.*
9. $(x \sqcup y)^+ = x^+ \sqcup y^+ x^+ \sqcup y^+$ *if* $xy = \bot$.
10. $\top(x \sqcap y) \sqcap \top(x \sqcap \overline{y}) = \bot$ *if x is injective.*

3 Orientations

In the remainder of this paper we model graphs using Kleene relation algebras. A (directed) graph is just an element of (the carrier set of) such an algebra. Graph x is undirected if x is symmetric: $x^\mathsf{T} = x$.

An *orientation* of undirected graph x is a directed graph y obtained by assigning a direction to each edge of x [8]. Algebraically this is formalised by

$$y \text{ is an orientation of } x \Leftrightarrow_{\text{def}} y \sqcap y^\mathsf{T} = \bot \land y \sqcup y^\mathsf{T} = x$$

expressing that y is asymmetric and its symmetric closure is x. Asymmetric requires that y has at most one directed edge between any two vertices; the second equation ensures y contains at least one directed edge between any two vertices connected by an edge in x. Graph x is *orientable* if it has an orientation y:

$$x \text{ is orientable} \Leftrightarrow_{\text{def}} \exists y : y \sqcap y^\mathsf{T} = \bot \land y \sqcup y^\mathsf{T} = x$$

It follows from this formalisation that every orientable graph is symmetric and irreflexive. We now consider the converse, namely, that every symmetric irreflexive element is orientable:

$$\forall x : x = x^\mathsf{T} \land x \sqsubseteq \overline{1} \Rightarrow \exists y : y \sqcap y^\mathsf{T} = \bot \land y \sqcup y^\mathsf{T} = x \tag{0}$$

The structure of this formula is similar to that of the axiom of choice given in Sect. 1; essentially a direction is chosen for each edge.

Formula (0) is independent of the axioms of Kleene relation algebras as witnessed by the following counterexample found by Nitpick [6]. The set $\{\bot, 1, \overline{1}, \top\}$ of relations over a two-element base set forms a Kleene relation algebra which is a subalgebra of the full algebra of relations. In this subalgebra, $\overline{1}$ is symmetric and irreflexive but not orientable.

We study two variants of orientations: one that admits loops and one that admits additional edges with an assigned direction.

- $y \sqcap y^\mathsf{T} \sqsubseteq 1 \land y \sqcup y^\mathsf{T} = x$ replaces asymmetric with antisymmetric in the definition of an orientation. This allows loops in x, which then must also occur in the orientation y. We call this a *loop-orientation*.
- $y \sqcap y^\mathsf{T} = \bot \land y \sqcup y^\mathsf{T} \sqsupseteq x$ requires the symmetric closure to contain x rather than to equal x. So y can contain extra edges, but at most one direction of each. We call this a *super-orientation*.
- $y \sqcap y^\mathsf{T} \sqsubseteq 1 \land y \sqcup y^\mathsf{T} \sqsupseteq x$ combines the two variants to obtain *loop-super-orientations*.

Definitions of loop-orientable, super-orientable and loop-super-orientable are derived for these variants similarly to orientable. Using these notions, we obtain several formulas equivalent to formula (0) as the following result shows.

Theorem 2. *The following eight properties are equivalent:*

1. *Every symmetric irreflexive element is orientable, that is, formula (0) holds.*
2. *Every symmetric element is loop-orientable.*
3. *Every irreflexive element is super-orientable.*
4. *Every element is loop-super-orientable.*
5. $\forall x : x = x^{\mathsf{T}} \Rightarrow \exists y : y \sqcap y^{\mathsf{T}} = x \sqcap 1 \wedge y \sqcup y^{\mathsf{T}} = x$.
6. $\forall x : x = x^{\mathsf{T}} \Rightarrow \exists y : y \sqcap y^{\mathsf{T}} \sqsubseteq x \sqcap 1 \wedge y \sqcup y^{\mathsf{T}} = x$.
7. $\forall x : \exists y : y \sqcap y^{\mathsf{T}} = x \sqcap 1 \wedge y \sqcup y^{\mathsf{T}} \sqsupseteq x$.
8. $\forall x : \exists y : y \sqcap y^{\mathsf{T}} \sqsubseteq x \sqcap 1 \wedge y \sqcup y^{\mathsf{T}} \sqsupseteq x$.

Theorems 2.2–2.4 show how the notions of loop-/super-orientation allow the assumptions of irreflexive/symmetric to be removed from formula (0).

The definition of an orientation generalises to the following useful ternary predicate:

$$S(x, y, z) \Leftrightarrow_{\mathrm{def}} y \sqcap y^{\mathsf{T}} = x \wedge y \sqcup y^{\mathsf{T}} = z$$

In words, the meet of y and y^{T} is x and their join is z. Both x and z need to be symmetric for $S(x, y, z)$ to hold, and $x \sqsubseteq y \sqsubseteq z$ follows, too. Hence, the intuitive meaning for undirected graphs x and z is:

- If an edge is in x and in z, it is also in y.
- If an edge is not in x and not in z, it is also not in y.
- If an edge is not in x but in z, exactly one direction of it is in y.

The following result gives consequences of this definition.

Theorem 3

1. $S(\bot, y, x)$ *if and only if* y *is an orientation of* x.
2. $S(1, y, x)$ *implies that* y *is a loop-orientation of* x.
3. $S(x \sqcap 1, y, x)$ *if and only if* y *is a loop-orientation of* x.
4. $S(y \sqcap 1, y, x)$ *if and only if* y *is a loop-orientation of* x.
5. $S(x, y, z)$ *if and only if* $y \sqcap y^{\mathsf{T}} = x \sqsubseteq z \wedge (y \sqcap \overline{y}^{\mathsf{T}}) \sqcup (y \sqcap \overline{y}^{\mathsf{T}})^{\mathsf{T}} = z \sqcap \overline{x}$.

Theorem 3.5 gives an alternative way to specify the ternary predicate. It requires $x \sqsubseteq z$, that x is the symmetric part of y, and that the difference $z \sqcap \overline{x}$ is the symmetric closure of the asymmetric part of y; see [21] for a study of the symmetric and asymmetric parts of a relation.

A special case of Theorem 3.1 is that $S(\bot, y, \overline{1})$ if and only if y is an orientation of $\overline{1}$. An orientation of the complete graph (without loops) $\overline{1}$ is also known as a *tournament* [8]. The existence of a tournament is equivalent to the conditions in Theorem 2 as the following result shows.

Theorem 4. *The following three properties are equivalent:*

1. *Formula (0) holds.*
2. $\exists y : S(\bot, y, \overline{1})$.
3. $\exists y : S(1, y, \top)$.

There are various ways of strengthening orientability. One is to require the orientation to be injective:

$$x \text{ is injectively orientable } \Leftrightarrow_{\text{def}} \exists y : y \sqcap y^{\mathsf{T}} = \bot \land y \sqcup y^{\mathsf{T}} = x \land yy^{\mathsf{T}} \sqsubseteq 1$$

Injectively orientable graphs correspond to graphs in which every component has at most one cycle, also known as pseudoforests [11,27]. They will be used in Sect. 4. A different strengthening requires orientations to be transitive [28]:

$$x \text{ is transitively orientable } \Leftrightarrow_{\text{def}} \exists y : y \sqcap y^{\mathsf{T}} = \bot \land y \sqcup y^{\mathsf{T}} = x \land yy \sqsubseteq y$$

Transitively orientable graphs, also known as comparability graphs, are the symmetric closures of strict orders. They will be used in Sect. 5.

4 Acyclicity of Undirected Graphs

In this section we discuss various ways to specify that an undirected graph x is acyclic. When justifying specifications informally, we implicitly assume that x is symmetric and irreflexive; we explicitly state such assumptions in theorems.

We present the specifications in order of increasing strength, give equivalent characterisations for most of them and study their relationships.

4.1. Our first specification requires that every orientation of x is acyclic (in the sense of directed graphs):

$$\forall y : y \sqcap y^{\mathsf{T}} = \bot \land y \sqcup y^{\mathsf{T}} = x \Rightarrow y^{+} \sqsubseteq \overline{1} \tag{1}$$

Intuitively, if x contained an undirected cycle then this cycle could be oriented and extended to an orientation of x that would not be acyclic. Conversely, if some orientation of x contained a cycle then the symmetric closure of this cycle would be an undirected cycle in x.

The following result shows an equivalent formulation of (1). It replaces $y^{+} \sqsubseteq \overline{1}$ with $y^{*} \sqcap y^{\mathsf{T}*} = 1$, which is equivalent for irreflexive y.

Theorem 5. *The following two properties are equivalent for any x:*

1. *x satisfies formula (1).*
2. $\forall y : y \sqcap y^{\mathsf{T}} = \bot \land y \sqcup y^{\mathsf{T}} = x \Rightarrow y^{*} \sqcap y^{\mathsf{T}*} = 1$.

4.2. Our second specification weakens the antecedent of formula (1) to asymmetric subsets of x:

$$\forall y : y \sqcap y^\mathsf{T} = \bot \wedge y \sqsubseteq x \Rightarrow y^+ \sqsubseteq \overline{1} \tag{2}$$

Every orientation of x clearly satisfies $y \sqsubseteq x$, so formula (2) implies formula (1). The converse implication holds for orientable elements according to the following result. It also gives equivalent formulations of (2).

Theorem 6. *The following three properties are equivalent for any symmetric x:*

1. *x satisfies formula (2).*
2. *$\forall y : y \sqcap y^\mathsf{T} = \bot \wedge y \sqcup y^\mathsf{T} \sqsubseteq x \Rightarrow y^+ \sqsubseteq \overline{1}$.*
3. *$\forall y : y \sqcap y^\mathsf{T} = \bot \wedge y \sqcup y^\mathsf{T} \sqsubseteq x \Rightarrow y^* \sqcap y^{\mathsf{T}*} = 1$.*

The last two of the above properties are equivalent for any x. Moreover,

4. *Formula (2) implies formula (1) for any x.*
5. *Formulas (2) and (1) are equivalent for any orientable x.*

A counterexample shows that formula (1) does not imply formula (2) for all symmetric irreflexive elements. The complex algebra $\mathrm{Cm}(G)$ of any group G is a relation algebra; see [12,19] for construction details. Moreover, $\mathrm{Cm}(G)$ is a Kleene relation algebra using $x^* = \bigcup_{i \in \mathbb{N}} x^i$. Consider $\mathrm{Cm}(\mathbb{Z}_4)$ where $\mathbb{Z}_4 = \{0, 1, 2, 3\}$ is the cyclic group of order 4. In $\mathrm{Cm}(\mathbb{Z}_4)$ the complex $x = \overline{1} = \{1, 2, 3\}$ satisfies formula (1) since x has no orientation as it is above symmetric atom $\{2\}$. But x is also above non-symmetric atom $y = \{1\}$ with $y^\mathsf{T} = \{3\}$ and $y^+ = \top = \mathbb{Z}_4$, whence x does not satisfy formula (2).

4.3. Our third specification avoids the reference to acyclic subgraphs. It requires that there is a unique way to sandwich x between a graph and its reflexive-transitive closure:

$$\forall y : y \sqsubseteq x \sqsubseteq y^* \Rightarrow y = x \tag{3}$$

Intuitively, if x contained an undirected cycle then one edge of this cycle could be removed without affecting reachability in the graph, so an element strictly below x would satisfy the antecedent. Conversely, if there was a y with $y \sqsubset x \sqsubseteq y^*$ then there would be an edge e in x that is not in y but in y^*, so there would be a path in y from the start vertex of e to its end vertex which together with e would form a cycle. The following result shows equivalent formulations of (3).

Theorem 7. *The following two properties are equivalent for any x:*

1. *x satisfies formula (3).*
2. *$\forall y : y \sqsubseteq x \wedge y^* = x^* \Rightarrow y = x$.*

The following two properties are equivalent for any x:

3. *$\forall y : y \sqsubseteq x \sqsubseteq y^+ \Rightarrow y = x$.*
4. *$\forall y : y \sqsubseteq x \wedge y^+ = x^+ \Rightarrow y = x$.*

All four of the above properties are equivalent for any irreflexive x.

4.4. Our fourth specification expresses the justification underlying formula (3) more directly:

$$\forall y : y \sqsubseteq x \Rightarrow x \sqcap y^* \sqsubseteq y \tag{4}$$

Intuitively, any edge e contained in both x and y^* must already be in y, otherwise the path obtained from y^* together with e would form a cycle. The following result shows that formulas (4) and (3) are equivalent and stronger than formula (2). It also gives further equivalent formulations of (4).

Theorem 8. *The following three properties are equivalent for any x:*

1. *x satisfies formula (4).*
2. $\forall y : y \sqsubseteq x \Rightarrow x \sqcap y^* = y$.
3. $\forall y : y \sqsubseteq x \Rightarrow y \sqcap (x \sqcap \overline{y})^* = \bot$.
4. *x satisfies formula (3).*

Moreover,

5. *Formula (4) implies formula (2) for any symmetric x.*
6. *Formulas (4) and (2) are equivalent for any symmetric irreflexive x if the following two axioms hold:*

$$\forall u : u \neq \bot \Rightarrow \exists v : v \textit{ is an arc} \wedge v \sqsubseteq u$$

$$\forall u : \forall v : u \textit{ is an arc} \wedge u \sqsubseteq v^* \Rightarrow \exists w : w \sqsubseteq v \wedge w \sqcap w^\mathsf{T} = \bot \wedge u \sqsubseteq w^*$$

The first of these axioms specifies that every non-empty graph contains an edge, which is similar to the point axiom [19,29]. The second of these axioms states that if the end vertex of an edge u is reachable from its start vertex using (directed) edges in v, then the same holds already in an asymmetric subset w of v. Intuitively, the asymmetric subset w is formed by the edges on the (directed) path from the start vertex of u to its end vertex.

A counterexample found by Nitpick shows that formula (2) does not imply formula (4) for all symmetric irreflexive elements. The set of symmetric complexes $\mathrm{SCm}(G) = \{x \in \mathrm{Cm}(G) \mid x = x^\mathsf{T}\}$ of a commutative group G forms a relation algebra which is a subalgebra of $\mathrm{Cm}(G)$ [12,16]. Since $\mathrm{SCm}(G)$ is closed under Kleene star, it also forms a Kleene relation algebra. In $\mathrm{SCm}(\mathbb{Z}_4)$ the complex $x = \overline{1} = \{1, 2, 3\}$ satisfies formula (2) because the only asymmetric complex is $\bot = \emptyset$. But x also contains atom $y = \{1, 3\}$ with $y^* = \top = \mathbb{Z}_4$, whence x does not satisfy formula (4).

4.5. Our fifth specification generalises the formulation given in Theorem 8.3. According to the latter formulation there cannot be an edge e in y such that there is a path from the source of e to its target using edges of x that are not in y. We now allow the edge e to be in y^*:

$$\forall y : y \sqsubseteq x \Rightarrow y^* \sqcap (x \sqcap \overline{y})^* = 1 \tag{5}$$

Intuitively, if there is a path in y, there cannot be a path from its start vertex to its end vertex using edges of x that are not in y, except if the start and end vertices coincide. Namely, if the start and end vertices were different, the two disjoint paths together would form a cycle. The following result shows that formula (5) is stronger than formula (4). It also gives equivalent formulations of (5).

Theorem 9. *The following six properties are equivalent for any x:*

1. *x satisfies formula (5).*
2. $\forall y : y \sqsubseteq x \Rightarrow y^* \sqcap (x \sqcap \overline{y})^+ \sqsubseteq 1$.
3. $\forall y : y \sqsubseteq x \Rightarrow y^+ \sqcap (x \sqcap \overline{y})^* \sqsubseteq 1$.
4. $\forall y : y \sqsubseteq x \Rightarrow y^+ \sqcap (x \sqcap \overline{y})^+ \sqsubseteq 1$.
5. $\forall y : \forall z : y \sqcap z = \bot \wedge y \sqcup z \sqsubseteq x \Rightarrow y^* \sqcap z^* = 1$.
6. $\forall y : \forall z : y \sqcap z = \bot \wedge y \sqcup z = x \Rightarrow y^* \sqcap z^* = 1$.

Moreover,

7. *Formula (5) implies formula (4) for any irreflexive x.*

The formulation in Theorem 9.6 is particularly conspicuous. If x is partitioned into y and z, then there cannot be a path from the same start vertex to the same end vertex in both partitions, except for the empty path if the start and end vertices coincide. The formulations in Theorems 9.5 and 9.6 generalise the formulations in Theorems 6.3 and 5.2, respectively, by replacing y^T with a new variable z.

A counterexample shows that formula (4) does not imply formula (5) for all symmetric irreflexive elements. Consider $\mathbb{Z}_{12} = \{0, 1, \dots, 10, 11\}$, the cyclic group of order 12. In $\mathrm{SCm}(\mathbb{Z}_{12})$, complex $x = \{2, 3, 9, 10\} \sqsubseteq \overline{1}$ satisfies formula (4) since only complexes $\bot$, $y_1 = \{2, 10\}$, $y_2 = \{3, 9\}$ and x are below x and

$$
\begin{aligned}
x \sqcap \bot^* &= x \sqcap 1 \sqsubseteq \overline{1} \sqcap 1 = \bot \\
x \sqcap y_1^* &= x \sqcap \{0, 2, 4, 6, 8, 10\} = y_1 \\
x \sqcap y_2^* &= x \sqcap \{0, 3, 6, 9\} = y_2 \\
x \sqcap x^* &= x
\end{aligned}
$$

But x does not satisfy formula (5) since

$$y_1^* \sqcap (x \sqcap \overline{y_1})^* = y_1^* \sqcap y_2^* = \{0, 2, 4, 6, 8, 10\} \sqcap \{0, 3, 6, 9\} = \{0, 6\} \neq \{0\} = 1$$

4.6. Our sixth specification asserts the existence of an orientation that is a forest. To this end, we strengthen the property of being injectively orientable, introduced at the end of Sect. 3, by replacing asymmetric with acyclic:

$$\exists y : y \sqcup y^\mathsf{T} = x \wedge y^+ \sqsubseteq \overline{1} \wedge yy^\mathsf{T} \sqsubseteq 1 \tag{6}$$

Note that $y^+ \sqsubseteq \overline{1}$ implies that y is asymmetric. With $y \sqcup y^\mathsf{T} = x$ it follows that y is an orientation of x. The properties acyclic and injective together are frequently used to specify forests with edges directed away from the roots of the component trees. Overall, the above property requires that there is a (directed) forest whose symmetric closure is x.

Whereas the previous specifications of acyclic graphs were universally quantified, formulation (6) is existentially quantified. The following result relates formula 6) to both the strongest and the weakest of the previous specifications, namely, formulas (5) and (1).

Theorem 10. *The following two properties are equivalent for any x:*

1. *x satisfies formula (6).*
2. *x is injectively orientable and satisfies formula (1).*

Moreover,

3. *Formula (6) implies formulas (1)–(5) for any x.*

The counterexample showing independence of formula (0) given in Sect. 3 also shows that formula (5) does not imply formula (6) for all symmetric irreflexive elements. In that algebra, $x = \overline{1}$ is an atom and satisfies the formulation in Theorem 9.6 since either $y = \bot$ or $z = \bot$ for any partition of x. But since x is not orientable, it does not satisfy formula (6) by Theorem 10.

We furthermore consider the following weakening of formula (6), which replaces the condition $y \sqcup y^\top = x$ with two of its consequences $y \sqsubseteq x$ and $x \sqsubseteq (y \sqcup y^\top)^*$:

$$x \text{ is spannable } \Leftrightarrow_{\text{def}} \exists y : y \sqsubseteq x \sqsubseteq (y \sqcup y^\top)^* \wedge y^+ \sqsubseteq \overline{1} \wedge yy^\top \sqsubseteq 1$$

This means that y no longer needs to contain a direction of every edge of x, but some edges can be entirely omitted provided their end vertices are still weakly connected in y. In other words, y is a spanning forest of x. A similar formalisation of spanning forests has been used in [14] for verifying the correctness of Kruskal's minimum spanning forest algorithm. The significance of being spannable for the present work is captured in the following result.

Theorem 11. *The following two properties are equivalent for any x:*

1. *x satisfies formula (6).*
2. *x is symmetric and spannable and satisfies formula (3).*

Moreover,

3. *$\overline{1}$ is spannable if a point exists.*

5 Transitive Orientations and Simple Paths

In Sect. 3 we have studied the existence of tournaments, that is, orientations of the complete graph without loops $\overline{1}$. In this section we additionally require orientations to be transitive.

Every orientation is asymmetric, and transitive asymmetric relations correspond to strict orders. Hence, the transitively orientable graphs are precisely the graphs of strict orders after ignoring edge directions. Applied to the complete graph without loops $\overline{1}$ this amounts to the existence of a strict total order.

Theorem 12. *The following two properties are equivalent for any x:*

1. *x is transitively orientable.*
2. *x is irreflexive and $\exists y : y \sqcup y^\mathsf{T} = x \wedge yy \sqsubseteq y$.*

In particular, the following two properties are equivalent:

3. *$\overline{1}$ is transitively orientable.*
4. *$\exists y : y \sqcup y^\mathsf{T} = \overline{1} \wedge yy \sqsubseteq y$.*

Moreover, each of the last two properties implies formula (0).

The following result gives additional equivalent properties.

Theorem 13. *The following five properties are equivalent:*

1. *$\overline{1}$ is transitively orientable.*
2. *$\exists y : y \sqcap y^\mathsf{T} \sqsubseteq 1 \wedge y \sqcup y^\mathsf{T} = \top \wedge yy \sqsubseteq y$.*
3. *$\exists y : y^+ \sqsubseteq \overline{1} \wedge y^* \sqcup y^{*\mathsf{T}} = \top$.*
4. *$\exists y : S(\bot, y^+, \overline{1})$.*
5. *$\exists y : S(1, y^*, \top)$.*

Theorem 13.2 is a translation of Theorem 12.4 to partial orders. The property in Theorem 13.3 requires that y is acyclic and unilaterally connected, in other words, between any two vertices there is exactly one (directed) path in y. Theorems 13.4 and 13.5 express this using the ternary predicate of Sect. 3.

We finally consider a special case of undirected acyclic graphs, namely those whose maximum degree is at most 2, that is, at most two edges are incident to each vertex. Every component of such a graph is a simple path [2]. To specify this we strengthen formula (6) by additionally requiring y to be univalent:

$$\exists y : y \sqcup y^\mathsf{T} = x \wedge y^+ \sqsubseteq \overline{1} \wedge yy^\mathsf{T} \sqsubseteq 1 \wedge y^\mathsf{T} y \sqsubseteq 1 \tag{7}$$

Intuitively, if the maximum degree of an acyclic undirected graph is at most 2, it can be oriented by choosing a directed simple path for each of its components. Conversely, if there is a vertex with at least 3 incident edges, any orientation will have at least two incoming or two outgoing edges at that vertex, making the orientation not injective or not univalent. Graphs with maximum degree 2 are not transitively orientable in general, but according to the following result their transitive closure (without loops) is transitively orientable.

Theorem 14

1. *Formula (7) implies formula (6) for any x.*
2. *$x^+ \sqcap \overline{1}$ is transitively orientable if x satisfies formula (7).*

6 Axioms and Algorithmic Proofs

In previous sections we have encountered two kinds of property. Properties such as being injectively/transitively orientable or being acyclic hold for some graphs but not for others. In contrast, properties such as formula (0), the two axioms in Theorem 8.6, and $\overline{1}$ being transitively orientable do not have free variables. Hence, they can serve as axioms that may or may not be assumed to hold in an algebraic setting. In this section we prove that these axioms hold under certain conditions. The conditions are somewhat restrictive from an algebraic perspective but nevertheless satisfied for many practical applications. Our focus is on the proof method which uses constructive algorithms.

For this section we assume that the given Kleene relation algebra is finite and the arc axiom holds, that is, every element except $\bot$ contains an arc:

$$\forall x : x \neq \bot \Rightarrow \exists y : y \text{ is an arc} \wedge y \sqsubseteq x$$

Finiteness is used to prove that algorithms terminate.

We first show that $\overline{1}$ is transitively orientable. To this end we use Szpilrajn's algorithm [32]. It constructs a total order that extends a given partial order. By applying this algorithm to the discrete partial order 1 we obtain the desired total order on the base set.

Partial correctness of Szpilrajn's algorithm has been proved in [3] using the automated theorem prover Prover9 [20]. We have transcribed the algorithm to Isabelle/HOL and proved its correctness using a Hoare-logic library [23,24], which we have extended to total correctness in previous work [14]. From the total-correctness proof we can extract the following result [28].

Theorem 15. *For every partial order p there is a total order t with $p \sqsubseteq t$.*

In particular, by setting $p = 1$ there exists a total order t, which is the condition in Theorem 13.2. Hence, $\overline{1}$ is transitively orientable by Theorem 13. Moreover, formula (0) holds by Theorem 12.

We next show that every symmetric element is spannable. To this end we use Kruskal's algorithm, which constructs a minimum spanning forest of an undirected graph [18]. We modify the algorithm so as to ignore edge weights, in which case it constructs a spanning forest. We have reused an existing specification and correctness proof of this algorithm from previous work [14]. The postcondition of the algorithm implies that the graph is spannable as per the definition in Sect. 4. The following result is a consequence of this.

Theorem 16. *Every symmetric element is spannable.*

We finally establish the second axiom given in Theorem 8.6.

Theorem 17. *Let x be an arc and let y be such that $x \sqsubseteq y^*$. Then there is an asymmetric z with $z \sqsubseteq y$ and $x \sqsubseteq z^*$.*

To construct the desired element z, we augment a standard breadth-first search algorithm [1] with a variable t:

input r, s	r = directed graph, s = start vertex
$t \leftarrow \bot$	t = constructed asymmetric element
$q \leftarrow s$	q = vertices reached so far
$p \leftarrow \overline{s} \sqcap r^\mathsf{T} s$	p = new vertices reached in the next step
while $p \neq \bot$ do	any new vertices?
$\quad t \leftarrow t \sqcup (r \sqcap q p^\mathsf{T})$	include all edges from current to new vertices
$\quad q \leftarrow q \sqcup p$	add new vertices to current set
$\quad p \leftarrow \overline{q} \sqcap r^\mathsf{T} p$	take another step
end	
output q, t	

As precondition we require that s is a vector. The loop invariant states that q is a vector, t is asymmetric, $t \sqsubseteq r$ and $t \sqsubseteq q = t^{\mathsf{T}*} s$ and $p = \overline{q} \sqcap r^\mathsf{T} q$. This allows us to prove the postcondition that t is asymmetric, $t \sqsubseteq r$ and $q = t^{\mathsf{T}*} s = r^{\mathsf{T}*} s$. Termination of the algorithm follows using the number of elements below $\overline{q} \sqcap r^{\mathsf{T}*} s$ as bound function. This reflects vertices that are reachable from s in r but have not been reached so far. From this total-correctness proof we can extract the following result.

Theorem 18. *For every element r and every vector s there is an asymmetric t with $t \sqsubseteq r$ and $t^{\mathsf{T}*} s = r^{\mathsf{T}*} s$.*

We can now show Theorem 17. Given an arc x and an element y with $x \sqsubseteq y^*$, we apply Theorem 18 using $r = y$ and vector $s = x\top$, obtaining an element t with the stated properties. The desired z is obtained as $z = t \sqcap y$. It is clearly below y and asymmetric as it is below the asymmetric t; moreover $x \sqsubseteq z^*$ follows from the assumptions and $t^{\mathsf{T}*} x \top = y^{\mathsf{T}*} x \top$.

7 Conclusion

In this paper we have used second-order properties expressed in relation algebras extended with a Kleene star to formalise that an undirected graph is acyclic in various ways. The formalisations are based on the concept of orientability, which we have therefore studied. We have also verified the correctness of constructive algorithms to validate several of the axioms.

The quantifiers used in second-order properties cause no issues for formal reasoning in Isabelle/HOL whose logic directly supports them as first-order formulas in relation algebras. Sledgehammer [5,26] can also be applied to such formulas and its integrated provers automatically find proofs in some cases. We should note that the formulas considered in this paper do not have complex nestings of quantifiers. In cases where steps are too big for automated provers, the quantifiers are easy to handle manually as most of them are at the outermost level. The integration with equational reasoning in the proof language Isar [34] is seamless; it might be further improved by automatically generating bits of boilerplate code to break down quantifiers depending on the structure of a formula. Isabelle/HOL would also support formalising the properties using second-order quantification over concrete relations, but we prefer working in relation algebras.

For simplicity we have presented all results in the framework of Kleene relation algebras. Our Isabelle/HOL theory shows that most results hold in more general structures, such as single-object bounded distributive allegories [10], Stone relation algebras and Stone-Kleene relation algebras [13,14]. A possible exception is the result that formula (3) implies formula (4), which we were able to prove only in Kleene relation algebras. The more general structures are useful for modelling weighted graphs. We will therefore apply the specifications of acyclic undirected graphs introduced in Sect. 4 to the verification of graph algorithms involving edge weights. Future work will consider the formalisation of further properties of graphs using higher-order formulas.

Acknowledgement. I thank Nicolas Robinson-O'Brien and the anonymous referees for helpful comments.

References

1. Berghammer, R.: Combining relational calculus and the Dijkstra-Gries method for deriving relational programs. Inf. Sci. **119**(3–4), 155–171 (1999)
2. Berghammer, R., Furusawa, H., Guttmann, W., Höfner, P.: Relational characterisations of paths. J. Log. Algebraic Meth. Program. **117**, 100590 (2020)
3. Berghammer, R., Struth, G.: On automated program construction and verification. In: Bolduc, C., Desharnais, J., Ktari, B. (eds.) MPC 2010. LNCS, vol. 6120, pp. 22–41. Springer, Heidelberg (2010). https://doi.org/10.1007/978-3-642-13321-3_4
4. Bird, R., Wadler, P.: Introduction to Functional Programming. Prentice Hall (1988)
5. Blanchette, J.C., Böhme, S., Paulson, L.C.: Extending Sledgehammer with SMT solvers. J. Autom. Reason. **51**(1), 109–128 (2013)
6. Blanchette, J.C., Nipkow, T.: Nitpick: a counterexample generator for higher-order logic based on a relational model finder. In: Kaufmann, M., Paulson, L.C. (eds.) ITP 2010. LNCS, vol. 6172, pp. 131–146. Springer, Heidelberg (2010). https://doi.org/10.1007/978-3-642-14052-5_11
7. Curry, H.B., Feys, R.: Combinatory Logic, vol. 1. North-Holland Publishing Company (1958)
8. Diestel, R.: Graph Theory, 3rd edn. Springer, Heidelberg (2005)
9. Dwinger, P.: Introduction to Boolean Algebras, 2nd edn. Physica-Verlag (1971)
10. Freyd, P.J., Ščedrov, A.: Categories, Allegories, North-Holland Mathematical Library, vol. 39. Elsevier Science Publishers (1990)
11. Gabow, H.N., Tarjan, R.E.: A linear-time algorithm for finding a minimum spanning pseudoforest. Inf. Process. Lett. **27**(5), 259–263 (1988)
12. Givant, S.: Introduction to Relation Algebras, Relation Algebras, vol. 1. Springer, Heidelberg (2017). https://doi.org/10.1007/978-3-319-65235-1
13. Guttmann, W.: An algebraic framework for minimum spanning tree problems. Theor. Comput. Sci. **744**, 37–55 (2018)
14. Guttmann, W.: Verifying minimum spanning tree algorithms with Stone relation algebras. J. Log. Algebraic Meth. Program. **101**, 132–150 (2018)

15. Guttmann, W.: Relational forests. Archive of Formal Proofs (2021). https://www.isa-afp.org/entries/Relational_Forests.html
16. Jipsen, P., Lukács, E.: Minimal relation algebras. Algebra Univers. **32**(2), 189–203 (1994)
17. Kozen, D.: A completeness theorem for Kleene algebras and the algebra of regular events. Inf. Comput. **110**(2), 366–390 (1994)
18. Kruskal, J.B., Jr.: On the shortest spanning subtree of a graph and the traveling salesman problem. Proc. Am. Math. Soc. **7**(1), 48–50 (1956)
19. Maddux, R.D.: Pair-dense relation algebras. Trans. Am. Math. Soc. **328**(1), 83–131 (1991)
20. McCune, W.: Prover9 and Mace4 (2005–2010). https://www.cs.unm.edu/~mccune/prover9/. Accessed 10 Aug 2021
21. Monjardet, B.: Axiomatiques et propriétés des quasi-ordres. Mathématiques et sciences humaines **63**, 51–82 (1978)
22. Ng, K.C.: Relation Algebras with Transitive Closure. Ph.D. thesis, University of California, Berkeley (1984)
23. Nipkow, T.: Winskel is (almost) right: towards a mechanized semantics textbook. Formal Aspects Comput. **10**(2), 171–186 (1998)
24. Nipkow, T.: Hoare logics in Isabelle/HOL. In: Schwichtenberg, H., Steinbrüggen, R. (eds.) Proof and System-Reliability, pp. 341–367. Kluwer Academic Publishers (2002)
25. Nipkow, T., Paulson, L.C., Wenzel, M.: Isabelle/HOL: A Proof Assistant for Higher-Order Logic. LNCS, vol. 2283. Springer, Heidelberg (2002). https://doi.org/10.1007/3-540-45949-9
26. Paulson, L.C., Blanchette, J.C.: Three years of experience with Sledgehammer, a practical link between automatic and interactive theorem provers. In: Sutcliffe, G., Ternovska, E., Schulz, S. (eds.) Proceedings of the 8th International Workshop on the Implementation of Logics, pp. 3–13 (2010)
27. Picard, J.C., Queyranne, M.: A network flow solution to some nonlinear 0–1 programming problems, with applications to graph theory. Networks **12**(2), 141–159 (1982)
28. Schmidt, G.: Relational Mathematics. Cambridge University Press (2011)
29. Schmidt, G., Ströhlein, T.: Relation algebras: concept of points and representability. Discret. Math. **54**(1), 83–92 (1985)
30. Schmidt, G., Ströhlein, T.: Relationen und Graphen. Springer, Heidelberg (1989). https://doi.org/10.1007/978-3-642-83608-4
31. Schönfinkel, M.: Über die Bausteine der mathematischen Logik. Math. Ann. **92**(3–4), 305–316 (1924)
32. Szpilrajn, E.: Sur l'extension de l'ordre partiel. Fundam. Math. **16**, 386–389 (1930)
33. Tarski, A.: On the calculus of relations. J. Symbolic Logic **6**(3), 73–89 (1941)
34. Wenzel, M.: Isar—a generic interpretative approach to readable formal proof documents. In: Bertot, Y., Dowek, G., Hirschowitz, A., Paulin, C., Théry, L. (eds.) TPHOLs 1999. LNCS, vol. 1690, pp. 167–183. Springer, Heidelberg (1999). https://doi.org/10.1007/3-540-48256-3_12

Relation-Algebraic Verification of Borůvka's Minimum Spanning Tree Algorithm

Walter Guttmann(✉) and Nicolas Robinson-O'Brien

Department of Computer Science and Software Engineering, University of Canterbury, Christchurch, New Zealand
walter.guttmann@canterbury.ac.nz, nic.robinson-obrien@pg.canterbury.ac.nz

Abstract. Previous work introduced a relation-algebraic framework for reasoning about weighted-graph algorithms. We use this framework to prove partial correctness of a sequential version of Borůvka's minimum spanning tree algorithm. This is the first formal proof of correctness for this algorithm. We also discuss new abstractions that make it easier to reason about weighted graphs.

Keywords: Algorithm verification · Formal methods · Kleene algebras · Relation algebras · Stone algebras · Weighted graphs

1 Introduction

The Minimum Spanning Tree (MST) problem is to find a subset of the edges of a graph that form a tree, connecting the graph's vertices, where the sum of the weights of the edges is minimal [32]. In 1926, Otakar Borůvka described the MST problem and gave an algorithm to solve it [6]. He was perhaps the first person to do so [12]. Borůvka's original paper is written in Czech; a translation can be found in [24]. Borůvka's MST algorithm was independently discovered by Choquet [8], Florek et al. [10], and Sollin [2]. Many textbooks do not treat Borůvka's MST algorithm with the same exposure as the algorithms of Prim [29] and Kruskal [21]; nevertheless, it is significant for its influence on running-time complexity improvements for MST algorithms [7,18,34].

Borůvka's MST algorithm computes a minimum spanning tree of a weighted, connected, undirected graph whose edge weights are distinct. The algorithm begins by initialising a forest with n trees, each containing a single vertex, where n is the number of vertices in the graph. While there is more than one tree in that forest, the following step is repeated. For each tree in the forest, find the edge in the graph with the smallest weight among all edges that leave the tree; all edges found in this way are added to the forest in this step.

A relation-algebraic framework for MST problems was introduced in [13] and has been used to formally verify Prim's [14] and Kruskal's [15] MST algorithms. In the present paper, we use this framework to formally verify a sequential version of Borůvka's MST algorithm. Its contributions are

U. Fahrenberg et al. (Eds.): RAMiCS 2021, LNCS 13027, pp. 225–240, 2021.
https://doi.org/10.1007/978-3-030-88701-8_14

- In Sect. 3 we define components of a graph in terms of a vector, representing a subset of vertices, and an equivalence, representing connectivity. We also introduce an operation, k, to select an arbitrary component of a graph. We show that this operation can be expressed in m-Kleene algebras.
- In Sect. 4 we present axioms for forests modulo an equivalence, a new abstraction that can represent a forest-like structure where clusters of vertices are conceptually collapsed to points forming a forest with the edges that connect them. A number of properties of this abstraction are also given. Additionally, we study paths between vertices in a forest modulo an equivalence and present a theorem for splitting such a path on one of its edges.
- In Sect. 5 we formalise Borůvka's MST algorithm in m-k-Stone-Kleene relation algebras. The formalisation uses the k operation.
- In Sect. 6 we discuss key invariants of our correctness proof of Borůvka's MST algorithm and highlight how we have used the abstractions introduced in previous sections. This is the first formal correctness proof of this algorithm.

Additionally, we have used Isabelle/HOL [27] to formally verify all results in this paper. The corresponding theories are available in the Archive of Formal Proofs [16] and proofs are omitted from this paper. The PDF version of this paper includes links to the relevant theorems and definitions, hosted online. Our proof of Borůvka's MST algorithm uses a Hoare-logic library and verification conditions are generated using a tactic of that library [25,26]. Further details of the correctness proof are described in [30].

There are other recent works verifying MST algorithms in Isabelle/HOL. For example, a functional version of Prim's algorithm was verified in [22] and an imperative version of Kruskal's algorithm was verified in [17]. These verifications use different frameworks than our work. For more examples and further related work see the survey [28] and [30].

2 Basic Definitions

In this section, we define the algebras that will be used in this paper. We are interested in algebras for reasoning about weighted graphs. Unweighted graphs have a straightforward representation as Boolean adjacency matrices so it makes sense that relation algebras, binary relations in particular, have been used to reason about graph algorithms [3,4,11,19,31]. Relation algebras can be generalised to Stone relation algebras to handle edge weights [13]. This is convenient since it does not involve additional structures to represent edge weights. Edge weights are typically numbered and lattices [1] provide a basis for comparing those weights.

Definition 1. *A* bounded distributive lattice, $(S, \sqcup, \sqcap, \bot, \top)$, *is a partial order, S, where for all $x, y, z \in S$ both a* join, $x \sqcup y$, *and a* meet, $x \sqcap y$ *exist and where $x \sqcap \bot = \bot$ and $x \sqcup \top = \top$, and finally where*

$$x \sqcup (y \sqcap z) = (x \sqcup y) \sqcap (x \sqcup z) \qquad x \sqcap (y \sqcup z) = (x \sqcap y) \sqcup (x \sqcap z)$$

Unfortunately, Boolean algebras cannot be used to represent edge weights as there is no suitable way to define a complement operation. The pseudo-complement operation of Stone algebras [5] weakens the complement axioms just enough to permit the inclusion of elements representing edge weights.

Definition 2. *A* Stone algebra, *$(S, \sqcup, \sqcap, ^{-}, \bot, \top)$, is a bounded distributive lattice, $(S, \sqcup, \sqcap, \bot, \top)$, with a pseudo-complement operation, $^{-}$, where for all $x, y \in S$*

$$\overline{x} \sqcup \overline{\overline{x}} = \top \qquad\qquad x \sqcap y = \bot \Leftrightarrow x \leq \overline{y}$$

The pseudo-complement $\overline{y}$ is the greatest element whose meet with y is $\bot$. If $x = \overline{\overline{x}}$ then x is said to be *regular*. If a Stone algebra has only regular elements then it is a Boolean algebra.

Stone relation algebras [13] include much of the structure we require from relation algebras [23,33] but without the restrictions of the complement operation of Boolean algebras.

Definition 3. *A Stone relation algebra, $(S, \sqcup, \sqcap, \cdot, ^{-}, ^{\top}, \bot, \top, 1)$, is a Stone algebra with operations composition, $\cdot$, and transpose, $^{\top}$, and a constant, 1, where for all $x, y, z \in S$*

$$\begin{aligned} (xy)z &= x(yz) & 1x &= x \\ (x \sqcup y)z &= xz \sqcup yz & \bot x &= \bot \\ (x^{\top})^{\top} &= x & \overline{\overline{xy}} &= \overline{\overline{x}}\,\overline{\overline{y}} \\ (xy)^{\top} &= y^{\top} x^{\top} & \overline{\overline{1}} &= 1 \\ (x \sqcup y)^{\top} &= x^{\top} \sqcup y^{\top} & xy \sqcap z &\leq x(y \sqcap x^{\top} z) \end{aligned}$$

Unless overridden with brackets, the operations have the precedence, from highest to lowest: $^{\top}$, $^{-}$, $\cdot$, $\sqcap$, $\sqcup$. Composition, $x \cdot y$, is often abbreviated to xy.

An element $x \in S$ is called *reflexive* if $1 \leq x$, *transitive* if $xx \leq x$, *symmetric* if $x = x^{\top}$, an *equivalence* if x is reflexive, transitive, and symmetric, a *vector* if $x\top = x$, *univalent* if $x^{\top}x \leq 1$, *injective* if $xx^{\top} \leq 1$, *surjective* if $1 \leq x^{\top}x$, *bijective* if x is injective and surjective, a *point* if x is a bijective vector and an *arc* if both $x\top$ and $x^{\top}\top$ are bijective.

For graphs, vectors are used to represent sets of vertices, points to represent a single vertex, and arcs to represent edges.

Stone-Kleene relation algebras combine Stone relation algebras with Kleene algebras to allow reasoning about reachability [13]. The unfold and induction axioms of the Kleene star are taken from [20].

Definition 4. *A* Stone-Kleene relation algebra, *$(S, \sqcup, \sqcap, \cdot, ^{-}, ^{\top}, ^{*}, \bot, \top, 1)$, is a Stone relation algebra, $(S, \sqcup, \sqcap, \cdot, ^{-}, ^{\top}, \bot, \top, 1)$, with an operation, *, where for all $x, y, z \in S$ the unfold and induction axioms hold*

$$\begin{aligned} 1 \sqcup xx^{*} &\leq x^{*} & z \sqcup xy \leq x &\Rightarrow zy^{*} \leq x \\ 1 \sqcup x^{*}x &\leq x^{*} & z \sqcup yx \leq x &\Rightarrow y^{*}z \leq x \end{aligned}$$

and additionally, $(\overline{\overline{x}})^{*} = \overline{\overline{x^{*}}}$.

We abbreviate xx^* as x^+. Furthermore, we call any $x \in S$ *acyclic* if $x^+ \leq \overline{1}$, and a *forest* if x is injective and acyclic.

Structure for reasoning about minimisation and aggregation is provided by m-Kleene algebras [14].

Definition 5. *An* m-Kleene algebra, $(S\,, \sqcup, \sqcap, \cdot\,, +, ^{-}, ^{\top}, ^{*}, s\,, m\,, \bot\,, \top\,, 1)$, *is a Stone-Kleene relation algebra,* $(S\,, \sqcup, \sqcap, \cdot\,, ^{-}, ^{\top}, ^{*}, \bot\,, \top\,, 1)$, *with operations addition,* $+$, *summation,* s, *and minimum selection,* m, *where for all* $x, y, z \in S$, *the summation properties are satisfied*

$$\begin{aligned} x \neq \bot \wedge s(x) \leq s(y) &\Rightarrow s(z) + s(x) \leq s(z) + s(y) \\ s(x) + s(\bot) &= s(x) \\ s(x) + s(y) &= s(x \sqcup y) + s(x \sqcap y) \\ s(x^{\top}) &= s(x) \end{aligned}$$

the linear property is satisfied

$$s(x) \leq s(y) \vee s(y) \leq s(x)$$

the minimum-selection properties are satisfied

$$\begin{aligned} m(x) &\leq \overline{\overline{x}} \\ x \neq \bot &\Rightarrow m(x) \text{ is an arc} \\ y \text{ is an arc} \wedge y \sqcap x \neq \bot &\Rightarrow s(m(x) \sqcap x) \leq s(y \sqcap x) \end{aligned}$$

and S *contains only finitely many regular elements.*

For reasoning about weighted graphs, we are interested in an instance of these algebras where the carrier set is comprised of square matrices whose entries are taken from the set of real numbers extended by $\bot$ and $\top$, the least and greatest elements respectively. This may be denoted as $\mathbb{R}'^{A \times A}$ where A is the index set of a square matrix and $\mathbb{R}' = \mathbb{R} \cup \{\bot, \top\}$.

In this model, an entry $\top$ in a matrix denotes an arc with unknown weight, $\bot$ the non-existence of an arc, and the real numbers arcs with weights corresponding to their values. Therefore, the regular elements (matrices over $\bot, \top$) describe the structure of graphs without weight information. The constant matrices $\bot$, $\top$ and 1 are defined as follows: $\bot_{ij} = \bot$ for all $i, j \in A$, $\top_{ij} = \top$ and

$$1_{ij} = \begin{cases} \top & \text{if } i = j, \\ \bot & \text{otherwise} \end{cases}$$

The operations $\sqcap$ and $\sqcup$ are the componentwise minimum and maximum, respectively. The binary $+$ operation on $\mathbb{R}'$ is the standard addition for real numbers and the maximum otherwise; for example, $\bot + \top = \top$. This operation is lifted to matrices, componentwise. The composition operation is defined as $(M \cdot N)_{ij} = \max_{k \in A} \min\{M_{ik}, N_{kj}\}$. The pseudo-complement on $\mathbb{R}'$ yields $\overline{x} = \bot$

for all $x \neq \bot$ and $\overline{\bot} = \top$. It is lifted componentwise to matrices. The $^\top$ operation is the usual transpose of matrices. The * operation describes reachability in a graph and is defined recursively using Conway's construction [9]:

$$\begin{pmatrix} a & b \\ c & d \end{pmatrix}^* = \begin{pmatrix} e^* & a^*bf^* \\ d^*ce^* & f^* \end{pmatrix} \quad \text{where} \quad \begin{pmatrix} e \\ f \end{pmatrix} = \begin{pmatrix} a \sqcup bd^*c \\ d \sqcup ca^*b \end{pmatrix}$$

The s-operation computes the sum of all weights in a matrix and is given by applying $+$ to all entries and storing the result in a fixed position of the returned matrix. The remaining entries in the resulting matrix are set to $\bot$. In this model, $s(\bot) = \bot$ and $\bot$ is the unit of $+$, however, there are models where neither holds [14]. The m-operation may be used for selecting an arc with minimum weight. When the input matrix contains at least one non-$\bot$ entry, the m-operation returns a matrix with $\top$ stored in the position corresponding to that of a minimum-weight arc and $\bot$ everywhere else. The result of $m(\bot)$ is $\bot$. This model is an m-Kleene algebra [13–15].

3 Component Selection

In graph theory there are notions of strongly-connected or weakly-connected components in a directed graph. We axiomatise an operation to select an arbitrary connected component. A component of a graph may be represented by a set of vertices as a vector.

Definition 6. *Let S be a Stone relation algebra and let $x, v \in S$. Then v represents* vector-classes *of x if x and v are regular, x is an equivalence, v is a vector, $xv \leq v$, and $v \neq \bot$. If v represents vector-classes of x and additionally, $vv^\top \leq x$ then v represents a* unique-vector-class *of x.*

A vector-class corresponds to one or more equivalence classes of x. The condition $xv \leq v$ ensures that v contains either all elements or no elements of each class. A unique-vector-class corresponds to one equivalence class. This can be used to represent the set of vertices of a particular component of a graph whose components are specified by an equivalence. The equivalence yielding the weakly-connected components of a graph, g, is obtained by taking the symmetric reflexive transitive closure, $x = (g \sqcup g^\top)^*$. For another example, the strongly-connected components of g are given by the equivalence $g^* \sqcap g^{\top *}$.

Definition 7. *A* k-Stone relation algebra, *$(S, \sqcup, \sqcap, \cdot, ^-, ^\top, k, \bot, \top, 1)$, is a Stone relation algebra, $(S, \sqcup, \sqcap, \cdot, ^-, ^\top, \bot, \top, 1)$, with a binary operation k, where for all $x, v \in S$ the element $k(x, v)$ is a regular vector and*

$$k(x,v) \leq \overline{\overline{v}} \tag{1}$$

$$k(x,v) \cdot k(x,v)^\top \leq x \tag{2}$$

$$x \cdot k(x,v) \leq k(x,v) \tag{3}$$

and, if v represents vector-classes of x then

$$k(x,v) \neq \bot \tag{4}$$

If v represents vector-classes of x the element $k(x, v)$ is a vector representing an arbitrary component that is connected according to x and contained in v. Axiom (1) expresses that the result of k is contained in the set of vertices we are selecting from, ignoring the weights. Axiom (2) makes any two vertices from the result of k connected in x. Axiom (3) expresses that the result of k is closed under being connected in x. This means that either all vertices of a component of x are included in the output of k, or none are. Axiom (4) requires that k returns a non-empty component if v represents vector-classes of x. If this is the case, the output of k represents a unique-vector-class.

Theorem 1. Let S be an m-Kleene algebra with a function k defined as

$$k(x, v) = \begin{cases} x \cdot m(v)\top & \text{if } v \text{ represents vector-classes of } x, \\ \bot & \text{otherwise} \end{cases}$$

Then S is a k-Stone relation algebra.

This particular implementation of k does not select an arbitrary component but rather a component containing a minimum-weight arc in v.

Definition 8. *An* m-k-Stone-Kleene relation algebra, $(S, \sqcup, \sqcap, \cdot, +, ^-, ^\top, ^*, s, m, k, \bot, \top, 1)$, *is an m-Kleene algebra,* $(S, \sqcup, \sqcap, \cdot, +, ^-, ^\top, ^*, s, m, \bot, \top, 1)$, *with a component selection operation, k, such that the reduct,* $(S, \sqcup, \sqcap, \cdot, ^-, ^\top, k, \bot, \top, 1)$, *is a k-Stone relation algebra.*

Previous work shows that matrices over $\mathbb{R}'$ form a model of m-Kleene algebras [13–15]. Every m-Kleene algebra is an m-k-Stone-Kleene relation algebra by Theorem 1. The correctness of Borůvka's MST algorithm will be proved in m-k-Stone-Kleene relation algebras, hence it holds in the weighted-graph model described in Sect. 2 and in many other models [14].

4 Forests Modulo an Equivalence

We generalise forests by giving axioms to treat a graph, d, as a forest modulo an equivalence, x. The arcs in d form a forest-like structure where each equivalence class of x forms a strongly-connected component. The intent of this abstraction is to provide an algebraic basis for reasoning about connectivity while forgetting about some of the structure of a graph.

4.1 Axioms and Properties

First, we define forests modulo an equivalence and give a number of properties for such structures.

Definition 9. *Let S be a Stone-Kleene relation algebra and let $x, d \in S$. Then, d is a* forest modulo x, *if x is an equivalence, xd is univalent, and*

$$x \sqcap dd^\top \leq 1 \qquad\qquad x \sqcap (xd)^+ \leq \bot$$

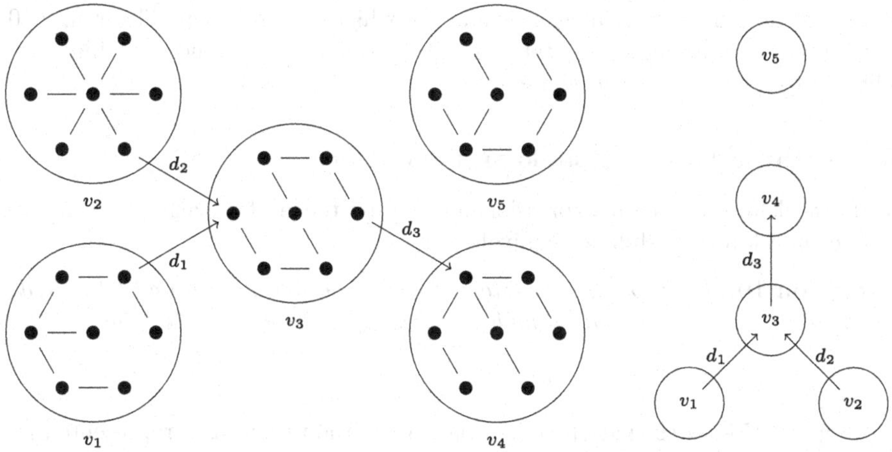

Fig. 1. An example of a forest modulo an equivalence (left) and a view of the quotient set of components (right). The equivalence classes are enclosed in circles and labelled v_1 to v_5. The arcs in d, labelled d_1 to d_3, form a forest modulo that equivalence.

These axioms describe a forest-like structure where the arcs in d are directed towards their respective root component(s). For example, the forest modulo an equivalence in Fig. 1 has two root components: v_4 and v_5. The special case of a forest modulo 1 is the transpose of a forest, that is, univalent and acyclic.

Theorem 2. Let S be a Stone-Kleene relation algebra and let $d, x \in S$ and let d be a forest modulo x. Then d is acyclic and univalent and

2.1 $d \sqcup d^\top \leq \overline{x}$
2.2 $dx \sqcup xd \leq \overline{x}$
2.3 $x(d \sqcup d^\top)x \leq \overline{x}$
2.4 $(dx)^+ \leq \overline{x}$
2.5 $(xd)^+ \leq \overline{x}$
2.6 $(xd)^* \sqcap (xd)^\top = \bot$
2.7 $d^\top xd \leq 1$
2.8 $(xd^\top)^+ xdxd^\top \leq (xd^\top)^+$
2.9 $(d^\top x)^*(xd)^* = (d^\top x)^* \sqcup (xd)^*$

Furthermore, let $a \in S$ be an arc and let $a \leq d$, then

2.10 $(d \sqcap \overline{a})^\top (xa\top) \leq \bot$
2.11 $(x(d \sqcap \overline{a}))^* xa\top = \big(x((d \sqcap \overline{a}) \sqcup (d \sqcap \overline{a})^\top)\big)^* xa\top$

Theorems 2.1 to 2.5 describe how d separates equivalence classes of x. For example, Theorem 2.4 states that taking steps $(dx)^+$ that involve one or more d-arcs leads to a different component. Theorem 2.6 follows from the acyclic-like structure of the forest modulo x. Theorem 2.7 is derivable from xd being univalent. Theorem 2.9 follows from x being an equivalence and xd being univalent. This theorem states that taking any number of steps backwards in the forest modulo x (away from the roots) followed by any number of steps forwards in the forest modulo x (towards the roots) is the same as going either forwards or backwards. Consider the situation if we take a step backwards between components of a forest modulo x without using an arc a, which is in d. Then, Theorem 2.10 states there is no sequence of steps we can take in the component we find ourselves in to next take a step along arc a.

4.2 Paths in Forests Modulo an Equivalence

Next, we define a general expression for the existence of a path between two vertices in a forest modulo an equivalence.

Definition 10. *Let S be an m-k-Stone-Kleene relation algebra and let $a, b, d, x \in S$. Then, $a \leadsto_x^d b$ holds if a and b are arcs, x is an equivalence, and*

$$a^\top \top \leq (xd)^* x b \top \tag{5}$$

Property (5) states that there is a path from the target of a, represented by the point $a^\top \top$, to the source of b, represented by the point $b\top$, in the forest d modulo x.

The following result states that for an arc e there is a path in $d \sqcup e$ between a and b if and only if there is either a path in d from a to b, or a path in d from a to e and one from e to b.

Theorem 3. Let S be an m-k-Stone-Kleene relation algebra where $a, b, e, d, x \in S$ are regular and let e be an arc and x be an equivalence. Then

$$a \leadsto_x^{d \sqcup e} b \Leftrightarrow a \leadsto_x^d b \vee \left(a \leadsto_x^d e \wedge e \leadsto_x^d b\right)$$

Theorem 3 allows us to split a path. An example use of this is given in Sect. 6.1.

5 Relational Formalisation of Borůvka's MST Algorithm

In this section, we formalise Borůvka's MST algorithm as a while-program, shown in Fig. 2. The variables of the program are elements of an m-k-Stone-Kleene relation algebra.

The input to the program is an undirected graph, g, modelled by a symmetric matrix. With the exception of g, all variables are regular elements. Graph g does not need to be connected. The algorithm constructs a directed minimum spanning forest f (of g) that is initialised as empty (line 2).

The outer while-loop executes until there are no arcs in g that could connect components of f (line 3). On lines 4 to 6, variables used by the inner while-loop are initialised. The forest h maintains a stable representation of f in each iteration of the inner while-loop. The vector j tracks the components still to be considered by the inner while-loop. The variable d tracks the arcs that have been added to f in each iteration of the inner while-loop. This variable is not required by the algorithm but is used in the correctness proof.

The inner while-loop exits when all components have been processed (line 7). As discussed in Sect. 7.1 of [15], for a directed graph, x, the weakly-connected

1 **input** g
2 $f \leftarrow \bot$
3 **while** $\overline{f^{\top *} f^{*}} \sqcap g \neq \bot$ **do**
4 $\quad d \leftarrow \bot$
5 $\quad h \leftarrow f$
6 $\quad j \leftarrow \top$
7 $\quad$ **while** $j \neq \bot$ **do**
8 $\quad\quad v \leftarrow k(c(h), j)$
9 $\quad\quad e \leftarrow m(v\overline{v}^{\top} \sqcap g)$
10 $\quad\quad$ **if** $e \leq \overline{f^{\top *} f^{*}}$ **then**
11 $\quad\quad\quad f \leftarrow f \sqcap \overline{e}^{\top}$
12 $\quad\quad\quad f \leftarrow (f \sqcap \overline{\top e f^{\top *}}) \sqcup (f \sqcap \top e f^{\top *})^{\top} \sqcup e$
13 $\quad\quad\quad d \leftarrow d \sqcup e$
14 $\quad\quad$ **end**
15 $\quad\quad j \leftarrow j \sqcap \overline{v}$
16 $\quad$ **end**
17 **end**
18 **output** f

Fig. 2. A relational formalisation of Borůvka's MST algorithm.

components are given by $(x \sqcup x^{\top})^{*}$. Furthermore, if x is a forest, this can be simplified to $c(x) = x^{\top *} x^{*}$. This is because, ignoring arc direction, two vertices are connected in a forest if there is a path from one vertex backwards in x, towards a root, and then forwards in x to the other vertex. An arbitrary component, $v = k(c(h), j)$, is selected from those that have not yet been considered (line 8). The k operation introduced in Definition 7 is used. The vector j represents the components not yet processed by the inner while-loop and the forest, h, represents f as it was when the current iteration of the outer loop started. The equivalence $c(h)$ describes the weakly-connected components of f, as they were at the start of the current iteration of the outer while-loop. The component v is then weakly connected in h and among those still to be processed by the inner while-loop. Since j is reduced by v at the end of each iteration of the inner while-loop and it continues until j is empty, every component of f is processed exactly once in each iteration of the outer while-loop.

A minimum-weighted arc, e, is selected from among the arcs whose source is in v and whose target is outside v (line 9). If e is not contained in a component of f (line 10) then f is updated (lines 11 and 12), otherwise, it is not. Before e is added to f, the algorithm ensures that any transpose of e, which may have been added in a previous iteration of the inner while-loop, is removed from f (line 11). The update on line 12 adds e to f and at the same time reverses certain arcs of f to maintain that f is a forest. These two updates give a new value for f, f', that is:

$$f' = \left(f \sqcap \overline{e}^{\top} \sqcap \overline{\top e (f \sqcap \overline{e}^{\top})^{\top *}}\right) \sqcup \left(f \sqcap \overline{e}^{\top} \sqcap \top e (f \sqcap \overline{e}^{\top})^{\top *}\right)^{\top} \sqcup e \qquad (6)$$

The variable d is updated to track the arcs that have been added to f in this iteration of the inner while-loop (line 13). The processed component is removed from j so that it is not considered in subsequent iterations of the inner while-loop (line 15). When the outer while-loop exits the algorithm terminates returning f (line 18) which contains the structure of a minimum spanning forest of g without weight information. The weighted version of the output may be obtained by taking the meet with g. The undirected version of the output can be obtained by taking the symmetric closure.

Borůvka's MST algorithm requires the input graph's arc weights to be distinct. Because our formalisation does not require this, we have added a check in the inner loop to ensure that no cycle is created (line 10). This check is also performed in the relational version of Kruskal's algorithm before adding an arc with minimal weight to the forest variable [15]. The relational version of Kruskal's algorithm iterates over the arcs of the graph while the inner while-loop of the algorithm we present iterates over component trees. This means that here we are often working with the properties of vectors. Both approaches keep track of the desired output by growing a minimum spanning forest, represented as a directed graph whose components are rooted directed trees. This structure is useful for maintaining that the output is injective and acyclic, properties used to conclude that the result is a forest. We select a minimum-weight arc whose source is in component v and whose target is outside v with $m(v\overline{v}^\top \sqcap g)$. This expression was used in [14] in a relational version of Prim's MST algorithm to select an arc with minimum weight that leaves a set of visited vertices.

The complexity of Eq. (6) results from representing f as a directed forest, which is also the approach taken in [15]. The advantage of this approach is that it is more simple to give a specification for being acyclic than if f was undirected.

6 Correctness Proof

In this section, we discuss the partial-correctness proof of the formalisation presented in Sect. 5. We work in, and our proof holds for any instance of, m-k-Stone-Kleene relation algebras. In particular, it holds for weighted matrices, $S = \mathbb{R}'^{A \times A}$.

We reuse the specification from [15] that was used to verify Kruskal's MST algorithm.

Definition 11. *Let S be an m-Kleene algebra where $f, g \in S$. Then, f is a* spanning forest *of g if f is a regular forest and*

$$f \leq \overline{\overline{g}}$$
$$\overline{\overline{g}}^* \leq c(f)$$

The spanning forest, f, is a minimum spanning forest *of g if for all $u \in S$ where u is a spanning forest of g, the following holds:*

$$s(f \sqcap g) \leq s(u \sqcap g)$$

Intuitively, a spanning forest is a maximal acyclic subset of arcs of a graph, that is, composed of spanning trees, one for each component of the graph. A minimum spanning forest is one where the sum of arc weights is minimal among all possible spanning forests of a graph.

Our correctness proof assumes that both while-loops terminate. In future work, we may eliminate this assumption by taking a similar approach as in [15]. Presently, under the assumption that both while-loops terminate, the following theorem gives the preconditions and invariants we use to establish the postcondition of the outer while-loop that f is a minimum spanning forest of g.

Theorem 4. Let S be an m-k-Stone-Kleene relation algebra and let $g \in S$ be symmetric. Then, the following invariant holds throughout Borůvka's MST algorithm:

4.1 g is symmetric;
4.2 f is a regular forest;
4.3 $f \leq \overline{\overline{g}}$, meaning that f is contained in g, ignoring arc weights;
4.4 there is a minimum spanning forest, w, of g, such that $f \leq w \sqcup w^\top$.

Establishing Theorems 4.1 to 4.3 at the start of the algorithm is easy. The variable g is symmetric as a result of the precondition; f is a regular forest since $\bot$ is regular, injective and acyclic; and $f \leq \overline{\overline{g}}$ since $\bot$ is the least element. We reuse the proof from [15] to establish Theorem 4.4.

In contrast to Prim's and Kruskal's algorithms, Borůvka's MST algorithm has a second while-loop that we must establish and maintain an invariant for.

Theorem 5. Let S be an m-k-Stone-Kleene relation algebra and let $j \in S$. Then, the following invariant holds throughout the inner while-loop of Borůvka's MST algorithm:

5.1 $g \neq \bot$, meaning that the graph has at least one arc;
5.2 d is regular;
5.3 j is a regular vector;
5.4 h is a regular forest;
5.5 $c(h)j = j$, meaning that j contains each component of h entirely or not at all;
5.6 d is a forest modulo $c(h)$;
5.7 $d\top \leq \overline{j}$, meaning that the sources of the arcs in d are not in the set of vertices still to be processed;
5.8 $f \sqcup f^\top = h \sqcup h^\top \sqcup d \sqcup d^\top$, meaning that, ignoring direction, f can be obtained by taking the join of h and d;
5.9 $\forall a, b : a \leadsto^{d}_{c(h)} b \wedge a \leq \overline{c(h)} \sqcap \overline{\overline{g}} \wedge b \leq d \Rightarrow s(b \sqcap g) \leq s(a \sqcap g)$, meaning that, for any arcs a and b, if there is a path from a to b in forest d modulo $c(h)$ and a is in the graph (ignoring weight) and is not contained in the components of h and b is contained in d then the weight of b is less than or equal to the weight of a.

The requirements of the invariant for the inner while-loop are also easy to establish owing to the values that the variables are initialised to.

Most of the work to maintain the invariants is to maintain the inner while-loop invariant since, aside from variable initialisation, the logic of the inner while-loop makes up the entirety of the outer while-loop.

The following result states the correctness of the algorithm.

Theorem 6. Let S be an m-k-Stone-Kleene relation algebra and let $g \in S$ be symmetric. Then, the following postcondition is established for Borůvka's MST algorithm: f is a minimum spanning forest of g.

In the remainder of this section we give two examples of how the invariant is maintained.

6.1 Maintaining Arc Weight Comparison in a Forest Modulo $c(h)$

To show that invariant 5.9 of Theorem 5 is maintained we must show that

$$a \leadsto^{d \sqcup e}_{c(h)} b \wedge a \leq \overline{c(h)} \sqcap \overline{\overline{g}} \wedge b \leq d \sqcup e \Rightarrow s(b \sqcap g) \leq s(a \sqcap g) \tag{7}$$

for any arcs a, b. To this end, we assume that invariant 5.9 holds (for the previous iteration of the inner while-loop) and that the antecedent of (7) is true.

Our proof is by case distinctions but within each case reasoning is algebraic. There are six cases to consider and we discuss one of these in more detail. We first make a case distinction where $b \neq e$ and $e \not\leq d$. Next, we use Theorem 3 to further split into two cases. The first is $a \leadsto^{d}_{c(h)} b$, that is, there is a path from a to b in d modulo $c(h)$, in which case we can conclude $s(b \sqcap g) \leq s(a \sqcap g)$ immediately from invariant 5.9 of Theorem 5. The second is $a \leadsto^{d}_{c(h)} e$ and $e \leadsto^{d}_{c(h)} b$, that is, there is a path in d modulo $c(h)$ from a to e and one from e to b.

We first treat the path from e to b. We have $e \leq \overline{c(h)} \sqcap \overline{\overline{g}}$, since e is contained in the graph and not contained in the components of h. Additionally, $b \leq d$, since e and b are arcs and $b \neq e$ and $e \not\leq d$ and $b \leq d \sqcup e$. Together with $e \leadsto^{d}_{c(h)} b$, it follows from invariant 5.9 of Theorem 5 that $s(b \sqcap g) \leq s(e \sqcap g)$.

Next, we treat the path from a to e. Either, the target of a is in the same component as the source of e or not. If it is then we have $a^\top \top \leq c(h)e\top$ and apply the following theorem.

Theorem 7. Let S be an m-k-Stone-Kleene relation algebra and let $a, e, g, v, x \in S$ where g is symmetric, v represents a unique-vector-class of x, $e = m(v\overline{v}^\top \sqcap g)$, a is an arc, $a \leq \overline{x} \sqcap \overline{\overline{g}}$, and $a^\top \top \leq xe\top$. Then, $s(e \sqcap g) \leq s(a \sqcap g)$.

This result allows us to show that the selected arc, e, that is outgoing from a component, v, must have a weight less than or equal to any other arc incoming to that component in d with respect to the forest modulo x. This is a consequence of m selecting a minimum-weighted arc.

To apply Theorem 7, we set $x = c(h)$ and $e = m(v\overline{v}^\top \sqcap g)$ where $v = k(c(h), j)$. Then, we have that g is symmetric and a is an arc from the invariant, and $c(h)$

is an equivalence. From the axioms of the k operation and since $j \neq \bot$, we know that v represents a unique-vector-class of $c(h)$. From the antecedent we have $a \leq \overline{c(h) \sqcap \overline{g}}$. Hence, in this case, $s(e \sqcap g) \leq s(a \sqcap g)$.

If the target of a is not in the same component of $c(h)$ as the source of e then we describe an arc, y, that is on the path from a to e in d modulo $c(h)$ and whose target is in the same component of $c(h)$ as the source of e. The arc y is defined as

$$y = d \sqcap \top e^\top c(h) \sqcap \left(c(h)d^\top\right)^* c(h)a^\top \top$$

The meet with d ensures that y is an arc between components of $c(h)$. The second part of this expression, $\top e^\top c(h)$, ensures that the target of y is in the same component of $c(h)$ as the source of e. The last part of this expression, $\left(c(h)d^\top\right)^* c(h)a^\top \top$, ensures that the source of y is reachable from the target of a by taking any number of steps in the forest d modulo $c(h)$. We can show that $s(y \sqcap g) \leq s(a \sqcap g)$ using invariant 5.9 of Theorem 5 in a similar manner as described above. Then, since we can apply Theorem 7 to show $s(e \sqcap g) \leq s(y \sqcap g)$, it follows that $s(e \sqcap g) \leq s(a \sqcap g)$.

Finally, the result for the path from a to e and the result for the path from e to b are combined to conclude that $s(b \sqcap g) \leq s(e \sqcap g) \leq s(a \sqcap g)$.

6.2 Extending f to a Minimum Spanning Forest

The key property of the invariant of the outer loop that must be maintained is that the forest, f, can be extended to a minimum spanning forest of the graph, g, ignoring arc direction, that is, there exists a minimum spanning forest, w, such that $f \leq w \sqcup w^\top$. We were able to reuse work from [15] in the maintenance of this invariant except the arc selected for replacement is changed.

If the arc added to f is not also in w then the added arc must replace an arc in w to ensure that w remains acyclic. In [15] the arc selected for replacement was the arc whose source was in the same component of f as the target of e. This arc does not suit our purposes because we do not have a convenient way to compare its weight with the weight of e. However, there is an easy comparison to be made between e and the arc, i, shown in Fig. 3, whose target is in the same component of f as the source of e. Namely, the weight of e is at least as small as the weight of i, since i is among those arcs that the algorithm chose e from with the minimum selection $m(v\overline{v}^\top \sqcap g)$.

Let $q = w \sqcap \top e w^{\top *}$, that is, the path from the root of w to the target of e and let $r = (w \sqcap \overline{q}) \sqcup q^\top$, that is, w with the path q reversed. Then, the desired forest, w', that extends f' is r, with i removed and e added. That is, $w' = (r \sqcap \overline{i}) \sqcup e$. The arc i is defined as

$$i = r \sqcap \overline{c(f)e\top} \sqcap \top e^\top c(f)$$

The meet with r limits i to only those arcs in the rooted directed forest w with the path q reversed. The second part of this expression, $\overline{c(f)e\top}$, specifies that the source of i cannot be in the same component of f as the source of e. Finally,

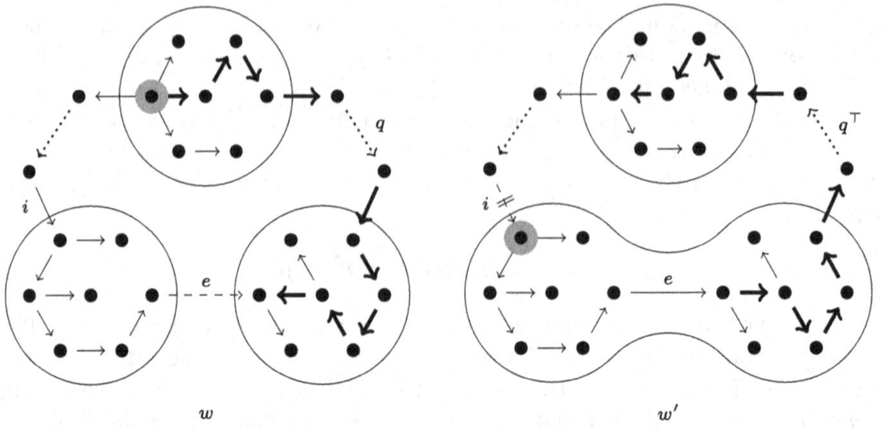

Fig. 3. Maintaining the invariant that f can be extended to a minimum spanning forest, w, before (left) and after (right) adding arc, e. The path, q, to the root of the rooted directed forest is reversed to maintain injectivity. The arc, i, whose target is in the same component of f as the source of e, is removed to maintain that w' is acyclic. The vertices enclosed in a circle denote a component, in f. The root of the rooted directed forest is highlighted grey.

the last part of the expression, $\top e^\top c(f)$, requires that the target of i is in the same component of f as the source of e.

We prove that these requirements uniquely identify an arc, i. After the update, the target of i becomes the root of w' in the component that e is in. Furthermore, we show that $s(e \sqcap g) \leq s(i \sqcap g)$ using Theorem 7.

7 Conclusion

We have formalised Borůvka's MST algorithm and proved its correctness. While we have benefited from the relation-algebraic framework introduced in [13] and the theorems subsequently developed in that framework, substantial additional proof work was required to complete our verification. To better structure the reasoning, we have axiomatised an operation, k, to select an arbitrary component. This axiomatisation uses a new definition for a component of a graph in terms of an equivalence, describing connectivity, and a vector, describing the subset of vertices we are selecting from. An implementation of k is given in m-Kleene algebras.

We have introduced a new abstraction, forests modulo an equivalence, that helps us to reason about forest-like structures of graphs by ignoring some arcs in a graph and focusing on others. The proof of our formalisation of Borůvka's MST algorithm applies this abstraction by considering the arcs connecting the components that are constructed by the inner while-loop of the algorithm as a forest modulo the components.

Much of our proof of Borůvka's MST algorithm used only the axioms of Stone-Kleene relation algebras and our proof holds for instances other than the weighted-graph model described in this paper. Different instances of m-Kleene algebras give rise to formalisations of various other algorithms, for example, the minimum bottleneck spanning tree problem [14]. Our proof holds for any instance that satisfies the axioms the proof is conducted in. This means Borůvka's MST algorithm is correct for various related MST problems.

Acknowledgement. We thank the anonymous referees for helpful comments.

References

1. Balbes, R., Dwinger, P.: Distributive Lattices. University of Missouri Press (1974)
2. Berge, C., Ghouila-Houri, A.: Programming, Games and Transportation Networks. Wiley, Methuen (1965)
3. Berghammer, R., von Karger, B., Wolf, A.: Relation-algebraic derivation of spanning tree algorithms. In: Jeuring, J. (ed.) MPC 1998. LNCS, vol. 1422, pp. 23–43. Springer, Heidelberg (1998). https://doi.org/10.1007/BFb0054283
4. Berghammer, R., Rusinowska, A., de Swart, H.: Computing tournament solutions using relation algebra and RelView. Eur. J. Oper. Res. **226**(3), 636–645 (2013)
5. Birkhoff, G.: Lattice Theory, Colloquium Publications, vol. XXV, 3rd edn., American Mathematical Society (1967)
6. Borůvka, O.: O jistém problému minimálním. Práce moravské přírodovědecké společnosti **3**(3), 37–58 (1926)
7. Chazelle, B.: A minimum spanning tree algorithm with inverse-Ackermann type complexity. J. ACM **47**(6), 1028–1047 (2000)
8. Choquet, G.: Étude de certains réseaux de routes. C. R. Hebd. Seances Acad. Sci. **206**, 310–313 (1938)
9. Conway, J.H.: Regular Algebra and Finite Machines. Chapman and Hall (1971)
10. Florek, K., Łukaszewicz, J., Perkal, J., Steinhaus, H., Zubrzycki, S.: Sur la liaison et la division des points d'un ensemble fini. Colloq. Math. **2**(3–4), 282–285 (1951)
11. Frias, M.F., Aguayo, N., Novak, B.: Development of graph algorithms with fork algebras. In: XIX Conferencia Latinoamericana de Informática, pp. 529–554 (1993)
12. Graham, R.L., Hell, P.: On the history of the minimum spanning tree problem. Ann. Hist. Comput. **7**(1), 43–57 (1985)
13. Guttmann, W.: Relation-algebraic verification of Prim's minimum spanning tree algorithm. In: Sampaio, A., Wang, F. (eds.) ICTAC 2016. LNCS, vol. 9965, pp. 51–68. Springer, Cham (2016). https://doi.org/10.1007/978-3-319-46750-4_4
14. Guttmann, W.: An algebraic framework for minimum spanning tree problems. Theoret. Comput. Sci. **744**, 37–55 (2018)
15. Guttmann, W.: Verifying minimum spanning tree algorithms with Stone relation algebras. J. Log. Algebraic Methods Program. **101**, 132–150 (2018)
16. Guttmann, W., Robinson-O'Brien, N.: Relational minimum spanning tree algorithms. Archive of Formal Proofs (2020). Formal proof development. https://isa-afp.org/entries/Relational_Minimum_Spanning_Trees.html
17. Haslbeck, M.P.L., Lammich, P., Biendarra, J.: Kruskal's algorithm for minimum spanning forest. Archive of Formal Proofs (2019). Formal proof development. https://isa-afp.org/entries/Kruskal.html

18. Karger, D.R., Klein, P.N., Tarjan, R.E.: A randomized linear-time algorithm to find minimum spanning trees. J. ACM **42**(2), 321–328 (1995)
19. Kehden, B., Neumann, F.: A relation-algebraic view on evolutionary algorithms for some graph problems. In: Gottlieb, J., Raidl, G.R. (eds.) EvoCOP 2006. LNCS, vol. 3906, pp. 147–158. Springer, Heidelberg (2006). https://doi.org/10.1007/11730095_13
20. Kozen, D.: A completeness theorem for Kleene algebras and the algebra of regular events. Inf. Comput. **110**(2), 366–390 (1994)
21. Kruskal, J.B., Jr.: On the shortest spanning subtree of a graph and the traveling salesman problem. Proc. Am. Math. Soc. **7**(1), 48–50 (1956)
22. Lammich, P., Nipkow, T.: Proof pearl: purely functional, simple and efficient priority search trees and applications to Prim and Dijkstra. In: Harrison, J., O'Leary, J., Tolmach, A. (eds.) 10th International Conference on Interactive Theorem Proving (ITP 2019). LIPIcs: Leibniz International Proceedings in Informatics, vol. 141, pp. 23:1–23:18. Schloss Dagstuhl – Leibniz-Zentrum für Informatik (2019)
23. Maddux, R.D.: Relation Algebras. Elsevier B.V. (2006)
24. Nešetřil, J., Milková, E., Nešetřilová, H.: Otakar Borůvka on minimum spanning tree problem – translation of both the 1926 papers, comments, history. Discret. Math. **233**(1–3), 3–36 (2001)
25. Nipkow, T.: Winskel is (almost) right: towards a mechanized semantics textbook. Formal Aspects Comput. **10**(2), 171–186 (1998)
26. Nipkow, T.: Hoare logics in Isabelle/HOL. In: Schwichtenberg, H., Steinbrüggen, R. (eds.) Proof and System-Reliability, pp. 341–367. Kluwer Academic Publishers (2002)
27. Nipkow, T., Paulson, L.C., Wenzel, M.: Isabelle/HOL: A Proof Assistant for Higher-Order Logic. LNCS, vol. 2283. Springer, Heidelberg (2002). https://doi.org/10.1007/3-540-45949-9
28. Nipkow, T., Eberl, M., Haslbeck, M.P.L.: Verified textbook algorithms. A biased survey. In: Hung, D.V., Sokolsky, O. (eds.) Automated Technology for Verification and Analysis, ATVA 2020. LNCS, vol. 12302, pp. 25–53. Springer, Heidelberg (2020). https://doi.org/10.1007/978-3-030-59152-6_2
29. Prim, R.C.: Shortest connection networks and some generalizations. Bell Syst. Tech. J. **36**(6), 1389–1401 (1957)
30. Robinson-O'Brien, N.: A formal correctness proof of Borůvka's minimum spanning tree algorithm. Master's thesis, University of Canterbury (2020). https://doi.org/10.26021/10196
31. Schmidt, G., Ströhlein, T.: Relations and Graphs. Springer, Heidelberg (1993). https://doi.org/10.1007/978-3-642-77968-8
32. Tarjan, R.E.: Data Structures and Network Algorithms, CBMS-NSF Regional Conference Series in Applied Mathematics, vol. 44. SIAM (1983)
33. Tarski, A.: On the calculus of relations. J. Symbolic Logic **6**(3), 73–89 (1941)
34. Yao, A.C.C.: An $O(|E| \log\log |V|)$ algorithm for finding minimum spanning trees. Inf. Process. Lett. **4**(1), 21–23 (1975)

Deciding FO-definability of Regular Languages

Agi Kurucz[1], Vladislav Ryzhikov[2(✉)], Yury Savateev[2,3], and Michael Zakharyaschev[2,3]

[1] King's College London, London, UK
agi.kurucz@kcl.ac.uk
[2] Birkbeck, University of London, London, UK
{vlad,yury,michael}@dcs.bbk.ac.uk
[3] HSE University, Moscow, Russia

Abstract. We prove that, similarly to known PSPACE-completeness of recognising $\mathsf{FO}(<)$-definability of the language $\boldsymbol{L}(\mathfrak{A})$ of a DFA $\mathfrak{A}$, deciding both $\mathsf{FO}(<,\equiv)$- and $\mathsf{FO}(<,\mathsf{MOD})$-definability (corresponding to circuit complexity in AC^0 and ACC^0) are PSPACE-complete. We obtain these results by first showing that known algebraic characterisations of FO-definability of $\boldsymbol{L}(\mathfrak{A})$ can be captured by 'localisable' properties of the transition monoid of $\mathfrak{A}$. Using our criterion, we then generalise the known proof of PSPACE-hardness of $\mathsf{FO}(<)$-definability, and establish the upper bounds not only for arbitrary DFAs but also for 2NFAs.

1 Introduction

This paper gives answers to some open questions related to finite automata, logic and circuit complexity. Research in this area goes back (at least) to the early1960 s when Büchi [8], Elgot [12] and Trakhtenbrot [28] showed that MSO(<) (monadic second-order) sentences over finite strict linear orders define exactly the class of regular languages.

$\mathsf{FO}(<)$-definable regular languages were proven to be the same as star-free languages [19], and their algebraic characterisation as languages with aperiodic syntactic monoids was obtained in [23]. Algebraic characterisations of FO-definability in other signatures, and circuit and descriptive complexity of regular languages were investigated in [3,4,26], which established an $\mathrm{AC}^0/\mathrm{ACC}^0/\mathrm{NC}^1$ trichotomy. In particular, the regular languages decidable in AC^0 are definable by $\mathsf{FO}(<,\equiv)$-sentences with unary predicates $x \equiv 0 \pmod n$; those in ACC^0 are definable by $\mathsf{FO}(<,\mathsf{MOD})$-sentences with quantifiers $\exists^n x\, \psi(x)$ checking whether the number of positions satisfying ψ is divisible by n; and all regular languages are definable in $\mathsf{FO}(\mathsf{RPR})$ with relational primitive recursion [11]; see Table 1.

The problem of deciding whether the language of a given DFA $\mathfrak{A}$ is $\mathsf{FO}(<)$-definable is known to be PSPACE-complete [7,10,25] (which is also a special case of general results on finite monoids [5,13]). As shown in [4], the algebraic criteria of Table 1 yield algorithms deciding whether a given regular language is in

U. Fahrenberg et al. (Eds.): RAMiCS 2021, LNCS 13027, pp. 241–257, 2021.
https://doi.org/10.1007/978-3-030-88701-8_15

Table 1. Definability, algebraic characterisations and circuit complexity of a regular language $\boldsymbol{L}$, where $M(\boldsymbol{L})$ is the syntactic monoid and $\eta_{\boldsymbol{L}}$ the syntactic morphism of $\boldsymbol{L}$.

Definability of $\boldsymbol{L}$	Algebraic characterisation of $\boldsymbol{L}$	Circuit complexity
$\mathsf{FO}(<)$	$M(\boldsymbol{L})$ is aperiodic	in AC^0
$\mathsf{FO}(<,\equiv)$	$\eta_{\boldsymbol{L}}$ is quasi-aperiodic	
$\mathsf{FO}(<,\mathsf{MOD})$	All groups in $M(\boldsymbol{L})$ are solvable	in ACC^0
$\mathsf{FO}(\mathsf{RPR})$	Arbitrary $M(\boldsymbol{L})$	in NC^1
Not in $\mathsf{FO}(<,\mathsf{MOD})$	$M(\boldsymbol{L})$ contains an unsolvable group	NC^1-hard

AC^0 and $\mathsf{FO}(<,\equiv)$-definable, or in ACC^0 and $\mathsf{FO}(<,\mathsf{MOD})$-definable, or NC^1-complete and is not $\mathsf{FO}(<,\mathsf{MOD})$-definable (unless $\mathrm{ACC}^0 = \mathrm{NC}^1$). However, these 'brute force' algorithms are not optimal, requiring the generation of the whole transition monoid of $\mathfrak{A}$, which can be of exponential size [14]. As far as we know, the precise complexity of these decision problems has remained open.

Our interest in the exact complexity of these problems is motivated by recent advances in ontology-based data access (OBDA) with linear time temporal logic *LTL* [1,2]. The classical (atemporal) OBDA paradigm [20,30] relies on a reduction of answering a query mediated by an ontology under the open-world semantics to evaluating a database query in a standard language such as SQL or its extension—that is, essentially, an extension of first-order logic—under the closed-world semantic. In the context of temporal OBDA, answering *LTL* ontology-mediated queries is equivalent to deciding certain regular languages given by an NFA or 2NFA of (possibly) exponential size, which gives rise to the circuit complexity and FO-definability problems for those languages. For further details the reader is referred to [22], which relies on the results we obtain below.

Our contribution in this paper is as follows. Let $\mathcal{L}$ be one of the languages $\mathsf{FO}(<,\equiv)$ or $\mathsf{FO}(<,\mathsf{MOD})$. First, using the algebraic characterisation results of [3,4,26], we obtain criteria for the $\mathcal{L}$-definability of the language $\boldsymbol{L}(\mathfrak{A})$ of any given DFA $\mathfrak{A}$ in terms of a limited part of the transition monoid of $\mathfrak{A}$ (Theorem 1). Then, by using our criteria and generalising the construction of [10], we show that deciding $\mathcal{L}$-definability of $\boldsymbol{L}(\mathfrak{A})$ for any minimal DFA $\mathfrak{A}$ is PSPACE-hard (Theorem 2). Finally, we apply our criteria to give a PSPACE-algorithm deciding $\mathcal{L}$-definability of $\boldsymbol{L}(\mathfrak{A})$ for not only any DFA but any 2NFA $\mathfrak{A}$ (Theorem 3).

2 Preliminaries

We begin by briefly reminding the reader of the basic algebraic and automata-theoretic notions required in the remainder of the paper.

2.1 Monoids and Groups

A *semigroup* is a structure $\mathfrak{S} = (S, \cdot)$ where $\cdot$ is an associative binary operation. Given $s, s' \in S$ and $n > 0$, we write s^n for $s \cdot \ldots \cdot s$ n-times, and often write ss'

for $s \cdot s'$. An element s in a semigroup $\mathfrak{S}$ is *idempotent* if $s^2 = s$. An element e in $\mathfrak{S}$ is an *identity* if $e \cdot x = x \cdot e = x$ for all $x \in S$. (It is easy to see that such an e is unique, if exists.) The identity element is clearly idempotent. A *monoid* is a semigroup with an identity element. For any element s in a monoid, we set $s^0 = e$. A monoid $\mathfrak{S} = (S, \cdot)$ is a *group* if, for any $x \in S$, there is $x^- \in S$—the *inverse of* x—such that $x \cdot x^- = x^- \cdot x = e$ (every element of a group has a unique inverse). A group is *trivial* if it has one element, and *nontrivial* otherwise.

Given two groups $\mathfrak{G} = (G, \cdot)$ and $\mathfrak{G}' = (G', \cdot')$, a map $h\colon G \to G'$ is a *group homomorphism from* $\mathfrak{G}$ *to* $\mathfrak{G}'$ if $h(g_1 \cdot g_2) = h(g_1) \cdot' h(g_2)$ for all $g_1, g_2 \in G$. (It is easy to see that any group homomorphism maps the identity of $\mathfrak{G}$ to the identity of $\mathfrak{G}'$ and preserves the inverses. The set $\{h(g) \mid g \in G\}$ is closed under $\cdot'$, and so is a group, the *image of* $\mathfrak{G}$ *under* h.) $\mathfrak{G}$ is a *subgroup of* $\mathfrak{G}'$ if $G \subseteq G'$ and the identity map id_G is a group homomorphism. Given $X \subseteq G$, the *subgroup of* $\mathfrak{G}$ *generated by* X is the smallest subgroup of $\mathfrak{G}$ containing X. The *order* $o_{\mathfrak{G}}(g)$ of an element g in $\mathfrak{G}$ is the smallest positive number n with $g^n = e$, which always exists. Clearly, $o_{\mathfrak{G}}(g) = o_{\mathfrak{G}}(g^-)$ and, if $g^k = e$ then $o_{\mathfrak{G}}(g)$ divides k. Also,

$$\text{if } g \text{ is a nonidentity element in a group } \mathfrak{G}, \text{ then } g^k \neq g^{k+1} \text{ for any } k. \tag{1}$$

A semigroup $\mathfrak{S}' = (S', \cdot')$ is a *subsemigroup* of a semigroup $\mathfrak{S} = (S, \cdot)$ if $S' \subseteq S$ and $\cdot'$ is the restriction of $\cdot$ to S'. Given a monoid $\boldsymbol{M} = (M, \cdot)$ and a set $S \subseteq M$, we say that S *contains the group* $\mathfrak{G} = (G, \cdot')$, if $G \subseteq S$ and $\mathfrak{G}$ is a subsemigroup of $\boldsymbol{M}$. Note that we do **not** require the identity of $\boldsymbol{M}$ to be in $\mathfrak{G}$, even if it is in S. If $S = M$, we also say that $\boldsymbol{M}$ *contains the group* $\mathfrak{G}$, or $\mathfrak{G}$ *is in* $\boldsymbol{M}$. We call a monoid $\boldsymbol{M}$ *aperiodic* if it does not contain any nontrivial groups.

Let $\mathfrak{S} = (S, \cdot)$ be a finite semigroup and $s \in S$. By the pigeonhole principle, there exist $i, j \geq 1$ such that $i + j \leq |S| + 1$ and $s^i = s^{i+j}$. Take the minimal such numbers, that is, let $i_s, j_s \geq 1$ be such that $i_s + j_s \leq |S| + 1$ and $s^{i_s} = s^{i_s + j_s}$ but $s^{i_s}, s^{i_s+1}, \dots, s^{i_s+j_s-1}$ are all different. Then clearly $\mathfrak{G}_s = (G_s, \cdot)$, where $G_s = \{s^{i_s}, s^{i_s+1}, \dots, s^{i_s+j_s-1}\}$, is a subsemigroup of $\mathfrak{S}$. It is easy to see that there is $m \geq 1$ with $i_s \leq m \cdot j_s < i_s + j_s \leq |S| + 1$, and so $s^{m \cdot j_s}$ is idempotent. Thus, for every element s in a semigroup $\mathfrak{S}$, we have the following:

$$\text{there is } n \geq 1 \text{ such that } s^n \text{ is idempotent;} \tag{2}$$

$$\mathfrak{G}_s \text{ is a group in } \mathfrak{S} \text{ (isomorphic to the cyclic group } \mathbb{Z}_{j_s}\text{);} \tag{3}$$

$$\mathfrak{G}_s \text{ is nontrivial iff } s^n \neq s^{n+1} \text{ for any } n. \tag{4}$$

Let $\delta\colon Q \to Q$ be a function on a finite set $Q \neq \emptyset$. For any $p \in Q$, the subset $\{\delta^k(p) \mid k < \omega\}$ with the obvious multiplication is a semigroup, and so we have:

$$\text{for every } p \in Q, \text{ there is } n_p \geq 1 \text{ such that } \delta^{n_p}(\delta^{n_p}(p)) = \delta^{n_p}(p); \tag{5}$$

$$\text{there exist } q \in Q \text{ and } n \geq 1 \text{ such that } q = \delta^n(q); \tag{6}$$

$$\text{for every } q \in Q, \text{ if } q = \delta^k(q) \text{ for some } k \geq 1, \text{ then there is } n,\ 1 \leq n \leq |Q|, \text{ with } q = \delta^n(q). \tag{7}$$

For a definition of *solvable* and *unsolvable* groups the reader is referred to [21]. Here, we only need the fact that any homomorphic image of a solvable group is solvable and the Kaplan–Levy criterion [16] (generalising Thompson's [27, Cor.3]) according to which a finite group $\mathfrak{G}$ is unsolvable iff it contains three elements a, b, c, such that $o_{\mathfrak{G}}(a) = 2$, $o_{\mathfrak{G}}(b)$ is an odd prime, $o_{\mathfrak{G}}(c) > 1$ and coprime to both 2 and $o_{\mathfrak{G}}(b)$, and abc is the identity element of $\mathfrak{G}$.

A one-to-one and onto function on a finite set S is called a *permutation on* S. The *order of a permutation* δ is its order in the group of all permutations on S (whose operation is composition, and its identity element is the identity permutation id_S). We use the standard cycle notation for permutations.

Suppose $\mathfrak{G}$ is a monoid of $Q \to Q$ functions, for some finite set $Q \neq \emptyset$. Let $S = \{q \in Q \mid e_{\mathfrak{G}}(q) = q\}$, where $e_{\mathfrak{G}}$ the identity element in $\mathfrak{G}$. For every function δ in $\mathfrak{G}$, let $\delta\lceil_S$ denote the restriction of δ to S. Then we have the following:

$$\mathfrak{G} \text{ is a group iff } \delta\lceil_S \text{ is a permutation on } S, \text{ for every } \delta \text{ in } \mathfrak{G}; \tag{8}$$

$$\text{if } \mathfrak{G} \text{ is a group and } \delta \text{ is a nonindentity element in it, then } \delta\lceil_S \neq \mathsf{id}_S \text{ and the order of the permutation } \delta\lceil_S \text{ divides } o_{\mathfrak{G}}(\delta). \tag{9}$$

2.2 Automata: DFAs, NFAs, 2NFAs

A *two-way nondeterministic finite automaton* is a quintuple $\mathfrak{A} = (Q, \Sigma, \delta, Q_0, F)$ that consists of an alphabet Σ, a finite set Q of states with a subset $Q_0 \neq \emptyset$ of initial states and a subset F of accepting states, and a transition function $\delta\colon Q \times \Sigma \to 2^{Q\times\{-1,0,1\}}$ indicating the next state and whether the head should move left (-1), right (1), or stay put. If $Q_0 = \{q_0\}$ and $|\delta(q, a)| = 1$, for all $q \in Q$ and $a \in \Sigma$, then $\mathfrak{A}$ is *deterministic*, in which case we write $\mathfrak{A} = (Q, \Sigma, \delta, q_0, F)$. If $\delta(q, a) \subseteq Q \times \{1\}$, for all $q \in Q$ and $a \in \Sigma$, then $\mathfrak{A}$ is a *one-way* automaton, and we write $\delta\colon Q \times \Sigma \to 2^Q$. As usual, DFA and NFA refer to one-way deterministic and non-deterministic finite automata, respectively, while 2DFA and 2NFA to the corresponding two-way automata. Given a 2NFA $\mathfrak{A}$, we write $q \to_{a,d} q'$ if $(q', d) \in \delta(q, a)$; given an NFA $\mathfrak{A}$, we write $q \to_a q'$ if $q' \in \delta(q, a)$. A *run* of a 2NFA $\mathfrak{A}$ is a word in $(Q\times\mathbb{N})^*$. A run $(q_0, i_0), \dots, (q_m, i_m)$ is a *run of* $\mathfrak{A}$ *on a word* $w = a_0 \dots a_n \in \Sigma^*$ if $q_0 \in Q_0$, $i_0 = 0$ and there exist $d_0, \dots, d_{m-1} \in \{-1, 0, 1\}$ such that $q_j \to_{a_j,d_j} q_{j+1}$ and $i_{j+1} = i_j + d_j$ for all j, $0 \leq j < m$. The run is *accepting* if $q_m \in F$, $i_m = n + 1$. $\mathfrak{A}$ *accepts* $w \in \Sigma^*$ if there is an accepting run of $\mathfrak{A}$ on w; the language $\boldsymbol{L}(\mathfrak{A})$ of $\mathfrak{A}$ is the set of all words accepted by $\mathfrak{A}$.

Given an NFA $\mathfrak{A}$, states $q, q' \in Q$, and $w = a_0 \dots a_n \in \Sigma^*$, we write $q \to_w q'$ if either $w = \varepsilon$ and $q' = q$ or there is a run of $\mathfrak{A}$ on w that starts with $(q_0, 0)$ and ends with $(q', n+1)$. We say that a state $q \in Q$ is *reachable* if $q' \to_w q$, for some $q' \in Q_0$ and $w \in \Sigma^*$.

Given a DFA $\mathfrak{A} = (Q, \Sigma, \delta, q_0, F)$ and a word $w \in \Sigma^*$, we define a function $\delta_w\colon Q \to Q$ by taking $\delta_w(q) = q'$ iff $q \to_w q'$. We also define an equivalence relation $\sim$ on the set $Q^r \subseteq Q$ of reachable states by taking $q \sim q'$ iff, for every $w \in \Sigma^*$, we have $\delta_w(q) \in F$ just in case $\delta_w(q') \in F$. We denote the $\sim$-class of q by $q/_\sim$, and let $X/_\sim = \{q/_\sim \mid q \in X\}$ for any $X \subseteq Q^r$. Define $\tilde{\delta}_w\colon Q^r/_\sim \to Q^r/_\sim$ by taking $\tilde{\delta}_w(q/_\sim) = \delta_w(q)/_\sim$. Then $\big(Q^r/_\sim, \Sigma, \tilde{\delta}, q_0/_\sim, (F \cap Q^r)/_\sim\big)$ is the *minimal DFA* whose language coincides with the language of $\mathfrak{A}$. Given a regular language $\boldsymbol{L}$, we denote by $\mathfrak{A}_{\boldsymbol{L}}$ the minimal DFA whose language is $\boldsymbol{L}$.

The *transition monoid of* a DFA $\mathfrak{A}$ is $M(\mathfrak{A}) = (\{\delta_w \mid w \in \Sigma^*\}, \cdot)$ with $\delta_v \cdot \delta_w = \delta_{vw}$, for any v, w. The *syntactic monoid* $M(\boldsymbol{L})$ *of* $\boldsymbol{L}$ is the transition monoid $M(\mathfrak{A}_{\boldsymbol{L}})$ of $\mathfrak{A}_{\boldsymbol{L}}$. The *syntactic morphism of* $\boldsymbol{L}$ is the map $\eta_{\boldsymbol{L}}$ from Σ^* to the domain of $M(\boldsymbol{L})$ defined by $\eta_{\boldsymbol{L}}(w) = \tilde{\delta}_w$. We call $\eta_{\boldsymbol{L}}$ *quasi-aperiodic* if $\eta_{\boldsymbol{L}}(\Sigma^t)$ is aperiodic for every $t < \omega$.

Suppose $\mathcal{L} \in \{\mathsf{FO}(<), \mathsf{FO}(<, \equiv), \mathsf{FO}(<, \mathsf{MOD})\}$. A language $\boldsymbol{L}$ over an alphabet Σ is $\mathcal{L}$-*definable* if there is an $\mathcal{L}$-sentence φ in the signature Σ, whose symbols are treated as unary predicates, such that, for any $w \in \Sigma^*$, we have $w = a_0 \dots a_n \in \boldsymbol{L}$ iff $\mathfrak{S}_w \models \varphi$, where $\mathfrak{S}_w$ is an FO-structure with domain $\{0, \dots, n\}$ ordered by $<$, in which $\mathfrak{S}_w \models a(i)$ iff $a = a_i$, for $0 \le i \le n$.

Table 1 summarises the known results that connect definability of a regular language $\boldsymbol{L}$ with properties of the syntactic monoid $M(\boldsymbol{L})$ and syntactic morphism $\eta_{\boldsymbol{L}}$ (see [4] for details) and with its circuit complexity under a reasonable binary encoding of $\boldsymbol{L}$'s alphabet (see, e.g., [7, Lemma 2.1]) and the assumption that $\mathrm{ACC}^0 \neq \mathrm{NC}^1$. We also remind the reader that a regular language is $\mathsf{FO}(<)$-definable iff it is star-free [26], and that $\mathrm{AC}^0 \subsetneq \mathrm{ACC}^0 \subseteq \mathrm{NC}^1$ [15,26].

3 Criteria of $\mathcal{L}$-definability

In this section, we show that the algebraic characterisations of FO-definability of $\boldsymbol{L}(\mathfrak{A})$ given in Table 1 can be captured by 'localisable' properties of the transition monoid of $\mathfrak{A}$, for any given DFA $\mathfrak{A}$. Note that Theorem 1 (*i*) was already observed in [25] and used in proving that $\mathsf{FO}(<)$-definability of $\boldsymbol{L}(\mathfrak{A})$ is PSPACE-complete [7,10,25]; while criteria (*ii*) and (*iii*) seem to be new.

Theorem 1. *For any DFA* $\mathfrak{A} = (Q, \Sigma, \delta, q_0, F)$, *the following criteria hold*:

- (*i*) $\boldsymbol{L}(\mathfrak{A})$ *is not* $\mathsf{FO}(<)$-*definable iff* $\mathfrak{A}$ *contains a nontrivial cycle, that is, there exist a word* $u \in \Sigma^*$, *a state* $q \in Q^r$, *and a number* $k \le |Q|$ *such that* $q \not\sim \delta_u(q)$ *and* $q = \delta_{u^k}(q)$;
- (*ii*) $\boldsymbol{L}(\mathfrak{A})$ *is not* $\mathsf{FO}(<, \equiv)$-*definable iff there are words* $u, v \in \Sigma^*$, *a state* $q \in Q^r$, *and a number* $k \le |Q|$ *such that* $q \not\sim \delta_u(q)$, $q = \delta_{u^k}(q)$, $|v| = |u|$, *and* $\delta_{u^i}(q) = \delta_{u^i v}(q)$, *for every* $i < k$;
- (*iii*) $\boldsymbol{L}(\mathfrak{A})$ *is not* $\mathsf{FO}(<, \mathsf{MOD})$-*definable iff there exist words* $u, v \in \Sigma^*$, *a state* $q \in Q^r$ *and numbers* $k, l \le |Q|$ *such that* k *is an odd prime*, $l > 1$ *and coprime to both* 2 *and* k, $q \not\sim \delta_u(q)$, $q \not\sim \delta_v(q)$, $q \not\sim \delta_{uv}(q)$ *and, for all* $x \in \{u, v\}^*$, *we have* $\delta_x(q) \sim \delta_{xu^2}(q) \sim \delta_{xv^k}(q) \sim \delta_{x(uv)^l}(q)$.

Proof. Throughout, we use the algebraic criteria of Table 1 for $\boldsymbol{L} = \boldsymbol{L}(\mathfrak{A})$. Thus, $M(\boldsymbol{L})$ is the transition monoid of the minimal DFA $\mathfrak{A}_{\boldsymbol{L}(\mathfrak{A})}$, whose transition function we denote by $\tilde{\delta}$.

(*i*) ($\Rightarrow$) Suppose $\mathfrak{G}$ is a nontrivial group in $M(\mathfrak{A}_{\boldsymbol{L}(\mathfrak{A})})$. Let $u \in \Sigma^*$ be such that $\tilde{\delta}_u$ is a nonidentity element in $\mathfrak{G}$. We claim that there is $p \in Q^r$ such that $\tilde{\delta}_{u^n}(p/_\sim) \neq \tilde{\delta}_{u^{n+1}}(p/_\sim)$ for any $n > 0$. Indeed, otherwise for every $p \in Q^r$ there is $n_p > 0$ with $\tilde{\delta}_{u^{n_p}}(p/_\sim) = \tilde{\delta}_{u^{n_p+1}}(p/_\sim)$. Let $n = \max\{n_p \mid p \in Q^r\}$. Then $\tilde{\delta}_{u^n} = \tilde{\delta}_{u^{n+1}}$, contrary to (1).

By (5), there is $m \geq 1$ with $\tilde{\delta}_{u^{2m}}(p/_\sim) = \tilde{\delta}_{u^m}(p/_\sim)$. Let $s/_\sim = \tilde{\delta}_{u^m}(p/_\sim)$. Then $s/_\sim = \tilde{\delta}_{u^m}(s/_\sim)$, and so the restriction of δ_{u^m} to the subset $s/_\sim$ of Q^r is an $s/_\sim \to s/_\sim$ function. By (6), there exist $q \in s/_\sim$ and $n \geq 1$ such that $(\delta_{u^m})^n(q) = q$. Thus, $\delta_{u^{mn}}(q) = q$, and so by (7), there is $k \leq |Q|$ with $\delta_{u^k}(q) = q$. As $s/_\sim \neq \tilde{\delta}_u(s/_\sim)$, we also have $q \not\sim \delta_u(q)$, as required.

(i) ($\Leftarrow$) Suppose the condition holds for $\mathfrak{A}$. Then there are $u \in \Sigma^*$, $q \in Q^r/_\sim$, and $k < \omega$ such that $q \neq \tilde{\delta}_u(q)$ and $q = \tilde{\delta}_{u^k}(q)$. Then $\tilde{\delta}_{u^n} \neq \tilde{\delta}_{u^{n+1}}$ for any $n > 0$. Indeed, otherwise we would have some $n > 0$ with $\tilde{\delta}_{u^n}(q) = \tilde{\delta}_{u^{n+1}}(q)$. Let i, j be such that $n = i \cdot k + j$ and $j < k$. Then

$$q = \tilde{\delta}_{u^k}(q) = \tilde{\delta}_{u^{(i+1)k}}(q) = \tilde{\delta}_{u^n u^{k-j}}(q) = \tilde{\delta}_{u^{n+1} u^{k-j}}(q) = \tilde{\delta}_{u^{(i+1)k} u}(q) = \tilde{\delta}_u(q).$$

So, by (3) and (4), $\mathfrak{G}_{\tilde{\delta}_u}$ is a nontrivial group in $M(\mathfrak{A}_{L(\mathfrak{A})})$.

(ii) ($\Rightarrow$) Let $\mathfrak{G}$ be a nontrivial group in $\eta_L(\Sigma^t)$, for some $t < \omega$, and let $u \in \Sigma^t$ be such that $\tilde{\delta}_u$ is a nonidentity element in $\mathfrak{G}$. As shown in the proof of *(i)* ($\Rightarrow$), there exist $s \in Q^r$ and $m \geq 1$ such that $s/_\sim \neq \tilde{\delta}_u(s/_\sim)$ and $s/_\sim = \tilde{\delta}_{u^m}(s/_\sim)$. Now let $v \in \Sigma^t$ be such that $\tilde{\delta}_v$ is the identity element in $\mathfrak{G}$, and consider δ_v. By (2), there is $\ell \geq 1$ such that δ_{v^ℓ} is idempotent. Then $\delta_{v^{2\ell-1}v^{2\ell}} = \delta_{v^{2\ell-1}}$. Thus, if we let $\bar{u} = uv^{2\ell-1}$ and $\bar{v} = v^{2\ell}$, then $|\bar{u}| = |\bar{v}|$ and $\delta_{\bar{u}^i} = \delta_{\bar{u}^i\bar{v}}$ for any $i < \omega$. Also, $\tilde{\delta}_{u^i} = \tilde{\delta}_{\bar{u}^i}$ for every $i \geq 1$, and so the restriction of $\delta_{\bar{u}^m}$ to $s/_\sim$ is an $s/_\sim \to s/_\sim$ function. By (6), there exist $q \in s/_\sim$ and $n \geq 1$ such that $(\delta_{\bar{u}^m})^n(q) = q$. Thus, $\delta_{\bar{u}^{mn}}(q) = q$, and so by (7), there is some $k \leq |Q|$ with $\delta_{\bar{u}^k}(q) = q$. As $s/_\sim \neq \tilde{\delta}_u(s/_\sim) = \tilde{\delta}_{\bar{u}}(s/_\sim)$, we also have $q \not\sim \delta_{\bar{u}}(q)$, as required.

(ii) ($\Leftarrow$) If the condition holds for $\mathfrak{A}$, then there exist $u, v \in \Sigma^*$, $q \in Q^r/_\sim$, and $k < \omega$ such that $q \neq \tilde{\delta}_u(q)$, $q = \tilde{\delta}_{u^k}(q)$, $|v| = |u|$, and $\tilde{\delta}_{u^i}(q) = \tilde{\delta}_{u^i v}(q)$, for every $i < k$. As $M(\mathfrak{A}_{L(\mathfrak{A})})$ is finite, it has finitely many subsets. So there exist $i, j \geq 1$ such that $\eta_L(\Sigma^{i|u|}) = \eta_L(\Sigma^{(i+j)|u|})$. Let z be a multiple of j with $i \leq z < i + j$. Then $\eta_L(\Sigma^{z|u|}) = \eta_L(\Sigma^{(z|u|)^2})$, and so $\eta_L(\Sigma^{z|u|})$ is closed under the composition of functions (that is, the semigroup operation of $M(\mathfrak{A}_{L(\mathfrak{A})})$). Let $w = uv^{z-1}$ and consider the group $\mathfrak{G}_{\tilde{\delta}_w}$ (defined above (2)–(4)). Then $G_{\tilde{\delta}_w} \subseteq \eta_L(\Sigma^{z|u|})$. We claim that $\mathfrak{G}_{\tilde{\delta}_w}$ is nontrivial. Indeed, we have $\tilde{\delta}_w(q) = \tilde{\delta}_{uv^{z-1}}(q) = \tilde{\delta}_u(q) \neq q$. On the other hand, $\tilde{\delta}_{w^k}(q) = \tilde{\delta}_{u^k}(q) = q$. By the proof of *(i)* ($\Leftarrow$), $\mathfrak{G}_{\tilde{\delta}_w}$ is nontrivial.

(iii) ($\Rightarrow$) Suppose $\mathfrak{G}$ is an unsolvable group in $M(\mathfrak{A}_{L(\mathfrak{A})})$. By the Kaplan–Levy criterion, $\mathfrak{G}$ contains three functions a, b, c such that $o_\mathfrak{G}(a) = 2$, $o_\mathfrak{G}(b)$ is an odd prime, $o_\mathfrak{G}(c) > 1$ and coprime to both 2 and $o_\mathfrak{G}(b)$, and $c \circ b \circ a = e_\mathfrak{G}$ for the identity element $e_\mathfrak{G}$ of $\mathfrak{G}$. Let $u, v \in \Sigma^*$ be such that $a = \tilde{\delta}_u$, $b = \tilde{\delta}_v$ and $c = (\tilde{\delta}_{uv})^-$, and let $k = o_\mathfrak{G}(\tilde{\delta}_v)$ and $r = o_\mathfrak{G}(c) = o_\mathfrak{G}(\tilde{\delta}_{uv})$. Then $r > 1$ and coprime to both 2 and k. Let $S = \{p \in Q^r/_\sim \mid e_\mathfrak{G}(p) = p\}$. As $\tilde{\delta}_x$ is $\mathfrak{G}$ for every $x \in \{u, v\}^*$, we have $e_\mathfrak{G} \circ \tilde{\delta}_x = \tilde{\delta}_x$. Thus,

$$\tilde{\delta}_{xu^2}(q) = \tilde{\delta}_{u^2}(\tilde{\delta}_x(q)) = e_\mathfrak{G}(\tilde{\delta}_x(q)) = (e_\mathfrak{G} \circ \tilde{\delta}_x)(q) = \tilde{\delta}_x(q), \quad \text{and}$$
$$\tilde{\delta}_{xv^k}(q) = \tilde{\delta}_{v^k}(\tilde{\delta}_x(q)) = e_\mathfrak{G}(\tilde{\delta}_x(q)) = (e_\mathfrak{G} \circ \tilde{\delta}_x)(q) = \tilde{\delta}_x(q), \quad \text{for every } q \in S.$$

Then, by (8), each of $\tilde{\delta}_u\restriction_S$, $\tilde{\delta}_v\restriction_S$ and $\tilde{\delta}_{uv}\restriction_S$ is a permutation on S. By (9), the order of $\tilde{\delta}_u\restriction_S$ is 2, the order of $\tilde{\delta}_v\restriction_S$ is k, and the order l of $\tilde{\delta}_{uv}\restriction_S$ is a > 1 divisor of r, and so it is coprime to both 2 and k. Also, we have $k, l \leq |S| \leq |Q|$. Further, for every x, if q is in S then $\tilde{\delta}_x(q) \in S$ as well. So we have

$$\tilde{\delta}_{x(uv)^l}(q) = \tilde{\delta}_{(uv)^l}(\tilde{\delta}_x(q)) = (\tilde{\delta}_{uv}\restriction_S)^l(\tilde{\delta}_x(q)) = \mathsf{id}_S(\tilde{\delta}_x(q)) = \tilde{\delta}_x(q), \quad \text{for all } q \in S.$$

It remains to show that there is $q \in S$ with $q \neq \tilde{\delta}_u(q)$, $q \neq \tilde{\delta}_u(q)$, and $q \neq \tilde{\delta}_{uv}(q)$. Recall that the length of any cycle in a permutation divides its order. First, we show there is $q \in S$ with $q \neq \tilde{\delta}_u(q)$ and $q \neq \tilde{\delta}_u(q)$. Indeed, as $\tilde{\delta}_u\restriction_S \neq \mathsf{id}_S$, there is $q \in S$ such that $\tilde{\delta}_u(q) = q' \neq q$. As the order of $\tilde{\delta}_u\restriction_S$ is 2, $\tilde{\delta}_u(q') = q$. If both $\tilde{\delta}_v(q) = q$ and $\tilde{\delta}_v(q') = q'$ were the case, then $\tilde{\delta}_{uv}(q) = q'$ and $\tilde{\delta}_{uv}(q') = q$ would hold, and so (qq') would be a cycle in $\tilde{\delta}_{uv}\restriction_S$, contrary to l being coprime to 2. So take some $q \in S$ with $\tilde{\delta}_u(q) = q' \neq q$ and $\tilde{\delta}_v(q) \neq q$. If $\tilde{\delta}_v(q') \neq q$ then $\tilde{\delta}_{uv}(q) \neq q$, and so q is a good choice. Suppose $\tilde{\delta}_v(q') = q$, and let $q'' = \tilde{\delta}_v(q)$. Then $q'' \neq q'$, as k is odd. Thus, $\tilde{\delta}_{uv}(q') \neq q'$, and so q' is a good choice.

(iii) ($\Leftarrow$) Suppose $u, v \in \Sigma^*$, $q \in Q^r$, and $k, l < \omega$ are satisfying the conditions. For every $x \in \{u, v\}^*$, we define an equivalence relation $\approx_x$ on $Q^r/_\sim$ by taking $p \approx_x p'$ iff $\tilde{\delta}_x(p) = \tilde{\delta}_x(p')$. Then we clearly have that $\approx_x \subseteq \approx_{xy}$, for all $x, y \in \{u, v\}^*$. As Q is finite, there is $z \in \{u, v\}^*$ such that $\approx_z = \approx_{zy}$ for all $y \in \{u, v\}^*$. Take such a z. By (2), $\tilde{\delta}_z^n$ is idempotent for some $n \geq 1$. We let $w = z^n$. Then $\tilde{\delta}_w$ is idempotent and we also have that

$$\approx_w = \approx_{wy} \quad \text{for all } y \in \{u, v\}^*. \tag{10}$$

Let $G_{\{u,v\}} = \{\tilde{\delta}_{wxw} \mid x \in \{u, v\}^*\}$. Then $G_{\{u,v\}}$ is closed under composition. Let $\mathfrak{G}_{\{u,v\}}$ be the subsemigroup of $M(\mathfrak{A}_{L(\mathfrak{A})})$ with universe $G_{\{u,v\}}$. Then $\tilde{\delta}_w = \tilde{\delta}_{w\varepsilon w}$ is an identity element in $\mathfrak{G}_{\{u,v\}}$. Let $S = \{p \in Q^r/_\sim \mid \tilde{\delta}_w(p) = p\}$. We show that

$$\text{for every } \tilde{\delta} \text{ in } \mathfrak{G}_{\{u,v\}},\ \tilde{\delta}\restriction_S \text{ is a permutation on } S, \tag{11}$$

and so $\mathfrak{G}_{\{u,v\}}$ is a group by (8). Indeed, take some $x \in \{u, v\}^*$. As $\tilde{\delta}_w(\tilde{\delta}_{wxw}(p)) = \tilde{\delta}_{wxww}(p) = \tilde{\delta}_{wxw}(p)$ for any $p \in Q^r/_\sim$, $\tilde{\delta}_{wxw}\restriction_S$ is an $S \to S$ function. Also, if $p, p' \in S$ and $\tilde{\delta}_{wxw}(p) = \tilde{\delta}_{wxw}(p')$ then $p \approx_{wxw} p'$. Thus, by (10), $p \approx_w p'$, that is, $p = \tilde{\delta}_w(p) = \tilde{\delta}_w(p') = p'$, proving (11).

We show that $\mathfrak{G}_{\{u,v\}}$ is unsolvable by finding an unsolvable homomorphic image of it. Let $R = \{p \in Q^r/_\sim \mid p = \tilde{\delta}_x(q) \text{ for some } x \in \{u, v\}^*\}$. We claim that, for every $\tilde{\delta}$ in $\mathfrak{G}_{\{u,v\}}$, $\tilde{\delta}\restriction_R$ is a permutation on R, and so the function h mapping every $\tilde{\delta}$ to $\tilde{\delta}\restriction_R$ is a group homomorphism from $\mathfrak{G}_{\{u,v\}}$ to the group of all permutations on R. Indeed, by (11), it is enough to show that $R \subseteq S$. Let $\overline{w} = \overline{z}_m \dots \overline{z}_1$, where $w = z_1 \dots z_m$ for some $z_i \in \{u, v\}$, $\overline{u} = u$ and $\overline{v} = v^{k-1}$. Since $\tilde{\delta}_x(q) = \tilde{\delta}_{x(u)^2}(q) = \tilde{\delta}_{x(v)^k}(q)$ for all $x \in \{u, v\}^*$, we obtain that

$$\begin{aligned}\tilde{\delta}_{yw\overline{w}}(q) = \ &\tilde{\delta}_{\overline{z}_{m-1}\dots\overline{z}_1}(\tilde{\delta}_{yz_1\dots z_m\overline{z}_m}(q)) = \tilde{\delta}_{\overline{z}_{m-1}\dots\overline{z}_1}(\tilde{\delta}_{yz_1\dots z_{m-1}}(q)) = \dots \\ &\dots = \tilde{\delta}_{\overline{z}_1}(\tilde{\delta}_{yz_1}(q)) = \tilde{\delta}_{xz_1\overline{z}_1}(q) = \tilde{\delta}_y(q), \quad \text{for all } y \in \{u, v\}^*. \end{aligned} \tag{12}$$

Now suppose $p \in R$, that is, $p = \tilde{\delta}_x(q)$ for some $x \in \{u, v\}^*$. Then, by (12),

$$\tilde{\delta}_w(p) = \tilde{\delta}_w(\tilde{\delta}_x(q)) = \tilde{\delta}_{xw}(q) = \tilde{\delta}_{xww\overline{w}}(q) = \tilde{\delta}_{xw\overline{w}}(q) = \tilde{\delta}_x(q) = p,$$

and so $p \in S$, as required.

Now let $\mathfrak{G}$ be the image of $\mathfrak{G}_{\{u,v\}}$ under h. We prove that $\mathfrak{G}$ is unsolvable by finding three elements a, b, c in it such that $o_{\mathfrak{G}}(a) = 2$, $o_{\mathfrak{G}}(b) = k$, $o_{\mathfrak{G}}(c)$ is coprime to both 2 and $o_{\mathfrak{G}}(b)$, and $c \circ b \circ a = \mathsf{id}_R$ (the identity element of $\mathfrak{G}$). So let $a = h(\tilde{\delta}_{wuw})$, $b = h(\tilde{\delta}_{wvw})$, and $c = h(\tilde{\delta}_{wuvw})^{-}$. Observe that, for every $x \in \{u,v\}^*$, $h(\tilde{\delta}_{wxw}) = \tilde{\delta}_x \restriction_R$, and so $c \circ b \circ a = \mathsf{id}_R$. Also, for any $\tilde{\delta}_x(q) \in R$, $a^2(\tilde{\delta}_x(q)) = (\tilde{\delta}_u \restriction_R)^2(\tilde{\delta}_x(q)) = \tilde{\delta}_{xu^2}(q) = \tilde{\delta}_x(q)$ by our assumption, so $a^2 = \mathsf{id}_R$. On the other hand, $q \in R$ as $\tilde{\delta}_\varepsilon(q) = q$, and $\mathsf{id}_R(q) = q \neq \tilde{\delta}_u(q)$ by assumption, so $a \neq \mathsf{id}_R$. As $o_{\mathfrak{G}}(a)$ divides 2, $o_{\mathfrak{G}}(a) = 2$ follows. Similarly, we can show that $o_{\mathfrak{G}}(b) = k$ (using that $\tilde{\delta}_{xv^k}(q) = \tilde{\delta}_x(q)$ for every $x \in \{u,v\}^*$, and $u \neq \tilde{\delta}_v(q)$). Finally (using that $\tilde{\delta}_{x(uv)^l}(q) = \tilde{\delta}_x(q)$ for every $x \in \{u,v\}^*$, and $u \neq \tilde{\delta}_{uv}(q)$), we obtain that $h(\tilde{\delta}_{wuvw})^l = \mathsf{id}_R$ and $h(\tilde{\delta}_{wuvw}) \neq \mathsf{id}_R$. Therefore, it follows that $o_{\mathfrak{G}}(c) = o_{\mathfrak{G}}\big(h(\tilde{\delta}_{wuvw})^{-}\big) = o_{\mathfrak{G}}\big(h(\tilde{\delta}_{wuvw})\big) > 1$ and divides l, and so coprime to both 2 and k, as required.

4 Deciding FO-definability: PSPACE-hardness

Kozen [18] showed that deciding whether the intersection of the languages recognised by a set of given deterministic DFAs is non-empty is PSPACE-complete. By carefully analysing Kozen's lower bound proof and using the criterion of Theorem 1 (*i*), Cho and Huynh [10] established that deciding $\mathsf{FO}(<)$-definability of $\boldsymbol{L}(\mathfrak{A})$ is PSPACE-hard, for any given minimal DFA $\mathfrak{A}$. We generalise their construction and use the criteria in Theorem 1 (*ii*)–(*iii*) to cover $\mathsf{FO}(<,\equiv)$- and $\mathsf{FO}(<,\mathsf{MOD})$-definability as well.

Theorem 2. *For any* $\mathcal{L} \in \{\mathsf{FO}(<), \mathsf{FO}(<,\equiv), \mathsf{FO}(<,\mathsf{MOD})\}$*, deciding* $\mathcal{L}$*-definability of the language* $\boldsymbol{L}(\mathfrak{A})$ *of a given minimal DFA* $\mathfrak{A}$ *is* PSPACE*-hard.*

Proof. Let $\boldsymbol{M}$ be a deterministic Turing machine that decides a language using at most $N = P_{\boldsymbol{M}}(n)$ tape cells on any input of size n, for some polynomial $P_{\boldsymbol{M}}$. Given such an $\boldsymbol{M}$ and an input $\boldsymbol{x}$, our aim is to define three minimal DFAs whose languages are, respectively, $\mathsf{FO}(<)$-, $\mathsf{FO}(<,\equiv)$-, and $\mathsf{FO}(<,\mathsf{MOD})$-definable iff $\boldsymbol{M}$ rejects $\boldsymbol{x}$, and whose sizes are polynomial in N and the size $|\boldsymbol{M}|$ of $\boldsymbol{M}$.

Suppose $\boldsymbol{M} = (Q, \Gamma, \gamma, \mathsf{b}, q_0, q_{acc})$ with a set Q of states, tape alphabet Γ with b for blank, transition function γ, initial state q_0 and accepting state q_{acc}. Without loss of generality we assume that $\boldsymbol{M}$ erases the tape before accepting, its head is at the left-most cell in an accepting configuration, and if $\boldsymbol{M}$ does not accept the input, it runs forever. Given an input word $\boldsymbol{x} = x_1 \dots x_n$ over Γ, we represent configurations $\mathfrak{c}$ of the computation of $\boldsymbol{M}$ on $\boldsymbol{x}$ by the N-long word written on the tape (with sufficiently many blanks at the end) in which the symbol y in the active cell is replaced by the pair (q, y) for the current state q. The accepting computation of $\boldsymbol{M}$ on $\boldsymbol{x}$ is encoded by a word $\sharp\, \mathfrak{c}_1 \,\sharp\, \mathfrak{c}_2 \,\sharp\, \dots \,\sharp\, \mathfrak{c}_{k-1} \,\sharp\, \mathfrak{c}_k \flat$ over the alphabet $\Sigma = \Gamma \cup (Q \times \Gamma) \cup \{\sharp, \flat\}$, with $\mathfrak{c}_1, \mathfrak{c}_2, \dots, \mathfrak{c}_k$ being the subsequent configurations. In particular, $\mathfrak{c}_1$ is the initial configuration on $\boldsymbol{x}$ (so it is of the form $(q_0, x_1)x_2 \dots x_n \mathsf{b} \dots \mathsf{b}$), and $\mathfrak{c}_k$ is the accepting configuration (so it is of the form $(q_{acc}, \mathsf{b})\mathsf{b} \dots \mathsf{b}$). As usual for this representation of computations, we may regard γ as a partial function from $\left(\Gamma \cup (Q \times \Gamma) \cup \{\sharp\}\right)^3$ to $\Gamma \cup (Q \times \Gamma)$ with $\gamma(\sigma^j_{i-1}, \sigma^j_i, \sigma^j_{i+1}) = \sigma^{j+1}_i$ for each $j < k$, where σ^j_i is the ith symbol of $\mathfrak{c}^j$.

Let $p_{M,x} = p$ be the first prime such that $p \geq N + 2$ and $p \not\equiv \pm 1 \pmod{10}$. By [6, Corollary 1.6], p is polynomial in N. Our first aim is to construct a $p+1$-long sequence $\mathfrak{A}_i$ of disjoint minimal DFAs over Σ. Each $\mathfrak{A}_i$ has size polynomial in N and $|\boldsymbol{M}|$, and it checks certain properties of an accepting computation on $\boldsymbol{x}$ such that $\boldsymbol{M}$ accepts $\boldsymbol{x}$ iff the intersection of the $\boldsymbol{L}(\mathfrak{A}_i)$ is not empty and consists of the single word encoding the accepting computation on $\boldsymbol{x}$.

We define each $\mathfrak{A}_i$ as an NFA, and assume that it can be turned to a DFA by adding a 'trash state' tr_i looping on itself with every $\sigma \in \Sigma$, and adding the missing transitions leading to tr_i. The DFA $\mathfrak{A}_0$ checks that an input starts with the initial configuration on $\boldsymbol{x}$ and ends with the accepting configuration:

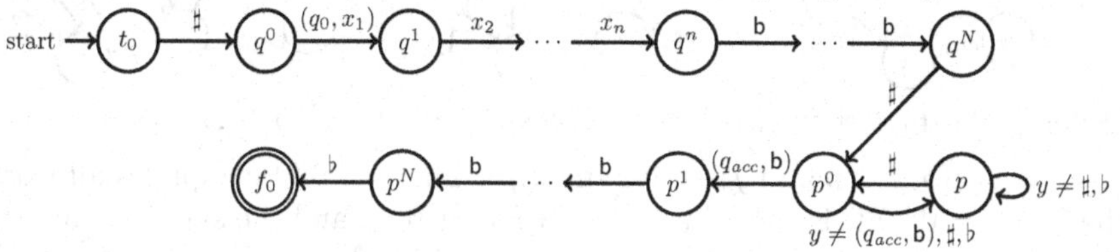

When $1 \leq i \leq N$, the DFA $\mathfrak{A}_i$ checks, for all $j < k$, whether the ith symbol of $\mathfrak{c}^j$ changes 'according to γ' in passing to $\mathfrak{c}^{j+1}$. The non-trash part of its transition function δ^i is as follows, for $1 < i < N$. (For $i = 1$ and $i = N$ some adjustments are needed.) For all $u, u', v, w, w', y, z \in \Gamma \cup (Q \times \Gamma)$,

$$\delta^i_\sharp(t_i) = q^0, \quad \delta^i_u(q^j) = q^{j+1}, \text{ for } j = 0, ..., i-3, \quad \delta^i_u(q^{i-2}) = r_u, \quad \delta^i_v(r_u) = r_{uv},$$
$$\delta^i_w(r_{uv}) = q^0_{\gamma(u,v,w)}, \quad \delta^i_y(q^j_z) = q^{j+1}_z, \text{ for } j = 0, ..., N-3,\ j \neq N-i-1,$$
$$\delta^i_\sharp(q^{N-i-1}_z) = q^{N-i}_z, \quad \delta^i_\flat(q^{N-i-1}_z) = f_i, \quad \delta^i_{u'}(q^{N-2}_z) = p_{u'z}, \quad \delta^i_z(p_{u'z}) = r_{u'z},$$

see below, where $z = \gamma(u, v, w)$ and $z' = \gamma(u', z, w')$:

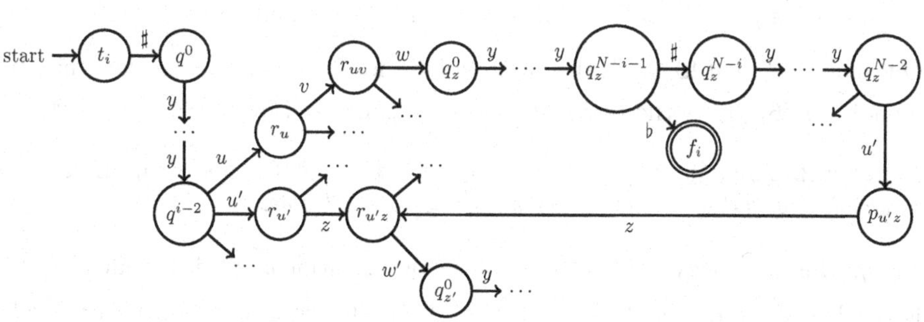

Finally, if $N + 1 \leq i \leq p$ then $\mathfrak{A}_i$ accepts all words over Σ with a single occurrence of $\flat$, which is the input's last character:

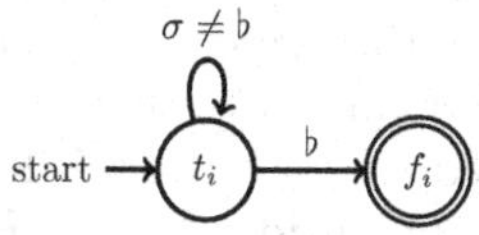

Note that $\mathfrak{A}_{p-1} = \mathfrak{A}_p$ as $p \geq N+2$. It is not hard to check that each of the $\mathfrak{A}_i$ is a minimal DFA that does not contain nontrivial cycles and the following holds:

Lemma 1. *$\boldsymbol{M}$ accepts $\boldsymbol{x}$ iff $\bigcap_{i=0}^{p} \boldsymbol{L}(\mathfrak{A}_i) \neq \emptyset$, in which case this language consists of a single word that encodes the accepting computation of $\boldsymbol{M}$ on $\boldsymbol{x}$.*

Next, we require three sequences of DFAs $\mathfrak{B}^p_<$, $\mathfrak{B}^p_\equiv$ and $\mathfrak{B}^p_{\mathsf{MOD}}$, where $p > 5$ is a prime number with $p \not\equiv \pm 1 \pmod{10}$; see the picture below for $p = 7$.

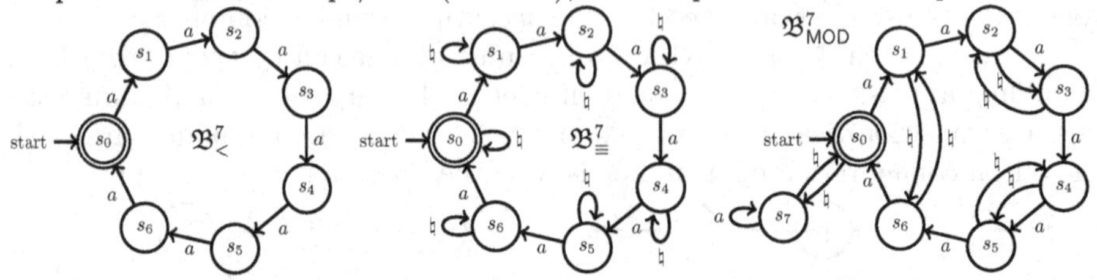

In general, the first sequence is $\mathfrak{B}^p_< = \big(\{s_i \mid i < p\}, \{a\}, \delta^{\mathfrak{B}^p_<}, s_0, \{s_0\}\big)$, where $\delta_a^{\mathfrak{B}^p_<}(s_i) = s_j$ if $i, j < p$ and $j \equiv i+1 \pmod{p}$. Then $\boldsymbol{L}(\mathfrak{B}^p_<)$ comprises all words of the form $(a^p)^*$, $\mathfrak{B}^p_<$ is the minimal DFA for $\boldsymbol{L}(\mathfrak{B}^p_<)$, and the syntactic monoid $M(\mathfrak{B}^p_<)$ is the cyclic group of order p (generated by the permutation $\delta_a^{\mathfrak{B}^p_<}$).

The second sequence is $\mathfrak{B}^p_\equiv = \big(\{s_i \mid i < p\}, \{a, \natural\}, \delta^{\mathfrak{B}^p_\equiv}, s_0, \{s_0\}\big)$, where $\delta_\natural^{\mathfrak{B}^p_\equiv}(s_i) = s_i$ and $\delta_a^{\mathfrak{B}^p_\equiv}(s_i) = s_j$ if $i, j < p$ and $j \equiv i+1 \pmod{p}$. One can check that $\boldsymbol{L}(\mathfrak{B}^p_\equiv)$ comprises all words of a's and $\natural$'s where the number of a's is divisible by p, $\mathfrak{B}^p_\equiv$ is the minimal DFA for this language, and $M(\mathfrak{B}^p_\equiv)$ is also the cyclic group of order p (generated by the permutation $\delta_a^{\mathfrak{B}^p_\equiv}$).

The third sequence is $\mathfrak{B}^p_{\mathsf{MOD}} = \big(\{s_i \mid i \leq p\}, \{a, \natural\}, \delta^{\mathfrak{B}^p_{\mathsf{MOD}}}, s_0, \{s_0\}\big)$, where

- $\delta_a^{\mathfrak{B}^p_{\mathsf{MOD}}}(s_p) = s_p$, and $\delta_a^{\mathfrak{B}^p_{\mathsf{MOD}}}(s_i) = s_j$ whenever $i, j < p$ and $j \equiv i+1 \pmod{p}$;
- $\delta_\natural^{\mathfrak{B}^p_{\mathsf{MOD}}}(s_0) = s_p$, $\delta_\natural^{\mathfrak{B}^p_{\mathsf{MOD}}}(s_p) = s_0$, and $\delta_\natural^{\mathfrak{B}^p_{\mathsf{MOD}}}(s_i) = s_j$ whenever $1 \leq i, j < p$ and $i \cdot j \equiv p-1 \pmod{p}$, that is, $j = -1/i$ in the finite field $\mathbb{F}_p$.

One can check that $\mathfrak{B}^p_{\mathsf{MOD}}$ is the minimal DFA for its language, and the syntactic monoid $M(\mathfrak{B}^p_{\mathsf{MOD}})$ is the permutation group generated by $\delta_a^{\mathfrak{B}^p_{\mathsf{MOD}}}$ and $\delta_\natural^{\mathfrak{B}^p_{\mathsf{MOD}}}$.

Lemma 2. *For any prime $p > 5$ with $p \not\equiv \pm 1 \pmod{10}$, the group $M(\mathfrak{B}^p_{\mathsf{MOD}})$ is unsolvable, but all of its proper subgroups are solvable.*

Proof. One can check that the order of the permutation $\delta_\natural^{\mathfrak{B}^p_{\mathsf{MOD}}}$ is 2, that of $\delta_a^{\mathfrak{B}^p_{\mathsf{MOD}}}$ is p, while the order of the inverse of $\delta_{\natural a}^{\mathfrak{B}^p_{\mathsf{MOD}}}$ is the same as the order of $\delta_{\natural a}^{\mathfrak{B}^p_{\mathsf{MOD}}}$, which is 3. So $M(\mathfrak{B}^p_{\mathsf{MOD}})$ is unsolvable, for any prime p, by the Kaplan–Levy criterion. To prove that all proper subgroups of $M(\mathfrak{B}^p_{\mathsf{MOD}})$ are solvable, we show that $M(\mathfrak{B}^p_{\mathsf{MOD}})$ is a subgroup of the *projective special linear group* $\mathrm{PSL}_2(p)$. If p is a prime with $p > 5$ and $p \not\equiv \pm 1 \pmod{10}$, then all proper subgroups of $\mathrm{PSL}_2(p)$ are solvable; see, e.g., [17, Theorem 2.1]. (So $M(\mathfrak{B}^p_{\mathsf{MOD}})$ is in fact isomorphic to the unsolvable group $\mathrm{PSL}_2(p)$.) Consider the set $P = \{0, 1, \ldots, p-1, \infty\}$ of all points of the projective line over the field $\mathbb{F}_p$. By identifying s_i with i for $i < p$, and s_p with ∞, we may regard the elements of $M(\mathfrak{B}^p_{\mathsf{MOD}})$ as $P \to P$ functions. The group $\mathrm{PSL}_2(p)$ consists of all $P \to P$ functions of the form $i \mapsto \frac{w \cdot i + x}{y \cdot i + z}$, where $w \cdot z - x \cdot y = 1$, with the field arithmetic of $\mathbb{F}_p$ extended by $i + \infty = \infty$ for any $i \in P$, $0 \cdot \infty = 1$ and $i \cdot \infty = \infty$ for $i \neq 0$. One can check that the two generators of $M(\mathfrak{B}^p_{\mathsf{MOD}})$ are in $\mathrm{PSL}_2(p)$: take $w = 1$, $x = 1$, $y = 0$, $z = 1$ for $\delta_a^{\mathfrak{B}^p_{\mathsf{MOD}}}$, and $w = 0$, $x = 1$, $y = p - 1$, $z = 0$ for $\delta_\natural^{\mathfrak{B}^p_{\mathsf{MOD}}}$.

Finally, we define three automata $\mathfrak{A}_<$, $\mathfrak{A}_\equiv$, $\mathfrak{A}_{\mathsf{MOD}}$ over the same tape alphabet $\Sigma_+ = \Sigma \cup \{a_1, a_2, \natural\}$, where a_1, a_2 are fresh symbols. We take, respectively, $\mathfrak{B}^p_<$, $\mathfrak{B}^p_\equiv$, $\mathfrak{B}^p_{\mathsf{MOD}}$ and replace each transition $s_i \to_a s_j$ in them by a fresh copy of $\mathfrak{A}_i$, for $i \leq p$, as shown in the picture below.

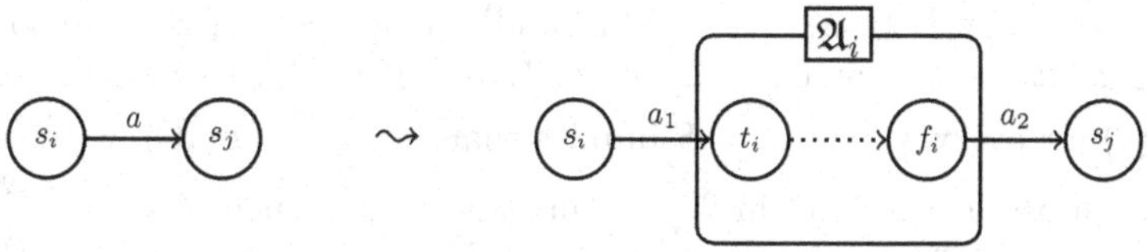

We make $\mathfrak{A}_<$, $\mathfrak{A}_\equiv$, $\mathfrak{A}_{\mathsf{MOD}}$ deterministic by adding a trash state *tr* looping on itself with every $y \in \Sigma_+$, and adding the missing transitions leading to *tr*. It follows that $\mathfrak{A}_<$, $\mathfrak{A}_\equiv$, and $\mathfrak{A}_{\mathsf{MOD}}$ are minimal DFAs of size polynomial in N, $|\boldsymbol{M}|$.

Lemma 3. *(i)* $\boldsymbol{L}(\mathfrak{A}_<)$ *is* $\mathsf{FO}(<)$*-definable iff* $\bigcap_{i=0}^p \boldsymbol{L}(\mathfrak{A}_i) = \emptyset$.
(ii) $\boldsymbol{L}(\mathfrak{A}_\equiv)$ *is* $\mathsf{FO}(<, \equiv)$*-definable iff* $\bigcap_{i=0}^p \boldsymbol{L}(\mathfrak{A}_i) = \emptyset$.
(iii) $\boldsymbol{L}(\mathfrak{A}_{\mathsf{MOD}})$ *is* $\mathsf{FO}(<, \mathsf{MOD})$*-definable iff* $\bigcap_{i=0}^p \boldsymbol{L}(\mathfrak{A}_i) = \emptyset$.

Proof. As $\mathfrak{A}_<, \mathfrak{A}_\equiv, \mathfrak{A}_{\mathsf{MOD}}$ are minimal, we can replace $\sim$ by $=$ in the conditions of Theorem 1. For the ($\Rightarrow$) directions, given some $w \in \bigcap_{i=0}^p \boldsymbol{L}(\mathfrak{A}_i)$, in each case we show how to satisfy the corresponding condition of Theorem 1: (i) take $u = a_1 w a_2$, $q = s_0$, and $k = p$; (ii) take $u = a_1 w a_2$, $v = \natural^{|u|}$, $q = s_0$, and $k = p$; (iii) take $u = \natural$, $v = a_1 w a_2$, $q = s_0$, $k = p$ and $l = 3$.

($\Leftarrow$) We show that the corresponding condition of Theorem 1 implies non-emptiness of $\bigcap_{i=0}^p \boldsymbol{L}(\mathfrak{A}_i)$. To this end, we define a $\Sigma_+^* \to \{a, \natural\}^*$ homomorphism by taking $h(\natural) = \natural$, $h(a_1) = a$, and $h(b) = \varepsilon$ for all other $b \in \Sigma_+$.

(i) and (ii): Let $\circ \in \{<, \equiv\}$ and suppose q is a state in $\mathfrak{A}^p_\circ$ and $u' \in \Sigma_+^*$ such that $q \neq \delta_{u'}^{\mathfrak{A}^p_\circ}(q)$ and $q = \delta_{(u')^k}^{\mathfrak{A}^p_\circ}(q)$ for some k. Let $S = \{s_0, s_1, \dots, s_{p-1}\}$. We claim that there exist $s \in S$ and $u \in \Sigma_+^*$ such that

$$s \neq \delta_u^{\mathfrak{A}^p_\circ}(s), \tag{13}$$

$$\delta_x^{\mathfrak{A}^p_\circ}(s) \in S, \quad \text{for every } x \in \{u\}^*. \tag{14}$$

Indeed, observe that none of the states along the cyclic $q \to_{(u')^k} q$ path Π in $\mathfrak{A}^p_\circ$ is *tr*. So there is some state along Π that is in S, as otherwise one of the $\mathfrak{A}_i$ would contain a nontrivial cycle. Therefore, u' must be of the form $w \natural^n a_1 w'$ for some $w \in \Sigma^*$, $n < \omega$ and $w' \in \Sigma_+^*$. It is easy to see that $s = \delta_{(u')^{k-1} w}^{\mathfrak{A}^p_\circ}(q)$ and $u = \natural^n a_1 w' w$ is as required in (13) and (14).

As $M(\mathfrak{B}^p_\circ)$ is a finite group, the set $\{\delta_{h(x)}^{\mathfrak{B}^p_\circ} \mid x \in \{u\}^*\}$ forms a subgroup $\mathfrak{G}$ in it (the subgroup generated by $\delta_{h(u)}^{\mathfrak{B}^p_\circ}$). We show that $\mathfrak{G}$ is nontrivial by finding a nontrivial homomorphic image of it. To this end, (14) implies that, for every $x \in \{u\}^*$, the restriction $\delta_x^{\mathfrak{A}^p_\circ}\restriction_{S'}$ of $\delta_x^{\mathfrak{A}^p_\circ}$ to the set $S' = \{\delta_y^{\mathfrak{A}^p_\circ}(s) \mid y \in \{u\}^*\}$ is an $S' \to S'$ function and $\delta_x^{\mathfrak{A}^p_\circ}\restriction_{S'} = \delta_{h(x)}^{\mathfrak{B}^p_\circ}\restriction_{S'}$. As $M(\mathfrak{B}^p_\circ)$ is a group of permutations on a set containing S', $\delta_{h(x)}^{\mathfrak{B}^p_\circ}\restriction_{S'}$ is a permutation of S', for every $x \in \{u\}^*$. Thus, $\{\delta_{h(x)}^{\mathfrak{B}^p_\circ}\restriction_{S'} \mid x \in \{u\}^*\}$ is a homomorphic image of $\mathfrak{G}$ that is nontrivial by (13).

As $\mathfrak{G}$ is a nontrivial subgroup of the cyclic group $M(\mathfrak{B}^p_\circ)$ of order p and p is a prime, $\mathfrak{G} = M(\mathfrak{B}^p_\circ)$. Then there is $x \in \{u\}^*$ with $\delta_{h(x)}^{\mathfrak{B}^p_\circ} = \delta_a^{\mathfrak{B}^p_\circ}$ (a permutation containing the p-cycle $(s_0 s_1 \dots s_{p-1})$ 'around' all elements of S), and so $S' = S$ and $x = \natural^n a_1 w a_2 w'$ for some $n < \omega$, $w \in \Sigma^*$, and $w' \in \Sigma^*_+$. As $n = 0$ when $\circ =<$ and $\delta_{\natural^n}^{\mathfrak{A}^p_=}(s)$ for every $s \in S$, $S' = S$ implies that $w \in \bigcap_{i=0}^{p-1} \boldsymbol{L}(\mathfrak{A}_i) = \bigcap_{i=0}^{p} \boldsymbol{L}(\mathfrak{A}_i)$.

(*iii*) Suppose q is a state in $\mathfrak{A}^p_{\mathsf{MOD}}$ and $u', v' \in \Sigma^*_+$ such that $q \neq \delta_{u'}^{\mathfrak{A}^p_{\mathsf{MOD}}}(q)$, $q \neq \delta_{v'}^{\mathfrak{A}^p_{\mathsf{MOD}}}(q)$, $q \neq \delta_{u'v'}^{\mathfrak{A}^p_{\mathsf{MOD}}}(q)$, and $\delta_x^{\mathfrak{A}^p_{\mathsf{MOD}}}(q) = \delta_{x(u')^2}^{\mathfrak{A}^p_{\mathsf{MOD}}}(q) = \delta_{x(v')^k}^{\mathfrak{A}^p_{\mathsf{MOD}}}(q) = \delta_{x(u'v')^l}^{\mathfrak{A}^p_{\mathsf{MOD}}}(q)$ for some odd prime k and number l that is coprime to both 2 and k. Take $S = \{s_0, s_1, \dots, s_p\}$. We claim that there exist $s \in S$ and $u, v \in \Sigma^*_+$ such that

$$s \neq \delta_u^{\mathfrak{A}^p_{\mathsf{MOD}}}(s), \; s \neq \delta_v^{\mathfrak{A}^p_{\mathsf{MOD}}}(s), \; s \neq \delta_{uv}^{\mathfrak{A}^p_{\mathsf{MOD}}}(s), \tag{15}$$

$$\delta_x^{\mathfrak{A}^p_{\mathsf{MOD}}}(s) \in S, \quad \text{for every } x \in \{u, v\}^*, \tag{16}$$

$$\delta_x^{\mathfrak{A}^p_{\mathsf{MOD}}}(s) = \delta_{xu^2}^{\mathfrak{A}^p_{\mathsf{MOD}}}(s) = \delta_{xv^k}^{\mathfrak{A}^p_{\mathsf{MOD}}}(s) = \delta_{x(uv)^l}^{\mathfrak{A}^p_{\mathsf{MOD}}}(s), \quad \text{for every } x \in \{u, v\}^*. \tag{17}$$

Indeed, by an argument similar to the one in the proof of (*i*) and (*ii*) above, we must have $u' = w_u \natural^n a_1 w'_u$ and $v' = w_v \natural^m a_1 w'_v$ for some $w_u, w_v \in \Sigma^*$, $n, m < \omega$ and $w'_u, w'_v \in \Sigma^*_+$. For every $x \in \{u, v\}^*$, as both $\delta_{xw_u}^{\mathfrak{A}^p_{\mathsf{MOD}}}(q)$ and $\delta_{xw_v}^{\mathfrak{A}^p_{\mathsf{MOD}}}(q)$ are in S, they must be the same state. Using this it is not hard to see that $s = \delta_{u'w_u}^{\mathfrak{A}^p_{\mathsf{MOD}}}(q)$, $u = \natural^n a_1 w'_u w_u$ and $v = \natural^m a_1 w'_v w_v$ are as required in (15)–(17).

As $M(\mathfrak{B}^p_{\mathsf{MOD}})$ is a finite group, the set $\{\delta_{h(x)}^{\mathfrak{B}^p_{\mathsf{MOD}}} \mid x \in \{u, v\}^*\}$ forms a subgroup $\mathfrak{G}$ in it (the subgroup generated by $\delta_{h(u)}^{\mathfrak{B}^p_{\mathsf{MOD}}}$ and $\delta_{h(v)}^{\mathfrak{B}^p_{\mathsf{MOD}}}$). We show that $\mathfrak{G}$ is unsolvable by finding an unsolvable homomorphic image of it. To this end, we let $S' = \{\delta_y^{\mathfrak{A}^p_{\mathsf{MOD}}}(s) \mid y \in \{u, v\}^*\}$. Then (16) implies that $S' \subseteq S$ and

$$\delta_{h(x)}^{\mathfrak{B}^p_{\mathsf{MOD}}}(s') = \delta_x^{\mathfrak{A}^p_{\mathsf{MOD}}}(s') \in S', \quad \text{for all } s' \in S \text{ and } x \in \{u, v\}^*, \tag{18}$$

and so the restriction $\delta_x^{\mathfrak{A}^p_{\mathsf{MOD}}}\restriction_{S'}$ of $\delta_x^{\mathfrak{A}^p_{\mathsf{MOD}}}$ to S' is an $S' \to S'$ function and $\delta_x^{\mathfrak{A}^p_{\mathsf{MOD}}}\restriction_{S'} = \delta_{h(x)}^{\mathfrak{B}^p_{\mathsf{MOD}}}\restriction_{S'}$. As $M(\mathfrak{B}^p_{\mathsf{MOD}})$ is a group of permutations on a set containing S', $\delta_{h(x)}^{\mathfrak{B}^p_{\mathsf{MOD}}}\restriction_{S'}$ is a permutation of S', for any $x \in \{u, v\}^*$. So $\{\delta_{h(x)}^{\mathfrak{B}^p_{\mathsf{MOD}}}\restriction_{S'} \mid x \in \{u, v\}^*\}$ is a homomorphic image of $\mathfrak{G}$ that is unsolvable by the Kaplan–Levy criterion: By (15), (17), and 2 and k being primes, the order of the permutation $\delta_{h(u)}^{\mathfrak{B}^p_{\mathsf{MOD}}}\restriction_{S'}$ is 2, the order of $\delta_{h(v)}^{\mathfrak{B}^p_{\mathsf{MOD}}}\restriction_{S'}$ is k, and the order of $\delta_{h(uv)}^{\mathfrak{B}^p_{\mathsf{MOD}}}\restriction_{S'}$ (which is the same as the order of its inverse) is a > 1 divisor of l, and so coprime to both 2 and k.

As $\mathfrak{G}$ is an unsolvable subgroup of $M(\mathfrak{B}^p_{\mathsf{MOD}})$, it follows from Lemma 2 that $\mathfrak{G} = M(\mathfrak{B}^p_{\mathsf{MOD}})$, and so $\{u, v\}^* \not\subseteq \natural^*$. We claim that $S' = S$ also follows. Indeed, let $x \in \{u, v\}^*$ be such that $\delta_{h(x)}^{\mathfrak{B}^p_{\mathsf{MOD}}} = \delta_a^{\mathfrak{B}^p_{\mathsf{MOD}}}$. As $|S'| \geq 2$ by (15), $s \in \{s_0, \dots, s_{p-1}\}$ must hold, and so $\{s_0, \dots, s_{p-1}\} \subseteq S'$ follows by (18). As there is $y \in \{u, v\}^*$ with $\delta_{h(y)}^{\mathfrak{B}^p_{\mathsf{MOD}}} = \delta_{\natural}^{\mathfrak{B}^p_{\mathsf{MOD}}}$, $s_p \in S'$ also follows by (18). Finally, as $\{u, v\}^* \not\subseteq \natural^*$, there is $x \in \{u, v\}^*$ of the form $\natural^n a_1 w a_2 w'$, for some $n < \omega$, $w \in \Sigma$ and $w' \in \Sigma_+^*$. As $S' = S$, $\delta_x^{\mathfrak{B}^p_{\mathsf{MOD}}}(s_i) \in S$ for every $i \leq p$, and so $w \in \bigcap_{i=0}^p \boldsymbol{L}(\mathfrak{A}_i)$.

Theorem 2 clearly follows from Lemmas 1 and 3.

5 Deciding $\mathcal{L}$-definability of 2NFAs in PSPACE

Using the criterion Theorem 1 (i), Stern [25] showed that deciding whether the language of any given DFA is $\mathsf{FO}(<)$-definable can be done in PSPACE. In this section, we also use the criteria of Theorem 1 to provide PSPACE-algorithms deciding whether the language of any given 2NFA is $\mathcal{L}$-definable, whenever $\mathcal{L} \in \{\mathsf{FO}(<), \mathsf{FO}(<, \equiv), \mathsf{FO}(<, \mathsf{MOD})\}$. Let $\mathfrak{A} = (Q, \Sigma, \delta, Q_0, F)$ be a 2NFA. Following [9], we first construct a(n exponential size) DFA $\mathfrak{A}'$ such that $\boldsymbol{L}(\mathfrak{A}) = \boldsymbol{L}(\mathfrak{A}')$. To this end, for any $w \in \Sigma^+$, we introduce four binary relations $\mathsf{b}_{lr}(w)$, $\mathsf{b}_{rl}(w)$, $\mathsf{b}_{rr}(w)$, and $\mathsf{b}_{ll}(w)$ on Q describing the *left-to-right*, *right-to-left*, *right-to-right*, and *left-to-left behaviour of* $\mathfrak{A}$ *on* w. Namely,

- $(q, q') \in \mathsf{b}_{lr}(w)$ if there is a run of $\mathfrak{A}$ on w from $(q, 0)$ to $(q', |w|)$;
- $(q, q') \in \mathsf{b}_{rr}(w)$ if there is a run of $\mathfrak{A}$ on w from $(q, |w| - 1)$ to $(q', |w|)$;
- $(q, q') \in \mathsf{b}_{rl}(w)$ if, for some $a \in \Sigma$, there is a run on aw from $(q, |aw| - 1)$ to $(q', 0)$ such that no $(q'', 0)$ occurs in it before $(q', 0)$;
- $(q, q') \in \mathsf{b}_{ll}(w)$ if, for some $a \in \Sigma$, there is a run on aw from $(q, 1)$ to $(q', 0)$ such that no $(q'', 0)$ occurs in it before $(q', 0)$.

For $w = \varepsilon$ (the empty word), we define the $\mathsf{b}_{ij}(w)$ as the identity relation on Q. Let $\mathsf{b} = (\mathsf{b}_{lr}, \mathsf{b}_{rl}, \mathsf{b}_{rr}, \mathsf{b}_{ll})$, where the b_{ij} are the behaviours of $\mathfrak{A}$ on some $w \in \Sigma^*$, in which case we can also write $\mathsf{b}(w)$, and let $\mathsf{b}' = \mathsf{b}(w')$, for some $w' \in \Sigma^*$. We define the composition $\mathsf{b} \cdot \mathsf{b}' = \mathsf{b}''$ with components b''_{ij} as follows. Let X and Y be the transitive closure of $\mathsf{b}'_{ll} \circ \mathsf{b}_{rr}$ and $\mathsf{b}_{rr} \circ \mathsf{b}'_{ll}$, respectively. Then we set:

$$\mathsf{b}''_{lr} = \mathsf{b}_{lr} \circ \mathsf{b}'_{lr} \cup \mathsf{b}_{lr} \circ X \circ \mathsf{b}'_{lr}, \qquad \mathsf{b}''_{rl} = \mathsf{b}'_{rl} \circ \mathsf{b}_{rl} \cup \mathsf{b}'_{rl} \circ Y \circ \mathsf{b}_{rl},$$
$$\mathsf{b}''_{rr} = \mathsf{b}'_{rr} \cup \mathsf{b}'_{rl} \circ Y \circ \mathsf{b}_{rr} \circ \mathsf{b}'_{lr}, \qquad \mathsf{b}''_{ll} = \mathsf{b}_{ll} \cup \mathsf{b}_{lr} \circ X \circ \mathsf{b}'_{ll} \circ \mathsf{b}_{rl}.$$

One can check that $\mathsf{b}'' = \mathsf{b}(ww')$. Define a DFA $\mathfrak{A}' = (Q', \Sigma, \delta', q'_0, F')$ by taking

$$Q' = \{(B_{lr}, B_{rr}) \mid B_{lr} \subseteq Q_0 \times Q,\ B_{rr} \subseteq Q \times Q\}, \quad q'_0 = (\{(q, q) \mid q \in Q_0\}, \emptyset),$$
$$F' = \{(B_{lr}, B_{rr}) \mid (q_0, q) \in B_{lr}, \text{ for some } q_0 \in Q_0 \text{ and } q \in F\},$$
$$\delta'_a((B_{lr}, B_{rr})) = (B'_{lr}, B'_{rr}), \text{ with } B'_{lr} = B_{lr} \circ X(a) \circ \mathsf{b}_{lr}(a),$$
$$B'_{rr} = B_{rr} \cup \mathsf{b}_{rl}(a) \circ Y(a) \circ \mathsf{b}_{lr}(a),$$

where $X(a)$ and $Y(a)$ are the reflexive and transitive closures of $\mathsf{b}_{ll}(a) \circ B_{rr}$ and $B_{rr} \circ \mathsf{b}_{ll}(a)$, respectively. It is not hard to see that, for any $w \in \Sigma^*$,

$$\delta'_w((B_{lr}, B_{rr})) = (B'_{lr}, B'_{rr}) \text{ iff } B'_{lr} = B_{lr} \circ X(w) \circ \mathsf{b}_{lr}(w) \text{ and } B'_{rr} = B_{rr} \cup \mathsf{b}_{rl}(w) \circ Y(w) \circ \mathsf{b}_{lr}(w), \quad (19)$$

where $X(w)$ and $Y(w)$ are the reflexive and transitive closures of $\mathsf{b}_{ll}(w) \circ B_{rr}$ and $B_{rr} \circ \mathsf{b}_{ll}(w)$, respectively. Also, one can show in a way similar to [24,29] that

$$\boldsymbol{L}(\mathfrak{A}) = \boldsymbol{L}(\mathfrak{A}'). \quad (20)$$

Next, we show that, even if the size of $\mathfrak{A}'$ is exponential in $\mathfrak{A}$, we can still use Theorem 1 to decide $\mathcal{L}$-definability of $\boldsymbol{L}(\mathfrak{A})$ in PSPACE:

Theorem 3. *For $\mathcal{L} \in \{\mathsf{FO}(<), \mathsf{FO}(<, \equiv), \mathsf{FO}(<, \mathsf{MOD})\}$, deciding $\mathcal{L}$-definability of $\boldsymbol{L}(\mathfrak{A})$, for any 2NFA $\mathfrak{A}$, is in* PSPACE.

Proof. Let $\mathfrak{A}'$ be the DFA defined above for the given 2NFA $\mathfrak{A}$. By Theorem 1 (*i*) and (20), $\boldsymbol{L}(\mathfrak{A})$ is not $\mathsf{FO}(<)$-definable iff there exist a word $u \in \Sigma^*$, a reachable state $q \in Q'$, and a number $k \leq |Q'|$ such that $q \not\sim \delta'_u(q)$ and $q = \delta'_{u^k}(q)$. We guess the required k in binary, q and a quadruple $\mathsf{b}(u)$ of binary relations on Q. Clearly, they all can be stored in polynomial space in $|\mathfrak{A}|$. To check that our guesses are correct, we first check that $\mathsf{b}(u)$ indeed corresponds to some $u \in \Sigma^*$. This is done by guessing a sequence $\mathsf{b}_0, \dots, \mathsf{b}_n$ of distinct quadruples of binary relations on Q such that $\mathsf{b}_0 = \mathsf{b}(u_0)$ and $\mathsf{b}_{i+1} = \mathsf{b}_i \cdot \mathsf{b}(u_{i+1})$, for some $u_0, \dots, u_n \in \Sigma$. (Any sequence with a subsequence starting after b_i and ending with b_{i+m}, for some i and m such that $\mathsf{b}_i = \mathsf{b}_{i+m}$, is equivalent, in the context of this proof, to the sequence with such a subsequence removed.) Thus, we can assume that $n \leq 2^{O(|Q|)}$, and so n can be guessed in binary and stored in PSPACE. So, the stage of our algorithm checking that $\mathsf{b}(u)$ corresponds to some $u \in \Sigma^*$ makes n iterations and continues to the next stage if $\mathsf{b}_n = \mathsf{b}(u)$ or terminates with an answer no otherwise. Now, using $\mathsf{b}(u)$, we compute $\mathsf{b}(u^k)$ by means of a sequence $\mathsf{b}_0, \dots, \mathsf{b}_k$, where $\mathsf{b}_0 = \mathsf{b}(u)$ and $\mathsf{b}_{i+1} = \mathsf{b}_i \cdot \mathsf{b}(u)$. With $\mathsf{b}(u)$ ($\mathsf{b}(u^k)$), we compute $\delta'_u(q)$ (respectively, $\delta'_{u^k}(q)$) in PSPACE using (19). If $\delta'_{u^k}(q) \neq q$, the algorithm terminates with an answer no. Otherwise, in the final stage of the algorithm, we check that $\delta'_u(q) \not\sim q$. This is done by guessing $v \in \Sigma^*$ such that $\delta'_v(q) = q_1$, $\delta'_v(\delta'_u(q)) = q_2$, and $q_1 \in F'$ iff $q_1 \notin F'$. We guess such a v (if exists) in the form of $\mathsf{b}(v)$ using an algorithm analogous to that for guessing u above.

By Theorem 1 (*ii*) and (20), $\boldsymbol{L}(\mathfrak{A})$ is not $\mathsf{FO}(<, \equiv)$-definable iff there there exist words $u, v \in \Sigma^*$, a reachable state $q \in Q'$, and a number $k \leq |Q'|$ such that $q \not\sim \delta'_u(q)$, $q = \delta'_{u^k}(q)$, $|v| = |u|$, and $\delta'_{u^i}(q) = \delta'_{u^i v}(q)$, for all $i < k$. We outline how to modify the algorithm for $\mathsf{FO}(<)$ above to check $\mathsf{FO}(<, \equiv)$-definability. First, we need to guess and check v in the form of $\mathsf{b}(v)$ in parallel with guessing and checking u in the form of $\mathsf{b}(u)$, making sure that $|v| = |u|$. For that, we guess a sequence of distinct pairs $(\mathsf{b}_0, \mathsf{b}'_0), \dots, (\mathsf{b}_n, \mathsf{b}'_n)$ such that the b_i are as above, $\mathsf{b}'_0 = \mathsf{b}(v_0)$ and $\mathsf{b}'_{i+1} = \mathsf{b}'_i \cdot \mathsf{b}(v_{i+1})$, for some $v_0, \dots, v_n \in \Sigma$. (Any such sequence with a subsequence starting after $(\mathsf{b}_i, \mathsf{b}'_i)$ and ending with $(\mathsf{b}_{i+m}, \mathsf{b}'_{i+m})$, for some

i and m such that $(\mathsf{b}_i, \mathsf{b}'_i) = (\mathsf{b}_{i+m}, \mathsf{b}'_{i+m})$, is equivalent to the sequence with that subsequence removed.) So $n \leq 2^{O(|Q|)}$. For each $i < k$, we can then compute $\delta'_{u^i}(q)$ and $\delta'_{u^i v}(q)$, using (19), and check whether whether they are equal.

Finally, by Theorem 1 (*iii*) and (20), $\boldsymbol{L}(\mathfrak{A})$ is not $\mathsf{FO}(<, \mathsf{MOD})$-definable iff there exist $u, v \in \Sigma^*$, a reachable state $q \in Q'$ and $k, l \leq |Q'|$ such that k is an odd prime, $l > 1$ and coprime to both 2 and k, $q \not\sim \delta'_u(q)$, $q \not\sim \delta'_v(q)$, $q \not\sim \delta'_{uv}(q)$, and $\delta'_x(q) \sim \delta'_{xu^2}(q) \sim \delta'_{xv^k}(q) \sim \delta'_{x(uv)^l}(q)$, for all $x \in \{u, v\}^*$. We start by guessing $u, v \in \Sigma^*$ in the form of $\mathsf{b}(u)$ and $\mathsf{b}(u)$, respectively. Also, we guess k and l in binary and check that k is an odd prime and l is coprime to both 2 and k. By (19), δ'_x is determined by $\mathsf{b}(x)$, for any $x \in \{u, v\}^*$. Thus, we can proceed as follows to verify that u, v, k and l are as required. We perform the following steps, for *each* quadruple b of binary relations on Q. First, we check whether $\mathsf{b} = \mathsf{b}(x)$, for some $x \in \{u, v\}^*$ (we discuss the algorithm for this below). If this is not the case, we construct the *next* quadruple b' and process it as this b. If it is the case, we compute all the states $\delta'_x(q)$, $\delta'_{xu^2}(q)$, $\delta'_{xv^k}(q)$, $\delta'_{x(uv)^l}(q)$, $\delta'_u(q)$, $\delta'_v(q)$, $\delta'_{uv}(q)$, and check their required (non)equivalences w.r.t. $\sim$, using the same method as for checking $\delta'_u(q) \not\sim q$ above. If they do not hold as required, our algorithm terminates with an answer no. Otherwise, we construct the *next* quadruple b' and process it as this b. When all possible quadruples b of binary relations of Q have been processed, the algorithm terminates with an answer yes.

Now, to check that a given quadruple b is equal to $\mathsf{b}(x)$, for some $x \in \{u, v\}^*$, we simply guess a sequence $\mathsf{b}_0, \dots, \mathsf{b}_n$ of quadruples of binary relations on Q such that $\mathsf{b}_0 = \mathsf{b}(w_0)$, $\mathsf{b}_n = \mathsf{b}$ and $\mathsf{b}_{i+1} = \mathsf{b}_i \cdot \mathsf{b}(w_{i+1})$, where $w_i \in \{u, v\}$. It follows from the argument above that it is enough to consider $n \leq 2^{O(|Q|)}$.

6 Further Research

The results obtained in this paper have been used for deciding the rewritability type of ontology-mediated queries (OMQs) given in linear temporal logic *LTL* [22]. As mentioned in the introduction, *LTL* OMQs can be simulated by automata. In the worst case, the automata are of exponential size, and deciding FO-rewritability of some OMQs may become ExpSpace-complete. On the other hand, there are natural and practically important fragments of *LTL* with automata of special forms whose FO-rewritability can be decided in PSpace, Π^p_2 or coNP. However, it remains to be seen whether the corresponding algorithms, even in the simplest case of $\mathsf{FO}(<)$-definability, are efficient enough for applications in temporal OBDA. Note that the problems considered in this paper are also relevant to the optimisation problem for recursive SQL queries.

Acknowledgements. This work was supported by UK EPSRC EP/S032282.

References

1. Artale, A., et al.: Ontology-mediated query answering over temporal data: a survey. In: Schewe, S., Schneider, T., Wijsen, J. (eds.) TIME, vol. 90, pp. 1–37 (2017)

2. Artale, A., et al.: First-order rewritability of ontology-mediated queries in linear temporal logic. Artif. Intell. **299**, 103536 (2021)
3. Barrington, D.: Bounded-width polynomial-size branching programs recognize exactly those languages in NC^1. J. Comput. Syst. Sci. **38**(1), 150–164 (1989)
4. Barrington, D., Compton, K., Straubing, H., Thérien, D.: Regular languages in NC^1. J. Comput. Syst. Sci. **44**(3), 478–499 (1992)
5. Beaudry, M., McKenzie, P., Thérien, D.: The membership problem in aperiodic transformation monoids. J. ACM **39**(3), 599–616 (1992)
6. Bennett, M., Martin, G., O'Bryant, K., Rechnitzer, A.: Explicit bounds for primes in arithmetic progressions. Illinois J. of Math. **62**(1–4), 427–532 (2018)
7. Bernátsky, L.: Regular expression star-freeness is PSPACE-complete. Acta Cybern. **13**(1), 1–21 (1997)
8. Büchi, J.R.: Weak second-order arithmetic and finite automata. Z. Math. Logik und Grundlagen der Math. **6**(1–6), 66–92 (1960)
9. Carton, O., Dartois, L.: Aperiodic two-way transducers and fo-transductions. In: Kreutzer, S. (ed.) CSL 2015, vol. 41 pp. 160–174 (2015)
10. Cho, S., Huynh, D.: Finite-automaton aperiodicity is PSPACE-complete. Theor. Comp. Sci. **88**(1), 99–116 (1991)
11. Compton, K., Laflamme, C.: An algebra and a logic for NC^1. Inf. Comput. **87**(1/2), 240–262 (1990)
12. Elgot, C.: Decision problems of finite automata design and related arithmetics. Trans. Am. Math. Soc. **98**, 21–51 (1961)
13. Fleischer, L., Kufleitner, M.: The intersection problem for finite monoids. In: Niedermeier, R., Vallée, B. (eds.) STACS 2018, vol. 96, pp. 1–14 (2018)
14. Holzer, M., König, B.: Regular languages, sizes of syntactic monoids, graph colouring, state complexity results, and how these topics are related to each other. Bull. EATCS **38**, 139–155 (2004)
15. Jukna, S.: Boolean Function Complexity. Advances and Frontiers, vol. 27. Springer, Heidelberg (2012). https://doi.org/10.1007/978-3-642-24508-4
16. Kaplan, G., Levy, D.: Solvability of finite groups via conditions on products of 2-elements and odd p-elements. Bull. Austr. Math. Soc. **82**(2), 265–273 (2010)
17. King, O.: The subgroup structure of finite classical groups in terms of geometric configurations. In: Webb, B. (ed.) Surveys in Combinatorics, Society Lecture Note Series, vol. 327, pp. 29–56. Cambridge University Press (2005)
18. Kozen, D.: Lower bounds for natural proof systems. In: Proceedings of FOCS 1977, pp. 254–266. IEEE Computer Society Press (1977)
19. McNaughton, R., Papert, S.: Counter-free automata. The MIT Press, Cambridge (1971)
20. Poggi, A., Lembo, D., Calvanese, D., De Giacomo, G., Lenzerini, M., Rosati, R.: Linking data to ontologies. J. Data Seman. **10**, 133–173 (2008)
21. Rotman, J.: An introduction to the theory of groups. Springer-Verlag, New York (1999). https://doi.org/10.1007/978-1-4612-4176-8
22. Ryzhikov, V., Savateev, Y., Zakharyaschev, M.: Deciding FO-rewritability of ontology-mediated queries in linear temporal logic. In: Combi, C., Eder, J., Reynolds, M. (eds.) TIME 2021, LIPIcs, pp. 6:1–7:15 (2021)
23. Schützenberger, M.: On finite monoids having only trivial subgroups. Inf. Control **8**(2), 190–194 (1965)
24. Shepherdson, J.C.: The reduction of two-way automata to one-way automata. IBM J. Res. Dev. **3**(2), 198–200 (1959)
25. Stern, J.: Complexity of some problems from the theory of automata. Inf. Control **66**(3), 163–176 (1985)

26. Straubing, H.: Finite Automata, Formal Logic, and Circuit Complexity. Birkhauser Verlag, Basel (1994)
27. Thompson, J.: Nonsolvable finite groups all of whose local subgroups are solvable. Bull. Amer. Math. Soc. **74**(3), 383–437 (1968)
28. Trakhtenbrot, B.: Finite automata and the logic of one-place predicates. Siberian Math. J. **3**, 103–131 (1962)
29. Vardi, M.: A note on the reduction of two-way automata to one-way atuomata. Inf. Process. Lett. **30**(5), 261–264 (1989)
30. Xiao, G., et al.: Ontology-based data access: a survey. In: Lang, J. (ed.) Proceedings of IJCAI 2018, pp. 5511–5519 (2018). ijcai.org

Relational Models for the Lambek Calculus with Intersection and Unit

Stepan L. Kuznetsov(✉)

Steklov Mathematical Institute of RAS, Moscow, Russia
sk@mi-ras.ru

Abstract. We consider the Lambek calculus extended with intersection (meet) operation. For its variant which does not allow empty antecedents, Andréka and Mikulás (1994) prove strong completeness w.r.t. relational models (R-models). Without the antecedent non-emptiness restriction, however, only weak completeness w.r.t. R-models (so-called square ones) holds (Mikulás 2015). Our goals are as follows. First, we extend the calculus with the unit constant, introduce a class of non-standard R-models for it, and prove completeness. This gives a simpler proof of Mikulás' result. Second, we prove that strong completeness does not hold. Third, we extend our weak completeness proof to the infinitary setting, to so-called iterative divisions (Kleene star under division).

Keywords: Lambek calculus · Relational semantics · Completeness

1 Introduction

We start with *the Lambek calculus* [10], formulated in a Gentzen-style sequent format. Lambek formulae are built from variables ($p, q, r, \ldots$) using three binary connectives: $\cdot$ (multiplication), $\backslash$ (left division), and $/$ (right division). The set of all formulae is denoted by Fm. Formulae are denoted by capital Latin letters. Capital Greek letters denote sequences of formulae; Λ stands for the empty sequence. Sequents are expressions of the form $\Pi \to B$. (Due to the non-commutative nature of the Lambek calculus, order in Π matters.) Here Π is called the antecedent and B the succedent of the sequent.

The axioms and inference rules of the original Lambek calculus [10], denoted by **L**, are as follows:

$$\frac{}{A \to A}\ Id \qquad \frac{\Pi \to A \quad \Gamma, A, \Delta \to C}{\Gamma, \Pi, \Delta \to C}\ Cut$$

$$\frac{\Pi \to A \quad \Gamma, B, \Delta \to C}{\Gamma, \Pi, A \backslash B, \Delta \to C}\ \backslash L \qquad \frac{A, \Pi \to B}{\Pi \to A \backslash B}\ \backslash R \qquad \frac{\Gamma, A, B, \Delta \to C}{\Gamma, A \cdot B, \Delta \to C}\ \cdot L$$

$$\frac{\Pi \to A \quad \Gamma, B, \Delta \to C}{\Gamma, B / A, \Pi, \Delta \to C}\ / L \qquad \frac{\Pi, A \to B}{\Pi \to B / A}\ / R \qquad \frac{\Pi \to A \quad \Delta \to B}{\Pi, \Delta \to A \cdot B}\ \cdot R$$

U. Fahrenberg et al. (Eds.): RAMiCS 2021, LNCS 13027, pp. 258–274, 2021.
https://doi.org/10.1007/978-3-030-88701-8_16

A distinctive feature of **L** is the $\Pi \neq \Lambda$ condition on rules $\backslash R$ and $/ R$, the so-called *Lambek's non-emptiness restriction.* This condition ensures that antecedents of all sequents in **L** derivations are non-empty ($\backslash R$ and $/ R$ are the only two rules which could possibly produce an empty antecedent).

Lambek's restriction is motivated by linguistic applications of the Lambek calculus [15, § 2.5]. From the logical point of view, however, it is also natural to consider a variant of **L** without this restriction [11]: in $\backslash R$ and $/ R$ now Π can be empty. This variant is called *the Lambek calculus allowing empty antecedents* and is usually denoted by $\mathbf{L}^*$. Throughout this paper, however, we use alternative notation, $\mathbf{L}^\Lambda$, in order to avoid conflict with Kleene star in Sect. 5.

It is important to keep in mind that **L** is not a conservative fragment of $\mathbf{L}^\Lambda$. Even if the sequent has a non-empty antecedent, empty antecedents could be necessary inside its derivation. An example is $(p \backslash p) \backslash q \to q$, which is derivable in $\mathbf{L}^\Lambda$, but not in **L**. Therefore, there is no easy way of translating results between **L** and $\mathbf{L}^\Lambda$, and certain properties of these systems differ, as we shall see below.

In this paper, we focus on *relational* semantics of the Lambek calculus.

Definition 1. *A relational model (R-model) is a triple* $\mathcal{M} = (W, U, v)$, *where* W *is a non-empty set,* $U \subseteq W \times W$ *is a transitive relation on* W *called the universal one, and* $v\colon \mathrm{Fm} \to \mathcal{P}(U)$ *is a valuation function mapping formulae to subrelations of* U. *The valuation function should obey the following conditions:*

$$v(A \cdot B) = v(A) \circ v(B) = \{(x, z) \mid \exists y \in W\ (x, y) \in v(A) \text{ and } (y, z) \in v(B)\};$$
$$v(A \backslash B) = v(A) \backslash_U v(B) = \{(y, z) \in U \mid \forall x \in W\ (x, y) \in v(A) \Rightarrow (x, z) \in v(B)\};$$
$$v(B / A) = v(B) /_U v(A) = \{(x, y) \in U \mid \forall z \in W\ (y, z) \in v(A) \Rightarrow (x, z) \in v(B)\}.$$

Definition 2. *An R-model* $\mathcal{M} = (W, U, v)$ *is a square one, if* $U = W \times W$.

Arbitrary R-models and square R-models form natural classes of models for **L** and $\mathbf{L}^\Lambda$ respectively. Let us define the truth condition of sequents in R-models.

Definition 3. *A sequent of the form* $A_1, \ldots, A_n \to B$ *is true in model* $\mathcal{M} = (W, U, v)$, *if* $v(A_1) \circ \ldots \circ v(A_n) \subseteq v(B)$. *For sequents with empty antecedents, truth is defined only in square R-models:* $\Lambda \to B$ *is true in* $\mathcal{M} = (W, W \times W, v)$, *if* $\delta = \{(x, x) \mid x \in W\} \subseteq v(B)$.

Let us also recall the general notion of *strong* soundness and completeness of a logic $\mathcal{L}$ (formulated as a sequent calculus) w.r.t. a class of models $\mathcal{K}$.

Definition 4. *Let* $\Pi \to B$ *and* $\mathcal{H}$ *be, respectively, a sequent and a set of sequents in the language of* $\mathcal{L}$. *The sequent* $\Pi \to B$ *semantically follows from* $\mathcal{H}$ *on the class of models* $\mathcal{K}$, *if for any model from* $\mathcal{K}$ *in which all sequents from* $\mathcal{H}$ *are true the sequent* $\Pi \to B$ *is also true. This is denoted by* $\mathcal{H} \vDash_{\mathcal{K}} \Pi \to B$.

Definition 5. *In the notations of the previous definition,* $\Pi \to B$ *syntactically follows from* $\mathcal{H}$ *in the logic* $\mathcal{L}$, *if* $\Pi \to B$ *is derivable in the calculus for* $\mathcal{L}$ *extended with sequents from* $\mathcal{H}$ *as extra axioms. This is denoted by* $\mathcal{H} \vdash_{\mathcal{L}} \Pi \to B$.

Definition 6. *The logic $\mathcal{L}$ is strongly sound w.r.t. the class of models $\mathcal{K}$, if $\mathcal{H} \vdash_{\mathcal{L}} \Pi \to B$ entails $\mathcal{H} \models_{\mathcal{K}} \Pi \to B$ for any $\Pi \to B$ and $\mathcal{H}$.*

Definition 7. *The logic $\mathcal{L}$ is strongly complete w.r.t. the class of models $\mathcal{K}$, if $\mathcal{H} \models_{\mathcal{K}} \Pi \to B$ entails $\mathcal{H} \vdash_{\mathcal{L}} \Pi \to B$ for any $\Pi \to B$ and $\mathcal{H}$.*

Notice that for substructural systems, like the Lambek calculus, strong soundness and completeness are significantly different from their more usual weak counterparts (that derivability of a sequent without any extra axioms yields its truth in all models from the given class, and the other way round). This is due to the fact that there is no deduction theorem available in these logics, and therefore formulae from $\mathcal{H}$ (even if $\mathcal{H}$ is finite) cannot be internalised into $\Pi \to B$.

One can easily check that the calculi $\mathbf{L}$ and $\mathbf{L}^{\Lambda}$ are strongly sound w.r.t. the corresponding classes of R-models: namely, all R-models for $\mathbf{L}$ and square R-models for $\mathbf{L}^{\Lambda}$. Strong completeness is non-trivial, and it was proved, for both calculi, by Andréka and Mikulás [1]:

Theorem 1 (Andréka, Mikulás 1994). *The calculus $\mathbf{L}$ is strongly complete w.r.t. the class of all R-models.*

Theorem 2 (Andréka, Mikulás 1994). *The calculus $\mathbf{L}^{\Lambda}$ is strongly complete w.r.t. the class of square R-models.*

The arguments used in [1] for proving Theorem 1 and Theorem 2, being similar, are yet not completely identical. The essential difference of the situations with and without Lambek's restriction gets revealed when one adds one more operation: intersection, or meet.

Remark 1. Adding the dual operation, join (union), immediately yields incompleteness [1,7], even in the weak sense, so we do not consider it. Indeed, in the presence of both meet and join, we get the distributivity law, which is not derivable in substructural logics like the Lambek calculus [16]. Moreover, unlike meet, with join alone, using Lambek divisions, one can formulate non-trivial corollaries of distributivity [7].

Intersection is axiomatized by the following rules:

$$\frac{\Gamma, A, \Delta \to C}{\Gamma, A \wedge B, \Delta \to C}\ \wedge L_1 \qquad \frac{\Gamma, B, \Delta \to C}{\Gamma, A \wedge B, \Delta \to C}\ \wedge L_2 \qquad \frac{\Pi \to A \quad \Pi \to B}{\Pi \to A \wedge B}\ \wedge R$$

In R-models, it is interpreted set-theoretically:

$$v(A \wedge B) = v(A) \cap v(B) = \{(x,y) \mid (x,y) \in v(A) \text{ and } (x,y) \in v(B)\}.$$

The corresponding calculi will be denoted by $\mathbf{L}\wedge$ and $\mathbf{L}^{\Lambda}\wedge$, depending on whether Lambek's restriction is imposed. One can easily check that such interpretation yields strong soundness for both systems.

As for completeness, in the presence of meet Lambek's restriction makes a significant difference. For $\mathbf{L}\wedge$, the argument of Andréka and Mikulás (Theorem 1) also works, as shown in the same article [1]:

Theorem 3 (Andréka, Mikulás 1994). *The calculus* $\mathbf{L}\wedge$ *is strongly complete w.r.t. the class of all R-models.*

For $\mathbf{L}^{\Lambda}\wedge$, in contrast, the reasoning of Andréka and Mikulás (Theorem 2) could not be easily extended. Later on, however, Mikulás [13,14] managed to modify the proof of Theorem 2 for $\mathbf{L}^{\Lambda}\wedge$—but this modification establishes only *weak* completeness (without hypotheses):

Theorem 4 (Mikulás 2015).[1] *If a sequent in the language of* $\cdot, \backslash, /, \wedge$ *is true in all square R-models, then it is derivable in* $\mathbf{L}^{\Lambda}\wedge$.

The results of the present paper are as follows.

1. We extend $\mathbf{L}^{\Lambda}\wedge$ with the explicit unit constant $\mathbf{1}$, introduce *non-standard* relational semantics for it (Sect. 2), and prove weak completeness (Sect. 3). Notice that the standard interpretation of the unit, $v(\mathbf{1}) = \delta$, does not give completeness. The reducts of our non-standard models to the language without $\mathbf{1}$, however, are standard R-models. Thus, we obtain a new, simpler proof of Mikulás' Theorem 4.
2. Mikulás [14, Remark 5.3] presents a series of potential counterexamples to strong completeness of $\mathbf{L}^{\Lambda}\wedge$ w.r.t. square R-models, but does not prove that they are indeed counterexamples. We prove (Sect. 4) that already the smallest non-trivial one of these examples indeed establishes failure of strong completeness.
3. We show that our proof of Theorem 4, unlike Mikulás' approaches, can be easily extended to infinite conjunctions. We consider (Sect. 5) a concrete interesting example of such conjunction, namely, so-called iterated divisions (Kleene star in the denominator of a division), and prove weak completeness w.r.t. square R-models.

2 Non-standard Models for Unit

Let us further extend $\mathbf{L}^{\Lambda}\wedge$ with the multiplicative unit constant $\mathbf{1}$, that is, an explicit constant for the neutral element of multiplication. The axiom and rule for $\mathbf{1}$, reflecting its neutrality, are as follows (see [12]), and the resulting calculus is denoted by $\mathbf{L}^{\Lambda}\wedge\mathbf{1}$.

$$\frac{\Gamma, \Delta \to C}{\Gamma, \mathbf{1}, \Delta \to C}\ \mathbf{1}L \qquad \frac{}{\Lambda \to \mathbf{1}}\ \mathbf{1}R$$

Notice that these rules exactly reflect neutrality of $\mathbf{1}$. Indeed, $A \cdot \mathbf{1} \to A$ is derived using $\mathbf{1}L$ (with $\cdot L$) and $A \to A \cdot \mathbf{1}$ is derived using $\mathbf{1}R$ via $\cdot R$. The rules for $\mathbf{1}$ are good sequent calculus rules in the sense of cut elimination, see [12].

[1] Here "Mikulás 2015" refers both to [13] and [14], which feature different proofs of Theorem 4.

This suggests interpreting $\mathbf{1}$ in square R-models as the diagonal relation:

$$v(\mathbf{1}) = \delta = \{(x,x) \mid x \in W\}.$$

We shall call this the standard interpretation of the unit.

Unfortunately, for the standard interpretation of the unit $\mathbf{L}^{\Lambda}\wedge\mathbf{1}$ does not enjoy completeness, even in the weak sense. A notable example of a sequent true in all standard square R-models, but not derivable in $\mathbf{L}^{\Lambda}\wedge\mathbf{1}$, is $\mathbf{1} \wedge F \wedge G \to (\mathbf{1} \wedge F) \cdot (\mathbf{1} \wedge G)$, given by Andréka and Mikulás [2]. For $F = G$, this is the contraction ("doubling") principle for formulae of the form $\mathbf{1} \wedge G$, that is, $\mathbf{1} \wedge G \to (\mathbf{1} \wedge G) \cdot (\mathbf{1} \wedge G)$. In the presence of contraction, even restricted to formulae of this specific form, the situation becomes quite complicated. In the view of the results of Chvalovský and Horčík [5] and Kanovich et al. [7] for closely related systems including such contraction principles, we conjecture that the complete system for standard square R-models (that is, the set of sequents true in all models of this class) is undecidable. (More precisely, it is probably at least Σ_1^0-hard, but maybe even higher in the complexity hierarchy.)

Another, independent counterexample to weak completeness w.r.t. standard models is $\mathbf{1} / (F / F) \to (\mathbf{1} / (F / F)) \cdot (\mathbf{1} / (F / F))$, given by Buszkowski [3]. This example uses division instead of intersection, and again it is a form of contraction. Thus, constructing an axiomatisation for the unit constant which is complete w.r.t. standard models, even if this is possible, is a non-trivial open question.

We overcome this issue by extending the class of models being considered, thus restoring completeness for the original system $\mathbf{L}^{\Lambda}\wedge\mathbf{1}$. The idea is as follows: while δ is the only neutral element for the set of *all* binary relations on W, for a set which includes only some relations (and does not include δ), the neutral element could be a different relation. This leads to the following definition.

Definition 8. *Let $\mathfrak{A} \subseteq \mathcal{P}(W \times W)$ be a family of binary relations over W, closed under $\circ$, $\backslash$, $/$, and $\cap$. Relation $\mathbf{1}_{\mathfrak{A}} \in \mathfrak{A}$ is called the $\mathfrak{A}$-unit, if $\mathbf{1}_{\mathfrak{A}} \circ R = R \circ \mathbf{1}_{\mathfrak{A}} = R$ for any $R \in \mathfrak{A}$.*

A standard algebraic argument shows that the $\mathfrak{A}$-unit, if it exists, is unique. Indeed, for another $\mathfrak{A}$-unit $\mathbf{1}'_{\mathfrak{A}} \in \mathfrak{A}$ we have $\mathbf{1}'_{\mathfrak{A}} = \mathbf{1}'_{\mathfrak{A}} \circ \mathbf{1}_{\mathfrak{A}} = \mathbf{1}_{\mathfrak{A}}$. However, $\mathbf{1}_{\mathfrak{A}}$ is not necessarily the diagonal relation $\delta = \{(x,x) \mid x \in W\}$, as the latter may be outside $\mathfrak{A}$.

Example 1. Let W be a non-empty set and let $W' = W \times \{1,2\}$. For each relation R on W let us define a relation R' as follows: $(x,i)R'(y,j)$, if xRy and $i \le j$. Let $\mathfrak{A}$ be the class of relations of the form R'. Then $\mathbf{1}_{\mathfrak{A}} = \delta' = \{((x,i),(x,j)) \mid i \le j\}$.

Lemma 1. *If $R \in \mathfrak{A}$ and $\mathbf{1}_{\mathfrak{A}}$ is the $\mathfrak{A}$-unit, then $\mathbf{1}_{\mathfrak{A}} \subseteq R$ if and only if $\delta \subseteq R$. In particular, $\delta \subseteq \mathbf{1}_{\mathfrak{A}}$.*

Proof. Let us first show that for any $R \in \mathfrak{A}$ we have $R = R / \mathbf{1}_{\mathfrak{A}}$. One inclusion is easy: $R \circ \mathbf{1}_{\mathfrak{A}} \subseteq R$ yields $R \subseteq R / \mathbf{1}_{\mathfrak{A}}$. For the other inclusion, we first notice that $(R / \mathbf{1}_{\mathfrak{A}}) \circ \mathbf{1}_{\mathfrak{A}} \subseteq R$ (this follows from $R / \mathbf{1}_{\mathfrak{A}} \subseteq R / \mathbf{1}_{\mathfrak{A}}$). Now, since $(R / \mathbf{1}_{\mathfrak{A}}) \in \mathfrak{A}$, we have $R / \mathbf{1}_{\mathfrak{A}} = (R / \mathbf{1}_{\mathfrak{A}}) \circ \mathbf{1}_{\mathfrak{A}} \subseteq R$. Using $R = R / \mathbf{1}_{\mathfrak{A}}$, we build a chain of equivalences: $\mathbf{1}_{\mathfrak{A}} \subseteq R \iff \delta \circ \mathbf{1}_{\mathfrak{A}} \subseteq R \iff \delta \subseteq R / \mathbf{1}_{\mathfrak{A}} \iff \delta \subseteq R$. □

Due to this lemma, we may keep the truth definition for sequents with empty antecedents the same. That is, we do not need to replace δ with $\mathbf{1}_{\mathfrak{A}}$.

Definition 9. *A non-standard square R-model with the unit is a structure* $\mathcal{M}^{\mathfrak{A}} = (W, \mathfrak{A}, \mathbf{1}_{\mathfrak{A}}, v)$, *where* W *is a non-empty set;* $\mathfrak{A} \subseteq \mathcal{P}(W \times W)$ *is a family of binary relations on* W, *closed under* $\circ$, $\backslash$, $/$, *and* $\cap$; $\mathbf{1}_{\mathfrak{A}}$ *is the* $\mathfrak{A}$*-unit;* $v \colon \mathrm{Fm} \to \mathfrak{A}$ *is a valuation function mapping formulae to relations from the family* $\mathfrak{A}$. *The valuation function should obey the conditions from Definition 1, with* $U = W \times W$, *and, additionally,* $v(\mathbf{1}) = \mathbf{1}_{\mathfrak{A}}$. *The truth of a sequent in a non-standard square R-model is defined exactly as in Definition 3.*

Proposition 1. *The calculus* $\mathbf{L}^{\Lambda}{\wedge}\mathbf{1}$ *is strongly sound w.r.t. the class of non-standard square R-models with the unit.*

Proof. As usual, we proceed by induction on the derivation. The interesting cases are $\mathbf{1}L$ and $\mathbf{1}R$, as others are copied from the standard strong soundness proof of $\mathbf{L}^{\Lambda}\wedge$ w.r.t. square R-models (without the unit). For $\mathbf{1}R$, we have to show that $\Lambda \to \mathbf{1}$ is true, that is, $\delta \subseteq v(\mathbf{1}) = \mathbf{1}_{\mathfrak{A}}$. This is a particular case of Lemma 1.

For $\mathbf{1}L$, we consider two cases. If both Γ and Δ are empty, then our induction hypothesis gives $\delta \subseteq v(C)$. By Lemma 1, this is equivalent to $\mathbf{1}_{\mathfrak{A}} \subseteq v(C)$ (recall that $v(C) \in \mathfrak{A}$), which is the truth of $\mathbf{1} \to C$. If, say, Δ is non-empty, then let D_1 be the first formula of Δ. By definition of the $\mathfrak{A}$-unit, we have $v(\mathbf{1}) \circ v(D_1) = \mathbf{1}_{\mathfrak{A}} \circ v(D_1) = v(D_1)$. Thus, interpretations of left-hand sides of the premise and the conclusion are identical. The case of non-empty Γ is symmetric. □

As for completeness, we prove only its weak version (Sect. 3). For strong completeness, there is a counterexample (Sect. 4).

3 Weak Completeness

We prove weak completeness of $\mathbf{L}^{\Lambda}{\wedge}\mathbf{1}$ w.r.t. the class of models defined in the previous section:

Theorem 5. *If a sequent (in the language of* $\backslash, /, \cdot, \wedge, \mathbf{1}$*) is true in all non-standard square R-models with the unit, then it is derivable in* $\mathbf{L}^{\Lambda}{\wedge}\mathbf{1}$.

Our proof follows the line of the proof of Theorem 1 (suprisingly, not Theorem 2, see Remark 3 below): we build a labelled graph with specific properties and use it to construct a universal model.

Throughout this section, $\vdash A \to B$ means "$A \to B$ is derivable in $\mathbf{L}^{\Lambda}{\wedge}\mathbf{1}$."

Lemma 2. *There exists a labelled directed graph* $G = (V, E, \ell)$, *where* $V \neq \varnothing$, $E \subseteq V \times V$, *and* $\ell \colon E \to \mathrm{Fm}$, *such that the following holds:*

1. *E is transitive;*
2. *E is reflexive and* $\ell(x, x) = \mathbf{1}$ *for any* $x \in V$;
3. *E is antisymmetric: if* $x \neq y$ *and* $(x, y) \in E$, *then* $(y, x) \notin E$;
4. *if* $(x, y) \in E$ *and* $(y, z) \in E$, *then* $\vdash \ell(x, z) \to \ell(x, y) \cdot \ell(y, z)$;

5. *if* $\vdash \ell(x,z) \to B \cdot C$, *then there exists such* $y \in V$ *that* $(x,y) \in E$, $(y,z) \in E$, $\vdash \ell(x,y) \to B$, *and* $\vdash \ell(y,z) \to C$;
6. *for any* $y \in V$ *and any formula* A *there exists such* $x \in V$ *that for any* $z \in V$ *if* $(y,z) \in E$, *then* $\ell(x,z) = A \cdot \ell(y,z)$;
7. *for any* $y \in V$ *and any formula* A *there exists such* $z \in V$ *that for any* $x \in V$ *if* $(x,y) \in E$, *then* $\ell(x,z) = \ell(x,y) \cdot A$.

Before proving Lemma 2, let us use it to establish Theorem 5.

Proof (of Theorem 5). Using graph G, we construct a *universal* non-standard square R-model $\mathcal{M}_0^{\mathfrak{A}} = (W, \mathfrak{A}, \mathbf{1}_{\mathfrak{A}}, v)$ in the following way:

$$W = V; \qquad v(A) = \{(x,y) \in E \mid \vdash \ell(x,y) \to A\};$$
$$\mathfrak{A} = \{v(A) \mid A \in \mathrm{Fm}\}; \qquad \mathbf{1}_{\mathfrak{A}} = v(\mathbf{1}).$$

Let us show that $\mathcal{M}_0^{\mathfrak{A}}$ is indeed a well-defined model.

Multiplication. If $(x,z) \in v(B \cdot C)$, then $\vdash \ell(x,z) \to B \cdot C$. By property 5 of graph G, there exists such y that $(x,y) \in v(B)$ and $(y,z) \in v(C)$. Therefore, $(x,z) \in v(B) \circ v(C)$. This establishes the inclusion $v(B \cdot C) \subseteq v(B) \circ v(C)$.

For the opposite inclusion, take $(x,y) \in v(B)$ and $(y,z) \in v(C)$. By transitivity, $(x,z) \in E$. By property 4 of G, $\vdash \ell(x,z) \to \ell(x,y) \cdot \ell(y,z)$. We derive $\ell(x,z) \to B \cdot C$ as follows:

$$\dfrac{\ell(x,z) \to \ell(x,y) \cdot \ell(y,z) \qquad \dfrac{\dfrac{\ell(x,y) \to B \quad \ell(y,z) \to C}{\ell(x,y), \ell(y,z) \to B \cdot C}\ {\cdot R}}{\ell(x,y) \cdot \ell(y,z) \to B \cdot C}\ {\cdot L}}{\ell(x,z) \to B \cdot C}\ \mathit{Cut}$$

Therefore, $(x,z) \in v(B \cdot C)$.

Division. Let $(y,z) \in v(A \setminus B)$, that is, $\vdash \ell(y,z) \to A \setminus B$. Take an arbitrary $x \in W$ such that $(x,y) \in v(A)$, that is, $\vdash \ell(x,y) \to A$. Now by transitivity $(x,z) \in E$, and $\ell(x,z) \to B$ is derived using two cuts, with $\ell(x,z) \to \ell(x,y) \cdot \ell(y,z)$ (property 4 of G) and $A \cdot (A \setminus B) \to B$. This establishes the inclusion $v(A \setminus B) \subseteq v(A) \setminus v(B)$.

For the opposite inclusion, take $(y,z) \in v(A) \setminus v(B)$ and apply property 6 to y and A. For the vertex $x \in W$ given by this property, we have $\ell(x,y) = A \cdot \ell(y,y) = A \cdot \mathbf{1}$ and $\ell(x,z) = A \cdot \ell(y,z)$. The first condition gives $(x,y) \in v(A)$ (because $\vdash A \cdot \mathbf{1} \to A$). Hence, $(x,z) \in v(B)$, i.e., $\vdash \ell(x,z) \to B$. Since $\ell(x,z) = A \cdot \ell(y,z)$, we may proceed as follows:

$$\dfrac{\dfrac{\dfrac{A \to A \quad \ell(y,z) \to \ell(y,z)}{A, \ell(y,z) \to A \cdot \ell(y,z)}\ {\cdot R} \qquad A \cdot \ell(y,z) \to B}{A, \ell(y,z) \to B}\ \mathit{Cut}}{\ell(y,z) \to A \setminus B}\ {\setminus R}$$

This establishes $(y,z) \in v(A \setminus B)$. Thus, we get $v(A) \setminus v(B) = v(A \setminus B)$.

The equality $v(B \,/\, A) = v(B) \,/\, v(A)$ is established symmetrically.

Intersection. We have $v(A \wedge B) = \{(x, y) \in E \mid \vdash \ell(x, y) \to A \wedge B\} = \{(x, y) \in E \mid \vdash \ell(x, y) \to A \text{ and } \vdash \ell(x, y) \to B\} = v(A) \cap v(B)$. In the second equality, the $\supseteq$ inclusion is by $\wedge R$, and the $\subseteq$ one is by cut with $A \wedge B \to A$ and $A \wedge B \to B$.

Unit. Here we have $v(\mathbf{1}) = \mathbf{1}_{\mathfrak{A}}$ by definition, and $\mathbf{1}_{\mathfrak{A}}$ is the $\mathfrak{A}$-unit. Indeed, since A is equivalent to $\mathbf{1} \cdot A$ and any relation in $\mathfrak{A}$ is of the form $v(A)$, we have $\mathbf{1}_{\mathfrak{A}} \circ v(A) = v(\mathbf{1}) \circ v(A) = v(\mathbf{1} \cdot A) = v(A)$. Similarly for $v(A) \circ \mathbf{1}_{\mathfrak{A}}$.

Now let us show that $\mathcal{M}_0^{\mathfrak{A}}$ is indeed a universal model, that is, a sequent is true in this model if and only if it is derivable in $\mathbf{L}^{\Lambda}\wedge\mathbf{1}$. The interesting direction is of course the "only if" one (the "if" direction is just weak soundness).

Moreover, we may consider only sequents of the form $\Lambda \to B$, since from $A_1, \ldots, A_n \to B$ one can derive $\Lambda \to A_n \setminus (A_{n-1} \setminus \ldots \setminus (A_1 \setminus B) \ldots)$, and *vice versa,* and by strong soundness these two sequents are true or false in $\mathcal{M}_0^{\mathfrak{A}}$ simultaneously.

Let $\Lambda \to B$ be true in $\mathcal{M}_0^{\mathfrak{A}}$, that is, $\delta \subseteq v(B)$. Take an arbitrary $x \in V$. We have $(x, x) \in v(B)$, that is, $\vdash \ell(x, x) \to B$. On the other hand, $\ell(x, x) = \mathbf{1}$ by property 2 of G. Applying cut with $\Lambda \to \mathbf{1}$ (axiom), we derive the desired sequent $\Lambda \to B$.

Existence of a universal model yields weak completeness: if a sequent is true in all models, then it is true in the universal one, and therefore derivable in the calculus. □

Now we finish our argument by proving Lemma 2. The spirit of this proof is the same as the central lemma of the proof of Theorem 1 by Andréka and Mikulás. However, for the step-by-step construction we use the countable schedule function, as in [8], which is sufficient for enumerating formulae, rather than consider abstract algebras of arbitrary cardinality, as in [1]; see Remark 2 below. The presentation of the proof closely follows the line of [8, Lemma 14]; the figures are adaptations of those by Andréka and Mikulás [1] to the reflexive situation.

Proof (of Lemma 2). We construct an increasing sequence of labelled graphs $G_n = (V_n, E_n, \ell_n)$, where each G_n is an induced subgraph of G_{n+1}. The countable set of vertices $V = \bigcup_{n=0}^{\infty} V_n$ is fixed before the procedure starts. Our aim is the union graph $G = (\bigcup_{n=0}^{\infty} V_n, \bigcup_{n=0}^{\infty} E_n, \bigcup_{n=0}^{\infty} \ell_n)$.

The desired properties 1–4 are *maintained* along the sequence, that is, they will hold for each G_n. In contrast, properties 5–7 are *achieved* only in the limit; each transition from G_n to G_{n+1} is a step towards satisfying one of these properties (in a particular case).

The initial graph G_0 is just a reflexive point with the required unit label on the loop: $G_0 = (\{\star\}, \{(\star, \star)\}, (\star, \star) \mapsto \mathbf{1})$. Properties 1–4 are trivially satisfied.

Each step is a transition of one of three types: for $t = 0, 1, 2$, a transition of type t is a step from G_{3i+t} to G_{3i+t+1}. In order to ensure that all necessary transitions are eventually performed, let us define two bijective schedule functions:

$$\sigma \colon \mathbb{N} \to (V \times \mathrm{Fm}) \times \mathbb{N}$$

$$\varsigma \colon \mathbb{N} \to (V \times V \times \mathrm{Fm} \times \mathrm{Fm}) \times \mathbb{N}$$

Here σ enumerates pairs of a (possible) vertex and a formula, and the second component (a natural number) ensures that each such pair is "visited" infinitely many times. The second function, ς, does the same for quadruples including two vertices and two formulae. Now let us define our transitions.

Transition of type 0, from G_{3i} to G_{3i+1}. Let $\sigma(i) = ((y, A), k)$. If $y \notin V_{3i}$, we skip: $G_{3i+1} = G_{3i}$. Otherwise we add a new vertex $x \in V - V_{3i}$ (such a vertex always exists, since V is countable and V_{3i} is finite) with a loop edge (x, x), $\ell(x, x) = \mathbf{1}$, and for each $z \in V_{3i}$, such that $(y, z) \in E_{3i}$, an edge (x, z) with $\ell(x, z) = A \cdot \ell(y, z)$. (In particular, we add an edge (x, y) with label $A \cdot \mathbf{1}$.)

Let us show that properties 1–4 keep valid for G_{3i+1}. Indeed, the new vertex x is reflexive, and the loop has the correct label $\mathbf{1}$. Antisymmetricity is also maintained: the new vertex x has no ingoing edges, except the loop.

Transitivity and property 4 are checked as follows. We have to verify that for any $x', y', z' \in V_{3i+1}$ if $(x', y') \in E_{3i+1}$ and $(y', z') \in E_{3i+1}$, then $(x', z') \in E_{3i+1}$ and $\vdash \ell(x', z') \to \ell(x', y') \cdot \ell(y', z')$. The interesting case is when at least one of these vertices is new (that is, not from V_{3i}). This means $x' = x$. If $y' = x$, we trivially get $(x', z') = (y', z') \in E_{3i+1}$, and $\vdash \ell(x', z') \to \ell(x', y') \cdot \ell(y', z')$, since $\ell(x', y') = \ell(x, x) = \mathbf{1}$. Now let y', z' be old vertices (from V_{3i}). Since (x, y') and (x, z') were added, edges (y, y') and (z, z') are in E_{3i}. Let $\ell(y, y') = B$, $\ell(y', z') = C$, and $\ell(y, z') = D$ (the latter edge exists by transitivity of G_{3i}). Then the picture is as follows (new edges are dashed):

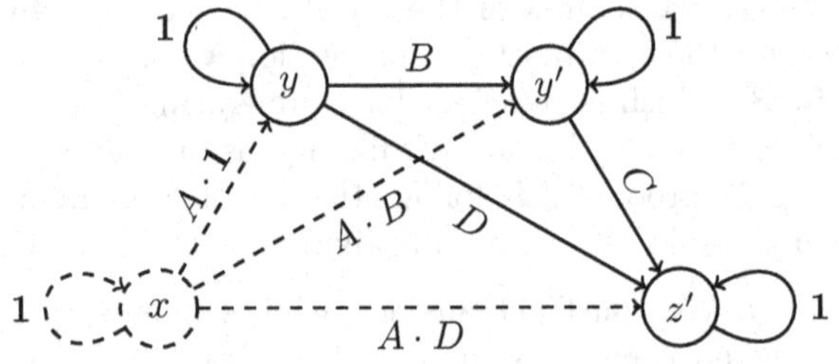

(Notice that some of the vertices y, y', z' could coincide; in this case, the corresponding edges are loops with label $\mathbf{1}$.)

We have indeed added the necessary edge (x, z'), and it remains to check that $\vdash \ell(x, z') \to \ell(x, y') \cdot \ell(y', z')$, that is, $A \cdot D \to (A \cdot B) \cdot C$. By associativity, we may replace $(A \cdot B) \cdot C$ with $A \cdot (B \cdot C)$, and then $A \cdot D \to A \cdot (B \cdot C)$ is derived from $D \to B \cdot C$ by applying $\cdot R$ and $\cdot L$. The sequent $D \to B \cdot C$ is derivable by property 4 of the old graph G_{3i}.

Transition of type 1, from G_{3i+1} to G_{3i+2}, is similar. Let $\sigma(i) = ((y, A), k)$. If $y \notin V_{3i}$, we skip, and otherwise add a new vertex z with its loop and for each $x \in V_{3i}$, if $(x, y) \in E_{3i}$, add an edge (x, z) with $\ell(x, z) = \ell(x, y) \cdot A$. As for type 0, properties 1–4 keep valid.

Transition of type 2, from G_{3i+2} to G_{3i+3}. Let $\varsigma(i) = ((x, z, B, C), k)$. If x or z is not in V_{3i+2} or if $\nvdash \ell(x, z) \to B \cdot C$, we skip. We also skip if $x = z$: in this case, we do not need to add a new vertex to satisfy property 5, see below.

Otherwise, we add a new vertex y, with its loop (y, y), $\ell(y, y) = \mathbf{1}$, and the following edges:

- edge (r, y), with $\ell(r, y) = \ell(r, x) \cdot B$, for each r such that $(r, x) \in E_{3i+2}$;
- edge (y, s), with $\ell(y, s) = C \cdot \ell(z, s)$, for each s such that $(z, s) \in E_{3i+2}$.

(In particular, we add edges (x, y) and (y, z) with labels $\mathbf{1} \cdot B$ and $C \cdot \mathbf{1}$ respectively.) The picture in this situation is as follows:

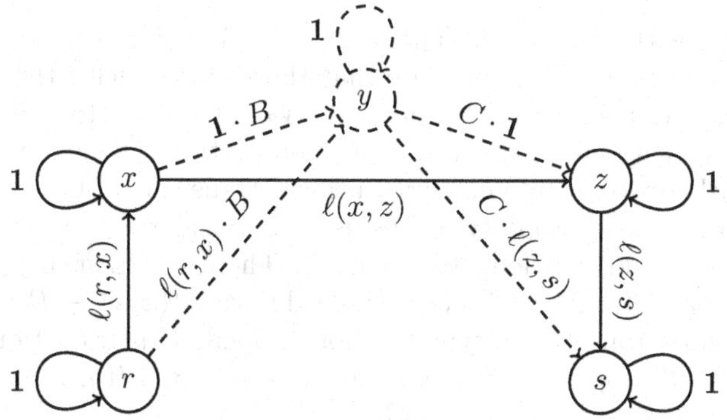

In this picture, it is possible that $r = x$, or $z = s$, or even both. In such a case, the corresponding edge is a loop with label $\mathbf{1}$. However, $x \neq z$ by assumption, and also $r \neq s$ (for any r, s in question). Indeed, if $r = s$, then by transitivity we get $(z, x) \in E_{3i+2}$, which violates antisymmetry of the old graph G_{3i+2}. Also, the new vertex y is a distinct one.

The $r \neq s$ condition yields antisymmetry of the new graph G_{3i+3}. Indeed, a possible violation of antisymmetry should involve the new vertex y, but then the other vertex should be r and s at the same time.

Reflexivity of the new graph, with $\mathbf{1}$ labels on the loops, is by construction.

Let us check transitivity and property 4. Take x', y', z' such that edges (x', y') and (y', z') belong to E_{3i+3}. The interesting case is when $x' \neq y'$ and $y' \neq z'$ (otherwise we just add a unit), and at least one of x', y', z' is the new vertex y. Moreover, by antisymmetry, which we have already proved, we have $x' \neq z'$.

Consider three cases.

Case 1: $x' = y$. Denote $s_1 = y'$ and $s_2 = z'$. Since $(y, s_1) \in E_{3i+3}$ and $s_1 \neq y$, we have $(z, s_1) \in E_{3i+2}$ and $\ell(y, s_1) = C \cdot \ell(z, s_1)$. (Possibly, $s_1 = z$.) For s_2, since it is also not y, we have $(s_1, s_2) \in E_{3i+2}$, and by transitivity of G_{3i+2} we get $(z, s_2) \in E_{3i+2}$. Therefore, $(y, s_2) \in E_{3i+3}$ and $\ell(y, s_2) = C \cdot \ell(z, s_2)$. Now by property 4 of the old graph we have $\vdash \ell(z, s_2) \to \ell(z, s_1) \cdot \ell(s_1, s_2)$, and via $\cdot R$, $\cdot L$, and associativity we obtain $\vdash C \cdot \ell(z, s_2) \to (C \cdot \ell(z, s_1)) \cdot \ell(s_1, s_2)$. This is the necessary sequent $\ell(y, s_2) \to \ell(y, s_1) \cdot \ell(y, s_2)$.

Notice that here antisymmetry is crucial: otherwise, we could have $z' = s_2 = y$ (i.e., s_1 plays both as s and r), in which case $\ell(y, s_2)$ would be $\mathbf{1}$, not $C \cdot \ell(z, s_2)$.

Case 2: $z' = y$. Considered symmetrically.

Case 3: $y' = y$. Denote $r = x'$ and $s = z'$; they are both distinct from y. We have $\ell(r, y) = \ell(r, x) \cdot B$ and $\ell(y, s) = C \cdot \ell(z, s)$. By shortcutting the path r–x–z–s using transitivity, we see that (r, s) is an edge of the old graph and $\vdash \ell(r, s) \to \ell(r, x) \cdot \ell(x, z) \cdot \ell(z, s)$. Now we recall that $\vdash \ell(x, z) \to B \cdot C$ by assumption and by cut and monotonicity conclude that $\vdash \ell(r, s) \to \ell(r, x) \cdot B \cdot C \cdot \ell(z, s)$. This is exactly (up to associativity) what we need: $\vdash \ell(r, s) \to \ell(r, y) \cdot \ell(y, s)$.

Our construction shows that properties 1–4 hold for each G_n. Therefore, they also hold for the limit graph G. Thus, it remains to show that G also enjoys properties 5–7.

Let us start with property 6. The vertex $y \in V$ belongs to V_n for some n. Using the bijectivity of σ, we conclude that there exists such i that $3i \geq n$ and $\sigma(i) = ((y, A), k)$ (for some k). Therefore, $y \in V_{3i}$, and at the transition of type 0 from G_{3i} to G_{3i+1} we added a vertex x, the properties of which are exactly the ones required. Property 7 is symmetric, using a transition of type 1.

Finally, let us prove property 5. Let us first consider the case where $x \neq z$. Again, vertices x and z belong to some V_n. There exists such i that $\varsigma(i) = ((x, z, B, C), k)$ and $3i+2 \geq n$. Since we indeed have $\vdash \ell(x, z) \to B \cdot C$ and $x \neq z$, the corresponding transition of type 2 is not skipped. This transition introduces y with the desired properties. Indeed, $\ell(x, y) = \mathbf{1} \cdot A$ and $\ell(y, z) = B \cdot \mathbf{1}$, which yields $\vdash \ell(x, y) \to B$ and $\vdash \ell(y, z) \to C$.

Now let $x = z$. Then $\ell(x, z) = \ell(x, x) = \mathbf{1}$, and we have $\vdash \mathbf{1} \to B \cdot C$. By cut with $\Lambda \to \mathbf{1}$ (axiom), we get $\vdash \Lambda \to B \cdot C$. Let us eliminate the cut rule in this proof.[2] The lowermost rule in the cut-free proof is nothing but $\cdot R$. Therefore, we get $\vdash \Lambda \to B$ and $\vdash \Lambda \to C$. In its turn, by $\mathbf{1}L$, this yields $\vdash \mathbf{1} \to B$ and $\vdash \mathbf{1} \to C$, or, in other words, $\vdash \ell(x, x) \to B$ and $\vdash \ell(x, x) \to C$. Thus, taking $y = x$ satisfies property 5. Notice that this is the only place in the proof where we cannot allow extra axioms from $\mathcal{H}$ and fail to prove strong completeness.[3] □

Remark 2. Andréka and Mikulás [1] in their proof of Theorem 1 use a more abstract algebraic framework: labels on graph edges are not formulae but elements of a residuated semi-lattice (that is, an algebraic model for **L**, see [6] for details). In other words, strong completeness appears as a corollary of a purely algebraic representation theorem. In our case, however, the representation theorem holds only for the Lindenbaum–Tarski algebra, which consists of equivalence classes of formulae. Indeed, the representation theorem for arbitrary algebras would have yielded strong completeness, which does not hold (see Sect. 4 below). Thus, it does not matter whether to use formulae (as we do) or elements of this algebra as labels.

---

[2] Cut elimination for $\mathbf{L}^{\Lambda}\wedge\mathbf{1}$ is standard. Below, in the proof of Theorem 6, we sketch the cut elimination proof for an extension of $\mathbf{L}^{\Lambda}\wedge\mathbf{1}$.

[3] This failure is actually even not due to the absence of cut elimination in the presence of $\mathcal{H}$. Indeed, one could just add $\mathbf{1} \to b \cdot c$ (b and c are variables) as an axiom, while $\mathbf{1} \to b$ and $\mathbf{1} \to c$ are not derivable. The extra axiom $\mathbf{1} \to b \cdot c$ can be reformulated as a good sequent calculus rule, see Sect. 4 for more details.

Remark 3. If one takes the reduct of a non-standard square R-model with the unit by removing the unit constant, the result is a square R-model in the usual sense. In particular, this holds for our universal model $\mathcal{M}_0^{\mathfrak{A}}$. Thus, we get Mikulás' Theorem 4 as a corollary of our Theorem 5. Labels in our proof are formulae, while Mikulás used filters, which are sets of formulae. The reason was that without an explicit unit constant there are incompatible formulae which should be labels of the same loop (e.g., $p \setminus p$ and $q \setminus q$ for different variables p and q). Using the explicit unit resolves this issue and makes things simpler.

4 Counterexample to Strong Completeness

Unlike the case with Lambek's restriction, for $\mathbf{L}^{\Lambda}\wedge$ strong completeness w.r.t. square R-models does not hold. By conservativity, this also yields failure of strong completeness of $\mathbf{L}^{\Lambda}\wedge\mathbf{1}$ w.r.t. the class of non-standard models defined in Sect. 2. A series of potential counterexamples to strong completeness was given by Mikulás [14]. Here we prove that the first one is indeed such a counterexample.

Theorem 6. *Let a, b, c, d be distinct variables. Then $a \setminus a \to b \cdot c \vDash_{\text{square R-models}} d \to d \cdot b \cdot \big((c \cdot b) \wedge (a \setminus a)\big) \cdot c$, but not $a \setminus a \to b \cdot c \vdash_{\mathbf{L}^{\Lambda}\wedge} d \to d \cdot b \cdot \big((c \cdot b) \wedge (a \setminus a)\big) \cdot c$. Therefore, $\mathbf{L}^{\Lambda}\wedge$ is not strongly complete w.r.t. square R-models.*

Proof. The first part (semantic entailment) is due to Mikulás [14, Remark 5.3]. We reproduce it here in order to keep this paper self-contained. Let us show that $(y, y) \in v(b \cdot ((c \cdot b) \wedge (a \setminus a)) \cdot c)$ for any $y \in W$. Then for any $(x, y) \in v(d)$ we shall have $(x, y) \in v(d \cdot b \cdot ((c \cdot b) \wedge (a \setminus a)) \cdot c)$.

We have $(y, y) \in v(a \setminus a)$, since $\delta \subseteq v(a) \setminus v(a)$ for any $v(a)$. Therefore, since $a \setminus a \to b \cdot c$ is true in $\mathcal{M}$, we get $(y, y) \in v(b) \circ v(c)$. This means that there exists such $z \in W$ that $(y, z) \in v(b)$ and $(z, y) \in v(c)$. In its turn, this gives $(z, z) \in v(c \cdot b)$; we also have $(z, z) \in v(a \setminus a)$, therefore $(z, z) \in v((c \cdot b) \wedge (a \setminus a))$.

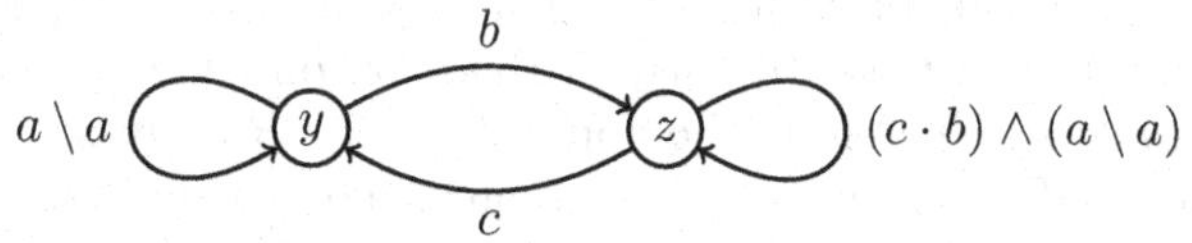

This yields $(y, y) \in v(b) \circ v((c \cdot b) \wedge (a \setminus a)) \circ v(c)$, q.e.d.

Now let us show that $d \to d \cdot b \cdot \big((c \cdot b) \wedge (a \setminus a)\big) \cdot c$ is not derivable from $a \setminus a \to b \cdot c$ in $\mathbf{L}^{\Lambda}\wedge$. We do it in a syntactic way. Suppose the contrary. Then derivability also holds in the larger calculus $\mathbf{L}^{\Lambda}\wedge\mathbf{1}$. In this derivation, let us substitute $\mathbf{1}$ for a and d. This yields derivability of $\mathbf{1} \to \mathbf{1} \cdot b \cdot \big((c \cdot b) \wedge (\mathbf{1} \setminus \mathbf{1})\big) \cdot c$ from $\mathbf{1} \setminus \mathbf{1} \to b \cdot c$.

Next, we notice that $\mathbf{1} \setminus \mathbf{1} \to b \cdot c$ is derivable from $\Lambda \to b \cdot c$ using $\mathbf{1}L$, $\mathbf{1}R$, and $\setminus L$. On the other hand, given in $\mathbf{1} \to \mathbf{1} \cdot b \cdot \big((c \cdot b) \wedge (\mathbf{1} \setminus \mathbf{1})\big) \cdot c$, we apply cut with $\Lambda \to \mathbf{1}$ (axiom) and $\mathbf{1} \cdot b \cdot \big((c \cdot b) \wedge (\mathbf{1} \setminus \mathbf{1})\big) \cdot c \to b \cdot \big((c \cdot b) \wedge \mathbf{1}\big) \cdot c$ (derivable in $\mathbf{L}^{\Lambda}\wedge\mathbf{1}$). This argument gives the following: $\Lambda \to b \cdot \big((c \cdot b) \wedge \mathbf{1}\big) \cdot c$ is derivable in $\mathbf{L}^{\Lambda}\wedge\mathbf{1}$, extended with $\Lambda \to b \cdot c$ as an extra axiom.

Let us introduce an auxiliary calculus $\mathbf{L}^{\Lambda}\wedge\mathbf{1}bc$, which is $\mathbf{L}^{\Lambda}\wedge\mathbf{1}$ extended with the following inference rule:

$$\frac{\Gamma, b, c, \Delta \to F}{\Gamma, \Delta \to F}\ bc$$

(Notice that here b and c are *concrete* variables, not meta-symbols.)

Adding this new rule to $\mathbf{L}^{\Lambda}\wedge\mathbf{1}$ is equivalent to adding $\Lambda \to b \cdot c$ as an axiom. Indeed, $\Lambda \to b \cdot c$ can be derived using the bc rule and the bc rule can be simulated using $\Lambda \to b \cdot c$ and cut.

The new calculus $\mathbf{L}^{\Lambda}\wedge\mathbf{1}bc$, however, enjoys cut elimination. The proof is standard (going back to Lambek's original paper [10]) and proceeds by nested induction: (1) on the complexity of the formula A being cut; (2) on the height of the derivation tree above the cut.

At each step we consider the lowermost rules in the derivations of the premises of cut. The only new situation here is when at least one of these rules is bc; other cases are standard.

For bc on the left, we propagate cut as follows:

$$\dfrac{\dfrac{\Pi', b, c, \Pi'' \to A}{\Pi', \Pi'' \to A}\ bc \qquad \Gamma, A, \Delta \to B}{\Gamma, \Pi', \Pi'', \Delta \to B}\ Cut \quad \rightsquigarrow \quad \dfrac{\dfrac{\Pi', b, c, \Pi'' \to A \qquad \Gamma, A, \Delta \to B}{\Gamma, \Pi', b, c, \Pi'', \Delta \to B}\ Cut}{\Gamma, \Pi', \Pi'', \Delta \to B}\ bc$$

For bc on the right:

$$\dfrac{\Pi \to A \qquad \dfrac{\Gamma', b, c, \Gamma'', A, \Delta \to B}{\Gamma', \Gamma'', A, \Delta \to B}\ bc}{\Gamma', \Gamma'', \Pi, \Delta \to B}\ Cut \quad \rightsquigarrow \quad \dfrac{\dfrac{\Pi \to A \qquad \Gamma', b, c, \Gamma'', A, \Delta \to B}{\Gamma', b, c, \Gamma'', \Pi, \Delta \to B}\ Cut}{\Gamma', \Gamma'', \Pi, \Delta \to B}\ bc$$

and similarly in the case when b, c appear in Δ.

Now we may suppose that $\Lambda \to b \cdot \big((c \cdot b) \wedge \mathbf{1}\big) \cdot c$ has a cut-free proof in $\mathbf{L}^{\Lambda}\wedge\mathbf{1}bc$. Let us track this proof from the goal sequent to the application of $\wedge R$ which introduces $(c \cdot b) \wedge \mathbf{1}$. Below this $\wedge R$ there are only two applications of $\cdot R$ and several ones of bc. Therefore, the application of $\wedge R$ derives $\Pi \to (c \cdot b) \wedge \mathbf{1}$ from $\Pi \to c \cdot b$ and $\Pi \to \mathbf{1}$, where Π is a sequence of b's and c's.

However, $\Pi \to \mathbf{1}$, if Π contains no connectives, is derivable only if Π is empty. Thus, we get derivability of $\Lambda \to c \cdot b$. Let us again track its derivation up to the application of $\cdot R$, which derives $\Phi, \Psi \to c \cdot b$ from $\Phi \to c$ and $\Psi \to b$. Here the sequence Φ, Ψ was obtained by several applications of the bc rule. Therefore, either Φ is empty, or the first element of Φ is b. On the other hand, $\Phi \to c$, where Φ is a sequence of b's and c's, is derivable only if $\Phi = c$. Contradiction. □

5 An Infinite Conjunction: Iterative Division

Now let us return to our weak completeness result (Sect. 3). As noted above (Remark 3), adding an explicit unit constant allowed us to label each edge in our graphs by just one formula, not a set of formulae.[4] Besides making the construction simpler, this has one more interesting consequence. Namely, we can easily extend our proof to the case of *infinite* conjunctions. In contrast, for Mikulás' proofs [13,14] this is not so easy, since the sets which are used as labels are filters, and filters are closed only under *finite* meets. Possibly, this issue could be overcome by modifying the notion of filter and making it closed under (specific kinds of) infinite meets, but our approach is much clearer and simpler.

In order to make our syntax simpler, we shall not consider arbitrary infinite conjunctions, since this would require development of an infinitary formula language. We concentrate on one important particular case, in which the formula language is finitary (while proofs could be infinitary). This particular case of infinitary conjunction is connected to the Kleene star.

In relational models, the Kleene star is the operation of taking the reflexive-transitive closure of a relation. Thus, it can be represented as an infinite union: $R^* = \delta \cup R \cup (R \circ R) \cup (R \circ R \circ R) \cup \ldots$ Adding a union-like connective, however, causes incompleteness issues connected with distributivity. A concrete corollary of the distributivity law, using meet and the Kleene star, which is not derivable without distributivity, is given in [9, Theorem 4.1].

When put under division, however, the infinite union turns into an infinite intersection: $S \,/\, R^* = S \cap (S \,/\, R) \cap ((S \,/\, R) \,/\, R) \cap \ldots$, and similarly for $R^* \setminus S$. Thus, instead of one unrestricted Kleene star, we consider two composite connectives: $A^* \setminus B$ and $B \,/\, A^*$. Following Sedlár [18], who introduced similar connectives in a non-associative setting and with positive iteration (Kleene plus) instead of Kleene star (due to Lambek's restriction), we call these connectives *iterative divisions*. Independently from Sedlár, such connectives were introduced in [8]. The system considered there is associative, but still has Lambek's restriction, so Kleene plus is used instead of Kleene star. In [8] it was proved that the Lambek calculus **L** extended with meet and iterative divisions is strongly complete w.r.t. the class of all R-models. In this paper, we shall prove a weak counterpart of that result for the system without Lambek's restriction.

An infinitary proof system for the Lambek calculus with Kleene star, or infinitary action logic, was introduced by Buszkowski and Palka [4,17]. We present a version of this system for iterative divisions, following [8]:

$$\frac{\Pi_1 \to A \quad \ldots \quad \Pi_n \to A \quad \Gamma, B, \Delta \to C}{\Gamma, \Pi_1, \ldots, \Pi_n, A^* \setminus B, \Delta \to C} \; {*}\backslash L,\ n \geq 0 \qquad \frac{\left(A^n, \Pi \to B\right)_{n=0}^{\infty}}{\Pi \to A^* \setminus B} \; {*}\backslash R$$

$$\frac{\Pi_1 \to A \quad \ldots \quad \Pi_n \to A \quad \Gamma, B, \Delta \to C}{\Gamma, B \,/\, A^*, \Pi_1, \ldots, \Pi_n, \Delta \to C} \; /{*}L,\ n \geq 0 \qquad \frac{\left(\Pi, A^n \to B\right)_{n=0}^{\infty}}{\Pi \to B \,/\, A^*} \; /{*}R$$

The system obtained by adding these rules to $\mathbf{L}^{\Lambda}\wedge\mathbf{1}$ will be denoted by $\mathbf{L}^{\Lambda}\wedge\mathbf{1}\mathrm{ItD}$. A version of this system with Lambek's restriction is undecidable (namely, Π_1^0-complete) [8]. For $\mathbf{L}^{\Lambda}\wedge\mathbf{1}\mathrm{ItD}$, we also conjecture Π_1^0-completeness, thus using infinitary proof machinery (omega-rules or similar) becomes inevitable.

[4] In the setting of Mikulaś [13,14], this formula generates a *principal* filter of all formulae which are "valid" on the edge.

In square R-models, the Kleene star is interpreted as the reflexive-transitive closure operation: $v(A^*) = (v(A))^*$. Thus, the interpretation of iterative divisions is as follows: $v(A^* \setminus B) = (v(A))^* \setminus v(B)$ and $v(B \,/\, A^*) = v(B) \,/ (v(A))^*$.

We extend the notion of non-standard square R-model with the unit (Definition 9) with this interpretation for iterative divisions. A routine check provides strong soundness. Notice that the usage of δ in the interpretation of the Kleene star does not conflict with the non-standard unit $\mathbf{1}_{\mathfrak{A}}$, since they are equivalent in the denominator: $R \,/\, \delta = R = R \,/\, \mathbf{1}_{\mathfrak{A}}$ (see the proof of Lemma 1).

Below we prove weak completeness. The strong one fails by Theorem 4. The reduct to the language without the unit yields "standard" square R-models with iterative divisions, thus we get soundness and weak completeness for them also.

Theorem 7. *If a sequent in the language with iterative divisions is true in all non-standard square R-models with the unit, the it is derivable in* $\mathbf{L}^{\Lambda}\wedge\mathbf{1}$ItD.

Proof. This extension of Theorem 5 is proved in the same way as we do in [8] for the case with Lambek's restriction. First, in Lemma 2 we replace the set of formulae used as labels by the one with iterative divisions. Thus, we get a new labelled graph G using the same step-by-step construction (that is, we do not need to re-prove Lemma 2). Next, the only thing we need to modify in the proof of Theorem 5 is to add one more case, iterative division, in the check that $\mathcal{M}_0^{\mathfrak{A}}$ is a well-defined model. Everything else remains the same.

Thus, we have to prove that $v(A^* \setminus B) = (v(A))^* \setminus v(B)$ and $v(B \,/\, A^*) = v(B) \,/ (v(A))^*$. We shall prove only the former, since the latter is symmetric. Let us first establish the $\subseteq$ inclusion.

Suppose that $(y, z) \in v(A^* \setminus B)$ and take an arbitrary $x \in W$ such that $(x, y) \in (v(A))^*$. Our aim is to show that $(x, z) \in v(B)$. The statement $(x, y) \in (v(A))^*$ means that there exists a number $n \geq 0$ and a sequence $x_0, x_1, \ldots, x_n \in W$ such that $x_0 = x$, $x_n = y$, and $(x_{i-1}, x_i) \in v(A)$ for each $i = 1, \ldots, n$. In particular, if $n = 0$, then we have $x = y$. For $n > 0$, we iterate property 4 of G and get $\vdash \ell(x, y) \to A^n$ and proceed as follows:

$$\dfrac{\ell(x,z) \to \ell(x,y) \cdot \ell(y,z) \qquad \dfrac{\dfrac{\ell(x,y) \to A^n \qquad \dfrac{\ell(y,z) \to A^* \setminus B \qquad \dfrac{\overbrace{A \to A \ \ldots}^{n \text{ times}} \quad B \to B}{A^n, A^* \setminus B \to B}\,{*}\backslash L}{A^n, \ell(y,z) \to B}\,\mathit{Cut}}{\ell(x,y), \ell(y,z) \to B}\,\mathit{Cut}}{\ell(x,y) \cdot \ell(y,z) \to B}\,{\cdot}L}{\ell(x,z) \to B}\,\mathit{Cut}$$

In the case of $n = 0$ we have $\ell(x, z) = \ell(y, z)$, and the sequent $\ell(y, z) \to B$ is derived using cut with $\ell(y, z) \to A^* \setminus B$ and the ${*}\backslash L$ rule with $n = 0$.

Now let us establish the $\supseteq$ inclusion. Suppose that $(y, z) \in (v(A))^* \setminus v(B)$. We need to show that $(y, z) \in v(A^* \setminus B)$, that is, $\vdash \ell(y, z) \to A^* \setminus B$. The latter is derived using the omega-rule ${*}\backslash R$ from the infinite series of sequents $\left(A^n, \ell(y, z) \to B\right)_{n=0}^{\infty}$. For $n = 0$, take $\delta \in (v(A))^*$ and conclude that $(y, z) \in$

$v(B)$, thus, $\vdash \ell(y,z) \to B$. For $n > 0$, we iterate property 6 of G and construct a sequence $x_0, x_1, \ldots, x_n$ such that $x_0 = y$, $(x_{i+1}, x_i) \in E$ and $\ell(x_{i+1}, t) = A \cdot \ell(x_i, t)$ for any t such that $(x_i, t) \in E$. Having $\ell(x_0, z) = \ell(y, z)$ and $\ell(x_0, y) = \ell(y, y) = \mathbf{1}$, by induction we get $\ell(x_n, z) = A^n \cdot \ell(y, z)$ and $\ell(x_n, y) = A^n \cdot \mathbf{1}$.

The latter yields $(x_n, y) \in v(A^n) = v(A) \circ \ldots \circ v(A) \subseteq (v(A))^*$. Thus, since $(y, z) \in (v(A))^* \setminus v(B)$, we have $(x_n, z) \in v(B)$, that is, $\vdash \ell(x_n, z) \to B$. Now the derivation of $A^n, \ell(y,z) \to B$ is as follows:

$$\dfrac{\dfrac{A^n \to A^n \quad \ell(y,z) \to \ell(y,z)}{A^n, \ell(y,z) \to \ell(x_n, z)}\,{\cdot}R \qquad \ell(x_n, z) \to B}{A^n, \ell(y,z) \to B}\,Cut$$

□

As a concluding remark, we note that this paper addresses semantical questions. For this reason, algorithmic complexity issues, namely, undecidability of the theory of standard R-models with the unit and Π_1^0-completeness of the calculus with iterated divisions $\mathbf{L}^{\Lambda}\wedge\mathbf{1ItD}$, are left open for future research. In the absence of strong completeness (see Sect. 4), the interesting question of finite axiomatisability of semantic entailment on square R-models, raised by Mikulás [14], is still open.

Acknowledgement. The author is grateful to Daniel Rogozin for fruitful discussions.

References

1. Andréka, H., Mikulás, S.: Lambek calculus and its relational semantics: completeness and incompleteness. J. Logic Lang. Inf. **3**(1), 1–37 (1994). https://doi.org/10.1007/BF01066355
2. Andréka, H., Mikulás, S.: Axiomatizability of positive algebras of binary relations. Algebra Univers. **66**, 7–34 (2011). https://doi.org/10.1007/s00012-011-0142-3
3. Buszkowski, W.: On the complexity of the equational theory of relational action algebras. In: Schmidt, R.A. (ed.) RelMiCS 2006. LNCS, vol. 4136, pp. 106–119. Springer, Heidelberg (2006). https://doi.org/10.1007/11828563_7
4. Buszkowski, W., Palka, E.: Infinitary action logic: complexity, models and grammars. Stud. Logica **89**(1), 1–18 (2008). https://doi.org/10.1007/s11225-008-9116-7
5. Chvalovský, K., Horčík, R.: Full Lambek calculus with contraction is undecidable. J. Symbolic Logic **81**(2), 524–540 (2016). https://doi.org/10.1017/jsl.2015.18
6. Galatos, N., Jipsen, P., Kowalski, T., Ono, H.: Residuated Lattices: An Algebraic Glimpse at Substructural Logics, Studies in Logic and the Foundations of Mathematics. Elsevier, Amsterdam (2007)
7. Kanovich, M., Kuznetsov, S., Scedrov, A.: Language models for some extensions of the Lambek calculus. Inf. Comput. 104760 (2021). https://doi.org/10.1016/j.ic.2021.104760
8. Kuznetsov, S.L., Ryzhkova, N.S.: A restricted fragment of the Lambek calculus with iteration and intersection operations. Algebra Logic **59**(2), 190–241 (2020). https://doi.org/10.1007/s10469-020-09586-9

9. Kuznetsov, S.: *-Continuity vs. induction: divide and conquer. In: AiML 2018. Advances in Modal Logic, vol. 12, pp. 493–510. College Publications, London (2018)
10. Lambek, J.: The mathematics of sentence structure. Am. Math. Mon. **65**, 154–170 (1958). https://doi.org/10.1080/00029890.1958.11989160
11. Lambek, J.: On the calculus of syntactic types. In: Jakobson, R. (ed.) Structure of Language and Its Mathematical Aspects, pp. 166–178. AMS (1961)
12. Lambek, J.: Deductive systems and categories II. Standard constructions and closed categories. In: Hilton, P.J. (ed.) Category Theory, Homology Theory and their Applications I. LNM, vol. 86, pp. 76–122. Springer, Heidelberg (1969). https://doi.org/10.1007/BFb0079385
13. Mikulás, S.: The equational theories of representable residuated semigroups. Synthese **192**, 2151–2158 (2015). https://doi.org/10.1007/s11229-014-0513-3
14. Mikulás, S.: Lower semilattice-ordered residuated semigroups and substructural logics. Stud. Logica **103**, 453–478 (2015). https://doi.org/10.1007/s11225-014-9574-z
15. Moot, R., Retoré, C.: The Logic of Categorial Grammars. LNCS, vol. 6850. Springer, Heidelberg (2012). https://doi.org/10.1007/978-3-642-31555-8
16. Ono, H., Komori, Y.: Logics without contraction rule. J. Symbolic Logic **50**(1), 169–201 (1985). https://doi.org/10.2307/2273798
17. Palka, E.: An infinitary sequent system for the equational theory of *-continuous action lattices. Fundam. Inform. **78**(2), 295–309 (2007)
18. Sedlár, I.: Iterative division in the distributive full non-associative Lambek calculus. In: Soares Barbosa, L., Baltag, A. (eds.) DALí 2019. LNCS, vol. 12005, pp. 141–154. Springer, Cham (2020). https://doi.org/10.1007/978-3-030-38808-9_9

Free Modal Riesz Spaces are Archimedean: A Syntactic Proof

Christophe Lucas[1(✉)] and Matteo Mio[2]

[1] ENS–Lyon, Lyon, France
christophe.lucas@ens-lyon.fr
[2] CNRS and ENS–Lyon, Lyon, France
matteo.mio@ens-lyon.fr

Abstract. We prove, using syntactical proof–theoretic methods, that free modal Riesz spaces are Archimedean. Modal Riesz spaces are Riesz spaces (real vector lattices) endowed with a positive linear 1–decreasing operator, and have found application in the development of probabilistic temporal logics in the field of formal verification. All our results have been formalised using the Coq proof assistant.

1 Introduction

Riesz spaces, also known as real vector lattices, are real vector spaces equipped with a lattice order ($\leq$) such that the vector space operations of addition and scalar multiplication are compatible with the order in the following sense: (1) if $x \leq y$ then $x + z \leq y + z$ and (2) if $x \leq y$ then $rx \leq ry$, for all $r \in \mathbb{R}_{\geq 0}$.

The simplest example of Riesz space is the linearly ordered vector space of real numbers $(\mathbb{R}, \leq)$ itself. More generally, for a given set X, the space of all functions $\mathbb{R}^X$ with operations and order defined pointwise is a Riesz space. If X carries some additional structure, such as a topology or a σ–algebra, then the spaces of continuous and measurable functions both constitute Riesz subspaces of $\mathbb{R}^X$. For this reason, the study of Riesz spaces originated at the intersection of functional analysis, algebra and measure theory and was pioneered in the 1930's by F. Riesz, G. Birkhoff, L. Kantorovich and H. Freudenthal among others. Today, the study of Riesz spaces constitutes a well–established field of research. We refer to [LZ71, JR77] as standard references.

An important class of Riesz spaces is given by Archimedean Riesz spaces. A Riesz space $(A, \leq)$ is Archimedean if, for any given pair of elements $a, b \in A$,

$$\big(\forall n \in \mathbb{N}.\ na \leq b\big) \Longrightarrow a \leq 0.$$

This work has been supported by the European Research Council (ERC) under the European Union's Horizon 2020 programme (CoVeCe, grant agreement No 678157), by the LABEX MILYON (ANR-10-LABX-0070) of Université de Lyon, within the program "Investissements d'Avenir" (ANR-11-IDEX- 0007) and by the French project ANR-20-CE48-0005 QuaReMe.

U. Fahrenberg et al. (Eds.): RAMiCS 2021, LNCS 13027, pp. 275–291, 2021.
https://doi.org/10.1007/978-3-030-88701-8_17

All the examples of Riesz spaces given above, given by collections of real valued functions, are Archimedean. For this reason the Archimedean property is of key importance in the theory of Riesz spaces.

It is well known that free Riesz spaces (i.e., free objects in the category of Riesz spaces and their homomorphisms) are Archimedean.

Modal Riesz spaces. In a series of recent works [MS17,MFM17,Mio18,FMM20] concerning the study and design of temporal logics for formal verification of probabilistic programs, the notion of *modal Riesz space* has been introduced as the algebraic semantics of the Riesz modal logic for Markov processes [FMM20].

A modal Riesz space (see Sect. 2.2) is a structure $(A, \leq, 1, \Diamond)$ where $(A, \leq)$ is a Riesz space, $1 \in A$ is a positive element $(1 \geq 0)$ and $\Diamond : A \to A$ is a unary operation which satisfies three axioms (see Fig. 2): linearity $(\Diamond(r_1 x + r_2 y) = r_1\Diamond(x)+r_2\Diamond(y))$, positivity (if $x \geq 0$ then $\Diamond(x) \geq 0$) and 1–decreasing $(\Diamond(1) \leq 1)$.

Examples of modal Riesz spaces are given in Sect. 2.2 and more can be found in [FMM20]. The class of modal Riesz spaces, being defined by a set of equations, constitutes a variety and thus free objects exist. In [FMM20, §6.3] the authors left open the following problem regarding modal Riesz spaces: is the free modal Riesz space on the empty set of generators[1] Archimedean? The main contribution of this paper is to give a general answer, covering any possible sets of generators, to this question.

Theorem 1. *Free modal Riesz spaces are Archimedean.*

Our Syntactic Proof. An interesting aspect of our proof is that it is syntactic and based on the proof–theoretic machinery of the hypersequent calculus **HMR** for modal Riesz spaces developed in [LM19,LM20]. One of the novel results, obtained in [LM20, Thm 4.13] using the **HMR** machinery, is the decidability of the equational theory of modal Riesz spaces. This work further illustrates, by proving Theorem 1, the general usefulness of the proof theory. We first reformulate the Archimedean property in terms of derivability in **HMR** and then prove it using proof–theoretic techniques based on the results from [LM20] (like, e.g., a form of cut–elimination). Our main technical result (Theorem 2) establishes that derivability in **HMR** is continuous, in an appropriate sense.

After a preliminary Sect. 2 consisting of technical background on (modal) Riesz spaces, and Sect. 3 summarising the main notions and results regarding the hypersequent calculus **HMR** from [LM20], our proof of Theorem 1 is presented in Sect. 4. To better present the argument, we first prove, using the sequence of steps outlined above, the known fact that free (non–modal) Riesz spaces are Archimedean. To this end, rather than **HMR**, we use its subsystem **HR** (also introduced in [LM20, §3] and presented in Sect. 3.1), which is sound and complete for the theory of (non–modal) Riesz spaces. Once this is done, we prove Theorem 1 tackling in Sect. 4.2 the additional complexity of modal Riesz spaces using the system **HMR**.

[1] The focus in [FMM20] is on the free Riesz space on the empty set of generators because it is the initial object in the category of modal Riesz spaces.

Coq formalisation. All our definitions and proofs have been formalised using the Coq proof assistant [Luc21]. See Sect. 2.3 for a detailed discussion.

2 Technical Background

In this section we present the basic definitions and results about Riesz spaces (Sect. 2.1), modal Riesz spaces (Sect. 2.2) and details about the Coq formalisation of the results of this work (Sect. 2.3).

2.1 Riesz Spaces

We refer to [LZ71, JR77] as standard references on the theory of Riesz spaces. The signature of Riesz spaces is given by $\Sigma_{\mathrm{RS}} = \{+, 0, \{r(\_)\}_{r\in\mathbb{R}}, \sqcup, \sqcap\}$ combining the signature of real vector spaces (addition, neutral element and scalar multiplication by reals) and of lattices (supremum and infimum). Given a set V, we denote with $\mathbf{T}_{\mathrm{RS}}(V)$ the set of Σ_{RS}–terms built from the set of atoms V. We use the letters ϕ and ψ to range over terms.

The class of Riesz spaces is the class of Σ_{RS}–algebras satisfying the axioms of Fig. 1, each of which can be expressed as universally quantified equations.

1. Axioms of real vector spaces:
   - Abelian groups: $x + (y + z) = (x + y) + z$, $x + y = y + x$, $x + 0 = x$, $x - x = 0$,
   - Axioms of scalar multiplication: $r_1(r_2x) = (r_1 \cdot r_2)x$, $1x = x$, $r(x + y) = (rx) + (ry)$, $(r_1 + r_2)x = (r_1x) + (r_2x)$,
2. Lattice axioms: (associativity) $x \sqcup (y \sqcup z) = (x \sqcup y) \sqcup z$, $x \sqcap (y \sqcap z) = (x \sqcap y) \sqcap z$, (commutativity) $z \sqcup y = y \sqcup z$, $z \sqcap y = y \sqcap z$, (absorption) $z \sqcup (z \sqcap y) = z$, $z \sqcap (z \sqcup y) = z$.
3. Compatibility axioms:
   - if $x \leq y$ then $x + z \leq y + z$, expressed equationally as: $(x \sqcap y) + z \leq y + z$,
   - if $x \leq y$ then $rx \leq ry$, for all $r \geq 0$, expressed equationally as: $r(x \sqcap y) \leq ry$, for all $r \geq 0$.

   where $x \leq y$ can be expressed by the equality $x \sqcap y = x$.

Fig. 1. Axioms of Riesz spaces.

Example 1. The Riesz space $(\mathbb{R}, +, 0, \max, \min)$ is a main example. Furthermore, for any set V, the collection of functions $\mathbb{R}^V$ $(f : V \to \mathbb{R})$ is a Riesz space where operations on functions are defined pointwise: e.g., $(f + g)(v) = f(v) + g(v)$. Subalgebras of $\mathbb{R}^V$ are, therefore, also Riesz spaces. For instance, if V is a topological space, the collection of continuous functions on $\mathbb{R}$ is a Riesz space.

Given two terms $\phi, \psi \in \mathbf{T}_{\mathrm{RS}}(V)$, we write $\phi \equiv_{\mathrm{RS}} \psi$ (or just $\phi \equiv \psi$ if clear form the context) if ϕ and ψ can be proved equal, in the usual apparatus of equational logic, from the axioms of Riesz spaces in Fig. 1.

Being definable purely by equations, the class of Riesz spaces is a variety in the sense of universal algebra. Therefore the category of Riesz spaces and their homomorphisms (functions preserving all Σ_{RS} operations) has free objects. Given a set V, we denote with $\mathbf{Free}_{\text{RS}}(V)$ the free Riesz space on the set V. The following definition and proposition are standard.

Definition 1 (Term algebra). *Given a set V, the term algebra $\mathbf{T}_{\text{RS}}(V)_{/\equiv}$ is the Riesz space whose elements are terms generated by V taken modulo the equivalence relation $\equiv_{\text{RS}}$, and operations defined on equivalence classes as: $[\phi]_\equiv + [\psi]_\equiv = [\phi+\psi]_\equiv$, $r[\phi]_\equiv = [r\phi]_\equiv$, $[\phi]_\equiv \sqcup [\psi]_\equiv = [\phi \sqcup \psi]_\equiv$, $[\phi]_\equiv \sqcap [\psi]_\equiv = [\phi \sqcap \psi]_\equiv$.*

Proposition 1. *For any set V, the free Riesz space $\mathbf{Free}_{\text{RS}}(V)$ and term Riesz space $\mathbf{T}_{\text{RS}}(V)_{/\equiv_{\text{RS}}}$ are isomorphic.*

We are now ready to define the Archimedean property of Riesz spaces (see, e.g., [LZ71, §22, Thm 22.2]).

Definition 2 (Archimedean Property). *A Riesz space A is* Archimedean *if, for any $a, b \in A$, it holds that: $\big(\forall n \in \mathbb{N}.\ na \leq b\big) \Longrightarrow a \leq 0$.*

The following result is well–known and follows from a theorem of Baker [Bak68, Thm 2.4] (see also [Ble73, Thm 2.3]) identifying the free Riesz space $\mathbf{Free}_{\text{RS}}(V)$ with a Riesz subspace of $\mathbb{R}^V \to \mathbb{R}$, and the following simple facts (see, e.g., [H.74, §1.15]): (i) the Riesz space $\mathbb{R}^X$ is Archimedean for any set X (so in particular for $X = \mathbb{R}^V$) and (ii) any Riesz subspace of an Archimedean Riesz space is Archimedean.

Proposition 2. *For any set V, the Riesz space $\mathbf{Free}_{\text{RS}}(V)$ is Archimedean.*

Syntactical conventions. We now introduce some convenient syntactical conventions. Rather than working with arbitrary scalar multiplications by $r \in \mathbb{R}$, it is often useful to introduce the derived *negation operator* $-\phi = (-1)\phi$ and restrict scalar multiplication only to strictly positive reals $r \in \mathbb{R}_{>0}$. Clearly this is not a restriction as one can, e.g., just rewrite $(-5)\phi$ to $-(5\phi)$ introducing the negation operator. Every Riesz term ϕ can be rewritten into a $\equiv_{\text{RS}}$-equivalent term ψ in *negation normal form* (NNF), where negation is only applied to variables, using the following valid equalities: $-(\phi \sqcap \psi) = (-\phi) \sqcup (-\psi)$, $-(\phi \sqcup \psi) = (-\phi) \sqcap (-\psi)$, $-(-\phi) = \phi$, $-(\phi+\psi) = (-\phi)+(-\psi)$, $-0 = 0$, $0\phi = 0$. We will use the capital letters A and B to range over Riesz terms in NNF, rather than ϕ and ψ. We write $\overline{A}$ for the NNF–term equivalent to the term $-A$. In particular, $\overline{x} = -x$. Note, therefore, that that terms in NNF can be seen as constructed, without negations, from the variables x and $\overline{x}$, with $x \in V$.

2.2 Modal Riesz Spaces

In this section we introduce the notion of modal Riesz space, a concept which has emerged as relevant in recent works [MS17, MFM17, Mio18, FMM20] concerning the study and design of temporal logics for formal verification of probabilistic programs.

The signature of modal Riesz spaces is given by $\Sigma_{\text{MRS}} = \Sigma_{\text{RS}} \cup \{1, \Diamond\}$ where Σ_{RS} is the signature of Riesz spaces, 1 is a constant symbol and $\Diamond$ is a unary function symbol (we will often omit the parenthesis on $\Diamond$, since it is a unary operator). Given a set V, we denote with $\mathbf{T}_{\text{MRS}}(V)$ the set of Σ_{MRS}–terms build from the set of generators V. Note that $\mathbf{T}_{\text{RS}}(V) \subsetneq \mathbf{T}_{\text{MRS}}(V)$ since $\Sigma_{\text{RS}} \subsetneq \Sigma_{\text{MRS}}$. We use the letters ϕ, ψ also to range over $\mathbf{T}_{\text{MRS}}(V)$.

Definition 3 (Modal Riesz spaces). *The class of modal Riesz spaces is the equationally defined class of Σ_{MRS}–algebras generated by the universally quantified equational axioms of Fig. 1 and the additional axioms of Fig. 2.*

4. Positivity of 1 : $0 \leq 1$.
   expressed equationally: $0 \sqcap 1 = 0$.
5. Modal axioms:
   - Linearity: $\Diamond(r_1 x + r_2 y) = r_1 \Diamond(x) + r_2 \Diamond(y)$.
   - Positivity: if $x \geq 0$ then $\Diamond(x) \geq 0$
     expressed equationally: $\Diamond(0 \sqcup x) \sqcap 0 = 0$.
   - 1-decreasing: $\Diamond(1) \leq 1$.
     expressed equationally: $\Diamond(1) \sqcap 1 = \Diamond(1)$.

Fig. 2. Additional axioms of modal Riesz spaces.

Example 2. A typical example of modal Riesz space is $M = (\mathbb{R}^n, \leq, 1^M, \Diamond^M)$, the n-dimensional vector space $\mathbb{R}^n$ with vectors ordered pointwise where 1^M is the constant 1 vector, and $\Diamond^M$ is a linear map $\mathbb{R}^n \to \mathbb{R}^n$, hence representable as a square matrix (also denoted $\Diamond^M$ with some abuse of notation), such that all entries $r_{i,j}$ are non–strictly positive (due to the positivity axiom) and where all the rows sum up to a value ≤ 1, i.e., for all $1 \leq i \leq n$ it holds that $\sum_{j=1}^{k} r_{i,j} \leq 1$ (due to the 1–decreasing axiom):

$$1^M = \begin{pmatrix} 1 \\ \vdots \\ 1 \end{pmatrix} \qquad \Diamond^M = \begin{pmatrix} r_{1,1} & r_{1,2} & \cdots & r_{1,n} \\ \vdots & \vdots & \ddots & \vdots \\ r_{n,1} & r_{n,2} & \cdots & r_{n,n} \end{pmatrix}$$

The modal Riesz space M can be seen as representing a discrete–time Markov chain, i.e., a probabilistic transition system having $\{1, \ldots, n\}$ states, where the probability of moving from state i to state j at the next step is $r_{i,j}$. The constraint $\sum_{j=1}^{k} r_{i,j} \leq 1$ represents the fact that there can be a nonzero probability of terminating the execution at the state i, thus not moving to any other state. These types of examples are the reason for the relevance of modal Riesz spaces to the axiomatisation of logics for expressing properties of probabilistic transition systems. In fact, the duality theory developed in [FMM20] shows that all Archimedean modal Riesz spaces with strong unit are topological generalisations of the example just presented and can be identified with discrete time Markov processes.

Remark 1. The choice of using the $\Diamond$ symbol for the unary operation of modal Riesz spaces might suggest the existence of a distinct De Morgan dual operator $\Box x = -\Diamond(-x)$. This is not the case since, due to linearity, $\Box x = \Diamond x$, i.e., $\Diamond$ is self dual. While using a different symbol such as $(\circ)$ might have been a better choice, we decided to stick to $\Diamond$ for backwards compatibility with previous works on modal Riesz spaces. Another source of potential ambiguity lies in the "modal" adjective itself. Of course other axioms for $\Diamond$ can be conceived (e.g., $\Diamond(x \sqcup y) = \Diamond(x) \sqcup \Diamond(y)$ instead of our $\Diamond(x+y) = \Diamond(x) + \Diamond(y)$, see, e.g., [DMS18]). Therefore different notions of modal Riesz spaces can be investigated, just like many types of classical modal logic exist (K, S4, S5, etc.). Once again, our choice of terminology is motivated by backwards compatibility with previous works.

We denote with $\equiv_{MRS}$ (or just $\equiv$, if clear from the context) the equivalence relation on $\mathbf{T}_{\mathrm{MRS}}(V)$ which equates modal Riesz terms that are provably equal from the axioms of Definition 3. Being equationally defined, the class of modal Riesz spaces is a variety in the sense of universal algebra. Therefore the category of modal Riesz spaces and their homomorphisms (functions preserving all Σ_{MRS} operations) has free objects. Note that every modal Riesz space is a Riesz space (since it satisfies all axioms of Fig. 1). Furthermore, any Riesz space R can be turned into a modal Riesz space by, e.g., defining $1^R = 0$ and $\Diamond^R = \mathrm{id}$, where $\mathrm{id} : R \to R$ is the identity map. Hence the notion of modal Riesz space is a conservative extension of that of Riesz space.

Given a set V, we denote with $\mathbf{Free}_{\mathrm{MRS}}(V)$ the free modal Riesz space on the set V and with $\mathbf{T}_{\mathrm{MRS}}(V)_{/\equiv_{\mathrm{MRS}}}$ the term algebra.

Proposition 3. *For any set V, the free modal Riesz space $\mathbf{Free}_{\mathrm{MRS}}(V)$ and the term modal Riesz space $\mathbf{T}_{\mathrm{MRS}}(V)_{/\equiv_{\mathrm{MRS}}}$ are isomorphic.*

The main result of this paper is Theorem 1, stating that $\mathbf{Free}_{\mathrm{MRS}}(V)$ is Archimedean. Our proof is presented in Sect. 4. This is a novel result and solves a problem left open [FMM20, §6.3]. We remark that free modal Riesz spaces can be rather complex objects. For instance, $\mathbf{Free}_{\mathrm{MRS}}(\emptyset)$ is not even finitely generated as a Riesz space [FMM20, §6]. For instance, the term $\Diamond^n 1$ can not equivalently be expressed by a Riesz combination of terms with $\Diamond$–depth (the maximum number of nested $\Diamond$ operators) lower than n.

Syntactical conventions. We extend the notion of negation normal form (NNF) from Riesz terms to modal Riesz space terms, taking in consideration the existence of the constant 1 in Σ_{RMS}. A modal Riesz term ϕ in $\mathbf{T}_{\mathrm{MRS}}(V)$ is in *negation normal form* (NNF) if the operator $(-)$ is only applied to atoms in V or the constant 1. Using the equality $(-\Diamond(\phi) = \Diamond(-\phi))$, every term ϕ in $\mathbf{T}_{\mathrm{MRS}}(V)$ is provably equal to a term in NNF. We will use the capital letters A and B to range over modal Riesz terms in NNF, rather than ϕ and ψ. We try to make it always clear if the term belong to $\mathbf{T}_{\mathrm{MRS}}(V)$ or just to $\mathbf{T}_{\mathrm{RS}}(V)$.

2.3 On the Coq Formalisation

All the results of this paper have been formalised using the Coq proof assistant and are publicly available [Luc21]. Throughout the paper, we refer to specific points of the formalisation by highlighting with a grey background either some portions of Coq code (as in the definition of `Axiom IPP` below) or by specifying the name of the lemma and its path as: `Repository [Luc21] : (Lemma) in Path.`

Our formalisation is based on the following mathematical notions and results.

1. The real numbers $\mathbb{R}$, functions on them ($+$, $\times$, etc.) and their basic properties.
2. The (strictly) positive real numbers $\mathbb{R}_{>0}$ with basic functions and properties.
3. The notion of polynomial expression, syntax and semantics.
4. Basic notions about limits of sequences of (tuples of) reals and (sequential) continuity of polynomial expressions.
5. The *infinitary pigeonhole principle*: for every sequence $u \in \mathbb{N}^{\mathbb{N}}$ bounded by some $m \in \mathbb{N}$ (i.e., $u_n < m$ for all n), there is a constant subsequence $(u_{\phi(n)})$ of u, i.e., there is $i \in [0..m[$ such that $u_{\phi(n)} = i$ for all n.
6. The *sequential compactness* of $\mathbb{R}$: if $u \in \mathbb{R}^{\mathbb{N}}$ is a sequence bounded by a lower bound $lb \in \mathbb{R}$ and a upper bound $ub \in \mathbb{R}$, then there is a subsequence $(u_{\phi(n)})$ of (u_n) and a real $l \in \mathbb{R}$ such that $\lim_{n \in \mathbb{N}} u_{\phi(n)} = l$.

Regarding (1), we use the default Coq implementation of real numbers $\mathbb{R}$. For (2), strictly positive reals are implemented as dependent pairs where the first element is the real number and the second element is a proof that this real is strictly positive. Operations and basic properties on $\mathbb{R}_{>0}$ are easily derived from those of $\mathbb{R}$ (standard library). For (3), polynomial expressions over the variables $\alpha_1, \ldots \alpha_n$ are simply defined by the grammar: $R, S := \alpha_i \mid r \in \mathbb{R} \mid R + S \mid RS$ (`Repository [Luc21] : (Poly : Type) in Utilities/polynomials.v.`) and interpreted as polynomial functions $P : \mathbb{R}^k \to \mathbb{R}$ as expected. Regarding (4), we use the `Coquelicot` library [BLM15] which provides definitions and results regarding uniform spaces (like $\mathbb{R}$), continuity, *etc.*. In particular we are able to derive the following statement.

Proposition 4 (Sequential continuity of Polynomial expressions). *Let R be a polynomial expression. For all $j \in [1..k]$, let $(t_{i,j})_{i \in \mathbb{N}} \in \mathbb{R}^{\mathbb{N}}$ and $t_j \in \mathbb{R}$ such that $\lim_{i \to +\infty} t_{i,j} = t_j$. Then $\lim_{i \to +\infty} R(t_{i,1}, ..., t_{i,k}) = R(t_1, ..., t_k)$.*

Proof. `Repository [Luc21] : (Poly_lim) in Utilities/pol_continuous.v.`

Finally, the infinitary pigeonhole principle (5) and sequential compactness (6) are stated as follows and assumed as axioms:

```
Axiom IPP : forall (u : nat → nat) m,
  (forall n, u n < m) →
  {' (phi, i) & prod (subseq_support phi) (*∃ phi,i,(subseq phi)∧ *)
              ((i < m) * (* (i < m) ∧ *)
              (forall n, u (phi n) = i))}. (*(∀ n, u_{phi n} = i) *)

...
```

```
Axiom SequentialCompactness : forall (u : nat → R) lb ub,
(forall n, prod (lb <= u n) (u n <= ub)) →
  {' (phi , l) & prod (subseq_support phi) (*∃  phi,l,(subseq phi)∧  *)
          (is_lim_seq (fun n ⇒ u (phi n)) l)}. (* (lim u_{phi n} = l) *)
```

Repository [Luc21]: `Utilities/R_complements.v`

3 Hypersequent Calculi

In this section we introduce a structural proof system called **HMR** from [LM20] (see also [LM19]) for the theory of modal Riesz spaces. We also discuss a subsystem of **HMR**, called **HR**, also introduced in [LM20, §3] for the theory of Riesz spaces. A proof system is called structural if it manipulates terms (or formulas) having a certain specific structure. For instance, Gentzen's sequent calculus **LK** [Gen34] manipulates and allows for the derivation of *sequents* S of the form $A_1, \dots A_n \vdash B_1, \dots, B_m$ which are interpreted as the Boolean term $(\!|S|\!) = (A_1 \wedge \dots \wedge A_n) \Rightarrow (B_1 \vee \dots \vee B_m)$. We say that **LK** is sound and complete for the theory of Boolean algebras because a sequent S is derivable in **LK** if and only if $(\!|S|\!) = \top$ is a valid identity in the theory of Boolean algebras.

In a similar way, the proof system **HMR** is structural as it manipulates structured terms G, called *hypersequents*, which are interpreted as modal Riesz space terms $(\!|G|\!)$. The system **HMR** is sound and complete with respect to the theory of modal Riesz spaces in the sense that G is derivable in **HMR** if and only if $(\!|G|\!) \geq 0$ (or $(\!|G|\!) \sqcap 0 = 0$, written equationally) is a valid identity in the theory of modal Riesz spaces. Similarly, the subsystem **HR** of **HMR**, only manipulating (non–modal) Riesz space terms, is sound and complete with respect to the theory of (non–modal) Riesz spaces.

The key advantage of working with structural proof systems, compared to non–structural deductive systems such as equational logic, appears from results such as the cut–elimination system (called CAN elimination theorem in the context of **H(M)R**), which greatly simplify the analysis of proofs.

We now procede with the formal definitions. We first present the subsystem **HR** (Sect. 3.1) and then the full system **HMR** (Sect. 3.2).

All definitions and results regarding **HR** and **HMR** have been formalised: Repository [Luc21]: `folders: /hr and /hmr` .

3.1 Hypersequent Calculus HR

In what follows, A and B range over Riesz terms in NNF (see end of Sect. 2.1) built from a set of variables V, ranged over by the letters x, y, z.

Definition 4 (Sequents and Hypersequents). *A* sequent *is a list of pairs* (r, A) *where* $r \in \mathbb{R}_{>0}$ *is a strictly positive real number and* A *is a term in NNF. The sequent* $\Gamma = ((r_1, A_1), ..., (r_n, A_n))$ *is written as:* $\vdash r_1.A_1, ..., r_n.A_n$. *The empty sequent is denoted by* $(\vdash)$. *A* hypersequent *is a nonempty list of sequents. The hypersequent* $G = [\Gamma_1, \Gamma_2, ..., \Gamma_n]$ *is written as:* $\vdash \Gamma_1 \mid \vdash \Gamma_2 \mid ... \mid \vdash \Gamma_n$.

We use the letters Γ, Δ and the letters G, H to range over sequents and hypersequents, respectively. Note that $(\vdash \Gamma)$ can, ambiguosly, denote both the sequent $\vdash \Gamma$ and the hypersequent $[\vdash \Gamma]$ consisting of only one sequent. The context should always determine which of the two interpretations is intended.

The proof system **HR** allows for the derivation of hypersequents using the axioms and deductive rules of Fig. 3. We write $\rhd_{\mathbf{HR}} G$ if the hypersequent G is derivable in the proof system **HR**. Before discussing the meaning of the rules and giving some examples, we define the interpretation of hypersequents and state the soundness and completeness of the proof system.

Definition 5 (Interpretation of Hypersequents). *We interpret sequents and hypersequents by Riesz terms as follows. A sequent $\Gamma = (\vdash r_1.A_1, \dots, r_n.A_n)$ is interpreted by the Riesz term $(\!|\Gamma|\!) = r_1 A_1 + \dots + r_n A_n$. In particular, for the empty sequent, $(\!|\vdash|\!) = 0$. A hypersequent $G = (\vdash \Gamma_1 \mid \cdots \mid \vdash \Gamma_n)$ is interpreted by the Riesz term $(\!|G|\!) = (\!|\vdash \Gamma_1|\!) \sqcup \dots \sqcup (\!|\vdash \Gamma_n|\!)$.*

Example 3. $(\!|\vdash 1.(x \sqcap y) \mid \vdash 1(\overline{x} \sqcup \overline{y}), 2.x|\!) = (1(x \sqcap y)) \sqcup (1(\overline{x} \sqcup \overline{y}) + 2x)$.

Lemma 1 (Soundness and Completeness [LM20, Thm 3.10 and 3.11]**).** *Let G be an hypersequent. Then $\rhd_{\mathbf{HR}} G$ if and only if $(\!|G|\!) \geq 0$ (or $(\!|G|\!) \sqcap 0 = 0$, written equationally) holds universally in all Riesz spaces.*

The meaning of most of the axioms and deductive rules of the hypersequent calculus **HR** is easy to grasp. For instance, the INIT rule allows to derive the empty sequent $(\vdash)$ and indeed this is a sound rule since $(\!|\vdash|\!) = 0$ and $(\!|\vdash|\!) \geq 0$. The contraction rule (C) reflects the idempotency of the lattice operation $\sqcup$, which is used to interpret the $(\mid)$ symbols of hypersequents. Similarly, the $(+)$–rule

Axiom:

$$\frac{}{\vdash}\ \text{INIT}$$

ID and CAN rules:

$$\frac{G \mid \vdash \Gamma}{G \mid \vdash \Gamma, \vec{r}.x, \vec{s}.\overline{x}}\ \text{ID}, \textstyle\sum r_i = \sum s_i \qquad \frac{G \mid \vdash \Gamma, \vec{s}.A, \vec{r}.\overline{A}}{G \mid \vdash \Gamma}\ \text{CAN}, \textstyle\sum r_i = \sum s_i$$

Structural rules:

$$\frac{G}{G \mid \vdash \Gamma}\ \text{W} \qquad \frac{G \mid \vdash \Gamma \mid \vdash \Gamma}{G \mid \vdash \Gamma}\ \text{C} \qquad \frac{G \mid \vdash \Gamma_1, \Gamma_2}{G \mid \vdash \Gamma_1 \mid \vdash \Gamma_2}\ \text{S} \qquad \frac{G \mid \vdash \Gamma_1 \quad G \mid \vdash \Gamma_2}{G \mid \vdash \Gamma_1, \Gamma_2}\ \text{M} \qquad \frac{G \mid \vdash r.\Gamma}{G \mid \vdash \Gamma}\ \text{T}$$

Logical rules:

$$\frac{G \mid \vdash \Gamma}{G \mid \vdash \Gamma, \vec{r}.0}\ 0 \qquad \frac{G \mid \vdash \Gamma, \vec{r}.A, \vec{r}.B}{G \mid \vdash \Gamma, \vec{r}.(A+B)}\ + \qquad \frac{G \mid \ \Gamma \vdash \Gamma, (s\vec{r}).A}{G \mid \ \Gamma \vdash \Gamma, \vec{r}.(sA)}\ \times$$

$$\frac{G \mid \vdash \Gamma, \vec{r}.A \mid \vdash \Gamma, \vec{r}.B}{G \mid \vdash \Gamma, \vec{r}.(A \sqcup B)}\ \sqcup \qquad \frac{G \mid \vdash \Gamma, \vec{r}.A \quad G \mid \vdash \Gamma, \vec{r}.B}{G \mid \vdash \Gamma, \vec{r}.(A \sqcap B)}\ \sqcap$$

Fig. 3. Inference rules of **HR** ([LM20]).

reflects the interpretation of commas in sequents as addition. In the T-rule, any sequent (in the hypersequent) can be multiplied by any positive scalar $r \in \mathbb{R}_{>0}$. This reflects the fact that if $rx \geq 0$ then $x \geq 0$, for every $r \in \mathbb{R}_{>0}$. Several rules adopt a vector notation ($\vec{r}$) to indicate that several terms are active in the rule. For example, the following is a valid instance of the rule ID:

$$\frac{G \mid \vdash \Gamma}{G \mid \vdash \frac{1}{2}.x, \frac{1}{3}.x, \frac{1}{6}.\overline{x}, \frac{2}{3}.\overline{x}} \text{ ID}, \quad \frac{1}{2} + \frac{1}{3} = \frac{1}{6} + \frac{2}{3}$$

because the proviso is satisfied. The ID rule expresses the fact that, since $\frac{1}{2}x + \frac{1}{3}x + \frac{1}{6}(-x) + \frac{2}{3}(-x) = 0$, the terms can be cancelled out. Note that the CAN rule has the same interpretation, but in the reverse direction. Finally, the rules M and S which, in various forms have appeared in the proof–theory literature (not necessarily related to Riesz spaces, see, e.g., [MOG09, Avr96, FR94]) do not have an equally simple interpretation, but are sound [LM20, Thm 3.10].

The key results regarding the hypersequent calculus **HR** from [LM20], which are relevant for this work, are stated as the following lemmas.

Lemma 2 (CAN–elimination [LM20, Thm 3.14]**).** *If a hypersequent G is derivable in **HR** then G has a **HR** derivation that does not use the CAN rule.*

Lemma 3 ([LM20, Thm 3.12])**.** *The rules $\{0, +, \sqcup, \sqcap, \times\}$ are invertible: if the conclusion of an instance of one of these rules is derivable, then all its premises are also derivable.*

An hypersequent G is called *atomic* if all terms A appearing in G are either variables or covariables, i.e., $A = x$ or $A = \overline{x}$, for $x \in V$.

Lemma 4 (λ-property for HR [LM20, Lemma 3.43]**).** *For all atomic hypersequents G formed using the variables and covariables $x_1, \overline{x_1}, \ldots, x_k, \overline{x_k}$ of the form $\vdash \Gamma_1 \mid \ldots \mid \vdash \Gamma_m$, where for each $i \in [1 \ldots m]$,*

$$\Gamma_i = \vec{r}_{i,1}.x_1, \ldots, \vec{r}_{i,k}.x_k, \vec{s}_{i,1}.\overline{x_1}, \ldots, \vec{s}_{i,k}.\overline{x_{i,k}}$$

*then G is derivable in **HR** if and only if there exist numbers $t_1, \ldots, t_m \in [0, 1]$, one for each sequent in G, such that:*

1. *there exists $i \in [1 \ldots m]$ such that $t_i = 1$, and*
2. *for every (co)variable $(x_j, \overline{x_j})$, it holds that: $\sum_{i=1}^{m} t_i(\sum \vec{r}_{i,j} - \sum \vec{s}_{i,j}) = 0$.*

It is important to appreciate how Lemma 4 reduces the derivability problem of atomic hypersequents in **HR** to the existence of a solution in a linear arithmetic problem. The derivability problem of arbitrary hypersequents can also be reduced to linear arithmetic by invoking, in an iterative fashion, Lemma 3 which allows to simplify the term–complexity of the considered hypersequents.

3.2 Hypersequent Calculus HMR

In this section we define the hypersequent calculus **HMR** from [LM20]. This is an extension of **HR** obtained by: (1) considering modal Riesz terms A, B (in NNF, see Sect. 2.2) rather than just (non-modal) Riesz terms, and (2) extending the set of rules of **HR** (Fig. 3) with the two additional rules of Fig. 3 dealing with the connectives $\{1, \Diamond\}$. The rule (1) is justified by the axiom $0 \leq 1$ of modal Riesz spaces (Definition 3). The ($\Diamond$) rule is justified by the the positivity and linearity of the $\Diamond$ operator as well as the axiom $\Diamond 1 \leq 1$ (see [LM20, §4.3]) (Fig. 4).

Additional rules:

$$\frac{G \mid \vdash \Gamma, \vec{r}.1, \vec{s}.\overline{1}}{G \mid \vdash \Gamma} \; 1, \textstyle\sum s_i \leq \sum r_i \qquad \frac{\vdash \Gamma, \vec{r}.1, \vec{s}.\overline{1}}{\vdash \Diamond\Gamma, \vec{r}.1, \vec{s}.\overline{1}} \; \Diamond, \textstyle\sum s_i \leq \sum r_i$$

Fig. 4. Additional inference rules of **HMR** ([LM20]).

We write $\rhd_{\mathbf{HMR}} G$ if the hypersequent G (involving modal Riesz terms $A, B, \ldots$) is derivable in the system **HMR**. By interpreting sequents and hypersequents as in Definition 5, the main results regarding **HR** extend to **HMR**: soundness, completeness, CAN–elimination and invertibility of the rules $\{0, +, \sqcup, \sqcap, \times\}$. Also a more sophisticated variant of the λ–property (Lemma 4) holds for **HMR** (see [LM20]), as we now state. A hypersequent G whose terms A are either atoms ($A = x$ or $A = \overline{x}$), or $A = 1$ or $A = \overline{1}$ or diamond–terms (i.e., $A = \Diamond B$, for some term B) is called a *basic hypersequent.*

Lemma 5 (λ-property of HMR [LM20, Lemma 4.44]**).** *For all basic hypersequents G formed using the variables and negated variables $x_1, \overline{x_1}, \ldots, x_k, \overline{x_k}$ of the form*

$$\vdash \Gamma_1, \Diamond\Delta_1, \vec{r'}_1.1, \vec{s'}_1.\overline{1} \mid \; ... \mid \vdash \Gamma_m, \Diamond\Delta_m, \vec{r'}_m.1, \vec{s'}_m.\overline{1}$$

*where $\Gamma_i = \vec{r}_{i,1}.x_1, ..., \vec{r}_{i,k}.x_k, \vec{s}_{i,1}.\overline{x_1}, ..., \vec{s}_{i,k}.\overline{x_{i,k}}$, for all $i \in [1 \ldots m]$, then G is derivable in **HMR** if and only if there exist numbers $t_1, ..., t_m \in [0, 1]$, one for each sequent in G, such that the following conditions hold:*

1. *there exists $i \in [1..m]$ such that $t_i = 1$,*
2. *for every (co)variable $(x_j, \overline{x_j})$ it holds that: $\sum_{i=1}^{m} t_i(\sum \vec{r}_{i,j} - \sum \vec{s}_{i,j}) = 0$*
3. $0 \leq \sum_{i=1}^{m} t_i(\sum \vec{r'}_i - \sum \vec{s'}_i)$,
4. *the following hypersequent (consisting of just one sequent) is derivable:* $\vdash t_1.\Delta_1, ..., t_m.\Delta_m, (t_1\vec{r'}_1).1, ..., (t_m\vec{r'}_m).1, (t_1\vec{s'}_1).\overline{1}, ..., (t_m\vec{s'}_m).\overline{1}$.

3.3 Parametrised Hypersequents

The hypersequents of **HMR** (and its subsystem **HR**) are built out of expressions of the form $(r.A)$ where $r > 0$ is a concrete real number (see Definition 4).

It is often useful, however, to state properties of parametrised families of hypersequents. For example, the hypersequent schema $(\vdash \alpha.x \mid\vdash \alpha.\overline{x})$, involving a variable α ranging over scalars, is derivable for all $\alpha > 0$.

Rather than just scalar variables (α, β), it is convenient to allow for even more general hypersequents schemas where in place of scalars we allow *polynomial expressions* over a certain number of variables α, β. We call such hypersequents *parametrised*. Given a parametrised hypersequent $G(\alpha_1, \ldots, \alpha_n)$, built using polynomial expressions R, S (see Sect. 2.3) involving scalar variables $\alpha_1, \ldots, \alpha_n$ we can obtain a concrete hypersequent (in the sense of Definition 4) $G(r_1, \ldots, r_n)$ by instantiating the scalar variables with concrete real numbers r_i and by evaluating the polynomial expressions as expected. Note, however, that since scalars r in expressions $(r.A)$ of concrete hypersequents are strictly positive real numbers, not all instantiations result in valid hypersequents. Therefore, when we write $G(r_1, \ldots, r_n)$, we implicitly mean that the substitution $[r_i/\alpha_i]$ results in a valid concrete hypersequent.

Example 4. $\vdash (\alpha_1 - 2\alpha_2).(x \sqcup y), (\alpha_1^2 - \frac{1}{2}).x$ is a parametrised hypersequent involving two scalar variables α_1 and α_2. The instance $[1/\alpha_1, -1/\alpha_2]$ results in the hypersequent $\vdash (3).(x \sqcup y), \frac{1}{2}.x$, and is therefore valid. The instance $[1/\alpha_1, 1/\alpha_2]$, instead, would result in $\vdash (-1).(x \sqcup y), \frac{1}{2}.x$ and is, therefore, not valid because (-1) is not a valid scalar.

Remark 2. Our main goal with the introduction of parametrised hypersequents is to express formally and schematically the last condition of the λ-property of **H(M)R** (Lemmas 4 and 5). To this end, even though scalars r in expressions $(r.A)$ of concrete hypersequents are strictly positive (Definition 4), it will be convenient to consider as valid also instances which results in the scalar 0. In this case, we use the convention $G \mid\vdash \Gamma, 0.A = G \mid\vdash \Gamma$, i.e., we remove every term that has a weight equal to 0.

4 Main Result – Proof of Theorem 1

In this section we present our main result, a syntactic proof–theoretical proof of Theorem 1: free modal Riesz spaces are Archimedean.

```
Lemma FreeMRS_archimedean : forall A B,
    (forall n, (INRpos n) *S A ≤ B) → (* forall n, (n+1)*A ≤ B *)
    A ≤ MRS_zero. (* A ≤ 0 *)
```

hmr_archimedean/archimedean.v

The same type of proof technique can also be used to prove the known fact that free (non–modal) Riesz spaces are Archimedean (Proposition 2).

As a first step, we express the Archimedean property for (modal) Riesz spaces as a derivability problem in the hypersequent calculus proof system **H(M)R**.

Lemma 6. *For any set V, the free (modal) Riesz space* $\mathbf{Free}_{(\mathrm{M})\mathrm{RS}}(V)$ *has the Archimedean property if and only if, for any (modal) Riesz terms A and B it holds that:* $(\forall n, \triangleright_{\boldsymbol{H(M)R}} \vdash 1.\overline{A}, \frac{1}{n}.B) \implies \triangleright_{\boldsymbol{H(M)R}} \vdash 1.\overline{A}$.

Proof. Recall that $\mathbf{Free}_{(\mathrm{M})\mathrm{RS}}(V)$ is isomorphic to $\mathbf{T}_{(\mathrm{M})\mathrm{RS}}(V)_{/\equiv}$ and therefore, by Definition 2, we have that $\mathbf{Free}_{(\mathrm{M})\mathrm{RS}}(V)$ is Archimedean if and only if the implication $(\forall n, n[A]_\equiv \leq [B]_\equiv) \Rightarrow [A]_\equiv \leq [0]_\equiv$ holds, for all (modal) Riesz terms A, B. Equivalently (using the identity $nx \leq y \Leftrightarrow -x + \frac{1}{n}y \geq 0$), $\mathbf{Free}_{(\mathrm{M})\mathrm{RS}}(V)$ is Archimedean if and only if $(\forall n, -[A]_\equiv + \frac{1}{n}[B]_\equiv \geq [0]_\equiv) \Rightarrow -[A]_\equiv \geq [0]_\equiv$holds. Finally, by the using the completeness and soundness of **H(M)R** (Theorem 1), this is equivalent to: $(\forall n, \rhd_{\mathbf{H(M)R}} \vdash 1.\overline{A}, \frac{1}{n}.B) \Rightarrow \rhd_{\mathbf{H(M)R}} \vdash 1.\overline{A}$. □

In order to establish the implication of Lemma 6 we prove a stronger result of independent interest about the hypersequent calculus **H(M)R**. This states that derivability in **H(M)R** is *continuous* in the sense that derivability preserves limits of scalars in hypersequents.

Theorem 2 (Continuity). *Let $G(\alpha_1, ..., \alpha_l)$ be a parametrized* ***H(M)R*** *hypersequent. Let $(s_{i,n}) \in \mathbb{R}^{\mathbb{N}}$ be a sequence of l–tuples of reals such that:*

1. *$G(s_{1,n}, ..., s_{l,n})$ is a valid instance of $G(\alpha_1, ..., \alpha_l)$, for all $n \in \mathbb{N}$, and*
2. *$\rhd_{\mathbf{H(M)R}} G(s_{1,n}, ..., s_{l,n})$ holds for all $n \in \mathbb{N}$,*
3. *For each $i \in \{1, \ldots, l\}$, the limit $\lim_{n \to +\infty} s_{i,n} = s_i$ exists.*

Then the limit instance $G(s_1, \ldots, s_l)$ is also valid and $\rhd_{\mathbf{H(M)R}} G(s_1, ..., s_l)$ holds.

Proof. Proofs for **HR** and **HMR** are presented in Sects. 4.1 and 4.2. □

The Archimedean property of free (modal) Riesz spaces is a direct corollary of Lemma 6 and Theorem 2, considering the parametrised hypersequent $\vdash 1.\overline{A}, \alpha_1.B$ and the sequence $s_{1,n} = \frac{1}{n}$, so that $s_1 = \lim_{n \to +\infty} \frac{1}{n} = 0$.

Corollary 1. $\mathbf{Free}_{(\mathrm{M})\mathrm{RS}}(V)$ *has the Archimedean property.*

4.1 Proof of Continuity for HR

As a first step, we prove that Theorem 2 holds for all **HR** hypersequents $G(\alpha_1, ..., \alpha_l)$ that are atomic, i.e., such that all terms appearing in G are either variables or covariables, i.e., $A = x$ or $A = \overline{x}$. Intuitively, this fact follows from the last point of Lemma 4, which reduces derivability of atomic hypersequents to the existence of a solution of a system of polynomial inequalities, and polynomials expressions are continuous.

Lemma 7 (Atomic continuity of HR). *The statement of Theorem 2 for* ***HR*** *holds for all atomic hypersequents $G(\alpha_1, ..., \alpha_l)$.*

Proof. `Repository [Luc21]: (HR_atomic_lim) in archimedean.v.` Let $G = \vdash \Gamma_1 \mid \ldots \mid \vdash \Gamma_m$ with $\Gamma_i = \vec{R}_{i,1}.x_1, ..., \vec{R}_{i,k}.x_k, \vec{S}_{i,1}.\overline{x_1}, ..., \vec{S}_{i,k}.\overline{x_k}$, where all the indexed expressions R, $\vec{R}$, S and $\vec{R}$ are polynomials over $\alpha_1, \ldots, \alpha_l$.

By assumption, each $G(s_{1,n}, ..., s_{l,n})$ is a valid instance (the evaluation of all polynomial expressions results in a strictly positive scalar or 0 scalar, see Remark 2). Furthermore, $G(s_{1,n}, ..., s_{l,n})$ is assumed to be derivable in **HR** which means, by Lemma 4, that there exist real numbers $t_{1,n}, ..., t_{m,n} \in [0, 1]$, such that:

1. there exists $i \in [1..m]$ such that $t_i = 1$, and
2. for every variable and covariable pair $(x_j, \overline{x_j})$, it holds that
$\sum_{i=1}^{m} t_{i,n}(\sum \overrightarrow{R}_{i,j}(s_{1,n}, ..., s_{l,n}) - \sum \overrightarrow{S}_{i,j}(s_{1,n}, ..., s_{l,n})) = 0$.

By the infinitary pigeon principle (see Sect. 2.3), there exists $i \in [1..m]$ such that $t_{i,n} = 1$ infinitely often and since $[0,1]^m$ is a compact space, by sequential compactness of $[0,1]$ (see Sect. 2.3), we can extract a subsequence $(t_{1,\sigma(j)}, ..., t_{m,\sigma(j)})_j$ converging to $(t_1, ..., t_m)$ with $t_{i,\sigma(j)} = 1$ for all j (and so $t_i = 1$). Finally, the identity $\sum_{i=1}^{m} t_i(\sum \overrightarrow{R}_{i,j}(s_1, ..., s_l) - \sum \overrightarrow{S}_{i,j}(s_1, ..., s_l)) = 0$ holds, because polynomial expressions are continuous and therefore preserve limits of converging sequences. Hence, we have that

1. there exists $i \in [1..m]$ such that $t_i = 1$, and
2. for every variable and covariable pair $(x_j, \overline{x_j})$, it holds that:
$\sum_{i=1}^{m} t_i(\sum \overrightarrow{R}_{i,j}(s_1, ..., s_l) - \sum \overrightarrow{S}_{i,j}(s_1, ..., s_l)) = 0$,

and, according to Lemma 4, this implies that $G(s_1, ..., s_l)$ is derivable. □

In order to conclude the proof of Theorem 2 for **HR**, we need to extend the result of Lemma 7 to arbitrary parametrised hypersequents $G(\alpha_1, \ldots, \alpha_l)$. This is done by showing that the continuity of parametrised hypersequents of a certain complexity can be reduced to the continuity of hypersequents of lower complexity, with the case of atomic hypersequent (Lemma 7) serving as base case. The main tool allowing this reduction is Lemma 3, which states that the logical rules of **HR** are invertible.

Definition 6 (Complexity). *The complexity of a sequent $\vdash \Gamma$, noted $|\vdash \Gamma|$, is the sum of all connectives $\{0, +, r(\_), \sqcup, \sqcap\}$ appearing in terms of Γ. The complexity of a (parametrized) hypersequent G, noted $|G|$, is a pair $(a, b) \in \mathbb{N}^2$ defined by: $a = \max_{\vdash \Gamma \in G} |\vdash \Gamma|$, the maximal complexity of sequents in G, and $b = |\{\vdash \Gamma \in G \mid |\vdash \Gamma| = a\}|$, the number of sequents in G with complexity a.*

Note that atomic hypersequents have complexity $|G| = (0, b)$. We are now ready to conclude the proof of Theorem 2.

Proof (general case). `Repository [Luc21]: (HR_lim) in archimedean.v.`
Let $G(\alpha_1, \ldots, \alpha_l)$ and $(s_{i,n}) \in \mathbb{R}^{\mathbb{N}}$ be as in the statement of Theorem 2. The proof goes by lexicographic induction on the complexity $|G|$.

If $|G| = (0, b)$, we can conclude with Lemma 7.

Otherwise $|G| = (a, b)$ for some $a, b > 1$. Hence G has the shape $G' \mid\vdash \Gamma, R.A$ where the complexity of $\vdash \Gamma, R.A$ is equal to a and A is a term with some outermost connective in $\{0, +, r(\_), \sqcup, \sqcap\}$. Here we only consider the case of $A = B \sqcup C$, the other cases being similar.

By assumption we know that, for all n, $\rhd_{\mathbf{HR}}(G' \mid \vdash \Gamma, R.(B \sqcup C))(s_{1,n}, ..., s_{l,n})$, for all tuples $(s_{1,n}, ..., s_{l,n})$. The invertibility of the $\sqcup$ rule (Lemma 3) implies that the following hypersequents, for each tuple $(s_{1,n}, ..., s_{l,n})$, are also derivable:

$$\rhd_{\mathbf{HR}}(G' \mid \vdash \Gamma, R.B \mid \vdash \Gamma, R.C)(s_{1,n}, ..., s_{l,n})$$

Note that the above hypersequents have complexity lower than $|G|$. Hence, by applying the induction hypothesis we obtain, by continuity, that

$$\rhd_{\mathbf{HR}}(G' \mid \vdash \Gamma, R.B \mid \vdash \Gamma, R.C)(s_1, ..., s_l)$$

holds. We can then conclude the argument by deriving the desired hypersequent as follows, by one application of the $\sqcup$ rule.

$$\frac{(G' \mid \vdash \Gamma, R.B \mid \vdash \Gamma, R.C)(s_1, ..., s_l)}{(G' \mid \vdash \Gamma, R.(B \sqcup C))(s_1, ..., s_l)} \sqcup \qquad \square$$

4.2 Proof of Continuity for HMR

The proof of Theorem 2 for **HMR** presented in this section has the same structure of the proof presented in Sect. 4.1 for **HR**. Namely, (1) we first prove a result similar to the atomic continuity Lemma 7 of Sect. 4.1 stating that Theorem 2 holds for certain "simple" hypersequents, and then (2) extend this result to arbitrary **HMR** hypersequents.

Regarding (1), the notion of "simple" is that of basic hypersequent associated with the λ–property of **HMR** (Lemma 5). Note, however, that unlike the corresponding λ–property of **HR** (Lemma 4), the statement of Lemma 5 reduces the derivability of basic **HMR** hypersequents not just to the existence of a solution in a system of polynomial inequalities, but also in terms of derivability of simpler (in terms of the number of $\Diamond$ operators in their terms) hypersequents. Hence, a slightly more sophisticated proof by induction on an appropriate notion of complexity of hypersequents is needed. Regarding (2), the key technical tool used (as in the proof of Sect. 4.1 for **HMR**) is the invertibility of the logical rules $\{0, +, r(\_), \sqcup, \sqcap\}$ of **HMR**, and is also based on an induction on the complexity of the hypersequent.

Definition 7 (Modal Depth and Outer Complexity). *The modal depth $\mathcal{D}(A)$ of a modal Riesz space term A is the maximum number of nested $\Diamond$'s in A, i.e., is defined inductively as: $\mathcal{D}(\Diamond B) = 1 + \mathcal{D}(B)$, $\mathcal{D}(A) = 0$ if $A \in \{x, \overline{x}, 1, \overline{1}, 0\}, \mathcal{D}(rB) = \mathcal{D}(B)$ and $\mathcal{D}(B \star C) = \max(\mathcal{D}(B), \mathcal{D}(C))$, for $\star \in \{+, \sqcup, \sqcap\}$.*

The outer complexity $\mathcal{O}(A)$ of a modal Riesz space term A is the total number of connectives $\{0, +, r(\_), \sqcup, \sqcap\}$ that do not appear under the scope of some $\Diamond$ in A, i.e., as: $\mathcal{O}(A) = 0$ if $A \in \{\Diamond B, x, \overline{x}, 1, \overline{1}\}$, $\mathcal{O}(0) = 1$, $\mathcal{O}(rB) = 1 + \mathcal{O}(B)$ and $\mathcal{O}(B \star C) = 1 + \mathcal{O}(B) + \mathcal{O}(C)$ for $\star \in \{+, \sqcup, \sqcap\}$.

Definition 8 (Hypersequent Complexity). *The modal depth and outer complexity of a sequent $\vdash \Gamma$ of the form $r_1.A_1, \ldots r_n.A_n$ are defined as: $\mathcal{D}(\vdash \Gamma) = \max_{i=1}^n \mathcal{D}(A_i)$ and $\mathcal{O}(\vdash \Gamma) = \sum_{i=1}^n \mathcal{O}(A_i)$. The complexity of a (parametrized) hypersequent G, noted $|G|$, is a triplet $(a, b, c) \in \mathbb{N}^3$ defined by: $a = \max_{\vdash \Gamma \in G} \mathcal{D}(\vdash \Gamma)$ is the maximum modal depth of any sequent in G, $b = \max_{\vdash \Gamma \in G} \mathcal{O}(\vdash \Gamma)$ is the maximum outer complexity of any sequent in G, and $c = |\{\vdash \Gamma \in G | \mathcal{O}(\vdash \Gamma) = b\}|$ is the number of sequents in G having outer complexity b.*

Note that hypersequents G with $|G| = (a, 0, c)$ are basic hypersequents (see Sect. 3.2). Furthermore note that if $|G| = (0, 0, c)$, then the hypersequent only contains terms of the form $\{x, \overline{x}, 1, \overline{1}\}$, and the statement of Lemma 5 simplifies (point 4 becomes trivial) and it reduces the **HMR** derivability of G to the solution of a system of polynomial equations.

We are now ready to present the proof of Theorem 2 for **HMR**.

Proof. `Repository [Luc21]: (HMR_lim) in archimedean.v.` Let $G(\alpha_1, \ldots, \alpha_l)$ and $(s_{i,n}) \in \mathbb{R}^{\mathbb{N}}$ be as in the statement of Theorem 2. We prove the result by lexicographic induction on $|G|$.

If $|G| = (0, 0, c)$, then Lemma 7 of Sect. 4.1 can be easily adapted, using the λ–property (Lemma 5) of **HMR**, to prove that $\triangleright_{\mathbf{HMR}} G(s_1, ..., s_l)$.

If $|G| = (a, b, c)$ with $b > 0$, following the same technique presented in Sect. 4.1 based on the invertibility of the logical rules $\{0, +, r(\_), \sqcup, \sqcap\}$, we can reduce the complexity of $|G|$ to some (a, b', c') with $b' < b$, apply the induction hypothesis and finally deduce that $\triangleright_{\mathbf{HMR}} G(s_1, ..., s_l)$.

Lastly, assume that $|G| = (a, 0, c)$, i.e., that G is a basic hypersequent and has the form $\vdash \Gamma_1, \Diamond\Delta_1, \overrightarrow{R'}_1.1, \overrightarrow{S'}_1.\overline{1} \mid \ldots \mid \vdash \Gamma_m, \Diamond\Delta_m, \overrightarrow{R}_m.1, \overrightarrow{S'}_m.\overline{1}$ where $\Gamma_i = \overrightarrow{R}_{i,1}.x_1, ..., \overrightarrow{R}_{i,k}.x_k, \overrightarrow{S}_{i,1}.\overline{x_1}, ..., \overrightarrow{S}_{i,k}.\overline{x_{i,k}}$. By assumption $G(s_{1,n}, ..., s_{l,n})$ has a proof for all n. Thus, by the λ–property (Lemma 5), for all n there exist numbers $t_{1,n}, ..., t_{m,n} \in [0, 1]$ such that the following conditions hold:

1. there exists $i \in [1..m]$ such that $t_{i,n} = 1$,
2. for every variable and covariable pair $(x_j, \overline{x_j})$, it holds that:
   $\sum_{i=1}^{m} t_{i,n}(\sum \overrightarrow{R}_{i,j}(s_{1,n}, ..., s_{l,n}) - \sum \overrightarrow{S}_{i,j}(s_{1,n}, ..., s_{l,n})) = 0$,
3. $0 \leq \sum_{i=1}^{m} t_{i,n}(\sum \overrightarrow{R'}_i(s_{1,n}, ..., s_{l,n}) - \sum \overrightarrow{S'}_i(s_{1,n}, ..., s_{l,n}))$,
4. the following hypersequent, consisting of only one sequent, is derivable:
   $(\vdash t_{1,n}.(\Delta_1, \overrightarrow{R'}_1.1, \overrightarrow{S'}_1.\overline{1}), , ..., t_{m,n}.(\Delta_m, \overrightarrow{R'}_m.1, t_{m,n}\overrightarrow{S'}_m).\overline{1})(s_{1,n}, ..., s_{l,n})$.

From point (4), by considering m additional scalar variables $\alpha_{l+1}, \ldots, \alpha_{l+m}$ and by defining $s_{l+i,n} = t_{i,n}$, the parametrised hypersequent H defined as $\vdash \alpha_{l+1}.(\Delta_1, \overrightarrow{R'}_1.1, \overrightarrow{S'}_1.\overline{1}), , ..., \alpha_{l+m}.(\Delta_m, \overrightarrow{R'}_m.1, t_{m,n}\overrightarrow{S'}_m.\overline{1})$ has a proof for all n. Note that $|H| < |G|$, due to the lower modal–depth complexity.

By assumption $\lim_{n \to +\infty} s_{i,n} = s_i$ exits for $i \in \{1, \ldots, l\}$ but the extended sequence $(s_1, \ldots, s_l, s_{l+1}, \ldots s_{l+m})$ might not have limits on the coordinates $l + 1, \ldots, l + m$. However, following the same argument of the proof of Lemma 7, by the infinitary pigeon principle and the sequential compactness of $[0, 1]^k$, we can extract a subsequence that converges on all coordinates and agrees with the existing limits s_i for $i \in \{1, \ldots, l\}$. Hence, by induction hypothesis, we can apply the continuity theorem to H and deduce that $H(s_1, ..., s_{l+m})$ has a proof.

Finally, in order to conclude the proof and prove that $\triangleright_{\mathbf{HMR}} G(s_1, ..., s_l)$, we apply once again the λ–property (Lemma 5) which states that $\triangleright_{\mathbf{HMR}} G(s_1, ..., s_l)$ is derivable if and only if the four points above hold, instantiated to $(s_1, ..., s_l)$. The fourth point has been established. The points (1–3) follow from the continuity of polynomial expressions as discussed in the proof of Lemma 7. $\square$

References

[Avr96] Avron, A.: The method of hypersequents in the proof theory of propositional non-classical logics. In: From Foundations to Applications: European Logic Colloquium, pp. 1–32. Oxford University Press (1996)

[Bak68] Baker, K.A.: Free vector lattices. Can. J. Math. **20**, 58–66 (1968)

[Ble73] Bleier, R.D.: Free vector lattices. Trans. Am. Math. Soc. **176**, 73–87 (1973)

[BLM15] Boldo, S., Lelay, C., Melquiond, G.: Coquelicot: a user-friendly library of real analysis for coq. Math. Comput. Sci. **9**(1), 41–62 (2014). https://doi.org/10.1007/s11786-014-0181-1

[DMS18] Denisa, D., Metcalfe, G., Schnüriger, L.: A real-valued modal logic. Log. Methods Comput. Sci. **14**(1), 10 (2018). https://doi.org/10.23638/LMCS-14(1:10)2018

[FMM20] Furber, R., Mardare, R., Mio, M.: Probabilistic logics based on riesz spaces. Logical Methods Comput. Sci. **16**(1), 6 (2020). https://doi.org/10.23638/LMCS-16(1:6)2020

[FR94] Fleury, A., Retoré, C.: The mix rule. Math. Struct. Comput. Sci. **4**(2), 273–285 (1994). https://doi.org/10.1017/S0960129500000451

[Gen34] Gentzen, G.: Untersuchungen über das logische Schließen. Math. Z. **39**, 405–431 (1934)

[H.74] Fremlin, D.H.: Topological Riesz Spaces and Measure Theory. Cambridge University Press, Cambridge (1974)

[JR77] De Jonge, E., Van Rooij, A.C.M.: Introduction to Riesz Spaces, vol. 78. Mathematical Centre Tracts, Amsterdam (1977)

[LM19] Lucas, C., Mio, M.: Towards a structural proof theory of probabilistic μ-calculi. In: Bojańczyk, M., Simpson, A. (eds.) FoSSaCS 2019. LNCS, vol. 11425, pp. 418–435. Springer, Cham (2019). https://doi.org/10.1007/978-3-030-17127-8_24

[LM20] Lucas, C., Mio, M.: Proof theory of Riesz spaces and modal Riesz spaces. In: Logical Methods in Computer Science (2021)

[Luc21] Lucas, C.: A complete Coq formalisation (2021). https://doi.org/10.5281/zenodo.4897186

[LZ71] Luxemburg, W.A.J., Zaanen, A.C.: Riesz Spaces, vol. 1. North-Holland Mathematical Library, Amsterdam (1971)

[MFM17] Mio, M., Furber, R., Mardare, R.: Riesz modal logic for Markov processes. In: 32nd ACM/IEEE Symposium on Logic in Computer Science (LICS), pp. 1–12. IEEE (2017). https://doi.org/10.1109/LICS.2017.8005091

[Mio18] Mio, M.: Riesz modal logic with threshold operators. In: 33rd Symposium on Logic in Computer Science (LICS), pp. 710–719. ACM (2018). https://doi.org/10.1145/3209108.3209118

[MOG09] Metcalfe, G., Olivetti, N., Gabbay, D.M.: Proof Theory for Fuzzy Logics. Applied Logic Series, vol. 36. Springer, Heidelberg (2009). https://doi.org/10.1007/978-1-4020-9409-5

[MS17] Mio, M., Simpson, A.: Łukasiewicz μ-calculus. Fundam. Inf. **150**(3–4), 317–346 (2017). https://doi.org/10.3233/FI-2017-1472

Polyadic Spaces and Profinite Monoids

Jérémie Marquès(✉)

Université Côte d'Azur, CNRS, LJAD (UMR 7351), Nice, France
jmarques@unice.fr

Abstract. Hyperdoctrines are an algebraization of first-order logic introduced by Lawvere in [11]. In [9], Joyal defines a polyadic space as the Stone dual of a Boolean hyperdoctrine. He also proposed to recover a polyadic space from a simpler core, its Stirling kernel. We generalize this here in order to adapt polyadic spaces to certain classes of first-order theories. We will see how these ideas can be applied to give a correspondence between some first-order theories with a linear order symbol and equidivisible profinite semigroup with open multiplication. The inspiration comes from the paper [6] of van Gool and Steinberg, where model theory is used to study pro-aperiodic monoids.

Keywords: Categorical logic · Profinite monoids · Logic on words

1 Introduction

In [6], van Gool and Steinberg use model theory to study free pro-aperiodic monoids. The starting point is the Schützenberger-McNaughton-Pappert theorem, which establishes that languages on an alphabet A recognized by aperiodic monoids are those given by first-order sentences in Büchi's logic on finite words [13]. In category theory, a standard construction is to encode a monoid M as a functor $n \mapsto M^n$ whose domain is the category of finite linear orders and order-preserving maps. When $M = M_A$ is the free pro-aperiodic monoid on an alphabet A, we will see, using categorical logic, how this functor also represents Büchi's logic.

In Sect. 2, we present Boolean hyperdoctrines in parallel with their pointwise duals, polyadic Stone spaces. Boolean hyperdoctrines are an algebraization of classical first-order logic introduced by Lawvere in [11]. They are a way of representing a first-order theory by a functor from finite sets to Boolean algebras, sending a finite set X of variables to the Lindenbaum-Tarski algebra of formulas whose free variables are in X. In [9], Joyal defines a polyadic space as the pointwise Stone dual of a Boolean hyperdoctrine, so that the Lindenbaum-Tarski algebras are replaced by type spaces. Intuitively speaking, a polyadic space is thus a functor sending a finite set X to a space of X-pointed models. See also [12], which mentions the link with cylindric and polyadic algebras. We

This project has received funding from the European Research Council (ERC) under the European Union's Horizon 2020 research and innovation program (grant agreement No. 670624).

U. Fahrenberg et al. (Eds.): RAMiCS 2021, LNCS 13027, pp. 292–308, 2021.
https://doi.org/10.1007/978-3-030-88701-8_18

will start by associating two functors to each first-order theory: the first is the Boolean hyperdoctrine representing it and the second is the dual polyadic space. Afterwards, functors arising in this way are characterized by two simple axioms (Definitions 2.6 and 2.7).

Joyal also proposed to recover with a free construction a polyadic space from a simpler core, its Stirling kernel. In Sect. 3, we introduce different notions of kernels of polyadic spaces and the use of left Kan extension to reconstruct polyadic spaces from their kernels. For instance, suppose that a theory T has a partial order symbol.[1] Since models of T come with a partial order, it makes sense to speak of X-pointed models when X is a finite poset. Thus, one can replace finite sets in the domain of the polyadic space representing T by finite posets. The resulting functor will then be a "kernel" of the polyadic space of T. We give a number of examples of kernels, but a general framework is still to be elaborated.

In the last section, Sect. 4, we apply these ideas to monoids and languages. We will see that the functor $n \mapsto M_A^n$ referred to above, with M_A the free pro-aperiodic monoid, is a kernel of the polyadic space associated to Büchi's logic on words. In our first main result, Theorem 4.3, we generalize this situation and characterize the profinite semigroups S such that the functor $n \mapsto S^n$ represents a first-order theory as those whose multiplication is open and equidivisible. Profinite semigroups with these two properties are studied for instance by Almeida, Costa, Costa and Zeitoun in [1]. In their setting, equidivisibility alone is less well-behaved. We show here that openness and equidivisibility are natural from the viewpoint of logic since they each correspond to one of the axioms of polyadic spaces (Definition 2.7). We also characterize in our second main result, Theorem 4.8, first-order theories arising from a profinite monoid as those whose models can be concatenated with some natural constraints. Thus, these theories are very special.

Finite sets will be written as natural numbers, so that n can also denote the set $\{1,\ldots,n\}$. We write β: **Set** $\to$ **Stone** for the Stone-Čech compactification functor, which is left adjoint to the forgetful functor **Stone** $\to$ **Set**. It is the usual Stone-Čech compactification restricted to discrete spaces. See [8, Chapter III, Sect. 2.1].

2 Hyperdoctrines and Polyadic Spaces

Boolean algebras are an algebraization of classical propositional logic. To get quantifiers in the picture, Lawvere remarked in [11] that they are adjoints to substitution of variables. We also get equality this way, and this gives a signature-independent algebraic account of classical first-order logic with equality. In this section, we explain how to arrive at this idea, using in parallel the dual notion of polyadic spaces introduced by Joyal in [9].

[1] More precisely, if there is a binary relational symbol $\leq$ subject to the axioms of partial orders.

2.1 Quantifiers as Adjoints

Let T be a first-order theory. We associate to it a functor $D_T\colon \mathbf{FinSet} \to \mathbf{Bool}$ defined as follows:

1. On objects, D_T sends a finite set n to the n-Lindenbaum-Tarski algebra of the theory, i.e. the Boolean algebra of first-order formulas on n free variables modulo equivalence according to T. The set n is the "context" of the formulas in $D_T(n)$. We will write these variables as $x_1,\ldots,x_n$, or $y_1,\ldots,y_n$ if several contexts are present.
2. On morphisms, if $f\colon n \to m$ is a function, then $D_T(f)\colon D_T(n) \to D_T(m)$ sends a formula $\varphi(x_1,\ldots,x_n)$ to the formula $\varphi(y_{f(1)},\ldots,y_{f(n)})$. That is, we substitute (in a capture-avoiding way) the variable x_i in $\varphi(x_1,\ldots,x_n)$ with the variable $y_{f(i)}$. In the special case where f is injective, this procedure only adds new variables to the context of the formula.

We will also use the notation $\varphi(\overline{x})$ for a formula on variables $\overline{x} = x_1,\ldots,x_n$, and write $\varphi(f(\overline{x}))$ for its image under $D_T(f)$.

A *Boolean hyperdoctrine* is a functor of the form D_T. A *polyadic space* is a functor $\mathbf{FinSet}^{\mathrm{op}} \to \mathbf{Stone}$ obtained by applying Stone duality pointwise to a hyperdoctrine. Let P_T be the polyadic space dual to D_T.

Remark 2.1. In model theory, $P_T(n)$ is usually called the space of (complete) *n-types* of T and is written $S_n(T)$ [7, Sect. 5.2]. It can be thought of as the set of n-pointed models of T modulo elementary equivalence. If $f\colon n \to m$ is a function, then $P_T(f)$ sends (the equivalence class of) an m-pointed model $(M, x_1,\ldots,x_m)$ to $(M, x_{f(1)},\ldots,x_{f(n)})$.

Example 2.2. A simple example that one can keep in mind is $P_T(n) = X^n$ for X a set. If X is infinite, we actually need to take the Stone-Čech compactification $P_T(n) = \beta(X^n)$ to get a polyadic space (otherwise, we only get a polyadic *set*, see Subsect. 3.3). The corresponding theory has one n-ary symbol for each subset of X^n, and theorems are formulas true for the obvious interpretation of these symbols on X.

Quantification. Consider the following introduction-elimination rule for existential quantification: $\exists\overline{x} : \varphi(\overline{x},\overline{y}) \vdash_{\overline{y}} \psi(\overline{y})$ if and only if $\varphi(\overline{x},\overline{y}) \vdash_{\overline{x},\overline{y}} \psi(\overline{y})$. (See [10, Part 2, Sect. 1].) It is actually an adjunction formula, and more generally, if $f\colon n \rightarrowtail m$ is an injection, then the left adjoint of $D_T(f)$ is given by existential quantification over the variables in m which are not in the image of f. Symmetrically, universal quantification gives the right adjoint of $D_T(f)$. Notice that these adjoints are not morphisms of Boolean algebras in general.

Let us translate this in terms of P_T. Let $h\colon A \to B$ be a morphism of Boolean algebras. Then h has a left adjoint $h_*\colon B \to A$ if and only if the dual map $\tilde{h}\colon \tilde{B} \to \tilde{A}$ is open. In this case, h_* corresponds to taking the direct image by $\tilde{h}$. So the condition we obtain on P_T is that $P_T(f)$ is open for each injection $f\colon n \rightarrowtail m$ in $\mathbf{FinSet}$.

Example 2.3. Getting back to our example $P_T(n) = X^n$, we see that existential quantification corresponds to projection. For instance, the two projections $X^2 \to X$ correspond to quantification along the two variables.

Equality. On the other hand, if $f : n \twoheadrightarrow m$ is a surjection, then $P_T(f) : P_T(m) \to P_T(n)$ is an injection whose image is the set of n-pointed models such that $x_i = x_j$ if $f(i) = f(j)$ (cf. Remark 2.1). Thus, $P_T(f)$ is open and dually, $D_T(f)$ has a left adjoint. The morphism $D_T(f) : D_T(n) \twoheadrightarrow D_T(m)$ realizes the quotient by the principal filter generated by $\bigwedge_{\substack{1 \le i,j \le n \\ f(i)=f(j)}} [x_i = x_j]$.

Example 2.4. Binary equality is given by the quotient $2 \twoheadrightarrow 1$. In the case of $P_T(n) = X^n$, the corresponding map is the diagonal inclusion $X \rightarrowtail X^2$.

2.2 The Beck–Chevalley Condition and Quasi-pullbacks

If we look at logic only syntactically, one hidden aspect is that substitution commutes with all other constructions: Boolean operations, quantification and equality. In order to express the commutativity of substitution and existential quantification, let X, Y, Z be finite sets and let $f : X \to Z$ be a function. Let $\varphi(\overline{x}, \overline{y})$ be a formula in $D_T(X \sqcup Y)$. We have two ways of building $\exists \overline{y} : \varphi(f(\overline{x}), \overline{y})$, as illustrated below on the left.

$$\begin{array}{ccc} \exists \overline{y} \colon \varphi(\overline{x}, \overline{y}) & \longmapsto & \exists \overline{y} \colon \varphi(f(\overline{x}), \overline{y}) \\ \uparrow & & \uparrow \\ \varphi(\overline{x}, \overline{y}) & \longmapsto & \varphi(f(\overline{x}), \overline{y}) \end{array} \qquad \begin{array}{ccc} D_T(X) & \xrightarrow{D_T(f)} & D_T(Z) \\ \exists \uparrow\downarrow & & \exists \uparrow\downarrow \\ D_T(X \sqcup Y) & \xrightarrow{D_T(f \sqcup \mathrm{id}_Y)} & D_T(Z \sqcup Y) \end{array}$$

In the diagram on the right, the plain arrows are the images by D_T of the canonical maps and the dashed arrows labeled by $\exists$ are the left adjoints. Thus, the condition we are interested in is that the square with the dashed arrows commutes. Using Definition 2.5 below, this condition says that D_T sends pushouts of an injection $(X \rightarrowtail X \sqcup Y)$ and an arbitrary function $(f : X \to Z)$ to Beck–Chevalley squares.

Definition 2.5. *We say that the commutative square of posets (1) satisfies the* Beck–Chevalley condition *(or is a Beck–Chevalley square) if u and v have left adjoints u_* and v_* such that the square (2) commutes.*

$$\begin{array}{ccc} A & \xrightarrow{f} & B \\ {\scriptstyle u}\downarrow & & \downarrow{\scriptstyle v} \\ C & \xrightarrow{g} & D \end{array} \quad (1) \qquad \begin{array}{ccc} A & \xrightarrow{f} & B \\ {\scriptstyle u_*}\uparrow & & \uparrow{\scriptstyle v_*} \\ C & \xrightarrow{g} & D \end{array} \quad (2) \qquad \begin{array}{ccc} S(D) & \xrightarrow{v^{-1}} & S(B) \\ {\scriptstyle u^{-1}}\downarrow & & \downarrow{\scriptstyle f^{-1}} \\ S(C) & \xrightarrow{g^{-1}} & S(A) \end{array} \quad (3)$$

The dual of the Beck–Chevalley condition is given by the back and forth conditions of the duality for operators [4]. Let $S : \mathbf{Bool}^{\mathrm{op}} \to \mathbf{Stone}$ be the

Stone dualization functor. Then (1) is Beck–Chevalley if and only if its dual (3) is a *quasi-pullback* or *quasi-cartesian*, meaning that for each $b \in S(B)$ and each $c \in S(C)$ such that $f^{-1}(b) = g^{-1}(c)$, there is some $d \in S(D)$ such that $v^{-1}(d) = b$ and $u^{-1}(d) = c$.

More generally, D_T sends *any* pushout square to a Beck–Chevalley square.

2.3 Definitions

We arrive at an algebraic definition of hyperdoctrines and dually of polyadic spaces.

Definition 2.6. *A* (Boolean) hyperdoctrine *is a functor* D: **FinSet** $\to$ **Bool** *such that:*

1. *For each function f between finite sets, $D(f)$ has a left adjoint.*
2. *D sends pushout squares to Beck–Chevalley squares.*

Definition 2.7. *A* polyadic (Stone) space *is a functor* P: **FinSet**^{op} $\to$ **Stone** *such that:*

1. *For each function f between finite sets, $P(f)$ is open.*
2. *P sends pushout squares in* **FinSet** *to quasi-pullbacks in* **Stone**.

Functors of the form D_T are axiomatized by Definition 2.6 (see [16]) and functors of the form P_T are axiomatized by Definition 2.7. The map $T \mapsto D_T$ forgets which are the base symbols of the signature, so that all we can recover from D_T is the Morleyization of T.

Definition 2.8. *A morphism of hyperdoctrines is a natural transformation whose naturality squares satisfy the Beck–Chevalley condition. Dually, a morphism of polyadic spaces is a natural transformation whose naturality squares are quasi-cartesian.*

Definition 2.9. *A* model *of a hyperdoctrine D is a morphism from D to some hyperdoctrine of the form $n \mapsto \mathscr{P}(X^n)$ as in Example 2.2. Dually, a model of a polyadic space P is a morphism of polyadic spaces $\beta(X^n) \to P(n)$.*

Models of D_T correspond to models of T: we interpret each relation definable over T as a subset of X. Since $\beta(X^n)$ is the free Stone space on X^n and using Proposition 2.11 below, models of a polyadic space P correspond to natural transformations $\alpha_n\colon X^n \to P(n)$ satisfying some weakening of Definition 2.8. This weakened condition says that if a n-tuple of the model satisfies some existential statement or an equality, then this is indeed concretely true in the model.

Definition 2.10. *In the commutative square on the left of (4) below, A and C are sets while B and D are Stone spaces with v continuous and open. We say that the square on the left of (4) is* weakly quasi-cartesian *if for each open $U \subseteq B$ and each $c \in C$ such that $g(c) \in v(U)$, there exists a witness $a \in A$ such that $f(a) \in U$ and $u(a) = c$.*

$$\begin{array}{ccc} A & \xrightarrow{f} & B \\ {\scriptstyle u}\downarrow & & \downarrow{\scriptstyle v} \\ C & \xrightarrow{g} & D \end{array} \qquad\qquad \begin{array}{ccc} \beta(A) & \longrightarrow & B \\ {\scriptstyle \beta(u)}\downarrow & & \downarrow{\scriptstyle v} \\ \beta(C) & \longrightarrow & D \end{array} \tag{4}$$

Proposition 2.11. *In the same situation as in Definition 2.10, the square on the left of (4) is weakly quasi-cartesian if and only if the square on the right of (4) (in* **Stone***) is quasi-cartesian.*

Remark 2.12. Let $\alpha\colon X^n \to P(n)$ be a natural transformation. Intuitively, the naturality squares of α are weakly quasi-cartesian if X has "all the points it should have to be a model." If we strengthen this condition and ask that the naturality squares of α are quasi-cartesian, we recover the notion of *ω-saturated model.*

Remark 2.13. Using the Yoneda lemma, a natural transformation $X^n \to P(n)$ can be identified with a point of $P(X) = \lim_{n \subseteq X \text{ finite}} P(n)$. The X-based models of P are then identified with a subspace of $P(X)$. This is used in Rasiowa and Sikorski's proof of Gödel's completeness theorem in [14]: they prove that if X is countable and each $P(n)$ has a countable basis, then this subspace of models is comeager. This also yields the omitting types theorem [7, Theorem 6.2.1].

3 Kernels and C-adic Spaces

3.1 Stirling Kernels

The idea of Stirling kernels is due to Joyal (private communication). Let T be a first-order theory. Let $P_T[n] \subseteq P_T(n)$ be the clopen subspace of models pointed by n distinct points. Then we can reconstruct $P_T(n)$ from $P_T[n]$ with the finite coproduct

$$P_T(n) = \coprod_{R \in \mathrm{Cong}(n)} P_T[n/R], \tag{5}$$

where $\mathrm{Cong}(n)$ is the set of equivalence relations on n. Each term $P_T[n/R]$ is identified with the clopen subset of $P_T(n)$ of n-pointed models whose points are equal if and only if they are R-equivalent.

Since restricting a pointing by n distinct points along an injection gives again distinct points, $n \mapsto P_T[n]$ is a functor $\mathbf{FinSetInj}^{\mathrm{op}} \to \mathbf{Stone}$ where **FinSetInj** is the category of finite sets and injections. This functor is called the *Stirling kernel* of $P_T(n)$, because the Stirling numbers of the second kind are the coefficients of $P_T[k]$ in the formula (5) (there is one copy of $P_T[k]$ for each equivalence relation of index k on n).

Another way of expressing the formula (5) is to say that $P_T(n)$ is the left Kan extension of $P_T[n]$ along the inclusion $\mathbf{FinSetInj}^{\mathrm{op}} \rightarrowtail \mathbf{FinSet}^{\mathrm{op}}$. For an introduction to Kan extensions, see [2, Sect. 3.7]. This also reconstructs the functoriality of $P_T(-)$. More concretely, let $f\colon n \to m$ be any function and let R be an equivalence relation on m. Let $f^{-1}(R)$ be the equivalence relation on n induced by R and f and let $\tilde{f}\colon n/f^{-1}(R) \to m/R$ be the corestriction of f. Then $P_T[m/R]$ is sent into $P_T[n/f^{-1}(R)]$ by $P_T(f)$ and this restriction coincides with $P_T[\tilde{f}]$.

In Theorem 3.4 below, we identify the conditions for a functor $\mathbf{FinSetInj}^{\mathrm{op}} \to \mathbf{Stone}$ to be a Stirling kernel. This provides another algebraization of first-order logic with equality. In some sense, this tells us that equality can be "freely adjoined".

In Definition 2.7, the first condition says that $P(f)$ is open and this is transferred to the Stirling kernel since $P[f]$ is a restrictions of $P(f)$ to a clopen subset. On the other hand, the second condition cannot be transferred so easily since the category **FinSetInj** does not admit pushouts. Let us give an example to illustrate that point.

Example 3.1. Consider the diagram $1 \leftarrow 0 \to 1$ in **FinSetInj**. The two minimal cocones over it are $1 \to 2 \leftarrow 1$ (the pushout in **FinSet**) and $1 \to 1 \leftarrow 1$. Since there is no injection $2 \to 1$, the pushout in **FinSet** is not a pushout in **FinSetInj** anymore. And indeed, the square (6) below is not a quasi-pullback: given two 1-pointed models (M, x) and (M', x') with $M \equiv M'$, it is possible that the only way to amalgamate them into a single 2-pointed model (N, a, b) with $(N, a) \equiv (M, x)$ and $(N, b) \equiv (M', x')$ is to take $a = b$. This means that instead of finding a point of $P_T[2]$ in the upper left corner of (6), we find a point of $P_T[1]$ in the upper left corner of (7).

$$\begin{array}{ccc} P_T[2] & \longrightarrow & P_T[1] \\ \downarrow & & \downarrow \\ P_T[1] & \longrightarrow & P_T[0] \end{array} \quad (6) \qquad \begin{array}{ccc} P_T[1] & \longrightarrow & P_T[1] \\ \downarrow & & \downarrow \\ P_T[1] & \longrightarrow & P_T[0] \end{array} \quad (7) \qquad \begin{array}{ccc} a & \xrightarrow{f} & b \\ {\scriptstyle g}\downarrow & & \downarrow{\scriptstyle u} \\ c & \xrightarrow{v} & d \end{array} \quad (8)$$

In order to take this into account, we need the following definitions.

Definition 3.2. *A functor* $P\colon \mathbf{C}^{\mathrm{op}} \to \mathbf{Set}$ *has the* amalgamation property *if for any diagram* $(f\colon a \to b, g\colon a \to c)$ *in* $\mathbf{C}$*, for any* $x \in P(b)$*,* $y \in P(c)$ *such that* $P(f)(x) = P(g)(y)$*, there exists a cocone as in (8) and a witness* $z \in P(d)$ *such that* $P(u)(z) = x$ *and* $P(v)(z) = y$*. Notice that* d *can also depend on* x *and* y*. A natural transformation is said to have the amalgamation property if each naturality square is quasi-cartesian. We use the same vocabulary for functors* $\mathbf{C}^{\mathrm{op}} \to \mathbf{Stone}$*. A natural transformation* $\alpha\colon P \to Q$ *has the* weak amalgamation property *if its naturality squares are weakly quasi-cartesian (Definition 2.10).*

We are now ready to generalize Definition 2.7 appropriately.

Definition 3.3. *Let* $\mathbf{C}$ *be a small category. A* $\mathbf{C}$-adic (Stone) space *is a functor* $P\colon \mathbf{C}^{\mathrm{op}} \to \mathbf{Stone}$ *with the amalgamation property and such that* $P(f)$ *is open for each arrow* f *in* $\mathbf{C}$*. A morphism of* $\mathbf{C}$*-adic spaces is a natural transformation with the amalgamation property.*

We will refer to $\mathbf{C}$ as the category of contexts. If $\mathbf{C}$ has pushouts, we recover the notion of Boolean hyperdoctrine since any cocone as in (8) factors through the pushout, and the amalgamation property says that pushout squares are sent to quasi-pullbacks. In particular, we get polyadic spaces with $\mathbf{C} = \mathbf{FinSet}$.

Theorem 3.4. *The functor* $[\mathbf{FinSetInj}^{\mathrm{op}}, \mathbf{Stone}] \to [\mathbf{FinSet}^{\mathrm{op}}, \mathbf{Stone}]$ *given by left Kan extension along* $\mathbf{FinSetInj}^{\mathrm{op}} \rightarrowtail \mathbf{FinSet}^{\mathrm{op}}$ *restricts to an equivalence between the category of* $\mathbf{FinSetInj}$*-adic spaces and the category of polyadic spaces.*

Given a first-order theory T, models of T can be defined in terms of $P_T[-]$ as a set X equipped with a natural transformation $\{\text{injections } n \rightarrowtail X\} \to P_T[n]$ with the weak amalgamation property.

3.2 Other Kernels

If we restrict our attention to smaller classes of first-order theories, it may be possible to consider more specialized kernels. We will only give here examples of kernels and no general definition.

Suppose T is a first-order theory with a distinguished linear order symbol. Let Δ_+ be the category of (possibly empty) finite linear orders and order-preserving maps. Then $P_T\langle n\rangle := \{\text{models of } T \text{ pointed by } n \text{ increasing points}\} \subseteq P_T(n)$ defines a functor $\Delta_+^{\mathrm{op}} \to \mathbf{Stone}$. Left Kan extension along the forgetful functor $\Delta_+ \to \mathbf{FinSet}$ yields an equivalence from the category of Δ_+-adic spaces to the category of polyadic spaces equipped with a linear order relation (morphisms need to respect this relation). Combining this with the idea of Stirling kernels, this category is again equivalent to the category of $\Delta_{+,\mathrm{inj}}$-adic spaces, where $\Delta_{+,\mathrm{inj}}$ is the category of finite linear orders and injective order-preserving maps.

An order is *bounded* if there is a greatest and a least element. Let Δ_{bound} be the category of finite bounded linear orders and bound-preserving order-preserving maps. Let $\Delta_{\mathrm{bound,inj}}$ be the subcategory of Δ_{bound} with only the injective morphisms. Then the Δ_{bound}-adic and $\Delta_{\mathrm{bound,inj}}$-adic spaces both correspond to first-order theories with a bounded linear order. The diagram (9) below summarizes the situation. Translation across different categories of contexts is done through left Kan extension along the arrows.

$$\begin{array}{ccccl}
\mathbf{FinSet} & \longleftarrow & \mathbf{FinSetInj} & & \text{first order theories} \\
\uparrow & & \uparrow & & \\
\Delta_+ & \longleftarrow & \Delta_{+,\mathrm{inj}} & & \text{with a linear order} \\
\uparrow & & \uparrow & & \\
\Delta_{\mathrm{bound}} & \longleftarrow & \Delta_{\mathrm{bound,inj}} & & \text{with a bounded linear order}
\end{array} \qquad (9)$$

As in the case of Stirling kernels, models are the intuitive ones: objects X with a natural transformation from $\{\text{maps } n \to X\}$ to $P(n)$ with the weak amalgamation property, where "objects" and "maps" are interpreted according to the situation. In general, a model of a **C**-adic space P is an ind-object $X \in \mathbf{Ind}(\mathbf{C})$ equipped with a natural transformation $X \to P$ with the weak amalgamation property. For an introduction to ind-objects, see [8, Chapter VI, Sect. 1].

Example 3.5. Here are some more examples of kernels. Items 5 and 6 will be used in Sect. 4.

1. The same thing works for ordered sets, graphs, pointed sets, sets with an equivalence relation, various notions of trees and forests,[2] etc.
2. If T is an algebraic theory and $\mathbf{Alg}^T_{\mathrm{fp}}$ is the category of finitely presented T-algebras, then $\mathbf{Alg}^T_{\mathrm{fp}}$-adic spaces correspond to first-order theories with a distinguished interpretation of T.
3. A $\mathbf{FinSet} \times \mathbf{FinSet}$-adic space is a theory in 2-sorted first order logic, or in other words a first-order theory with a distinguished unary relational symbol R (one sort is the interpretation of R and the other one is its complement). Taking the left Kan extension along $+\colon \mathbf{FinSet} \times \mathbf{FinSet} \to \mathbf{FinSet}$ forgets the distinguished unary relational symbol.
4. A pointed bounded linear order is a pair of bounded linear orders, so a $\Delta^2_{\mathrm{bound}}$-adic space is a Δ_{bound}-adic space with a distinguished constant. The translation is made via a left Kan extension along the functor $\boxplus\colon \Delta^2_{\mathrm{bound}} \to \Delta_{\mathrm{bound}}$ joining two bounded linear orders by gluing their endpoints.
5. If $P, Q\colon \Delta^{\mathrm{op}}_{\mathrm{bound}} \to \mathbf{Stone}$ are two Δ_{bound}-adic spaces, then $(a, b) \mapsto P(a) \times Q(b)$ is the $\Delta^2_{\mathrm{bound}}$-adic space whose models are bounded linear orders with a distinguished point, a P-structure on the lower half and a Q-structure on the upper half.
6. If P is a Δ_{bound}-adic space, then $(a, b) \mapsto P(a \boxplus b)$ is the $\Delta^2_{\mathrm{bound}}$-adic space whose models are pointed models of P. Actually, precomposition by $\boxplus$ defines a functor from Δ_{bound}-adic spaces to $\Delta^2_{\mathrm{bound}}$-adic spaces and it is the right adjoint to left Kan extension along $\boxplus$.
7. On the other hand, it is *false* that if P is a **FinSet**-adic space, then $(a, b) \mapsto P(a + b)$ is the $\mathbf{FinSet}^2$-adic space whose models are models of P with a distinguished subset, as it was the case for Δ_{bound}.

3.3 Some Properties of C-adic Spaces

Instead of **C**-adic spaces, we can consider **C**-adic *sets*. They are functors $\mathbf{C}^{\mathrm{op}} \to \mathbf{Set}$ with the amalgamation property. Under some hypothesis on **C**, they can be compactified to give **C**-adic spaces.

Proposition 3.6. *Let* **C** *be a small category. For any two arrows* $f\colon a \to b$, $g\colon a \to c$ *with a common domain, let* $\mathbf{C}_{f,g}$ *be the category of cocones over the diagram composed of* f *and* g*. Suppose that for each* $\mathbf{C}_{f,g}$*, there is a weakly initial finite set of objects* $T_{f,g}$*, i.e. such that each other object admits an arrow from an element of* $T_{f,g}$*. Then post-composition by* $\beta\colon \mathbf{Set} \to \mathbf{Stone}$ *gives a functor from* **C***-adic sets to* **C***-adic Stone spaces.*

[2] Trees and forests are considered as special ordered sets for the notion to be first-order.

Remark 3.7. The hypothesis of Property 3.6 above is also the condition for the duals of **C**-adic spaces to be a multisorted algebraic variety, with one sort for each object of **C**. A consequence of this algebraic presentation is that products of hyperdoctrines are computed pointwise and dually coproducts of **C**-adic spaces are computed pointwise. Another consequence is that subalgebras are again hyperdoctrines, but this is true more generally as seen in the proposition below.

Proposition 3.8. *Let* **C** *be a small category. Let* $P, Q\colon \mathbf{C}^{\mathrm{op}} \to \mathbf{Stone}$*. Suppose* P *is a* **C***-adic space. Let* $\alpha\colon P \to Q$ *be a natural transformation that is pointwise surjective. If* α *has the amalgamation property, then* Q *is a* **C***-adic space.*

Remark 3.9. A *polyadic subspace* of a polyadic space P is a morphism of polyadic spaces $Q \to P$ which is pointwise injective. If **C** has an initial object 0, then polyadic sub-spaces of P are classified by closed subsets of $P(0)$. In particular, P cannot be written as a non-trivial coproduct if and only if $P(0) = 1$, i.e. if the theory is "complete." If P is a **C**-adic *set* instead of a **C**-adic Stone space, then we can maximally decompose P as a coproduct indexed by points of $P(0)$.

Gödel's Completeness. We can generalize the statement of Gödel's completeness theorem to **C**-adic spaces. We will say that **C** satisfies Gödel's completeness theorem if for any **C**-adic space P, for any $X \in \mathbf{Ind}(\mathbf{C})$ and any natural transformation $\alpha\colon X \to P$, we can extend this "pre-model" to a model, i.e. find an ind-object $Y \in \mathbf{Ind}(\mathbf{C})$, a natural transformation $\tilde{\alpha}\colon Y \to P$ with the weak amalgamation property and a natural transformation $f\colon X \to Y$ such that $\alpha = \tilde{\alpha} \circ f$. When **C** has pushouts, the proof of the classical Gödel theorem with Henkin's models can be adapted. But this is not enough to cover all the cases where it is true, since it can sometimes be deduced from other categories of contexts through the use of kernels. For instance, none of the categories of Diagram (9) admit pushouts except for **FinSet**, but Gödel's completeness theorem for these categories can be deduced from **FinSet**.

When Gödel's completeness theorem holds, it allows for a description of morphisms in terms of models. Given a **C**-adic space P, we say that a family of clopen subsets of the spaces $P(c)$ is *generating* if for each $c \in \mathbf{C}$, each clopen subset of $P(c)$ can be expressed from the given family of clopen subsets using the Boolean operations together with direct and inverse images by maps of the form $P(f)$. If $\mathbf{C} = \mathbf{FinSet}$ and if $P = P_T$ is the polyadic space associated to a first-order theory T, then the clopen subsets corresponding to the symbols of the signature of T are generating.

Proposition 3.10. *Let* **C** *be a category satisfying Gödel's completeness theorem. Let* P *and* Q *be two* **C***-adic spaces. Then a morphism* $P \to Q$ *is given by a way of transforming, for any* $X \in \mathbf{Ind}(\mathbf{C})$*,* P*-structures on* X *into* Q*-structures on* X *such that for any clopen subset* $\varphi \subseteq Q(c)$*, there exists a clopen subset* $F(\varphi) \subseteq P(c)$ *defining the same subset of* $X(c)$ *for each* P*-model* X*. Moreover, it is enough to check this condition on a generating family of clopen subsets.*

Remark 3.11. This way of viewing morphisms of polyadic spaces respects composition. In particular, an isomorphism $P \cong Q$ is given by the data of a bijection between P-structures on X and Q-structures on X for each ind-object X, such that each subset of $X(c)$ definable over Q is definable over P (uniformly in X) and conversely. Beth's definability theorem, which is a consequence of Gödel's completeness theorem in this setting, says that this converse is automatically satisfied: any morphism $P \to Q$ of $\mathbf{C}$-adic spaces inducing a bijection between P-models and Q-models is an isomorphism.

4 Application to Logic on Words

4.1 Correspondence Between Theories and Monoids

As we saw in Subsect. 3.2, a first-order theory with a bounded linear order symbol is a special functor $\Delta_{\text{bound}}^{\text{op}} \to \mathbf{Stone}$. Looking at a bounded linear order as a lattice, the duality between finite lattices and finite posets specializes to a duality between Δ_{bound} and Δ_+. Hence, functors $\Delta_{\text{bound}}^{\text{op}} \to \mathbf{Stone}$ correspond to functors $\Delta_+ \to \mathbf{Stone}$ by precomposing with this equivalence. Functors $\Delta_+ \to \mathbf{Stone}$ can also be used to encode profinite monoids S, meaning that S is also a Stone space with continuous multiplication. The functor associated to S will be called P_S. It sends n to S^n, the injection $0 \to 1$ to the neutral element $e\colon 1 \cong S^0 \to S^1$ and the unique surjection $2 \to 1$ to the multiplication $S^2 \to S$.

Before we continue, we should introduce some notation. When we want to emphasize the ambient category, we will write the object of Δ_+ with n elements as $\underline{n}$, and the object of Δ_{bound} with n elements as $\underbracket{n}$. For convenience, we will also write $\underbracket{n} \mapsto P_S(\underbracket{n})$ for the functor obtained by precomposing $\underline{n} \mapsto P_S(\underline{n})$ with the equivalence $\Delta_{\text{bound}}^{\text{op}} \cong \Delta_+$. We will write $\underline{n} + \underline{m}$ for the linear order obtained by concatenating $\underline{n}$ and $\underline{m}$. The dual of this operation is $\underbracket{n} \boxplus \underbracket{m} = \underbracket{n - 1 + m}$, obtained by merging the last point of $\underbracket{n}$ with the first point of $\underbracket{m}$.

The condition on P_S to encode a monoid is that it is *monoidal*, meaning that $P_S(\underline{0}) = 1$ and that it comes equipped with an isomorphism $P_S(\underline{n}+\underline{m}) \cong P_S(\underline{n}) \times P_S(\underline{m})$ satisfying with some commutativity conditions, or dually an isomorphism $P_S(\underbracket{n} \boxplus \underbracket{m}) \cong P_S(\underbracket{n}) \times P_S(\underbracket{m})$. If we replace the duality $\Delta_+ \cong \Delta_{\text{bound}}^{\text{op}}$ with the duality $\Delta_{+,\text{surj}} \cong \Delta_{\text{bound,inj}}^{\text{op}}$, we get the same thing but with semigroups instead of monoids. Left Kan extension along $\Delta_{+,\text{surj}} \rightarrowtail \Delta_+$ freely adjoins a neutral element. Note that a monoidal functor is obtained by adding some structure to a functor, so that we will speak of monoidal functor structures.

The two questions raised by this connection are:

1. When does a monoid correspond to a theory with a bounded linear order symbol? More precisely, under which conditions on S is P_S a Δ_{bound}-adic space? (Theorem 4.3.)
2. When does a theory with a bounded linear order symbol correspond to a monoid? More precisely, given a Δ_{bound}-adic space P, how can we understand a monoidal functor structure on P from the viewpoint of logic? (Theorem 4.8.)

From Monoids to Theories. Let S be a semigroup. The amalgamation property of $P_S \colon \underline{n} \mapsto S^n$ applied to the diagram $\underline{2} \to \underline{1} \leftarrow \underline{2}$ gives the definition below. The three cases correspond to the three minimal cones over $\underline{2} \to \underline{1} \leftarrow \underline{2}$ in $\Delta_{+,\mathrm{surj}}$.

Definition 4.1. *A monoid or semigroup S is* equidivisible *if for any $a, b, x, y \in S$ such that $ab = xy$, either $(a, b) = (x, y)$ or there is a $k \in S$ such that $ak = x$ and $ky = b$, or $xk = a$ and $kb = y$. In the case of monoids, the definition can be simplified by taking $k = e$ when $(a, b) = (x, y)$.*

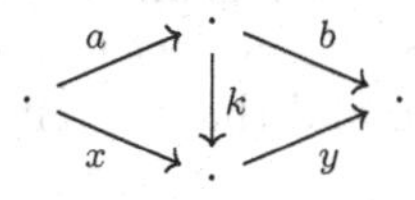

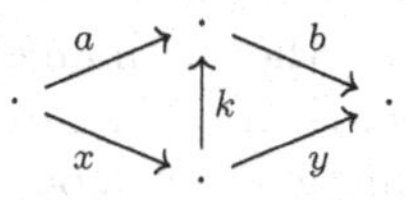

As we see in the next proposition, equidivisibility is enough to recover the full amalgamation property.

Proposition 4.2. *Let S be a semigroup. Then P_S has the amalgamation property if and only if S is equidivisible. If S is a monoid, then P_S has the amalgamation property if and only if S is obtained by freely adjoining an identity to an equidivisible semigroup.*

As a consequence, we obtain the following theorem.

Theorem 4.3. *Let S be a profinite semigroup. Then P_S is a $\Delta_{\mathrm{bound,inj}}$-adic space if and only if S is equidivisible and its multiplication is open.*

Let S be a profinite monoid. Then P_S is a Δ_{bound}-adic space if and only if S is obtained from a profinite equidivisible semigroup with open multiplication by freely adjoining an isolated neutral element.

Let S be a profinite monoid as in Theorem 4.3. Here is a description of what is a P_S-model, or S-model for short.

Proposition 4.4. *Let X be a bounded linear order. Think of X as a category with* $\mathrm{Hom}(x, y) = 1$ *if $x \leq y$ and 0 otherwise. Think of S as a category with one object. Then a structure of S-model on X is given by a functor $F \colon X \to S$ such that, writing $F(x, y)$ for the image of the unique arrow $x \to y$ when it exists, we have $F(x, y) = e \iff x = y$ and for any clopen subset $U, V \subseteq S$ and any $x \leq y$ in X,*

$$F(x, y) \in UV \iff \exists z \in [x, y] : F(x, z) \in U \wedge F(z, y) \in V. \tag{10}$$

Remark 4.5. If we replace (10) by the following, we get the notion of ω-saturated model (cf. Remark 2.12).

$$F(x, y) = uv \iff \exists z \in [x, y] : F(x, z) = u \wedge F(z, y) = v.$$

Remark 4.6. There is also a natural axiomatization of this theory. Each clopen subset $U \subseteq S$ corresponds to a binary relational symbol on X that can be satisfied by two points x, y only if $x \leq y$. We write $U(x, y)$ for this symbol. Here are the axioms.

1. All the axioms for bounded linear orders.
2. $U(x,y) \implies x \leq y$
3. $[U \cap V](x,y) \iff U(x,y) \wedge V(x,y)$ and similarly for all Boolean operations.
4. $\{e\}(x,y) \iff x = y$
5. $[UV](x,y) \iff \exists z : U(x,z) \wedge V(z,y)$

The set of axioms 3 ensures that each couple of points determines a prime filter on the dual Boolean algebra of S and thus a point of S. Predicates on n ordered free variables (i.e. under the condition $x_1 \leq \cdots \leq x_n$) are identified to clopen subsets of S^{n+1}. An n-type of this theory can be described as an ordering of the n variables together with an element $(a_0, \ldots, a_n) \in S^{n+1}$, subject to the relation that if $a_i = e$, then we can exchange the ith variable and the $(i+1)$th variable in the ordering without modifying the n-type.

From Theories to Monoids. Let's now turn to the second question: Given a Δ_{bound}-adic space P, how can we understand a monoidal functor structure on P from the viewpoint of logic? We will first study the structure transferred from S to P_S. Let S be a profinite monoid such that P_S is a Δ_{bound}-adic space. Proposition 4.7 below shows that the monoid structure of S underlies a monoid structure on the class of S-models.

Given two bounded linear orders X and Y, let $X \boxplus Y$ be their concatenation obtained by gluing $\max X$ and $\min Y$. If $F\colon X \to S$ and $G\colon Y \to S$ are two S-models, we define $F \boxplus G\colon X \boxplus Y \to S$ by

$$(F \boxplus G)(x,y) = \begin{cases} F(x,y) & \text{if } x, y \in X, \\ G(x,y) & \text{if } x, y \in Y, \\ F(x, \max X)G(\min Y, y) & \text{if } x \in X,\ y \in Y. \end{cases}$$

Proposition 4.7. *Let S be a profinite monoid such that P_S is a Δ_{bound}-adic space. Let $F\colon X \to S$ and $G\colon Y \to S$ be two S-models. Then $F \boxplus G$ is also an S-model.*

Proof. We need to prove that if $(F \boxplus G)(x,y) \in UV$ with $U, V \subseteq S$ clopen subsets, $x \in X$ and $y \in Y$, then there is some z in X or some z in Y such that $(F \boxplus G)(x,z) \in U$ and $(F \boxplus G)(z,y) \in V$. Let $a := F(x, \max X)$ and $b := G(\min Y, y)$. We know that $ab = uv$ for some $u \in U$ and $v \in V$. Because of the amalgamation property, this means that, without loss of generality, there is some $k \in S$ such that $kb = v$ and $uk = a$. We deduce that $a \in U(V/b)$. Hence, there exists some $z \in X$ such that $F(x,z) \in U$ and $F(z, \max X) \in V/b$. We get $(F \boxplus G)(x,z) \in U$ and $(F \boxplus G)(z,y) = F(z, \max X)G(\min Y, y) \in V$ as desired. □

This operation of concatenation can be explained abstractly as follows. Let P, Q be two Δ_{bound}-adic spaces. The theory whose models are pairs of models, one of P and one of Q, is represented by the functor $P \times Q\colon \Delta_{\text{bound}} \times \Delta_{\text{bound}} \to$ **Stone** sending (n,m) to $P(n) \times Q(m)$. As said in example 3.5, left Kan extension along $\boxplus\colon \Delta_{\text{bound}} \times \Delta_{\text{bound}} \to \Delta_{\text{bound}}$ (the functor of concatenation by gluing the endpoints) translates a $\Delta_{\text{bound}} \times \Delta_{\text{bound}}$-adic space as a Δ_{bound}-adic space by forgetting the distinguished point. We will write $P * Q$ for the left Kan extension of $P \times Q$ along $\boxplus$. It is called the *Day convolution* of P and Q and we have the coend formula $[P * Q](n) = \int^{(a,b)} P(a) \times Q(b) \times \mathrm{Hom}(n, a \boxplus b)$.

As in Example 3.5 again, if P is a Δ_{bound}-adic space, then the $\Delta_{\text{bound}} \times \Delta_{\text{bound}}$-adic space $(a, b) \mapsto P(a \boxplus b)$ represents the theory whose models are pointed models of P. Let P_* be the left Kan extension of this functor along $\boxplus$. Then P_* is the theory P with a distinguished constant freely added and the counit $P_* \to P$ forgets this new constant. A monoidal functor structure on P is given by an isomorphism $P(a) \times P(b) \cong P(a \boxplus b)$, i.e. an isomorphism $P * P \cong P_*$. By composing $P * P \cong P_* \to P$, we get the concatenation of Proposition 4.7.

If X and Y are two bounded linear orders, thought of as ind-objects $\Delta^{\text{op}}_{\text{bound}} \to$ **Set**, then their Day convolution $X * Y$ is $X \boxplus Y$. To concatenate two models $X \to P$ and $Y \to P$, we compose $X * Y \to P * P \to P$.

Recall from Proposition 3.10 and Remark 3.11 (using Beth's definability theorem) that an isomorphism between Δ_{bound}-adic spaces $P(a) \times P(b) \cong P(a \boxplus b)$ can be seen at the level of models. The theorem below makes this more explicit in the present case.

Theorem 4.8. *Let T be a first-order theory with a bounded linear order symbol. Let $P\colon \Delta^{\text{op}}_{\text{bound}} \to$ **Stone** be the corresponding Δ_{bound}-adic space. Suppose that T has exactly one model of cardinality 1, or in other words that $P(\underset{\vdash\!\dashv}{1}) = 1$. Then monoidal functor structures on P are in correspondence with families indexed by pointed bounded linear orders (X, p) of bijections between T-structures on X and pairs of T-structures on the two segments $\{x \in X \mid x \leq p\}$ and $\{x \in X \mid x \geq p\}$, subject to the following conditions:*

1. *The induced concatenation of T-models is associative, or equivalently given a bipointed T-model (X, a, b), the two induced T-structures on $[a, b] \subseteq X$ are equal.*
2. *The unique model of cardinality 1 acts as a neutral element for concatenation.*
3. *Formulas can be relativized to segments: given any formula $\varphi(x_1, \ldots, x_n)$ on n ordered variables ($x_1 \leq \cdots \leq x_n$), there is some formula $\psi(x_1, \ldots, x_n, p)$ on $n+1$ ordered variables such that an $(n+1)$-pointed model $(X, x_1, \ldots, x_n, p)$ of P with $x_1 \leq \cdots \leq x_n \leq p$ satisfies ψ if and only if $(\{x \in X \mid x \leq p\}, x_1, \ldots, x_n)$ satisfies φ. Symmetrically for $p \leq x_1 \leq \cdots \leq x_n$.*

Moreover, it is enough to check the last condition on the symbols of the signature of T other than the order.

Example 4.9. As an example of a profinite monoid built using Theorem 4.8, we can take T to be simply the theory of bounded linear orders. Since the only symbol of the signature is the order, there is almost nothing to check. This monoid is uncountable [15, Corollary 6.17]. If T is instead the theory of successor ordinals, then the monoid becomes countable and has been described in [3, Definition 39].

4.2 Application to Logic on Words

Logic on Words. The goal now is to give a presentation of logic on words from the viewpoint of polyadic spaces. Let A be a finite alphabet. Since A^* is equidivisible, the functor $\underline{n} \mapsto [A^*]^n \in \mathbf{Set}$ has the amalgamation property, meaning that it is a Δ_{bound}-adic set. By Property 3.6, the Stone-Čech compactification $\underline{n} \mapsto \beta([A^*]^n)$ is a Δ_{bound}-adic space whose dual is $\underline{n} \mapsto \mathscr{P}([A^*]^n)$. We do not get a profinite monoid this way, but we will obtain some equidivisible profinite monoids as *quotients* of $\beta([A^*]^n)$ in Corollary 4.10.

Let us describe the associated logic. Each finite word over A corresponds to a model whose points are positions between two letters, before the first or after the last one, so that a word with k letters gives a model with $k+1$ elements. A predicate with n free variables is, in the most general possible way, a subset of all the n-pointed finite models. Looking at the corresponding Δ_{bound}-adic space, since $\underline{n}$-pointed words can be identified with $[A^*]^{n-1}$, a predicate in context $\underline{n}$ is a subset of $[A^*]^{n-1}$.

As seen in Remark 3.9, each Δ_{bound}-adic set can be decomposed uniquely as a coproduct of irreducible Δ_{bound}-adic sets P with $P(\underline{2}) = 1$. In the present case, this decomposition gives that $\underline{n} \mapsto [A^*]^n$ is the coproduct of all its finite models. This means that a morphism of Δ_{bound}-adic spaces from $\beta([A^*]^n)$ to $P(\underline{n})$ is a way of interpreting each finite word with k letters as a model of P with $k+1$ points.

The Logic Associated to a Boolean Algebra of Languages. Let $B \subseteq \mathscr{P}(A^*)$ be a Boolean algebra of languages and make the following assumptions.

1. The morphism $\mathscr{P}(A^*) \to \mathscr{P}(A^* \times A^*)$ dual to multiplication $A^* \times A^* \to A^*$ sends $B \subseteq \mathscr{P}(A^*)$ into $B \sqcup B \subseteq \mathscr{P}(A^* \times A^*)$.[3] This is equivalent to B being included in regular languages and stable by division on the left and on the right, see [5].
2. B is stable under concatenation and contains $\{\varepsilon\}$ where ε is the empty word.

Under condition 1, the whole family $B^{\sqcup n} \subseteq \mathscr{P}([A^*]^n)$ defines a subfunctor and the Stone dual M of B inherits the structure of a profinite monoid, the multiplication being dual to $B \to B \sqcup B$. Condition 2 ensures that the $B^{\sqcup n} \subseteq \mathscr{P}([A^*]^n)$ are stable under the action of the left adjoints. Since hyperdoctrines are algebraic systems (cf. Remark 3.7), the subalgebra $B^{\sqcup n}$ is again a hyperdoctrine. We can also dualize and use Proposition 3.8 instead: quotients of polyadic spaces are still polyadic spaces.

Corollary 4.10. *Under conditions 1 and 2, the dual M of B is an equidivisible profinite monoid with open multiplication, whose neutral element is the only invertible and is topologically isolated. In particular, the free profinite monoid on A is equidivisible.*

---

[3] The set $B \sqcup B \subseteq \mathscr{P}(A^* \times A^*)$ is the Boolean subalgebra generated by sets of the form $L \times A^*$ and $A^* \times L$ where $L \in B$. It is also the coproduct of two copies of B in the category of Boolean algebras.

By Theorem 4.3, each such Boolean algebra has an associated first-order logic whose predicates in context $\underline{n}$ are the Boolean algebra $B^{\sqcup(n-1)} \subseteq \mathscr{P}([A^*]^{n-1})$. The conditions 1 and 2 ensure that these subsets are stable under the constructions of first-order logic. Here is an example to illustrate more concretely the roles of both conditions.

Example 4.11. Suppose L and L' are two languages in B. Let $[L](-,-)$ be the 2-ary predicate associated to L (formally, it is $A^* \times L \times A^*$). Let $\bot$, $\top$ be the constants for the minimum and the maximum. For instance, $[L](\bot,\top)$ is the formula saying that the whole word is in L. The formula

$$\varphi := \forall x : [L](x,\top) \implies \exists y : y \geq x \wedge [L'](\bot,y)$$

expresses that for each suffix in L, there is a prefix in L' ending at least at the start of the suffix. Let us show that the language associated to φ is in B. First, we need to place the formula $[L'](\bot,y)$ in a context where there is one more variable $x \leq y$. This is done using condition 1: there exists a finite list $(U_i,V_i)_{i=1}^n \subseteq B \times B$ such that we can rewrite $[L'](\bot,y)$ as $\bigvee_{i=1}^n [U_i](\bot,x) \wedge [V_i](x,y)$. We then apply the quantifier $\exists y$ using condition 2 and combine with "$[L](x,\top) \implies$", this gives $[\overline{L}](x,\top) \vee \bigvee_{i=1}^n [U_i](\bot,x) \wedge [V_iA^*](x,\top)$. In order to apply the universal quantifier, we negate the formula, put it in normal disjunctive form, apply the existential quantification and negate again to obtain the final formula

$$\bigcap_{s:n\longrightarrow\{0,1\}} \overline{\left(\bigcap_{i\in s^{-1}(0)} \overline{U_i}\right)\left(\overline{L\cap \bigcap_{i\in s^{-1}(1)} \overline{V_iA^*}}\right)}.$$

Example 4.12. Here are some Boolean algebras satisfying conditions 1 and 2.

1. The Boolean algebra of all regular languages. The associated logic is Büchi's monadic second-order logic on words.
2. The Boolean algebra of star-free languages. In [6], van Gool and Steinberg apply model theory to study the free pro-aperiodic monoid A. It is built as a submonoid of another monoid $\Lambda(A)$, which is constructed in our setting by applying Theorem 4.8 to the theory of discrete orders with endpoints whose pairs of consecutive points are labeled by A. Since each finite word is in particular a discrete labeled order, we have a morphism of polyadic spaces $\beta([A^*]^n) \to \Lambda(A)^n$ and the free pro-aperiodic monoid is the image of this morphism. We can interpret [6, Lemma 3.4, p. 13] as saying that this image is a polyadic subspace.
3. The Boolean algebra of languages of generalized star-height n.
4. Other examples can be found in [1, Sect. 3], where one can find a characterization of the pseudovarieties recognizing Boolean algebras of languages satisfying condition 2 (condition 1 is automatically satisfied).

Acknowledgments. This work has been done under the supervision of Mai Gehrke and André Joyal.

References

1. Almeida, J., et al.: The linear nature of pseudowords. In: Publicacions Matemàtiques, vol. 63, pp. 361–422, July 2019. ISSN: 0214-1493. https://doi.org/10.5565/publmat6321901
2. Borceux, F.: Handbook of Categorical Algebra 1 - Basic Category Theory. Cambridge University Press, Cambridge (1994). https://doi.org/10.1017/CBO9780511525858
3. Doner, J.E., Mostowski, A., Tarski, A.: The elementary theory of well-odering - a metamathematical study. In: Logic Colloquium 1977. Studies in Logic and the Foundations of Mathematics, vol. 96, pp. 1–54. Elsevier (1978). https://doi.org/10.1016/S0049-237X(08)71988-8
4. Gehrke, M.: Stone duality, topological algebra, and recognition. J. Pure Appl. Algebra **220**(7), 2711–2747 (2016)
5. Gehrke, M., Grigorieff, S., Pin, J.É.: A topological approach to recognition. In: Abramsky, S., Gavoille, C., Kirchner, C., Meyer auf der Heide, F., Spirakis, P.G. (eds.) ICALP 2010. LNCS, vol. 6199, pp. 151–162. Springer, Heidelberg (2010). https://doi.org/10.1007/978-3-642-14162-1_13
6. van Gool, S.J., Steinberg, B.: Pro-aperiodic monoids via saturated models. Israel J. Math. **234**(1), 451–498 (2019). https://doi.org/10.1007/s11856-019-1947-6
7. Hodges, W.: A Shorter Model Theory. Cambridge University Press, Cambridge (1997). ISBN: 0521587131
8. Johnstone, P.: Stone Spaces, vol. xxi, p. 370. Cambridge University Press, Cambridge (1982). ISBN: 0521238935
9. Joyal, A.: Polyadic spaces and elementary theories. In: Notices of the American Mathematical Society, April 1971
10. Lambek, J., Scott, P.J.: Introduction to Higher Order Categorical Logic. Cambridge University Press, Cambridge (1986). ISBN: 0521246652
11. Lawvere, F.W.: Adjointness in foundations. In: Dialectica (1969). http://citeseerx.ist.psu.edu/viewdoc/summary?doi=10.1.1.386.6900
12. Marquis, J.-P., Reyes, G.E.: The history of categorical logic: 1963–1977. In: Sets and Extensions in the Twentieth Century. Handbook of the History of Logic, pp. 689–800. Elsevier (2012). https://doi.org/10.1016/B978-0-444-51621-3.50010-4
13. Pin, J.-E.: Logic on words. In: Current trends in theoretical computer science, pp. 254–273. World Scientific Publishing (2001)
14. Rasiowa, H., Sikorski, R.: Proof of the completeness theorem of Gödel. In: Fundamenta Mathematicae **37**(1), 193–200 (1950). http://eudml.org/doc/213213
15. Rosenstein, J.G.: Linear Orderings, vol. xvii, p. 487. Academic Press, New York (1981). ISBN: 0125976801
16. Seely, R.A.G.: Hyperdoctrines, natural deduction and the beck condition. Math. Logic Q. **29**(10), 505–542 (1983). https://doi.org/10.1002/malq.19830291005

Time Warps, from Algebra to Algorithms

Sam van Gool[1], Adrien Guatto[1], George Metcalfe[2(✉)], and Simon Santschi[2]

[1] IRIF, Université de Paris, Paris, France
{vangool,guatto}@irif.fr
[2] Mathematical Institute, University of Bern, Bern, Switzerland
{george.metcalfe,simon.santschi}@math.unibe.ch

Abstract. Graded modalities have been proposed in recent work on programming languages as a general framework for refining type systems with intensional properties. In particular, continuous endomaps of the discrete time scale, or *time warps*, can be used to quantify the growth of information in the course of program execution. Time warps form a complete residuated lattice, with the residuals playing an important role in potential programming applications. In this paper, we study the algebraic structure of time warps, and prove that their equational theory is decidable, a necessary condition for their use in real-world compilers. We also describe how our universal-algebraic proof technique lends itself to a constraint-based implementation, establishing a new link between universal algebra and verification technology.

Keywords: Residuated lattices · Universal algebra · Decision procedures · Graded modalities · Type systems · Programming languages

1 Introduction

Program types are almost as old as programs themselves. Their initial role was to allow compilers to determine data sizes at compilation time, e.g., distinguishing machine integers from double precision numbers [1]. Type system research has developed tremendously since these humble beginnings, benefiting from close connections to logic [16]. For example, dependent types are expressive enough to serve as specification languages for program results [24,25].

Another line of research into type systems aims to classify not only *what* programs compute, but also *how* they do so. Such type systems describe the *effect* of a program—e.g., which parts of memory it modifies [19]—or the *resources* it requires—e.g., how long it takes to run [12]. Recently, *graded modalities* [7,8] have emerged as a unified setting for describing effect- and resource-annotated types. A graded modality $\Box$ allows programmers to form a new type $\Box_f A$ from a type A and a *grading* f. The meaning of $\Box_f A$ depends on the system at hand, but can generally be understood as a modification of A that includes the behavior prescribed by f.

Supported by Swiss National Science Foundation grant 200021_165850.

U. Fahrenberg et al. (Eds.): RAMiCS 2021, LNCS 13027, pp. 309–324, 2021.
https://doi.org/10.1007/978-3-030-88701-8_19

In many cases, gradings come equipped with an ordered algebraic structure that is relevant for programming applications. Most commonly, they form a monoid whose binary operation corresponds to a notion of composition such that $\Box_{gf}A$ is related to $\Box_f\Box_g A$. It is also often the case that gradings can be ordered by some sort of *precision* ordering along which the graded modality acts contravariantly. That is, we have a *generic* program of type $\Box_g A \to \Box_f A$ if $f \leq g$, allowing us to freely move from more to less precise types. As a consequence, the structure of this ordering is reflected by the operations available on types; for example, when the infimum of f and g exists, it permits the conversion of two values of types $\Box_f A$ and $\Box_g B$ into a single value of type $\Box_{f\wedge g}(A \times B)$.

The additional flexibility and descriptive power gained by adopting graded modalities in a programming language comes at a price, however. The language implementation must now be able to manipulate gradings in various ways; in particular, it should be able to decide the ordering between gradings in order to distinguish between well-typed and ill-typed programs. In this paper, we address this issue for a specific class of gradings known as *time warps*: sup-preserving functions on $\omega^+ = \omega \cup \{\omega\}$, or, equivalently, monotonic functions $f\colon \omega^+ \to \omega^+$ satisfying $f(0) = 0$ and $f(\omega) = \bigvee\{f(n) \mid n \in \omega\}$ [14]. Informally, time warps describe the growth of data along program execution. In this setting, any type A describes a family of sets $(A_n)_{n\in\omega}$, where A_n is the set of values classified by A at execution step n. The type $\Box_f A$ classifies the set of values of $A_{f(n)}$ at step n. This typing discipline generalizes a long line of works on programming languages for embedded systems [5] and type theories with modal recursion operators [2,22].

Let us denote the set of time warps by $\mathscr{W}$. Then $\langle \mathscr{W}, \circ, id\rangle$ is a monoid, where $fg := f \circ g$ denotes the composition of $f, g \in \mathscr{W}$, and id is the identity function. Moreover, equipping $\mathscr{W}$ with the pointwise order, defined by

$$f \leq g \;:\Longleftrightarrow\; f(p) \leq g(p) \text{ for all } p \in \omega^+,$$

yields a complete distributive lattice $\langle \mathscr{W}, \wedge, \vee\rangle$ satisfying, for all $f, g_1, g_2, h \in \mathscr{W}$,

$$f(g_1 \vee g_2)h = fg_1h \vee fg_2h \quad\text{and}\quad f(g_1 \wedge g_2)h = fg_1h \wedge fg_2h,$$

with a least element $\bot$ that maps all $p \in \omega^+$ to 0, and a greatest element $\top$ that maps all $p \in \omega^+\backslash\{0\}$ to ω. Note that the operation $\circ$ is a double quasi-operator on this lattice in the sense of [10,11], and that the structure $\langle \mathscr{W}, \wedge, \vee, \circ, id\rangle$ belongs to the family of unital quantales of sup-preserving functions on a complete lattice studied in [23].

The monoidal structure of time warps plays the expected role in programming applications. In particular, $\Box_{gf}A$ and $\Box_f\Box_g A$ are isomorphic, as are $\Box_{id}A$ and A. However, time warps also admit further additional algebraic structure of interest for programming. Since they are sup-preserving, there exist binary operations $\backslash, /$ on $\mathscr{W}$, called *residuals*, satisfying for all $f, g, h \in \mathscr{W}$,

$$f \leq h/g \iff fg \leq h \iff g \leq f\backslash h.$$

From a programming perspective, residuals play a role similar to that of weakest preconditions in deductive verification. The type $\Box_{h/g}A$ can be seen as the *most general* type B such that $\Box_h A$ can be sent generically to $\Box_g B$. Similarly, $f\backslash h$ is the most general (largest) time warp f' such that $\Box_h A$ can be sent generically to $\Box_{f'}\Box_f A$. Such questions arise naturally when programming in a modular way [14], justifying the consideration of residuated structure in gradings.

The algebraic structure $\mathbf{W} = \langle \mathscr{W}, \wedge, \vee, \circ, \backslash, /, id, \bot, \top \rangle$, referred to here as the *time warp algebra*, belongs to the family of (bounded) residuated lattices, widely studied as algebraic semantics for substructural logics [3,9,20]. The main goal of this paper is to prove the following theorem, a necessary condition for the use of time warps in real-world compilers:

Theorem 1. *The equational theory of the time warp algebra* $\mathbf{W}$ *is decidable.*

A *time warp term* is a member of the term algebra over a countably infinite set of variables of the algebraic language with binary operation symbols $\wedge, \vee, \circ, \backslash, /$, and constant symbols $id, \bot, \top$, and a *time warp equation* consists of an ordered pair of terms s, t, denoted by $s \approx t$. Let $s \le t$ denote the equation $s \wedge t \approx s$, noting that $\mathbf{W} \models s \approx t$ if, and only if, $\mathbf{W} \models s \le t$ and $\mathbf{W} \models t \le s$, and, by residuation, $\mathbf{W} \models s \le t$ if, and only if, $\mathbf{W} \models id \le t/s$. Clearly, to prove Theorem 1, it will suffice to provide an algorithm that decides $\mathbf{W} \models id \le t$ for any time warp term t.

Overview of the Proof of Theorem 1

We prove Theorem 1 by describing an algorithm with the following behavior:

Input. A time warp term t in the variables $x_1, \dots, x_k$.
Output. If $\mathbf{W} \models id \le t$, the algorithm returns '`Valid`'; if $\mathbf{W} \not\models id \le t$, the algorithm returns '`Invalid` at $(\hat{f}_1, \dots, \hat{f}_k, p)$' for some $p \in \omega^+$ and finite descriptions $\hat{f}_1, \dots, \hat{f}_k$ of time warps $f_1, \dots, f_k$, such that $[\![t]\!](p) < p$, where $[\![t]\!]$ is the time warp obtained from t by mapping each x_i to f_i.

We now give a high-level overview of the three main steps of the algorithm; the details and the proof of its correctness will occupy us for the rest of the paper.

I. Pre-processing into a Disjunction of Basic Terms. In Sect. 2, we show how to effectively obtain for any time warp term t, a time warp term

$$t' := \bigwedge_{i=1}^{m} \bigvee_{j=1}^{n_i} t_{i,j},$$

such that $\mathbf{W} \models t \approx t'$, where each $t_{i,j}$ is a *basic term*, constructed using $\circ$, id, $\bot$, and the defined operations $s^\ell := id/s$, $s^\mathsf{r} := s\backslash id$, and $s^\circ := \top\backslash s$ (Theorem 9). Since $\mathbf{W} \models id \le t$ if, and only if, $\mathbf{W} \models id \le \bigvee_{j=1}^{n_i} t_{i,j}$ for each $i \in \{1, \dots, m\}$, our task is reduced to giving an algorithm with the required behavior for terms of the form $t_1 \vee \dots \vee t_n$, where each t_i is a basic term. Once we have an algorithm that solves this case, we can run it for each of the m conjuncts of t' in turn, returning '`Invalid` at $(\hat{f}_1, \dots, \hat{f}_k, p)$' whenever this is the result of one of these runs, and otherwise '`Valid`'.

II. Finitary Characterization Through Diagrams. The crucial step in our algorithm is the finitary characterization of 'potential counterexamples' for an equation of the form $id \leq t_1 \vee \cdots \vee t_n$, where each t_i is a basic term. Our main tool for providing these finitary characterizations is the notion of a *diagram*.[1]

Let us give an example to illustrate the basic idea. To falsify the equation $id \leq xyx^\ell \vee y^\ell$ in $\mathbf{W}$, it suffices to find time warps f_x and f_y, and an element $p \in \omega^+$, such that $(f_x \circ f_y \circ f_x^\ell)(p) < p$ and $f_y^\ell(p) < p$. Although time warps are, as functions on ω^+, infinite objects, only *finitely many* of the values of f_x and f_y are relevant for falsifying the equation. Moreover, an upper bound for the number of values required for such a counterexample can be computed. The condition $(f_x \circ f_y \circ f_x^\ell)(p) < p$ is 'unravelled' by stating that there exist $\alpha_1, \alpha_2, \alpha_3 \in \omega^+$ such that $\alpha_3 < p$, where $\alpha_1 := f_x^\ell(p)$, $\alpha_2 := f_y(\alpha_1)$, and $\alpha_3 := f_x(\alpha_2)$. More formally, using a 'time variable' κ to refer to the value p, we build a finite *sample set* $\Gamma_1 \supseteq \{\kappa, x^\ell[\kappa], y[x^\ell[\kappa]], x[y[x^\ell[\kappa]]]\}$, where Γ_1 is 'saturated' with extra conditions used to describe, e.g., the behavior of f_x^ℓ at relevant values. Similarly, we obtain a finite saturated sample set $\Gamma_2 \supseteq \{\kappa, y^\ell[\kappa]\}$ for the condition $f_y^\ell(p) < p$. The problem of deciding if there exists a counterexample to $id \leq xyx^\ell \vee y^\ell$ then becomes the problem of deciding if there exists a suitable function $\delta \colon \Gamma_1 \cup \Gamma_2 \to \omega^+$ satisfying $\delta(x[y[x^\ell[\kappa]]]) < \delta(\kappa)$ and $\delta(y^\ell[\kappa]]]) < \delta(\kappa)$. In particular, δ should determine partial sup-preserving functions $\hat{f}_x$ and $\hat{f}_y$ on ω^+ satisfying $\hat{f}_x(\delta(\alpha)) = \delta(x[\alpha])$ for all $x[\alpha] \in \Gamma_1 \cup \Gamma_2$, and $\hat{f}_y(\delta(\alpha)) = \delta(y[\alpha])$ for all $y[\alpha] \in \Gamma_1 \cup \Gamma_2$.

Clearly, not every function δ from a saturated sample set to ω^+ extends to a valuation in $\mathbf{W}$; e.g., if $\delta(\kappa) = 0$, then we must also have $\delta(x[\kappa]) = 0$. Moreover, although time warp equations in the residual-free language can be decided by considering an algebra of sup-preserving functions on a *finite* totally ordered set, this is not the case for the full language.[2] Section 3 develops a general theory that precisely characterizes the functions—called *diagrams*—that extend to valuations and can be used to falsify a given equation. This allows us to prove that there exists a counterexample to $id \leq t_1 \vee \cdots \vee t_n$ if, and only if, there exists a diagram $\delta \colon \Gamma \to \omega^+$ satisfying $\delta(\kappa) > \delta(t_i[\kappa])$ for each $i \in \{1, \ldots, n\}$, where Γ is the finite saturated sample set extending $\{t_1[\kappa], \ldots, t_n[\kappa]\}$ (Theorem 31).

III. Encoding as a Satisfiability Query. In the last step of the algorithm, described in Sect. 4, we use the decidability of the satisfiability problem in the first-order logic of natural numbers with the natural ordering and successor. More precisely, we show that the existence of a diagram in Theorem 31 can be encoded as an existential first-order sentence in that signature. Concretely, our algorithm constructs a quantifier-free formula which is satisfiable in the structure $(\mathbb{N}, \leq, S, 0)$ if, and only if, there exists a diagram as specified by Theorem 31.

[1] The name 'diagram' recalls a similar concept used to prove the decidability of the equational theory of lattice-ordered groups in [15].

[2] Indeed, the equational theory of the time warp algebra without residuals coincides with the equational theory of distributive lattice-ordered monoids [6], but an elegant (finite) axiomatization of the equational theory in the full language is not known.

Moreover, a satisfying assignment can be converted into a valuation into **W** that provides a counterexample to the equation $id \leq t_1 \vee \cdots \vee t_n$.

2 A Normal Form for Time Warps

The main aim of this section is to provide a normal form for time warp terms. Our first step is to provide a more precise description of the left and right residuals of time warps. Note that to prove that two time warps are equal, it suffices to show that they coincide on every non-zero natural number, since for any time warp f, it is always the case that $f(0) = 0$ and $f(\omega) = \bigvee\{f(n) \mid n \in \omega\}$.

Lemma 2. *For any time warps f, g and $p \in \omega^+$,*

$$\text{(a)}\ \ (f\backslash g)(p) = \begin{cases} 0 & \textit{if } p = 0; \\ \bigvee\{q \in \omega^+ \mid f(q) \leq g(p)\} & \textit{if } p \in \omega\backslash\{0\}; \\ \bigvee\{q \in \omega^+ \mid (\exists m \in \omega)(f(q) \leq g(m))\} & \textit{if } p = \omega \end{cases}$$

$$\text{(b)}\ \ (g/f)(p) = \bigwedge g[\{q \in \omega^+ \mid p \leq f(q)\}].$$

Proof. (a) Let h denote the function defined by cases on the right of the equation. Clearly, h is monotonic and satisfies $h(0) = 0$ and $h(\omega) = \bigvee\{h(n) \mid n \in \omega\}$, so h is a time warp. Moreover, since f preserves arbitrary joins, $fh \leq g$, and hence $h \leq f\backslash g$. For the converse, just observe that for any $n \in \omega\backslash\{0\}$, since $f((f\backslash g)(n)) \leq g(n)$, also $(f\backslash g)(n) \leq h(n)$. So $h = f\backslash g$.

(b) Let h be the function defined by $h(p) := \bigwedge g[\{q \in \omega^+ \mid p \leq f(q)\}]$. Clearly, h is monotonic and satisfies $h(0) = 0$ and $h(\omega) = \bigvee\{h(n) \mid n \in \omega\}$, so h is a time warp. Moreover, $hf \leq g$, and hence $h \leq g/f$. For the converse, let $n \in \omega\backslash\{0\}$. If $q \in \omega^+$ satisfies $n \leq f(q)$, then $(g/f)(n) \leq (g/f)(f(q)) \leq g(q)$, and hence $(g/f)(n) \leq \bigwedge g[\{q \in \omega^+ \mid n \leq f(q)\}] = h(n)$. So $h = g/f$. □

Next, we show that residuals of time warps distribute over joins and meets.

Lemma 3. *For any time warps f, g, h,*

(a) $f\backslash(g \wedge h) = (f\backslash g) \wedge (f\backslash h)$ (e) $(g \wedge h)/f = (g/f) \wedge (h/f)$
(b) $(g \wedge h)\backslash f = (g\backslash f) \vee (h\backslash f)$ (f) $f/(g \wedge h) = (f/g) \vee (f/h)$
(c) $f\backslash(g \vee h) = (f\backslash g) \vee (f\backslash h)$ (g) $(g \vee h)/f = (g/f) \vee (h/f)$
(d) $(g \vee h)\backslash f = (g\backslash f) \wedge (h\backslash f)$ (h) $f/(g \vee h) = (f/g) \wedge (f/h)$.

Proof. Parts (a), (d), (e), and (h) hold in any residuated lattice (see, e.g., [3]). For (b), consider any $n \in \omega\backslash\{0\}$. Using Lemma 2(a),

$$\begin{aligned} ((g \wedge h)\backslash f)(n) &= \bigvee\{q \in \omega^+ \mid (g \wedge h)(q) \leq f(n)\} \\ &= \bigvee\{q \in \omega^+ \mid g(q) \leq f(n) \text{ or } h(q) \leq f(n)\} \\ &= \bigvee\{q \in \omega^+ \mid g(q) \leq f(n)\} \vee \bigvee\{q \in \omega^+ \mid h(q) \leq f(n)\} \\ &= ((g\backslash f) \vee (h\backslash f))(n). \end{aligned}$$

For (f), consider any $n \in \omega\backslash\{0\}$. Using Lemma 2(b),

$$\begin{aligned}(f/(g \wedge h))(n) &= \bigwedge f[\{q \in \omega^+ \mid n \leq (g \wedge h)(q)\}] \\ &= \bigwedge f[\{q \in \omega^+ \mid n \leq g(q) \text{ and } n \leq h(q)\}] \\ &= \bigwedge f[\{q \in \omega^+ \mid n \leq g(q)\}] \vee \bigwedge f[\{q \in \omega^+ \mid n \leq h(q)\}] \\ &= ((g/f) \vee (h/f))(n).\end{aligned}$$

Parts (c) and (g) are proved similarly. □

It follows from Lemma 3 that every time warp term is equivalent to a meet of joins of terms constructed using the operations $\circ$, $\backslash$, $/$, id, $\bot$, and $\top$. However, we can take this simplification process one step further by expressing the residuals of time warps in terms of their restrictions to certain unary operations.

Definition 4. For any time warp f, let

$$f^\ell := id/f, \quad f^{\mathsf{r}} := f\backslash id, \quad \text{and} \quad f^\circ := \top\backslash f.$$

Lemma 5. *For any time warps f, g,*

(a) $f\backslash g = f^{\mathsf{r}} g \vee (\top f)^{\mathsf{r}} \vee g^\circ$
(b) $g/f = g f^\ell \vee (f^\ell)^\circ$.

Proof. For (a), note first that clearly $f^{\mathsf{r}} g \vee (\top f)^{\mathsf{r}} \vee g^\circ \leq f\backslash g$. For the converse, consider any $n \in \omega\backslash\{0\}$. If $g(n) = 0$, then, by Lemma 2(a),

$$(f\backslash g)(n) = \bigvee\{q \in \omega^+ \mid f(q) \leq 0\} = \bigvee\{q \in \omega^+ \mid \top f(q) \leq id(n)\} = (\top f)^{\mathsf{r}}(n).$$

If $g(n) \in \omega\backslash\{0\}$, then, by Lemma 2(a),

$$(f\backslash g)(n) = \bigvee\{q \in \omega^+ \mid f(q) \leq g(n)\} = \bigvee\{q \in \omega^+ \mid f(q) \leq id(g(n))\} = f^{\mathsf{r}}(g(n)).$$

Finally, if $g(n) = \omega$, then, by Lemma 2(a),

$$(f\backslash g)(n) = \bigvee\{q \in \omega^+ \mid f(q) \leq \omega)\} = \omega = \bigvee\{q \in \omega^+ \mid \top(q) \leq \omega)\} = g^\circ(n).$$

So $f\backslash g = f^{\mathsf{r}} g \vee (\top f)^{\mathsf{r}} \vee g^\circ$.

For (b), note first that clearly $g f^\ell \vee (f^\ell)^\circ \leq g/f$. For the converse, consider any $n \in \omega\backslash\{0\}$. If $\{q \in \omega^+ \mid n \leq f(q)\} = \emptyset$, then, by Lemma 2(b),

$$(g/f)(n) = \bigwedge g[\emptyset] = \omega = ((f^\ell)^\circ)(n).$$

Otherwise, $\{q \in \omega^+ \mid n \leq f(q)\} \neq \emptyset$ and, by Lemma 2(b),

$$(g/f)(n) = \bigwedge g[\{q \in \omega^+ \mid n \leq f(q)\}] = g(\bigwedge id[\{q \in \omega^+ \mid n \leq f(q)\}]) = (g f^\ell)(n).$$

So $g/f = g f^\ell \vee (f^\ell)^\circ$. □

To gain a better understanding of these defined unary operations, we observe that Lemma 2 yields for any $n \in \omega\backslash\{0\}$,

$$
\begin{aligned}
f^{\circ}(n) &= \max\{m \in \omega^{+} \mid \omega \leq f(n)\} \\
f^{\mathsf{r}}(n) &= \max\{m \in \omega^{+} \mid f(m) \leq n\} \\
f^{\ell}(n) &= \bigwedge\{m \in \omega^{+} \mid n \leq f(m)\}.
\end{aligned}
$$

The following lemmas collect some simple consequences of these observations.

Lemma 6. *For any time warp f and $n \in \omega\backslash\{0\}$,*

$$
\begin{aligned}
f^{\circ}(n) = 0 &\iff f(n) < \omega \\
f^{\circ}(n) = \omega &\iff f(n) = \omega \\
f^{\circ}(\omega) = 0 &\iff f(k) < \omega \text{ for all } k \in \omega \\
f^{\circ}(\omega) = \omega &\iff f(k) = \omega \text{ for some } k \in \omega.
\end{aligned}
$$

Lemma 7. *For any time warp f, $n \in \omega\backslash\{0\}$, and $m \in \omega$,*

$$
\begin{aligned}
f^{\mathsf{r}}(n) = m &\iff f(m) \leq n < f(m+1) \\
f^{\mathsf{r}}(n) = \omega &\iff f(\omega) \leq n \\
f^{\mathsf{r}}(\omega) = m &\iff f(m+1) = \omega \text{ and } f^{\mathsf{r}}(k) = m \text{ for some } k \in \omega \\
f^{\mathsf{r}}(\omega) = \omega &\iff f(\omega) < \omega \text{ or } (f(\omega) = \omega \text{ and } \forall k \in \omega : f(k) < \omega).
\end{aligned}
$$

Lemma 8. *For any time warp f, $n \in \omega\backslash\{0\}$, and $m \in \omega$,*

$$
\begin{aligned}
f^{\ell}(n) = m &\iff f(m-1) < n \leq f(m) \\
f^{\ell}(n) = \omega &\iff f(\omega) < n \\
f^{\ell}(\omega) = m &\iff f(m) = \omega \text{ and } f^{\ell}(k) = m \text{ for some } k \in \omega \\
f^{\ell}(\omega) = \omega &\iff f(\omega) < \omega \text{ or } (f(\omega) = \omega \text{ and } \forall k \in \omega : f(k) < \omega).
\end{aligned}
$$

Note also that $\top = \bot^{\ell}$. We call a time warp term *basic* if it is constructed using only $\circ$, *id*, $\bot$, and the defined operations $t^{\ell} := id/t$, $t^{\mathsf{r}} := t\backslash id$, and $t^{\circ} := \top\backslash t$. Our normal form theorem now follows, using Lemma 5 to remove residuals from a time warp term, then Lemma 3 and other distributivity properties of $\mathbf{W}$ to push out meets and joins, preserving equivalence in $\mathbf{W}$ at every step.

Theorem 9. *There is an effective procedure that given any time warp term t, produces positive integers $m, n_1, \ldots, n_m$ and a set of basic time warp terms $\{t_{i,j} \mid 1 \leq i \leq m;\ 1 \leq j \leq n_i\}$ satisfying $\mathbf{W} \models t \approx \bigwedge_{i=1}^{m} \bigvee_{j=1}^{n_i} t_{i,j}$.*

Corollary 10. *The equational theory of $\mathbf{W}$ is decidable if, and only if, there exists an effective procedure that decides for any finite non-empty set of basic time warp terms $\{t_1, \ldots, t_n\}$ if $\mathbf{W} \models id \leq t_1 \vee \cdots \vee t_n$.*

We conclude this section by introducing a further notion that will be useful for providing finitary characterizations of time warps.

Definition 11. For any time warp f, let

$$\text{last}(f) := \bigwedge\{p \in \omega^+ \mid f(p) = f(\omega)\}.$$

Observe that $\text{last}(f) < \omega$ if, and only if, f is eventually constant, i.e., increases a finite number of times, and that $\text{last}(f)$ can be defined equivalently in the language of time warps as $(f^\ell f)(\omega)$. For future reference, we record the following easy consequences of this definition.

Lemma 12. *For any time warps f, g,*

(a) $\text{last}(fg) = \omega \iff (\text{last}(f) = \omega$ *and* $\text{last}(g) = \omega)$
(b) $\text{last}(f) = \omega \iff \text{last}(f^{\mathsf{r}}) = \omega \iff \text{last}(f^\ell) = \omega$.

3 Diagrams

In this section, we define diagrams as finitary characterizations of 'potential counterexamples' for equations of the form $id \leq t_1 \vee \cdots \vee t_n$, where each t_i is a basic time warp term. This definition is obtained by considering relevant properties of time warps assigned to variables in a refuting valuation, and it therefore follows easily that if $\mathbf{W} \not\models id \leq t_1 \vee \cdots \vee t_n$, then there exists a suitable refuting diagram. The more challenging direction is to show that every refuting diagram extends to a refuting valuation witnessing $\mathbf{W} \not\models id \leq t_1 \vee \cdots \vee t_n$.

Note first that, using Theorem 9, we may without loss of generality express validity in $\mathbf{W}$ using a simplified language where the restricted residuals are taken as fundamental operations. Let $\mathscr{T}_V$ be a countably infinite set of *term variables*, with elements denoted by x, y, z, etc.

Definition 13. A *basic term* belongs to the grammar

$$\mathscr{T} \ni t, u ::= x \mid tu \mid t^{\mathsf{o}} \mid t^\ell \mid t^{\mathsf{r}} \mid id \mid \bot.$$

We also define valuations and interpretations explicitly for basic terms.

Definition 14. A *valuation* θ is a map $\mathscr{T}_V \to \mathscr{W}$. The *interpretation* of a basic term t under θ, denoted by $[\![t]\!]_\theta$, is the time warp defined inductively by

$$[\![x]\!]_\theta := \theta(x), \qquad [\![tu]\!]_\theta := [\![t]\!]_\theta [\![u]\!]_\theta, \qquad [\![t^\star]\!]_\theta := [\![t]\!]_\theta^\star \text{ for } \star \in \{\mathsf{o}, \ell, \mathsf{r}\}.$$

Corollary 10 tells us that the equational theory of $\mathbf{W}$ is decidable if, and only if, there exists an effective procedure that decides, for any finite set of basic terms T, if there exists a valuation θ and $p \in \omega^+$ such that $[\![t]\!]_\theta(p) < p$ for all $t \in T$. To refer to this element p, we let $\mathscr{I}_V$ be a countably infinite set of *time variables* containing elements denoted by κ, κ', etc., noting that in fact only one time variable will be required for the proofs in this paper. We now define a new language of 'samples' that will be used to refer to values considered in a diagram.

Definition 15. A *sample* belongs to the grammar (where t is any basic term)

$$\mathscr{I} \ni \alpha ::= \kappa \mid t[\alpha] \mid \mathsf{s}(\alpha) \mid \mathsf{p}(\alpha) \mid \mathsf{last}(t).$$

Although samples are purely syntactic, the notation is indicative of their intended meaning. Given an initial sample set $\{t_1[\kappa], \ldots, t_n[\kappa]\}$, obtained from the equation $id \leq t_1 \vee \cdots \vee t_n$, the idea is to 'saturate' this set by adding further samples required to describe the existence of a counterexample.

Definition 16. A sample set Δ is called *saturated* if whenever $\alpha \in \Delta$ and $\alpha \rightsquigarrow \beta$, also $\beta \in \Delta$, where $\rightsquigarrow$ is the relation between samples defined by

$$\begin{aligned} t[\alpha] &\rightsquigarrow \alpha & t^{\circ}[\alpha] &\rightsquigarrow t[\alpha] \\ \mathsf{s}(\alpha) &\rightsquigarrow \alpha & t^{\mathsf{r}}[\alpha] &\rightsquigarrow t[t^{\mathsf{r}}[\alpha]], t[\mathsf{s}(t^{\mathsf{r}}[\alpha])] \\ \mathsf{p}(\alpha) &\rightsquigarrow \alpha & t^{\ell}[\alpha] &\rightsquigarrow t[t^{\ell}[\alpha]], t[\mathsf{p}(t^{\ell}[\alpha])] \\ tu[\alpha] &\rightsquigarrow t[u[\alpha]] & t[\alpha] &\rightsquigarrow t[\mathsf{last}(t)]. \end{aligned}$$

The *saturation* of a sample set Δ is

$$\Delta^{\rightsquigarrow} := \{\beta \mid \exists \alpha \in \Delta, \alpha \rightsquigarrow^{*} \beta\},$$

where $\rightsquigarrow^{*}$ denotes the reflexive transitive closure of $\rightsquigarrow$.

A proof of the following result can be found in [13, Appendix A.1].

Lemma 17. *The saturation of a finite sample set is finite.*

Let us fix, until after Definition 25, a saturated sample set Δ.

Definition 18. A *Δ-prediagram* is a map $\delta \colon \Delta \to \omega^{+}$.

We now give a list of conditions for a Δ-prediagram to be a *Δ-diagram.*

Definition 19. For $p \in \omega^{+}$, let

$$p \ominus 1 := \begin{cases} p-1 & \text{if } p \in \omega \setminus \{0\} \\ p & \text{if } p \in \{0, \omega\} \end{cases}, \qquad p \oplus 1 := \begin{cases} p+1 & \text{if } p \in \omega \\ p & \text{if } p = \omega \end{cases}.$$

Definition 20. A Δ-prediagram δ is called *structurally-sound* if

$$\begin{aligned} \forall t[\alpha], t[\beta] \in \Delta,\ \delta(\alpha) \leq \delta(\beta) &\Rightarrow \delta(t[\alpha]) \leq \delta(t[\beta]) && (1)\\ \forall t[\alpha] \in \Delta,\ \delta(\alpha) = 0 &\Rightarrow \delta(t[\alpha]) = 0 && (2)\\ \forall \mathsf{p}(\alpha) \in \Delta,\ \delta(\mathsf{p}(\alpha)) &= \delta(\alpha) \ominus 1 && (3)\\ \forall \mathsf{s}(\alpha) \in \Delta,\ \delta(\mathsf{s}(\alpha)) &= \delta(\alpha) \oplus 1 && (4)\\ \forall t[\alpha] \in \Delta,\ \delta(\mathsf{last}(t)) \leq \delta(\alpha) &\Leftrightarrow \delta(t[\alpha]) = \delta(t[\mathsf{last}(t)]) && (5)\\ \forall t[\mathsf{last}(t)] \in \Delta,\ \delta(\mathsf{last}(t)) = \omega &\Rightarrow \delta(t[\mathsf{last}(t)]) = \omega. && (6) \end{aligned}$$

Definition 21. A Δ-prediagram δ is called *logically-sound* if

$$\begin{aligned} \forall id[\alpha] \in \Delta,\ \delta(id[\alpha]) &= \delta(\alpha) && (7)\\ \forall \bot[\alpha] \in \Delta,\ \delta(\mathsf{last}(\bot)) &= 0 && (8)\\ \forall tu[\alpha] \in \Delta,\ \delta(tu[\alpha]) &= \delta(t[u[\alpha]]) && (9)\\ \forall tu[\mathsf{last}(tu)] \in \Delta,\ \delta(\mathsf{last}(tu)) = \omega &\Rightarrow \delta(\mathsf{last}(t)) = \delta(\mathsf{last}(u)) = \omega. && (10) \end{aligned}$$

Definition 22. A Δ-prediagram δ is called *o-sound* if

$$\forall t^{\mathsf{o}}[\alpha] \in \Delta,\ \delta(t^{\mathsf{o}}[\alpha]) = 0 \text{ or } \delta(t^{\mathsf{o}}[\alpha]) = \omega \tag{11}$$

$$\forall t^{\mathsf{o}}[\alpha] \in \Delta,\ \delta(\alpha) < \omega \Rightarrow (\delta(t^{\mathsf{o}}[\alpha]) = \omega \Leftrightarrow \delta(t[\alpha]) = \omega) \tag{12}$$

$$\forall \mathsf{last}(t^{\mathsf{o}}) \in \Delta,\ \delta(\mathsf{last}(t^{\mathsf{o}})) < \omega \tag{13}$$

$$\forall t[\alpha], t^{\mathsf{o}}[\mathsf{last}(t^{\mathsf{o}})] \in \Delta,\ (\delta(t^{\mathsf{o}}[\mathsf{last}(t^{\mathsf{o}})]) < \omega \text{ and } \delta(\alpha) < \omega) \Rightarrow \delta(t[\alpha]) < \omega. \tag{14}$$

Definition 23. A Δ-prediagram δ is called *r-sound* if

$$\forall t[t^{\mathsf{r}}[\alpha]] \in \Delta,\ \delta(t[t^{\mathsf{r}}[\alpha]]) \leq \delta(\alpha) \tag{15}$$

$$\forall t^{\mathsf{r}}[\alpha] \in \Delta,\ (0 < \delta(\alpha) < \omega \text{ and } \delta(t^{\mathsf{r}}[\alpha]) < \omega) \Rightarrow \delta(\alpha) < \delta(t[\mathsf{s}(t^{\mathsf{r}}[\alpha])]) \tag{16}$$

$$\forall t^{\mathsf{r}}[\mathsf{last}(t^{\mathsf{r}})] \in \Delta,\ \delta(\mathsf{last}(t^{\mathsf{r}})) = \omega \Rightarrow \delta(\mathsf{last}(t)) = \omega \tag{17}$$

$$\forall t^{\mathsf{r}}[\mathsf{last}(t^{\mathsf{r}})] \in \Delta,\ \delta(t^{\mathsf{r}}[\mathsf{last}(t^{\mathsf{r}})]) < \omega \Rightarrow \delta(t[\mathsf{s}(t^{\mathsf{r}}[\mathsf{last}(t^{\mathsf{r}})])]) = \omega. \tag{18}$$

Definition 24. A Δ-prediagram δ is called *ℓ-sound* if

$$\forall t[t^{\ell}[\alpha]] \in \Delta,\ \delta(t^{\ell}[\alpha]) < \omega \Rightarrow \delta(\alpha) \leq \delta(t[t^{\ell}[\alpha]]) \tag{19}$$

$$\forall t^{\ell}[\alpha] \in \Delta,\ (0 < \delta(\alpha) < \omega \text{ and } \delta(t^{\ell}[\alpha]) < \omega) \Rightarrow \delta(t[\mathsf{p}(t^{\ell}[\alpha])]) < \delta(\alpha) \tag{20}$$

$$\forall t[t^{\ell}[\alpha]] \in \Delta,\ (\delta(\alpha) < \omega \text{ and } \delta(t^{\ell}[\alpha]) = \omega) \Rightarrow \delta(t[t^{\ell}[\alpha]]) < \delta(\alpha) \tag{21}$$

$$\forall t^{\ell}[\mathsf{last}(t^{\ell})] \in \Delta,\ \delta(\mathsf{last}(t^{\ell})) = \omega \Rightarrow \delta(\mathsf{last}(t)) = \omega \tag{22}$$

$$\forall t^{\ell}[\mathsf{last}(t^{\ell})] \in \Delta,\ \delta(t^{\ell}[\mathsf{last}(t^{\ell})]) < \omega \Rightarrow \delta(t[t^{\ell}[\mathsf{last}(t^{\ell})]]) = \omega. \tag{23}$$

Definition 25. A Δ-prediagram δ is called a *Δ-diagram* if it is structurally sound, logically sound, *o*-sound, ℓ-sound, and r-sound.

It follows from the next proposition that any counterexample to the validity of an equation in $\mathbf{W}$ restricts to a finite diagram witnessing this failure. More precisely, if $\mathbf{W} \not\models id \leq t_1 \vee \cdots \vee t_n$, where each t_i is a basic term, and Δ is the saturation of the sample set $\{t_1[\kappa], \ldots, t_n[\kappa]\}$, then there exists a Δ-diagram δ satisfying $\delta(\kappa) > \delta(t_i[\kappa])$ for each $i \in \{1, \ldots, n\}$.

Proposition 26. *Let T be a set of basic terms, κ a time variable, and Δ the saturation of the sample set $\{t[\kappa] \mid t \in T\}$. Then for any valuation θ and $p \in \omega^+$, there exists a Δ-diagram δ such that $\delta(\kappa) = p$ and $\delta(t[\kappa]) = [\![t]\!]_\theta(p)$ for all $t \in T$.*

Proof. We define the map $\delta\colon \Delta \to \omega^+$ recursively by

$$\begin{aligned} \delta(\kappa) &:= p \\ \forall t[\alpha] \in \Delta,\quad \delta(t[\alpha]) &:= [\![t]\!]_\theta(\delta(\alpha)) \\ \forall \mathsf{last}(t) \in \Delta,\ \delta(\mathsf{last}(t)) &:= \mathrm{last}([\![t]\!]_\theta) \\ \forall \mathsf{p}(\alpha) \in \Delta,\quad \delta(\mathsf{p}(\alpha)) &:= \delta(\alpha) \ominus 1 \\ \forall \mathsf{s}(\alpha) \in \Delta,\quad \delta(\mathsf{s}(\alpha)) &:= \delta(\alpha) \oplus 1. \end{aligned}$$

The map δ is well-defined since $\alpha \in \Delta$ if, and only if, there exist samples $\alpha_1, \ldots, \alpha_n$ such that $\alpha_1 = t[\kappa]$ for some $t \in T$, $\alpha_n = \alpha$, and $\alpha_j \rightsquigarrow \alpha_{j+1}$ for each $j \in \{1, \ldots, n-1\}$. So δ is a Δ-prediagram. A proof that δ is a Δ-diagram—i.e., that δ satisfies conditions conditions (1) to (23)—may be found in [13, Appendix A.2]. □

We now turn our attention to proving that every Δ-diagram δ extends to a valuation θ satisfying $[\![t]\!]_\theta(\delta(\alpha)) = \delta(t[\alpha])$ for all $t[\alpha] \in \Delta$. First, we use δ to define a partial sup-preserving function $\lfloor t \rfloor_\delta$ for each basic term t.

Definition 27. For any Δ-diagram δ and basic term t, let

$$\lfloor t \rfloor_\delta := \{(\delta(\alpha), \delta(t[\alpha])) \mid t[\alpha] \in \Delta\}.$$

A time warp f *extends* $\lfloor t \rfloor_\delta$ if $f(i) = j$ for all $(i, j) \in \lfloor t \rfloor_\delta$, and *strongly extends* $\lfloor t \rfloor_\delta$ if also

$$\text{either } \lfloor t \rfloor_\delta = \emptyset \text{ or } (\lfloor t \rfloor_\delta \neq \emptyset \text{ and } \delta(\mathsf{last}(t)) = \omega \implies \mathsf{last}(f) = \omega).$$

Lemma 28. *There exists an effective procedure that produces for any finite Δ-diagram δ and term variable x, an algorithmic description of a time warp f that strongly extends $\lfloor x \rfloor_\delta$.*

Proof. If $\lfloor x \rfloor_\delta = \emptyset$, then any time warp strongly extends it, so assume $\lfloor x \rfloor_\delta \neq \emptyset$. By (1), $\lfloor x \rfloor_\delta$ can be considered as a partial map from ω^+ to ω^+. Moreover, since Δ is saturated, and, by (5), $\delta(x[\mathsf{last}(x)]) \geq j$ for all $(i, j) \in \lfloor x \rfloor_\delta$, we have $(\delta(\mathsf{last}(x)), \delta(x[\mathsf{last}(x)])) \in \lfloor x \rfloor_\delta$.

Let $X := \lfloor x \rfloor_\delta \cup \{(0, 0), (\omega, \delta(x[\mathsf{last}(x)]))\}$. This is still a partial map by (2) and (5). For each $i \in \omega$, there exists a unique pair $(i_1, j_1), (i_2, j_2) \in X$ such that $i_1 \leq i < i_2$ and there is no $(i_3, j_3) \in X$ with $i_1 < i_3 < i_2$, and we define

$$f(i) := \min(j_2, j_1 \oplus (i - i_1)),$$

where $n \oplus m := \min\{\omega, n + m\}$. Let also $f(\omega) := \delta(x[\mathsf{last}(x)])$).

Clearly f is monotonic. It extends $\lfloor x \rfloor_\delta$, since $i = i_1 < \omega$ implies $f(i_1) = \min(j_2, j_1) = j_1$. In particular, $f(0) = 0$. To confirm that f is a time warp, it remains to show that $f(\omega) = \bigvee\{f(i) \mid i \in \omega\}$. If $\delta(x[\mathsf{last}(x)]) = f(\omega) < \omega$, then, by (6), $\delta(\mathsf{last}(x)) < \omega$ and, by monotonicity, $f(i) = f(\omega)$ for each $i \geq \delta(\mathsf{last}(x))$ and $f(\omega) = f(\delta(\mathsf{last}(x))) = \bigvee\{f(i) \mid i \in \omega\}$. If $f(\omega) = \omega$, then for each $j \in \omega$, there exists an $i \in \omega$ such that $f(i) > j$, and hence $\bigvee_{i<\omega} f(i) = \omega = f(\omega)$.

Finally, suppose that $\delta(\mathsf{last}(x)) = \omega$. Then (6) yields $(\omega, \omega) \in \lfloor x \rfloor_\delta$ and for any $(i, j) \in \lfloor x \rfloor_\delta$, if $i \in \omega$, then also $j \in \omega$. Hence, $\mathsf{last}(f) = \omega$, by the definition of f. So f strongly extends $\lfloor x \rfloor_\delta$. □

Lemma 29. *For every basic term t, valuation θ, and Δ-diagram δ, if $\theta(x)$ strongly extends $\lfloor x \rfloor_\delta$ for every term variable x, then $[\![t]\!]_\theta$ strongly extends $\lfloor t \rfloor_\delta$.*

Proof. By induction on t. The case $t = x$ is immediate and the other cases follow by a series of lemmas proved in [13, Appendix A.2], and the induction hypothesis. □

The next proposition is then a direct consequence of Lemmas 28 and 29.

Proposition 30. *There is an effective procedure that produces for any finite Δ-diagram δ, an algorithmic description of a valuation θ satisfying $[\![t]\!]_\theta(\delta(\alpha)) = \delta(t[\alpha])$ for all $t[\alpha] \in \Delta$.*

We are now ready to establish the main theorem of this section.

Theorem 31. *Let $t_1, \ldots, t_n$ be basic terms, κ a time variable, and Δ the saturation of the sample set $\{t_1[\kappa], \ldots, t_n[\kappa]\}$. Then $\mathbf{W} \not\models id \leq t_1 \vee \cdots \vee t_n$ if, and only if, there exists a Δ-diagram δ such that $\delta(\kappa) > \delta(t_i[\kappa])$ for all $i \in \{1, \ldots, n\}$.*

Proof. Suppose first that $\mathbf{W} \not\models id \leq t_1 \vee \cdots \vee t_n$. Then there exist a valuation θ and $p \in \omega^+$ such that $p = id(p) > [\![t_i]\!]_\theta(p)$ for all $i \in \{1, \ldots, n\}$. Hence, by Proposition 26, there exists a Δ-diagram δ such that $\delta(\kappa) = p > [\![t_i]\!]_\theta(p) = \delta(t_i[\kappa])$ for all $i \in \{1, \ldots, n\}$.

Now suppose that there exists a Δ-diagram δ such that $\delta(\kappa) > \delta(t_i[\kappa])$ for all $i \in \{1, \ldots, n\}$. Then, by Proposition 30, there exists a valuation θ such that $[\![t_i]\!]_\theta(\delta(\kappa)) = \delta(t_i[\kappa])$ for all $i \in \{1, \ldots, n\}$. So $id(\delta(\kappa)) = \delta(\kappa) > [\![t_i]\!]_\theta(\delta(\kappa))$ for all $i \in \{1, \ldots, n\}$. Hence $\mathbf{W} \not\models id \leq t_1 \vee \cdots \vee t_n$. □

4 Decidability via Logic

Let $t_1, \ldots, t_n$ be basic terms, κ a time variable, and Δ the saturation of the sample set $\{t_1[\kappa], \ldots, t_n[\kappa]\}$. Our aim in this section is to express the existence of a Δ-diagram witnessing $\mathbf{W} \not\models id \leq t_1 \vee \ldots \vee t_n$, as stated in Theorem 31, via an existential sentence over the natural numbers with the ordering and successor relations. Since the first-order theory of this structure is decidable, it follows that the equational theory of $\mathbf{W}$ is decidable, concluding the proof of Theorem 1.

Note that in the logic encoding, we will no longer allow ω as a value for the variables. The theoretical reason why this is possible is that the ordinal ω^+ admits a first-order (even quantifier-free) interpretation in ω. However, we will avoid relying upon such model-theoretic generalities here and just give the necessary concrete definitions.

Our construction of a first-order formula ϕ encoding the existence of a Δ-diagram uses the samples in Δ as variables and proceeds in two steps:

1. We define a formula ψ with variables in Δ, intended to be interpreted in ω^+, using the order relation symbol $\preceq$, the successor relation symbol $\mathcal{S}$, and two further unary relation symbols $\mathcal{O}$ and $\mathcal{I}$, where the intended interpretations of $\mathcal{O}(x)$ and $\mathcal{I}(x)$ are "$x = \omega$" and "$x = 0$", respectively.
2. We obtain ϕ by eliminating the symbols $\mathcal{O}$ and $\mathcal{I}$ from ψ and re-interpreting $\preceq$ and $\mathcal{S}$ using an encoding of ω^+ in the structure $(\mathbb{N}, \leq, S, 0)$.

Let τ be the relational first-order signature with two binary relation symbols $\preceq$ and $\mathcal{S}$, and two unary relation symbols $\mathcal{O}$ and $\mathcal{I}$. We consider ω^+ as a τ-structure by defining $\preceq^{\omega^+}$ to be the natural ordering of ω^+, $\mathcal{S}^{\omega^+} := \{(n, n+1) \mid n \in \omega\} \cup \{(\omega, \omega)\}$, $\mathcal{I}^{\omega^+} := \{0\}$, and $\mathcal{O}^{\omega^+} := \{\omega\}$. Note that a Δ-prediagram is a valuation of the variables in Δ in this structure.

We define ψ by translating the defining properties of being a Δ-diagram into quantifier-free formulas of first-order logic in the signature τ with variables from Δ. In the following definition, the symbols $\curlywedge$ and $\curlyvee$ denote the logical connectives 'and' and 'or', respectively, and the notation $a \prec b$ is shorthand for $a \preceq b \curlywedge \neg(b \preceq a)$. Note also that ψ is well-defined, since Δ is finite by Lemma 17.

Definition 32. Let ψ be the first-order quantifier-free τ-formula

$$\curlywedge(\mathsf{struct} \cup \mathsf{log} \cup \mathsf{bounds} \cup \mathsf{right} \cup \mathsf{left} \cup \mathsf{fail}),$$

where the first five sets, corresponding to Definitions 20–24 in the definition of a diagram, and fail, expressing the failure of $id \leq t_1 \vee \ldots \vee t_n$ in $\mathbf{W}$ at the time variable κ, are defined as follows:

$$\begin{aligned}
\mathsf{struct} := {} & \{\alpha \preceq \beta \Rightarrow t[\alpha] \preceq t[\beta] \mid t[\alpha], t[\beta] \in \Delta\} \cup \\
& \{\mathcal{I}(\alpha) \Rightarrow \mathcal{I}(t[\alpha]) \mid t[\alpha] \in \Delta\} \cup \\
& \{\mathcal{S}(\mathsf{p}(\alpha), \alpha) \curlyvee (\mathcal{I}(\mathsf{p}(\alpha)) \curlywedge \mathcal{I}(\alpha)) \mid \mathsf{p}(\alpha) \in \Delta\} \cup \\
& \{\mathcal{S}(\alpha, \mathsf{s}(\alpha)) \mid \mathsf{s}(\alpha) \in \Delta\} \cup \\
& \{\mathsf{last}(t) \preceq \alpha \Leftrightarrow t[\alpha] = t[\mathsf{last}[t]] \mid t[\alpha] \in \Delta\} \cup \\
& \{\mathcal{O}(\mathsf{last}(t)) \Rightarrow \mathcal{O}(t[\mathsf{last}(t)]) \mid t[\mathsf{last}(t)] \in \Delta\}
\end{aligned}$$

$$\begin{aligned}
\mathsf{log} := {} & \{id[\alpha] = \alpha \mid id[\alpha] \in \Delta\} \cup \\
& \{\mathcal{I}(\mathsf{last}(\bot)) \mid \bot[\alpha] \in \Delta\} \cup \\
& \{tu[\alpha] = t[u[\alpha]] \mid tu[\alpha] \in \Delta\} \cup \\
& \{\mathcal{O}(\mathsf{last}(tu)) \Rightarrow (\mathcal{O}(\mathsf{last}(t)) \curlywedge \mathcal{O}(\mathsf{last}(u))) \mid tu[\mathsf{last}(tu)] \in \Delta\}
\end{aligned}$$

$$\begin{aligned}
\mathsf{bounds} := {} & \{\mathcal{I}(t^{\circ}[\alpha]) \curlyvee \mathcal{O}(t^{\circ}[\alpha]) \mid t^{\circ}[\alpha] \in \Delta\} \cup \\
& \{\neg\mathcal{O}(\alpha) \Rightarrow (\mathcal{O}(t^{\circ}[\alpha]) \Leftrightarrow \mathcal{O}(t[\alpha])) \mid t^{\circ}[\alpha] \in \Delta\} \cup \\
& \{\neg\mathcal{O}(\mathsf{last}(t^{\circ})) \mid \mathsf{last}(t^{\circ}) \in \Delta\} \cup \\
& \{(\neg\mathcal{O}(t^{\circ}[\mathsf{last}(t^{\circ})]) \curlywedge \neg\mathcal{O}(\alpha)) \Rightarrow \neg\mathcal{O}(t[\alpha]) \mid t[\alpha], t^{\circ}[\mathsf{last}(t^{\circ})] \in \Delta\}
\end{aligned}$$

$$\begin{aligned}
\mathsf{right} := {} & \{t[t^{\mathsf{r}}[\alpha]] \preceq \alpha \mid t[t^{\mathsf{r}}[\alpha]] \in \Delta\} \cup \\
& \{(\neg\mathcal{I}(\alpha) \curlywedge \neg\mathcal{O}(\alpha) \curlywedge \neg\mathcal{O}(t^{\mathsf{r}}[\alpha]) \Rightarrow \alpha \prec t[\mathsf{s}(t^{\mathsf{r}}[\alpha])] \mid t[\mathsf{s}(t^{\mathsf{r}}[\alpha])] \in \Delta\} \cup \\
& \{\mathcal{O}(\mathsf{last}(t^{\mathsf{r}})) \Rightarrow \mathcal{O}(\mathsf{last}(t)) \mid t^{\mathsf{r}}[\mathsf{last}(t^{\mathsf{r}})] \in \Delta\} \cup \\
& \{\neg\mathcal{O}(t^{\mathsf{r}}[\mathsf{last}(t^{\mathsf{r}})]) \Rightarrow \mathcal{O}(t[\mathsf{s}(t^{\mathsf{r}}[\mathsf{last}(t^{\mathsf{r}})])]) \mid t[\mathsf{s}(t^{\mathsf{r}}[\mathsf{last}(t^{\mathsf{r}})])] \in \Delta\}
\end{aligned}$$

$$\begin{aligned}
\mathsf{left} := {} & \{\neg\mathcal{O}(t^{\ell}[\alpha]) \Rightarrow \alpha \preceq t[t^{\ell}[\alpha]] \mid t[t^{\ell}[\alpha]] \in \Delta\} \cup \\
& \{(\neg\mathcal{I}(\alpha) \curlywedge \neg\mathcal{O}(\alpha) \curlywedge \neg\mathcal{O}(t^{\ell}[\alpha])) \Rightarrow t[\mathsf{p}(t^{\ell}[\alpha])] \prec \alpha \mid t[\mathsf{p}(t^{\ell}[\alpha])] \in \Delta\} \cup \\
& \{(\neg\mathcal{O}(\alpha) \curlywedge \mathcal{O}(t^{\ell}[\alpha])) \Rightarrow t[t^{\ell}[\alpha]] \prec \alpha \mid t[t^{\ell}[\alpha]] \in \Delta\} \cup \\
& \{\mathcal{O}(\mathsf{last}(t^{\ell})) \Rightarrow \mathcal{O}(\mathsf{last}(t)) \mid t^{\ell}[\mathsf{last}(t^{\ell})] \in \Delta\} \cup \\
& \{\neg\mathcal{O}(\{t^{\ell}[\mathsf{last}(t^{\ell})]\} \Rightarrow \mathcal{O}(t[t^{\ell}[\mathsf{last}(t^{\ell})]]) \mid t^{\ell}[\mathsf{last}(t^{\ell})] \in \Delta\}
\end{aligned}$$

$$\mathsf{fail} := \{t_i[\kappa] \prec \kappa \mid 1 \leq i \leq n\}.$$

The next lemma then follows directly from the definition of a Δ-diagram.

Lemma 33. *Let $\delta\colon \Delta \to \omega^+$ be a Δ-prediagram. Then $\omega^+, \delta \models \psi$ if, and only if, δ is a Δ-diagram such that $\delta(t_i[\kappa]) < \delta(\kappa)$ for each $i \in \{1, \dots, n\}$.*

Theorem 31 and Lemma 33 together show that $\mathbf{W} \not\models id \leq t_1 \vee \ldots \vee t_n$ if, and only, if ψ is satisfiable in ω^+. We could therefore conclude the proof of Theorem 1 at this point by appealing to classical decidability results on the first-order theory of ordinals [17]. Instead, however, we show explicitly how to interpret the τ-structure ω^+ inside the standard model $(\mathbb{N}, \leq, S, 0)$, which is more commonly available in satisfiability solvers.

Consider the first-order signature σ with two binary relation symbols $\leq$ and S, and one constant symbol 0, and let $\mathbb{N}$ denote the σ-structure based on the natural numbers, where $\leq^{\mathbb{N}}$ is the usual order, $S^{\mathbb{N}} := \{(n, n+1) \mid n \in \mathbb{N}\}$, and $0^{\mathbb{N}} := 0$. The following definition and lemma contain the crucial observation needed for encoding τ-formulas over ω^+ into σ-formulas over $\mathbb{N}$.[3]

Definition 34. Define the bijection $\iota\colon \mathbb{N} \to \omega^+$ by $\iota(0) := \omega$, and $\iota(n) := n - 1$ for each $n \in \omega \setminus \{0\}$.

For any valuation $w\colon \Delta \to \mathbb{N}$, let $\hat{w}\colon \Delta \to \omega^+$ denote the function defined by $\hat{w}(x) := \iota(w(x))$. Note that the map $w \mapsto \hat{w}$ is a bijection between $\mathbb{N}^\Delta$ and $(\omega^+)^\Delta$, since ι is a bijection.

Lemma 35. *Let χ be a quantifier-free τ-formula. Define χ' to be the quantifier-free σ-formula obtained from χ by making the following symbolic substitutions for every occurrence of an atomic formula in χ:*

(i) *$\mathcal{O}(x)$ is replaced by $x = 0$*
(ii) *$\mathcal{I}(x)$ is replaced by $S(0, x)$*
(iii) *$\mathcal{S}(x, y)$ is replaced by $(x = 0 \curlywedge y = 0) \curlyvee (\neg(x = 0) \curlywedge S(x, y))$*
(iv) *$x \preceq y$ is replaced by $y = 0 \curlyvee (\neg(x = 0) \curlywedge x \leq y)$.*

Then, for any valuation $w\colon \Delta \to \mathbb{N}$, $\mathbb{N}, w \models \chi'$ if, and only if, $\omega^+, \hat{w} \models \chi$.

Proof. By induction on the complexity of χ. The induction step is immediate, and the atomic cases essentially follow from the definitions; we just show the proof for $x \preceq y$ as an example. For any valuation w, we have $\omega^+, \hat{w} \models x \preceq y$ if, and only if, $\hat{w}(y) = \omega$ or $(\hat{w}(x) \neq \omega$ and $\hat{w}(x) \leq \hat{w}(y))$ in ω^+. Using the definition of $\hat{w}$, this is equivalent to $w(y) = 0$ or $(w(x) \neq 0$ and $w(x) \leq w(y))$ in $\mathbb{N}$, that is, $\mathbb{N}, w \models y = 0 \curlyvee (\neg(x = 0) \curlywedge x \leq y)$. □

Finally, we define our quantifier-free σ-formula ϕ encoding the non-validity of $id \leq t_1 \vee \cdots \vee t_n$ in $\mathbf{W}$.

Definition 36. Let $\phi := \psi'$, the σ-formula obtained from the τ-formula ψ (Definition 32) by performing the replacements in Lemma 35.

[3] We thank Thomas Colcombet for suggesting this idea.

We are now ready to put everything together.

Theorem 37. *The time warp equation $id \leq t_1 \vee \cdots \vee t_n$ is valid in* $\mathbf{W}$ *if, and only if, the quantifier-free σ-formula ϕ is unsatisfiable in $\mathbb{N}$. Moreover, any valuation $w\colon \Delta \to \mathbb{N}$ such that $\mathbb{N}, w \models \phi$ effectively yields a valuation θ of the time warp variables occurring in $t_1 \vee \cdots \vee t_n$ such that $\mathbf{W}, \theta \models id \not\leq t_1 \vee \cdots \vee t_n$.*

Proof. By Theorem 31, the equation $id \leq t_1, \vee \cdots \vee t_n$ is not valid in $\mathbf{W}$ if, and only if, there exists a Δ-diagram δ such that $\delta(t_i[\kappa]) < \delta(\kappa)$ for all $i \in \{1, \dots, n\}$. By Lemma 33, the latter is equivalent to the existence of a valuation $v\colon \Delta \to \omega^+$ such that $\omega^+, v \models \psi$. By Lemma 35, the latter is in turn equivalent to the existence of a valuation $w\colon \Delta \to \mathbb{N}$ such that $\mathbb{N}, w \models \phi$.

For the second claim, we retrace our steps. If $w\colon \Delta \to \mathbb{N}$ is a valuation such that $\mathbb{N}, w \models \phi$, define the function $\delta\colon \Delta \to \omega^+$ by $\delta(\alpha) := \iota(w(\alpha))$ for $\alpha \in \Delta$. By Lemma 33, δ is a Δ-diagram such that $\delta(t_i[\kappa]) < \delta(\kappa)$ for each $i \in \{1, \dots, n\}$. By Proposition 30, δ effectively yields a valuation θ that falsfies $id \leq t_1 \vee \cdots \vee t_n$. □

Theorem 1 follows now directly from Theorem 37 and the decidability of the first-order theory of $\mathbb{N}$ (see, e.g., [17]).

Concluding Remark. The proof of Theorem 37, together with the normal form results of Sect. 2, provides a decision procedure for the equational theory of the time warp algebra, as explained in Sect. 1. We are currently in the process of implementing this decision procedure in a software tool. This tool is written in the OCaml functional programming language [18] and uses the Z3 theorem prover [21] to decide the satisfiability of the final logic formula. Our experiments with a preliminary implementation for basic time warp terms have been encouraging so far, and we hope to integrate a full version in a compiler for graded modalities. From a complexity perspective, the most challenging issue is to deal with the potentially very large saturated sample sets and corresponding logic formulas produced by time warp equations. We therefore intend to consider encodings of the decision problem for time warps using alternative, possibly more efficient, data structures supported by the Z3 theorem prover, such as *arrays*, as suggested by a referee of this paper (see [4]).

References

1. Backus, J.W., et al.: The FORTRAN automatic coding system. In: Astrahan, M.M. (ed.) Proceedings of IRE-AIEE-ACM 1957 (Western), pp. 188–198. ACM (1957)
2. Birkedal, L., Møgelberg, R.E., Schwinghammer, J., Støvring, K.: First steps in synthetic guarded domain theory: step-indexing in the topos of trees. Log. Methods Comput. Sci. **8**(4), 55–64 (2012)
3. Blount, K., Tsinakis, C.: The structure of residuated lattices. Int. J. Algebr. Comput. **13**(4), 437–461 (2003)
4. Bradley, A.R., Manna, Z., Sipma, H.B.: What's decidable about arrays? In: Emerson, E.A., Namjoshi, K.S. (eds.) VMCAI 2006. LNCS, vol. 3855, pp. 427–442. Springer, Heidelberg (2005). https://doi.org/10.1007/11609773_28

5. Caspi, P., Pouzet, M.: Synchronous Kahn networks. In: Proceedings of ICFP 1996, pp. 226–238. ACM (1996)
6. Colacito, A., Galatos, N., Metcalfe, G., Santschi, S.: From distributive ℓ-monoids to ℓ-groups, and back again (2021). https://arxiv.org/pdf/2103.00146
7. Fujii, S., Katsumata, S., Melliès, P.-A.: Towards a formal theory of graded monads. In: Jacobs, B., Löding, C. (eds.) FoSSaCS 2016. LNCS, vol. 9634, pp. 513–530. Springer, Heidelberg (2016). https://doi.org/10.1007/978-3-662-49630-5_30
8. Gaboardi, M., Katsumata, S.Y., Orchard, D., Breuvart, F., Uustalu, T.: Combining effects and coeffects via grading. ACM SIGPLAN Not. **51**(9), 476–489 (2016)
9. Galatos, N., Jipsen, P., Kowalski, T., Ono, H.: Residuated Lattices: An Algebraic Glimpse at Substructural Logics. Elsevier, Amsterdam (2007)
10. Gehrke, M., Priestley, H.: Canonical extensions of double quasioperator algebras: an algebraic perspective on duality for certain algebras with binary operations. J. Pure Appl. Algebra **209**(1), 269–290 (2007)
11. Gehrke, M., Priestley, H.: Duality for double quasioperator algebras via their canonical extensions. Stud. Logica. **86**(1), 31–68 (2007)
12. Ghica, D.R., Smith, A.I.: Bounded linear types in a resource semiring. In: Shao, Z. (ed.) ESOP 2014. LNCS, vol. 8410, pp. 331–350. Springer, Heidelberg (2014). https://doi.org/10.1007/978-3-642-54833-8_18
13. van Gool, S., Guatto, A., Metcalfe, G., Santschi, S.: Time warps, from algebra to algorithms (with appendix) (2021). https://arxiv.org/abs/2106.06205
14. Guatto, A.: A generalized modality for recursion. In: Dawar, A., Grädel, E. (eds.) Proceedings of LICS 2018, pp. 482–491. ACM (2018)
15. Holland, W., McCleary, S.: Solvability of the word problem in free lattice-ordered groups. Houston J. Math. **5**(1), 99–105 (1979)
16. Howard, W.A.: The formulae-as-types notion of construction. In: Curry, H., Hindley, B., Roger, S.J., Jonathan, P. (eds.) To H. B. Curry: Essays on Combinatory Logic, Lambda Calculus, and Formalism, pp. 479–490. Academic Press (1980)
17. Läuchli, H., Leonard, J.: On the elementary theory of linear order. Fund. Math. **59**, 109–116 (1966)
18. Leroy, X., Doligez, D., Frisch, A., Garrigue, J., Rémy, D., Vouillon, J.: The OCaml system release 4.12 (2021). https://ocaml.org/releases/4.12/htmlman/index.html
19. Lucassen, J., Gifford, D.: Polymorphic effect systems. In: Proceedings of POPL 1988, pp. 47–57. ACM (1988)
20. Metcalfe, G., Paoli, F., Tsinakis, C.: Ordered algebras and logic. In: Hosni, H., Montagna, F. (eds.) Uncertainty and Rationality, vol. 10, pp. 1–85. Publications of the Scuola Normale Superiore di Pisa (2010)
21. de Moura, L., Bjørner, N.: Z3: an efficient SMT solver. In: Ramakrishnan, C.R., Rehof, J. (eds.) TACAS 2008. LNCS, vol. 4963, pp. 337–340. Springer, Heidelberg (2008). https://doi.org/10.1007/978-3-540-78800-3_24
22. Nakano, H.: A modality for recursion. In: Proceedings of LICS 2000, pp. 255–266. IEEE (2000)
23. Santocanale, L.: The involutive quantaloid of completely distributive lattices. In: Fahrenberg, U., Jipsen, P., Winter, M. (eds.) RAMiCS 2020. LNCS, vol. 12062, pp. 286–301. Springer, Cham (2020). https://doi.org/10.1007/978-3-030-43520-2_18
24. The Agda Development Team: The Agda Dependently-Typed Programming Language (2021). https://wiki.portal.chalmers.se/agda/Main/HomePage
25. The Coq Development Team: The Coq Proof Assistant (2021). https://coq.inria.fr

On Algebra of Program Correctness and Incorrectness

Bernhard Möller[1], Peter O'Hearn[2,3](✉), and Tony Hoare[4]

[1] Universität Augsburg, Augsburg, Germany
[2] Facebook, London, UK
[3] University College London, London, UK
[4] University of Cambridge, Cambridge, UK

Abstract. Variants of Kleene algebra have been used to provide foundations of reasoning about programs, for instance by representing Hoare Logic (HL) in algebra. That work has generally emphasised program correctness, i.e., proving the absence of bugs. Recently, Incorrectness Logic (IL) has been advanced as a formalism for the dual problem: proving the presence of bugs. IL is intended to underpin the use of logic in program testing and static bug finding. Here, we use a Kleene algebra with diamond operators and countable joins of tests, which embeds IL, and which also is complete for reasoning about the image of the embedding. Next to embedding IL, the algebra is able to embed HL, and allows making connections between IL and HL specifications. In this sense, it unifies correctness and incorrectness reasoning in one formalism.

1 Introduction

1.1 Context

My basic mistake was to set up proof in opposition to testing, where in fact both of them are valuable and mutually supportive ways of accumulating evidence of the correctness and serviceability of programs. T. Hoare [17]

Beginning with fundamental work of Kozen and others, variants of Kleene algebra have been used as foundations for program logics. Typically, a translation is given from a logic, such as Hoare Logic (HL, [16]), into the algebra [22]. This approach has been extended to other program logics such as modal [11] and concurrency logics [18]. Work has generally emphasised correctness, i.e., proving the absence of bugs. While that is a worthy ideal, significant programs are often not wholly free of bugs and may never be as they continue evolving. So much attention and energy is spent in engineering practice on testing and other methods of finding specific bugs, rather than proving that none can ever occur.

Despite the practical importance of bug finding, theoretical research on reasoning about programs has concentrated to a much greater extent on correctness. Testing and verification are sometimes even seen in opposition to one another. The third author described his regret for this in the quotation above, from a retrospective in 2009, 40 years after the appearance of Hoare's Logic. In recent

U. Fahrenberg et al. (Eds.): RAMiCS 2021, LNCS 13027, pp. 325–343, 2021.
https://doi.org/10.1007/978-3-030-88701-8_20

years he and the second author have turned attention to theories of testing and static analysis as ways of showing program incorrectness, and the second author introduced Incorrectness Logic (IL) as a dual formalism to HL, oriented to proving the presence of bugs rather than their absence [25]. It was shown in [25] that IL can be used to represent a variety of bug finding approaches, ranging from traditional testing, to symbolic execution [8], to compositional analyses which use logic to summarise the effect of a program component [14].

The present paper deals with representing IL in Kleene algebra. We don't repeat the motivations for IL or the examples illustrating it, and instead refer the reader to the developments in [25,26]. Our interest here is to extend Kleene algebra's successful treatment of program correctness to encompass incorrectness as well. The first aim is to pinpoint the algebraic properties needed to represent IL; secondly we want to represent both IL and HL in the same algebra. This second aim replaces the opposition described in the quotation at the beginning with theoretical unification. We are building on work on the unifying role of Kleene algebra in connecting denotational and operational semantics and program logics for correctness (e.g., [18,19]); we are adding incorrectness to the picture here. Note that both HL and IL deal with partial correctness, not with termination. Other limitations of our results are mentioned at the end of the paper.

1.2 Technical Approach

IL uses an under-approximate triple [25,27], dual to Hoare's triple:

$$\text{(IL)}\ \ [p]\ c\ [q] \Leftrightarrow_{df} q \subseteq \mathsf{sp}(c,p) \qquad\qquad \text{(HL)}\ \ \{p\}\ c\ \{q\} \Leftrightarrow_{df} q \supseteq \mathsf{sp}(c,p)$$

Here, $\mathsf{sp}(c,p)$ is the strongest postcondition: in a binary relation model, c is a relation, p a set of states, and $\mathsf{sp}(c,p)$ the image of the restriction of c to p. In the sequel we abbreviate pre/postcondition by just pre/post. The Hoare triple $\{p\}\ c\ \{q\}$ stipulates that the post q be a superset, an *over-approximation*, of the states reachable via c from p, while the under-approximate triple $[p]\ c\ [q]$ requires that q be a subset, an *under-approximation*.

The terminology over/under-approximation comes from automatic program analysis. The method of Floyd [13] associates an assertion describing a superset of the reachable states with each program point, and this corresponds to Hoare triples (or more generally to abstract interpretations [10]). Over-approximation may lead to false positives (bug claims that are not true) in program analysis, but not false negatives (missed bugs). Dually, an analysis computing an under-approximation at each program point avoids false positives but may suffer from false negatives. Under-approximate analysis is performed by testing and symbolic bug-finding tools. In such an analysis there would be an under-approximate triple relating the program start state with any given program point.

It is well known that $\mathsf{sp}(c,p)$ can be seen as a backwards diamond modality in the sense of dynamic logic. So, it is natural to employ a Modal Kleene Algebra [12] to represent the under-approximate triple algebraically. This has the pleasant consequence that the Hoare triple, which is usually defined in Kleene algebra without recourse to $\mathsf{sp}(c,p)$, enjoys a description that can be connected at once to its under-approximate cousin in a way that formalises aspects of testing and verification as mutually supportive ways of obtaining evidence (see Theorem 4.1 and Theorem 4.5). In addition to connecting over- and under-approximate triples, we also study a version of IL as in [25] in which assertions are embedded as statements within programs in such a way that their violation signals an error.

We start from one of Kozen's variants of Kleene algebra [21], an idempotent semiring (equivalently, an ordered monoid with all finite joins, cf. Definition 2.1) with an additional operator * satisfying two unfolding and two induction axioms. Modal Kleene Algebra enriches that with diamond operators [12]. This indeed gives us a way to interpret all IL proof rules except an infinitary proof principle which is used to obtain a completeness theorem for under-approximate reasoning:

$$\frac{\forall n \in \mathbb{N} : [p_n]\ a\ [p_{n+1}]}{[p_0]\ a^*\ [\bigvee_{n\in\mathbb{N}} p_n]} \qquad (\mathit{Iteration})$$

To model this rule we need countable joins of tests, where a test is a complemented algebra element below the unit of sequential composition. We show that Kleene Algebra with diamonds and countable joins of tests is sound and (relatively) complete for under-approximate triples. A version of this infinitary principle is actually sound in Hoare logic but usually not stated because loop invariants provide a complete reasoning technique for over-approximation. Loop invariants are not complete for under-approximation.

Our technical development begins in Sect. 2 with the algebraic framework, viz. Modal Kleene algebra with countable suprema of tests (CTC algebras). In Sect. 3 we prove soundness and completeness of a proof system for under-approximate triples over CTC algebras. Section 4 connects under-approximation to incorrectness by showing how under-approximate triples can be used to disprove Hoare triples, as well as treating a language with embedded assertions with *error* and *ok* information. In Sect. 5 we use the algebra to present another variant of incorrectness reasoning based on backwards rather than forwards under-approximation. Section 6 concludes.

2 Modal Kleene Algebra

Throughout the paper we refer to a particular example, the "relation model", which is an algebra where the carrier $A = P(S \times S)$ is the set of binary relations on a set S. $a \cdot b$ denotes the composition in diagrammatic order (sequential composition) of relations, $+$ denotes their union, 1 is the identity relation, 0 the empty relation, and a^* is the reflexive-transitive closure of a. If $p \subseteq S$ then $\mathsf{sp}(a,p)$ is the image $\{s' \mid \exists s \in p : (s,s') \in a\}$. The relation model is a Boolean quantale (details below) where a^* is a certain least fixed-point. Some Boolean quantales, like the algebras of relations and of sets of graph paths, satisfy the algebraic properties we need to interpret IL, thus giving us a wide range of models, but we seek to identify lesser structure that supports interpretation.

2.1 Idempotent Semirings, Tests and Diamonds

Definition 2.1
1. An *idempotent semiring*, briefly I-semiring, is a structure $(A, +, \cdot, 0, 1)$ such that $(A, +, 0)$ is a commutative monoid with idempotent addition, $(A, \cdot, 1)$ is a monoid, multiplication distributes from both sides over addition and 0 is an *annihilator* for multiplication, that is, $0 \cdot a = 0 = a \cdot 0$ for all $a \in A$.
2. Every I-semiring can be partially ordered by setting $a \leq b \Leftrightarrow_{df} a + b = b$. Then $+$ and $\cdot$ are isotone w.r.t. $\leq$ and 0 is the least element. This makes A an upper semilattice with join operator $+$ and least element 0. If existing, the least upper bound (lub) of a subset $B \subseteq A$ is denoted by $\bigsqcup B$. With this, $a + b = \bigsqcup \{a, b\}$. For uniformity we write $\bigsqcup_{i<k} a_i$ instead of $\sum_{i<k} a_i$.

Definition 2.2. A *test* in an I-semiring is an element p that has a complement $\neg p$ relative to the multiplicative unit 1, namely $p + \neg p = 1$ and $p \cdot \neg p = 0 = \neg p \cdot p$. The set of all tests in A is denoted by $\mathsf{test}(A)$. A is called *countably test-complete (CTC)* if every countable subset of $\mathsf{test}(A)$ has a lub. (This is equivalent to stipulating that every countable chain of tests has a lub.)

The complement $\neg p$ is unique when it exists. The composition $p \cdot q$ of tests represents logical conjunction and is the meet of p and q. Symmetrically, $p + q$ represents disjunction and is the join. Finally, $p \leq q$ represents implication.

Using tests we can axiomatise the modal operators diamond and box. For IL we only need the backward diamond. For $a \in A$ and $p \in \mathsf{test}(A)$ the test $\langle a|p$ characterises the set of states that can be reached from p in a single a-step, i.e., the image of p under a. We use the following axiomatization of diamond.

Definition 2.3. A *backward diamond semiring* is a structure $(A, +, \cdot, 0, 1, \langle\, |)$ such that $(A, +, \cdot, 0, 1)$ is an I-semiring and $\langle\, | : A \times \mathsf{test}(A) \to \mathsf{test}(A)$ is an operator satisfying the axioms

$$\langle a|q \leq p \Leftrightarrow q \cdot a \leq a \cdot p \qquad (bdia1) \qquad\qquad \langle a \cdot b|q = \langle b|(\langle a|p) \qquad (bdia2)$$

The *backward box* is the De Morgan dual of the diamond: $[a|p =_{df} \neg\langle a|\neg q$.

(*bdia1*) implies that diamond is additive in both arguments, while (*bdia2*) stipulates that it is multiplicative in its first argument and hence preserves composition. In the relation model the diamond $\langle a|p$ corresponds to $\mathsf{sp}(a, p)$.

Our presentation uses a direct axiomatisation of backward modalities. An alternative is to axiomatise a codomain operator $^{\urcorner} : A \to \mathsf{test}(A)$ and then define $\langle a|p = (p \cdot a)^{\urcorner}$. Conversely, one can define codomain in terms of diamond as $a^{\urcorner} =_{df} \langle a|1$. Adopting a definition based on Kleene Algebra with (Co)Domain [12] or based on diamonds is just a presentational choice.

Definition 2.4. A *modal semiring* is a structure $(A, +, \cdot, 0, 1, \langle\, |, \langle\, |)$ such that $(A, +, \cdot, 0, 1, \langle\, |)$ is a backward diamond semiring and the *forward diamond operator* $|\,\rangle : A \times \mathsf{test}(A) \to \mathsf{test}(A)$ satisfies the (dual) axiom

$$|a\rangle q \leq p \Leftrightarrow a \cdot q \leq p \cdot a \qquad (fdia1)$$

The test $|a\rangle q$ algebraically represents the inverse image of q under a. It has been shown in [12] (Cor. 5.8) that (*bdia1*), (*bdia2*) and (*fdia1*) imply multiplicativity of forward diamond, i.e., $|a \cdot b\rangle q = |a\rangle(|b\rangle p)$ (*fdia2*). The *forward box* is $|a]p =_{df} \neg|a\rangle\neg q$; it corresponds to the weakest liberal precondition $\mathsf{wlp}(a, q)$.

The following property is fundamental for our completeness results; see [23] for more details. Moreover, forward diamonds have good other use later.

Lemma 2.5. *Assume a modal semiring.*
1. *We have Galois connections* $\langle a|p \leq q \Leftrightarrow p \leq |a]q$ *and* $|a\rangle p \leq q \Leftrightarrow p \leq [a|q$.
2. *As lower adjoints of Galois connections the diamonds preserve all existing lubs in their second argument.*

We discuss some related axiomatisations. The purely relational *monotype factor* of [3] coincides with the forward box and wlp. The *dynamic negation* of [20] in dynamic relation algebras (reducts of relation algebras) coincides with the complement of domain. In Boolean quantales, the domain operator was defined via a Galois connection in [5]. Since here we strive for a maximally general algebraic basis, we have decided for an axiomatisation equivalent to that in [11,12].

2.2 Iteration and Kleene Algebra

Next we represent arbitrary finite iteration by an additional operator $^* : A \to A$.

Definition 2.6. An I-semiring with star is called a *Kleene algebra* if * satisfies

$$1 + a \cdot a^* \leq a^* \qquad 1 + a^* \cdot a \leq a^* \qquad (\textit{Star Unfold})$$
$$b + a \cdot c \leq c \Rightarrow a^* \cdot b \leq c \qquad b + c \cdot a \leq c \Rightarrow b \cdot a^* \leq c \qquad (\textit{Star induction})$$

The following property (by an easy induction on n) is essential for a number of rules of IL.

Lemma 2.7. *In an I-semiring with star and one of the star unfold axioms, the element* a^* *is an upper bound of the sets* $\{a^n \mid n \in \mathbb{N}\}$ *and* $\{a^{\leq n} \mid n \in \mathbb{N}\}$, *where* $a^{\leq i} =_{df} \sum_{j<i} a^j$.

Lemma 2.8 [12]. *In a Kleene algebra with backward diamond one has, without any further assumptions, the rules*

$$p + \langle a|\langle a^*|p \leq \langle a^*|p \qquad (\textit{Diamond Star Unfold})$$
$$p + \langle a|q \leq q \Rightarrow \langle a^*|p \leq q\ . \qquad (\textit{Diamond Star Induction})$$

Dual rules hold for the forward diamond.

Definition 2.9. A *quantale* (e.g. [28]) is an I-semiring in which the order $\leq$ induces a complete lattice and $\cdot$ preserves arbitrary lubs in both arguments. It is *Boolean* if its complete lattice is a Boolean algebra.

Lemma 2.10 [12]. *Any Boolean quantale is a CTC Kleene Algebra (cf. Definition 2.2) and admits pre-diamonds satisfying (bdia1) and (fdia1).*

3 Under-Approximation and Over-Approximation

It is standard that one can encode while programs into Kleene algebra. Given test p and algebra elements a, b, we can represent while p do a as $(p \cdot a)^* \cdot \neg p$, if p then a else b as $p \cdot a + \neg p \cdot b$, $a;b$ as $a \cdot b$ and the identity *skip* as 1. 0 is equivalent to while true do *skip*. Throughout this section, by a *program* we mean a Kleene algebra element generated from 0, 1 and a set of atomic commands and arbitrary tests using $+$, $\cdot$ and *. (In the next section we will need to distinguish programs from algebra when working with expressions that map less directly to algebra.)

3.1 Under-Approximate Triples

Definition 3.1. We assume a CTC Kleene algebra A with backward diamond.
1. An *under-approximate (or IL)* triple is a formula $[p]\ a\ [q]$, where $p, q \in \mathsf{test}(A)$ and $a \in A$ is a program.
2. $[p]\ a\ [q]$ is *valid* in A, in signs $\models [p]\ a\ [q]$, iff $q \leq \langle a|p$.
3. $[p]\ a\ [q]$ is *provable* iff it can be derived using the rules in Fig. 1.

Divergence
$$\frac{}{[p]\ 0\ [0]}$$

Skip
$$\frac{}{[p]\ 1\ [p]}$$

Assume
$$\frac{q \in \mathsf{test}(A)}{[p]\ q\ [p \wedge q]}$$

Choice
$$\frac{\exists i \in \{1,2\} : [p]\ a_i\ [q]}{[p]\ a_1 + a_2\ [q]}$$

Sequencing
$$\frac{[p]\ a\ [r] \quad [r]\ b\ [q]}{[p]\ a \cdot b\ [q]}$$

Iteration
$$\frac{\forall n \in \mathbb{N} : [p_n]\ a\ [p_{n+1}]}{[p_0]\ a^*\ [\bigvee_{n \in \mathbb{N}} p_n]}$$

Disjunction
$$\frac{[p_1]\ a\ [q_1] \qquad [p_2]\ a\ [q_2]}{[p_1 \vee p_2]\ a\ [q_1 \vee q_2]}$$

Consequence
$$\frac{p \leq p' \quad [p]\ a\ [q] \quad q' \leq q}{[p']\ a\ [q']}$$

Atom
$$\frac{a \in \mathit{Atoms}}{[p]\ a\ [\langle a|p]}$$

Fig. 1. Proof rules for under-approximation

As in [22] we are dealing with a "propositional" program logic: rules involving variables are left out. Some of these, such as an axiom for assignment statements

$$\frac{}{[p]\ x := e\ [\exists x' : p[x'/x] \wedge x' = e[x'/x]]}$$

could be covered (in a particular model) by the axiom for atoms in Fig. 1, which requires that the strongest post be present for each atomic command. Others, such as the frame rule for variable mutation and substitution rules, would require additional inference rules.

An example triple is $[\, x = 0\,]\ (x := x+1)^*\ [\, x \geq 0\,]$, saying that execution of the loop can result in x taking on any non-negative integer. This property can be proven using (*Iteration*) with $p_n = (n \geq 0 \wedge x = n)$. The triple $[\, x = n\,]\ x := x+1\ [\, x = n+1\,]$ used in the premise of the rule shows n decreasing in the backwards direction (hence the name Backwards Variant for this rule in [25]). The post can be written as an infinite disjunction, or finitely using a quantifier: $[\, x = 0\,]\ (x := x+1)^*\ [\, \exists n : n \geq 0 \wedge x = n\,]$. This is where the CTC assumption has its place; without it such countable disjunctions are not well defined. The assertion $x \geq 0$ is the strongest post, and this pre/post pair actually gives us a valid Hoare triple as well. But we can shrink the post using (*Consequence*) to go below the strongest post, to obtain $[\, x = 0\,]\ (x := x+1)^*\ [\, x \geq 1\,]$: the positive integers are a valid under-aproximation.

We record this example as `loop1()` in Fig. 4, where `presumes` is used for the pre-assertion in an under-approximate triple, and `achieves` for the post-assertion. In a program analysis tool we would not expect the human to specify the `presumes` and `achieves`, or the variant p; they would (hopefully) be inferred.

It can be helpful to contrast these rules with those from the over-approximate logic HL (see Fig. 3). The rule for divergence has "false" as the post for under-approximation, where "true" is the post in HL, the choice rule has an $\exists$ in the premise where HL has $\forall$, the consequence rule uses $\leq$ where over-approximation uses $\geq$, and the iterate rule uses a possibly infinite disjunction while for over-approximation we can use loop invariance with no need for infinitary constructs.

It is worth noting that $\leq$ of the algebra is used in the proof rules only between tests (assertions), and not general algebra elements (commands). In program logic this is done because one expects to have a reasonable way to decide $\leq$ between tests, by means of a theorem prover or an abstract interpreter, but deciding $\leq$ between commands has been less common (at least, historically). However, rules involving $\leq$ between commands can be handy, especially for metatheory.

Isotony in Command
$$\frac{[p]\ a\ [q] \qquad a \leq a'}{[p]\ a'\ [q]}$$

Full Disjunction
$$\frac{[p]\ a\ [\bigvee_{i<k} q_i]}{\bigwedge_{i<k} ([p]\ a\ [q_i])}$$

False Post
$$\frac{}{[p]\ a\ [0]}$$

Unfold I
$$\frac{}{[p]\ a^*\ [p]}$$

Unfold II
$$\frac{[p]\ a^* \cdot a\ [q]}{[p]\ a^*\ [q]}$$

Unfold III
$$\frac{[p]\ a \cdot a^*\ [q]}{[p]\ a^*\ [q]}$$

Bounded Unrolling
$$\frac{\bigwedge_{i<k} ([p]\ a^i\ [q_i])}{[p]\ a^*\ [\bigvee_{i<k} q_i]}$$

Fig. 2. Further Proof Rules for Under-Approximation

One such is the rule (*Isotony in Command*) from Fig. 2. (In contrast, Hoare triples are anti-isotone in their command.) Isotony together with (*Star Unfold*) of Kleene algebra justifies the three (*Unfold*) rules (all of which are unsound in HL).

It turns out that these are what is called in logic *admissible* rules: if the premises are derivable from the given rules then so is the conclusion (but there might not be a single direct derivation from the premises to the conclusion). O'Hearn [25] included the first and second rules as they allow a direct derivation of a rule for bounded model checking. De Vries and Koutavas [27] avoided unrolling rules, and a justification for this decision is a completeness theorem (where all true triples can be derived, if not all true proof *rules*).

The two further admissible but non-derivable rules (*Full Disjunction*) and (*False Post*) deal with choice. Note that in the premise of (*Full Disjunction*) we use semiring lub (sum), whereas in the conclusion we use meta-conjunction.

Theorem 3.2. (Soundness and Completeness). *Assume a modal CTC semiring with star.*

1. *With star unfold, the rules in Figs. 1 and 2 are sound (preserve validity).*
2. *If also star induction holds then the rules in Fig. 1 are complete, i.e., every valid triple is provable.*

Note that these results do not require any form of $*$-continuity.

Proof [Iteration Soundness]. Here is the proof for the iteration rule. First we show by induction on n that $p_n \leq \langle a^* | p_0$ for all $n \in \mathbb{N}$.

- For $n = 0$, by $1 \leq a^*$ with isotony of diamond, $p_0 = \langle 1 | p_0 \leq \langle a^* | p_0$.
- Assume $p_n \leq \langle a^* | p_0$. Then by the rule premise with the definition of under-approximate triples, the induction hypothesis with isotony of diamond, multiplicativity of diamond and $a \cdot a^* \leq a^*$ with isotony of diamond:

$$p_{n+1} \leq \langle a | p_n \leq \langle a | \langle a^* | p_0 = \langle a \cdot a^* | p_0 \leq \langle a^* | p_0$$

Hence $\langle a^* | p_0$ is an upper bound of all the p_n and by the definition of a lub we obtain $\bigvee_{n \in \mathbb{N}} p_n \leq \langle a^* | p_0$, which is the conclusion of the rule.

The proofs for the other rules are not difficult, and do not use CTC at all. □

The completeness result (proof overleaf) is sometimes termed "relative" completeness because it uses a proof theory with potential incomputable elements that we assume oracles for deciding or writing: in our case $\leq$ queries and infinite disjunctions. Also, an assumption that posts $\langle a | p$ be expressible is sidestepped by allowing arbitrary tests instead of those built following a specific syntax.

The completeness argument is standard: first we show that triples for strongest posts are provable. This, in turn, relies on a characterization of the strongest post for iteration, which is where induction principles from Kleene algebra are used.

Lemma 3.3. *In a modal CTC Kleene algebra all elements a and tests p satisfy* $\langle a^* | p = \bigvee_{n \in \mathbb{N}} \langle a^n | p$.

Proof ($\geq$). This is immediate from $a^n \leq a^*$ for all $n \in \mathbb{N}$ (Lemma 2.7), isotony of diamond and the definition of lubs.

($\leq$) We use (*Diamond Star Induction*) for $q =_{df} \bigvee_{n\in\mathbb{N}} \langle a^n|p$. The premise of that rule is by lattice algebra equivalent to $p \leq q \wedge \langle a|q \leq q$. The first conjunct is true, since $p = \langle 1|p = \langle a^0|p$. For the second conjunct we calculate by the definition of q, diamond preserving existing lubs (Lm. 2.5.1), multiplicativity of diamond, laws of powers, lattice algebra and definition of q:

$$\langle a|q = \langle a|(\bigvee_{n\in\mathbb{N}} \langle a^n|p) = \bigvee_{n\in\mathbb{N}} \langle a|\langle a^n|p = \bigvee_{n\in\mathbb{N}} \langle a^n \cdot a|p = \bigvee_{n\in\mathbb{N}} \langle a^{n+1}|p \leq \bigvee_{k\in\mathbb{N}} \langle a^k|p = q$$

□

Lemma 3.4. *In a modal CTC Kleene algebra the triple* $[\,p\,]\ a\ [\,\langle a|p\,]$ *is provable.*

Proof. By induction on the generation of the program a. For atomic a the claim holds by the proof rule for atoms.

- For choice, additivity of diamond yields $[\,p\,]\ a+b\ [\,\langle a+b|p\,] \Leftrightarrow [\,p\,]\ a+b\ [\,\langle a|p + \langle b|p\,]$. Now, by (*Disjunction*), (*Choice*) twice and the induction hypothesis for a, b:

$$\begin{aligned}[\,p\,]\ a+b\ [\,\langle a|p+\langle b|p\,] &\dashv [\,p\,]\ a+b\ [\,\langle a|p\,]\ \wedge\ [\,p\,]\ a+b\ [\,\langle b|p\,] \\ &\dashv [\,p\,]\ a\ [\,\langle a|p\,]\ \wedge\ [\,p\,]\ b\ [\,\langle b|p\,] \dashv \mathsf{TRUE}\end{aligned}$$

- For composition, multiplicativity of diamond yields $[\,p\,]\ a\cdot b\ [\,\langle a\cdot b|p\,] \Leftrightarrow [\,p\,]\ a\cdot b\ [\,\langle b|\langle a|p\,]$. Now, by (*Sequencing*) and induction hypothesis for a, b:

$$[\,p\,]\ a\cdot b\ [\,\langle b|\langle a|p\,] \dashv [\,p\,]\ a\ [\,\langle a|p\,] \wedge [\,\langle a|p\,]\ b\ [\,\langle b|\langle a|p\,] \dashv \mathsf{TRUE}$$

- For iteration, by the induction hypothesis for a, all triples $[\,p_n\,]\ a\ [\,p_{n+1}\,]$ are derivable, since $p_{n+1} = \langle a|p_n$. Therefore, (*Iteration*) yields the triple $[\,p_0\,]\ a^*\ [\,\bigvee_{n\in\mathbb{N}} p_n\,]$, which is equivalent to $[\,p_0\,]\ a^*\ [\,\langle a^*|p_0\,]$ by Lemma 3.3.

□

Completeness then follows since from $[\,p\,]\ a\ [\,\langle a|p\,]$ we can shrink the post using (*Consequence*) to obtain that $[\,p\,]\ a\ [\,q\,]$ is provable for any $q \leq \langle a|p$.

3.2 Over-Approximate Triples

We have established that the algebra provides a faithful representation of under-approximate reasoning. In this section we briefly indicate how the prior strength of Kleene algebra for correctness (over-approximation) is maintained.

Definition 3.5

1. An *over-approximate* triple is a formula $\{p\}\ a\ \{q\}$, where $p, q \in \mathsf{test}(A)$ and $a \in A$ is a program.
2. $\{p\}\ a\ \{q\}$ is *valid* in A, in signs $\models \{p\}\ a\ \{q\}$, just if $q \geq \langle a|p$.
3. $\{p\}\ a\ \{q\}$ is *provable* iff it can be derived using the rules in Fig. 3.

Theorem 3.6 (Soundness and Completeness) *[23]. The proof rules in Fig. 3 are sound and complete in any modal Kleene algebra.*

Divergence	*Skip*	*Assume*
$\dfrac{}{\{p\}\ 0\ \{1\}}$	$\dfrac{}{\{p\}\ 1\ \{p\}}$	$\dfrac{q \in \mathsf{test}(A)}{\{p\}\ q\ \{p \wedge q\}}$
Choice	*Sequencing*	*Iteration*
$\dfrac{\forall i \in \{1,2\}:\ \{p\}\ a_i\ \{q\}}{\{p\}\ a_1 + a_2\ \{q\}}$	$\dfrac{\{p\}\ a\ \{r\} \quad \{r\}\ b\ \{q\}}{\{p\}\ a \cdot b\ \{q\}}$	$\dfrac{\{p\}\ a\ \{p\}}{\{p\}\ a^*\ \{p\}}$
Disjunction	*Consequence*	*Atom*
$\dfrac{\{p_1\}\ a\ \{q_1\} \qquad \{p_2\}\ a\ \{q_2\}}{\{p_1 \vee p_2\}\ a\ \{q_1 \vee q_2\}}$	$\dfrac{p \geq p' \quad \{p\}\ a\ \{q\} \quad q' \geq q}{\{p'\}\ a\ \{q'\}}$	$\dfrac{a \in \mathit{Atoms}}{\{p\}\ a\ \{\langle a \vert p\}}$

Fig. 3. Proof Rules for Over-Approximation

Revisiting our earlier example, $\{x = 0\}\ (x := x + 1)^*\ \{x \geq 1\}$ is not valid but $\{x = 0\}\ (x := x + 1)^*\ \{x \geq -1\}$ is, as can be proven by selecting $x \geq -1$ as the loop invariant, together with use of the Consequence rule with the implication from $x = 0$ to $x \geq -1$. $x \geq -42$ is strictly over-approximate, and $[x = 0]\ (x := x + 1)^*\ [x \geq -1]$ can't be proven because (*Consequence*) reverses the implications between tests and the implication from $x = 0$ to $x \geq -1$ is in the wrong direction for under-approximate reasoning.

In contrast to our result for under-approximation (Theorem 3.2(1)), we do not need the CTC hypothesis to show soundness here: the iteration rule is based on loop invariants rather than an infinite disjunction. But, we do need to have a Kleene algebra requirement for soundness, where the under-approximate case (Theorem 3.2(1)) does not: star induction implies the over-approximate iteration rule. These differences underline that the under-approximate theory is not obtained from the over-approximate theory at once by appeal to order duality.

The completeness proof for Theorem 3.6 follows the same pattern as our earlier completeness result: we establish by generation induction on a that $\{p\}\ a\ \{\langle a|p\}$ is derivable and then apply the rule of consequence. An algebraic relative completeness result for HL was given already in [23] using forwards diamonds, and a proof with backwards diamonds is possible and omitted. (Note that the under-approximate triple does not admit a backwards predicate transformer semantics [25], and that is why we have used backwards diamonds in the present paper.)

This result extends at once to CTC algebras. We have included it to emphasise: in modal CTC Kleene algebras, we have sound and complete representations of both under-approximate and over-approximate triples, in the same algebra.

Aside: The Ideal Formulation. We have axiomatised the strongest post via a backwards diamond. There are other algebraic encodings which avoid modalities, as in a generalized representation of $\{b\}$ a $\{c\}$ as $b \cdot a \leq c$, where b and c are not required to be tests [18]. Here, $\leq$ judges approximation between entire programs and not just tests. A similar under-approximate triple is obtained by defining $[b]$ a $[c]$ as $b \cdot a \geq c$. Let's call this the *generalized under-approximate triple.*

To connect the generalized and classical triples we represent pres and posts in terms of the programs b and c. If p is a test and our semilattice has a greatest element $\top$, then $\top \cdot p$ is the *test ideal* for p. Intuitively, $\top \cdot p$ represents a program that can do anything but, if it terminates, must leave p being true at the end. In the relation model, it is the relation that maps any input state to every state in p. Using test ideals, we can define under- and over-approximate triples as

$$\models [p]\ a\ [q] \Leftrightarrow_{df} \top \cdot p \cdot a \geq \top \cdot q \qquad \models \{p\}\ a\ \{q\} \Leftrightarrow_{df} \top \cdot p \cdot a \leq \top \cdot q$$

The right hand sides of the equivalences are generalized triples, with $\top \cdot p$ and $\top \cdot q$ for b and c. Let us call these the *ideal* interpretations of the two triples, and the ones in rest of the paper the *modal* interpretations. The ideal and modal interpretations agree in the concrete relation model ([25], Fact 11). We chose to work with modalities as their direct connection to the official definition of the under-approximate triple seemed natural, but we emphasize that it is possible to develop a full treatment of IL based on ideals rather than modalities.

Another encoding represents $\{p\}$ a $\{q\}$ as $p \cdot a \cdot \neg q = 0$: assuming p then executing a cannot contradict q [22]. We are not aware of a similar interpretation, avoiding ideals and modalities, for the under-approximate triple. **End of Aside**

4 Incorrectness

An under-approximate triple $[p]$ a $[q]$ on its own tells us nothing about whether a program is incorrect: it just tells us a subset of what can happen. In the terminology of program testing, such a triple gives us information that is similar to a "test case", with the generalization that if p and q are assertions describing multiple program states then a single triple can cover more than a single input or output. A test case is not yet a test, as it has no way to judge violation: we also need a "test oracle", which tells us if a run is considered erroneous.

4.1 Disproving Hoare Triples

Suppose we are given a Hoare triple $\{p\}$ c $\{q\}$, but we are not told whether it is valid. We can consider such a triple as "putative", a test oracle (or a specification): given an input/output pair of states the oracle says "no" if the input state satisfies p and the output state doesn't satisfy q.

```
int x;

void loop1()
/*  presumes: [true],    achieves: [ok: x>=0]     */
  { x = 0;
    Kleene-star{    /* p(n) = (x==n) */
      x = x+1;
  } }

void specloop1()
/*  requires: {true},    ensures: {x != 42,000,000} */
   { loop1();
   }

void testloop1()
   { loop1();
     assert(x != 42,000,000);
    }

 void loop2()
  { x = 0;
     Kleene-star{
         assert(x != 42,000,000);
         x = x+1;
   } } }
```

Fig. 4. Iteration Examples

See Fig. 4 for an example. In `specloop1()` we use the keywords `requires` and `ensures` to indicate the pre and post in a putative Hoare triple. The `presumes` and `ensures` assertions from `loop1()` are those of an under-approximate triple, and give us a way to prove falsity of the Hoare triple: the `achieves` assertion in `loop1()` says that x can be any positive integer, and this is incompatible with the `ensures` assertion in `specloop1()`. Note also that if we were to change `specloop1()` by replacing `42,000,000` with `86,000,000` then we could re-use the `achieves` assertion in `loop1()` to give us another disproof.

These ideas on testing and oracles/specification are captured in the following result, which refers to interpretations of triples in modal Kleene algebra.

Theorem 4.1. $\not\models \{p\}\ a\ \{q\} \Leftrightarrow \exists p', q' : p' \leq p \wedge q' \not\leq q \wedge \models [p']\ a\ [q']$.

Proof. ($\Rightarrow$) By the definitions, $\not\models \{p\}\ a\ \{q\} \Leftrightarrow \langle a|p \not\leq q$ and $\models [p]\ a\ [\langle a|p]$. Hence we may choose $p' = p$ and $q' = \langle a|p$.
($\Leftarrow$) By generalised contraposition, the definitions and isotony of diamond with transitivity of $\leq$,

$$\begin{aligned}
&((p' \leq p \wedge q' \not\leq q \wedge \models [p']\ a\ [q']) \Rightarrow \not\models \{p\}\ a\ \{q\}) \Leftrightarrow \\
&((p' \leq p \wedge \models [p']\ a\ [q'] \wedge \models \{p\}\ a\ \{q\}) \Rightarrow q' \leq q) \Leftrightarrow \\
&((p' \leq p \wedge q' \leq \langle a|p' \wedge \langle a|b \leq q) \Rightarrow q' \leq q) \Leftrightarrow \mathsf{TRUE}
\end{aligned}$$
□

Informally, if an execution of a lands outside the post, then the post can't hold. In concrete program testing in the relation model, p' and q' denote singleton sets. But the theorem also covers cases where we use assertions to cover many states at once, as in `loop1()`. Connecting back to the discussion of false positives and under-approximation in the Introduction: if we were to attempt to apply Theorem 4.1, if q' did not under-approximate the reachable states we might get incorrect suggested disproofs; false positives. Also, under-approximation furnishes information for verification: when we shouldn't try to verify. Conversely, if a Hoare triple holds, we needn't try to use testing to falsify it.

These remarks connecting verification and testing are obvious intuitively in concrete models [25]. The point of stating this theorem is as a sanity check that the algebra faithfully represents their connection.

4.2 Error Statements and Embedded Assertions

Hoare triples provide a popular form of specifications/test oracles, especially in theoretical work. An even more popular method, in widespread use across industry in program testing, is embedded "assert" statements. When an assert fails it indicates program error and execution halts. This gives us a facility similar to the post in a Hoare triple, as in `testloop1()` in Fig. 4. But, an assert statement does not need to be in post position, and in this sense is more flexible than pre/post specs. A concocted example is `loop2()` from Fig. 4. A more realistic example is testing code in Fig. 5, taken from the open source code of the toolkit `OpenSSL`. The call to `test_dtls1_heartbleed_excessive_plaintext_length()` in the `main()` program eventually calls `dtls1_write_bytes()` after executing other instructions and, as we can see from the comment, the embedded `assert` statement catches a bug which was present in earlier versions of the software. The `assert` statement functions like the post in a Hoare triple, in that its failure indicates a program error, but it has further statements following it: it is a program statement, and it need not be placed at the end of a function.

```
int main(int argc, char *argv[])
        { ....
            /* The following test causes an assertion failure at
             * ssl/d1_pkt.c:dtls1_write_bytes() in versions prior to
                 1.0.1g: */
            (OPENSSL_VERSION_NUMBER >= 0x1000107fL ?
             test_dtls1_heartbleed_excessive_plaintext_length()
           ....
int dtls1_write_bytes(SSL *s, int type, const void *buf, int len)
        {
        int i;
        OPENSSL_assert(len <= SSL3_RT_MAX_PLAIN_LENGTH);
        s->rwstate=SSL_NOTHING;
        i=do_dtls1_write(s, type, buf, len, 0);
        return i;
        }}
```

Fig. 5. Test code from `heartbeat_test.c` in `openssl1.0.1h`

Turning back to theory, these assert statements are distinct from the "tests" used in Kleene algebra. The latter corresponds more to the "assume" statement in programming language theory: $assume(p)$ simply discards the current program path when p is false but does not halt execution, whereas $assert(p)$ results in abnormal program termination. Hence, c is not executed in $assert(p); c$ when p fails. Abnormal termination is itself different from the 0 of Kleene algebra, which means program divergence, since it is equivalent to while true do $skip$ (or $1^*; \neg 1$).

We approach the use of embedded assertions to specify incorrectness following Incorrectness Logic, which itself follows the lead of C and other programming languages. A special instruction *error*, distinct from 1 and 0, is used to cause abnormal termination before a program's end is reached; assert statements can be described using a combination of tests/Booleans and *error*.

We consider the following grammar of commands,

$$c ::= \text{atom} \mid skip \mid diverge \mid error \mid p \mid c + c \mid c; c \mid c^*$$

and set $assert(p) =_{def} (p; error) + \neg p = \text{if } p \text{ then } error \text{ else } skip$. Although *error* does not correspond directly to a Kleene algebra element, we can do a semantics of a language with *error* by representing a program as a pair (a, e) of Kleene algebra expressions. The a component describes what happens in executions where no errors are raised, while the e component describes what happens when errors do occur. In terms of the relation model, a program denotes a pair of relations $(a, e) \in P(S \times S) \times P(S \times S)$, and this model is equivalent to a treatment of exceptions in denotational semantics via the isomorphism with $P(S \times (S + S))$.

The mapping of such program expressions to pairs of Kleene algebra elements is given in Fig. 6. This semantics, when specialised to the relation model, is that of [25]. Notice how neither sequencing nor iteration map pointwise to their Kleene cousins $\cdot$ and *: Sequencing uses "short circuiting", where upon an error in the first operand execution halts and does not continue with the second, while iteration allows an error to happen on a final execution of a loop body, after some number of normally terminating executions. Finally, note that $[\![error]\!]$ is distinct from $[\![diverge]\!] = (0, 0) = [\![\texttt{while}\ true\ \texttt{do}\ skip]\!]$.

$$\begin{array}{lcllcl}
[\![skip]\!] & = & (1, 0) & [\![diverge]\!] & = & (0, 0) \\
[\![error]\!] & = & (0, 1) & [\![p]\!] & = & (p, 0) \\
[\![c_1 + c_2]\!] & = & \multicolumn{4}{l}{\big([\![c_1]\!]\mathsf{ok} + [\![c_2]\!]\mathsf{ok}\,,\ [\![c_1]\!]\mathsf{er} + [\![c_2]\!]\mathsf{er}\big)} \\
[\![c_1; c_2]\!] & = & \multicolumn{4}{l}{\big([\![c_1]\!]\mathsf{ok} \cdot [\![c_2]\!]\mathsf{ok}\,,\ [\![c_1]\!]\mathsf{er} + ([\![c_1]\!]\mathsf{ok} \cdot [\![c_2]\!]\mathsf{er}\big)} \\
[\![c^*]\!] & = & \multicolumn{4}{l}{\big(([\![c]\!]\mathsf{ok})^*\,,\ ([\![c]\!]\mathsf{ok})^* \cdot [\![c]\!]\mathsf{er}\big)} \\
[\![atom]\!] & \in & A \times A & \multicolumn{3}{l}{\text{(assumed given)}}
\end{array}$$

Fig. 6. Semantics for Programs with Embedded Errors; $[\![c]\!] \in A \times A$

4.3 Incorrectness Logic

We are now in a position to formulate a general algebraic form of the program logic from [25]. This is based on distinguishing two post-assertion forms, one $[\,p\,]\ c\ [\,\mathsf{ok} : q\,]$ for normal termination and the other $[\,p\,]\ c\ [\,\mathsf{er} : q\,]$ for erroneous.

Definition 4.2

1. An *under-approximate triple with error information* is a formula $[p]\,c\,[\mathsf{ok}:q]$ or $[p]\,c\,[\mathsf{er}:q]$, where $p, q \in \mathsf{test}(A)$ and c is a program.
2. (a) $[p]\,c\,[\mathsf{ok}:q]$ is *valid*, in signs $\models [p]\,c\,[\mathsf{ok}:q]$, iff $q \leq \langle fst[\![c]\!] | p$.
   (b) $[p]\,c\,[\mathsf{er}:q]$ is *valid*, in signs $\models [p]\,c\,[\mathsf{er}:q]$, iff $q \leq \langle snd[\![c]\!] | p$.

$$\textit{Error:er}\qquad \frac{}{[p]\ \textit{error}\ [\mathsf{er}:p]}$$

$$\textit{Error:ok}\qquad \frac{}{[p]\ \textit{error}\ [\mathsf{ok}:0]}$$

$$\textit{Divergence:er}\qquad \frac{}{[p]\ \textit{diverge}\ [\mathsf{er}:0]}$$

$$\textit{Skip:er}\qquad \frac{}{[p]\ \textit{skip}\ [\mathsf{er}:0]}$$

$$\textit{Assume:er}\qquad \frac{q \in \mathsf{test}(A)}{[p]\ q\ [\mathsf{er}:0]}$$

$$\textit{Short Circuit:er}\qquad \frac{[p]\ c_1\ [\mathsf{er}:q]}{[p]\ c_1;c_2\ [\mathsf{er}:q]}$$

$$\textit{Sequencing:er}\qquad \frac{[p]\ c_1\ [\mathsf{ok}:q] \qquad [q]\ c_2\ [\mathsf{er}:r]}{[p]\ c_1;c_2\ [\mathsf{er}:r]}$$

$$\textit{Iterate:er}\qquad \frac{[p]\ c^*;c\ [\mathsf{er}:q]}{[p]\ c^*\ [\mathsf{er}:q]}$$

Fig. 7. Under-approximate Proof Rules for Embedded Errors

For the proof rules, first we translate all of the earlier proof rules for under-approximate triples $[p]\ c\ [q]$ into the $[p]\ c\ [\mathsf{ok}:q]$ form.

Definition 4.3. For each rule in Fig. 1 define a corresponding proof rule for $[p]\ c\ [\mathsf{ok}:q]$ triples by replacing $\cdot$ with $;$, 0 with *diverge*, 1 with *skip* and add strongest post axioms for atomic commands. Also, add (*Choice*) and (*Disjunction*) with er and ok conclusions. Call the resulting set of rules *Fig.* 1 *translated.*

Theorem 4.4 (Soundness and Completeness). *Assume a modal CTC semiring with star.*

1. With star unfold, the rules in Figs. 7 plus Fig. 1 translated are sound.
2. If also star induction holds then every true triple is provable.

There is a departure from [25]: we include the rule (*Iterate:er*) for er conclusions only. The soundness of this rule follows from the semantics of iteration in the error case. If we included this rule for ok conclusions as well (as in [25]), then we would appeal to the Kleene law $c^*;c \leq c^*$ plus isotony in command for soundness, and this would require stronger assumptions than Theorem 4.4(1). From the point of view of provability it is possible to leave out the stronger rule for ok conclusions, as the backwards variant rule lets us prove all true ok conclusions. The proof of completeness of each error case is straightforward and omitted.

An example which utilises (*Iterate:er*) is `loop2()` from Fig. 4. The assertion eventually fails after the loop body is successfully executed `42,000,000` times, and then upon the next iteration abnormal termination occurs. We leave as an exercise for the reader to supply the assertions formalizing this argument.

4.4 Over-Approximating Errors

It is possible to formulate an over-approximate system for reasoning about errors as well. We define $\{p\}\ c\ \{\mathsf{ok} : q\}$ to be valid iff $q \geq \langle fst[\![c]\!]|p$, and $\{p\}\ c\ \{\mathsf{er} : q\}$ to be valid iff $q \geq \langle snd[\![c]\!]|p$. Such triples can be used to show that a program doesn't raise an error (when the er conclusion is empty), or that it doesn't terminate normally (when the ok conclusion is empty).

As before there are properties in the algebra connecting testing and verification. Specifically, if 0 over-approximates the error states, then no non-0 under-approximate error conclusion is possible (we needn't test for it); and, if a non-0 under-approximates error, then some errors must occur (we needn't verify).

Theorem 4.5. *1.* $\models \{p\}\ c\ \{\mathsf{er} : 0\} \Leftrightarrow \forall q : (\models [p]\ c\ [\mathsf{er} : q] \Rightarrow q = 0)$.
2. $\exists q :\ \models [p]\ c\ [\mathsf{er} : q]$ *and* $q \neq 0 \Leftrightarrow\ \not\models \{p\}\ c\ \{\mathsf{er} : 0\}$.

We don't give a full treatment of proof theory for this extension of HL, but it is worth contrasting some rules with those for under-approximation. First, sequencing again takes into account short-circuit evaluation, but now must consider the short-circuit and normal cases together top achieve over-approximation.

$$\frac{\{p\}\ c_1\ \{\mathsf{er} : r_1\} \qquad \{p\}\ c_1\ \{\mathsf{ok} : q\} \qquad \{q\}\ c_2\ \{\mathsf{er} : r_2\}}{\{p\}\ c_1; c_2\ \{\mathsf{er} : r_1 \vee r_2\}}$$

Second, iteration cannot (however implicitly) make use of $c^*; c \leq c^*$ plus isotony in command, because in over-approximate logic isotony fails (anti-isotony holds). Instead, even in the error case, can make use of a loop invariant (cf. [9]).

$$\frac{\{p\}\ c\ \{\mathsf{ok} : p\} \quad \{p\}\ c\ \{\mathsf{er} : p\}}{\{p\}\ c^*\ \{\mathsf{er} : p\}}$$

5 Backwards Under-Approximation and Incorrectness

A broadly similar technical development can be done if we replace the strongest post, or the image of a relation, by the inverse image. We obtain a triple $[p\}\ a\ [q\}$ where p under-approximates the states obtained executing a backwards from q.

More precisely, in the relation model define the weakest possible pre of a relation, $\mathsf{wpp}(a, q) =_{df} \{s \mid \exists s'.(s, s') \in a \wedge s' \in q\}$. Then we can define

$$\models [p\}\ a\ [q\} \ \Leftrightarrow_{df}\ p \subseteq \mathsf{wpp}(q, a)$$

Another way to describe $[p\}\ a\ [q\}$ is by saying that every state in p can reach some state in q via a [27]. Note that wpp is neither Dijkstra's weakest precondition, nor the weakest liberal precondition. This triple is different from $\{p\}\ a\ \{q\}$, because a still might land outside of q when executing in the forwards direction.

This yields a logic for reasoning about backwards under-approximation. We don't develop the logic in full, but mention several salient points.

1. Algebraically, $\mathsf{wpp}(q, a)$ is represented by the forward diamond $|q\rangle a$ from Definition 2.4, and we define $[p\}\ a\ [q\} \Leftrightarrow_{df} p \leq |a\rangle q$.
2. The iteration rule becomes

$$\frac{\forall n \in \mathbb{N} : [p_{n+1}\}\ a\ [p_n\}}{[\bigvee_{n\in\mathbb{N}} p_n\}\ a^*\ [p_0\}}$$

   This is in the form of variant rules for total correctness [1] except that the triple does not ensure termination on every path or that no path leads outside the post. It is similar to a rule for diamonds in first-order dynamic logic [15].
3. The assignment axiom is now the backwards-running one

$$\frac{}{[q[e/x]\}\ x := e\ [q\}}$$

4. Disproving a Hoare triple can still be done but requires a different approach than Theorem 4.1: landing outside q falsifies a Hoare triple.

$$\not\models \{p\}\ a\ \{q\} \Leftrightarrow \exists p', q' : 0 \neq p' \leq p \wedge q' \cdot q = 0 \wedge \models [p'\}\ a\ [q'\}$$

Kleene algebra can serve as a foundation for reasoning about backwards under-approximation; e.g., must transitions [4] and reachability witnesses [2].

6 Conclusion and Outlook

Kleene algebra encapsulates basic principles of simple imperative programs, and many works have shown how these principles can provide a foundation for program verification. In this paper we have studied how Kleene algebra can also be used as a foundation for reasoning about the presence of bugs, as in (static or dynamic) program testing. This has the potential to significantly expand the application area of Kleene algebra. While logics for verification have received voluminous treatment in theoretical research, the practical impact of program testing currently dwarfs that of verification: it is much more widely deployed in engineering practice, helping the creation of almost all software products.

In this paper we looked at two specific theories, Hoare Logic (HL) and Incorrectness Logic (IL). The main difference between them is that HL (like verification of safety properties, generally) is a formalism of over-approximation, where IL (like testing generally) rests on under-approximation. We used Modal Kleene algebra to express the under-approximate triple directly. As a happy consequence, the power of Kleene algebra to describe HL is preserved, and properties can be stated linking HL and IL specifications. In this sense, the modal Kleene algebra of the current paper can be said to unify HL and IL.

Our technical results pertain to the specific theories of HL and IL, not to the entire broader informal concepts of correctness and incorrectness. The results here concern safety properties and not liveness properties or hyperproperties. Also, HL and IL are basic theories, and the extension of our results to further programming features is not obvious. For example, Concurrent Kleene Algebra [18] through its exchange law supports a proof rule for over-approximate and not under-approximate reasoning about concurrent processes, and reversing the law furnishes a rule for under-approximation, but how to treat correctness and incorrectness in one algebra for concurrency is not immediately obvious. For another, sequential separation logic relies on an interpretation of triples which avoids memory errors, where the incorrectness version does not use such an interpretation [26]; again, they do not connect *at once* as HL and IL do here.

A contribution of the paper has been to pinpoint properties relevant to completeness of under-approximate reasoning (test completeness), when Kleene properties are and are not needed, and that $*$-continuity is not. These observations should persist into extensions to other programming features, whether or not the exact form of unification achieved here can survive extension.

To conclude, under-approximation and incorrectness have been under-studied in foundational theory, and there is much to be learnt about them and their relation to correctness theories (see [7] for an example of recent theoretical learnings in the area between them, albeit without algebra). It is our hope that some of the problems in this area will be taken up by others.

Acknowledgements. Helpful comments were provided by Jules Desharnais, Roland Glück, Mark Harman and the anonymous referees.

References

1. Apt, K.: Ten years of Hoare's logic: a survey – part 1. ACM TOPLAS **3**(4), 431–483 (1981)
2. Asadi, A., Chatterjee, K., Fu, H., Goharshady, A., Mahdavi, M.: Polynomial reachability witnesses via Stellensätze. In: PLDI (2021)
3. Backhouse, R., van der Woude, J.: Demonic operators and monotype factors. Math. Struct. Comp. Sci. **3**, 417–433 (1993)
4. Ball, T., Kupferman, O., Yorsh, G.: Abstraction for falsification. In: CAV, pp. 67–81 (2005)
5. Brunn, T., Möller, B., Russling, M.: Layered graph traversals and Hamiltonian path problems – an algebraic approach. Tech. Rep. 1997–08, Institute of Computer Science, University of Augsburg, December 1997. Revision in Jeuring, J. (ed.): Math. Prog. Constr. LNCS 1422, 96–121. Springer (1998)
6. Brookes, S., O'Hearn, P.W.: Concurrent Separation Logic. ACM SIGLOG News **3**(3), 47–65 (2016)
7. Bruni, R., Giacobazzi, R., Gori, R., Ranzato, F.: A logic for locally complete abstract interpretations. In: LICS (2021)
8. Cadar, C., Sen, K.: Symbolic execution for software testing: three decades later. CACM **65**(2), 82–90 (2013)
9. Clint, M., Hoare, T.: Program proving: jumps and functions. Acta Informatica **1**, 214–224 (1972)
10. Cousot, P., Cousot, R.: Abstract interpretation: a unified lattice model for static analysis of programs by construction or approximation of fixpoints. In: POPL, pp. 238–252 (1977)

11. Desharnais, J., Möller, B., Struth, G.: Modal Kleene algebra and applications – a survey. J. Rel. Meth. Comp. Sci. **1**, 93–131 (2004)
12. Desharnais, J., Möller, B., Struth, G.: Kleene algebra with domain. ACM TOCL **7**, 798–833 (2006)
13. Floyd, R.W.: Assigning meanings to programs. Mathematical aspects of computer science. In: Proceedings of Symposium on Applied Mathematics. vol. 19, pp. 19–32. AMS (1967)
14. Godefroid, P.: Compositional dynamic test generation. In: POPL, pp 47–54 (2007)
15. Harel, D. (ed.): First-Order Dynamic Logic. LNCS, vol. 68. Springer, Heidelberg (1979). https://doi.org/10.1007/3-540-09237-4
16. Hoare, C.A.R.: An axiomatic basis for computer programming. CACM **12**(10), 576–580 (1969)
17. Hoare, T.: Retrospective: an axiomatic basis for computer programming. CACM **52**(10), 30–32 (2009)
18. Hoare, T., Möller, B., Struth, G., Wehrman, I.: Concurrent Kleene algebra and its foundations. J. Log. Algebr. Program **80**(6), 266–296 (2011)
19. Hoare, T., van Staden, S.: In praise of algebra. Formal Aspects Comput. **24**(4–6), 423–431 (2012)
20. Hollenberg, M.: An equational axiomatization of dynamic negation and relational composition. J. Logic Lang. Inf. **6**, 381–401 (1997)
21. Kozen, D.: A completeness theorem for Kleene algebras and the algebra of regular events. Inf. Comp. **110**, 366–390 (1994)
22. Kozen, D.: On Hoare logic and Kleene algebra with tests. ACM TOCL **1**(1), 60–76 (2000)
23. Möller, B., Struth, G.: Algebras of modal operators and partial correctness. TCS **351**, 221–239 (2006)
24. O'Hearn, P.W.: Separation logic. Commun. ACM **62**(2), 86–95 (2019)
25. O'Hearn, P.W.: Incorrectness logic. PACML(POPL) **4**, 10:1–10:32 (2020)
26. Raad, A., Berdine, J., Dang, H.-H., Dreyer, D., O'Hearn, P.W., Villard, J.: Local reasoning about the presence of bugs: incorrectness separation logic. CAV **2**, 225–252 (2020)
27. de Vries, E., Koutavas, V.: Reverse Hoare logic. In: Barthe, G., Pardo, A., Schneider, G. (eds.) SEFM 2011. LNCS, vol. 7041, pp. 155–171. Springer, Heidelberg (2011). https://doi.org/10.1007/978-3-642-24690-6_12
28. Rosenthal, K.: Quantales and their applications. In: Pitman Research Notes in Math, vol. 234 (1990)

Computing Least and Greatest Fixed Points in Absorptive Semirings

Matthias Naaf(✉)

RWTH Aachen University, Aachen, Germany
naaf@logic.rwth-aachen.de

Abstract. We present two methods to algorithmically compute both least and greatest solutions of polynomial equation systems over absorptive semirings (with certain completeness and continuity assumptions), such as the tropical semiring. Both methods require a polynomial number of semiring operations, including semiring addition, multiplication and an infinitary power operation.

Our main result is a closed-form solution for least and greatest fixed points based on the fixed-point iteration. The proof builds on the notion of (possibly infinite) derivation trees; a careful analysis of the shape of these trees allows us to collapse the fixed-point iteration to a linear number of steps. The second method is an iterative symbolic computation in the semiring of generalized absorptive polynomials, largely based on results on Kleene algebras.

Keywords: Fixed-point computation · Absorptive semirings · Semiring provenance

1 Introduction

A recent line of research on semiring provenance analysis for databases [5,11,12], logic [4,9] and games [10] has identified the class of absorptive, commutative semirings as an appropriate domain for provenance semantics of fixed-point logics [4] and games with fixed-point semantics, such as Büchi or parity games. The underlying idea is to replace the Boolean evaluation of formulae by computations in certain semirings. From this point of view, a formula is essentially a polynomial expression over some semiring, and fixed-point formulae evaluate to least or greatest solutions of polynomial equation systems. To guarantee the existence and meaningfulness (when interpreted as provenance information) of these fixed points, one assumes that the semiring is equipped with a natural order that is a complete lattice (for the existence) and that the semiring is absorptive, that is, $1 + a = 1$ for all elements a. Absorption guarantees a duality of the semiring operations in the sense that addition is increasing, with least element 0, while multiplication is decreasing, with greatest element 1, and it is this property that leads to meaningful provenance information of greatest fixed points [4].

U. Fahrenberg et al. (Eds.): RAMiCS 2021, LNCS 13027, pp. 344–361, 2021.
https://doi.org/10.1007/978-3-030-88701-8_21

This raises the question how one can (efficiently) compute least and greatest solutions of polynomial equation systems over such semirings. The textbook approach is the fixed-point iteration: start by setting all indeterminates to the smallest (or greatest) semiring value, then repeatedly evaluate the equations to obtain new values for all indeterminates. In the Boolean setting, this terminates in at most n steps on n indeterminates (due to monotonicity), but we are also interested in larger and especially infinite semirings such as the tropical semiring[1] $\mathbb{T} = (\mathbb{R}_{\geq 0} \cup \{\infty\}, \min, +_{\mathbb{R}}, \infty, 0)$. Several techniques have been developed to compute least solutions. For ω-continuous semirings (where suprema exist and are compatible with the semiring operations), Hopkins and Kozen [14] have defined a faster iteration scheme based on differentials and more recently, Esparza, Kiefer and Luttenberger [6] have used this idea to generalize Newton's method to ω-continuous semirings. This works surprisingly well for a wide variety of semirings (in fact, their results for idempotent semirings subsume our result for least fixed points). Gondran and Minoux [8] use quasi-inverses of elements and matrices to compute least solutions of linear systems and univariate polynomial equations over dioids. This applies to absorptive semirings, where elements have the trivial quasi-inverse $a^* = 1$ and hence quasi-inverses of matrices always exist.

Our goal is to complement the results in [6,14] by also computing *greatest* solutions, as our motivation stems from semiring provenance where both least and greatest fixed points are considered. To this end, we work with absorptive, *fully* continuous semirings (requiring continuity for both suprema and infima).

Example. Consider the following graph whose edges are annotated by cost values in the tropical semiring. A natural example of a greatest fixed point is the minimal cost of an infinite path. This corresponds to the greatest solution of the equation system given on the right, where each node is represented by an indeterminate and costs appear as coefficients (notice that the right-hand sides are indeed polynomial expressions in terms of the semiring operations).

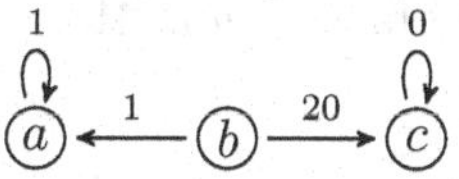

$$
\begin{aligned}
X_a &= 1 +_{\mathbb{R}} X_a \\
X_b &= \min(1 +_{\mathbb{R}} X_a, 20 +_{\mathbb{R}} X_c) \\
X_c &= 0 +_{\mathbb{R}} X_c
\end{aligned}
$$

When we speak of *least* or *greatest* solutions, we always refer to the natural order of the semiring. In the case of the tropical semiring, this is the inverse of the standard order, so $\infty <_{\mathbb{T}} 20 <_{\mathbb{T}} 1 <_{\mathbb{T}} 0$. While the least solution of the above system is trivially $X_a = X_b = X_c = \infty$, the fixed-point iteration for the greatest solution is infinite:

$$
\begin{pmatrix}0\\0\\0\end{pmatrix} \mapsto \begin{pmatrix}1\\1\\0\end{pmatrix} \mapsto \begin{pmatrix}2\\2\\0\end{pmatrix} \mapsto \begin{pmatrix}3\\3\\0\end{pmatrix} \mapsto \cdots \mapsto \begin{pmatrix}20\\20\\0\end{pmatrix} \mapsto \begin{pmatrix}21\\20\\0\end{pmatrix} \mapsto \begin{pmatrix}22\\20\\0\end{pmatrix} \mapsto \begin{pmatrix}23\\20\\0\end{pmatrix} \mapsto \cdots
$$

and converges to the greatest solution: $X_a = \infty$, $X_b = 20$ and $X_c = 0$.

Main Result. The essential idea to compute such solutions is that greatest fixed points are composed of two parts: a cyclic part that is repeated indefinitely

[1] We use $+_{\mathbb{R}}$ for the addition on $\mathbb{R}$ to distinguish it from the semiring operation $+$.

(the loop at a or c) and a reachability part to get to the cycle (the edges from b). As both parts can consist of at most n nodes, all information we need is already present after n steps of the fixed-point iteration; we can use this information to abbreviate the iteration. The formal proof of this observation is based on (infinite) derivation trees, inspired by the derivation trees in the analysis of Newton's method [6] and infinite strategy trees in [4]. We show that these trees provide an alternative description of the fixed-point iteration; a careful analysis of the shape of the derivation trees then leads to our main result:

Theorem 1. *Let F be the operator induced by a polynomial equation system in n indeterminates over an absorptive, fully-continuous, commutative semiring. We can compute in a polynomial number of semiring operations:*

- *the least solution: $F^n(\mathbf{0})$,*
- *the greatest solution: $F^n(\,F^n(\mathbf{1})^\infty\,)$.*

Here, a^∞ is the *infinitary power* operation $a^\infty := \prod_{n \in \mathbb{N}} a^n$ which is well-defined (and usually easy to compute) in the absorptive semirings we consider. For instance, in the tropical semiring we have $0^\infty = 0$ and $a^\infty = \infty$ for $a \neq 0$.

Symbolic Approach. Our second approach is a technique to eliminate indeterminates one by one, based on the work of Hopkins and Kozen on Kleene algebras [14]. We apply their symbolic approach to the semiring $\mathbb{S}^\infty[\boldsymbol{X}]$ of generalized absorptive polynomials, which is perhaps the most relevant semiring for provenance analysis with fixed points, and extend it to include greatest solutions.

Outline. Section 2 introduces the problem setting, in particular the relevant class of semirings, as well as derivation trees. Section 3 establishes the connection between derivation trees and the fixed-point iteration, and Sect. 4 builds on this concept to prove our main result. The symbolic approach for absorptive polynomials is briefly outlined in Sect. 5. Due to space reasons, several proofs have been omitted. All details can be found in the full version [16].

2 Preliminaries

2.1 Polynomial Equation Systems

Throughout the paper, we fix a finite set $\boldsymbol{X} = \{X_1, \dots, X_\ell\}$ of ℓ pairwise different indeterminates. We represent monomials over $\boldsymbol{X}$ as mappings $m\colon \boldsymbol{X} \to \mathbb{N}$. We use bold symbols to denote tuples: $\mathbf{a} = (a_1, \dots, a_\ell)$, in particular $\mathbf{0} = (0, \dots, 0)$ and $\mathbf{1} = (1, \dots, 1)$. To simplify the presentation, we avoid numbered indices and instead index tuples by $\boldsymbol{X}$. That is, for a tuple $\mathbf{a} = (a_1, \dots, a_\ell)$ and an indeterminate $X \in \boldsymbol{X}$, we write $\mathbf{a}_X$ for the entry a_i such that $X_i = X$.

Definition 2. A *polynomial* P over a semiring $(K, +, \cdot, 0, 1)$ and indeterminates $\boldsymbol{X}$ is a finite formal sum of the form $P = \sum_{i=1}^{k} c_i \cdot m_i$, where the m_i are pairwise different monomials over $\boldsymbol{X}$ and $c_i \in K \setminus \{0\}$ are arbitrary coefficients.

Abusing notation, we write $m \in P$ if there is an i with $m = m_i$, and $c \cdot m \in P$ if additionally $c = c_i$. We may write $P(X_1, \ldots, X_\ell)$ to make the indeterminates explicit. Then, $P(a_1, \ldots, a_\ell) \in K$ is the semiring value obtained by instantiating each indeterminate X_i by $a_i \in K$ and evaluating the resulting expression in K.

Definition 3. A *polynomial equation system* $\mathcal{E}$ over a semiring K and indeterminates $\boldsymbol{X} = \{X_1, \ldots, X_\ell\}$ is a family of equations $\mathcal{E}\colon \big(X_i = P_i(X_1, \ldots, X_\ell)\big)_{1 \le i \le \ell}$ with polynomials P_i over $\boldsymbol{X}$ and K.

We associate with $\mathcal{E}$ the operator $F_\mathcal{E}\colon K^\ell \to K^\ell$ defined by $F_\mathcal{E}(a_1, \ldots, a_\ell)_X = P_X(a_1, \ldots, a_\ell)$, for $X \in \boldsymbol{X}$. The least (greatest) solution to $\mathcal{E}$ is thus the least (greatest) fixed point of $F_\mathcal{E}$. We drop the index if $\mathcal{E}$ is clear from the context.

Notice that these are quadratic systems, with the number of equations equal to the number of indeterminates. We recall the example from the introduction in the tropical semiring (where semiring addition is min and semiring multiplication is $+_\mathbb{R}$). Using $\boldsymbol{X} = \{X_a, X_b, X_c\}$, we refer to the polynomial equation system as $(X = P_X)_{X \in \boldsymbol{X}}$. For example, P_{X_b} is the polynomial $\min(1 +_\mathbb{R} X_a, 20 +_\mathbb{R} X_c)$ consisting of the two coefficient-monomial pairs $1 +_\mathbb{R} X_a$ and $20 +_\mathbb{R} X_c$.

2.2 Semirings

Definition 4. A (commutative) *semiring* is an algebraic structure $(K, +, \cdot, 0, 1)$, with $0 \neq 1$, such that $(K, +, 0)$ and $(K, \cdot, 1)$ are commutative monoids, $\cdot$ distributes over $+$, and $0 \cdot a = a \cdot 0 = 0$. It is *idempotent* if $a + a = a$ and *absorptive* if $1 + a = 1$, for all $a \in K$.

In an idempotent semiring K, the *natural order* $\le_K$ is the partial order with $a \le_K b$ if $a + b = b$, for $a, b \in K$. We drop the index if K is clear from the context.

All semirings considered in this paper are commutative and absorptive (except for $\mathbb{N}^\infty$ below). Absorption (also called *0-closed* or *bounded* [15]) implies idempotence and is equivalent to 1 being the $\le_K$-maximal element and to multiplication being decreasing, i.e., $ab \le_K a$ for all $a, b \in K$ (dually to increasing addition).

To guarantee the existence of fixed points, we further require that the natural order is a complete lattice[2] so that suprema $\bigsqcup$ and infima $\bigsqcap$ always exist (with respect to $\le_K$). In addition, we make a continuity assumption stating that the semiring operations commute with the lattice operations on chains (a chain is a totally ordered set). This is crucial for most of our proofs, but does not seem to be a strong restriction in practice: all natural examples of complete-lattice semirings we are aware of are in fact also fully continuous (a notable exception are binary relations with union and composition, but the latter is not commutative).

[2] In idempotent semirings, it is equivalent to only assume suprema/infima of chains [4].

Definition 5. An idempotent semiring K is *fully continuous* if $\leq_K$ is a complete lattice and for all $a \in K$, all nonempty chains $C \subseteq K$ and $\circ \in \{+, \cdot\}$,

$$\bigsqcup(a \circ C) = a \circ \bigsqcup C \quad \text{and} \quad \bigsqcap(a \circ C) = a \circ \bigsqcap C.$$

A homomorphism $h\colon K_1 \to K_2$ on fully-continuous semirings is *fully continuous* if $h(\bigsqcup C) = \bigsqcup h(C)$ and $h(\bigsqcap C) = \bigsqcap h(C)$, for all nonempty chains $C \subseteq K_1$.

Fully-continuous semirings are similar to quantales. More precisely, every absorptive, fully-continuous semiring K induces a quantale $(K, \bigsqcup, \cdot)$ with the top element 1 as unit (see [16] for details). The main difference is that we additionally require compatibility of semiring operations with infima (but only of chains). Another related concept is that of topological dioids in [8] which requires that both operations are compatible with suprema of countable chains.

Since multiplication is decreasing by absorption, powers of an element a form a descending chain $1 \geq a \geq a^2 \geq \dots$ whose infimum we denote by a^∞.

Definition 6. In an absorptive, fully-continuous semiring K, we define the *infinitary power* of $a \in K$ as $a^\infty := \bigsqcap_{n \in \mathbb{N}} a^n$, and $\mathbf{a}^\infty := (a_1^\infty, \dots, a_\ell^\infty)$ for tuples.

Using continuity of multiplication, one can easily verify the properties $(ab)^\infty = a^\infty b^\infty$, $(a^n)^\infty = a^\infty$ and $(a+b)^\infty = a^\infty + b^\infty$ (see [3] for details). We remark that it is usually quite easy to compute the infinitary power. One can further define infinite sum and product operations on families $(a_i)_{i \in I}$ over K with arbitrary index set I. Summation is simply defined as supremum; products can be defined as infimum over finite subproducts (see [3, Appendix]). Here we only need infinite products over finite domain $\{a_i \mid i \in I\}$ (as the polynomials we consider have finitely many coefficients), which are commutative, associative and commute with fully-continuous homomorphisms and the infinitary power.

Fully-continuous homomorphisms further preserve fixed points of monotone functions, in particular least and greatest solutions of polynomial systems:

Lemma 7. *Let $h\colon K_1 \to K_2$ be a fully-continuous homomorphism on absorptive, fully-continuous semirings. Let $\mathcal{E}\colon (X_i = P_i)_{1 \leq i \leq n}$ be a polynomial equation system over K_1. Let $h(\mathcal{E})\colon (X_i = h(P_i))_{1 \leq i \leq n}$ result from $\mathcal{E}$ by applying h to all coefficients. Then, $\mathbf{lfp}(F_{h(\mathcal{E})}) = h(\mathbf{lfp}(F_\mathcal{E}))$ and $\mathbf{gfp}(F_{h(\mathcal{E})}) = h(\mathbf{gfp}(F_\mathcal{E}))$.*

Examples. Some examples of absorptive, fully-continuous semirings are:

- The *Boolean semiring* $\mathbb{B} = (\{0,1\}, \vee, \wedge, 0, 1)$ is the habitat of logical truth.
- $\mathbb{T} = (\mathbb{R}^\infty_{\geq 0}, \min, +_\mathbb{R}, \infty, 0)$ is the *tropical* semiring used for cost computations.
- The *Viterbi* semiring $\mathbb{V} = ([0,1], \max, \cdot, 0, 1)$ is isomorphic to $\mathbb{T}$ and can be used to model confidence scores.
- The *Łukasiewicz* semiring $\mathbb{L} = ([0,1], \max, \star, 0, 1)$ with $a \star b = \max(0, a+b-1)$, used in many-valued logics.

- The *min-max* semiring on a totally ordered set $(A, \leq)$ with least element a and greatest element b is the semiring $(A, \max, \min, a, b)$.
- The semiring of *generalized absorptive polynomials* $\mathbb{S}^\infty[\boldsymbol{X}]$, defined below.

We write $\mathbb{N}^\infty$ for the semiring of natural numbers extended by a special element ∞ (with $n \cdot \infty = n + \infty = \infty$, for $n \neq 0$). It is neither absorptive nor idempotent (but fully continuous w.r.t. the standard order on natural numbers).

Absorptive Polynomials. The most important absorptive, fully-continuous semiring, both from a provenance perspective and for our proofs, is the semiring of *(generalized[3]) absorptive polynomials* $\mathbb{S}^\infty[\boldsymbol{X}]$. We briefly summarize its definition and key properties from [4]. Given a finite set $\boldsymbol{X}$ of indeterminates, a (generalized) monomial over $\boldsymbol{X}$ is a mapping $m \colon \boldsymbol{X} \to \mathbb{N}^\infty$ (here we also allow the exponent ∞), multiplication adds exponents and the neutral element is $1 \colon X \mapsto 0$. We say that a monomial m_1 *absorbs* m_2, denoted $m_1 \succeq m_2$, if $m_1(X) \leq m_2(X)$ for all $X \in \boldsymbol{X}$ (notice that absorption is the *inverse* of the pointwise order on the exponents). In order to mimic the algebraic property of absorption, polynomials are *antichains* of monomials (which are always finite). Addition and multiplication are defined as usual, but we drop monomials that are absorbed after each operation. For example, $(XY^2 + X^2Y) \cdot X^\infty = X^\infty Y^2 + X^\infty Y = X^\infty Y$.

Definition 8. The semiring $(\mathbb{S}^\infty[\boldsymbol{X}], +, \cdot, 0, 1)$ of *(generalized) absorptive polynomials* consists of all antichains of monomials (w.r.t. absorption). We write 0 for the empty antichain and 1 for the antichain $\{1\}$. Given $P, Q \in \mathbb{S}^\infty[X]$, define

$$P + Q = \mathsf{Maximals}(P \cup Q), \quad P \cdot Q = \mathsf{Maximals}\{m_1 \cdot m_2 \mid m_1 \in P, m_2 \in Q\},$$

where $\mathsf{Maximals}(M)$ denotes the set of $\succeq$-maximal monomials in M.

This semiring is fully continuous, with $\bigsqcup S = \sum S = \mathsf{Maximals}(\bigcup S)$ for sets S, and absorptive. Moreover, $\mathbb{S}^\infty[\boldsymbol{X}]$ is the most general such semiring, as made explicit in the following universal property. Together with Lemma 7, this is a fruitful tool to simplify reasoning about all absorptive, fully-continuous semirings.

Theorem 9 (universal property, [4]**).** *Every mapping $h \colon \boldsymbol{X} \to K$ into an absorptive, fully-continuous semiring K uniquely extends to a fully-continuous semiring homomorphism $h \colon \mathbb{S}^\infty[\boldsymbol{X}] \to K$ (by means of polynomial evaluation).*

For our technical results, we need the following observations based on [3].

Lemma 10 ([3,16])**.** *Let $S \subseteq \mathbb{S}^\infty[\boldsymbol{X}]$ and $P \in \mathbb{S}^\infty[\boldsymbol{X}]$. Then,*

1. $P \cdot \sum S = \sum\{P \cdot Q \mid Q \in S\}$, and

---

[3] $\mathbb{S}^\infty[\boldsymbol{X}]$ generalizes the semiring $\mathrm{Sorp}(\boldsymbol{X})$ of absorptive polynomials in [5] by adding the exponent ∞ (which is needed to have *fully*-continuous homomorphisms in Theorem 9). We only use $\mathbb{S}^\infty[\boldsymbol{X}]$ in this paper and hence drop *generalized* in the following.

2. $(\sum S)^\infty = \sum\{Q^\infty \mid Q \in S\}$, *and*
3. $h(\sum S) = \sum h(S)$, *if* $h\colon \mathbb{S}^\infty[\boldsymbol{X}] \to K$ *is a fully-continuous homomorphism.*

Lemma 11 ([3]). *Let* $(P_i)_{i\in\mathbb{N}}$ *be a descending* ω*-chain with* $P_i \in \mathbb{S}^\infty[\boldsymbol{X}]$*. Then,*

$$\prod_{i\in\mathbb{N}} P_i = \bigsqcup\Big\{\prod_{i\in\mathbb{N}} m_i \;\Big|\; \begin{array}{l}(m_i)_{i\in\mathbb{N}} \text{ is a descending } \omega\text{-chain}\\ \text{of monomials with } m_i \in P_i\end{array}\Big\}.$$

To clearly distinguish between indeterminates in polynomial equation systems and absorptive polynomials, we often use the indeterminate set $\boldsymbol{A} = \{A_1, \dots, A_k\}$ for the latter, in particular when we use values from $\mathbb{S}^\infty[\boldsymbol{A}]$ as coefficients.

2.3 Derivation Trees

Inspired by the analysis of Newton's method [6], we use derivation trees to describe the behaviour of polynomial equation systems. For the intuition behind this notion, think of a polynomial system as a formal grammar: The indeterminates are the nonterminal symbols, coefficients the terminal symbols, and each monomial in P_X gives rise to a production rule for X. We essentially consider derivation trees of this grammar in the usual sense, except that we ignore the order of children (we use commutative semirings) and allow infinite derivations.

Definition 12. A *derivation tree* $T = (V, E, \mathsf{var}, \mathsf{yd})$ over a semiring K and indeterminates $\boldsymbol{X}$ is a (possibly infinite) tree (V, E) with node labelings $\mathsf{var}\colon V \to \boldsymbol{X}$ and $\mathsf{yd}\colon V \to K$, the *yield* of v. We say that T is *from* X if for the root ε, we have $\mathsf{var}(\varepsilon) = X$. For convenience, we often write $v \in T$ instead of $v \in V$ and refer to v with $\mathsf{var}(v) = X$ as an occurrence of X in T.

We associate with each node the monomial $\mathsf{mon}(v) = \prod_{w\in vE} \mathsf{var}(w)$ composed of its children's indeterminates. We say that T is *compatible with the system* $(X = P_X)_{X\in\boldsymbol{X}}$ if for each node, $\mathsf{yd}(v) \cdot \mathsf{mon}(v) \in P_{\mathsf{var}(v)}$. The set of all derivation trees from X that are compatible with the system $\mathcal{E}$ is denoted $\mathcal{T}(\mathcal{E}, X)$.

Given an equation system $\mathcal{E}\colon (X = P_X)_{X\in\boldsymbol{X}}$ and an indeterminate X, a derivation tree $T \in \mathcal{T}(\mathcal{E}, X)$ first chooses from the equation $X = P_X$ a monomial $\mathsf{mon}(\varepsilon)$ together with its coefficient $\mathsf{yd}(\varepsilon)$. On the next level, it then makes analogous choices for all indeterminates occurring in $\mathsf{mon}(\varepsilon)$, where the exponent specifies how often an indeterminate occurs. The leaves v of such a derivation tree (if they exist) have $\mathsf{mon}(v) = 1$ and correspond to absolute coefficients in one of the equations. See Fig. 1 for an example. We define the yield of an entire tree as the combined yield of all nodes.

Definition 13. The *yield* of a derivation tree $T = (V, E, \mathsf{var}, \mathsf{yd})$ over K is the (possibly infinite) product $\mathsf{yd}(T) = \prod_{v\in V} \mathsf{yd}(v) \in K$.

Recall that we assume K to be absorptive and fully continuous, hence infinite products are well-defined. Moreover, we can rearrange the product $\prod_v \mathsf{yd}(v)$ over all nodes to group together nodes with the same monomial label (and hence the same yield). We can thus compare the yields of two trees simply by counting occurrences of monomials.

Definition 14. Let $\mathcal{E}$ be an equation system over $\boldsymbol{X}$, let $Y \in \boldsymbol{X}$ and $m \in P_Y$. For derivation trees $T \in \mathcal{T}(\mathcal{E}, X)$, we define

$$|T|_{m,Y} = \big|\{v \in T \mid \mathsf{mon}(v) = m, \mathsf{var}(v) = Y\}\big| \in \mathbb{N}^\infty$$

as the number of occurrences of $m \in P_Y$ in T. Notice that we use pairs (m, Y) to unambiguously refer to $m \in P_Y$, as m may also occur in other polynomials of $\mathcal{E}$.

Lemma 15 (yield comparison). *Given a polynomial system $\mathcal{E}\colon (X = P_X)_{X\in \boldsymbol{X}}$ over an absorptive, fully-continuous semiring and trees $T, T' \in \mathcal{T}(\mathcal{E}, X)$,*

- *if $|T|_{m,Y} \geq |T'|_{m,Y}$ for all $m \in P_Y$, $Y \in \boldsymbol{X}$, then $\mathsf{yd}(T) \leq \mathsf{yd}(T')$,*
- *if $|T|_{m,Y} = 0$ implies $|T'|_{m,Y} = 0$ for all m, Y, then $\mathsf{yd}(T)^\infty \leq \mathsf{yd}(T')^\infty$.*

3 Derivation Trees and the Fixed-Point Iteration

As a first step towards our main result, this section shows that we can express least and greatest solutions in terms of the yields of derivation trees. Notice that a single derivation tree does not correspond to a solution of the equation system, but only to (the derivation of) a single term in the solution. We thus consider the sum over all derivation trees.

For least solutions, this was already shown (for a slightly different notion of derivation trees) in [6]. Here we are mostly concerned with the proof for greatest solutions, as this is much more involved due to the trees being infinite.

Theorem 16. *Let K be an absorptive, fully-continuous semiring. Let $\mathcal{E}\colon (X = P_X)_{X\in \boldsymbol{X}}$ be a polynomial equation system over K. Then for each $X \in \boldsymbol{X}$,*

$$\mathbf{lfp}(F_\mathcal{E})_X = \sum_{\substack{T\in\mathcal{T}(\mathcal{E},X),\\ T \text{ is finite}}} \mathsf{yd}(T), \qquad \mathbf{gfp}(F_\mathcal{E})_X = \sum_{T\in\mathcal{T}(\mathcal{E},X)} \mathsf{yd}(T).$$

We recall that summation is equivalent to supremum (in idempotent semirings). Here and in the following, we use summation in reminiscence of the general, non-idempotent case (cf. [6]) and only switch to supremum as needed. Towards a proof, we first observe that it suffices to prove Theorem 16 for the most general semiring $K = \mathbb{S}^\infty[\boldsymbol{A}]$. That is, with the coefficients being absorptive polynomials (not to be confused with the polynomials of $\mathcal{E}$). For any other semiring, we can

$$
\begin{aligned}
X_1 &= aX_1 + bX_2X_3\\
X_2 &= cX_1^2\\
X_3 &= d
\end{aligned}
$$

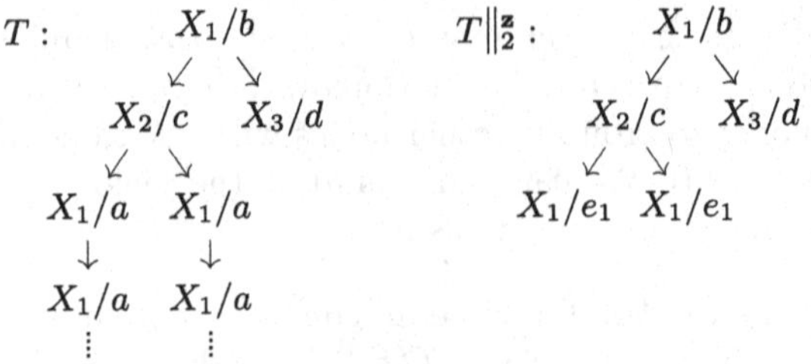

Fig. 1. A derivation tree T and its $(2, \mathbf{e})$-truncation for a sample equation system, with node labels $\mathsf{var}(v)/\mathsf{yd}(v)$. The trees have yield $\mathsf{yd}(T) = a^\infty bcd$ and $\mathsf{yd}(T\|_2^{\mathbf{e}}) = e_1^2 bcd$.

first abstract the equations by replacing coefficients with pairwise different indeterminates in $\boldsymbol{A}$, then apply the theorem and undo the abstraction (see [16]). This works due to the universal property of $\mathbb{S}^\infty[\boldsymbol{A}]$ and Lemmas 7 and 10.

For the remaining section, we thus fix a polynomial system $\mathcal{E}\colon (X = P_X)_{X \in \boldsymbol{X}}$ over $K = \mathbb{S}^\infty[\boldsymbol{A}]$ and consider the induced operator F. The proof proceeds by inductively relating the steps of the fixed-point iterations $F^n(\mathbf{0})$ and $F^n(\mathbf{1})$ to prefixes of derivation trees. These prefixes are defined by simply cutting off the trees at a certain depth and modifying the yield of nodes at the cut-off depth (eventually, we will use yield 0 for the least and 1 for the greatest fixed point).

Definition 17. Let $T = (V, E, \mathsf{var}, \mathsf{yd}) \in \mathcal{T}(\mathcal{E}, X)$, $n \in \mathbb{N}$ and $\mathbf{b} \in K^\ell$. Let $V_{\leq n} \subseteq V$ be the nodes at depth $\leq n$. We define the $(n, \mathbf{b})$-*truncation* of T as

$$T\|_n^{\mathbf{b}} := (V_{\leq n},\, E \cap V_{\leq n}^2,\, \mathsf{var},\, \mathsf{yd}'), \quad \mathsf{yd}'(v) = \begin{cases} \mathbf{b}_{\mathsf{var}(v)}, & v \text{ at depth } n,\\ \mathsf{yd}(v), & \text{otherwise.} \end{cases}$$

This defines a derivation tree (compatible with $\mathcal{E}$ except for its leaves) and we define $\mathsf{mon}(v)$ and $\mathsf{yd}(T\|_n^{\mathbf{b}})$ as in Definitions 12 and 13 (cf. Fig. 1).

The following, mostly technical lemma establishes the general connection between truncations of derivation trees and the fixed-point iteration. It follows by an inductive argument, relying on the inductive nature of derivation trees and on the distributivity of semiring operations to express the yield (see [16]).

Lemma 18 (tree iteration). *Let $K = \mathbb{S}^\infty[\boldsymbol{A}]$ and $\mathbf{b} \in K^\ell$. Then, $F^n(\mathbf{b})_X = \sum_{T \in \mathcal{T}(\mathcal{E}, X)} \mathsf{yd}(T\|_n^{\mathbf{b}})$, for all $n \in \mathbb{N}$, $X \in \boldsymbol{X}$.*

To prove Theorem 16, all that is left to do is to consider the supremum of the iteration $F^n(\mathbf{0})$ and the corresponding tree truncations, and dually the infimum of $F^n(\mathbf{1})$. For the infimum, one last obstacle needs to be resolved: We must show that whenever we pick for each n some n-truncation, their infimum can still be realized as yield of an actual (infinite) tree, even if we pick a different tree to truncate for each n. A similar observation has been used for strategy trees of model-checking games in [4], where it was called *puzzle lemma*. The proof in our setting is similar, but slightly simpler (due to a simpler notion of truncations). The rough idea is to obtain T' by repeating certain parts of the tree $T_n\|^{\mathbf{1}}_n$ for a sufficiently large n (see [3,16] for details).

Lemma 19 (puzzle lemma [4]**).** *Let $X \in \boldsymbol{X}$ and $K = \mathbb{S}^\infty[\boldsymbol{A}]$. Let $(T_n)_{n\in\mathbb{N}}$ be a family of trees $T_n \in \mathcal{T}(\mathcal{E}, X)$ such that their yields $\mathsf{yd}(T_n\|^{\mathbf{1}}_n)$ form a descending chain. Then there is a tree $T' \in \mathcal{T}(\mathcal{E}, X)$ with $\mathsf{yd}(T') \geq \prod_n \mathsf{yd}(T_n\|^{\mathbf{1}}_n)$.*

With this taken care of, we can prove that the sum of all (finite) derivation trees gives the least and greatest solutions.

Proof (of Theorem 16*).* We first observe that F is fully continuous, as it is defined by polynomial expressions over a fully-continuous semiring. By Kleene's fixed-point theorem, we can thus express its least (or greatest) fixed point as supremum of $F^n(\mathbf{0})$ (or infimum of $F^n(\mathbf{1})$) over $n \in \mathbb{N}$. By idempotence, sums coincide with suprema, so for the least solution we immediately obtain:

$$\mathbf{lfp}(F)_X = \bigsqcup_{n\in\mathbb{N}} F^n(\mathbf{0})_X \overset{(18)}{=} \bigsqcup_{n\in\mathbb{N}} \Big(\sum_{T\in\mathcal{T}(\mathcal{E},X)} \mathsf{yd}(T\|^{\mathbf{0}}_n)\Big) = \sum_{T\in\mathcal{T}(\mathcal{E},X)} \Big(\bigsqcup_{n\in\mathbb{N}} \mathsf{yd}(T\|^{\mathbf{0}}_n)\Big).$$

Now observe that $\mathsf{yd}(T\|^{\mathbf{0}}_n) = \mathsf{yd}(T)$ if T has height $< n$, otherwise $\mathsf{yd}(T\|^{\mathbf{0}}_n) = 0$. Hence $\bigsqcup_n \mathsf{yd}(T\|^{\mathbf{0}}_n) = \mathsf{yd}(T)$ if T is finite and 0 otherwise.

It remains to consider the greatest solution. We apply Lemma 11 to express the infimum in $\mathbb{S}^\infty[\boldsymbol{A}]$ as a supremum:

$$\begin{aligned}\mathbf{gfp}(F)_X &= \prod_{n\in\mathbb{N}} F^n(\mathbf{1})_X \overset{(18)}{=} \prod_{n\in\mathbb{N}} \Big(\sum_{T\in\mathcal{T}(\mathcal{E},X)} \mathsf{yd}(T\|^{\mathbf{1}}_n)\Big) \\ &\overset{(11)}{=} \bigsqcup \Big\{ \prod_{n\in\mathbb{N}} y_n \;\Big|\; \begin{array}{l}(y_n)_{n\in\mathbb{N}} \text{ is a descending chain of monomials} \\ \text{with } y_n = \mathsf{yd}(T_n\|^{\mathbf{1}}_n) \text{ for some } T_n \in \mathcal{T}(\mathcal{E}, X)\end{array}\Big\} \\ &\overset{(19)}{=} \bigsqcup \Big\{ \mathsf{yd}(T') \;\Big|\; T' \in \mathcal{T}(\mathcal{E}, X)\Big\} = \sum_{T\in\mathcal{T}(\mathcal{E},X)} \mathsf{yd}(T).\end{aligned}$$

In the last line, we apply the puzzle lemma. This gives us for each monomial chain $(y_n)_{n\in\mathbb{N}}$ an infinite tree T' with $\mathsf{yd}(T') \geq \prod_n y_n$. Conversely, each tree T' induces the monomial chain defined by $y_n = \mathsf{yd}(T'\|^{\mathbf{1}}_n)$. It is easy to see that this chain has infimum $\mathsf{yd}(T')$, so we have equality. □

4 Closed Form Solution

This section is devoted to the proof of our main result:

Theorem 1. *Let K be an absorptive, fully-continuous semiring. Let $\mathcal{E}\colon (X = P_X)_{X\in\boldsymbol{X}}$ be a polynomial equation system over K and $\boldsymbol{X} = \{X_1, \ldots, X_\ell\}$ with induced operator $F_{\mathcal{E}}\colon K^\ell \to K^\ell$. Then,*

$$\mathbf{lfp}(F_{\mathcal{E}}) = F^\ell_{\mathcal{E}}(\mathbf{0}), \quad \mathbf{gfp}(F_{\mathcal{E}}) = F^\ell_{\mathcal{E}}(\, F^\ell_{\mathcal{E}}(\mathbf{1})^\infty\,).$$

Towards the proof, we again fix a polynomial equation system $\mathcal{E}\colon (X = P_X)_{X \in \boldsymbol{X}}$ over an absorptive, fully-continuous semiring K with induced operator F. Recall that ℓ denotes the number of equations (and indeterminates) of $\mathcal{E}$. Our strategy is to prove that we can always find derivation trees of a certain shape, and that the yield of all other derivation trees is absorbed by these trees.

4.1 Deterministic Derivation Trees

Definition 20. A derivation tree $T = (V, E, \mathsf{var}, \mathsf{yd})$ is said to be *deterministic* if $\mathsf{mon}(v)$ depends only on $\mathsf{var}(v)$. That is, the indeterminate labels of the (unordered) children of a node v are determined by the node's indeterminate label so that the relation $\{(\mathsf{var}(v), \mathsf{mon}(v)) \mid v \in V\}$ is a function.

To reason about $\mathbf{gfp}(F)$, we must reason about and construct infinite derivation trees. This is straight-forward for deterministic trees.

Lemma 21 (deterministic construction). *Let $\boldsymbol{X}_0 \subseteq \boldsymbol{X}$. For each $X \in \boldsymbol{X}_0$, let m_X be a monomial with $m_X \in P_X$ such that all indeterminates occurring in m_X are contained in $\boldsymbol{X}_0$. Then for each $X \in \boldsymbol{X}_0$, there is a deterministic tree $T \in \mathcal{T}(\mathcal{E}, X)$ with $\mathsf{var}(v) \in \boldsymbol{X}_0$ and $\mathsf{mon}(v) = m_{\mathsf{var}(v)}$ for all nodes $v \in T$.*

Proof (sketch). Starting from the root $\mathsf{var}(\varepsilon) = X$, define the (possibly infinite) tree T inductively by repeatedly adding to each leaf v child nodes according to $m_{\mathsf{var}(v)}$, always maintaining the desired property for all inner nodes. □

It is easy to see that deterministic trees are uniquely defined by their prefix up to depth $\ell - 1$, as at depth ℓ each path must either end or start to repeat (we only have ℓ indeterminates). Once we consider the infinitary power $\mathsf{yd}(T)^\infty$, it only matters which coefficients c occur in T, but not how often, since $(c^n)^\infty = c^\infty$ for all $n > 0$. This leads to the following simple but essential observations.

Lemma 22. *If $T \in \mathcal{T}(\mathcal{E}, X)$ is deterministic, every indeterminate that occurs in T also occurs in the truncation $T\|^{1}_{\ell-1}$. It follows that $\mathsf{yd}(T)^\infty = \mathsf{yd}(T\|^{1}_{\ell})^\infty$.*

Corollary 23. *For each $T \in \mathcal{T}(\mathcal{E}, X)$, there is a deterministic tree $T' \in \mathcal{T}(\mathcal{E}, X)$ such that $\mathsf{yd}(T\|^{1}_{\ell})^\infty \leq \mathsf{yd}(T')^\infty \leq \mathsf{yd}(T')$.*

Proof (sketch). Choose any way to determinize $T\|^{1}_{\ell}$ by Lemma 21 using only monomials appearing in $T\|^{1}_{\ell}$. This is always possible, as $T\|^{1}_{\ell}$ contains at most ℓ indeterminates and hence every path must contain a repetition or end in a leaf (cf. [16]). The inequalities hold by Lemma 15 (yield comparison) and absorption. □

4.2 Constructing Simple Trees

The main insight behind Theorem 1 is that when we sum over the yield of all derivation trees, it suffices to consider trees of a particular shape corresponding to our intuition from the introduction: These trees consist of an arbitrary prefix up to (at most) depth ℓ (the reachability part), followed by deterministic trees (the cyclic part). See Fig. 2d for an illustration.

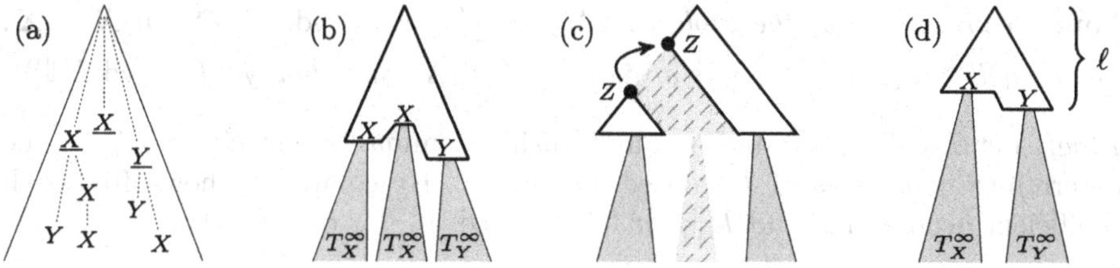

Fig. 2. Illustration of the construction steps in the proof of Lemma 24.

Lemma 24 (main lemma). *For each* $T \in \mathcal{T}(\mathcal{E}, X)$*, there is a derivation tree* $T' \in \mathcal{T}(\mathcal{E}, X)$ *such that all subtrees rooted at depth* ℓ *in* T' *are deterministic and use only monomials* $m \in P_Y$ *(with* $Y \in \boldsymbol{X}$*) that occur infinitely often in* T*. Moreover,* $|T'|_{m,Y} \leq |T|_{m,Y}$ *for all* $m \in P_Y, Y \in \boldsymbol{X}$*.*

Proof. Let $\boldsymbol{X}_\infty \subseteq \boldsymbol{X}$ be the set of indeterminates that occur infinitely often in T (may be empty). We write $V_X = \{v \in T \mid \mathsf{var}(v) = X\}$ for the set of nodes labeled X. For each $X \in \boldsymbol{X}_\infty$, the set V_X is infinite. As the polynomial P_X is finite, there must thus be infinitely many $v \in V_X$ with the same monomial $\mathsf{mon}(v)$. For each $X \in \boldsymbol{X}_\infty$, choose such an infinitely often occurring monomial $m_X \in P_X$. Using Lemma 21, we obtain for each $X \in \boldsymbol{X}_\infty$ a deterministic tree $T_X^\infty \in \mathcal{T}(\mathcal{E}, X)$ such that for all $v \in T_X^\infty$: $\mathsf{var}(v) \in \boldsymbol{X}_\infty$ and $\mathsf{mon}(v)$ occurs infinitely often in T.

Let W be the set of earliest occurrences of $\boldsymbol{X}_\infty$ in T (cf. Fig. 2a). Now let S be the tree that results from T by replacing the subtree at each $v \in W$ with the tree $T_{\mathsf{var}(v)}^\infty$ (cf. Fig. 2b). The tree S is almost of the desired shape, but the trees T_X^∞ may be rooted at depth $> \ell$. To fix this, we consider the prefix up to the subtrees T_X^∞ and eliminate all repetitions of indeterminates within the prefix. As all indeterminates in the prefix occur only finitely often, we can eliminate repetitions by replacing each first occurrence of an indeterminate Z by a last occurrence of Z within the prefix (cf. Fig. 2c). We refer to [16] for a more formal construction and correctness proof.

To see why this results in a tree T' with $|T'|_{m,Y} \leq |T|_{m,Y}$ for all m, Y, recall that the trees T_X^∞ only use monomials that occur infinitely often in T. All further construction steps only remove parts of the tree, thus the number of occurrences is not increased and the inequality holds. □

4.3 Proof of the Main Result

We relate infinite trees of this shape to the expression $F^\ell(F^\ell(\mathbf{1})^\infty)$. The deterministic trees rooted at depth ℓ correspond to the inner term $F^\ell(\mathbf{1})^\infty$, relying on Lemma 22 to ensure that ℓ applications of F suffice. The outer applications of F then correspond to the prefix on which we impose no further restrictions (except that it has height at most ℓ). The following rather technical lemma formalizes this intuition.

Lemma 25. *Let* $\mathbf{b}$ *be the tuple with* $\mathbf{b}_X = \sum_{T \in \mathcal{T}(\mathcal{E},X)} \mathsf{yd}(T\|_\ell^{\mathbf{1}})^\infty$ *for* $X \in \boldsymbol{X}$. *For each* $T \in \mathcal{T}(\mathcal{E}, X)$, *there is a tree* $T' \in \mathcal{T}(\mathcal{E}, X)$ *such that* $\mathsf{yd}(T) \leq \mathsf{yd}(T'\|_\ell^{\mathbf{b}})$.

Proof. Let $T \in \mathcal{T}(\mathcal{E}, X)$ and obtain T' using Lemma 24. Let $S_1, \ldots, S_k$ be the deterministic subtrees of T' rooted at depth ℓ. By comparing how often each coefficient occurs in T and $T'\|_\ell^{\mathbf{1}}$, it follows (see [16] for details) that

$$\mathsf{yd}(T) \leq \mathsf{yd}(T'\|_\ell^{\mathbf{1}}) \cdot \prod_{i=1}^{k} \mathsf{yd}(S_i)^\infty$$

Now let $v_1, \ldots, v_k \in T'$ be the root nodes of the deterministic subtrees $S_1, \ldots, S_k$. By Lemma 22, $\mathsf{yd}(S_i)^\infty = \mathsf{yd}(S_i\|_\ell^{\mathbf{1}})^\infty \leq \mathbf{b}_{\mathsf{var}(v_i)}$, and thus

$$\mathsf{yd}(T) \leq \mathsf{yd}(T'\|_\ell^{\mathbf{1}}) \cdot \prod_{i=1}^{k} \mathbf{b}_{\mathsf{var}(v_i)} = \mathsf{yd}(T'\|_\ell^{\mathbf{b}}).$$

□

We are now ready to prove our main result. The statement on the least solution follows rather directly from our earlier considerations. For greatest fixed points, the previous lemma already proves the difficult direction.

Proof (of Theorem 1**).** It suffices to consider the case $K = \mathbb{S}^\infty[\boldsymbol{A}]$ (so that Lemmas 10 and 18 apply), as the general statement follows with Lemma 7. We first consider the least solution. It is clear by monotonicity of F that $F^\ell(\mathbf{0})_X \leq \mathbf{lfp}(F)_X$. By Theorem 16 and Lemma 18, it thus suffices to prove

$$\mathbf{lfp}(F)_X = \sum_{\substack{T \in \mathcal{T}(\mathcal{E},X) \\ T \text{ finite}}} \mathsf{yd}(T) \overset{!}{\leq} \sum_{T \in \mathcal{T}(\mathcal{E},X)} \mathsf{yd}(T\|_\ell^{\mathbf{0}}) = F^\ell(\mathbf{0})_X.$$

Let $T \in \mathcal{T}(\mathcal{E}, X)$ be finite and obtain T' by Lemma 24. As T is finite, no monomials can occur infinitely often. Hence T' has no subtrees rooted at depth ℓ and is thus of height $< \ell$. But then, $\mathsf{yd}(T) \leq \mathsf{yd}(T') = \mathsf{yd}(T'\|_\ell^{\mathbf{0}}) \leq F^\ell(\mathbf{0})_X$.

For the greatest solution, we know that $\mathbf{gfp}(F)_X = \sum_{T \in \mathcal{T}(\mathcal{E},X)} \mathsf{yd}(T)$. On the other hand, Lemma 18 (tree iteration) entails

$$(F^\ell(\mathbf{1})_X)^\infty = \Big(\sum_{T \in \mathcal{T}(\mathcal{E},X)} \mathsf{yd}(T\|_\ell^{\mathbf{1}}) \Big)^\infty \overset{(10)}{=} \sum_{T \in \mathcal{T}(\mathcal{E},X)} \mathsf{yd}(T\|_\ell^{\mathbf{1}})^\infty.$$

Let $\mathbf{b} = F^\ell(\mathbf{1})^\infty$. Applying Lemma 18 again gives

$$F^\ell(F^\ell(\mathbf{1})^\infty)_X = F^\ell(\mathbf{b})_X = \sum_{T \in \mathcal{T}(\mathcal{E},X)} \mathsf{yd}(T\|_\ell^{\mathbf{b}}).$$

The direction $\mathbf{gfp}(F) \leq F^\ell(F^\ell(\mathbf{1})^\infty)$ follows immediately from Lemma 25. For the other direction, let $T \in \mathcal{T}(\mathcal{E}, X)$. Let $v_1, \ldots, v_k$ be the nodes at depth ℓ in T. By distributivity (Lemma 10), we get

$$\begin{aligned}
\mathsf{yd}(T\|_\ell^{\mathsf{b}}) &= \mathsf{yd}(T\|_\ell^{1}) \cdot \prod_{1\leq i\leq k} \mathsf{b}_{\mathsf{var}(v_i)} \\
&\overset{\text{dist.}}{=} \mathsf{yd}(T\|_\ell^{1}) \cdot \sum \Big\{ \prod_{1\leq i\leq k} \mathsf{yd}(S_i\|_\ell^{1})^\infty \mid S_i \in \mathcal{T}(\mathcal{E}, \mathsf{var}(v_i)) \text{ for all } i \Big\} \\
&\overset{(23)}{\leq} \mathsf{yd}(T\|_\ell^{1}) \cdot \sum \Big\{ \prod_{1\leq i\leq k} \mathsf{yd}(S_i') \mid S_i' \in \mathcal{T}(\mathcal{E}, \mathsf{var}(v_i)) \text{ for all } i \Big\} \\
&\overset{\text{dist.}}{=} \sum \Big\{ \mathsf{yd}(T) \mid T \in \mathcal{T}(\mathcal{E}, X) \Big\} = \mathbf{gfp}(F)_X .
\end{aligned}$$

□

Using this result, we can compute least and, most importantly, greatest solutions of polynomial equation systems in a polynomial number of semiring operations (including the infinitary power). Notice that, although the proof relied on $\mathbb{S}^\infty[\boldsymbol{A}]$, the computation happens only in the semiring we consider.

Example 26. Recall the equations $X_a = 1 +_\mathbb{R} X_a$, $X_b = \min(1 +_\mathbb{R} X_a, 20 +_\mathbb{R} X_c)$ and $X_c = 0 +_\mathbb{R} X_c$ from the introduction. Notice that the one-element of the tropical semiring is the real value 0. Using Theorem 1, we collapse the infinite fixed-point iteration to

$$\begin{pmatrix}0\\0\\0\end{pmatrix} \overset{F}{\longmapsto} \begin{pmatrix}1\\1\\0\end{pmatrix} \overset{F}{\longmapsto} \begin{pmatrix}2\\2\\0\end{pmatrix} \overset{F}{\longmapsto} \begin{pmatrix}3\\3\\0\end{pmatrix} \overset{\infty}{\longmapsto} \begin{pmatrix}\infty\\\infty\\0\end{pmatrix} \overset{F}{\longmapsto} \begin{pmatrix}\infty\\20\\0\end{pmatrix} \circlearrowleft^{F}$$

and obtain the expected solution. In this example, one iteration of F would actually suffice (instead of $\ell = 3$ iterations), since cycles have length one (see the graph in the introduction). In general, all ℓ steps are required (see [16]). ⌟

5 Symbolic Computation

This section briefly describes our second approach focused specifically on polynomial equation systems over the semiring $\mathbb{S}^\infty[\boldsymbol{A}]$. To this end, we adapt results of Hopkins and Kozen on Kleene[4] algebras [14] which include absorptive semirings. They express least solutions using symbolic derivatives of polynomials and we show how we can adapt this idea to greatest solutions.

It is convenient to slightly reformulate our problem setting: Instead of a system $\mathcal{E}\colon (X = P_X)_{X\in\boldsymbol{X}}$ with polynomials P_X over $\boldsymbol{X}$ and coefficients $\mathbb{S}^\infty[\boldsymbol{A}]$, we now regard P_X as an absorptive polynomial $P_X \in \mathbb{S}^\infty[\boldsymbol{A} \cup \boldsymbol{X}]$. This allows a

[4] We note that absorptive, fully-continuous semirings also satisfy the additional axioms of Cohen's ω-algebras [2] by setting $a^* = 1$ and $a^\omega = a^\infty$. However, these axioms seem too weak to axiomatize the operation a^∞; we discuss an alternative in [16].

more uniform treatment of indeterminate elimination and it is easy to see that it does not affect the solutions. To simplify notation, we write $\mathbb{S}^\infty[\boldsymbol{A}, X]$ for $\mathbb{S}^\infty[\boldsymbol{A} \cup \{X\}]$. Recall that for $P(X) \in \mathbb{S}^\infty[\boldsymbol{A}, X]$, the polynomial $P(a) \in \mathbb{S}^\infty[\boldsymbol{A}]$ results from P by replacing X with $a \in \mathbb{S}^\infty[\boldsymbol{A}]$.

5.1 Solutions in One Dimension

We first show how least and greatest solutions of a single equation $X = P(X)$ can be computed using the derivative of P.

Definition 27. Let $P(X) \in \mathbb{S}^\infty[\boldsymbol{A}, X]$. We denote the *partial derivative* of P with respect to X as P' (leaving X implicit) and define it inductively by $X' = 1$, $Y' = 0$ for $Y \in \boldsymbol{A}$, and for $P(X), Q(X) \in \mathbb{S}^\infty[\boldsymbol{A}, X]$,

$$(PQ)' = P' \cdot Q + P \cdot Q', \quad (P+Q)' = P' + Q', \quad (P^\infty)' = P^\infty \cdot P'.$$

Hopkins and Kozen prove that these partial derivatives satisfy the classical chain rule and a version of Taylor's theorem. Both proofs also apply to our setting, with straight-forward adaptions to handle the infinitary power. The least solution is then $P'(P(0))^* \cdot P(0)$, which in our setting is equal to $P(0)$ and can in fact be derived directly from absorption, without derivatives. However, using derivatives allows us to also express greatest solutions (see [16] for details):

Theorem 28. *Let $P(X) \in \mathbb{S}^\infty[\boldsymbol{A}, X]$. Then $X = P(X)$ has the least solution $P(0)$ and the greatest solution $P(0) + P'(1)^\infty$ in $\mathbb{S}^\infty[\boldsymbol{A}]$.*

5.2 Solutions of Larger Systems

To solve larger systems, we iteratively solve single equations to remove indeterminates one by one. The reason this works is the uniformity of Theorem 28: Both solutions hold under all possible instantiations as a consequence of the universal property of $\mathbb{S}^\infty[\boldsymbol{A}]$, see [16] for details. One elimination step works as follows:

Theorem 29. *Consider the equation system $\mathcal{E}$: $X = P(X,Y), Y = Q(X,Y)$ with $P, Q \in \mathbb{S}^\infty[\boldsymbol{A}, X, Y]$. Let further*

$$\begin{aligned} H(Y) &= \mathbf{gfp}(X \mapsto P(X,Y)) && \in \mathbb{S}^\infty[\boldsymbol{A}, Y], \\ b &= \mathbf{gfp}(Y \mapsto Q(H(Y),Y)) && \in \mathbb{S}^\infty[\boldsymbol{A}]. \end{aligned}$$

Then $(H(b), b)$ is the greatest solution of $\mathcal{E}$.

Here, we write $\mathbf{gfp}(X \mapsto P(X,Y))$ for the greatest solution of the equation $X = P(X,Y)$ by Theorem 28 (treating Y as coefficient). The approach for least solutions is completely symmetric.

We can apply this second technique to semirings other than $\mathbb{S}^\infty[\boldsymbol{A}]$ by first performing a symbolic abstraction. That is, we replace all coefficients by pairwise different indeterminates from $\boldsymbol{A}$, then compute the solution and apply the reverse instantiation (which preserves solutions).

Example 30. Recall our example in the tropical semiring $\mathbb{T}$. By replacing coefficients with indeterminates x, y, z, we obtain the equation system on the right.

$$\begin{aligned} X_a &= 1 +_{\mathbb{R}} X_a & & X_a = x \cdot X_a \\ X_b &= \min(1 +_{\mathbb{R}} X_a, 20 +_{\mathbb{R}} X_c) & \rightsquigarrow \quad & X_b = x \cdot X_a + y \cdot X_c \\ X_c &= 0 +_{\mathbb{R}} X_c & & X_c = z \cdot X_c \end{aligned}$$

We solve the system over $\mathbb{S}^\infty[X_a, X_b, X_c, x, y, z]$ by the symbolic approach:

- $\mathbf{gfp}(X_a \mapsto x \cdot X_a) = 0 + x^\infty = x^\infty$ (by Theorem 28)
- $\mathbf{gfp}(X_b \mapsto x \cdot x^\infty + y \cdot X_c) = x^\infty + y \cdot X_c$ (we first instantiate X_a by x^∞)
- $\mathbf{gfp}(X_c \mapsto z \cdot X_c) = z^\infty$

The greatest solution is thus $X_a = x^\infty$, $X_b = x^\infty + yz^\infty$, $X_c = z^\infty$. Applying the reverse substitution, we get the expected solution $(\infty, 20, 0)$ in $\mathbb{T}$.

Usually, the closed-form solution in Theorem 1 is preferable, as we can work directly in the target semiring. The symbolic technique is best suited to compute solutions in $\mathbb{S}^\infty[\boldsymbol{A}]$, which is of interest for semiring provenance analysis.

Remark 31. For least solutions, the one-dimensional solution in Theorem 28 in fact implies the solution in Theorem 1, as shown in [7, Prop. 28]. It seems an interesting question if a similar connection holds for greatest solutions, i.e., can the solution $F^\ell(F^\ell(\mathbf{1})^\infty)$ for ℓ equations be derived from the solution $P(0) + P'(1)^\infty$ of a single equation by algebraic methods, without derivation trees?

6 Conclusion

We have presented two methods to compute least and, most importantly, greatest solutions of polynomial equation systems over absorptive, fully-continuous semirings. Both methods require only polynomially many applications of the semiring operations and the infinitary power, in terms of the number of equations.

While we assume full continuity mostly to guarantee the existence of both kinds of solutions, absorption is a strong assumption that leads to a particularly simple way of computing solutions. Our motivation to consider absorptive semirings comes from semiring provenance of fixed-point logics, where our methods can directly be applied to compute provenance information, for example of Büchi games or formulae of least fixed-point logic LFP.

The first method, and our main result, is a closed-form solution that works in any absorptive, fully-continuous semiring and is as easy as computing the standard fixed-point iteration with an added application of the infinitary power. To prove the correctness for greatest solutions, we extended the notion of derivation trees used in the analysis of Newton's method [6] to infinite trees. Derivation trees provide an intuitive tool to understand the fixed-point iteration, but require somewhat involved arguments and constructions to properly handle infinite trees.

Our main technical contribution is that it suffices to consider trees of a particular shape resembling the solution term $F^{\ell}(F^{\ell}(\mathbf{1})^{\infty})$, intuitively corresponding to a reachability prefix with infinitely repeating deterministic subtrees. For the second method, we applied results on least solutions over Kleene algebras [14] specifically to the semiring of generalized absorptive polynomials, and extended these results by similar observations for greatest solutions.

Comparing the two proofs, we see that the symbolic approach has a simpler algebraic proof (cf. [16]) raising the question whether we can avoid the constructions of infinite trees in our main proof in favor of algebraic arguments. A further direction for future work is to study systems of nested fixed points over absorptive semirings. Recently, quasipolynomial-time algorithms have been developed to solve such systems in the Boolean case [1] or over finite lattices [13]. With the simple computation based on the fixed-point iteration, absorptive semirings might be a candidate to further increase the applicability of these algorithms.

Acknowledgements. I would like to thank the anonymous reviewers for their helpful comments and for suggesting related concepts, in particular [2,8] and Remark 31.

References

1. Arnold, A., Niwiński, D., Parys, P.: A quasi-polynomial black-box algorithm for fixed point evaluation. In: Baier, C., Goubault-Larrecq, J. (eds.) 29th EACSL Annual Conference on Computer Science Logic (CSL 2021). Leibniz International Proceedings in Informatics (LIPIcs), vol. 183, pp. 9:1–9:23. Dagstuhl (2021). https://doi.org/10.4230/LIPIcs.CSL.2021.9
2. Cohen, E.: Separation and reduction. In: Backhouse, R., Oliveira, J.N. (eds.) MPC 2000. LNCS, vol. 1837, pp. 45–59. Springer, Heidelberg (2000). https://doi.org/10.1007/10722010_4
3. Dannert, K., Grädel, E., Naaf, M., Tannen, V.: Generalized absorptive polynomials and provenance semantics for fixed-point logic. arXiv: 1910.07910 [cs.LO] (2019). https://arxiv.org/abs/1910.07910, full version of [4]
4. Dannert, K., Grädel, E., Naaf, M., Tannen, V.: Semiring provenance for fixed-point logic. In: Baier, C., Goubault-Larrecq, J. (eds.) 29th EACSL Annual Conference on Computer Science Logic (CSL 2021). Leibniz International Proceedings in Informatics (LIPIcs), vol. 183, pp. 17:1–17:22. Dagstuhl (2021). https://doi.org/10.4230/LIPIcs.CSL.2021.17
5. Deutch, D., Milo, T., Roy, S., Tannen, V.: Circuits for datalog provenance. In: Proceedings of 17th International Conference on Database Theory ICDT, pp. 201–212. OpenProceedings.org (2014). https://doi.org/10.5441/002/icdt.2014.22
6. Esparza, J., Kiefer, S., Luttenberger, M.: Newtonian program analysis. J. ACM **57**(6), 33 (2010). https://doi.org/10.1145/1857914.1857917
7. Ghilardi, S., Gouveia, M.J., Santocanale, L.: Fixed-point elimination in the intuitionistic propositional calculus. ACM Trans. Comput. Log. **21**(1), 4:1–4:37 (2019). https://doi.org/10.1145/3359669
8. Gondran, M., Minoux, M.: Graphs, Dioids and Semirings: New Models and Algorithms. Operations Research/Computer Science Interfaces, vol. 41. Springer, Boston (2008). https://doi.org/10.1007/978-0-387-75450-5

9. Grädel, E., Tannen, V.: Semiring provenance for first-order model checking. arXiv:1712.01980 [cs.LO] (2017). https://arxiv.org/abs/1712.01980
10. Grädel, E., Tannen, V.: Provenance analysis for logic and games. Mosc. J. Comb. Number Theory **9**(3), 203–228 (2020). https://doi.org/10.2140/moscow.2020.9.203
11. Green, T., Karvounarakis, G., Tannen, V.: Provenance semirings. In: Principles of Database Systems PODS, pp. 31–40 (2007). https://doi.org/10.1145/1265530.1265535
12. Green, T., Tannen, V.: The semiring framework for database provenance. In: Proceedings of PODS, pp. 93–99. ACM (2017). https://doi.org/10.1145/3034786.3056125
13. Hausmann, D., Schröder, L.: Computing nested fixpoints in quasipolynomial time. arXiv:1907.07020 [cs.CC] (2019). https://arxiv.org/abs/1907.07020
14. Hopkins, M., Kozen, D.: Parikh's theorem in commutative Kleene algebra. In: Proceedings of 14th Symposium on Logic in Computer Science, pp. 394–401. IEEE (1999). https://doi.org/10.1109/LICS.1999.782634
15. Mohri, M.: Semiring frameworks and algorithms for shortest-distance problems. J. Autom. Lang. Comb. **7**(3), 321–350 (2002). https://doi.org/10.25596/jalc-2002-321
16. Naaf, M.: Computing least and greatest fixed points in absorptive semirings. arXiv: 2106.00399 [cs.LO] (2021). https://arxiv.org/abs/2106.00399, full version of this paper

A Variety Theorem for Relational Universal Algebra

Chad Nester(✉)

Tallinn University of Technology, Tallinn, Estonia

Abstract. We consider an analogue of universal algebra in which generating symbols are interpreted as relations. We prove a variety theorem for these relational algebraic theories, in which we find that their categories of models are precisely the definable categories. The syntax of our relational algebraic theories is string-diagrammatic, and can be seen as an extension of the usual term syntax for algebraic theories.

1 Introduction

Universal algebra is the study of what is common to algebraic structures, such as groups and rings, by algebraic means. The central idea of universal algebra is that of a *theory*, which is a syntactic description of some class of structures in terms of generating symbols and equations involving them. A *model* of a theory is then a set equipped with a function for each generating symbol in a way that satisfies the equations. There is a further notion of *model morphism*, and together the models and model morphisms of a given theory form a category. These categories of models are called *varieties*. Much of classical algebra can be understood as the study of specific varieties. For example, group theory is the study of the variety of groups, which arises from the theory of groups in the manner outlined above.

A given variety will in general arise as the models of more than one theory. A natural question to ask, then, is when two theories present the same variety. To obtain a satisfying answer to this question it is helpful to adopt a more abstract perspective. Theories become categories with finite products, models become functors, and model morphisms become natural transformations. Our reward for this shift in perspective is the following answer to our question: two theories present equivalent varieties in case they have equivalent idempotent splitting completions. Thus, from a certain point of view universal algebra is the study of categories with finite products.

This point of view has developed into *categorical* universal algebra. For any sort of categorical structure we can treat categories with that structure as theories, functors that preserve it as models, and natural transformations thereof as model morphisms. The aim is then to figure out what sort of categories arise as

This research was supported by the ESF funded Estonian IT Academy research measure (project 2014-2020.4.05.19-0001).

U. Fahrenberg et al. (Eds.): RAMiCS 2021, LNCS 13027, pp. 362–377, 2021.
https://doi.org/10.1007/978-3-030-88701-8_22

models and model morphisms of this kind – that is, to determine the appropriate notion of variety. For example, if we take categories with finite limits to be our theories, then varieties correspond to locally finitely presentable categories [2].

The familiar syntax of classical algebra – consisting of *terms* built out of variables by application of the generating symbols – is inextricably bound to finite product structure. In leaving finite products behind for more richly-structured settings, categorical universal algebra also leaves behind much of the syntactic elegance of its classical counterpart. While methods of specifying various sorts of theory (categories with structure) exist, these are often cumbersome, lacking the intuitive flavour of classical universal algebra.

The present paper concerns an analogue of classical universal algebra in which the generating symbols are understood as *relations* instead of functions. The role of classical terms is instead played by string diagrams, and categories with finite products become cartesian bicategories of relations in the sense of [10] – an idea that first appears in [7]. This allows us to present relational algebraic theories in terms of generators and equations, in the style of classical universal algebra. In fact, this approach to syntax for relational theories extends the classical syntax for algebraic theories, which admits a similar diagrammatic presentation.

Our development is best understood in the context of recent work on partial algebraic theories [11], in which the string-diagrammatic syntax for algebraic theories is modified to capture partial functions. This modification of the basic syntax coincides with an increase in the expressive power of the framework, corresponding roughly to the equalizer completion of a category with finite products [8]. The move to relational algebraic theories involves a further modification of the string-diagrammatic syntax, corresponding roughly to the regular completion of a category with finite limits [9]. Put another way, in [11] the (string-diagrammatic) syntax for algebraic theories is extended to express a certain kind of equality, and the resulting terms denote partial functions. In this paper, we further extend the string-diagrammatic syntax to express existential quantification, and the resulting terms denote relations.

Contributions. The central contribution of this paper is a variety theorem characterizing the categories that arise as the models and model morphisms of some relational algebraic theory (Theorem 48). Specifically, we will see that these are precisely the *definable* categories of [19]. As a consequence we obtain that two relational algebraic theories present the same definable category if and only if splitting the partial equivalence relations in each yields equivalent categories (Theorem 49). We illustrate the use of our framework with a number of examples, including the theory of regular semigroups [16] and the theory of effectoids [24]. Lemma 10 is also novel, and we consider it to be a minor contribution

Related Work. The study of universal algebra began with the work of Birkhoff [6]. A few decades later, Lawvere introduced the categorical perspective in his doctoral thesis [22]. A modern account of universal algebra from the categorical perspective is [3]. A highlight of this account is the variety theorem for algebraic theories [1], which our variety theorem for relational algebraic theories is explicitly modelled on. An important result in categorical algebra is Gabriel-Ulmer duality [15], which tells us that if we consider categories with finite limits as our notion of algebraic theory, then the corresponding notion of variety is that of a locally finitely presentable category [2]. Our development relies on the related notion of a definable category [19,20], which recently arose in the development of an analogue of Gabriel-Ulmer duality for regular categories.

We use cartesian bicategories of relations [10] as our notion of relational algebraic theory. Our development relies on several results from the theory of allegories [14], in which cartesian bicategories of relations coincide with the notion of a unitary pre-tabular allegory. We also make use of the theory of regular and exact completions [9]. Of course, all of this relies on the theory of regular and exact categories [5]. The idea of using string diagrams as terms in more general notions of algebraic theories is relatively recent, and relies on the work of Fox [13]. The present paper can be considered a generalisation of recent work on partial theories [11] to include relations. The idea to treat cartesian bicategories of relations as theories with models in the category of sets and relations originally appeared in [7], although no variety theorem is provided therein.

Organization and Prerequisites. In Sect. 2 we introduce categories of abstract relations. In Sect. 3 we give the definition of a relational algebraic theory, and provide a number of examples. Section 4 contains the proof of the variety theorem. We assume familiarity with category theory, including regular categories [5], string diagrams for monoidal categories [17] and their connection to algebraic theories [3], and some 2-category theory [18]. We will behave as though all monoidal categories are *strict* monoidal categories, justifying this behaviour in the usual way by appealing to the coherence theorem for monoidal categories [21].

2 The Algebra of Relations

In the context of algebraic theories, finite product structure serves as an algebra of functions. In this section, we consider an analogous algebra of relations. There are two perspectives from which to consider this algebra of relations: As internal relations in a regular category, or through cartesian bicategories of relations. The two perspectives are very closely related, and we require both: it is through regular categories that our development connects to the wider literature on categorical algebra, but our syntax for relational theories will be the string-diagrammatic syntax for cartesian bicategories of relations.

To begin, we recall the category Rel of sets and relations, which will serve as the universe of models for relational theories in the same way that the category Set of sets and functions is the universe of models for classical algebraic theories.

Definition 1. *The category* Rel *has sets as objects, with arrows* $f : X \to Y$ *given by binary relations* $f \subseteq X \times Y$*. The composite of arrows* $f : X \to Y$, $g : Y \to Z$ *is defined by* $fg = \{(x, z) \mid \exists y \in Y.(x, y) \in f \wedge (y, z) \in g\}$*, and the identity relation on* X *is* $\{(x, x) \mid x \in X\}$.

2.1 Categories of Internal Relations

In any regular category we can construct an abstract analogue of Definition 1. Instead of sub*sets* $R \subseteq A \times B$, we represent relations as sub*objects* $R \rightarrowtail A \times B$. This approach to categorifying the theory of relations has a relatively long history [14], and integrates well with standard categorical logic due to the ubiquity of regular categories there.

Definition 2. *Let* $\mathbb{C}$ *be a regular category. The associated category of* internal relations, $\mathsf{Rel}(\mathbb{C})$, *is defined as follows:*

objects *are objects of* $\mathbb{C}$
arrows $r : A \to B$ *are jointly monic spans* $r = \langle f, g\rangle : R \rightarrowtail A \times B$ *modulo equivalence as subobjects of* $A \times B$. *That is,* $r : R \rightarrowtail A \times B$ *and* $r' : R' \rightarrowtail A \times B$ *are equivalent (and thus define the same arrow of* $\mathsf{Rel}(\mathbb{C})$*) in case there exists an isomorphism* $\alpha : R \to R'$ *such that* $\alpha r' = r$.
composition *of two arrows* $r : A \to B$ *and* $s : B \to C$ *given respectively by* $\langle f, g\rangle : R \rightarrowtail A \times B$ *and* $\langle h, k\rangle : S \rightarrowtail B \times C$ *is defined by first constructing the pullback of* h *along* g, *pictured below on the left. This defines an arrow* $\langle h'f, g'k\rangle : R \times_B S \to A \times C$. *The composite* $rs : A \to C$ *is defined to be the monic part of the image factorization of this arrow, pictured below on the right.*

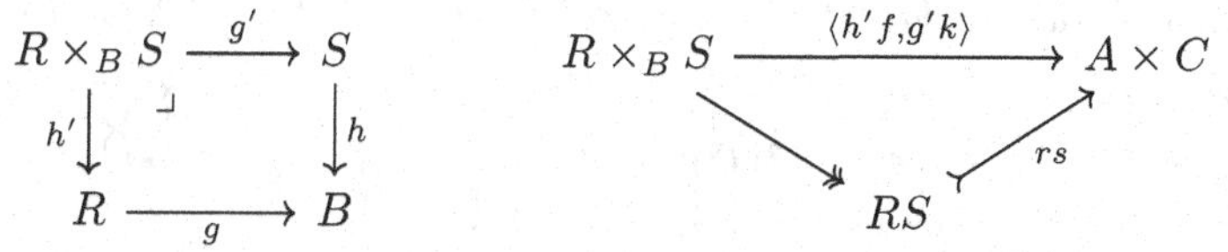

identities $1_A : A \to A$ *are are given by diagonal maps* $\Delta_A : A \rightarrowtail A \times A$.

Example 3. Set is a regular category, and the category of internal relations in Rel(Set) is precisely the usual category of sets and relations Rel.

2.2 Cartesian Bicategories of Relations

It is difficult to work with relations internal to a regular category directly. Routine calculations often involve complex interaction between pullbacks and image factorizations, and this quickly becomes intractable. A much more tractable setting for working with relations is provided by cartesian bicategories of relations, which admit a convenient graphical syntax.

Cartesian bicategories of relations are defined in terms of commutative special frobenius algebras, which provide the basic syntactic scaffolding of our approach:

Definition 4. *Let* $\mathbb{X}$ *be a symmetric strict monoidal category. A* commutative special frobenius algebra *in* $\mathbb{X}$ *is a 5-tuple* $(X, \delta_X, \mu_X, \varepsilon_X, \eta_X)$, *as in*

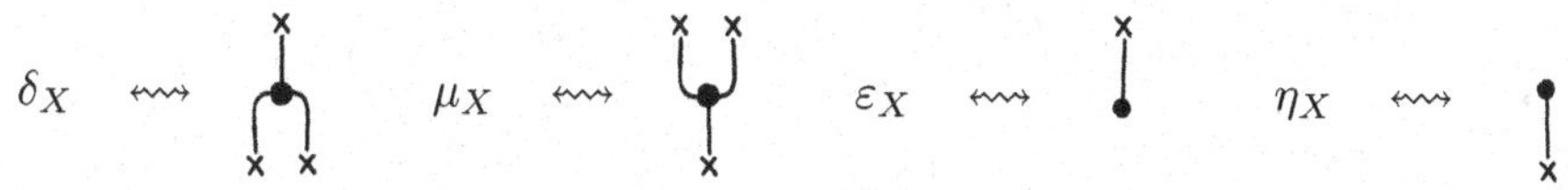

such that

(i) $(X, \delta_X, \varepsilon_X)$ *is a commutative comonoid:*

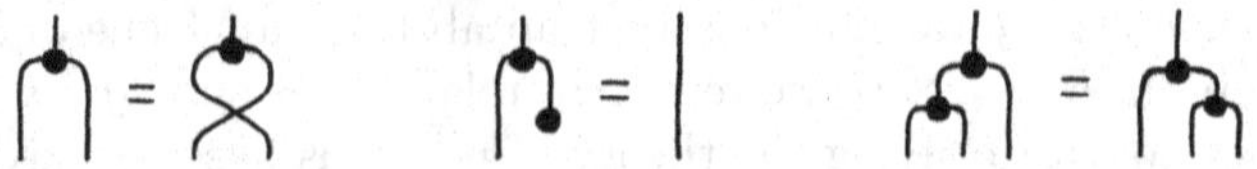

(ii) (X, μ_X, η_X) *is a commutative monoid:*

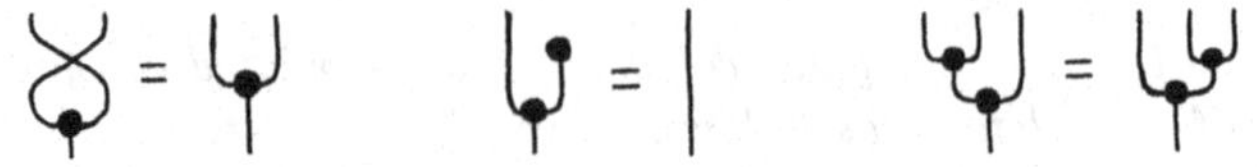

(iii) μ_X *and* δ_X *satisfy the special and frobenius equations:*

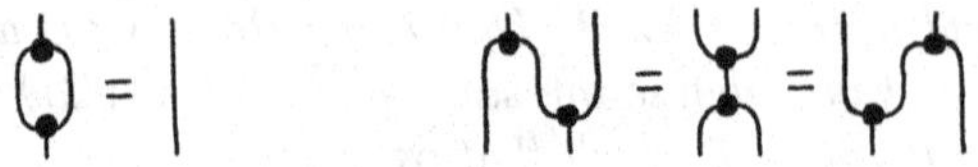

An intermediate notion is that of a hypergraph category, in which objects are coherently equipped with commutative special frobenius algebra structure:

Definition 5. *A symmetric strict monoidal category* $\mathbb{X}$ *is called a* hypergraph category *[12] in case:*

(i) Each object X *of* $\mathbb{X}$ *is equipped with a commutative special frobenius algebra.*
(ii) The frobenius algebra structure is coherent, i. e., for all X, Y *we have:*

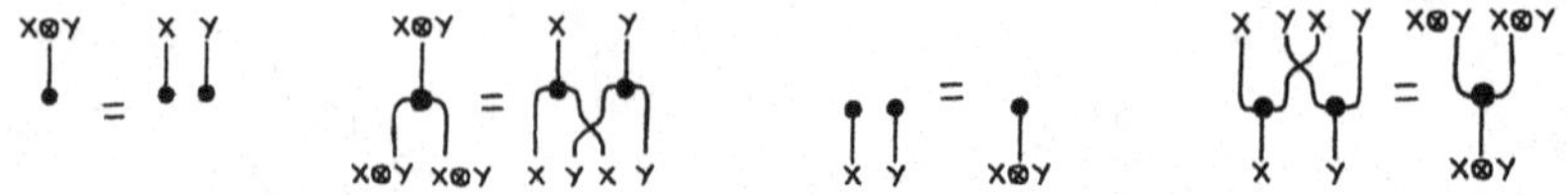

Now a cartesian bicategory of relations is a hypergraph category enjoying certain additional structure:

Definition 6. *A* cartesian bicategory of relations *[10] is a poset-enriched hypergraph category* $\mathbb{X}$ *such that:*

(i) The comonoid structure is lax natural. *That is, for all arrows* f *of* $\mathbb{X}$*:*

(ii) Each of the frobenius algebras satisfy:

Example 7. The category Rel is a cartesian bicategory of relations with

$$\delta_X = \{(x, (x, x)) \mid x \in X\} \qquad \mu_X = \{((x, x), x) \mid x \in X\}$$

$$\varepsilon_X = \{(x, *) \mid x \in X\} \qquad \eta_X = \{(*, x) \mid x \in X\}$$

where $*$ is the unique element of the singleton set $I = \{*\}$.

Example 8. If $\mathbb{C}$ is a regular category then $\mathsf{Rel}(\mathbb{C})$ is a cartesian bicategory of relations with $X \otimes Y = X \times Y$, $I = 1$, and

$$\delta_X = \langle 1_X, \Delta_X \rangle : X \rightarrowtail X \times (X \times X) \qquad \mu_X = \langle \Delta_X, 1_X \rangle : X \rightarrowtail (X \times X) \times X$$

$$\varepsilon_X = \langle 1_X, !_X \rangle : X \rightarrowtail X \times 1 \qquad \eta_X = \langle !_X, 1_X \rangle : X \rightarrowtail 1 \times X$$

Where Δ_X is the diagonal morphism and $!_X$ is the unique morphism into the terminal object 1 of $\mathbb{C}$.

Cartesian bicategories of relations admit meets of hom-sets:

Lemma 9 ([7]). *Every cartesian bicategory of relations has meets of parallel arrows, with $f \cap g$ for $f, g : X \to Y$ defined by*

Further, the meet determines the poset-enrichment in that $f \leq g \Leftrightarrow f \cap g = f$.

We point out this allows for a much simpler presentation, as in:

Lemma 10. *A hypergraph category $\mathbb{X}$ is a cartesian bicategory of relations if and only if for each arrow f:*

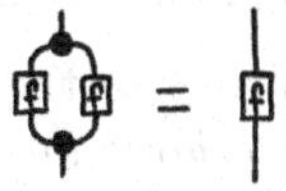

We will require a 2-category of cartesian bicategories of relations in our development. Our notion of 1-cell is a structure-preserving functor as in:

Definition 11. *A* morphism *of cartesian bicategories of relations $F : \mathbb{X} \to \mathbb{Y}$ is a strict monoidal functor that preserves the frobenius algebra structure:*

$$F(\delta_X) = \delta_{FX} \qquad F(\mu_X) = \mu_{FX} \qquad F(\varepsilon_X) = \varepsilon_{FX} \qquad F(\eta_X) = \eta_{FX}$$

and the correct sort of 2-cell turns out to be a *lax* natural transformation:

Definition 12. *Let $\mathbb{X}, \mathbb{Y}$ be cartesian bicategories of relations, and let $F, G : \mathbb{X} \to \mathbb{Y}$ be morphisms thereof. Then a* lax transformation *$\alpha : F \to G$ consists of an $\mathbb{X}_0$-indexed family of arrows $\alpha_X : F(X) \to G(X)$ such that for each arrow $f : X \to Y$ of $\mathbb{X}$ we have $F(f)\alpha_Y \leq \alpha_X G(f)$ in $\mathbb{Y}$.*

Definition 13. *Let* RAT *be the 2-category with cartesian bicategories of relations as 0-cells, their morphisms as 1-cells, and lax transformations as 2-cells.*

An important class of arrows in a cartesian bicateory of relations are the *maps*, which should be thought of as those relations that happen to be functions.

Definition 14 (Maps). *An arrow $f : X \to Y$ in a cartesian bicategory of relations is called:*

(i) simple *in case the equation below on the left holds.*
(ii) total *in case the equation below on the right holds.*

(iii) A map *in case it is both simple and total.*

The maps of a cartesian bicategory of relations always form a subcategory $\mathsf{Map}(\mathbb{X})$. For example, $\mathsf{Map}(\mathsf{Rel}) \cong \mathsf{Set}$. More generally:

Theorem 15 ([14]). *For $\mathbb{C}$ a regular category, there is an equivalence of categories $\mathbb{C} \simeq \mathsf{Map}(\mathsf{Rel}(\mathbb{C}))$.*

Remarkably, the components of lax transformations are always maps:

Lemma 16 ([7]). *If $\mathbb{X}, \mathbb{Y}$ are cartesian bicategories of relations, $F, G : \mathbb{X} \to \mathbb{Y}$ are morphisms thereof and $\alpha : F \to G$ is a lax transformation, then each component $\alpha_X : FX \to GX$ of α is necessarily a map.*

3 Relational Algebraic Theories

In this section we define relational algebraic theories along with the models and model morphisms, and consider a number of examples.

Definition 17. [7] *A* relational algebraic theory *is a cartesian bicategory of relations. A* model *of a relational algebraic theory $\mathbb{X}$ is a morphism of cartesian bicategories of relations $F : \mathbb{X} \to \mathsf{Rel}$. A* model morphism *$\alpha : F \to G$ is a lax transformation.*

It is convenient to present relational algebraic theories somewhat informally in terms of string-diagrammatic generators and (in)equations between them, with the structure of a cartesian bicategory of relations implicitly present. A more formal account would proceed in terms of monoidal equational theories, from which the cartesian bicategory of relations giving the associated relational algebraic theory may be freely constructed [7].

Example 18 (Sets). The relational algebraic theory with no generators and no equations has sets as models and functions as model morphisms (see Lemma 16), and so the associated category of models is Set.

Example 19 (Posets). Consider the relational theory with a single generator (below left) which is required to be reflexive, transitive, and antisymmetric:

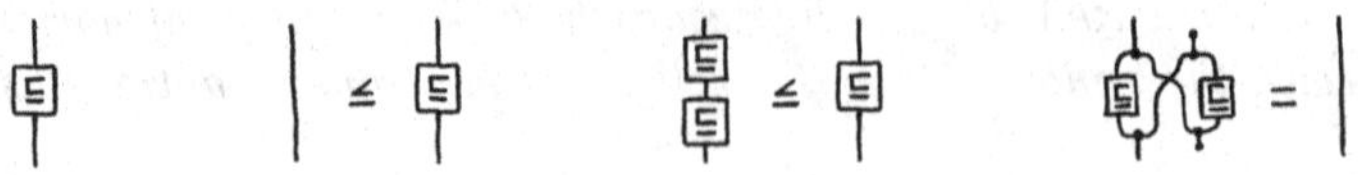

The associated category of models is the category of posets and monotone maps.

Example 20 (Nonempty Sets). Consider the relational theory with no generating symbols and a single equation:

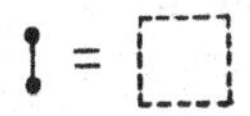

Models of the associated relational algebraic theory are sets X such that the generating equation is satisfied in Rel:

$$\eta_X \varepsilon_X = \{(*,*)\} = \square_I$$

where η_X and ε_X are defined as in Definition 1. If we calculate the relational composite, we find that:

$$\eta_X \varepsilon_X = \{(*,*) \mid \exists x \in X.(*,x) \in \eta_X \wedge (x,*) \in \varepsilon_X\} = \{(*,*) \mid \exists x \in X\}$$

and so models are nonempty sets. The theory of nonempty sets contains no generating morphisms, and so model morphisms are simply functions. Contrast this to the category of *pointed* sets, in which morphisms must preserve the point.

Example 21 (Regular Semigroups). A *semigroup* is a set equipped with an associative binary operation, denoted by juxtaposition. A semigroup S is *regular* [16] in case

$$\forall a \in S. \exists x \in S. axa = a$$

The relational theory of semigroups has a single generating symbol (below left) which is required to be simple, total, and associative:

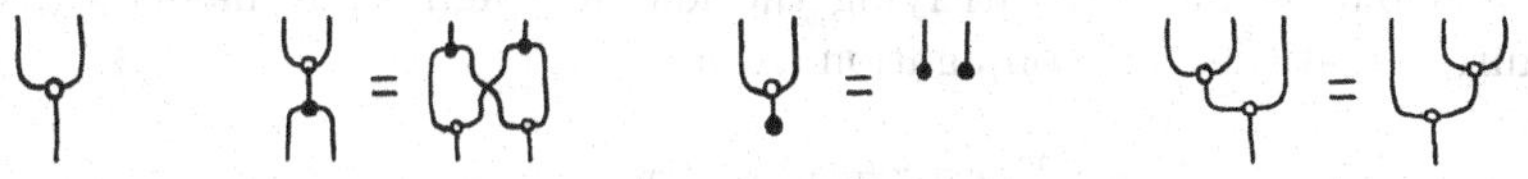

To capture the regular semigroups we include the following equation:

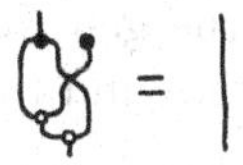

The associated category of models is the category of regular semigroups and semigroup homomorphisms.

Example 22 (Effectoids). An *effectoid* [24] is a set A equipped with a unary relation $\not\in \mapsto \_ \subseteq A$, a binary relation $\_ \preceq \_ \subseteq A \times A$, and a ternary relation $\_ ; \_ \mapsto \_ \subseteq A \times A \times A$ satisfying:

(Identity) For all $a, a' \in A$,

$$\exists x \in A.(\not\in \mapsto x) \wedge (x ; a \mapsto a') \Leftrightarrow a \preceq a' \Leftrightarrow \exists y \in A.(\not\in \mapsto y) \wedge (a ; y \mapsto a')$$

(Associativity) For all $a, b, c, d \in A$,

$$\exists x.(a ; b \mapsto x) \wedge (x ; c \mapsto d) \Leftrightarrow \exists y.(b ; c \mapsto y) \wedge (a ; y \mapsto d)$$

(Reflexive Congruence 1) For all $a \in A$, $a \preceq a$.
(Reflexive Congruence 2) For all $a, a' \in A$, $(\not\in \mapsto a) \wedge (a \preceq a') \Rightarrow (\not\in \mapsto a')$
(Reflexive Congruence 3) For all $a, b, c \in A$, $\exists x.(a\,; b \mapsto x) \wedge (x \preceq c) \Rightarrow (a\,; b \preceq c)$

To obtain a relational theory of effectoids, we ask for three generating symbols corresponding respectively to the unary, binary, and ternary relation:

Then the identity and associativity axioms become:

And the reflexive congruence axioms become:

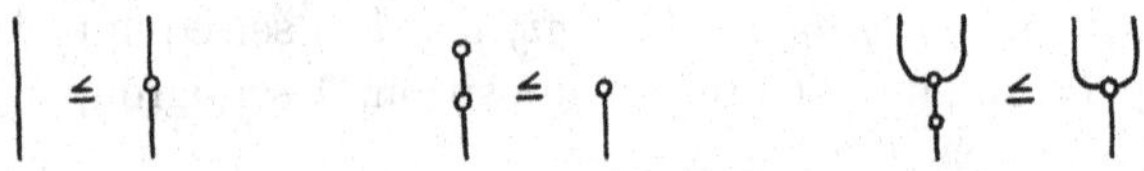

The models of this relational theory are precisely the effectoids.

Example 23 (Generalized Separation Algebras). A *generalized separation algebra* [4] is a partial monoid satisfying the left and right cancellativity axioms, which further satisfies the conjugation axiom:

$$\forall x, y.(\exists z.x \circ z = y) \Leftrightarrow (\exists w.w \circ x = y)$$

To capture generalized separation algebras as a relational algebraic theory, we require two generating symbols in the generating monoidal equational theory, corresponding to the monoid operation and the unit:

Both are required to be simple, and the unit is required to be total:

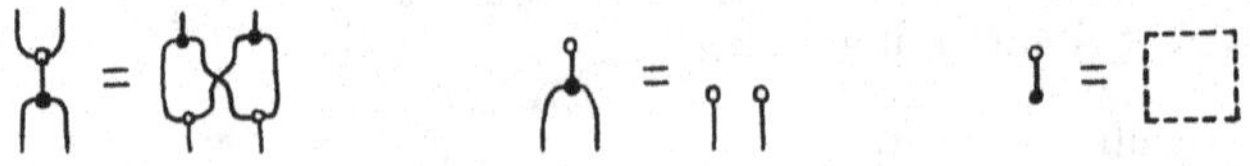

The associativity and unitality axioms become:

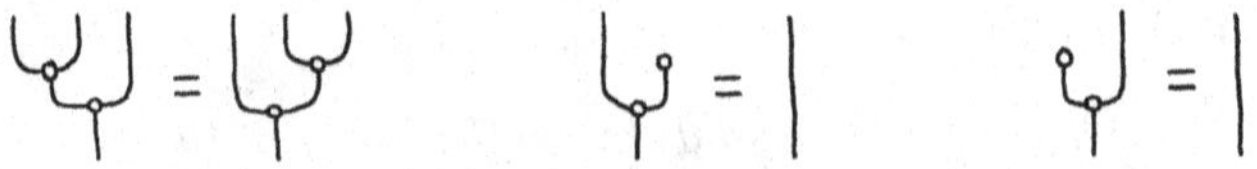

Now, define upside-down versions of the generators as in:

Then left cancellativity, right cancellativity, and conjugation are, respectively:

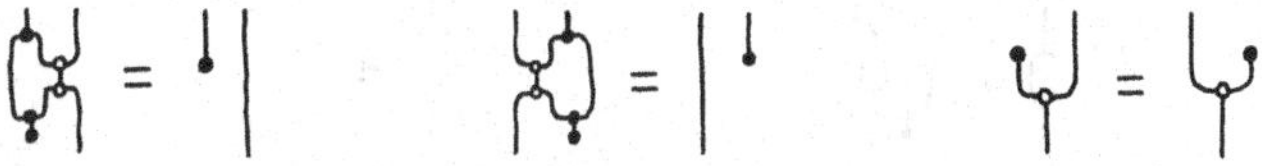

The corresponding category of models is the category of generalized separation algebras and partial monoid homomorphisms.

Example 24 (Algebraic Theories). Let $\mathbb{X}$ be an algebraic theory, and let $(\mathbb{X}_{\mathsf{eq}})_{\mathsf{reg/lex}}$ be the regular completion of $\mathbb{X}$ [8,9]. $\mathsf{Rel}((\mathbb{X}_{\mathsf{eq}})_{\mathsf{reg/lex}})$ is a relational algebraic theory. Further, its models and model morphisms (as a relational algebraic theory) coincide with the models and model morphisms of $\mathbb{X}$ (as an algebraic theory). Conversely, if $\mathbb{X}$ is a relational algebraic theory, then the maps of $\mathbb{X}$ form a subcategory $\mathsf{Map}(\mathbb{X})$. $\mathsf{Map}(\mathbb{X})$ has finite products, and so defines an algebraic theory in the usual sense. Further, the notions of model and model morphism for relational algebraic theories restrict to the usual notions for algebraic theories on the category of maps.

Example 25 (Essentially Algebraic Theories). An *essentially algebraic theory* [23] is (among many equivalent presentations) a category $\mathbb{X}$ with finite limits. Models are the finite-limit preserving functors $\mathbb{X} \to \mathsf{Set}$, and model morphisms are natural transformations. For $\mathbb{X}$ an essentially algebraic theory let $\mathbb{X}_{\mathsf{reg/lex}}$ be the regular completion of $\mathbb{X}$ [9]. Then $\mathsf{Rel}(\mathbb{X}_{\mathsf{reg/lex}})$ is a relational algebraic theory. Further, its models and model morphisms (as a relational algebraic theory) coincide with the models and model morphisms of $\mathbb{X}$ (as an essentially algebraic theory). Conversely, if $\mathbb{X}$ is a relational algebraic theory then the simple maps of $\mathbb{X}$ are a partial algebraic theory in the sense of [11] – which turn out to be equivalent to essentially algebraic theories. The notions of model and model morphism for relational theories restrict to the corresponding notions for partial theories.

4 The Variety Theorem

In this section we prove the variety theorem for relational algebraic theories. We do this in phases: first we introduce some necessary terminology concerning classes of idempotents, and recall some details of the idempotent splitting completion. Next, we make the relationship between bicategories of relations and regular categories precise. We then show how the situation extends to include exact categories, this being necessary because exactness is the difference between regular categories and definable categories. Finally, we introduce definable categories, which end up being the varieties of our relational theories. This is structured so that the variety theorem follows immediately. We end by showing precisely when two relational theories present the same definable category.

4.1 Flavours of Idempotent Splitting

We begin by introducing some important kinds of arrow in a relational theory:

Definition 26. *An arrow $f : A \to A$ of a relational algebraic theory is called reflexive in case $1 \leq f$,* coreflexive *in case $f \leq 1$, a* partial equivalence relation *in case it is symmetric and transitive as in:*

and is called an equivalence relation *if it is reflexive, symmetric, and transitive.*

Notice in particular that every partial equivalence relation is idempotent, that every coreflexive arrow is a partial equivalence relation, and that every equivalence relation is a partial equivalence relation. We also recall the idempotent splitting completion relative to a class of idempotents in a category:

Definition 27. *Let $\mathbb{X}$ be a category, and let $\mathcal{E}$ be a collection of idempotents in $\mathbb{X}$. Define a category $\mathsf{Split}_{\mathcal{E}}(\mathbb{X})$ in which objects are pairs (X, a) where X is a object of $\mathbb{X}$ and $a : X \to X$ is in $\mathcal{E}$, and arrows $f : (X, a) \to (Y, b)$ are arrows $f : X \to Y$ of $\mathbb{X}$ such that $afb = f$. Composition is composition in $\mathbb{X}$, and identities are given by $a = 1_{(X,a)} : (X, a) \to (X, a)$.*

Every member of $\mathcal{E}$ splits in $\mathsf{Split}_{\mathcal{E}}(\mathbb{X})$. It turns out that splitting partial equivalence relations works well with cartesian bicategories of relations:

Proposition 28 ([14]). *If $\mathbb{X}$ is a relational algebraic theory and $\mathcal{E}$ is a class of partial equivalence relations in $\mathbb{X}$, then $\mathsf{Split}_{\mathcal{E}}(\mathbb{X})$ is a relational algebraic theory.*

4.2 Tabulation and Regular Categories

We begin our exposition of the correspondence between regular categories and relational algebraic theories by recalling the notion of tabulation [10]. Intuitively, a tabulation of an arrow represents it as a subobject in the category of maps.

Definition 29. *A* tabulation *of an arrow $f : X \to Y$ in a relational algebraic theory $\mathbb{X}$ consists of a pair of maps (h, k) such that the equation below on the left holds in $\mathbb{X}$, and the map below on the right is monic in $\mathsf{Map}(\mathbb{X})$:*

$\mathbb{X}$ is tabular *in case every arrow of $\mathbb{X}$ admits a tabulation. Further, define $\mathsf{RAT}_{\mathsf{tab}}$ to be the full 2-subcategory of RAT (Definition 13) on the tabular 0-cells.*

The category of maps of a tabular relational algebraic theory is regular, and conversely the category of internal relations in a regular category is tabular:

Proposition 30. *Let* REG *be the 2-category of regular categories, regular functors, and natural transformation. Then:*

(i) If $\mathbb{X}$ *is a tabular relational algebraic theory then* $\mathsf{Map}(\mathbb{X})$ *is regular. This extends to a 2-functor* $\mathsf{Map} : \mathsf{RAT}_{\mathsf{tab}} \to \mathsf{REG}$.
(ii) If $\mathbb{C}$ *is a regular category, then* $\mathsf{Rel}(\mathbb{C})$ *is tabular. This extends to a 2-functor* $\mathsf{Rel} : \mathsf{REG} \to \mathsf{RAT}_{\mathsf{tab}}$.

Tabular relational theories and regular categories are thus interchangeable:[1]

Theorem 31. *There is an equivalence of 2-categories* $\mathsf{Map} : \mathsf{RAT}_{\mathsf{tab}} \simeq \mathsf{REG} : \mathsf{Rel}$.

Finally, any relational theory can be made tabular by splitting the coreflexives:

Proposition 32. *Let* $\mathbb{X}$ *be a relational algebraic theory, and let* cor *be the collection of coreflexives in* $\mathbb{X}$. *Then* $\mathbb{X}$ *is tabular if and only if every member of* cor *splits. In particular,* $\mathsf{Split}_{\mathsf{cor}}(\mathbb{X})$ *is always tabular. This extends to a 2-adjunction* $\mathsf{Split}_{\mathsf{cor}} : \mathsf{RAT} \dashv \mathsf{RAT}_{\mathsf{tab}} : U$ *where* U *is the evident forgetful functor.*

4.3 Effectivity and Exact Categories

We begin by recalling the closely related notions of effectivity and exactness:

Definition 33 ([14]). *A relational algebraic theory* $\mathbb{X}$ *is* effective *in case all partial equivalence relations in* $\mathbb{X}$ *split. Let* $\mathsf{RAT}_{\mathsf{eff}}$ *be the full 2-subcategory of* RAT *on the effective 0-cells.*

Definition 34 ([9]). *A regular category* $\mathbb{C}$ *is* exact *in case* $\mathsf{Rel}(\mathbb{C})$ *is effective. Let* EX *be the full 2-subcategory of* REG *on the exact 0-cells.*

It is straightforward to verify that Theorem 31 restricts to the effective case:

Proposition 35. *If* $\mathbb{X}$ *is an effective relational algebraic theory, then* $\mathsf{Map}(\mathbb{X})$ *is exact. Conversely, if* $\mathbb{C}$ *is an exact category, then* $\mathsf{Rel}(\mathbb{C})$ *is effective. This extends to an equivalence of 2-categories* $\mathsf{Map} : \mathsf{RAT}_{\mathsf{eff}} \simeq \mathsf{EX} : \mathsf{Rel}$.

Splitting equivalence relations makes tabular relational theories effective:

Proposition 36. *Let* $\mathbb{X}$ *be a tabular relational algebraic theory, and let* eq *be the collection of equivalence relations in* $\mathbb{X}$. *Then* $\mathsf{Split}_{\mathsf{eq}}(\mathbb{X})$ *is effective. This extends to a 2-adjunction* $\mathsf{Split}_{\mathsf{eq}}\mathsf{RAT}_{\mathsf{tab}} \dashv \mathsf{RAT}_{\mathsf{eff}} : U$ *where* U *is the evident forgetful functor.*

We may therefore give the exact completion of a regular category as follows:

---

[1] We note that we restrict our attention to the 0- and 1-cells then this is proven in [10]. Our contribution is to extend this to include 2-cells.

Proposition 37 ([9,20]). *If* $\mathbb{C}$ *is regular, define the* exact completion *of* $\mathbb{C}$ *by*

$$\mathbb{C}_{\mathsf{ex/reg}} = \mathsf{Map}(\mathsf{Split}_{\mathsf{eq}}(\mathsf{Rel}(\mathbb{X})))$$

Then $\mathbb{C}_{\mathsf{ex/reg}}$ *is exact. This extends to a 2-adjunction* $\mathsf{ex/reg} : \mathsf{REG} \dashv \mathsf{EX} : U$ *where* U *is the evident forgetful functor.*

We summarize the relationship of regularity and exactness to relational theories:

Corollary 38. *The following diagram of left 2-adjoint commutes:*

$$\begin{array}{ccc} \mathsf{RAT}_{\mathsf{tab}} & \xrightarrow[\sim]{\mathsf{Map}} & \mathsf{REG} \\ {\scriptstyle \mathsf{Split}_{\mathsf{eq}}}\downarrow & & \downarrow{\scriptstyle \mathsf{ex/reg}} \\ \mathsf{RAT}_{\mathsf{eff}} & \xrightarrow[\mathsf{Map}]{\sim} & \mathsf{EX} \end{array}$$

where the arrows marked with $\sim$ *are part of a 2-equivalence.*

Similarly, splitting partial equivalence relations allows us to summarize the role of the idempotent splitting completion:

Proposition 39. *Write* per *to denote the collection of partial equivalence relations in a relational algebraic theory. There is a 2-adjunction* $\mathsf{Split}_{\mathsf{per}} : \mathsf{RAT} \dashv \mathsf{RAT}_{\mathsf{eff}} : U$ *where* U *is the evident forgetful functor. Further, for any relational algebraic theory* $\mathbb{X}$, *we have* $\mathsf{Split}_{\mathsf{per}}(\mathbb{X}) \simeq \mathsf{Split}_{\mathsf{eq}}(\mathsf{Split}_{\mathsf{cor}}(\mathbb{X}))$, *and so the following diagram of left 2-adjoints commutes:*

$$\begin{array}{ccc} \mathsf{RAT} & \xrightarrow{\mathsf{Split}_{\mathsf{cor}}} & \mathsf{RAT}_{\mathsf{tab}} \\ & {\scriptstyle \mathsf{Split}_{\mathsf{per}}}\searrow & \downarrow{\scriptstyle \mathsf{Split}_{\mathsf{eq}}} \\ & & \mathsf{RAT}_{\mathsf{eff}} \end{array}$$

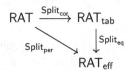

Proof. The proof that $\mathsf{Split}_{\mathsf{per}}$ defines a 2-functor which is left adjoint to the forgetful 2-functor is straightforward, and similar to Proposition 32. A proof that $\mathsf{Split}_{\mathsf{per}}(\mathbb{X}) \simeq \mathsf{Split}_{\mathsf{eq}}(\mathsf{Split}_{\mathsf{cor}}(\mathbb{X}))$ can be found in [14, 2.169], it follows immediately that our diagram of left 2-adjoints commutes.

4.4 Definable Categories

The final idea involved in our variety theorem is that of a definable category [19]. Definable categories come from categorical universal algebra. If we take regular categories as our notion of theory, regular functors into Set as our notion of model, and natural transformations as our model morphisms, then definable categories are the corresponding varieties. We follow the exposition of [20], and in particular we formulate definable categories via finite injectivity classes:

Definition 40 (Finite Injectivity Class). *Let $h : A \to B$ be an arrow of $\mathbb{X}$. Then an object C of $\mathbb{X}$ is said to be* h-injective *in case the function of hom-sets $\mathbb{X}(h, C) : \mathbb{X}(B, C) \to \mathbb{X}(A, C)$ defined by $X(h, C)(f) = hf$ is injective. If M is a finite set of arrows in $\mathbb{X}$, write $\mathsf{inj}(M)$ for the full subcategory on the objects C of $\mathbb{X}$ that are h-injective for each $h \in M$. We say that each $\mathsf{inj}(M)$ is a* finite injectivity class *in $\mathbb{X}$.*

Definable categories are defined relative to an ambient locally finitely presentable category. It is an open problem to give a free-standing characterization [19].

Definition 41. *A category is said to be* definable *if it arises as a finite injectivity class in some locally finitely presentable category. If $\mathbb{X}$ and $\mathbb{Y}$ are definable categories, a functor $F : \mathbb{X} \to \mathbb{Y}$ is called an* interpretation *in case it preserves products and directed colimits. Let DEF be the 2-category with definable categories as 0-cells, interpretations as 1-cells, and natural transformations as 2-cells.*

From any definable category we can obtain an exact category by considering its interpretations into Set.

Proposition 42 ([20])**.** *If $\mathbb{X}$ is a definable category then the functor category $\mathsf{DEF}(\mathbb{X}, \mathsf{Set})$ is an exact category. This extends to a 2-functor $\mathsf{DEF}(\_, \mathsf{Set}) : \mathsf{DEF}^{\mathsf{op}} \to \mathsf{EX}$.*

Similarly, for any regular category the associated category of regular functors into Set is definable.

Proposition 43 ([20])**.** *If $\mathbb{C}$ is a regular category then the functor category $\mathsf{REG}(\mathbb{C}, \mathsf{Set})$ is definable. This extends to a 2-functor $\mathsf{REG}(\_, \mathsf{Set}) : \mathsf{REG} \to \mathsf{DEF}^{\mathsf{op}}$.*

If the category in question is exact, then considering interpretations of the resulting definable category into Set yields the original exact category. This lifts to the 2-categorical setting.

Proposition 44 ([20])**.** *There is an adjunction of 2-categories $\mathsf{REG}(-, \mathsf{Set}) : \mathsf{REG} \dashv \mathsf{DEF}^{\mathsf{op}} : \mathsf{DEF}(-, \mathsf{Set})$ which specializes to an equivalence of 2-categories $\mathsf{REG}(-, \mathsf{Set}) : \mathsf{EX} \simeq \mathsf{DEF}^{\mathsf{op}} : \mathsf{DEF}(-, \mathsf{Set})$.*

This gives another way to describe the exact completion of a regular category:

Proposition 45 ([20])**.** *If $\mathbb{C}$ is regular then $\mathbb{C}_{\mathsf{ex/reg}} \simeq \mathsf{DEF}(\mathsf{REG}(\mathbb{C}, \mathsf{Set}), \mathsf{Set})$.*

Thus, we may summarize the relationship between definable, regular, and exact categories as follows:

Corollary 46 ([20, **Sect. 9,10**])**.** *The following diagram of left 2-adjoints commutes.*

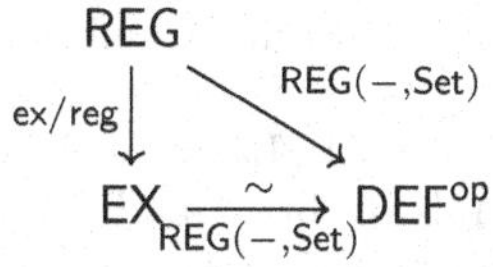

where the arrow marked with $\sim$ is part of a 2-equivalence.

The ingredients of our variety theorem for relational algebraic theories are now assembled. Together, Proposition 39, Corollary 38, and Corollary 46 give:

Corollary 47. *There following diagram of left 2-adjoints commutes:*

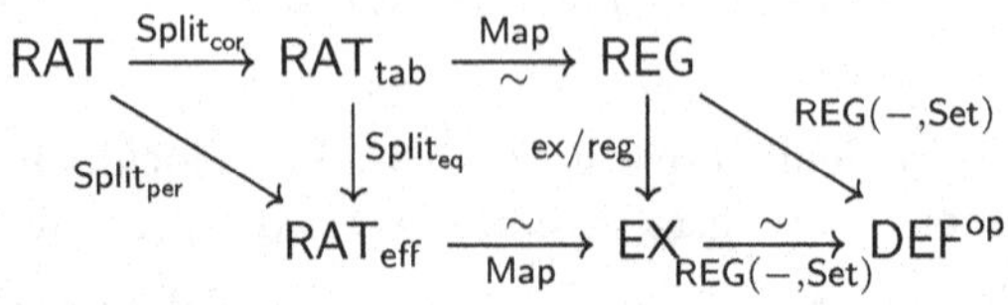

where the arrows marked with $\sim$ *are part of a 2-equivalence.*

Now our variety theorem is an immediate consequence of Corollary 47:

Theorem 48. *There is an adjunction of 2-categories* $\mathsf{Mod} : \mathsf{RAT} \dashv \mathsf{DEF}^{\mathsf{op}} : \mathsf{Th}$

It may not be immediately clear what this tells us about the category of models and model morphisms of a relational algebraic theory, so let us briefly discuss. Consider an arbitrary relational algebraic theory $\mathbb{X}$. Our universe of models Rel is tabular, so models of $\mathbb{X}$ and models of $\mathsf{Split}_{\mathsf{cor}}(\mathbb{X})$ are the same thing since the image of any coreflexive in $\mathbb{X}$ already splits in Rel. Then the category of models of $\mathbb{X}$ and model morphisms thereof is $\mathsf{RAT}_{\mathsf{tab}}(\mathsf{Split}_{\mathsf{cor}}(\mathbb{X}), \mathsf{Rel})$. When we transport this across the 2-equivalence $\mathsf{Map} : \mathsf{RAT}_{\mathsf{tab}} \xrightarrow{\sim} \mathsf{REG}$ it becomes $\mathsf{REG}(\mathsf{Map}(\mathsf{Split}_{\mathsf{cor}}(\mathbb{X})), \mathsf{Set})$, a definable category. Thus, categories of models and model morphisms of regular algebraic theories are definable categories.

Now, Set is exact, so Rel is effective, which means that much like the models of $\mathbb{X}$ and $\mathsf{Split}_{\mathsf{cor}}(\mathbb{X})$, the models of $\mathbb{X}$ and $\mathsf{Split}_{\mathsf{per}}(\mathbb{X})$ are the same. We have shown that $\mathsf{RAT}_{\mathsf{eff}} \simeq \mathsf{EX} \simeq \mathsf{DEF}^{\mathsf{op}}$, and so the question of when two relational algebraic theories generate the same category of models and model morphisms can be answered as follows:

Theorem 49. *Two relational algebraic theories* $\mathbb{X}$ *and* $\mathbb{Y}$ *present equivalent definable categories if and only if* $\mathsf{Split}_{\mathsf{per}}(\mathbb{X})$ *and* $\mathsf{Split}_{\mathsf{per}}(\mathbb{Y})$ *are equivalent.*

Compare this to the case of algebraic theories, in which two theories present the same variety in case splitting *all* idempotents yields equivalent categories [1].

References

1. Adámek, J., Lawvere, F.W., Rosický, J.: On the duality between varieties and algebraic theories. Algebra Univers. **49**(1), 35–49 (2003)
2. Adamek, J., Rosický, J.: Locally Presentable and Accessible Categories. London Mathematical Society Lecture Note Series. Cambridge University Press, Cambridge (1994)
3. Adámek, J., Rosický, J., Vitale, E.M., Lawvere, F.W.: Algebraic Theories: A Categorical Introduction to General Algebra. Cambridge Tracts in Mathematics, Cambridge University Press, Cambridge (2010)

4. Alexander, S., Jipsen, P., Upegui, N.: On the structure of generalized effect algebras and separation algebras. In: Desharnais, J., Guttmann, W., Joosten, S. (eds.) RAMiCS 2018. LNCS, vol. 11194, pp. 148–165. Springer, Cham (2018). https://doi.org/10.1007/978-3-030-02149-8_10
5. Barr, M.: Exact categories. In: Barr, M., Grillet, P.A., van Osdol, D.H. (eds.) Exact Categories and Categories of Sheaves. LNM, vol. 236, pp. 1–120. Springer, Heidelberg (1971). https://doi.org/10.1007/BFb0058580
6. Birkhoff, G.: On the structure of abstract algebras. In: Mathematical Proceedings of the Cambridge Philosophical Society, vol. 31, pp. 433–454. Cambridge University Press, Cambridge (1935)
7. Bonchi, F., Pavlovic, D., Sobociński, P.: Functorial semantics for relational theories (2017). arXiv:1711.08699
8. Bunge, M., Carboni, A.: The symmetric topos. J. Pure Appl. Algebra **105**(3), 233–249 (1995)
9. Carboni, A., Vitale, E.M.: Regular and exact completions. J. Pure Appl. Algebra **125**(1), 79–116 (1998)
10. Carboni, A., Walters, R.F.C.: Cartesian bicategories I. J. Pure Appl. Algebra **49**(1), 11–32 (1987)
11. Di Liberti, I., Loregian, F., Nester, C., Sobociński, P.: Functorial semantics for partial theories. Proc. ACM Program. Lang. **5**(POPL), 1–28 (2021)
12. Fong, B., Spivak, D.I.: Hypergraph categories (2019)
13. Fox, T.: Coalgebras and cartesian categories. Comm. Algebra **4**(7), 665–667 (1976)
14. Freyd, P.J., Scedrov, A.: Categories, Allegories (1990)
15. Gabriel, P.: Unzerlegbare Darstellungen. I. Manuscripta Math. **6**, 71–103 (1972); correction, ibid. **6**, 309 (1972)
16. Green, J.A.: On the structure of semigroups. Ann. Math. 163–172 (1951)
17. Joyal, A., Street, R.: The geometry of tensor calculus I. Adv. Math. **88**(1), 55–112 (1991)
18. Kelly, G.M., Street, R.: Review of the elements of 2-categories. In: Kelly, G.M. (ed.) Category Seminar. LNM, vol. 420, pp. 75–103. Springer, Heidelberg (1974). https://doi.org/10.1007/BFb0063101
19. Kuber, A., Rosický, J.: Definable categories. J. Pure Appl. Algebra **222**(5), 1006–1025 (2018)
20. Lack, S., Tendas, G.: Enriched regular theories. J. Pure Appl. Algebra **224**(6), 106268 (2020)
21. Mac Lane, S.: Categories for the Working Mathematician. Springer, New York (1971). https://doi.org/10.1007/978-1-4612-9839-7
22. Lawvere, F.W.: Functorial Semantics of Algebraic Theories: And, Some Algebraic Problems in the Context of Functorial Semantics of Algebraic Theories (1963)
23. Palmgren, E., Vickers, S.J.: Partial horn logic and cartesian categories. Ann. Pure Appl. Logic **145**(3), 314–353 (2007)
24. Tate, R.: The sequential semantics of producer effect systems. In: Proceedings of the 40th Annual ACM SIGPLAN-SIGACT Symposium on Principles of Programming Languages, pp. 15–26 (2013)

On Tools for Completeness of Kleene Algebra with Hypotheses

Damien Pous[1], Jurriaan Rot[2(✉)], and Jana Wagemaker[2]

[1] CNRS, LIP, ENS de Lyon, Lyon, France
[2] Radboud University, Nijmegen, The Netherlands
jrot@cs.ru.nl

Abstract. In the literature on Kleene algebra, a number of variants have been proposed which impose additional structure specified by a theory, such as Kleene algebra with tests (KAT) and the recent Kleene algebra with observations (KAO), or make specific assumptions about certain constants, as for instance in NetKAT. Many of these variants fit within the unifying perspective offered by *Kleene algebra with hypotheses*, which comes with a canonical language model constructed from a given set of hypotheses. For the case of KAT, this model corresponds to the familiar interpretation of expressions as languages of guarded strings.

A relevant question therefore is whether Kleene algebra together with a given set of hypotheses is complete with respect to its canonical language model. In this paper, we revisit, combine and extend existing results on this question to obtain tools for proving completeness in a modular way. We showcase these tools by reproving completeness of KAT and KAO, and prove completeness of a new variant of KAT where the collection of tests only forms a distributive lattice.

1 Introduction

Kleene algebras (KA) [8,17] are algebraic structures involving an iteration operation, Kleene star, corresponding to reflexive-transitive closure in relational models and to language iteration in language models. Its axioms are complete w.r.t. relational models and language models [3,18,28], and the resulting equational theory is decidable via automata algorithms (in fact, PSPACE-complete [29]).

These structures were later extended in order to deal with common programming constructs. For instance, Kleene algebras with tests (KAT) [22], which combine Kleene algebra and Boolean algebra, make it possible to represent the control flow of while programs. Kleene star is used for while loops, and Boolean tests are used for the conditions of such loops, as well as the conditions in if-then-else statements. Again, the axioms of KAT are complete w.r.t. appropriate classes of models, and its equational theory remains in PSPACE. Proving so is

An extended version of this abstract, with proofs, may be found on HAL at https://hal.archives-ouvertes.fr/hal-03269462/ [31]. This work has been supported by the ERC (CoVeCe, grant No 678157) and by the LABEX MILYON (ANR-10-LABX-0070), within the program ANR-11-IDEX-0007.

U. Fahrenberg et al. (Eds.): RAMiCS 2021, LNCS 13027, pp. 378–395, 2021.
https://doi.org/10.1007/978-3-030-88701-8_23

non-trivial: Kozen's proof reduces completeness of KAT to completeness of KA, via a direct syntactic transformation on terms.

Another extension is Concurrent Kleene algebra (CKA) [13], where a binary operator for parallelism is added. The resulting theory is characterised by languages of pomsets rather than languages of words, and is ExpSpace-complete [6]. Trying to have both tests and concurrency turned out to be non-trivial, and called for yet another notion: Kleene algebras with observations (KAO) [15], which are again complete w.r.t. appropriate models, and decidable.

When used in the context of program verification, e.g., in a proof assistant, such structures make it possible to write algebraic proofs of correctness, and to mechanise some of the steps: when two expressions e and f representing two programs happen to be provably equivalent in KA, KAT, or KAO, one does not need to provide a proof, one can simply call a certified decision procedure [4,30]. However, this is often not enough [1,12,26]: most of the time, the expressions e and f are provably equal only under certain assumptions on their constituants. For instance, to prove that $(a+b)^*$ and a^*b^* are equal, one may have to use that in the considered instance, we have $ba = ab$. In other words, one would like to prove equations under some assumptions, to have algorithms for the Horn theory of Kleene algebra and its extensions rather than just their equational theories.

Unfortunately, those Horn theories are typically undecidable [21,24], even with rather restricted forms of hypotheses (e.g., commutation of two letters, as in the above example). Nevertheless, important and useful classes of hypotheses can be 'eliminated', by reducing to the plain and decidable case of the equational theory. This is for instance the case of *Hoare hypotheses* [23], of the shape $e = 0$, which make it possible to encode Hoare triples for partial correctness in KAT.

In some cases, one wants to exploit hypotheses about specific constituants (e.g., a and b in the above example). In other situations, one wants to exploit assumptions on the whole structure. For instance, in commutative Kleene algebra [5,8,33], one assumes that the product is commutative everywhere.

Many of these extensions of Kleene algebra (KAT, KAO, commutative KA, specific hypotheses) fit into the generic framework of Kleene algebra with hypotheses [10], providing in each case a canonical model in terms of closed languages.

We show that we recover standard models in this way, and we provide tools to establish completeness and decidability of such extensions, in a modular way. The key notion is that of *reduction* from one set of hypotheses to another. We summarise existing reductions and we provide a toolbox for combining those reductions together. We use this toolbox in order to obtain new and modular proofs of completeness for KAT and KAO, as well as for the fragment of KAT where tests are only assumed to form a distributive lattice.

Note however that there are Kleene algebra extensions like *action algebras* [32] or *action lattices* [20], which do not seem to fit into the framework of Kleene algebra with hypotheses: it is not clear how to interpret the additional operations as letters with additional structure.

2 Kleene Algebra, Hypotheses, Closures

A *Kleene algebra* [8,19] is a tuple $(K, +, \cdot, ^*, 0, 1)$ such that $(K, +, \cdot, 0, 1)$ is an idempotent semiring, and * is a unary operator on K such that for all $x, y \in K$ the following axioms are satisfied:

$$1 + x \cdot x^* \leq x^* \qquad x + y \cdot z \leq z \Rightarrow y^* \cdot x \leq z \qquad x + y \cdot z \leq y \Rightarrow x \cdot z^* \leq y$$

There, as later in the paper, we write $x \leq y$ as a shorthand for $x + y = y$. Given the idempotent semiring axioms, $\leq$ is a partial order in every Kleene algebra, and all operations are monotone w.r.t. that order.

We let e, f range over regular expressions over an alphabet Σ, defined by:

$$e, f ::= e + f \mid e \cdot f \mid e^* \mid 0 \mid 1 \mid a \in \Sigma$$

We write $\mathcal{T}(\Sigma)$ for the set of such expressions, or simply $\mathcal{T}$ when the alphabet is clear from the context. Given alphabets Σ and Γ, a function $h \colon \Sigma \to \mathcal{T}(\Gamma)$ extends uniquely into a homomorphism $h \colon \mathcal{T}(\Sigma) \to \mathcal{T}(\Gamma)$, which we refer to as the *homomorphism generated by h*. As usual, every regular expression e gives rise to a language $[\![e]\!] \in \mathcal{P}(\Sigma^*)$. Given two regular expressions, we moreover write $\mathsf{KA} \vdash e = f$ when $e = f$ is derivable from the axioms of Kleene algebra. (Equivalently, when the equation $e = f$ holds universally, in all Kleene algebras.)

The central theorem of Kleene algebra is the following:

Theorem 2.1 (Soundness and Completeness of KA [3,18,28]**).** *For all $e, f \in \mathcal{T}$, we have $\mathsf{KA} \vdash e = f$ if and only if $[\![e]\!] = [\![f]\!]$.*

As a consequence, the equational theory of Kleene algebras is decidable.

Our goal is to extend this result to the case where we have additional hypotheses on some of the letters of the alphabet, or axioms restricting the behaviour of certain operations. Those are represented by sets of *inequations*, i.e., pairs (e, f) of regular expressions written $e \leq f$ for the sake of clarity. Given a set H of such inequations, we write $\mathsf{KA}_H \vdash e \leq f$ when the inequation $e \leq f$ is derivable from the axioms of Kleene algebra and the hypotheses in H (similarly for equations). By extension, we write $\mathsf{KA}_H \vdash H'$ when $\mathsf{KA}_H \vdash e \leq f$ for all $e \leq f$ in H'.

Note that we consider letters of the alphabet as constants rather than variables. In particular, while we have $\mathsf{KA}_{ba \leq ab} \vdash (a + b)^* \leq a^*b^*$, we do not have $\mathsf{KA}_{ba \leq ab} \vdash (a + c)^* \leq a^*c^*$. Formally, we use a notion of derivation where there is no substitution rule, and where we have all instances of Kleene algebra axioms as axioms. When we want to consider hypotheses that are universally valid, it suffices to use all their instances. For example, to define commutative Kleene algebra, we simply use the infinite set $\{ef \leq fe \mid e, f \in \mathcal{T}\}$.

We associate a canonical language model to KA with a set of hypotheses H, defined by *closure* under H. For $u, v \in \Sigma^*$ and $L \subseteq \Sigma^*$, let $uLv \triangleq \{uxv \mid x \in L\}$.

Definition 2.2 (H-closure). Let H be a set of hypotheses and $L \subseteq \Sigma^*$ a language. The *H-closure of L*, denoted as $\mathrm{cl}_H(L)$, is the smallest language containing L s.t. for all $e \leq f \in H$ and $u, v \in \Sigma^*$, if $u[\![f]\!]v \subseteq \mathrm{cl}_H(L)$, then $u[\![e]\!]v \subseteq \mathrm{cl}_H(L)$.

Fixpoint theory makes it possible to characterise the H-closure of a language L as the least (pre)fixpoint of the function $\mathrm{st}'_{H,L}(X) = \mathrm{st}_H(X) \cup L$, where

$$\mathrm{st}_H(X) = \bigcup\{u[\![e]\!]v \mid e \leq f \in H, u, v \in \Sigma^*, u[\![f]\!]v \subseteq X\}.$$

This least fixpoint can be characterised more explicitly by transfinite iteration: we have $\mathrm{cl}_H(L) = \bigcup_\alpha \mathrm{st}_H^\alpha(L)$ where $\mathrm{st}_H^{\alpha+1}(L) = \mathrm{st}_H(\mathrm{st}_H^\alpha(L))$ for every ordinal α, and $\mathrm{st}_H^\lambda(L) = \bigcup_{\alpha<\lambda} \mathrm{st}_H^\alpha(L)$ for every limit ordinal λ.

This notion of closure gives a *closed interpretation* of regular expressions, $\mathrm{cl}_H([\![-]\!])$, for which KA_H is sound:

Theorem 2.3 ([10, **Theorem 2**]). *If* $\mathsf{KA}_H \vdash e = f$, *then* $\mathrm{cl}_H([\![e]\!]) = \mathrm{cl}_H([\![f]\!])$.

In the sequel, we shall prove the converse implication, completeness, for specific choices of H: we say that KA_H is *complete* if for all expressions e, f:

$$\mathrm{cl}_H([\![e]\!]) = \mathrm{cl}_H([\![f]\!]) \text{ implies } \mathsf{KA}_H \vdash e = f\,.$$

We could hope that completeness always holds, notably because the notion of closure is invariant under inter-derivability of the considered hypotheses, as a consequence of the following lemma:

Lemma 2.4 ([16, **Lemma 4.10**]). *Let H and H' be sets of hypotheses such that* $\mathsf{KA}_H \vdash H'$. *Then* $\mathrm{cl}_{H'} \subseteq \mathrm{cl}_H$.

Unfortunately, there are concrete instances for which KA_H is known not to be complete. For instance, there is a finitely presented monoid (thus a finite set H_0 of equations) such that $\{(e, f) \mid \mathrm{cl}_{H_0}([\![e]\!]) = \mathrm{cl}_{H_0}([\![f]\!])\}$ is not r.e. [25, Theorem 1]. Since derivability in KA_H is r.e. as soon as H is, KA_{H_0} cannot be complete.

Before turning to techniques for proving completeness, let us describe the closed interpretation of regular expressions for two specific choices of hypotheses.

Let us consider first *commutative Kleene algebra*, obtained as explained in the Introduction using the set $\{ef \leq fe \mid e, f \in \mathcal{T}(\Sigma)\}$. Under Kleene algebra axioms, this set is equiderivable with its restriction to letters, $C = \{ab \leq ba \mid a, b \in \Sigma\}$ (a consequence of [1, Lemma 4.4]).

The associated closure can be characterised as follows:

$$\mathrm{cl}_C(L) = \{w \in \Sigma^* \mid \exists v \in L.\, |w|_x = |v|_x \text{ for all } x \in \Sigma\}$$

where $|w|_x$ denotes the number of occurences of x in w. Thus, $w \in \mathrm{cl}_C(L)$ if it is a permutation of some word in L.

This semantics matches precisely the one used in [8] for commutative Kleene algebra: there, a function $[\![-]\!]_c \colon \mathcal{T}(\Sigma) \to \mathcal{P}(\mathbb{N}^\Sigma)$ interprets regular expressions as subsets of $\mathbb{N}^\Sigma$, whose elements are thought of as "commutative words": these assign to each letter the number of occurrences, but there is no order of letters. Let $q \colon \mathcal{P}(\Sigma^*) \to \mathcal{P}(\mathbb{N}^\Sigma)$, $q(L) = \{\lambda x.|w|_x \mid w \in L\}$; this map computes the *Parikh image* of a given language L, that is, the set of multisets representing occurences of letters in words in L. Then this semantics is characterised by $[\![-]\!]_c = q \circ [\![-]\!]$.

One may observe that $[\![-]\!]_c = q(\mathrm{cl}_C([\![-]\!]))$, since cl_C only adds words to a language which have the same number of occurences of each letter as some word which is already there. Conversely, we have $\mathrm{cl}_C([\![-]\!]) = q'([\![-]\!]_c)$, where $q' \colon \mathcal{P}(\mathbb{N}^\Sigma) \to \mathcal{P}(\Sigma^*)$, $q'(L) = \{w \mid p \in L,\ \forall x \in \Sigma, |w|_x = p(x)\}$. As a consequence, we have $[\![e]\!]_c = [\![f]\!]_c$ if and only if $\mathrm{cl}_C([\![e]\!]) = \mathrm{cl}_C([\![f]\!])$.

From there, we can easily deduce from the completeness result in [8, Chapter 11, Theorem 4], attributed to Pilling (see also [5]), that KA_C is complete.

Let us now consider a single hypothesis: $D = \{ab \leq 0\}$ for some letters a and b. The D-closure of a language L consists of those words that either belong to L, or contain ab as a subword. As a consequence, we have $\mathrm{cl}_D([\![e]\!]) = \mathrm{cl}_D([\![f]\!])$ if and only if $[\![e]\!]$ and $[\![f]\!]$ agree on all words not containing the pattern ab.

In this example, we can easily obtain decidability and completeness of KA_D. Indeed, consider the function $r \colon \mathcal{T}(\Sigma) \to \mathcal{T}(\Sigma)$, $r(e) = e + \Sigma^* ab \Sigma^*$. For all e, we have $\mathsf{KA}_D \vdash e = r(e)$, and $\mathrm{cl}_D([\![e]\!]) = [\![r(e)]\!]$. As a consequence, we have

$$
\begin{aligned}
& \mathrm{cl}_D([\![e]\!]) = \mathrm{cl}_D([\![f]\!]) \\
\Leftrightarrow\quad & [\![r(e)]\!] = [\![r(f)]\!] \\
\Leftrightarrow\quad & \mathsf{KA} \vdash r(e) = r(f) && \text{(Theorem 2.1)} \\
\Rightarrow\quad & \mathsf{KA}_D \vdash e = f
\end{aligned}
$$

The first step above establishes decidability of the closed semantics; the following ones reduce the problem of completeness for KA_D to that for KA alone, which is known to hold. By soundness (Theorem 2.3), the last line implies the first one, so that these conditions are all equivalent.

This second example exploits and illustrate a simple instance of the framework we design in the sequel to prove completeness of various sets of hypotheses.

3 Reductions

As illustrated above, the overall strategy is to reduce completeness of KA_H, for a given set of hypotheses H, to completeness of Kleene algebra. The core idea is to provide a map r from expressions to expressions, which incorporates the hypotheses H in the sense that $[\![r(e)]\!] = \mathrm{cl}_H([\![e]\!])$, and such that $r(e)$ is provably equivalent to e under the hypotheses H. This idea leads to the unifying notion of *reduction*, developed in [10,16,25].

Definition 3.1 (Reduction). Assume $\Gamma \subseteq \Sigma$ and let H, H' be sets of hypotheses over Σ and Γ respectively. We say that H *reduces to* H' if $\mathsf{KA}_H \vdash H'$ and there exists a map $r \colon \mathcal{T}(\Sigma) \to \mathcal{T}(\Gamma)$ such that for all $e \in \mathcal{T}(\Sigma)$,

1. $\mathsf{KA}_H \vdash e = r(e)$, and
2. $\mathrm{cl}_H([\![e]\!]) \cap \Gamma^* = \mathrm{cl}_{H'}([\![r(e)]\!])$.

We often refer to such a witnessing map r itself as a reduction. Generalising the above example, we obtain the key property of reductions:

Theorem 3.2. *Suppose H reduces to H'. If $\mathsf{KA}_{H'}$ is complete, then so is KA_H.*

Proof. Let r be the map for the reduction from H to H'. For all $e, f \in \mathcal{T}(\Sigma)$,

$$\begin{array}{rlr} & \mathrm{cl}_H([\![e]\!]) = \mathrm{cl}_H([\![f]\!]) & \\ \Rightarrow & \mathrm{cl}_{H'}([\![r(e)]\!]) = \mathrm{cl}_{H'}([\![r(f)]\!]) & (r \text{ a reduction (item 2)}) \\ \Rightarrow & \mathsf{KA}_{H'} \vdash r(e) = r(f) & (\text{completeness of } H') \\ \Rightarrow & \mathsf{KA}_H \vdash r(e) = r(f) & (\mathsf{KA}_H \vdash H') \\ \Rightarrow & \mathsf{KA}_H \vdash e = f & (r \text{ a reduction (item 1)}) \end{array}$$ □

An important case is when $H' = \emptyset$: given a reduction from H to $\emptyset$, Theorem 3.2 gives completeness of KA_H, by completeness of KA. Such reductions are what we ultimately aim for. However, in the examples later in this paper, these reductions are composed of smaller ones, which do make use of intermediate hypotheses. Section 3.2 contains general techniques for combining reductions.

While we focus on completeness in this paper, note that reductions can also be used to prove decidability. More precisely, if KA'_H is complete and decidable, and H reduces to H' via a computable reduction r, then KA_H is decidable.

The following result from [16] (cf. Remark 3.5) gives a sufficient condition for the existence of a reduction. This is useful for reductions where the underlying map r is a homomorphism.

Lemma 3.3. *Assume $\Gamma \subseteq \Sigma$ and let H, H' be sets of hypotheses over Σ and Γ respectively, such that $\mathsf{KA}_H \vdash H'$. If there exists a homomorphism $r\colon \mathcal{T}(\Sigma) \to \mathcal{T}(\Gamma)$ such that:*

1. *For all $a \in \Gamma$, we have $\mathsf{KA} \vdash a \leq r(a)$.*
2. *For all $a \in \Sigma$, we have $\mathsf{KA}_H \vdash a = r(a)$.*
3. *For all $e \leq f \in H$, we have $\mathsf{KA}_{H'} \vdash r(e) \leq r(f)$.*

then H reduces to *H'.*

Example 3.4. We consider KA together with a global "top element" $\top$ and the axiom $e \leq \top$. To make this precise in Kleene algebra with hypotheses, we assume an alphabet Σ with $\top \in \Sigma$, and take the set of hypotheses $H_\top = \{e \leq \top \mid e \in \mathcal{T}(\Sigma)\}$. Then $\mathrm{cl}_{H_\top}(L)$ contains those words obtained from a word $w \in L$ by replacing every occurence of $\top$ in w by arbitrary words in Σ^*.

We claim that H reduces to $\emptyset$. To this end, define the homomorphism $r\colon \mathcal{T}(\Sigma) \to \mathcal{T}(\Sigma)$ by $r(\top) = \Sigma^*$ (where we view Σ as an expression consisting of the sum of its elements) and $r(a) = a$ for $a \in \Sigma$ with $a \neq \top$. Each of the conditions of Lemma 3.3 is now easy to check. Thus r is a reduction, so by Theorem 3.2, $\mathsf{KA}_{H_\top}$ is complete.

Note that this implies completeness w.r.t. validity of equations in all (regular) language models, where $\top$ is interpreted as the largest language: indeed, the closed semantics $\mathrm{cl}_{H_\top}([\![-]\!])$ is generated by such a model.

At the end of Sect. 2, we discussed commutative KA as an instance of Kleene algebra with hypotheses H. While KA_H is complete in that case, there is no reduction from H to $\emptyset$, as cl_H does not preserve regularity. Indeed, $\mathrm{cl}_H([\![(ab)^*]\!]) = \{w \mid |w|_a = |w|_b\}$ which is not regular. The completeness proof in [5,8] is self-contained, and does not rely on completeness of KA.

Remark 3.5. The idea to use two sets of hypotheses in Definition 3.1 is from [16], where reductions are defined slightly differently: the alphabet is fixed (that is, $\Sigma = \Gamma$), and the last condition is instead defined as $\mathrm{cl}_H([\![e]\!]) = \mathrm{cl}_H([\![f]\!]) \Rightarrow \mathrm{cl}_{H'}([\![r(e)]\!]) = \mathrm{cl}_{H'}([\![r(f)]\!])$. An extra notion of *strong* reduction is then introduced, which coincides with our definition if $\Sigma = \Gamma$. By allowing a change of alphabet, we do not need to distinguish reductions and strong reductions. Lemma 3.3 is in [16, Lemma 4.23], adapted here to the case with two alphabets (this is taken care of in *loc. cit.* by assuming $\mathrm{cl}_{H'}$ preserves languages over Γ).

3.1 Basic Reductions

The following result collects several sets of hypotheses for which we have reductions to $\emptyset$. These mostly come from the literature. They form basic building blocks used in the more complex reductions that we present in the examples below.

Lemma 3.6. *Each of the following sets of hypotheses reduce to the empty set (of hypotheses over Σ).*

(i) $\{u_i \leq w_i \mid i \in I\}$ *with* $u_i, w_i \in \Sigma^*$ *and* $|u_i| \leq 1$ *for all* $i \in I$
(ii) $\{1 \leq \sum_{a \in S_i} a \mid i \in I\}$ *with each* $S_i \subseteq \Sigma$ *finite*
(iii) $\{e \leq 0\}$ *for* $e \in \mathcal{T}(\Sigma)$
(iv) $\{ea \leq a\}$ *and* $\{ae \leq a\}$ *for* $a \in \Sigma$, $e \in \mathcal{T}(\Sigma \backslash \{a\})$

Proof. (i) is [25, Theorem 2]. (The result mentions equations, but in the proof only the relevant inequations are used.) (ii) is [10, Proposition 6]. (iii) is basically due to [7], but since it is phrased differently there we include a proof in [31]. Hypotheses of a similar form as (iv) are studied in the setting of Kleene algebra with tests in [12], we include a proof in [31]. □

Note that Item iii above covers finite sets of hypotheses of the form $\{e_i \leq 0\}_{i \in I}$, as these can be encoded as the single hypothesis $\sum_{i \in I} e_i \leq 0$.

3.2 Compositional Reductions

The previous subsection gives reductions to the empty set for single equations. However, in the examples we often start with a *collection* of hypotheses of different shapes, which we wish to reduce to the empty set. Therefore, we now discuss a few techniques for combining reductions.

Throughout this section, for sets of hypotheses $H_1, \ldots, H_n$ we often denote the associated closure by cl_i instead of cl_{H_i}, $\mathrm{cl}_{i,j}$ instead of $\mathrm{cl}_{H_i \cup H_j}$ and $\mathrm{cl}_{i \ldots j}$

instead of $\mathrm{cl}_{\bigcup_{i\leq k\leq j} H_k}$. Similarly, we write st_i instead of st_{H_i} etc. First, there are the basic observations that reductions compose (Lemma 3.7) and that equiderivable sets of hypotheses always reduce to each other, via the identity (Lemma 3.8).

Lemma 3.7. *Let H_1, H_2 and H_3 be sets of hypotheses. If H_1 reduces to H_2 and H_2 reduces to H_3 then H_1 reduces to H_3.*

Lemma 3.8. *Let H_1, H_2 be sets of hypotheses over a common alphabet. If $\mathsf{KA}_{H_1} \vdash H_2$ and $\mathsf{KA}_{H_2} \vdash H_1$ then H_1 and H_2 reduce to each other.*

The following useful lemma allows to combine reductions by union. Its assumptions allow to compose the reductions sequentially. A similar lemma is formulated in the setting of bi-Kleene algebra in [14, Lemma 4.46].

Lemma 3.9. *Let $H_1, \ldots, H_n, H$ be sets of hypotheses over Σ, with $n \geq 1$. If H_i reduces to H for all i, and $\mathrm{cl}_{1\ldots n} = \mathrm{cl}_n \circ \cdots \circ \mathrm{cl}_1$, then $\bigcup_{i\leq n} H_i$ reduces to H.*

The next lemma is useful to show the second requirement in Lemma 3.9.

Lemma 3.10. *Let $H_1, \ldots, H_n$ be sets of hypotheses, such that $\mathrm{cl}_i \circ \mathrm{cl}_j \subseteq \mathrm{cl}_j \circ \mathrm{cl}_i$ for all i, j with $i < j$. Then $\mathrm{cl}_{1\ldots n} = \mathrm{cl}_n \circ \cdots \circ \mathrm{cl}_1$.*

The condition $\mathrm{cl}_i \circ \mathrm{cl}_j \subseteq \mathrm{cl}_j \circ \mathrm{cl}_i$ is equivalent to $\mathrm{cl}_{i,j} = \mathrm{cl}_j \circ \mathrm{cl}_i$. With that formulation, Lemma 3.10 is stated in bi-Kleene algebra as [14, Lemma 4.50].

We now proceed with several lemmas that help proving $\mathrm{cl}_1 \circ \mathrm{cl}_2 \subseteq \mathrm{cl}_2 \circ \mathrm{cl}_1$. In particular, these allow to use the "one-step closure" st_H from the fixed point characterisation of cl_H (below Definition 2.2), for cl_1 in Lemma 3.11 and additionally for cl_2 in Lemma 3.12, assuming further conditions.

Lemma 3.11. *For all H_1, H_2, if $\mathrm{st}_1 \circ \mathrm{cl}_2 \subseteq \mathrm{cl}_2 \circ \mathrm{cl}_1$, then $\mathrm{cl}_1 \circ \mathrm{cl}_2 \subseteq \mathrm{cl}_2 \circ \mathrm{cl}_1$.*

Lemma 3.12. *Let H_1, H_2 be sets of hypotheses such that the right-hand sides of inequations in H_1 are all words. If $\mathrm{st}_1 \circ \mathrm{st}_2 \subseteq \mathrm{cl}_2 \circ \mathrm{st}_1^{=}$ then $\mathrm{cl}_1 \circ \mathrm{cl}_2 \subseteq \mathrm{cl}_2 \circ \mathrm{cl}_1$.*

We return to hypotheses of the form $H_1 = \{e \leq 0\}$. For any term e and with H_2 an arbitrary set of hypotheses, we have $\mathrm{cl}_1 \circ \mathrm{cl}_2 \subseteq \mathrm{cl}_2 \circ \mathrm{cl}_1$. As a consequence, Lemma 3.6(iii) extends to the following result, which shows we can always get rid of finite sets of hypotheses of the form $e \leq 0$. A similar result, in terms of Horn formulas and in the context of KAT, is shown in [11].

Lemma 3.13. *For a set of hypotheses H and term e, $H \cup \{e \leq 0\}$ reduces to H.*

4 Kleene Algebra with Tests

In this section we apply the machinery from the previous sections to obtain a modular completeness proof for Kleene algebra with tests [27].

A *Kleene algebra with tests (KAT)* is a Kleene algebra X containing a Boolean algebra L such that the meet of L coincides with the product of X, the join of X coincides with the sum of X, the top element of L is the multiplicative identity of X, and the bottom elements of X and L coincide.

Syntactically, we fix two finite sets Σ and Ω of primitive actions and primitive tests. We denote the set of Boolean expressions over alphabet Ω by $\mathcal{T}_{\mathsf{BA}}$:

$$\phi, \psi ::= \phi \vee \psi \mid \phi \wedge \psi \mid \neg\phi \mid \bot \mid \top \mid o \in \Omega$$

We write $\mathsf{BA} \vdash \phi = \phi'$ when this equation is derivable from Boolean algebra axioms [2,9], and similarly for inequations.

We let α, β range over *atoms*: elements of the set $\mathsf{At} \triangleq 2^{\Omega}$. Those may be seen as valuations for Boolean expressions, or as complete conjunctions of literals: α is implicitly seen as the Boolean formula $\bigwedge_{\alpha(o)=1} o \wedge \bigwedge_{\alpha(o)=0} \neg o$. They form the atoms of the Boolean algebra generated by Ω. We write $\alpha \models \phi$ when ϕ holds under the valuation α. A key property of Boolean algebras is that for all atoms α and formulas ϕ, we have

$$\alpha \models \phi \Leftrightarrow \mathsf{BA} \vdash \alpha \leq \phi \qquad \text{and} \qquad \mathsf{BA} \vdash \phi = \bigvee_{\alpha \models \phi} \alpha$$

The KAT terms over alphabets Σ and Ω are the regular expressions over the alphabet $\Sigma + \mathcal{T}_{\mathsf{BA}}$: $\mathcal{T}_{\mathsf{KAT}} \triangleq \mathcal{T}(\Sigma + \mathcal{T}_{\mathsf{BA}})$. We write $\mathsf{KAT} \vdash e = f$ when this equation is derivable from the axioms of KAT, and similarly for inequations.

The standard interpretation of KAT associates to each term a language of guarded strings. A *guarded string* is a sequence of the form $\alpha_0 a_0 \alpha_1 a_1 \ldots a_{n-1}\alpha_n$ with $a_i \in \Sigma$ for all $i < n$, and $\alpha_i \in \mathsf{At}$ for all $i \leq n$. We write $\mathcal{GS}$ for the set $\mathsf{At} \times (\Sigma \times \mathsf{At})^*$ of such guarded strings. Now, the interpretation $\mathcal{G}\colon \mathcal{T}(\Sigma + \mathcal{T}_{\mathsf{BA}}) \to 2^{\mathcal{GS}}$ is defined as the homomorphic extension of the assignment $\mathcal{G}(a) = \{\alpha a \beta \mid \alpha, \beta \in \mathsf{At}\}$ for $a \in \Sigma$ and $\mathcal{G}(\phi) = \{\alpha \mid \alpha \models \phi\}$ for $\phi \in \mathcal{T}_{\mathsf{BA}}$, where for sequential composition of guarded strings the *coalesced product* is used. The coalesced product of guarded strings $u\alpha$ and βv is defined as $u\alpha v$ if $\alpha = \beta$ and undefined otherwise.

Theorem 4.1 (**[27, Theorem 8]**). *For all* $e, f \in \mathcal{T}_{\mathsf{KAT}}$*, we have* $\mathsf{KAT} \vdash e = f$ *iff* $\mathcal{G}(e) = \mathcal{G}(f)$.

We now reprove this result using Kleene algebra with hypotheses. We start by defining the additional axioms of KAT as hypotheses.

Definition 4.2. We write bool for the set of all instances of Boolean algebra axioms over $\mathcal{T}_{\mathsf{BA}}$ and glue for the following set of hypotheses relating the Boolean algebra connectives to the Kleene algebra ones

$$\mathsf{glue} = \{\phi \wedge \psi = \phi \cdot \psi,\ \phi \vee \psi = \phi + \psi \mid \phi, \psi \in \mathcal{T}_{\mathsf{BA}}\} \cup \{\bot = 0,\ \top = 1\}$$

We then define $\mathsf{kat} = \mathsf{bool} \cup \mathsf{glue}$.

(Note that all these equations are actually understood as double inequations.)

We prove completeness of $\mathsf{KA}_{\mathsf{kat}}$ in Sect. 4.2 below, by constructing a suitable reduction. Recall that this means completeness w.r.t. the interpretation $\mathrm{cl}_{\mathsf{kat}}([\![-]\!])$ in terms of closed languages. Before proving completeness of $\mathsf{KA}_{\mathsf{kat}}$, we compare it to the classical completeness (Theorem 4.1). First note that $\mathsf{KA}_{\mathsf{kat}}$ contains the same axioms as Kleene algebra with tests, so that provability in $\mathsf{KA}_{\mathsf{kat}}$ and KAT coincide: $\mathsf{KA}_{\mathsf{kat}} \vdash e = f$ iff $\mathsf{KAT} \vdash e = f$. Comparing the interpretation $\mathrm{cl}_{\mathsf{kat}}([\![-]\!])$ to the guarded string interpretation $\mathcal{G}$ is slightly more subtle, and is the focus of the next subsection.

4.1 Relation to Guarded String Interpretation

To relate the guarded string model and the model obtained with closure under kat, we first develop the following lemmas.

The key step consists in characterising the strings that are present in the closure of a language of guarded strings (Lemma 4.3) below. First observe that a guarded string may always be seen as a word over the alphabet $\Sigma + \mathcal{T}_{\mathsf{BA}}$. Conversely, a word over the alphabet $\Sigma + \mathcal{T}_{\mathsf{BA}}$ can always be decomposed as a sequence $\phi_0 a_0 \cdots \phi_{n-1} a_{n-1} \phi_n$ where $a_i \in \Sigma$ for all $i < n$ and each ϕ_i is a possibly empty sequence of Boolean expressions. We let ϕ range over such sequences, and we write $\overline{\phi}$ for the conjunction of the elements of ϕ.

Lemma 4.3. *Let L be a language of guarded strings. We have*

$$\begin{aligned} & \phi_0 a_0 \cdots \phi_{n-1} a_{n-1} \phi_n \in \mathrm{cl}_{\mathsf{kat}}(L) \\ \Leftrightarrow \quad & \forall\, (\alpha_i)_{i \le n}, \quad (\forall i \le n, \alpha_i \models \overline{\phi_i}) \quad \Rightarrow \quad \alpha_0 a_0 \cdots \alpha_{n-1} a_{n-1} \alpha_n \in L \end{aligned}$$

Then we show that the kat-closures of $[\![e]\!]$ and $\mathcal{G}(e)$ coincide:

Lemma 4.4. *For all KAT expressions e, $\mathrm{cl}_{\mathsf{kat}}([\![e]\!]) = \mathrm{cl}_{\mathsf{kat}}(\mathcal{G}(e))$.*

Let $\mathcal{GS}$ be the set of all guarded strings. We also have:

Lemma 4.5. *For all KAT expressions e, $\mathcal{G}(e) = \mathrm{cl}_{\mathsf{kat}}([\![e]\!]) \cap \mathcal{GS}$.*

As an immediate consequence of these two lemmas, we can finally relate the guarded strings languages semantics to the kat-closed languages one:

Corollary 4.6. *Let $e, f \in \mathcal{T}_{\mathsf{kat}}$. We have $\mathcal{G}(e) = \mathcal{G}(f) \Leftrightarrow \mathrm{cl}_{\mathsf{kat}}([\![e]\!]) = \mathrm{cl}_{\mathsf{kat}}([\![f]\!])$.*

4.2 Completeness

To prove completeness of the closed language model wrt kat, we proceed as follows. First, we reduce the hypotheses in kat to a simpler set of axioms: by putting the Boolean expressions into normal forms via the atoms, we can get rid of the hypotheses in bool [31]. We do not remove the hypotheses in glue directly: we transform them into the following hypotheses about atoms:

$$\mathsf{atom} = \{\alpha \cdot \beta \le 0 \mid \alpha, \beta \in \mathsf{At},\ \alpha \ne \beta\} \cup \{\alpha \le 1 \mid \alpha \in \mathsf{At}\} \cup \{1 \le \sum_{\alpha \in \mathsf{At}} \alpha\}$$

We thus first show that kat reduces to atom. Second, we use results from Sects. 3.1 and 3.2 to reduce from atom to $\emptyset$, to obtain completeness of $\mathsf{KA}_{\mathsf{kat}}$.

Let $r\colon \mathcal{T}(\Sigma + \mathcal{T}_{\mathsf{BA}}) \to \mathcal{T}(\Sigma + \mathsf{At})$ be the homomorphism defined by

$$r(x) = \begin{cases} a & x = a \in \Sigma \\ \sum_{\alpha \models \phi} \alpha & x = \phi \in \mathcal{T}_{\mathsf{BA}} \end{cases}$$

We show below that r yields a reduction from kat to atom, using Lemma 3.3. In the sequel, we use atom_1, atom_2 and atom_3, or simply 1, 2, 3, to denote the three families of inequations in atom.

Lemma 4.7. *The homomorphism r yields a reduction from* kat *to* atom.

Proof. We use Lemma 3.3. We first need to show $\mathsf{KA}_{\mathsf{kat}} \vdash \mathsf{atom}$: for $\alpha, \beta \in \mathsf{At}$ with $\alpha \neq \beta$, we have the following derivations in $\mathsf{KA}_{\mathsf{kat}}$

$$\alpha \cdot \beta = \alpha \wedge \beta = \bot = 0 \qquad \alpha \leq \top = 1 \qquad 1 = \top = \bigvee_{\alpha \models \top} \alpha = \sum_{\alpha \in \mathsf{At}} \alpha$$

Now for $a \in \Sigma + \mathsf{At}$, we have $a = r(a)$ (syntactically): if $a = \mathsf{a} \in \Sigma$, then $r(\mathsf{a}) = \mathsf{a}$; if $a = \alpha \in \mathsf{At}$, then $r(\alpha) = \sum_{\alpha \models \alpha} \alpha = \alpha$. The first condition about r is thus satisfied, and it suffices to verify the second condition about r for $\phi \in \mathcal{T}_{\mathsf{BA}}$. In this case, we have we have $\mathsf{KA}_{\mathsf{kat}} \vdash r(\phi) = \sum_{\alpha \models \phi} \alpha = \bigvee_{\alpha \models \phi} \alpha = \phi$. The third and last condition is proven in [31]. □

Now we must reduce atom to the empty set. We can immediately get rid of atom_1: by Lemma 3.13, atom reduces to $\mathsf{atom}_{2,3}$. For atom_2 and atom_3, we have individiual reductions to the empty set via Lemma 3.6(i) and (ii), respectively. We combine those reductions via Lemma 3.9, by showing that their corresponding closures can be organised as follows:

Lemma 4.8. *We have* $\mathrm{cl}_{\mathsf{atom}_{2,3}} = \mathrm{cl}_{\mathsf{atom}_3} \circ \mathrm{cl}_{\mathsf{atom}_2}$

Proof. We simply write 2 and 3 for atom_2 and atom_3. Since the right-hand sides of atom_2 are words, by Lemma 3.12 and Lemma 3.10, it suffices to prove $\mathrm{st}_2 \circ \mathrm{st}_3 \subseteq \mathrm{cl}_3 \circ \mathrm{st}_2$. We actually prove $\mathrm{st}_2 \circ \mathrm{st}_3 \subseteq \mathrm{st}_3 \circ \mathrm{st}_2$.

Assume $w \in \mathrm{st}_2(\mathrm{st}_3(L))$ for some language L. Hence, $w = u\alpha v$ for some atom α and words u, v such that and $uv \in \mathrm{st}_3(L)$. In turn, we uv must be equal to $u'v'$ for some words u', v' such that for all atoms β, $u'\beta v' \in L$. By symmetry, we may assume $|u| \leq |u'|$, i.e., $u' = uw$, $v = wv'$ for some word w. In this case, we have $uw\beta v' \in L$ for all β, whence $u\alpha w\beta v' \in \mathrm{st}_2(L)$ for all β, whence $u\alpha v = u\alpha wv' \in \mathrm{st}_3(\mathrm{st}_2(L))$, as required. □

Putting everything together, we finally obtain completeness of $\mathsf{KA}_{\mathsf{kat}}$.

Theorem 4.9. *For all $e, f \in \mathcal{T}_{\mathsf{KAT}}$, $\mathrm{cl}_{\mathsf{kat}}([\![e]\!]) = \mathrm{cl}_{\mathsf{kat}}([\![f]\!])$ implies $\mathsf{KA}_{\mathsf{kat}} \vdash e = f$.*

Proof. kat reduces to atom (lemma 4.7), which reduces to $\mathsf{atom}_{2,3}$ by Lemma 3.13. The latter set reduces to the empty set by Lemma 4.8, Lemma 3.6 and Lemma 3.9. Thus kat reduces to the empty set, and we conclude via completeness of Kleene algebra (Theorem 2.1) and Theorem 3.2. □

5 Kleene Algebra with Observations

A *Kleene algebra with Observations (KAO)* is a Kleene algebra which also contains a Boolean algebra, but the connection between the Boolean algebra and the Kleene algebra is different than for KAT: instead of having the axiom $\phi \wedge \psi = \phi \cdot \psi$ for all $\phi, \psi \in \mathcal{T}_{\mathsf{BA}}$, we only have $\phi \wedge \psi \leq \phi \cdot \psi$ [15]. This system was introduced to allow for concurrency and tests in a Kleene algebra framework, because associating $\phi \cdot \psi$ and $\phi \wedge \psi$ in a concurrent setting is no longer appropriate: $\phi \wedge \psi$ is one event, where we instantenously test whether both ϕ and ψ are true, while $\phi \cdot \psi$ performs first the test ϕ, and then ψ, and possibly other things can happen between those tests in another parallel thread. Hence, the behaviour of $\phi \wedge \psi$ should be included in $\phi \cdot \psi$, but they are no longer equivalent. (Note that even if we add the axiom $1 = \top$, in which case we have that $\phi \cdot \psi$ is below both ψ and ϕ, this is not enough to collapse $\phi \cdot \psi$ and $\phi \wedge \psi$, because $\phi \cdot \psi$ need not be an element of the Boolean algebra.)

Algebraically this constitutes a small change, and an ad-hoc completeness proof is in [15]. Here we show how to obtain completeness within our framework. We also show how to add the additional and natural axiom $1 = \top$, which is not present in [15], and thereby emphasise the modular aspect of the approach.

Similar to KAT, we add the additional axioms of KAO to KA as hypotheses. The additional axioms of KAO are the axioms of Boolean algebra and the axioms specifying the interaction between the two algebras. The KAO-terms are the same as the KAT-terms: regular expression over the alphabet $\Sigma + \mathcal{T}_{\mathsf{BA}}$.

Definition 5.1. We define the set of hypotheses $\mathsf{kao} = \mathsf{bool} \cup \mathsf{glue}'$, where

$$\mathsf{glue}' = \{\phi \wedge \psi \leq \phi \cdot \psi,\ \phi \vee \psi = \phi + \psi \mid \phi, \psi \in \mathcal{T}_{\mathsf{BA}}\} \cup \{\bot = 0\}$$

We prove completeness with respect to the closed interpretation under hypotheses: $\mathrm{cl}_{\mathsf{kao}}(\llbracket - \rrbracket)$. As shown below, this also implies completeness for the language model presented in [15]. We take similar steps as for KAT:

1. Reduce kao to the simpler set of axioms $\mathsf{contr} = \{\alpha \leq \alpha \cdot \alpha \mid \alpha \in \mathsf{At}\}$, where $\mathsf{At} = 2^{\Omega}$ is the set of atoms, as in Sect. 4.
2. Use results from Sect. 3.1 to reduce contr to the empty set.

For the first step, we use the same homomorphism r as for KAT.

Lemma 5.2. *The homomorphism r yields a reduction from* kao *to* contr.

Proof. Like for Lemma 4.7, we use Lemma 3.3. We show $\mathsf{KA}_{\mathsf{kao}} \vdash \mathsf{contr}$: for $\alpha \in \mathsf{At}$, we have $\mathsf{KA}_{\mathsf{kao}} \vdash \alpha = \alpha \wedge \alpha \leq \alpha \cdot \alpha$. The first and second condition about r are obtained like in the KAT case: the glueing equations for $\wedge$ were not necessary there. The third and last condition is proven in [31]. □

Theorem 5.3. *For all $e, f \in \mathcal{T}_{\mathsf{KAT}}$, $\mathrm{cl}_{\mathsf{kao}}(\llbracket e \rrbracket) = \mathrm{cl}_{\mathsf{kao}}(\llbracket f \rrbracket)$ implies $\mathsf{KA}_{\mathsf{kao}} \vdash e = f$.*

Proof. kao reduces to contr (Lemma 5.2), which reduces to $\emptyset$ by Lemma 3.6(i), as both α and $\alpha \cdot \alpha$ are words and α is a word of length 1. □

The semantics in [15] actually corresponds to $\mathrm{cl}_{\mathsf{contr}}([\![r(-)]\!])$ rather than $\mathrm{cl}_{\mathsf{kao}}([\![-]\!])$. But these are equivalent: kao reducing to contr via r (the proof of Theorem 3.2 actually establishes that when H reduces to H' via r and $\mathsf{KA}_{H'}$ is complete, we have $\mathrm{cl}_H([\![e]\!]) = \mathrm{cl}_H([\![f]\!])$ iff $\mathrm{cl}_{H'}([\![r(e)]\!]) = \mathrm{cl}_{H'}([\![r(f)]\!])$ for all e, f).

Because we set up KAO in a modular way, we can now easily extend it with the extra axiom $\top = 1$. Combining the proofs that r is a reduction from kat to atom and from kao to contr, we can easily see that r is also a reduction from $\mathsf{kao} \cup \{\top = 1\}$ to $\mathsf{contr} \cup \mathsf{atom}_2 \cup \mathsf{atom}_3$. To obtain completeness, it thus suffices to explain how to combine the closures w.r.t. contr, atom_2, and atom_3.

Lemma 5.4. *We have* $\mathrm{cl}_2 \circ \mathrm{cl}_{\mathsf{contr}} \subseteq \mathrm{cl}_{\mathsf{contr}} \circ \mathrm{cl}_2$ *and* $\mathrm{cl}_{\mathsf{contr}} \circ \mathrm{cl}_3 \subseteq \mathrm{cl}_3 \circ \mathrm{cl}_{\mathsf{contr}}$

Theorem 5.5. $\mathsf{KA}_{\mathsf{kao} \cup \{\top = 1\}}$ *is complete.*

Proof. Because we know $\mathrm{cl}_{2,3} = \mathrm{cl}_3 \circ \mathrm{cl}_2$ (Lemma 4.8) and hence $\mathrm{cl}_2 \circ \mathrm{cl}_3 \subseteq \mathrm{cl}_3 \circ \mathrm{cl}_2$, we can use Lemma 3.10 with Lemma 5.4 to deduce $\mathrm{cl}_{\mathsf{contr},2,3} = \mathrm{cl}_3 \circ \mathrm{cl}_{\mathsf{contr}} \circ \mathrm{cl}_2$, and obtain completeness via Lemma 3.9 and the fact that contr, 2 and 3 all reduce to the empty set. □

6 Kleene Algebra with Positive Tests

In KAT, tests are assumed to form a Boolean algebra. Here we study the structure obtained by assuming that they only form a distributive lattice. A *Kleene algebra with positive tests (KAPT)* is a Kleene algebra X containing a lattice L such that the meet of L coincides with the product of X, the join of X coincides with the sum of X, and all elements of L are below the multiplicative identity of X. (We discuss the variant where we have a bounded lattice at the end, see Remark 6.6). Since the product distributes over sums in X, L must be a distributive lattice. Also note that there might be elements of X below 1 that do not belong to L.

As before, we fix two finite sets Σ and Ω of primitive actions and primitive tests. Then we consider regular expressions over the alphabet $\Sigma + \mathcal{T}_{\mathsf{DL}}$, where $\mathcal{T}_{\mathsf{DL}}$ is the set of lattice expressions over Ω: expressions built from elements of Ω and two binary connectives $\vee$ and $\wedge$.

We write dl for the set of all instances of distributive lattice axioms over $\mathcal{T}_{\mathsf{DL}}$ [9], and we set $\mathsf{kapt} \triangleq \mathsf{dl} \cup \mathsf{glue}''$ where

$$\mathsf{glue}'' \triangleq \{\phi \wedge \psi = \phi \cdot \psi,\ \phi \vee \psi = \phi + \psi \mid \phi, \psi \in \mathcal{T}_{\mathsf{DL}}\} \cup \{\phi \leq 1 \mid \phi \in \mathcal{T}_{\mathsf{DL}}\}$$

Like for Boolean algebras, the free distributive lattice over Ω is finite and can be described easily. An *atom* α is a non-empty subset of Ω, and we write At for the set of such atoms as before. However, while an atom $\{a, b\}$ of Boolean algebra was implicitly interpreted as the term $a \wedge b \wedge \neg c$ (when $\Omega = \{a, b, c\}$), the same atom in the context of distributive lattices is implicitly interpreted as the term $a \wedge b$—there are no negative literals in distributive lattices. Again similarly to the case of Boolean algebras, the key property for atoms in distributive lattices is the following: for all atoms α and formulas ϕ, we have

$$\alpha \models \phi \Leftrightarrow \mathsf{DL} \vdash \alpha \leq \phi \qquad \text{and} \qquad \mathsf{DL} \vdash \phi = \bigvee_{\alpha \models \phi} \alpha$$

Like for KAT, such a property makes it possible to reduce kapt to the following set of equations on the alphabet $\Sigma + \mathsf{At}$.

$$\mathsf{atom}' \triangleq \{\alpha \cdot \beta = \alpha \cup \beta \mid \alpha, \beta \in \mathsf{At}\} \cup \{\alpha \leq 1 \mid \alpha \in \mathsf{At}\}$$

(Note that in the right-hand side of the first equation, $\alpha \cup \beta$ is a single atom, whose implicit interpretation is $\alpha \wedge \beta$.)

Lemma 6.1. *There is a reduction from* kapt *to* atom′, *witnessed by the homomorphism* $r \colon \mathcal{T}(\Sigma + \mathcal{T}_{\mathsf{DL}}) \to \mathcal{T}(\Sigma + \mathsf{At})$ *defined by*

$$r(x) = \begin{cases} a & x = a \in \Sigma \\ \sum_{\alpha \models \phi} \alpha & x = \phi \in \mathcal{T}_{\mathsf{DL}} \end{cases}$$

As a consequence, in order to get decidability and completeness for KAPT (i.e., kapt), it suffices to reduce atom′ to the empty set. Let us number the three kinds of inequations that appear in this set:

$$1{=}\{\alpha{\cup}\beta \leq \alpha{\cdot}\beta \mid \alpha, \beta \in \mathsf{At}\} \quad 2{=}\{\alpha{\cdot}\beta \leq \alpha{\cup}\beta \mid \alpha, \beta \in \mathsf{At}\} \quad 3{=}\{\alpha \leq 1 \mid \alpha \in \mathsf{At}\}$$

Lemma 3.6(i) gives reductions to the empty set for 1 and 3, but so far we have no reduction for 2. We actually do not know if there is a reduction from 2 to the empty set. Instead, we establish a reduction from 2 together with 3 to 3 alone.

Lemma 6.2. *There is a reduction from 2,3 to 3, witnessed by the homomorphism* $r \colon \mathcal{T}(\Sigma + \mathsf{At}) \to \mathcal{T}(\Sigma + \mathsf{At})$ *defined by*

$$r(x) = \begin{cases} a & x = a \in \Sigma \\ \sum\{\alpha_1 \cdot \ldots \cdot \alpha_n \mid \alpha = \bigcup_{i \leq n} \alpha_i,\ i \neq j \Rightarrow \alpha_i \neq \alpha_j\} & x = \alpha \in \mathsf{At} \end{cases}$$

(Note that the above reduction requires 3 in its target, and cannot be extended directly into a reduction from 1,2,3 to 3: $r(\alpha \cup \beta) \leq r(\alpha \cdot \beta)$ cannot be proved in KA_3—take $\alpha = \{a\}$, $\beta = \{b\}$, then ba is a term in $r(\alpha \cup \beta)$ which is not provably below $ab = r(\alpha \cdot \beta)$.)

Composed with the existing reduction from 3 to the empty set (Lemma 3.6(i)), we thus have a reduction from 2,3 to the empty set. It remains to combine this reduction to the one from 1 to the empty set (Lemma 3.6(i) again). To this end, we would like to use Lemma 3.9, which simply requires us to prove that the closure $\mathrm{cl}_{\mathsf{atom}'} = \mathrm{cl}_{1,2,3}$ is equal either to $\mathrm{cl}_1 \circ \mathrm{cl}_{2,3}$ or to $\mathrm{cl}_{2,3} \circ \mathrm{cl}_1$. Unfortunately, this is not the case. To see this, suppose we have two atomic tests a and b. For the first option, consider the singleton language $\{ab\}$ (a word consisting of two atoms); we have $ba \in \mathrm{cl}_{1,2,3}(\{ab\})$ (because $(a \wedge b) \in \mathrm{cl}_1(\{ab\})$, and then using cl_2) but $ba \notin \mathrm{cl}_1(\mathrm{cl}_{2,3}(\{ab\}))$. For the second option, consider the singleton language $\{a\}$; we have $(a \wedge b) \in \mathrm{cl}_{1,2,3}(\{a\})$, because $ab \in \mathrm{cl}_3(\{a\})$, but $(a \wedge b) \notin \mathrm{cl}_{2,3}(\mathrm{cl}_1(\{a\}))$ because $\mathrm{cl}_1(\{a\})$ is just $\{a\}$, and $\mathrm{cl}_{2,3}$ does not make it possible to forge conjunctions.

In order to circumvent this difficulty, we use a fourth family of equations:

$$4 = \{\alpha \cup \beta \leq \alpha \mid \alpha, \beta \in \mathsf{At}\}$$

These axioms are immediate consequences of 1 and 3. Therefore, 1, 2, 3 reduces to 1, 2, 3, 4. Moreover they consist of 'letter-letter' inequations, which are covered by Lemma 3.6(i): 4 reduces to the empty set. We shall further prove that $\mathrm{cl}_{1,2,3,4} = \mathrm{cl}_4 \circ \mathrm{cl}_3 \circ \mathrm{cl}_2 \circ \mathrm{cl}_1$ and $\mathrm{cl}_{2,3} = \mathrm{cl}_3 \circ \mathrm{cl}_2$, so that Lemma 3.9 applies to obtain a reduction from 1,2,3,4 to the empty set.

Lemma 6.3. *We have the following inclusions of functions:*
(i) $\mathrm{st}_1 \circ \mathrm{st}_2 \subseteq \mathrm{st}_2 \circ \mathrm{st}_1 + \mathrm{id}$
(ii) $\mathrm{st}_1 \circ \mathrm{st}_3 \subseteq \mathrm{st}_3 \circ \mathrm{st}_1 + \mathrm{st}_4$
(iii) $\mathrm{st}_1 \circ \mathrm{st}_4 \subseteq \mathrm{st}_4 \circ \mathrm{st}_1$
(iv) $\mathrm{st}_2 \circ \mathrm{st}_3 \subseteq \mathrm{st}_3 \circ \mathrm{st}_2 + \mathrm{st}_3 \circ \mathrm{st}_3$
(v) $\mathrm{st}_2 \circ \mathrm{st}_4 \subseteq \mathrm{st}_4 \circ \mathrm{st}_2 + \mathrm{st}_4 \circ \mathrm{st}_4 \circ \mathrm{st}_2$
(vi) $\mathrm{st}_3 \circ \mathrm{st}_4 \subseteq \mathrm{st}_4 \circ \mathrm{st}_3$

Lemma 6.4. *We have* $\mathrm{cl}_{1,2,3,4} = \mathrm{cl}_4 \circ \mathrm{cl}_3 \circ \mathrm{cl}_2 \circ \mathrm{cl}_1 = \mathrm{cl}_4 \circ \mathrm{cl}_{2,3} \circ \mathrm{cl}_1$.

Proof. We use Lemma 3.10 and Lemma 3.12 repeatedly, on (combinations of) the inclusions provided by Lemma 6.3. See the proof in [31]. □

Theorem 6.5. $\mathsf{KA}_{\mathsf{kapt}}$ *reduces to the empty set, and is complete and decidable.*

Proof. kapt reduces to atom' by Lemma 6.1, which in turn reduces to 1, 2, 3, 4 by Lemma 3.8. The latter is composed of three sets of hypotheses, 1, 4, and 2, 3. All three of them reduce to the empty set: the first two by Lemma 3.6(i), and the third one by Lemma 6.2. These three reductions can be composed together by Lemma 3.9 and Lemma 6.4. □

Remark 6.6. The case of Kleene algebras containing a *bounded* distributive lattice, with extremal elements $\bot$ and $\top$ coinciding with 0 and 1, may be obtained as follows. Allow the empty atom $\emptyset$ in At (interpreted as $\top$), and add the inequation $5 = \{1 \leq \emptyset\}$ to atom'. Lemma 6.1 extends easily, and we have a reduction from 5 to the empty set (Lemma 3.6(i)). Therefore it suffices to find how to combine cl_5 with the other closures. We have $\mathrm{cl}_{1,2,3,4,5} = \mathrm{cl}_5 \circ \mathrm{cl}_4 \circ \mathrm{cl}_3 \circ \mathrm{cl}_2 \circ \mathrm{cl}_1$ (see [31]), so that we can conclude that the equational theory of Kleene algebras with a bounded distributive lattice is complete and decidable.

7 Related Work

There is a range of papers on completeness and decidability of Kleene algebra together with specific forms of hypotheses, starting with [7]. The general case of Kleene algebra with hypotheses, and reductions to prove completeness, has been studied recently in [10,16,25]. The current paper combines and extends these results, and thereby aims to provide a comprehensive overview and a showcase of how to apply these techniques to concrete case studies (KAT, KAO and the new theory KAPT). Below, we discuss each of these recent works in more detail.

Kozen and Mamouras [25] define the canonical language model for KA with a set of hypotheses in terms of rewriting systems, as well as reductions and their role in completeness, and provide reductions for equations of the form $1 = w$ and $a = w$ (cf. Lemma 3.6). Their general results cover completeness results which instantiate to KAT and NetKAT. In fact, the assumptions made in their technical development are tailored towards these cases; for instance, their assumption $\alpha\beta \leq \bot$ (in Assumption 2) rules out KAPT. The current paper focuses more on generality and how to construct reductions in a modular way.

Doumane et al. [10] also define reductions, with an emphasis on (un)decidability. In particular, they cover hypotheses of the form $1 \leq \sum_{a \in S} a$ (cf. Lemma 3.6). A first step towards modularity may also be found in [10, Proposition 3].

Kappé et al. [16] study hypotheses on top of bi-Kleene algebra, where the canonical interpretation is based on pomset languages, and ultimately prove completeness of *concurrent Kleene algebra with observations*; many of the results there apply to the word case as well. We follow this paper for the basic definitions and results for the general theory of Kleene algebra with hypotheses, with a small change in the actual definition of a reduction (Remark 3.5). Compositionality in the sense of Sect. 3.2 is treated in Kappé's PhD thesis [14]. We extend these results with Lemmas 3.11, 3.12, which simplify the work needed to combine hypotheses. Further, we highlight the word case in this paper (as opposed to the pomset languages in concurrent Kleene algebra), by showcasing several examples.

References

1. Angus, A., Kozen, D.: Kleene algebra with tests and program schematology. Technical Report TR2001-1844, CS Dpt., Cornell University, July 2001
2. Birkhoff, G., Bartee, T.C.: Modern Applied Algebra. McGraw-Hill (1970)
3. Boffa, M.: Une remarque sur les systèmes complets d'identités rationnelles. Informatique Théorique et Applications **24**, 419–428 (1990)
4. Braibant, T., Pous, D.: An efficient coq tactic for deciding kleene algebras. In: Kaufmann, M., Paulson, L.C. (eds.) ITP 2010. LNCS, vol. 6172, pp. 163–178. Springer, Heidelberg (2010). https://doi.org/10.1007/978-3-642-14052-5_13
5. Brunet, P.: A note on commutative Kleene algebra. CoRR, abs/1910.14381 (2019)
6. Brunet, P., Pous, D., Struth, G.: On decidability of concurrent Kleene algebra. In: CONCUR, vol. 85 of LIPIcs, pp. 24:1–24:16. Schloss Dagstuhl (2017)
7. Cohen, E.: Hypotheses in Kleene algebra. Tech. rep., Bellcore, N.J. (1994)
8. Conway, J.: Regular Algebra and Finite Machines. Chapman and Hall mathematics series. Dover Publications, Incorporated (2012)
9. Davey, B., Priestley, H.: Introduction to Lattices and Order. Cambridge University Press (1990)

10. Doumane, A., Kuperberg, D., Pous, D., Pradic, P.: Kleene algebra with hypotheses. In: Bojańczyk, M., Simpson, A. (eds.) FoSSaCS 2019. LNCS, vol. 11425, pp. 207–223. Springer, Cham (2019). https://doi.org/10.1007/978-3-030-17127-8_12
11. Hardin, C.: Modularizing elimination of r = 0 in Kleene algebra. In: LMCS, vol. 1, no. 3 (2005)
12. Hardin, C., Kozen, D.: On the elimination of hypotheses in Kleene algebra with tests. Technical report TR2002-1879, CS Dpt., Cornell University, October 2002
13. Hoare, T., Möller, B., Struth, G., Wehrman, I.: Concurrent Kleene algebra and its foundations. J. Logic Algebraic Program. **80**(6), 266–296 (2011)
14. Kappé, T.: Concurrent Kleene Algebra: Completeness and Decidability. PhD thesis, UCL, London (2020)
15. Kappé, T., Brunet, P., Rot, J., Silva, A., Wagemaker, J., Zanasi, F.: Kleene algebra with observations. In: CONCUR, pp. 41:1–41:16 (2019)
16. Kappé, T., Brunet, P., Silva, A., Wagemaker, J., Zanasi, F.: Concurrent Kleene algebra with observations: from hypotheses to completeness. In: FoSSaCS 2020. LNCS, vol. 12077, pp. 381–400. Springer, Cham (2020). https://doi.org/10.1007/978-3-030-45231-5_20
17. Kleene, S.C.: Representation of events in nerve nets and finite automata. In: Automata Studies, pp. 3–41. Princeton University Press (1956)
18. Kozen, D.: A completeness theorem for Kleene Algebras and the algebra of regular events. In: LICS, pp. 214–225. IEEE Computer Society (1991)
19. Kozen, D.: A completeness theorem for Kleene algebras and the algebra of regular events. I&C **110**(2), 366–390 (1994)
20. Kozen, D.: On action algebras. In: Logic and Information Flow, pp. 78–88. MIT Press (1994)
21. Kozen, D.: Kleene algebra with tests and commutativity conditions. In: Margaria, T., Steffen, B. (eds.) TACAS 1996. LNCS, vol. 1055, pp. 14–33. Springer, Heidelberg (1996). https://doi.org/10.1007/3-540-61042-1_35
22. Kozen, D.: Kleene algebra with tests. Trans. Program. Lang. Syst. **19**(3), 427–443 (1997)
23. Kozen, D.: On Hoare logic and Kleene algebra with tests. ACM Trans. Comput. Log. **1**(1), 60–76 (2000)
24. Kozen, D.: On the complexity of reasoning in Kleene algebra, p. 179. I&C (2002)
25. Kozen, D., Mamouras, K.: Kleene algebra with equations. In: Esparza, J., Fraigniaud, P., Husfeldt, T., Koutsoupias, E. (eds.) ICALP 2014. LNCS, vol. 8573, pp. 280–292. Springer, Heidelberg (2014). https://doi.org/10.1007/978-3-662-43951-7_24
26. Kozen, D., Patron, M.-C.: Certification of compiler optimizations using Kleene algebra with tests. In: Lloyd, J., et al. (eds.) CL 2000. LNCS (LNAI), vol. 1861, pp. 568–582. Springer, Heidelberg (2000). https://doi.org/10.1007/3-540-44957-4_38
27. Kozen, D., Smith, F.: Kleene algebra with tests: completeness and decidability. In: van Dalen, D., Bezem, M. (eds.) CSL 1996. LNCS, vol. 1258, pp. 244–259. Springer, Heidelberg (1997). https://doi.org/10.1007/3-540-63172-0_43
28. Krob, D.: Complete systems of B-rational identities. T.C.S. **89**(2), 207–343 (1991)
29. Meyer, A., Stockmeyer, L.J.: The equivalence problem for regular expressions with squaring requires exponential space. In: SWAT, pp. 125–129. IEEE (1972)
30. Pous, D.: Kleene algebra with tests and coq tools for while programs. In: Blazy, S., Paulin-Mohring, C., Pichardie, D. (eds.) ITP 2013. LNCS, vol. 7998, pp. 180–196. Springer, Heidelberg (2013). https://doi.org/10.1007/978-3-642-39634-2_15
31. Pous, D., Rot, J., Wagemaker, J.: On tools for completeness of Kleene algebra with hypotheses. Extended version of the present paper, with proofs, available on HAL at https://hal.archives-ouvertes.fr/hal-03269462/ (2021)

32. Pratt, V.: Action logic and pure induction. In: van Eijck, J. (ed.) JELIA 1990. LNCS, vol. 478, pp. 97–120. Springer, Heidelberg (1991). https://doi.org/10.1007/BFb0018436
33. Redko, V.: On the algebra of commutative events. Ukrain. Mat **16**, 185–195 (1964)

Skew Metrics Valued in Sugihara Semigroups

Luigi Santocanale(✉)

Laboratoire d'Informatique et des Systèmes, UMR 7020,
Aix-Marseille Université, CNRS, Marseille, France
luigi.santocanale@lis-lab.fr

Abstract. We consider skew metrics (equivalently, transitive relations that are tournaments, linear orderings) valued in Sugihara semigroups on autodual chains. We prove that, for odd chains and chains without a unit, skew metrics classify certain tree-like structures that we call perfect augmented plane towers. When the chain is finite and has cardinality $2K+1$, skew metrics on a set X give rise to perfect rooted plane trees of height K whose frontier is a linear preorder of X.

Any linear ordering on X gives rise to an ordering on the set of its skew metrics valued in an arbitrary involutive residuated lattice Q. If Q satisfies the mix rule, then this poset is most often a lattice. We study this lattice for $X = \{1, \ldots, n\}$ and Q the Sugihara monoid on the chain of cardinality $2K+1$. We give a combinatorial model of this lattice by describing its covers as moves on a space of words coding perfect augmented plane trees. Using the combinatorial model, we develop enumerative considerations on this lattice.

1 Introduction

Linear orders and trees are fundamental structures in Computer Science and Mathematics. We might consider linear orders using some object of truth values different from the classical two-element Boolean algebra. The theory of linear orders in an intuitionistic setting intrinsically suffers from the lack of a well-behaved negation; a striking consequence of this is the existence of different types of intuitionistic ordinals [11,24]. However, when the object of truth values is an involutive residuated lattice or a Girard quantale, negation is again fully operative, generalized linear orders are easily axiomatized, and a rich theory can be developed, capable to generalize non trivial results on classical linear orders.

For such object of truth values, linear orders can be equivalently defined either as some kind of metric valued in the quantale where the symmetry property of the distance is replaced by skewness—that is, we require $\delta(y,x) = \delta(x,y)^*$ where $(-)^*$ is the negation—or as transitive relations on the quantale obeying this law, analogous to the requirement that a binary relation is a tournament.

As part of a general investigation of these objects, see [9,20], we investigate in this paper skew metrics valued in Sugihara monoids and, more generally, in

U. Fahrenberg et al. (Eds.): RAMiCS 2021, LNCS 13027, pp. 396–412, 2021.
https://doi.org/10.1007/978-3-030-88701-8_24

Sugihara semigroups. Research on Sugihara monoids can be traced back to [5] and constitutes nowadays a quite active domain, see e.g. [6,7]. More importantly, Sugihara semigroups arise as the unique idempotent involutive residuated lattice structure that can be given to an autodual chain. Linear orders on the Sugihara chain with three elements—that is, linear preorders or pseudo-permutations—have already been investigated [2,14], partly motivated from complexity issues related to the representation of temporal reasoning [25]. The importance of linear preorders in relation with the combinatorics and geometry of Coxeter groups and hyperplane arrangements was remarked already in [2] and has been once more emphasized in [4]. For us, it is the canonicity of Sugihara semigroups and the use of these structures in combinatorics that motivates further investigations of linear orders on arbitrary Sugihara semigroups.

We focus in this paper on autodual chains C for which either the positive cone C^+ has no least element, or satisfying $C^+ \cap C^- \neq \emptyset$, C^- being the negative cone. We claim that the cases left can be easily studied from the present research. We show that skew metrics on a set X valued in the Sugihara semigroup on C are in bijection with some tree-like objects that we call augmented perfect towers and can be neatly described as functors from the poset C^+ to the category of linearly ordered sets with few additional properties and structure. In particular all the maps involved in such a functor T are surjective and, moreover, a cone from X to T (in the category of sets) is given, reflecting the fact that leaves are labeled by subsets of X.

Once this correspondence is established, we further study the case where $X = \{1, \ldots, n\}$ and $C = \{-K, \ldots, -1, 0, 1, \ldots, K\}$, in which case augmented towers are indeed rooted plane trees, that are perfect (meaning that each branch has equal length) of height $K + 1$ and leaves are labelled by subsets of X, so the frontier forms an ordered partition of X. When X is linearly ordered, skew metrics defined on X can be ordered and most often this ordering yields a lattice. For such choice of X and C, we describe the poset of skew metrics and, by representing trees as words, we determine covers of this poset as moves or rewrites on these words.

Relying on the combinatorial description of skew metrics as words, we give enumerative results on these combinatorial objects and these posets such as determining the size and the length.

Let us mention that the combinatorial model obtained is unavoidably close to the one of [2,14]. In these works the combinatorial model is given and the algebra of Sugihara monoids is mostly used for proving the lattice property of pseudo-permutations. Here the flow has opposite direction: the algebraic framework is given, and the problem, solved, is to instantiate the algebra into the combinatorics.

In this work converge our previous research on the lattice structures that arise from linear orders [9,20–22] and research on the algebraic structures of logic that are in use in combinatorics, see e.g. [18,19]. W.r.t. the first line of research, a main advance here is the recognition of the primary role of the notion of skew metric or linear order on an involutive residuated lattice, compared to closed/open constructions, and its framing in a relational setting. W.r.t. the second line of research, we identify via [8] a connection of Sugihara monoids to the combinatorics of hyperplane arrangements, thus witnessing once more the value of algebraic structures of logic in the realm of combinatorics.

The paper is structured as follows. In Sect. 2 we recall the definition of involutive residuated lattices, of Sugihara semigroups, state their canonicity and few properties needed in the rest of the paper. In Sect. 3 we develop and exemplify the elementary theory of skew metrics. In Sect. 4 we characterize skew metrics as perfect augmented towers. In Sect. 5 we recall results on the ordering on the set of skew metrics valued in an arbitrary involutive residuated lattice. In Sect. 6 we give a combinatorial model of this poset in the case $X = \{1, \ldots, n\}$ and Q is the Sugihara monoid on a chain of size $2K + 1$. We conclude in Sect. 7 with enumerative results concerning these posets.

2 Sugihara Semigroups on Autodual Chains

In this section we recall elementary facts on Sugihara semigroups, which are involutive residuated lattices on autodual chains. We take the view that residuated lattices might not have units, as indeed it will be the case for many autodual chains that we consider.

Definition 1. *An* involutive residuated lattice *is a structure* $\langle L, \wedge, \vee, \otimes, (-)^* \rangle$ *such that* $\langle L, \wedge, \vee \rangle$ *is a lattice,* $\otimes$ *is a semigroup operation on* L *compatible with the order,* $(-)^*$ *is an antitone involution such that*

$$x \otimes y \leq z \quad \textit{iff} \quad y \otimes z^* \leq x^* \quad \textit{iff} \quad z^* \otimes x \leq y^* . \tag{1}$$

We call the relations in (1) the shift relations. Let us define

$$x \oplus y := (y^* \otimes x^*)^* , \qquad x \backslash y := x^* \oplus y , \qquad x / y := x \oplus y^* .$$

It is easy to see that the shift relations are equivalent to asking that the semigroup operation is residuated in both variables:

$$x \otimes y \leq z \quad \text{iff} \quad y \leq x \backslash z \quad \text{iff} \quad x \leq z / y , \qquad \text{for each } x, y, z \in L.$$

By an *autodual chain* we mean a totally ordered set C coming with an antitone involution $(-)^* : C \to C$. For C such an autodual chain we define the absolute value as expected: $|x| := \max(x, x^*)$, for each $x \in C$. Notice then that we also have $|x|^* = \min(x, x^*)$. If we let

$$C^+ := \{ x \in C \mid x^* \leq x \} , \qquad C^- := \{ x \in C \mid x \leq x^* \} ,$$

then $C = C^+ \cup C^-$, and $C^+ \cap C^-$ is either empty, or it is the singleton containing the unique fixed point of $(-)^*$. On C we can define the following two operations:

$$x \otimes y := \begin{cases} x , & |y| < |x| , \\ y , & |x| < |y| , \\ \min(x, y) , & |x| = |y| , \end{cases} \qquad x \oplus y := \begin{cases} x , & |y| < |x| , \\ y , & |x| < |y| , \\ \max(x, y) , & |x| = |y| . \end{cases} \tag{2}$$

These operations are associative, commutative, idempotent, and dual. Moreover, they satisfy the mix rule, meaning that $x \otimes y \leq x \oplus y$, for each $x, y \in C$.

Example 2. On the chain $\{-1, 0, 1\}$ these two idempotent semigroup structures are as follows:

$\otimes$	-1	0	1
-1	-1	-1	-1
0	-1	0	1
1	-1	1	1

$\oplus$	-1	0	1
-1	-1	-1	1
0	-1	0	1
1	1	1	1

Proposition 3. *For any autodual chain C, the structure $\langle C, \min, \max, \otimes, (-)^*\rangle$ is an involutive residuated lattice.*

The next statement, possibly part of the folklore, witnesses the canonicity of this semigroup structure.

Proposition 4. *If C is an autodual chain, then there is exactly one idempotent semigroup operation $\otimes$ on C making $\langle C, \min, \max, \otimes, (-)^*\rangle$ into an involutive residuated lattice.*

This canonical semigroup structure is known as the *Sugihara monoid* on the autodual chain C, see e.g. [7]. Indeed, units are usually considered in involutive residuated lattices, so we characterise next when such a semigroup structure has a unit.

Lemma 5. *The semigroup structure $\otimes$ has a unit ι if and only if the set C^+ has a greatest lower bound $\bigwedge C^+$. In either case, we have $\iota = \bigwedge C^+$.*

As a consequence of the lemma, each auto-dual chain C has at most one unital idempotent involutive residuated lattice structure on it, and exactly one if C is a complete chain.

In the following, we let $\mathbf{3}$ be the chain $\{-1, 0, 1\}$, which we consider with its Sugihara semigroup structure. For C an autodual chain and $k \in C^+$, we define $\chi_k : C \to \mathbf{3}$ as follows:

$$\chi_k(x) := \begin{cases} 1\,, & k < x\,, \\ 0\,, & k^* \leq x \leq k\,, \\ -1\,, & x < k^*\,. \end{cases} \tag{3}$$

The map χ_k is monotone, thus a lattice homomorphism. Let us remark that χ_k is not a semigroup homomorphism since, for example, if x, y, k are such that $x^* < y < k^* \leq k < y^* < x$, then $\chi_k(x \otimes y) = \chi_k(x) = 1$, while $\chi_k(x) \otimes \chi_k(y) = 1 \otimes -1 = -1$. Yet, χ_k satisfies the two properties stated in the following proposition, relevant for the considerations to come.

Proposition 6. *For each $x, y \in C$, we have*

$$\chi_k(x^*) = \chi_k(x)^*\,, \qquad \chi_k(x) \otimes \chi_k(y) \leq \chi_k(x \otimes y)\,.$$

3 Skew Metrics Valued in an Involutive Residuated Lattice

In this section, we let $Q = \langle Q, \wedge, \vee, \otimes, (-)^* \rangle$ be a fixed involutive residuated lattice. For a set X, we let $\Delta_X := \{ (x,x) \mid x \in X \}$.

Definition 7. *A* Q-relation *on* X *is a map* $f : X^2 \setminus \Delta_X \to Q$. *A* Q*-relation* f *is*

- transitive *if* $f(x,y) \otimes f(y,z) \leq f(x,z)$,
- cotransitive *if* $f(x,z) \leq f(x,y) \oplus f(y,z)$,
- skew *if* $f(x,y) = f(y,x)^*$,

for each pairwise distinct $x, y, z \in X$.

The reader might be surprised about our choice of the domain of a Q-relation. Indeed, we could have defined a Q-relation as a map $f : X^2 \to Q$ and, for example, said it is reflexive if $1 \leq f(x,x)$, so a reflexive and transitive Q-relation is nothing else than a category enriched over Q, see [13]. However, we shall insist on the last property, skewness.[1] If f is skew and also defined on Δ_X, then $f(x,x) = f(x,x)^*$, for each $x \in X$, that is, the duality coming from Q has at least one fixed point. Moreover, if we ask the relation $1 \leq f(x,x)$ to hold, then $1 \leq f(x,x) = f(x,x)^* \leq 1^*$. This leaves out many involutive residuated lattices that either do not have units or, for example, for which $1^* < 1$. Our choice is therefore dictated by the aim to consider the largest number of examples. On the other hand, if Q has a unit 1 such that $1^* = 1$, we can freely assume that f is defined on the entire X^2 with $f(x,x) = 1$, for each $x \in X$.

Lemma 8. *If a* Q*-relation is skew, then it is transitive if and only if it is cotransitive.*

Definition 9. *A* skew Q-metric *(or* skew metric*, if* Q *is understood) on* X *is a cotransitive skew* Q*-relation.*

By the previous lemma, a skew metric is transitive. We prefer the name skew metric (to cotransitive skew Q-relation), since the conditions

$$f(x,z) \leq f(x,y) \oplus f(y,z)\,, \qquad\qquad f(y,x) = f(x,y)^*\,,$$

satisfied by a skew metric suggest that f is a distance where symmetry of a distance is being replaced by skewness, see e.g. [12,15]. The next examples explain why skew metrics are generalized linear orders.

Example 10. Let **2** be the two element Boolean algebra. Skew **2**-metrics on X bijectively correspond to (strict) linear orders on X. Indeed, consider a function $f : X^2 \setminus \Delta_X \to \mathbf{2}$ and define $R_f := \{ (x,y) \mid f(x,y) = 1 \}$. Then $f(x,y) \wedge f(y,z) \leq f(x,z)$ holds iff R_f is transitive, $f(x,y) \leq \neg f(y,x)$ holds iff R_f is antisymmetric (where $\neg$ stands for Boolean complement), and $f(x,y) \geq \neg f(y,x)$ holds iff R_f is total (or linear). Indeed, skew **2**-relations correspond to tournaments.

[1] In [12] a property analogous to skewness is considered. In this work the star operation appearing in the relation $f(y,x) = f(x,y)^*$ is *monotone*.

Example 11. This example is the most relevant for the following. A linear preorder on a set X is a transitive relation R which is total: for each x, y, xRy or yRx. Let **3** be the Sugihara monoid on the chain $\{-1, 0, 1\}$. Skew **3**-metrics on X bijectively correspond to linear preorders on X via the mapping sending $f : X^2 \setminus \Delta(X) \to \mathbf{3}$ to $R_f := \{(x, y) \mid f(x, y) \geq 0\}$. Again, $f(x, y) \otimes f(y, x) \leq f(x, z)$ yields transitivity of R_f, while $f(x, y) = f(y, x)^*$, that is, $f(x, y) + f(y, x) = 0$, yields totality of R_f. Let us remark that, in turn, linear preorders bijectively correspond to ordered partitions of the set X. We can directly define an ordered partition of X from a skew metric $f : X^2 \setminus \Delta_X \to \mathbf{3}$ as follows. Say that $x \sim_0 y$ if $x = y$ or $f(x, y) = 0$. Then $\sim_0$ is an equivalence relation, so the blocks of the partition are the equivalence classes of $\sim_0$. If $x \sim_0 x'$ and $y \sim_0 y'$, then $f(x', y') = f(x, y)$, as witnessed by the following computation: $f(x, y) = f(x', x) \otimes f(x, y) \otimes f(y, y') \leq f(x', y') \leq f(x', x) \oplus f(x, y) \oplus f(y, y') = f(x, y)$. That is, we can define f on the set of equivalence classes $X/{\sim_0}$ and then the map from the quotient $f : (X/\sim_0)^2 \setminus \Delta_{(X/\sim_0)} \to \mathbf{2}$ yields a total ordering on the blocks.

Example 12. It was shown in [20] that if X is finite, then skew metrics valued in the involutive residuated lattice of sup-preserving maps from the unit interval $[0, 1]$ bijectively correspond to images of continuous (in the topological sense) maps $[0, 1] \to [0, 1]^X$ that are isotone, and preserve the endpoints. Alternatively, they correspond to maximal chains in the cube $[0, 1]^X$.

The statements below, whose proofs are straightforward, illustrate the elementary algebra that can be developed around skew metrics.

Lemma 13. *Let $f : X^2 \setminus \Delta_X \to Q$ be a skew metric. If $g : Y \to X$ is injective, then $f \circ g : Y^2 \setminus \Delta_Y \to Q$ is also a skew metric. If Q is commutative, then $f^* : X^2 \setminus \Delta_X \to Q$, defined by $f^*(x, y) := f(y, x)$, is a skew metric.*

Definition 14. *A* monoidal map *from an involutive residuated lattice Q to an involutive residuated lattice Q' is a function $h : Q \to Q'$ such that $h(x^*) = h(x)^*$ and $h(x) \otimes h(y) \leq h(x \otimes y)$.*

Lemma 15. *If $f : X^2 \setminus \Delta_X \to Q$ is a skew metric and $h : Q \to Q'$ is a monoidal function, then $h \circ f : X^2 \setminus \Delta_X \to Q'$ is a skew metric.*

Remark 16. Let C be an autodual chain and $k \in C^+$. As we have seen in Proposition 6, the map $\chi_k : C \to \mathbf{3}$ defined in Eq. (3) is monoidal. According to Lemma 15, $\chi_k \circ f : X^2 \setminus \Delta_X \to \mathbf{3}$ is a skew metric, for each skew metric $f : X^2 \setminus \Delta_X \to C$.

4 Augmented Plane Towers

We characterise in this section the combinatorial objects arising from skew metrics valued in Sugihara semigroups that are either unitless or odd. Let us make these notions precise.

Definition 17. *An autodual chain C is* even *if $C^+ \cap C^- = \emptyset$ and, otherwise, it is odd. We say that C is* interesting *if either C is even and $\bigwedge C^+$ does not exist (so the idempotent semigroup structure on C is unitless) or C is odd.*

The reader can easily verify that a *finite* autodual chain is odd if and only if it has odd cardinality and, otherwise, it is even. In the following we fix an interesting autodual chain C. Observe that if C is finite, then it is odd. Let us use LinOrd to denote the category of linearly ordered sets and order preserving maps, and $U : \mathsf{LinOrd} \longrightarrow \mathsf{Set}$ to denote the forgetful functor from this category to the category of sets and functions. For the next definition, recall that a poset can be regarded as a category whose objects are the elements of the poset and for which there is exactly one arrow between two elements x, y when $x \leq y$.

Definition 18. *A* plane tower *is a functor $T : C^+ \longrightarrow \mathsf{LinOrd}$. For X any set, an* augmented plane tower *on X is a pair (τ, T) with T a plane tower and $\tau : X \longrightarrow U \circ T$ a cone.*

We spell out what the definition means. A plane tower T is a pair of collections $\{ T_k \mid k \in C^+ \}$ and $\{ T_{j,k} \mid j, k \in C^+, j \leq k \}$. For each $k \in C^+$, T_k is a linearly ordered set; for $j, k \in C^+$ and $j \leq k$, $T_{j,k} : T_j \longrightarrow T_k$ is an order preserving map. These data satisfy the following constraints: $T_{k,k}$ is the identity and, for $j \leq k \leq u$, $T_{j,u} = T_{k,u} \circ T_{j,k}$. A cone $\tau : X \longrightarrow U \circ T$ is a collection $\{ \tau_k \mid k \in C^+ \}$ of functions such that $\tau_k : X \to T_k$ and, for $j \leq k$, $\tau_k = T_{j,k} \circ \tau_j$. Let us insist on the fact that X is just a set, it is not linearly ordered, while all the sets T_k are linearly ordered.

Remark 19. For T a plane tower, let $El(T)$ be the poset whose elements are pairs (k, x) with $x \in T_k$ and for which $(k_1, x_1) \leq (k_2, x_2)$ if $k_2 \geq k_1$ and $x_2 = T_{k_1,k_2}(x_1)$. It easily verified that the C^+ is dually well-founded (e.g. Noetherian) if and only if $El(T)$ is a tree in the sense of set theory—that is, each downset $\downarrow (k, x) = \{ (k', x') \mid k \leq k', x' = T_{k,k'}(x) \}$ is well-ordered. Thus, if C^+ is dually well-founded and in particular if C^+ is finite, then we call T a *plane tree* instead of a plane tower.

Definition 20. *An augmented plane tower (τ, T) is* perfect *if each map τ_i is surjective. An augmented plane tower is* complete *if, for each $x, y \in X$, the set $Eg(x, y) := \{ k \in C^+ \mid \tau_k(x) = \tau_k(y) \}$ has a least element.*[2]

It is an elementary exercise to verify that if an augmented plane tower (τ, T) is *perfect*, then also the maps $T_{j,k} : T_j \to T_k$, $j \leq k$, are surjective. The following lemma exemplifies some consequences of these conditions.

Lemma 21. *If (τ, T) is a perfect complete augmented tower from X, then, whenever $k = \bigwedge J$, the canonical map $T_k \longrightarrow \lim_{j \in J} T_j$ is injective.*

[2] For finite binary trees, the adjectives perfect, full, and complete have precise yet distinct meanings. We adopt the wording perfect for a (non necessarily binary) tree (or a tower) all of whose branches have equal length. The wording complete refers here to a completeness property of the poset C^+.

Next, for a skew metric $f : X^2 \setminus \Delta_X \rightarrow C$, we give the following definitions:

$$x \sim_k^f y \text{ if } x = y \text{ or } |f(x,y)| \leq k\,, \qquad [x]_k^f := \{ y \in X \mid y \sim_k^f x \}\,.$$
$$T_k^f := \{ [x]_k^f \mid x \in X \}\,, \qquad [x]_k^f <_k^f [y]_k^f \text{ if } k < f(x,y)\,.$$

Notice that, for $j \leq k$, we have $[x]_j^f \subseteq [x]_k^f$ and so, for such j, k, we can define

$$T_{j,k}^f([x]_j^f) := [x]_k^f\,, \qquad \tau_k^f(x) := [x]_k^f\,.$$

Proposition 22. *$t^f := (\tau^f, T^f)$ is a perfect and complete augmented plane tower from X.*

Proof. Quite obviously T^f is a functor and τ^f is a cone from X to T_k^f.

Every set T_k^f is linearly ordered by $<_k^f$, since this ordering is induced by the skew metric $\chi_k \circ f : X \rightarrow \mathbf{3}$, see Example 11 and Remark 16. Moreover, for $j, k \in C^+$ and $j \leq k$, $j < f(x,y)$ and $|f(x,y)| \not\leq k$ imply $k < f(x,y)$. That is, for such j, k, $[x]_j^f <_j^f [y]_j^f$ implies $[x]_k^f \leq_k^f [y]_k^f$, so we can take as codomain of the functor T^f the category of linearly ordered sets and isotone functions. The maps τ_k^f are surjective, so t^f is perfect. For completeness, observe that $\bigwedge\{ k \in C^+ \mid [x]_k^f = [y]_k^f \} = |f(x,y)|$, since by definition $[x]_k^f = [y]_k^f$ if and only if $|f(x,y)| \leq k$. □

For $t = (\tau, T)$ a complete augmented tower, we set

$$\delta_t(x,y) := \bigwedge\{ k \in C^+ \mid \tau_k(x) = \tau_k(y) \}\,.$$

Notice that δ_t is an ultrametric on X valued in C^+, meaning that, for each $x, y, z \in X$, $\delta_t(x,y) = \delta_t(y,x)$ and $\delta_t(x,z) \leq \max(\delta_t(x,y), \delta_t(y,z))$. If C^+ has a least element 0, then the condition $\delta_t(x,y) = 0$ implies $x = y$ holds if and only if the map $\tau_0 : X \rightarrow T_0$ is injective.

Let now $t = (\tau, T)$ be a perfect and complete augmented plane tower. Observe that if $k < \delta_t(x,y)$, then we have either $\tau_k(x) < \tau_k(y)$, or $\tau_k(y) < \tau_k(x)$, but not both. Moreover, if $k, k' < \delta_t(x,y)$ and $\tau_k(x) < \tau_k(y)$, then $\tau_{k'}(x) < \tau_{k'}(y)$ as well. Indeed, if $\tau_k(x) < \tau_k(y)$ and $\tau_{k'}(y) < \tau_{k'}(x)$, then, for $K = \max(k, k')$, $\tau_K(x) \leq \tau_K(y)$ and $\tau_K(y) \leq \tau_K(x)$, thus $\tau_K(x) = \tau_K(y)$ with $K < \delta_t(x,y)$, a contradiction. Therefore, we define

$$\varsigma_t(x,y) := \begin{cases} -1\,, & \text{if, for some } k < \delta_t(x,y),\ \tau_k(y) < \tau_k(x)\,, \\ 1\,, & \text{otherwise}\,. \end{cases}$$

We define then $f_t : X^2 \setminus \Delta_X \rightarrow C$ by

$$f_t(x,y) := \varsigma_t(x,y) \cdot \delta_t(x,y)\,,$$

where the action of $\{ -1, 1 \}$ on C is as expected: $1 \cdot k = k$, and $-1 \cdot k = k^*$. Accordingly, we use the notation $-k$ as equivalent to k^*.

Proposition 23. *f_t so defined is a skew-metric on C.*

Proof. Firsty we argue that, for $x, y, z \in X$ arbitrary pairwise distinct, $f_t(x,y) \otimes f_t(y,z) \leq f_t(x,z)$. Recalling that $f_t(x,y) \otimes f_t(y,z) \in \{ f_t(x,y), f_t(y,z) \}$, we suppose that $f_t(x,y) \otimes f_t(y,z) = f_t(x,y)$ (if $f_t(x,y) \otimes f_t(y,z) = f_t(y,z)$, then the argument is similar). Under this assumption, we have either (i) $\delta_t(x,y) > \delta_t(y,z)$, or (ii) $\delta_t(x,y) = \delta_t(y,z)$ and $\varsigma_t(x,y) = -1$. Suppose (i). Since $\delta_t(x,y) > \delta_t(y,z)$, then $\delta_t(x,z) = \delta_t(x,y)$: indeed, $\delta_t(x,z) \leq \max(\delta_t(x,y), \delta_t(y,z)) = \delta_t(x,y)$, and $\delta_t(x,y) \leq \max(\delta_t(x,z), \delta_t(y,z))$ implies $\delta_t(x,y) \leq \delta_t(x,z)$. Therefore, in order to show that $f_t(x,y) \leq f_t(x,z)$, we need to argue that $\varsigma_t(x,y) = 1$ implies $\varsigma_t(x,z) = 1$. Assume therefore that $\varsigma_t(x,y) = 1$ and let $k = \delta_t(y,z)$; we have then $\tau_k(x) < \tau_k(y) = \tau_k(z)$, so $\varsigma_t(x,z) = 1$. We suppose now (ii), that is, $\delta_t(x,y) = \delta_t(y,z)$ and $\varsigma_t(x,y) = -1$. If $\varsigma_t(x,z) = 1$, then we obviously have $f_t(x,y) \leq f_t(x,z)$. Thus, we can assume that $\varsigma_t(x,z) = -1$. Considering that $\delta_t(x,z) \leq \max(\delta_t(x,y), \delta_t(y,z)) = \delta_t(x,y)$, then we immediately have $f_t(x,y) = -1 \cdot \delta_t(x,y) \leq -1 \cdot \delta_t(x,z) = f_t(x,z)$.

Next, we argue that $f_t(y,x) = f_t(x,y)^*$. Clearly, we have $\delta_t(x,y) = \delta_t(y,x)$. If $\varsigma_t(x,y) = -1$, then, for some $k < \delta_t(x,y)$, $\tau_k(y) < \tau_k(x)$, thus $\varsigma_t(y,x) = 1$ and $f_t(y,x) = -f_t(x,y)$. Suppose, therefore, that $\varsigma_t(x,y) = 1$, so $\tau_k(x) < \tau_k(y)$ for all $k < \delta_t(x,y)$. If $\delta_t(x,y)$ is not the least element of C^+, then we deduce $\varsigma_t(y,x) = -1$, so $f_t(y,x) = -f_t(x,y)$. Otherwise, $\delta_t(x,y)$ is the least element of C^+ and therefore $\varsigma_t(x,y) = \varsigma_t(y,x) = 1$, but also $C^+ \cap C^- = \{ \delta_t(x,y) \}$, since we assume that C is interesting, thus odd if C^+ has a least element. Then $f_t(x,y) = \delta_t(x,y) = -\delta_t(x,y) - \delta_t(y,x) = -f_t(y,x)$. □

Proposition 24. *For $f : X^2 \setminus \Delta_X \to C$ a skew metric, we have $f_{t^f} = f$.*

Proof. As we already observed, $\delta_{t^f}(x,y) = |f(x,y)|$. Moreover, $\varsigma_{t^f}(x,y) = -1$ iff for some $k < |f(x,y)|$ we have $[y]^f_k <^f_k [x]^f_k$, where the last inequality is equivalent, by definition, to $k < f(y,x)$. Now saying that, for some $k \in C^+$, $k < f(y,x)$, that is, $f(x,y) < -k$, is equivalent to saying that $f(x,y)$ is strictly negative, i.e. $f(x,y) \in C^- \setminus C^+$. Then, if $f(x,y)$ is strictly negative, then $f(x,y) = -|f(x,y)| = \varsigma_{t^f}(x,y) \cdot \delta_{t^f}(x,y) = f_{t^f}(x,y)$. If $f(x,y)$ is positive, then $\varsigma_{t^f}(x,y) = 1$, $\delta_{t^f}(x,y) = f(x,y)$, and again $f(x,y) = \varsigma_{t^f}(x,y) \cdot \delta_{t^f}(x,y) = f_{t^f}(x,y)$. □

Let T, T' be two plane towers. Recall that a natural transformation $\alpha : T \to T'$ is a collection $\{ \alpha_k : T_k \to T'_k \mid k \in C^+ \}$ such that α_k is order preserving and such that, for $j \leq k$, $T'_{j,k} \circ \alpha_j = \alpha_k \circ T_{j,k}$. Such a natural transformation is a natural isomorphism if each α_k has an order preserving inverse. We say that $(\tau, T), (\tau', T')$ are isomorphic if there is such a natural isomorphism $\alpha : T \to T'$ such that $\tau' = \alpha \circ \tau$, that is, $\tau'_k = \alpha_k \circ \tau_k$ for each $k \in C^+$.

Proposition 25. *If $t = (\tau, T)$ is a perfect and complete augmented plane tower, then t^{f_t} is naturally isomorphic to t.*

Proof. Observe that $x \sim_k^{f_t} y$ iff $|f_t(x,y)| = \delta_{t^{f_t}}(x,y) \leq k$ iff $\tau_k(x) = \tau_k(y)$. That is, the equivalence relation $\sim_k^{f_t}$ is the kernel of τ_k. Since moreover τ_k is surjective, the function α_k sending $[x]_k^{f_t}$ to $\tau_k(x)$ is bijection from $T_k^{f_t}$ to T_k.

Something more can be said: $[x]_k^{f_t} <_k^{f_t} [y]_k^{f_t}$ iff $k < f_t(x,y)$, iff $\tau_k(x) < \tau_k(y)$. Therefore α_k is a bijective embedding of posets, whence an invertible map in the category of linear orders and isotone maps.
It is also obvious that $\tau = \alpha \circ \tau^{f_t}$, since this relation amounts to $\alpha_k(\tau_k^{f_t}(x)) = \alpha_k([x]_k^{f_t}) = \tau_k(x)$. From this relation and surjectivity of $\tau_k^{f_t}$ it also follows that $\alpha : T^{f_t} \to T$ is natural, which can be verified by inferring commutativity of the inner square from commutativity of the outer triangle in the diagram on the right.

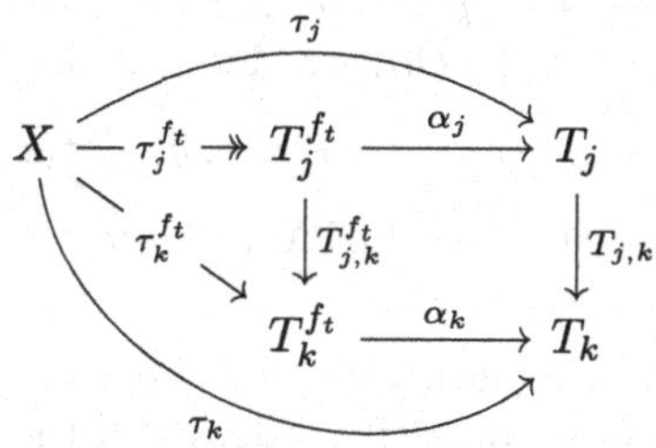

□

5 The Posets of Skew Metrics

In this section we consider again an arbitrary involutive residuated lattice Q and recall more advanced algebraic properties of skew metrics valued in Q. More precisely, we pinpoint that skew metrics can be ordered and that most often this ordering is a lattice.

Observe that it is not interesting to order skew metrics pointwise. For example, if $f, g : X^2 \setminus \Delta_X \to Q$ and $f(x,y) < g(x,y)$, then $g(y,x) = g(x,y)^* < f(x,y)^* = f(y,x)$. That is, a pointwise ordering is necessarily discrete (all the elements are incomparable). We can get a more interesting ordering if we assume that X is totally ordered. In this case, we let $\mathcal{I}_X := \{ (x,y) \in X^2 \setminus \Delta_X \mid x < y \}$ and also introduce the following concept:

Definition 26. *A map $f : \mathcal{I}_X \to Q$ is* clopen *if, whenever $x < y < z \in X$,*

$$f(x,y) \otimes f(y,z) \leq f(x,z) \leq f(x,y) \oplus f(y,z). \tag{4}$$

We let $\mathrm{Clop}_X(Q)$ *be the set of clopen maps $f : \mathcal{I}_X \to Q$.*

Roughly speaking (and up to a choice of a total ordering on X) clopen maps and skew metrics are the same kind of objects, as stated below:

Proposition 27. *Every clopen map $f : \mathcal{I}_X \to Q$ extends uniquely to a skew metric $f : X^2 \setminus \Delta_X \to Q$. Therefore, every total order on X determines a bijection from the set* $\mathrm{Clop}_X(Q)$ *to the set of skew metrics on X valued in Q.*

The set $\mathrm{Clop}_X(Q)$ can be ordered pointwise in a non trivial way. It was argued in [20, see Theorem 21] that if $X = \{1, \ldots, n\}$ and Q is any involutive residuated lattice satisfying the mix rule, then $\mathrm{Clop}_X(Q)$ is a lattice. By inspecting the proof of this result, it is not difficult to generalize it as follows:

Theorem 28. *Let X be a totally ordered set. If Q is an involutive residuated lattice satisfying the mix rule, and every interval of X is finite or Q is complete as a lattice, then* $\mathrm{Clop}_X(Q)$ *is a lattice.*

As we have insisted on skew metrics, we can rephrase the previous statement in terms of skew metrics.

Theorem 29. *Let X be a totally ordered set, let Q be an involutive residuated lattice satisfying the mix rule. Let* $\mathrm{SMet}_X(Q)$ *be the set of skew metrics on X valued in Q. Order* $\mathrm{SMet}_X(Q)$ *as follows:*

$$f \leq g \;\; \textit{iff} \;\; g(x,y) \leq f(x,y)\,, \textit{ for each } x,y \textit{ such that } x < y\,. \qquad (5)$$

If every interval of X is finite or Q is complete as a lattice, then $\mathrm{SMet}_X(Q)$ *is a lattice.*

Let us remark that, for coherence with existing literature, we are considering in (5) the opposite ordering of the pointwise ordering on the restriction to $\mathcal{I}_X$.

Example 30. Suppose $X = \{1,\ldots,n\}$. If $Q = \mathbf{2}$, then $\mathrm{SMet}_X(Q)$ is the lattice all permutations of X, known as the Permutohedron or the weak Bruhat order on the symmetric group, see e.g. [3,10,22]. If $Q = \mathbf{3}$ is the Sugihara semigroup on the three element chain, then $\mathrm{SMet}_X(Q)$ is isomorphic to the lattice of pseudo-permutations of X, see [2,14,21]. For Q the lattice of sup-preserving maps from the chain $[0,1]$ to itself, the poset $\mathrm{Clop}_X(Q)$—and therefore $\mathrm{SMet}_X(Q)$—was studied in [9,20]. We study in the next section the lattice $\mathrm{SMet}_X(Q)$ for Q a Sugihara monoid on an odd finite chain.

Remark 31. For X a finite total order, the construction sending Q to $\mathrm{SMet}_X(Q)$ can be made into a limit preserving functor from the category of involutive residuated lattices satisfying the mix rule into the category of lattices, see [20]. As a consequence, given that a Sugihara monoid on the finite even chain $\mathbf{2k}$ can be embedded in the Sugihara monoid on the odd chain $\mathbf{2k+1}$, the lattice $\mathrm{SMet}_X(\mathbf{2k})$ can be described as a sublattice of $\mathrm{SMet}_X(\mathbf{2k+1})$. For this reason we have given priority to the investigation of the lattices of the form $\mathrm{SMet}_X(\mathbf{2k+1})$.

6 The Poset of Augmented Plane Trees

In this section we study the ordered set $\mathrm{SMet}_X(Q)$, with $X = \{1,\ldots,n\}$ and Q the Sugihara monoid on the finite chain of size $2K+1$, denoted henceforth by $\mathrm{SMet}_{n,K}$. The aim is to give a combinatorial model of this poset, by describing its covers as moves (i.e. elementary transformations or rewrite rules) on a set of combinatorial objects, in the spirit of [16]. Since skew metrics correspond—under the bijection described in Sect. 4—to $K+1$-level plane trees whose leaves are labelled by subsets of X, these subsets forming a partition of X,[3] we should describe the ordering directly on this kind of objects. However, mostly for compactness, we prefer to code trees as words and handle the latter.

[3] These objects are called $K+1$-level labeled linear rooted trees with n leaves on The On-Line Encyclopedia of Integer Sequences [1], cf. Fig. 2.

Coding trees as words. We fix n and K and consider disjoint alphabets $\Sigma_0 := \{1, \ldots, n\}$ and $\Sigma_1 := \{|^1, \ldots, |^K\}$. We think of Σ_1 as an alphabet of walls of distinct heights. If $w \in (\Sigma_0 \cup \Sigma_1)^*$, then the walls from Σ_1 "split w into blocks". More precisely if w' is obtained from w by erasing letters from Σ_1, then w' can subdivided into blocks of contiguous letters from Σ_0. We define the set $Tw(n, K)$ (of tree-words) as the set of words w over the alphabet $\Sigma_0 \cup \Sigma_1$ satisfying the following conditions:

(i) The blocks of w are non-empty. That is, there are no contiguous walls, and walls do appear neither in first nor in last position.
(ii) The word obtained from w by erasing the walls is a permutation.
(iii) If two letters $x, y \in \Sigma_0$ are in the same block of w and $x < y$, then x appears on the left of y in w.

Example 32. The word $2|^2 13|^1 4$ belongs to $Tw(4, 2)$. The words $2|^2 134|^1$ and $2|^2|^3 13|^1 4$ violate the first constraint. The word $2|^2 23|^1 4$ violates the second constraint. The word $2|^2 31|^1 4$ violates the third constraint.

We take for granted that a word in $Tw(n, K)$ codes a perfect plane tree of height K augmented from $\{1, \ldots, n\}$, see Fig. 1 for examples. Yet, a few remarks are due. By identifying a word $w \in Tw(n, K)$ with the augmented tree it codes, we have

$$\delta_w(x, y) = \max\{\, k \mid \text{the symbol } |^k \text{ separates } x \text{ from } y \text{ in } w \,\},$$

and, for x, y such that $1 \leq x < y \leq n$,

$$\varsigma_w(x, y) = -1 \;\text{ iff }\; y \text{ appears before } x \text{ in the permutation underlying } w\,,$$

where the latter relation is a consequence of the fact that letters belonging to the same block appear in increasing order.

Positive and Negative Walls, Enabled Walls. For $w \in Tw(n, K)$ and $k \in \{1, \ldots, K\}$, let $w\uparrow_k$ be the word in $Tw(n, K)$ obtained by first erasing all the walls $|^j \in \Sigma_1$ with $j < k$, and then by reorganising contiguous blocks so to satisfy the third constraint. For example, $(1|^1 3|^2 2)\uparrow_2 = 13|^2 2$. For $w \in Tw(n, K)$ and an occurrence of a letter $|^k \in \Sigma_1$ in w, the *left* (resp., right) *scope* of this occurrence is the block on its left (resp., right) in $w\uparrow_k$; we say that such an occurrence is *positive* (resp., *negative*) if all the letters in its left scope are smaller (resp., greater) than those in its right scope. If such an occurrence is either positive or negative, then we say that it is enabled. For example, by $1\underline{|^2}3|^1 2$ is positive, $2|^1 3\underline{|^2}2$ is negative, $2|^3 3\underline{|^2}2$ is neither negative nor positive, so it is not enabled.

Moves. An *erosion move* replaces a positive occurrence of a wall $|^k$ by $|^{k-1}$, if $k > 1$, or deletes it, if $k = 1$. For $k = 1$, we call such an instance of an erosion move a *join move*, for obvious reasons. Dually, a *build move* occurs when a negative occurrence of a wall $|^k$ (with $0 < k < K$) is replaced by $|^{k+1}$. We can also consider the case when $k = 0$, which amounts to (i) inserting a wall $|^1$, thus splitting a block into two new blocks and then (ii) swapping the relative positions of these two new blocks. We call this a *split move*. Moves are illustrated in Fig. 1.

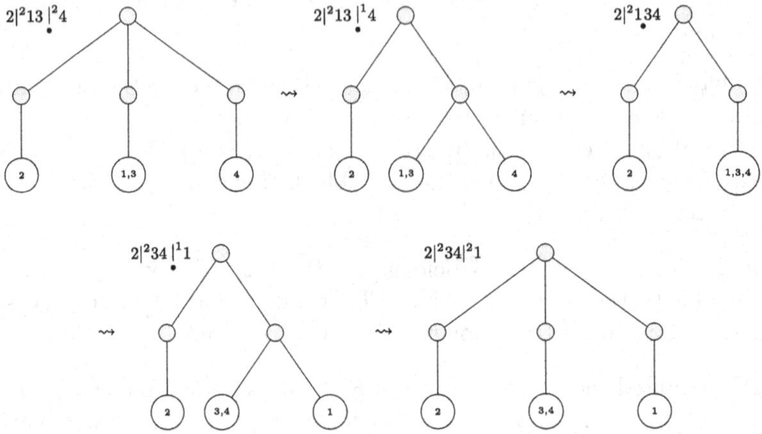

Fig. 1. Erosion, join, split, build moves, exemplified in the order

As from Eq. (5), we order $Tw(n, K)$ by saying that $w \leq u$ if, whenever $x < y$, $f_u(x, y) \leq f_w(x, y)$, where f_w, f_u are the skew metrics corresponding to w and u, respectively.

Theorem 33. *A word u is an upper cover of w in the poset $Tw(n, K)$ if and only if u is obtained from w by one of these moves.*

The theorem is an immediate consequence of the following two propositions, together with the straightforward observation that distinct moves from the same word yield incomparable words.

Proposition 34. *For each $w, u \in Tw(n, K)$, if u can be obtained from w by any of these moves, then $w < u$.*

Proof. Let $w, u \in Tw(n, K)$ be as stated, we need to show that, for each $(x, y) \in \mathcal{I}_X$, $f_u(x, y) \leq f_w(x, y)$, and $f_u(x, y) < f_w(x, y)$ for some $(x, y) \in \mathcal{I}_X$. Notice that if x, y are on the opposite scopes of the wall $|^k$ whose value k is being decreased or increased, then $\delta_u(x, y) \neq \delta_w(x, y)$. Therefore, we shall show that, for all $(x, y) \in \mathcal{I}_X$, $f_u(x, y) \leq f_w(x, y)$.

If the relative positions of x, y are not changed, that is, if $\varsigma_u(x, y) = \varsigma_w(x, y)$, and if $\delta_u(x, y) = \delta_w(x, y)$, then $f_u(x, y) = f_w(x, y)$. If $\varsigma_u(x, y) \neq \varsigma_w(x, y)$, then this happens with a split move and in this case we also have $\delta_w(x, y) \neq \delta_u(x, y)$. We suppose therefore that $\delta_w(x, y) \neq \delta_u(x, y)$.

If $\delta_u(x, y) < \delta_w(x, y)$, then u is obtained from w by erosion of a positive occurrence of a wall $|^k$ with $k := \delta_w(x, y)$. Then, x, y appear in the scopes of this wall, and since this occurrence is positive, x appear on the left and y appears on the right of the wall. Thus, we have $\varsigma_w(x,y) = \varsigma_u(x,y) = 1$ and $f_u(x,y) < f_w(x,y)$.

If $\delta_u(x,y) > \delta_w(x,y)$, then u is obtained from w by a build move of a negative occurrence of a wall $|^k$ with $k := \delta_w(x,y)$. Let us suppose first that $k > 0$. Thus, x,y appear in the scopes of this occurrence and, since this occurrence is negative, x appears in the right scope and y in the left scope. Thus we have $\varsigma_w(x,y) = \varsigma_w(x,y) = -1$, $\delta_u(x,y) = k+1$, and therefore $f_u(x,y) < f_w(x,y)$. Let us suppose finally that $k = 0$, that is, u is the result of a split move, so $\delta_w(x,y) = 0$ and $\delta_u(x,y) = 1$. Therefore, $f_w(x,y) = 0$, x,y belong to the same block of w, while x,y are separated by a wall $|^1$ in u, x being on its right scope and y being on its left scope. We have therefore $\varsigma_u(x,y) = -1$, $\delta_u(x,y) = 1$, and therefore $f_u(x,y) = -1 < 0 = f_w(x,y)$. □

Proposition 35. *If $w < u$, then there exists w', obtained from w by one of these moves, such that $w' \leq u$.*

Proof. For this proof, recall that $x \leq_0^w y$ if and only if $0 \leq f_w(x,y)$. Even if this is just a preorder, but we can still define standard notions, such as the closed interval $[x,y]_0^w := \{ z \mid x \leq_0^w z \leq y \}$, using which, the block of x is $[x]_0^w := [x,x]_0^w$.

Since $w < u$, the set $A := \{ (x,y) \in \mathcal{I}_X \mid f_u(x,y) < f_w(x,y) \}$ is nonempty. Consider a pair $(x,y) \in A$ minimizing the function δ_w on A. Moreover, among all such pairs, choose (x,y) such that the cardinality of $[x,y]_0^w \cup [y,x]_0^w$ is minimum. We suppose firstly that $\delta_w(x,y) = 0$, that is, x,y belong to the same block $[x]_0^w$. We split this block in two so to obtain w' with $f_u \leq f_{w'}$. To achieve this, consider that the restriction of the equivalence relation $\sim_0^u$ to $[x]_0^w$ splits it into blocks, say $b_1, \ldots, b_m$. If $x', y' \in [x]_0^w$, $x' < y'$, and $x' \in b_i, y' \in b_j$ with $i \neq j$, then $\varsigma_u(x',y') = -1$. This is a consequence of $\delta_u(x',y') > 0$ (since $[x']_0^u \neq [y']_0^u$) and $f_u(x',y') \leq f_w(x',y') = f_w(x,y) = 0$. Therefore, we can order the blocks so that $b_i < b_j$ if, for some $y' \in b_i$, $x' \in b_j$, $x' < y'$. Without loss of generality, we can assume that $b_1 < b_2 < \ldots < b_m$. Therefore, if we let $s := \mathrm{card}(b_m)$, then we can split $[x]_0^w$ at position s to obtain w'. We have $f_{w'}(x',y') = -1$ for each $(x',y') \in \mathcal{I}_X$ with $x' \in b_m$ and $y' \in [x]_0 \setminus b_m$ and, otherwise, $f_{w'}(x',y') = f_w(x',y')$. This shows that $f_u \leq f_{w'}$.

Suppose now that $\delta_w(x,y) > 0$ and let $k := \delta_w(x,y)$. By minimality of the cardinal of $[x,y]_0^w \cup [y,x]_0^w$, it follows that there is at most one wall $|^k$ separating x from y in $w \uparrow_k$. We claim that this occurrence is enabled. Indeed, take x', y' on the opposite scopes of the wall and observe that $\max(\delta_w(x,x'), \delta_w(y,y')) < \delta_w(x,y) = \delta_w(x',y')$. From this, it follows that

$$f_w(x',y') = f_w(x,x') \otimes f_w(x',y') \otimes f_w(y',y) \leq f_w(x,y)\,,$$

and, dually, $f_w(x,y) \leq f_w(x',y')$. Thus $f_w(x,y) = f_w(x',y')$ and, consequently, $\varsigma_w(x',y') = \varsigma_w(x,y)$. Therefore, if $\varsigma_w(x,y) = 1$, that is, if $x \leq_0^w y$, the occurrence of $|^k$ is positive, and if $\varsigma_w(x,y) = -1$ (i.e. $y \leq_0^w x$), the occurrence of $|^k$ is negative. Suppose that the occurrence of $|^k$ is positive. We have $f_{w'}(x',y') =$

$k-1$, for each $(x',y') \in \mathcal{I}_X$ with x' in the left scope of this wall and y' in its right scope and, otherwise, $f_{w'}(x',y') = f_w(x',y')$. This shows that $f_u \leq f_{w'}$. A similar argument shows that $f_u \leq f_{w'}$ if the occurrence of $|^k$ is negative. □

7 Enumerative Considerations

Several enumerative questions concerning the lattices $\mathrm{SMet}_{n,K}$ may be answered via the combinatorial model. We can determine the length of the posets $\mathrm{SMet}_{n,K}$, that is, the length of a longest chain. It is easily seen that a chain cannot have length greater than $2K\frac{n(n-1)}{2}$ and we claim that this is the length of some chain. We construct such a chain in $\mathrm{SMet}_{n+1,K}$ by concatenating a longest chain $\mathrm{SMet}_{n,K}$ with n sequences of $2K$ moves switching contiguous letters, as suggested below:

$$1|^K2\ldots|^Kn|^Kn+1 \leadsto^* n|^K\ldots 2|^K1|^Kn+1$$
$$\leadsto^{2K} n|^K\ldots 2|^Kn+1|^K1 \leadsto^{(n-1)2K} n+1|^Kn|^Kn-1\ldots|^Kn+1\,.$$

Letting $\ell_{n,K}$ be the length of such a sequence, we have the recurrence $\ell_{1,K} = 0$ and $\ell_{n+1,K} = \ell_{n+1,K} + 2nK$, yielding $\ell_{n,K} = Kn(n-1)$. Notice that a minimal sequence of moves from the bottom to the top elements of this poset has length $2K(n-1)$, so in particular these posets are not ranked.

The cardinalities $f(n,K) := \mathrm{card}(\mathrm{SMet}_{n,K})$ can be computed by

$$f(n,K) \overset{(a)}{=} \sum_{i=1}^{n} i!\left\{ {n \atop i} \right\} K^{i-1} \overset{(b)}{=} \sum_{i=0}^{n-1} \left\langle {n \atop i} \right\rangle K^i(K+1)^{n-1-i}\,,$$

where in these equalities $\left\{ {n \atop i} \right\}$ is the Stirling number of the second kind, counting the number of partitions of an n-element set into i blocks, while $\left\langle {n \atop i} \right\rangle$ is the Eulerian number, counting the number of permutations of n-elements with i descent positions.[4] Both formulas for $f(n,K)$ immediately follows from the correspondence with words in $Tw(n,K)$ given in the previous section. Equality (a) can be understood as follows. The number $j!\left\{ {n \atop j} \right\}$ counts the number of ordered partitions of an n-element set into j blocks and K^{j-1} counts the ways we can assign heights to the separating walls. Equality (b) stems from a well-known relation between ordered partitions and permutations. It can be read out as follows: given a permutation with i descent positions, we construct a word in $Tw(n,K)$ by (i) inserting a wall at each descent position and choosing an height for it in K different ways, (ii) for the other $n-1-i$ positions, either we do not insert a wall or we insert a wall and assign it an height, resulting in $K+1$ choices. Inspecting the values of the function $f(n,K)$ on The On-Line Encyclopedia of Integer

4 A descent position in a permutation $\sigma_1\sigma_1\ldots\sigma_n$ is an index $i \in \{1,\ldots,n-1\}$ such that $\sigma_i > \sigma_{i+1}$. The numbers $\left\langle {n \atop i} \right\rangle$ can be easily computed via the alternating formula $\left\langle {n \atop i} \right\rangle = \sum_{j=0}^{i}(-1)^j\binom{n+1}{j}(k+1-j)^n$, see e.g. [17].

n/K	1	2	3	4	5	6	7	OEIS
2	3	5	7	9	11	13	15	
3	13	37	73	121	181	253	337	*A003154*
4	75	365	1015	2169	3971	6565	10095	*A193252*
5	541	4501	17641	48601	108901	212941	378001	
6	4683	66605	367927	1306809	3583811	8288293	16984815	
7	47293	1149877	8952553	40994521	137595781	376372333	890380177	
OEIS	*A000670*	*A050351*	*A050352*	*A050353*				

Fig. 2. Cardinalities of $\mathrm{SMet}_{n,K}$

Sequences [1], see Fig. 2, we came across the reference [8]. This work, which also pinpoints the recursion $f(1, K) = 1$, $f(n + 1, K) = 1 + K \sum_{i=1}^{n} f(i, K)$, allows to establish a connection between skew metrics on Sugihara monoids and the geometry of hyperplane arrangements, see e.g. [23], a connection already known for the Sugihara monoid **3**, see e.g. [2,4]. It is proved in [8] that $f(n, K)$ is the number of maximal elements of the intersection poset of the affine braid arrangement $\{ H_{i,j,k} \mid 1 \leq i < j \leq n, -K \leq k \leq K \}$, with $H_{i,j,k}$ being the affine hyperplane of equation $x_j = x_i + k$. Sugihara monoids (and, more generally, involutive residuated lattices, as argued in [20]) therefore appear to have a pervasive role in this realm of geometry and in the related combinatorics. It is still a long way towards making this role fully explicit, but surely it is a research path that we want to pursue.

References

1. OEIS Foundation Inc.: The On-Line Encyclopedia of Integer Sequences (2021). http://oeis.org
2. Boulier, F., Hivert, F., Krob, D., Novelli, J.: Pseudo-permutations II: geometry and representation theory. In: Cori, R., Mazoyer, J., Morvan, M., Mosseri, R. (eds.) DM-CCG 2001, volume AA of DMTCS Proceedings, pp. 123–132 (2001)
3. Caspard, N., Santocanale, L., Wehrung, F., Grätzer, G., Wehrung, F.: Permutohedra and Associahedra. In: Lattice Theory: Special Topics and Applications, pp. 215–286. Springer, Cham (2016). https://doi.org/10.1007/978-3-319-44236-5_7
4. Dermenjian, A., Hohlweg, C., Pilaud, V.: The facial weak order and its lattice quotients. Trans. Amer. Math. Soc. **370**(2), 1469–1507 (2018)
5. Dunn, M.J.: Algebraic completeness results for R-mingle and its extensions. J. Symb. Log. **35**(1), 1–13 (1970)
6. Fussner, W., Galatos, N.: Categories of models of R-mingle. Ann. Pure Appl. Logic **170**(10), 1188–1242 (2019)
7. Galatos, N., Raftery, J.: A category equivalence for odd Sugihara monoids and its applications. J. Pure Appl. Algebra **216**(10), 2177–2192 (2012)
8. Gill, R.: The number of elements in a generalized partition semilattice. Discret. Math. **186**(1), 125–134 (1998)
9. Gouveia, M.J., Santocanale, L.: MIX $\star$-autonomous Quantales and the continuous weak order. In: Desharnais, J., Guttmann, W., Joosten, S. (eds.) RAMiCS 2018. LNCS, vol. 11194, pp. 184–201. Springer, Cham (2018). https://doi.org/10.1007/978-3-030-02149-8_12

10. Guilbaud, G.T., Rosenstiehl, P.: Analyse algébrique d'un scrutin. Mathématiques et Sciences Humaines **4**(9–33) (1963)
11. Joyal, A., Moerdijk, I.: Algebraic Set Theory. London Mathematical Society Lecture Note Series. Cambridge University Press, Cambridge (1995)
12. Kabil, M., Pouzet, M., Rosenberg, I.G.: Free monoids and generalized metric spaces. Eur. J. Comb. **80**, 339–360 (2019)
13. Kelly, G.M.: Basic concepts of enriched category theory. Repr. Theory Appl. Categ. (10), vi+137 (2005)
14. Krob, D., Latapy, M., Novelli, J.-C., Phan, H.-D., Schwer, S.: Pseudo-permutations I: first combinatorial and lattice properties. In: FPSAC 2001 (2001)
15. Lawvere, F.W.: Metric spaces, generalized logic and closed categories. Rendiconti del Seminario Matematico e Fisico di Milano **XLIII**, 135–166 (1973)
16. Newman, M.H.A.: On theories with a combinatorial definition of "equivalence". Ann. Math. **43**(2), 223–243 (1942)
17. Petersen, T.K.: Eulerian Numbers. Birkhäuser/Springer, New York (2015). https://doi.org/10.1007/978-1-4939-3091-3
18. Santocanale, L.: On discrete idempotent paths. In: Mercaş, R., Reidenbach, D. (eds.) WORDS 2019. LNCS, vol. 11682, pp. 312–325. Springer, Cham (2019). https://doi.org/10.1007/978-3-030-28796-2_25
19. Santocanale, L.: The involutive quantaloid of completely distributive lattices. In: Fahrenberg, U., Jipsen, P., Winter, M. (eds.) RAMiCS 2020. LNCS, vol. 12062, pp. 286–301. Springer, Cham (2020). https://doi.org/10.1007/978-3-030-43520-2_18
20. Santocanale, L., Gouveia, M.J.: The continuous weak order. J. Pure Appl. Algebra **225**, 106472 (2021)
21. Santocanale, L., Wehrung, F., Grätzer, G., Wehrung, F.: Generalizations of the permutohedron. In: Lattice Theory: Special Topics and Applications, pp. 287–397. Springer, Cham (2016). https://doi.org/10.1007/978-3-319-44236-5_8
22. Santocanale, L., Wehrung, F.: The equational theory of the weak Bruhat order on finite symmetric groups. J. Eur. Math. Soc. **20**(8), 1959–2003 (2018)
23. Stanley, R.P.: An introduction to hyperplane arrangements. In: Lecture notes, IAS/Park City Mathematics Institute (2004)
24. Taylor, P.: Intuitionistic sets and ordinals. J. Symb. Log. **61**(3), 705–744 (1996)
25. Vilain, M., Kautz, H., van Beek, P.: Constraint propagation algorithms for temporal reasoning: a revised report. In: Weld, D.S., de Kleer, J. (eds.) Readings in Qualitative Reasoning About Physical Systems, pp. 373–381. Morgan Kaufmann (1990)

Computing Distributed Knowledge as the Greatest Lower Bound of Knowledge

Carlos Pinzón[4], Santiago Quintero[3], Sergio Ramírez[4], and Frank Valencia[1,2](✉)

[1] CNRS-LIX, École Polytechnique de Paris, Palaiseau, France
[2] Pontificia Universidad Javeriana-Cali, Cali, Colombia
[3] LIX, École Polytechnique de Paris, Palaiseau, France
[4] INRIA Saclay, Palaiseau, France

Abstract. Let L be a distributive lattice and $\mathcal{E}(L)$ be the set of join endomorphisms of L. We consider the problem of finding $f \sqcap_{\mathcal{E}(L)} g$ given L and $f, g \in \mathcal{E}(L)$ as inputs. (1) We show that it can be solved in time $O(n)$ where $n = |L|$. The previous upper bound was $O(n^2)$. (2) We characterize the standard notion of distributed knowledge of a group as the greatest lower bound of the join-endomorphisms representing the knowledge of each member of the group. (3) We show that deciding whether an agent has the distributed knowledge of two other agents can be computed in time $O(n^2)$ where n is the size of the underlying set of states. (4) For the special case of $S5$ knowledge, we show that it can be decided in time $O(n\alpha_n)$ where α_n is the inverse of the Ackermann function.

Keywords: Distributive knowledge · Join-endomorphims · Lattice algorithms

1 Introduction

Structures involving a lattice L and its set of join-endomorphisms $\mathcal{E}(L)$ are ubiquitous in computer science. For example, in *Mathematical Morphology (MM)* [3], a well-established theory for the analysis and processing of geometrical structures founded upon lattice theory, join-endomorphisms correspond to one of its fundamental operations: *dilations*. In this and many other areas, lattices are used as rich abstract structures that capture the fundamental principles of their domain of application.

We believe that devising efficient algorithms in the abstract realm of lattice theory could be of great utility: We may benefit from many representation results and identify general properties that can be exploited in the particular domain of application of the corresponding lattices. In fact, we will use *distributivity* and *join-irreducibility* to reduce significantly the time and space needed to solve particular lattice problems. In this paper we focus on algorithms for the meet of join-endomorphisms.

This work has been partially supported by the ECOS-NORD project FACTS (C19M03).

U. Fahrenberg et al. (Eds.): RAMiCS 2021, LNCS 13027, pp. 413–432, 2021.
https://doi.org/10.1007/978-3-030-88701-8_25

We shall begin with a maximization problem: *Given a lattice* L *of size* n *and* $f, g \in \mathcal{E}(L)$, *find the greatest lower bound* $h = f \sqcap_{\mathcal{E}(L)} g$. Notice that the input is L not $\mathcal{E}(L)$. Simply taking $h(a) = f(a) \sqcap_L g(a)$ for all $a \in L$ does not work because the resulting h may not even be a join-endomorphism. Previous lower bounds for solving this problem are $O(n^3)$ for arbitrary lattices and $O(n^2)$ for distributive lattices [22]. We will show that this problem can actually be solved in $O(n)$ for distributive lattices.

Distributed knowledge [15] corresponds to knowledge that is distributed among the members of a group, without any of its members necessarily having it. This notion can be used to analyse the implications of the knowledge of a community if its members were to combine their knowledge, hence its importance. We will show that distributed knowledge can be seen as the meet of the join-endomorphisms representing the knowledge of each member of a group.

The standard structures in economics for multi-agent knowledge [23] involve a set of *states (or worlds)* Ω and a *knowledge operator* $K_i : \mathcal{P}(\Omega) \rightarrow \mathcal{P}(\Omega)$ describing the events, represented as subsets of Ω, that an agent i knows. The event of i knowing the event E is $K_i(E) = \{\omega \in \Omega \mid \mathcal{R}_i(\omega) \subseteq E\}$ where $\mathcal{R}_i \subseteq \Omega^2$ is the accessibility relation of i and $\mathcal{R}_i(\omega) = \{\omega' \mid (\omega, \omega') \in \mathcal{R}_i\}$. The event of having distributed knowledge of E by i and j is $D_{\{i,j\}}(E) = \{\omega \in \Omega \mid \mathcal{R}_i(\omega) \cap \mathcal{R}_j(\omega) \subseteq E\}$ [7].

Knowledge operators are join-endomorphisms of $L = (\mathcal{P}(\Omega), \supseteq)$. Intuitively, the lower an agent i (its knowledge function) is placed in $\mathcal{E}(L)$, the "wiser" (or more knowledgeable) the agent is. We will show that $D_{\{i,j\}} = K_i \sqcap_{\mathcal{E}(L)} K_j$, i.e., $D_{\{i,j\}}$ can be viewed as the *least* knowledgeable agent that is *wiser* than both i and j.

We also consider the following decision problem: *Given the knowledge of agents* i, j, m, *decide whether* m *has the distributed knowledge of* i *and* j, *i.e.*, $K_m = D_{\{i,j\}}$. The knowledge of an agent k can be represented by $K_k : \mathcal{P}(\Omega) \rightarrow \mathcal{P}(\Omega)$. If available it can also be represented, exponentially more succinctly, by $\mathcal{R}_k \subseteq \Omega^2$. In the first case the problem reduces to checking whether $K_m = K_i \sqcap_{\mathcal{E}(L)} K_j$. In the second the problem reduces to $\mathcal{R}_m = \mathcal{R}_i \cap \mathcal{R}_j$ and this can be done in $O(n^2)$ where $n = |\Omega|$.

Nevertheless, we show that even without the accessibility relations, if inputs are the knowledge operators, represented as arrays, the problem can be still be solved in $O(n^2)$. We obtain this result using tools from lattice theory to exponentially reduce the number of tests on the knowledge operators (arrays) needed to decide the problem.

Furthermore, if the inputs are the accessibility relations and they are equivalences (hence they can be represented as partitions), we show that the problem can be solved basically in *linear time*: More precisely, in $O(n\alpha_n)$ where α_n is an extremely slow growing function; the inverse of the Ackermann function. It is worth noticing that if accessibility relations can be represented as partitions, the structures are known as Aumann structures [2] and they characterize a standard notion of knowledge called $S5$ [7].

To prove the $O(n\alpha_n)$ bound we show a new result of independent interest using a Disjoint-Set data structure [8]: The intersection of two *partitions* of a set of size n can be computed in $O(n\alpha_n)$. This result may have applications beyond *knowledge*, particularly in domains where Disjoint-Set is typically used; e.g., *given two undirected graphs G_1 and G_2 with the same nodes, find an undirected graph G_3 such that two nodes are connected in it iff they are connected in both G_1 and G_2.*

Contributions and Organization. The main contributions are the following:

1. We prove that for distributive lattices of size n, the meet of join-endomorphisms can be computed in time $O(n)$. Previous upper bound was $O(n^2)$.
2. We show that distributed knowledge of a given group can be viewed as the meet of the join-endomorphisms representing the knowledge of each member of the group.
3. We show that the problem of whether an agent has the distributed knowledge of two other can be decided in time $O(n^2)$ where $n = |\Omega|$.
4. If the agents' knowledge can be represented as partitions, the problem in (3) can be decided in $O(n\alpha_n)$. To obtain this we provide a procedure, interesting in its own right, that computes the intersection of two *partitions* of a set of size n in $O(n\alpha_n)$.

The above results are given in Sects. 3 and 6. For conducting our study, in the intermediate sections (Sects. 4 and 5) we will adapt some representation and duality results (e.g., Jónsson-Tarski duality [17]) to our structures. Some of these results are part of the folklore in lattice theory but for completeness we provide simple proofs of them. We also provide experimental results for the above-mentioned effective procedures.

2 Notation, Definitions and Elementary Facts

We list facts and notation used throughout the paper. We index joins, meets, and orders with their corresponding poset but often omit the index when it is clear from the context.

Partially Ordered Sets and Lattices. A poset L is a *lattice* iff each finite nonempty subset of L has a supremum and infimum in L. It is a *complete lattice* iff each subset of L has a supremum and infimum in L. A poset L is *distributive* iff for every $a, b, c \in L$, $a \sqcup (b \sqcap c) = (a \sqcup b) \sqcap (a \sqcup c)$. We write $a \| b$ to denote that a and b are incomparable in the underlying poset. A *lattice of sets* is a set of sets ordered by inclusion and closed under finite unions and intersections. A *powerset lattice* is a lattice of sets that includes all the subsets of its top element.

Definition 1 (Downsets, Covers, Join-irreducibility [5]**).** *Let L be a lattice and $a, b \in L$. We say b is covered by a, written $b \prec a$, if $b \sqsubset a$ and there is no $c \in L$ s.t., $b \sqsubset c \sqsubset a$. The* down-set (up-set) *of a is $\downarrow a \stackrel{\text{def}}{=} \{b \in L \mid b \sqsubseteq a\}(\uparrow a \stackrel{\text{def}}{=}$

$\{b \in L \mid b \sqsupseteq a\}$), *and the set of elements* covered *by a is* $\downarrow^1 a \stackrel{\text{def}}{=} \{b \mid b \prec a\}$. *An element* $c \in L$ *is said to be* join-irreducible *if* $c = a \sqcup b$ *implies* $c = a$ *or* $c = b$. *If L is finite, c is* join-irreducible *if* $|\downarrow^1 c| = 1$. *The set of all join-irreducible elements of L is* $\mathcal{J}(L)$ *and* $\downarrow^{\mathcal{J}} c \stackrel{\text{def}}{=} \downarrow c \cap \mathcal{J}(L)$.

Posets of Maps. A map $f : X \to Y$ where X and Y are posets is *monotonic (or order-preserving)* if $a \sqsubseteq_X b$ implies $f(a) \sqsubseteq_Y f(b)$ for every $a, b \in X$. We say that f *preserves* the join of $S \subseteq X$ iff $f(\bigsqcup S) = \bigsqcup\{f(c) \mid c \in S\}$. A *self-map* on X is a function $f : X \to X$. If X and Y are posets, we use $\langle X \to Y \rangle$ to denote the poset of monotonic functions from X to Y. The functions in $M = \langle X \to Y \rangle$ are ordered pointwise: i.e., $f \sqsubseteq_M g$ iff $f(a) \sqsubseteq_Y g(a)$ for every $a \in X$.

Definition 2 (Join-endomorphisms and $\mathcal{E}(L)$). *Let L be a lattice. We say that a self-map is a* (bottom preserving) join-endomorphism *iff it preserves the join of every finite subset of L. Define* $\mathcal{E}(L)$ *as the set of all join-endomorphisms of L. Furthermore, given* $f, g \in \mathcal{E}(L)$, *define* $f \sqsubseteq_{\mathcal{E}} g$ *iff* $f(a) \sqsubseteq g(a)$ *for every* $a \in L$.

Proposition 3 ([5,12]). *Let L be a lattice.*

P.1 $f \in \mathcal{E}(L)$ *iff* $f(\bot) = \bot$ *and* $f(a \sqcup b) = f(a) \sqcup f(b)$ *for all* $a, b \in L$.
P.2 *If* $f \in \mathcal{E}(L)$ *then f is monotonic.*
P.3 *If L is a complete lattice, then* $\mathcal{E}(L)$ *is a complete lattice.*
P.4 $\mathcal{E}(L)$ *is a complete distributive lattice iff L is a complete distributive lattice.*

P.5 *If L is finite and distributive,* $\mathcal{E}(L) \cong \langle \mathcal{J}(L) \to L \rangle$.
P.6 *If L is a finite lattice,* $e = \bigsqcup_L \{c \in \mathcal{J}(L) \mid c \sqsubseteq e\}$ *for every* $e \in L$.
P.7 *If L is finite and distributive,* $f \in \mathcal{E}(L)$ *iff* $(\forall e \in L)\ f(e) = \bigsqcup\{f(e') \mid e' \in \downarrow^{\mathcal{J}} e\}$.

We shall use these posets in our examples: $\bar{\mathbf{n}}$ is $\{1, \ldots, n\}$ with the order $x \sqsubseteq y$ iff $x = y$ and $\mathbf{M}_n \stackrel{\text{def}}{=} (\bar{\mathbf{n}}_\bot)^\top$ is the lattice that results from adding a top and bottom to $\bar{\mathbf{n}}$.

3 Computing the Meet of Join-Endomorphisms

Join-endomorphisms and their meet arise as fundamental computational operations in computer science. We therefore believe that the problem of computing these operations in the abstract realm of lattice theory is a relevant issue: We may identify general properties that can be exploited in all instances of these lattices.

In this section, we address the problem of computing the meet of join endomorphisms. Let us consider the following *maximization* problem.

Problem 4. Given a lattice L of size n and two join-endomorphisms $f, g : L \to L$, find the *greatest* join-endomorphism $h : L \to L$ below both f and g: i.e., $h = f \sqcap_{\mathcal{E}(L)} g$.

Notice that the lattice $\mathcal{E}(L)$, which could be exponentially bigger than L [22], is not an input to the problem above. It may not be immediate how to find h; e.g., see the endomorphism h in Fig. 1a for a small lattice of four elements. A *naive approach* to find $f \sqcap_{\mathcal{E}(L)} g$ could be to attempt to compute it pointwise by taking $h(a) = f(a) \sqcap_L g(a)$ for every $a \in L$. Nevertheless, the somewhat appealing equation

$$(f \sqcap_{\mathcal{E}(L)} g)\,(a) = f(a) \sqcap_L g(a) \tag{1}$$

does not hold in general, as illustrated in the lattices $\mathbf{M}_2$ and $\mathbf{M}_3$ in Fig. 1b and 1c.

A general approach in [22] for arbitrary lattices shows how to find h in Problem 4 by successive approximations $\sigma^0 \sqsupset \sigma^1 \sqsupset \cdots \sqsupset \sigma^i$, starting with some self-map σ^0 known to be smaller than both f and g, and greater than h; while keeping the invariant $\sigma^i \sqsupseteq h$. The starting point is the naive approach above: $\sigma^0(a) = f(a) \sqcap g(a)$ for all $a \in L$. The approach computes decreasing upper bounds of h by correcting in σ^i the image under σ^{i-1} of some values $b, c, b \sqcup c$ violating the property $\sigma^{i-1}(b) \sqcup \sigma^{i-1}(c) = \sigma^{i-1}(b \sqcup c)$. The correction satisfies $\sigma^{i-1} \sqsupset \sigma^i$ and maintains the invariant $\sigma^i \sqsupseteq h$. This approach eventually finds h in $O(n^3)$ basic lattice operations (binary meets and joins).

Recall that in finite distributive lattices, and more generally in co-Heyting algebras [21], the subtraction operator $\ominus$ is uniquely determined by the *Galois connection* $b \sqsupseteq c \ominus a$ iff $a \sqcup b \sqsupseteq c$. Based on the following proposition it was shown in [22] that if the only basic operations are joins or meets, h can be computed in $O(n^3)$ of them, but if we also allow subtraction as a basic operation, the bound can be improved to $O(n^2)$.

Proposition 5 ([22]). *Let L be a finite distributive lattice. Let $h = f \sqcap_{\mathcal{E}(L)} g$. Then (1) $h(c) = \bigsqcap_L \{f(a) \sqcup g(b) \mid a \sqcup b \sqsupseteq c\}$, and (2) $h(c) = \bigsqcap_L \{f(a) \sqcup g(c \ominus a) \mid a \in \downarrow c\}$.*

Nevertheless, it turns out that we can partly use Eq. 1 to obtain a better upper bound. The following lemma states that Eq. 1 holds if L is distributive and $a \in \mathcal{J}(L)$.

Lemma 6. *Let L be a finite distributive lattice and $f, g \in \mathcal{E}(L)$. Then the following equation holds: $(f \sqcap_{\mathcal{E}(L)} g)\,(a) = f(a) \sqcap_L g(a)$ for every $a \in \mathcal{J}(L)$.*

Proof. From Proposition 5, $(f \sqcap_{\mathcal{E}(L)} g)(a) = \bigsqcap\{f(a') \sqcup g(a \ominus a') \mid a' \in \downarrow a\}$. Note that since $a \in \mathcal{J}(L)$ if $a' \in \downarrow a$ then $a \ominus a' = a$ when $a \neq a'$, and $a \ominus a' = \bot$ when $a = a'$. Then, $\{f(a') \sqcup g(a \ominus a') \mid a' \in \downarrow a\} = \{f(a') \sqcup g(a \ominus a') \mid a' \sqsubset a\} \cup \{f(a) \sqcup g(\bot)\} = \{f(a') \sqcup g(a) \mid a' \sqsubset a\} \cup \{f(a)\} = \{f(a') \sqcup g(a) \mid \bot \sqsubset a' \sqsubset a\} \cup \{f(a), g(a)\}$. By absorption, we know that $(f(a') \sqcup g(a)) \sqcap g(a) = g(a)$. Finally, using properties of $\sqcap$, $(f \sqcap_{\mathcal{E}(L)} g)(a) = \bigsqcap(\{f(a') \sqcup g(a) \mid \bot \sqsubset a' \sqsubset a\} \cup \{f(a), g(a)\}) = \bigsqcap\{f(a') \sqcup g(a) \mid \bot \sqsubset a' \sqsubset a\} \sqcap f(a) \sqcap g(a) = f(a) \sqcap g(a)$. □

It is worth noting the Lem.6 may not hold for non-distributive lattices. This is illustrated in Fig. 1c with the archetypal non-distributive lattice $\mathbf{M}_3$. Suppose that f and g are given as in Fig. 1c. Let $h = f \sqcap_{\mathcal{E}(L)} g$ with $h(a) = f(a) \sqcap g(a)$

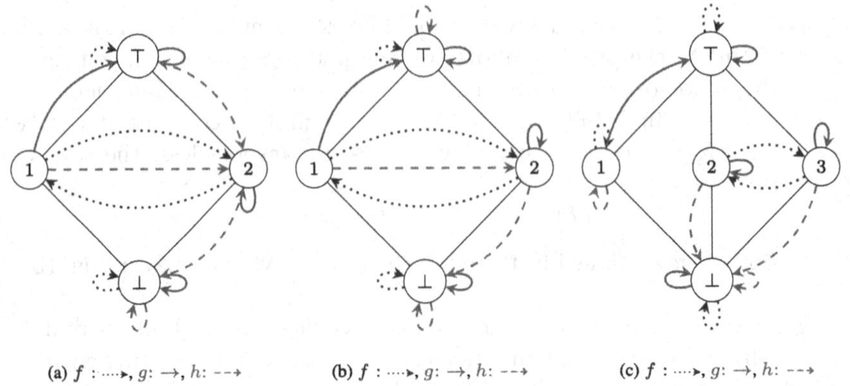

Fig. 1. (a) $h = f \sqcap_{\mathcal{E}(L)} g$. (b) $h(a) \stackrel{\text{def}}{=} f(a) \sqcap g(a)$ for $a \in \mathbf{M}_2$ is not in $\mathcal{E}(\mathbf{M}_2)$: $h(1 \sqcup 2) \neq h(1) \sqcup h(2)$. (c) Any $h : \mathbf{M}_3 \to \mathbf{M}_3$ s.t. $h(a) = f(a) \sqcap g(a)$ for $a \in \mathcal{J}(\mathbf{M}_3)$ is not in $\mathcal{E}(\mathbf{M}_3)$: $h(\top) = h(1 \sqcup 2) = h(1) \sqcup h(2) = 1 \neq \bot = h(2) \sqcup h(3) = h(2 \sqcup 3) = h(\top)$.

for all $a \in \{1, 2, 3\} = \mathcal{J}(\mathbf{M}_3)$. Since h is a join-endomorphism, we would have $h(\top) = h(1 \sqcup 2) = h(1) \sqcup h(2) = 1 \neq \bot = h(2) \sqcup h(3) = h(2 \sqcup 3) = h(\top)$, a contradiction.

Lemma 6 and P.7 lead us to the following characterization of meets over $\mathcal{E}(L)$.

Theorem 7. *Let L be a finite distributive lattice and $f, g \in \mathcal{E}(L)$. Then $h = f \sqcap_{\mathcal{E}(L)} g$ iff h satisfies*

$$h(a) = \begin{cases} f(a) \sqcap_L g(a) & \text{if } a \in \mathcal{J}(L) \text{ or } a = \bot \\ h(b) \sqcup_L h(c) & \text{if } b, c \in \downarrow^1 a \text{ with } b \neq c \end{cases} \tag{2}$$

Proof. The only-if direction follows from Lem. 6 and Proposition P.7. For the if-direction, suppose that h satisfies Eq. 2. If $h \in \mathcal{E}(L)$ the result follows from Lemma 6 and P.7. To prove $h \in \mathcal{E}(L)$ from P.7 it suffices to show

$$h(e) = \bigsqcup \{h(e') \mid e' \in \downarrow^{\mathcal{J}} e\} \tag{3}$$

for every $e \in L$. From Eq. 2 and since f and g are monotonic, h is monotonic. If $e \in \mathcal{J}(L)$ then $h(e') \sqsubseteq h(e)$ for every $e' \in \downarrow^{\mathcal{J}} e$. Therefore, $\bigsqcup\{h(e') \mid e' \in \downarrow^{\mathcal{J}} e\} = h(e)$. If $e \notin \mathcal{J}(L)$, we proceed by induction. Assume Eq. 3 holds for all $a \in \downarrow^1 e$. By definition, $h(e) = h(b) \sqcup h(c)$ for any $b, c \in \downarrow^1 e$ with $b \neq c$. Then, we have $h(b) = \bigsqcup\{h(e') \mid e' \in \downarrow^{\mathcal{J}} b\}$ and $h(c) = \bigsqcup\{h(e') \mid e' \in \downarrow^{\mathcal{J}} c\}$. Notice that $e' \in \downarrow^{\mathcal{J}} b$ or $e' \in \downarrow^{\mathcal{J}} c$ iff $e' \in \downarrow^{\mathcal{J}} (b \sqcup c)$, since L is distributive. Thus, $h(e) = h(b) \sqcup h(c) = \bigsqcup\{h(e') \mid e' \in \downarrow^{\mathcal{J}} (b \sqcup c)\} = \bigsqcup\{h(e') \mid e' \in \downarrow^{\mathcal{J}} e\}$ as wanted. □

We conclude this section by stating the time complexity $O(n)$ to compute h in the above theorem. As in [22], the time complexity is determined by the number of basic binary lattice operations (i.e., meets and joins) performed during execution.

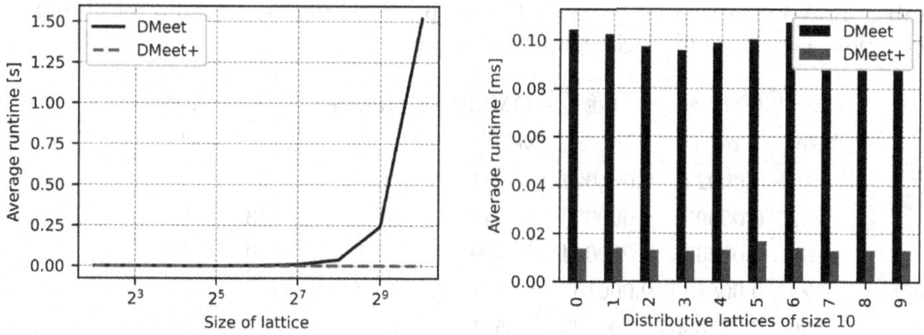

(a) Powerset lattices. (b) Arbitrary distributive lattices.

Fig. 2. Comparison between an implementation of Proposition 5 (DMEET) and Theorem 7 (DMEET+).

Corollary 8. *Given a distributive lattice L of size n, and functions $f, g \in \mathcal{E}(L)$, the function $h = f \sqcap_{\mathcal{E}(L)} g$ can be computed in $O(n)$ binary lattice operations.*

Proof. If $a \in \mathcal{J}(L)$ then from Theorem 7, $h(a)$ can be computed as $f(a) \sqcap g(a)$. If $a = \bot$ then $h(a)$ is $\bot$. If $a \notin \mathcal{J}(L)$ and $a \neq \bot$, we pick any $b, c \in \downarrow^1 a$ such that $b \neq c$ and compute $h(a)$ recursively as $h(b) \sqcup h(c)$ by Theorem 7. We can use a lookup table to keep track of the values of $a \in L$ for which $h(a)$ has been computed, starting with all $a \in \mathcal{J}(L)$. Since $h(a)$ is only computed once for each $a \in L$, either as a meet for elements in $\mathcal{J}(L)$ or as a join otherwise, we only perform n binary lattice operations.

Experimental Results. Now we present some experimental results comparing the average runtime between the previous algorithm in [22] based on Proposition 5, referred to as DMEET, and the proposed algorithm in Theorem 7, called DMEET+.

Figure 2 shows the average runtime of each algorithm, from 100 runs with a random pair of join-endomorphisms. For Fig. 2a, we compared each algorithm against powerset lattices of sizes between 2^2 and 2^{10}. For Fig. 2b, 10 random distributive lattices of size 10 were selected. In both cases, all binary lattice operation are guaranteed a complexity in $O(1)$ to showcase the quadratic nature of DMEET compared to the linear growth of DMEET+. The time reduction from DMEET to DMEET+ is also reflected in a reduction on the number of $\sqcup$ and $\sqcap$ operations performed as illustrated in Table 1. For DMEET+, given a distributive lattice L of size n, $\#\sqcap = |\mathcal{J}(L)|$ and $\#\sqcup = |L| - |\mathcal{J}(L)| - 1$ ($\bot$ is directly mapped to $\bot$).

Table 1. Average runtime in seconds over powerset lattices. Number of $\sqcup$ and $\sqcap$ operations performed for each algorithm.

Size	DMEET Time [s]	DMEET+ Time [s]	DMEET #$\sqcup$	DMEET+ #$\sqcup$	DMEET #$\sqcap$	DMEET+ #$\sqcap$
16	0.000246	0.000024	81	11	81	4
32	0.000971	0.000059	243	26	243	5
64	0.002659	0.000094	729	57	729	6
128	0.008735	0.000163	2187	120	2187	7
256	0.038086	0.000302	6561	247	6561	8
512	0.244304	0.000645	19683	502	19683	9
1024	1.518173	0.001468	59049	1013	59049	10

4 A Representation of Join-Irreducible Elements of $\mathcal{E}(L)$

In this section we state a characterization of the join-irreducible elements of the lattice of join-endomorphisms $\mathcal{E}(L)$. We use it to prove a representation result for join-endomorphisms. Some of these results may be part of the folklore in lattice theory, our purpose here is to identify and use them as technical tools in the following section.

The following family of functions can be used to represent $\mathcal{J}(\mathcal{E}(L))$.

Definition 9. *Let L be a lattice and $a, b \in \mathcal{J}(L)$. Let $f_{a,b} : L \to L$ be given by $f_{a,b}(x) \stackrel{\text{def}}{=} b$ if $x \in \uparrow a$, otherwise $f_{a,b}(x) \stackrel{\text{def}}{=} \bot$.*

It is easy to verify that $f_{a,b}(\bot) = \bot$. On the other hand, for every $c, d \in L$, $f_{a,b}(c \sqcup d) = f_{a,b}(c) \sqcup f_{a,b}(d)$ follows from the fact that $a \in \mathcal{J}(L)$ and by cases on $c \sqcup d \in \uparrow a$ and $c \sqcup d \notin \uparrow a$. Thus, from P.1 we know that $f_{a,b}$ is a join-endomorphism, and from P.2 it is monotone. Therefore, $f_{a,b}\restriction_{\mathcal{J}(L)} \in \langle \mathcal{J}(L) \to L \rangle$. In addition, we point out the following rather technical lemma that gives us way to construct from a function $g \in \langle \mathcal{J}(L) \to L \rangle$, a function $h \in \langle \mathcal{J}(L) \to L \rangle$ covered by g. The proof is given in the technical report available at https://hal.archives-ouvertes.fr/hal-03323638.

Lemma 10. *Let L be a finite lattice. Let $g \in \langle \mathcal{J}(L) \to L \rangle$, $x_0 \in \mathcal{J}(L)$ and $y_0 \in L$ be such that $y_0 \in \downarrow^1 g(x_0)$ and $g(x) \sqsubseteq y_0$ for all $x \sqsubset x_0$. Define $h : \mathcal{J}(L) \to L$ as $h(x) \stackrel{\text{def}}{=} y_0$ if $x = x_0$ else $h(x) \stackrel{\text{def}}{=} g(x)$. Then h is monotonic and g covers h.*

We proceed to characterize the join-irreducible elements of the lattice $\mathcal{E}(L)$. The next lemma, together with P.6, tell us that every join-endomorphism in $\mathcal{E}(L)$ can be expressed solely as a join of functions of the form $f_{a,b}$ defined in Definition 9.

Lemma 11. *Let L be a finite distributive lattice. For any join-endomorphism $f \in \mathcal{E}(L)$, f is join-irreducible iff $f = f_{a,b}$ for some $a, b \in \mathcal{J}(L)$.*

Proof. For notational convenience let $M = \langle \mathcal{J}(L) \to L \rangle$. From P.5 it suffices to prove: $g \in M$ is join-irreducible in M iff $g = g_{a,b}$ for some $a, b \in \mathcal{J}(L)$ where $g_{a,b} = f_{a,b}\restriction_{\mathcal{J}(L)}$. We use the following immediate consequence of Lemma 10.

Property ($\star$): Let $g \in M$, $x_1, x_2 \in \mathcal{J}(L)$ and $y_1, y_2 \in L$, be such that for each $i \in \{1,2\}$, $y_i \in \downarrow^1 g(x_i)$ and $g(x) \sqsubseteq y_i$ for all $x \sqsubset x_i$. If $x_1 \neq x_2$ or $y_1 \neq y_2$, then there are two distinct functions $g_1, g_2 \in M$ that are covered by g in M.

1. For the only-if direction, let $X = \{x \in \mathcal{J}(L) \mid g(x) \neq \bot\}$ and $Y = \{g(x) \mid x \in X\}$. If $X = \emptyset$, then $g(x) = \bot$ for all $x \in \mathcal{J}(L)$, in which case g is not join-irreducible in M. Thus, necessarily, $X \neq \emptyset$ and $Y \neq \emptyset$. Let us now prove that: (a) X has a minimum element $a \in \mathcal{J}(L)$ with $g(a) \in \mathcal{J}(L)$, and (b) $Y = \{g(a)\}$.
   (a) Let $x_1, x_2 \in X$ be minimal elements in X. For each $i \in \{1,2\}$, let $y_i \in \downarrow^1 g(x_i)$. Since x_i is minimal, it follows that $g(x) = \bot$ for all $x \sqsubset x_i$. From ($\star$) and the fact that g is join-irreducible, we have $x_1 = x_2$ and $y_1 = y_2$. Thus, X has a minimum element. We refer to such element as a. Furthermore, $|\downarrow^1 g(a)| = 1$, i.e. $g(a) \in \mathcal{J}(L)$.
   (b) Let $Y^* = Y \setminus \{g(a)\}$. For the sake of contradiction, suppose $Y^* \neq \emptyset$. Let $y \in Y^*$ be a minimal element and $x^* \in X$ be a minimal of $X^* = \{x \in X \mid g(x) = y\}$. Since $a \sqsubset x^*$ and $y \neq g(a)$, we have $g(a) \sqsubset g(x^*) = y$. Then there is at least one $z \in \downarrow^1 y$ such that $g(a) \sqsubseteq z \sqsubset y$. Since g is monotonic, $Im(g) = \{\bot\} \cup Y$ and y is minimal in Y^*, for all $x \sqsubset x^*$, we have $g(x) \in \{\bot, g(a)\}$. Therefore, $g(x) \sqsubseteq z$ for all $x \sqsubset x^*$. From ($\star$), with $x_1 = a$, $x_2 = x^*$, $y_1 \in \downarrow^1 g(a)$ and $y_2 = z$, it follows that g is not join-irreducible in M, a contradiction.

   Monotonicity of g and (a)-(b), imply $Im(g) = \{\bot, b\}$ with $b = g(a)$. Thus $g = g_{a,b}$.
2. We prove that $g = g_{a,b}$ has a unique cover in M. Let c be the only cover of b. Define $g^*: \mathcal{J}(L) \to L$ as $g^*(x) = c$ if $x = a$ else $g^*(x) = g(x)$. From Lemma 10, it follows that $g^* \in M$ and $g_{a,b}$ covers g^* in M. It suffices that for any $h \in M$ with $h \sqsubset_M g_{a,b}$, $h \sqsubseteq_M g^*$ holds. Take any such $h \in M$. Since $h(a) \neq b$, $h(a) \sqsubset b$. Thus $h(a) \sqsubseteq c$, so $h(a) \sqsubseteq g^*(a)$. Indeed, for any $x \neq a$, $h(x) \sqsubset g(x) = g^*(x)$. Then $h \sqsubseteq_M g^*$. □

We conclude with a corollary of Lemma 11 that provides a representation theorem for join-endomorphism on distributive lattices. We will use this result in the next section.

Corollary 12. *Let L be a finite distributive lattice and let $f \in \mathcal{E}(L)$. Then $f = F_R$ where $R = \{(a,b) \in \mathcal{J}(L)^2 \mid a \sqsubseteq f(b)\}$ and $F_R : L \to L$ is the function given by $F_R(c) \stackrel{\text{def}}{=} \bigsqcup\{a \in \mathcal{J}(L) \mid (a,b) \in R \text{ and } c \sqsupseteq b \text{ for some } b \in \mathcal{J}(L)\}$.*

Proof. From P.6 $f = \bigsqcup_{\mathcal{E}(L)} \{g \in \mathcal{J}(\mathcal{E}(L)) \mid g \sqsubseteq_{\mathcal{E}} f\}$. Thus,

$$
\begin{aligned}
f(c) &= \Big(\bigsqcup\nolimits_{\mathcal{E}(L)} \{g \in \mathcal{J}(\mathcal{E}(L)) \mid g \sqsubseteq_{\mathcal{E}} f\}\Big)(c) = \bigsqcup\{g(c) \mid g \in \mathcal{J}(\mathcal{E}(L)) \text{ and } g \sqsubseteq_{\mathcal{E}} f\} \\
&= \bigsqcup\{f_{b,a}(c) \mid (b,a) \in \mathcal{J}(L)^2 \text{ and } f_{b,a} \sqsubseteq_{\mathcal{E}} f\} \qquad \text{(Lemma 11)} \\
&= \bigsqcup\{f_{b,a}(c) \mid (b,a) \in \mathcal{J}(L)^2 \text{ and } a \sqsubseteq f(b)\} \\
&= \bigsqcup\{a \in \mathcal{J}(L) \mid (b,a) \in \mathcal{J}(L)^2, a \sqsubseteq f(b) \text{ and } c \sqsupseteq b \text{ for some } b \in \mathcal{J}(L)\} \\
&= \bigsqcup\{a \in \mathcal{J}(L) \mid (b,a) \in R \text{ and } c \sqsupseteq b \text{ for some } b \in \mathcal{J}(L)\} = F_R(c)
\end{aligned}
$$

5 Distributive Lattices and Knowledge Structures

In this section, we introduce some knowledge structures from economics [2,23] and relate them to distributive lattices by adapting fundamental duality results between modal algebras and frames [17]. We will use these structures and their relation to distributive lattices in the algorithmic results in the next section. We use the term *knowledge* to encompass various epistemic concepts including $S5$ knowledge and belief [7].

Definition 13 ([23]). *A (finite)* Knowledge Structure (KS) *for a set of* agents $\mathcal{A}$ *is a tuple* $(\Omega, \{K_i\}_{i\in\mathcal{A}})$ *where* Ω *is a finite set and each* $K_i : \mathcal{P}(\Omega) \to \mathcal{P}(\Omega)$ *is given by* $K_i(E) = \{\omega \in \Omega \mid \mathcal{R}_i(\omega) \subseteq E\}$ *where* $\mathcal{R}_i \subseteq \Omega^2$ *and* $\mathcal{R}_i(\omega) = \{\omega' \mid (\omega, \omega') \in \mathcal{R}_i\}$.

The elements $\omega \in \Omega$ and the subsets $E \subseteq \Omega$ are called *states* and *events*, resp. We refer to K_i and $\mathcal{R}_i$ as the *knowledge operator* and the *accessibility relation* of agent i.

The notion of *event* may be familiar to some readers from probability theory; for example the event *"public transportation is suspended"* corresponds the set of states at which public transportation is suspended. An event E *holds at* ω if $\omega \in E$. Thus Ω, the event that holds at every ω, corresponds to true in logic, union of events corresponds to disjunction, intersection to conjunction, and complementation in Ω to negation. We use $\overline{E}$ for $\Omega \setminus E$. We write $E \Rightarrow F$ for the event $\overline{E} \cup F$ which corresponds to classic logic implication. We say that E *entails* F if $E \subseteq F$. The event of i *knowing* E is $K_i(E)$.

The following properties hold for all events E and F of any KS $(\Omega, \{K_i\}_{i\in\mathcal{A}})$. It is easy to see that ($\mathfrak{K}1$) $K_i(\Omega) = \Omega$, i.e., agents know the event that holds at every state, i.e., Ω. A distinctive property of knowledge is ($\mathfrak{K}2$) $K_i(E) \cap K_i(F) = K_i(E \cap F)$; i.e., if an agent knows two events, she knows their conjunction. In fact, $\mathfrak{K}2$ implies ($\mathfrak{K}3$) $(K_i(E) \cap K_i(E \Rightarrow F)) \subseteq K_i(F)$. This property expresses modus ponens for knowledge. Other property implied by $\mathfrak{K}2$ is that knowledge is monotonic: ($\mathfrak{K}4$) if $E \subseteq F$ then $K_i(E) \subseteq K_i(F)$, i.e., agents know the consequences of their knowledge.

An agent i is *wiser* (or *more knowledgeable*) than j iff $K_j(E) \subseteq K_i(E)$ for every event E; i.e., if j knows E so does i.

Aumann Structures. Aumann structures are the standard event-based formalism in economics and decision theory [7] for reasoning about knowledge. *A (finite) Aumann structure (AS) is a KS where all the accessibility relations are equivalences.*[1] The intended notion of knowledge of AS is $S5$; i.e., the knowledge captured by Prop.$\mathfrak{K}1$-$\mathfrak{K}2$ and the following three fundamental properties which hold for any AS: ($\mathfrak{K}5$) $K_i(E) \subseteq E$, ($\mathfrak{K}6$) $K_i(E) \subseteq K_i(K_i(E))$, and ($\mathfrak{K}7$) $\overline{K_i(E)} \subseteq K_i(\overline{K_i(E)})$. The first says that if an agents knows E then E cannot be false, the second and third state that agents know what they know and what they do not know.

Extended KS. We now introduce a simple extension of KS that will allow us to give a uniform presentation of our results.

Definition 14 (EKS). *A tuple* $(\Omega, \mathcal{S}, \{K_i\}_{i \in \mathcal{A}})$ *is said to be an* extended knowledge structure (EKS) *if (1)* $(\Omega, \{K_i\}_{i \in \mathcal{A}})$ *is a KS, and (2)* $\mathcal{S}$ *is a subset of* $\mathcal{P}(\Omega)$ *that contains* Ω *and it is closed under union, intersection and application of* K_i *for every* $i \in \mathcal{A}$.

Notation. Given an underlying EKS $(\Omega, \mathcal{S}, \{K_i\}_{i \in \mathcal{A}})$ and $f : \mathcal{P}(\Omega) \to \mathcal{P}(\Omega)$ we shall use $\tilde{f}$ for the function $f\lceil_{\mathcal{S}} : \mathcal{S} \to \mathcal{P}(\Omega)$, i.e., $\tilde{f}(E) = f(E)$ for every $E \in \mathcal{S}$. Because of the closure properties of $\mathcal{S}$, for every $i \in \mathcal{A}$ we have $\widetilde{K_i} : \mathcal{S} \to \mathcal{S}$.

Notice that the AS and, in general KS, are EKS where $\mathcal{S} = \mathcal{P}(\Omega)$. Also Kripke frames [7] can be viewed as EKS with $\mathcal{S} = \mathcal{P}(\Omega)$. Other structures not discussed in this paper such as set algebras with operators (SOS) [24] and general frames [4] can be represented as EKSs where $\mathcal{S}$ is required to be closed under complement.

5.1 Extended KS and Distributive Lattices

The knowledge operators of an EKS are join-endomorphisms on a distributive lattice. This is an easy consequence of $\mathfrak{K}1$ and $\mathfrak{K}2$, and the closure properties of EKS. The next proposition tells us that the *wiser* the agent, the lower that (its knowledge operator) is placed in the corresponding lattice.

Proposition 15. *Let* $(\Omega, \mathcal{S}, \{K_i\}_{i \in \mathcal{A}})$ *be an EKS. Then* $L = (\mathcal{S}, \supseteq)$ *is a distributive lattice and for each* $i \in \mathcal{A}$, $\widetilde{K_i} \in \mathcal{E}(L)$.

Conversely, the join-endomorphisms of distributive lattices correspond to knowledge operators of EKS. Recall that every distributive lattice is isomorphic to (the dual of) a lattice of sets. The next proposition is an adaptation to finite distributive lattices of Jónsson-Tarski duality for general-frames and boolean algebras with operators [17].

[1] The presentation of AS [2] uses a partition $\mathcal{P}_i = \{\mathcal{R}_i(\omega) \mid \omega \in \Omega\}$ of Ω and $K_i(E)$ is equivalently defined as $\{\omega \in \Omega \mid \mathcal{P}_i(\omega) \subseteq E\}$ where $\mathcal{P}_i(\omega)$ is the cell of $\mathcal{P}_i$ containing ω.

Proposition 16. *Let L be dual to a finite lattice of sets with a family $\{f_i \in \mathcal{E}(L)\}_{i \in I}$. Then $(\Omega, \mathcal{S}, \{K_i\}_{i \in I})$ is an EKS where $\mathcal{S} = L, \Omega = \bot_L$, and for every $i \in I$, $\mathcal{R}_i = \{(\omega, \omega') \in \Omega^2 \mid$ for all $E \in \mathcal{S}, \omega \in f_i(E)$ implies $\omega' \in E\}$. Also, for $i \in I$, $\widetilde{K}_i = f_i$.*

Proof. Notice that $L = \mathcal{S}$ is closed under union and intersection since L is the dual of a lattice of sets. Showing $\widetilde{K}_i = f_i$ also proves that $\mathcal{S}$ is closed under K_i. Recall that $\widetilde{K}_i(E) = K_i(E)$ for each $E \in \mathcal{S}$. Thus, it remains to prove $K_i(E) = f_i(E)$ for all $E \in \mathcal{S}$. From $\mathfrak{K}1$ and the fact that f_i is a join-endomorphism, $K_i(E) = f_i(E) = \Omega$ for $E = \Omega$. Hence, choose an arbitrary $E \neq \Omega$. First suppose that $\tau \in f_i(E)$. From the definition of $\mathcal{R}_i$ if $(\tau, \tau') \in \mathcal{R}_i$, $\tau' \in E$. Hence $\mathcal{R}_i(\tau) \subseteq E$, so $\tau \in K_i(E)$.

Now suppose that $\tau \in K_i(E)$ but $\tau \notin f_i(E)$. From $\tau \in K_i(E)$ we obtain:

$$\textit{for all } \tau' \in \Omega \textit{ if } (\tau, \tau') \in \mathcal{R}_i \textit{ then } \tau' \in E. \tag{4}$$

From the assumption $\tau \notin f_i(E)$ and the monotonicity of join-endomorphisms (P.2):

$$\textit{for every } F \in \mathcal{S} \textit{ if } F \subseteq E \textit{ then } \tau \notin f_i(F). \tag{5}$$

Let $X = \{E' \in \mathcal{S} \mid \tau \in f_i(E')\}$. If $X = \emptyset$ then from the definition of $\mathcal{R}_i$ we conclude $\mathcal{R}_i(\tau) = \Omega$ which contradicts (4) since $E \neq \Omega$. If $X \neq \emptyset$ take $S = \bigcap X$. Since f_i is a join-endomorphism, it distributes over intersection (i.e., the join in L), we conclude $\tau \in f(S)$. Thus, if $S \subseteq E$ we obtain a contradiction with (5). If $S \nsubseteq E$ then there exists $\tau' \in S$ such that $\tau' \notin E$. From the definition of S, $\tau' \in E'$ for each E' such that $\tau \in f_i(E')$. But this implies $(\tau, \tau') \in \mathcal{R}_i$ and $\tau' \notin E$, a contradiction with (4). □

Nevertheless, we can use our general characterization of join endomorphisms in the previous section (Corollary 12) to obtain a simpler relational construction for join endomorphisms of powerset lattices (boolean algebras). Unlike the construction in Proposition 16, this characterization of $\mathcal{R}_i$ does not appeal to universal quantification.

Proposition 17. *Let L be dual to a finite powerset lattice with a family $\{f_i \in \mathcal{E}(L)\}_{i \in I}$. Let $(\Omega, \{K_i\}_{i \in I})$ be the KS where $\Omega = \bot_L$ and $\mathcal{R}_i = \{(\omega, \omega') \mid \omega \in \overline{f_i(\ \overline{\{\omega'\}}\ )}\ \}$. Then, for every $i \in \mathcal{A}$, $K_i = f_i$.*

Proof. Since L is dual to a powerset lattice, $\sqcup = \cap$, $\sqsubseteq = \supseteq$, and $\mathcal{J}(L) = \{\overline{\{\tau\}} \mid \tau \in \Omega\}$. Let $Q = \{(\overline{\{\sigma\}}, \overline{\{\tau\}}) \mid (\sigma, \tau) \in \mathcal{R}_i\}$. Notice that for every $(\overline{\{\sigma\}}, \overline{\{\tau\}}) \in Q$, we have $\sigma \in \overline{f_i(\ \overline{\{\tau\}}\ )}$. Equivalently, $\{\sigma\} \subseteq \overline{f_i(\ \overline{\{\tau\}}\ )}$ and $f_i(\ \overline{\{\tau\}}\ ) \subseteq \overline{\{\sigma\}}$. Therefore, from Corollary 12, it follows that for every $E \in L$, $f_i(E) = \bigcap\{\overline{\{\sigma\}} \in \mathcal{J}(L) \mid (\overline{\{\sigma\}}, \overline{\{\tau\}}) \in Q_i$ and $E \subseteq \overline{\{\tau\}}$ for some $\overline{\{\tau\}} \in \mathcal{J}(L)\}$. We complete the proof as follows:

$$
\begin{aligned}
f_i(E) &= \bigcap\{\overline{\{\sigma\}} \in \mathcal{J}(L) \mid \exists \overline{\{\tau\}} \in \mathcal{J}(L) : ((\overline{\{\sigma\}}, \overline{\{\tau\}}) \in Q \text{ and } E \subseteq \overline{\{\tau\}})\} \\
&= \bigcap\{\overline{\{\sigma\}} \in \mathcal{J}(L) \mid \neg\forall \overline{\{\tau\}} \in \mathcal{J}(L) : ((\overline{\{\sigma\}}, \overline{\{\tau\}}) \in Q \implies E \not\subseteq \overline{\{\tau\}})\} \\
&= \bigcap\{\Omega \setminus \{\sigma\} \in \mathcal{J}(L) \mid \neg\forall \tau \in \Omega : ((\sigma, \tau) \in \mathcal{R}_i \implies \tau \in E)\} \\
&= \bigcap\{\Omega \setminus \{\sigma\} \in \mathcal{J}(L) \mid \neg(\mathcal{R}_i(\sigma) \subseteq E)\} \\
&= \Omega \setminus \{\sigma \in \Omega \mid \neg(\mathcal{R}_i(\sigma) \subseteq E)\} = \{\sigma \in \Omega \mid \mathcal{R}_i(\sigma) \subseteq E\} = K_i(E)
\end{aligned}
$$

We conclude this section by pointing out that accessibility relations can be obtained from knowledge operators. We can use Proposition 17 to compute $\mathcal{R}_i$ from K_i. For AS we can obtain the equivalence class $\mathcal{R}_i(\omega)$ directly from K_i.

Corollary 18. *Let $\mathcal{K} = (\Omega, \{K_i\}_{i \in \mathcal{A}})$ be a KS. Then (1) $\mathcal{R}_i = \{(\omega, \omega') \mid \omega \in \overline{K_i(\overline{\{\omega'\}})}\}$. (2) If $\mathcal{K}$ is an AS then $\mathcal{R}_i(\omega) = \overline{K_i(\overline{\{\omega\}})}$ for every $\omega \in \Omega$.*

6 Distributed Knowledge

The notion of *distributed knowledge* represents the information that two or more agents may have as a group but not necessarily individually. Intuitively, it is what someone who knows what each agent, in a given group, knows. As described in [7], while common knowledge can be viewed as what "any fool" knows, distributed knowledge can be viewed as what a "wise man" would know.

Let $(\Omega, \{K_i\}_{i \in \mathcal{A}})$ be a KS and $i, j \in \mathcal{A}$. The *distributed knowledge* of i and j is represented by $D_{\{i,j\}} : \mathcal{P}(\Omega) \to \mathcal{P}(\Omega)$ defined as $D_{\{i,j\}}(E) = \{\omega \in \Omega \mid \mathcal{R}_i(\omega) \cap \mathcal{R}_j(\omega) \subseteq E\}$ where $\mathcal{R}_i$ and $\mathcal{R}_j$ are the accessibility relations for i and j.

The following property captures the notion of distributed knowledge by relating group to individual knowledge: ($\mathfrak{K}8$) $(K_i(E) \cap K_j(E \Rightarrow F)) \subseteq D_{\{i,j\}}(F)$. It says that if one agents knows E and the other knows that E implies F, together they have the distributed knowledge of F even if neither agent knew F.

Example 19. Let E be the event "Bob's *boss is working from home*" and F be the event "*public transportation is suspended*". Suppose that agent Alice knows that Bob's boss is working from home (i.e., $K_A(E)$), and that agent Bob knows that his boss works from home only when public transportation is suspended (i.e., $K_B(E \Rightarrow F)$). Thus, if they told each other what they knew, they would have distributed knowledge of F (i.e., $D_{\{A,B\}}(F)$). Indeed, $K_A(E) \cap K_B(E \Rightarrow F)$ entails $D_{\{A,B\}}(F)$ from $\mathfrak{K}8$.

A self-explanatory property relating individual and distributed knowledge is ($\mathfrak{K}9$) $K_i(E) \subseteq D_{\{i,j\}}(E)$. Furthermore, the above basic properties of knowledge Prop.$\mathfrak{K}1$-$\mathfrak{K}2$ also hold if we replace the K_i with $D_{\{i,j\}}$: Intuitively, distributed knowledge is knowledge. Indeed, imagine an agent m that combines i and j's knowledge by having an accessibility relation $\mathcal{R}_m = \mathcal{R}_i \cap \mathcal{R}_j$. In this case we would have $K_m = D_{\{i,j\}}$. Therefore, any KS may include distributed knowledge as one of its knowledge operators. For simplicity, we are considering distributed knowledge of two agents but this can be easily extended to arbitrary groups of agents. E.g. if $K_m = D_{\{i,j\}}$ then $D_{\{k,m\}}$ represents the distributed knowledge of three agents i, j and k.

6.1 The Meet of Knowledge.

In Sect. 5.1 we identified knowledge operators and join endomorphisms. We now show that the notion of distributed knowledge corresponds exactly to the meet of the knowledge operators in the lattice of all join-endomorphisms in $(\mathcal{S}, \supseteq)$.

Theorem 20. *Let $(\Omega, \mathcal{S}, \{K_i\}_{i \in \mathcal{A}})$ be an EKS and let L be the lattice $(\mathcal{S}, \supseteq)$. Let us suppose that $K_m = D_{\{i,j\}}$ for some $i, j, m \in \mathcal{A}$. Then $\widetilde{K}_m = \widetilde{K}_i \sqcap_{\mathcal{E}(L)} \widetilde{K}_j$.*

Proof. Let us assume $K_m = D_{\{i,j\}}$. Then from the closure properties of $\mathcal{S}$, we have $\widetilde{D}_{\{i,j\}} = \widetilde{K}_m : \mathcal{S} \to \mathcal{S}$. Let $f = \widetilde{K}_i \sqcap_{\mathcal{E}(L)} \widetilde{K}_j$. (Recall that the order relation $\sqsubseteq_L$ over L is reversed inclusion $\supseteq$, joins are intersections and meets are unions.)

From Proposition 19, for every $E \in \mathcal{S}$, $D_{\{i,j\}}(E) \sqsubseteq_L K_i(E), K_j(E)$. Thus $\widetilde{D}_{\{i,j\}}$ is a lower bound of both $\widetilde{K}_i$ and $\widetilde{K}_j$ in $\mathcal{E}(L)$, so $\widetilde{D}_{\{i,j\}} \sqsubseteq_{\mathcal{E}(L)} f$.

To prove $f \sqsubseteq_{\mathcal{E}(L)} \widetilde{D}_{\{i,j\}}$, take $\tau \in \widetilde{D}_{\{i,j\}}(E) = D_{\{i,j\}}(E)$ for an arbitrary $E \in \mathcal{S}$. By definition of $D_{\{i,j\}}$, we have

$$\mathcal{R}_i(\tau) \cap \mathcal{R}_j(\tau) \subseteq E. \tag{6}$$

From Proposition 5

$$f(E) = \bigcup \{K_i(F) \cap K_j(H) \mid F, H \in \mathcal{S} \text{ and } F \cap H \subseteq E\} \tag{7}$$

Take $F = \mathcal{R}_i(\tau)$ and $H = \mathcal{R}_j(\tau)$, from (6), $F \cap H \subseteq E$. By definition of knowledge operator, $\tau \in K_i(F)$ and $\tau \in K_j(H)$. From (7), $\tau \in f(E)$. Thus $f \sqsubseteq_{\mathcal{E}(L)} \widetilde{D}_{\{i,j\}}$. □

The theorem above allows us to characterize an agent m having the distributed knowledge of i and j as the *least knowledgeable* agent wiser than both i and j. In the next section we consider the decision problem of whether a given m indeed has the distributed knowledge of i and j.

6.2 The Distributed Knowledge Problem

In what follows, let $(\Omega, \{K_i\}_{i \in \mathcal{A}})$ be a KS and let $n = |\Omega|$. Let us now consider the following decision: *Given the knowledge of agents i, j, m, decide whether m has the distributed knowledge of i and j, i.e., $K_m = D_{\{i,j\}}$.*

The input for this problem is the knowledge of the agents and it can be represented using either knowledge operators K_i, K_j, K_m or accessibility relations $\mathcal{R}_i, \mathcal{R}_j, \mathcal{R}_m$. For each representation, the algorithm that solves the problem $K_m = D_{\{i,j\}}$ can be implemented differently. For the first representation, it follows from Theorem 20 that $K_m = D_{\{i,j\}}$ holds if and only if $K_m = K_i \sqcap_{\mathcal{E}(L)} K_j$ where $L = (\mathcal{P}(\Omega), \supseteq)$. For the second one, we can verify $\mathcal{R}_m = \mathcal{R}_i \cap \mathcal{R}_j$ instead. Indeed, as stated in Corollary 18, one representation can be obtained from the other, hence an alternative solution for the decision problem is to translate the input from the given representation into the other one before solving.

Accessibility relations represent knowledge much more compactly than knowledge operators because the former are relations on Ω^2 while the latter are relations on $\mathcal{P}(\Omega)^2$. For this reason, it would seem in principle that the algorithm for handling the knowledge operator would be slower by several orders of magnitude. Nevertheless, we can use our lattice theoretical results from previous sections to show that this is not necessarily the case, thus it is worth considering both types of representations.

From Knowledge Operators. We wish to determine $K_m = D_{\{i,j\}}$ by establishing whether $K_m = K_i \sqcap_{\mathcal{E}(L)} K_j$ where $L = (\mathcal{P}(\Omega), \supseteq)$. Let us assume the following bitwise representation of knowledge operators. The states in Ω are numbered as $\omega_1, \ldots, \omega_n$. Each event E is represented as a number $\#E \in [0..2^n - 1]$ whose binary representation has its k-th bit set to 1 iff $\omega_k \in E$. Each input knowledge operator K_i is represented as an array $\mathtt{K}_i$ of size 2^n that stores $\#K_i(E)$ at position $\#E$, i.e., $\mathtt{K}_i[\ \#E\ ] = \#K_i(E)$.

From Lemma 6, $K_m = K_i \sqcap_{\mathcal{E}(L)} K_j$ iff $K_m(E) = K_i(E) \cup K_j(E)$ for every join-irreducible element E in L. Notice that $E \in \mathcal{J}(L)$ iff E has the form $\overline{\{\omega_k\}}$ for some $\omega_k \in \Omega$. Moreover, $\#\overline{\{\omega_k\}} = (2^n - 1) - 2^k$. These facts lead us to the following result.

Theorem 21. *Given the arrays $\mathtt{K}_i, \mathtt{K}_j, \mathtt{K}_m$ where $i, j, m \in I$, there is an effective procedure that can decide $K_m = D_{\{i,j\}}$ in time $O(n^2)$ where $n = |\Omega|$.*

Proof. Let $L = (\mathcal{P}(\Omega), \supseteq)$. We have $K_m = D_{\{i,j\}}$ iff $K_m = K_i \sqcap_{\mathcal{E}(L)} K_j$ (Theorem 20) iff $K_m(E) = K_i(E) \cup K_j(E)$ for every $E \in \mathcal{J}(L)$ (Lemma 6). Furthermore, $E \in \mathcal{J}(L)$ iff $E = \overline{\{\omega\}}$ for some $\omega \in \Omega$. Then we can conclude that $E \in \mathcal{J}(L)$ iff $\#E = (2^n - 1) - 2^k$ for some $k \in [0..n-1]$. Therefore, $K_m = D_{\{i,j\}}$ iff for every $k \in [0..n-1]$

$$\mathtt{K}_m[\ p_k\ ] \ = \ \mathtt{K}_i[\ p_k\ ] \ \mid \ \mathtt{K}_j[\ p_k\ ] \tag{8}$$

where $p_k = (2^n - 1) - 2^k$ and $\mid$ is the OR operation over the bitwise representation of $\mathtt{K}_i[\ p_k\ ]$ and $\mathtt{K}_i[\ p_k\ ]$. For each $k \in [0..n-1]$, the equality test and the OR operation in Eq. 8 can be computed in $O(n)$. Hence the total cost is $O(n^2)$. $\square$

From Accessibility Relations. A very natural encoding for accessibility relations is to use a binary $n \times n$ matrix. If the input is encoded using three matrices $\mathtt{M}_i, \mathtt{M}_j$ and $\mathtt{M}_m$, we can test whether $\mathcal{R}_m = \mathcal{R}_i \cap \mathcal{R}_j$ (a proxy for $K_m = D_{\{i,j\}}$) in $O(n^2)$ by checking pointwise if $\mathtt{M}_m[a, b] = \mathtt{M}_i[a, b] \cdot \mathtt{M}_j[a, b]$.

It suggests that for *AS* we can use a different encoding and check $\mathcal{R}_m = \mathcal{R}_i \cap \mathcal{R}_j$ practically in *linear time*: More precisely in $O(\alpha_n n)$ where α_n is the inverse of the Ackermann function[2]. The key point is that the relations of AS are equivalences so they can be represented as *partitions*. The proof of the following result, which is interesting in its own right, shows an $O(n\alpha_n)$ procedure for deciding $\mathcal{R}_m = \mathcal{R}_i \cap \mathcal{R}_j$.

Theorem 22. *Let $\mathcal{R}_1, \mathcal{R}_2, \mathcal{R}_3 \subseteq \Omega^2$ be equivalences over a set Ω of $n = |\Omega|$ elements. There is an $O(\alpha_n n)$ algorithm for the following problem:*

> ***Input:*** *Each $\mathcal{R}_i$ in partition form, i.e. an array of disjoint arrays of elements of Ω, whose concatenation produces Ω. This is readable in $O(n)$.*
> ***Output:*** *Boolean answer to whether $\mathcal{R}_3 = \mathcal{R}_1 \cap \mathcal{R}_2$.*

Proof. We use the Disjoint-Sets data structure [8] whose details are included in the technical report https://hal.archives-ouvertes.fr/hal-03323638. We can view a disjoint-set as a function $r : I \to I$ that satisfies $r \circ r = r$ and can be evaluated at a particular index in $O(\alpha_n)$. The element $r(i)$ corresponds to the class representative of i for each $i \in I$, so that $i \sim_r j$ if and only if $r(i) = r(j)$.

If we let r_i denote a disjoint-set for $\mathcal{R}_i$ for each $i \in \{1, 2, 3\}$, and we let q denote the disjoint-set for $\mathcal{R}_1 \cap \mathcal{R}_2$, then the problem can be divided into computing the disjoint-set q in $O(n\alpha_n)$ and verifying whether $\sim_q = \sim_{r_3}$ also in $O(n\alpha_n)$. To organize these claims, let us consider the following algorithm descriptions.

> **Intersection.** Takes two disjoint-sets r_1 and r_2, and produces a disjoint-set q such that $i \sim_q j$ iff $i \sim_{r_1} j$ and $i \sim_{r_2} j$.
> **Canonical.** Takes a disjoint-set r and produces another $\hat{r}$ with $\sim_r = \sim_{\hat{r}}$, but such that $\hat{r}(i) \leq i$ for all $i \in I$.
> **Equality.** Takes two disjoint-sets r_1, r_2 and determines if $i \sim_{r_1} j$ iff $i \sim_{r_2} j$ for all $i, j \in I$. This problem is reduced simply to checking if $\hat{r}_1 = \hat{r}_2$.

We proceed to show that Algorithms 1 and 2 compute q and $\hat{r}$ (in array form) in $O(n\alpha_n)$. The complexity follows from the fact that they must read the input function(s) pointwise and all other operations are linear. It remains to show correctness only.

The array $\mathtt{g}$ in Algorithm 1 is any version of the inverse image of $\mathtt{f}$, i.e. $\mathtt{f}[\mathtt{g}[y]] = y$ for every $y \in Im(\mathtt{f})$. This guarantees $\mathtt{f} \circ \mathtt{g} \circ \mathtt{f} = \mathtt{f}$ and hence $\mathtt{q} \circ \mathtt{q} = \mathtt{g} \circ \mathtt{f} \circ \mathtt{g} \circ \mathtt{f} = \mathtt{g} \circ \mathtt{f} = \mathtt{q}$. Moreover, for any $i, j \in I$, $\mathtt{q}[i] = \mathtt{q}[j]$ iff $\mathtt{g}[\mathtt{f}[i]] = \mathtt{g}[\mathtt{f}[j]]$ by definition; iff $\mathtt{f}[i] = \mathtt{f}[j]$ because $\mathtt{f}$ is injective; iff $r_1(i) = r_1(j)$ and $r_2(i) = r_2(j)$; iff $i \sim_{r_1} j$ and $i \sim_{r_2} j$.

Regarding Algorithm 2, for all $i \in I$, $i \sim t[r(i)]$, thus $r(i) = r(t[r(i)])$. This is, $r = r \circ \mathtt{t} \circ r$. Thus, $\hat{\mathtt{r}} \circ \hat{\mathtt{r}} = \mathtt{t} \circ r \circ \mathtt{t} \circ r = \mathtt{t} \circ r = \hat{\mathtt{r}}$. Moreover, for any $i, j \in I$, $i \sim j$ iff $r(i) = r(j)$; iff $t[r(i)] = t[r(j)]$ since $\mathtt{t}$ is injective on J; iff $\hat{\mathtt{r}}[i] = \hat{\mathtt{r}}[j]$ by definition.

[2] $\alpha_n \stackrel{\text{def}}{=} \min\{k : A(k,k) \geq n\}$, where A is the Ackermann function. The growth of α_n is negligible in practice, e.g., $\alpha_n = 4$ for $n = 2^{2^{2^{65536}}} - 3$.

Algorithm 1. Intersection of disjoint sets in $O(n\alpha_n)$

procedure INTERSECTION(r_1, r_2)
  Let $\mathtt{f} : I \to I \times I$ be an array
  For each $i \in I$ do
    $\mathtt{f}[i] \leftarrow (r_1(i), r_2(i))$
  Let $\mathtt{g} : Im(\mathtt{f}) \to I$ be a hash map
  For each $i \in I$ do $\mathtt{g}[\mathtt{f}[i]] \leftarrow i$
  Let $\mathtt{q} : I \to I$ be an array
  For each $i \in I$ do $\mathtt{q}[i] \leftarrow \mathtt{g}[\mathtt{f}[i]]$
  return q

Algorithm 2. Equality of disjoint sets in $O(n\alpha_n)$

procedure CANONICAL(r)
  (Comment) $J \stackrel{\text{def}}{=} \{r(i) : i \in I\}$.
  Let $\mathtt{t} : J \to I$ be a hash map.
  For each $i \in I$ do $t[r(i)] \leftarrow r(i)$.
  For each $i \in I$ do
    $t[r(i)] \leftarrow \min(t[r(i)], i)$
  Let $\hat{\mathtt{r}} : I \to I$ be an array
  For each $i \in I$ do $\hat{\mathtt{r}}[i] \leftarrow t_{r(i)}$
  return $\hat{\mathtt{r}}$

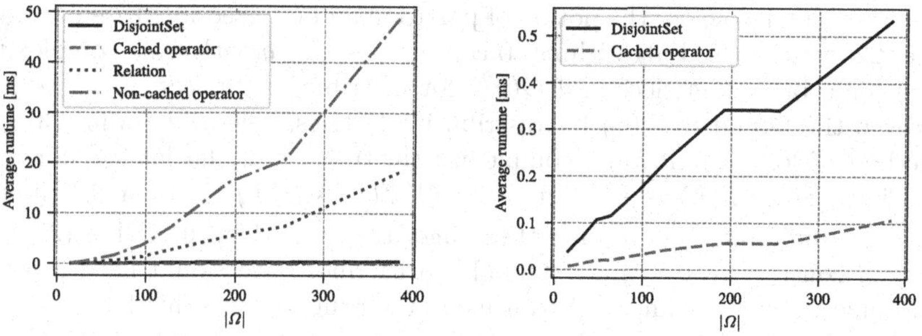

Fig. 3. Runtime comparison of several algorithms that solve the distributed knowledge problem.

Experimental Results. Figure 3 shows the average runtime (100 random executions) of the four algorithms listed below for the distributed knowledge problem. Fixing the number of elements $n = |\Omega|$ elements, the input for each execution consisted of three randomly generated partitions P_i, P_j and P_m. The first two are generated independently and uniformly over the set of all possible partitions of n elements. The third, P_m, corresponds with 50% probability to the intersection of the relations of the first two, and to a different but very similar partition otherwise, so as to increase the problem difficulty.

1. The "Cached operator" algorithm is the one described in Theorem 21. It assumes that the input knowledge operators can be evaluated in $O(1)$ at any join-irreducible input $E \subseteq \Omega$. Its complexity is $O(n^2)$, because bit-mask operations are linear w.r.t. the number of bits. However, this is compensated heavily in practice by the speed of bit-masking operations, at least for the sizes depicted.
2. The "Disjoint set" algorithm is the one described in Theorem 22 ($O(n\alpha_n)$). It takes the accessibility relations in partition form as input.
3. The "Relation" algorithm ($O(n^2)$) takes as input the accessibility relations in the form of $n \times n$ binary matrices, and simply verifies if the pointwise-and matches.

4. The "Non-cached operator" ($O(n^2)$) algorithm is that of the "Cached operator" when the cost of evaluating $K_i(\cdot)$ is taken into account. It shows that although the "Cached operator" algorithm is very fast, its speed depends heavily on the assumption that the knowledge operators are pre-computed.

7 Concluding Remarks and Related Work

We have used some standard tools from lattice theory to characterize the notion of distributed knowledge and provide efficient procedures to compute the meet of join-endomorphisms. Furthermore, we provide an algorithm to compute the intersection of *partitions* of a set of size n in $O(n\alpha_n)$. As illustrated in the introduction, this algorithm may have applications for graph connected components and other domains where the notion of partition and intersection arise naturally.

In [22] we proposed algorithms to compute $f \sqcap_{\mathcal{E}(L)} g$ with time complexities $O(n^3)$ for arbitrary lattices and $O(n^2)$ for distributive lattices. Here we have improved the bound to $O(n)$ for distributive lattices. The authors in [13] gave a method of logarithmic time complexity (in the size of the lattice) for meet operations. Since $\mathcal{E}(L)$ is isomorphic to $\mathcal{O}(\mathcal{J}(L) \times \mathcal{J}(L)^{op})$ for a distributive lattice L, finding $f \sqcap_{\mathcal{E}(L)} g$ with their algorithm would be in $O(\log_2(2^{n^2})) = O(n^2)$ in contrast to our linear bound. Furthermore, we would need a lattice isomorphic to $\mathcal{E}(L)$ to find $f \sqcap_{\mathcal{E}(L)} g$ using their algorithm. This lattice can be exponentially bigger than L [22] which is the input to our algorithm. We also provided experimental results illustrating the performance of our procedures. We followed the work in [16] for generating random distributive lattices.

The finite representation results we used in Sects. 4 and 5 to obtain our main results are adaptations from standard results from duality theory. Jónsson and Tarski [17,19] originally presented an extension of boolean algebras with operators (BAO), called canonical extensions, provided with some representation theorems. Roughly speaking, the representation theorems state that (1) every relation algebra is isomorphic to a complete and atomic relation algebra and (2) every boolean algebra with operators is isomorphic to a complex algebra that is complete and atomic. The idea behind this result, as was presented later by Kripke in [20], basically says that the operators can be recovered from certain binary relations and vice versa. Another approach to this duality was given by Goldblatt [11] where it is stated that the variety of normal modal algebras coincides with the class of subalgebras defined on the class of all frames. Canonical extensions have been useful for the development of duality and algebra. Jónsson proved an important result for modal logic in [18] and the authors of [6,9,10] have generalized canonical extensions for BAOs to distributive and arbitrary bounded lattices and posets.

Distributed knowledge was introduced in [15] and various axiomatization and expressiveness for it have been provided, e.g., in [1,14]. In terms of computational complexity, the satisfiability problem for epistemic logic with distributed knowledge ($S5^D$) has been shown to be PSPACE-complete [7]. Nevertheless, we are not aware of any lattice theoretical characterization of distributed knowledge nor algorithms to decide if an agent has the distributed knowledge of others.

Acknowledgments. We are indebted to the anonymous referees and editors of RAMiCS 2021 for helping us to improve the overall quality of the paper. We thank Luigi Santocanale for his constructive comments.

References

1. Ågotnes, T., Wáng, Y.N.: Resolving distributed knowledge. Artif. Intell. **252**, 1–21 (2017)
2. Aumann, R.J.: Agreeing to disagree. Ann. Stat. **4**, 1236–1239 (1976)
3. Bloch, I., Heijmans, H., Ronse, C.: Mathematical morphology. In: Handbook of Spatial Logics. pp. 857–944. Springer Netherlands (2007). https://doi.org/10.1007/978-1-4020-5587-4_14
4. Chagrov, A., Zakharyaschev, M.: Modal Logic, vol. 35. Oxford University Press, New York (1997)
5. Davey, B.A., Priestley, H.A.: Introduction to Lattices and Order, 2nd edn. Cambridge University Press, Cambridge (2002)
6. Dunn, J.M., Gehrke, M., Palmigiano, A.: Canonical extensions and relational completeness of some substructural logics. J. Symbolic Logic **70**(3), 713–740 (2005)
7. Fagin, R., Halpern, J.Y., Moses, Y., Vardi, M.Y.: Reasoning about Knowledge, 4th edn. MIT press, Cambridge (1995)
8. Galler, B.A., Fisher, M.J.: An improved equivalence algorithm. Commun. ACM **7**(5), 301–303 (1964)
9. Gehrke, M., Harding, J.: Bounded lattice expansions. J. Algebra **238**(1), 345–371 (2001)
10. Gehrke, M., Jónsson, B.: Bounded distributive lattice expansions. Mathematica Scandinavica **94**(1), 13–45 (2004). http://www.jstor.org/stable/24493402
11. Goldblatt, R.: Varieties of complex algebras. Ann. Pure Appl. Logic **44**(3), 173–242 (1989). https://doi.org/10.1016/0168-0072(89)90032-8
12. Grätzer, G., Schmidt, E.: On the lattice of all join-endomorphisms of a lattice. Proc. Am. Math. Soc. **9**, 722–722 (1958)
13. Habib, M., Nourine, L.: Tree structure for distributive lattices and its applications. Theor. Comput. Sci. **165**(2), 391–405 (1996)
14. Hakli, R., Negri, S.: Proof theory for distributed knowledge. In: Sadri, F., Satoh, K. (eds.) CLIMA 2007. LNCS (LNAI), vol. 5056, pp. 100–116. Springer, Heidelberg (2008). https://doi.org/10.1007/978-3-540-88833-8_6
15. Halpern, J.Y., Moses, Y.: Knowledge and common knowledge in a distributed environment. J. ACM **37**(3), 549–587 (1990)
16. Jipsen, P., Lawless, N.: Generating all finite modular lattices of a given size. Algebra Universalis **74**(3), 253–264 (2015)
17. Jónnson, B., Tarski, A.: Boolean algebras with operators, Part II. Am. J. Math. **74**(1), 127–162 (1952). http://www.jstor.org/stable/2372074
18. Jónsson, B.: On the canonicity of Sahlqvist identities. Studia Logica **53**(4), 473–491 (1994)
19. Jónsson, B., Tarski, A.: Boolean algebras with operators. Part I. Am. J. Mathematics **73**(4), 891–939 (1951), http://www.jstor.org/stable/2372123

20. Kripke, S.A.: A completeness theorem in modal logic. J. Symbolic Logic **24**(1), 1–14 (1959)
21. McKinsey, J.C.C., Tarski, A.: On closed elements in closure algebras. Ann. Math. **47**(1), 122–162 (1946). http://www.jstor.org/stable/1969038
22. Quintero, S., Ramirez, S., Rueda, C., Valencia, F.: Counting and computing join-endomorphisms in lattices. In: Fahrenberg, U., Jipsen, P., Winter, M. (eds.) RAMiCS 2020. LNCS, vol. 12062, pp. 253–269. Springer, Cham (2020). https://doi.org/10.1007/978-3-030-43520-2_16
23. Samet, D.: Agreeing to disagree: the non-probabilistic case. Games Econ. Behav. **69**(1), 169–174 (2010). https://doi.org/10.1016/j.geb.2008.09.032
24. Samet, D.: S5 knowledge without partitions. Synthese **172**(1), 145–155 (2010)

Relational Sums and Splittings in Categories of *L*-fuzzy Relations

Michael Winter(✉)

Department of Computer Science, Brock University, St. Catharines, ON L2S 3A1, Canada
mwinter@brocku.ca

Abstract. Dedekind categories and similar structures provide a suitable framework to reason about binary relations in an abstract setting. Arrow categories extend this theory by certain operations and axioms so that additional aspects of *L*-fuzzy relations become expressible. In particular, arrow categories allow to identify crisp relations among all relations. On the other hand, the new operations and axioms in arrow categories force the category to be uniform, i.e., to be within a particular subclass of Dedekind categories. As an extension, arrow categories inherit constructions from Dedekind categories such as the definition of relational sums and splittings. However, these constructions are usually modified in arrow categories by requiring that certain relations are additionally crisp. This additional crispness requirement and the fact that the category is uniform raises a general question about these constructions in arrow categories. When can we guarantee the existence of the construction with and without the additional requirement of crispness in the given arrow category or an extension thereof? This paper provides a complete answer to this complex question for the two constructions mentioned.

1 Introduction

Allegories and Dedekind categories, in particular, provide a suitable framework to reason about binary relations [1,3–5]. Typically one is interested in an allegory that provides some additional constructions. For example, the relational sum is an abstract version of the disjoint union of two sets and is based on the two injections ι and κ mapping elements from the two separate objects into the sum. This construction can be used to model computations on two separated parts of the input by one relation. Another construction of interest is a splitting. This construction can be used to model a subset of a given domain as well as the set of equivalence classes induced by an equivalence relation.

Beside set-theoretic binary relations also so-called *L*-fuzzy relations establish a Dedekind category, i.e., Dedekind categories cover also relations that use elements from a complete Heyting algebra *L* as truth or membership values instead of the Boolean values true and false. However, the theory of Dedekind categories does not allow to characterize crisp relations among *L*-fuzzy relations, i.e., those relations that only rely on the smallest element 0 (false) and the greatest element 1 (true) of *L* [8]. Because of this

M. Winter—The author gratefully acknowledges support from the Natural Sciences and Engineering Research Council of Canada (283267).

U. Fahrenberg et al. (Eds.): RAMiCS 2021, LNCS 13027, pp. 433–447, 2021.
https://doi.org/10.1007/978-3-030-88701-8_26

reason arrow categories were introduced. These categories add two arrow operations to theory returning the support and kernel of a L-fuzzy relation. Similar to Dedekind categories one is typically interested in an arrow category that provides some additional constructions such as the relational sum and splittings mentioned above. However, the original definition of a relational sum does not require that the corresponding injections ι and κ are crisp. Even though relational sums are unique up to isomorphism we may have two relational sums in one arrow category, one with crisp injections and one with non-crisp injections. This is possible because the notion of crispness is not invariant under isomorphisms [8]. Now the following question arises. If an arrow category has a relational sum for two objects, does it also have a relational sum of these objects with crisp injections? And if not, is there an extension of the category that will have such a crisp version? The same question can also be asked for a splitting of a crisp partial equivalence relation.

The construction of a splitting raises an additional question. Arrow categories model the so-called fixed basis approach to fuzziness, i.e., all relations in a given arrow category use the same lattice of truth values. Algebraically this implies that arrow categories are uniform, i.e., the arrow category seen as a Dedekind category satisfies an additional axiom preventing that some partial equivalence relations split. It would be interesting to characterize those partial equivalence relations that may split in arrow categories. In this paper we will provide a complete answer to the questions mentioned above.

The paper is organized as follows. In Sect. 2 we will recall the basic concepts from categories, Dedekind categories, and arrow categories. Section 3 will cover the relational sum. We will show that an arrow category may have a relational sum without providing a crisp version thereof. In addition, we will show that any arrow category can be embedded into a category of matrices that provides a relational sum with crisp injections. In Sect. 4 we will first investigate the class of partial equivalence relation that may split in an arrow category. In order to do so we will introduce the notion of a pseudo-crisp relation. After that we proceed similar to Sect. 3 and show that an arrow category may have a splitting of a crisp partial equivalence relation without providing a crisp version thereof. Finally, we will show that any arrow category can be embedded into a larger category that provides a crisp splitting of a crisp partial equivalence relation. Section 5 will outline some future work.

2 Mathematical Preliminaries

In this section we want to provide the mathematical notions used in this paper. Therefore we recall some basic notions from category theory and introduce Dedekind and arrow categories [1,8,9].

We will write $R : A \to B$ to indicate that a morphism R of a category C has source A and target B. Composition and the identity morphism are denoted by ; and $\mathbb{I}_A$, respectively. We will use composition from left to right, i.e., $Q;R$ means first Q and then R.

Definition 1. *A Dedekind category $\mathcal{R}$ is a category satisfying the following:*

1. *For all objects A and B the collection $\mathcal{R}[A, B]$ is a complete Heyting algebra. Meet, join, the implication operation, the induced ordering, the least and the greatest element are denoted by* $\sqcap, \sqcup, \rightarrow, \sqsubseteq, \perp\!\!\!\perp_{AB}, \top\!\!\!\top_{AB}$, *respectively.*
2. *There is a monotone operation* $\breve{}$ *(called converse) mapping a relation* $Q : A \to B$ *to* $Q^{\smile} : B \to A$ *such that for all relations* $Q : A \to B$ *and* $R : B \to C$ *the following holds:* $(Q;R)^{\smile} = R^{\smile};Q^{\smile}$ *and* $(Q^{\smile})^{\smile} = Q$.
3. *For all relations* $Q : A \to B, R : B \to C$ *and* $S : A \to C$ *the modular law* $(Q;R) \sqcap S \sqsubseteq Q;(R \sqcap (Q^{\smile};S))$ *holds.*
4. *For all relations* $R : B \to C$ *and* $S : A \to C$ *there is a relation* $S/R : A \to B$ *(called the left residual of S and R) such that for all* $X : A \to B$ *the following holds:* $X;R \sqsubseteq S \Leftrightarrow X \sqsubseteq S/R$.

Given a complete Heyting algebra L, an L-fuzzy relation R between two sets A and B is a function $A \times B \to L$ assigning to each pair of A and B elements a degree of membership in R. A meet and a join operation on L-fuzzy relations can be defined component-wise using the corresponding operations on L. Together with the usual converse operation and composition defined by

$$(Q;R)(a,c) = \bigsqcup_{b \in B} Q(a,b) \sqcap R(b,c),$$

we obtain the Dedekind category **Rel**(L) of sets and L-fuzzy relations.

If the sets A and B are finite we will present relations Q between A and B often as matrices labeled by the elements from A (rows) and B (columns). The entries of the matrix are taken from L so that an entry u in the row labeled a and column labeled b simply indicates $Q(a,b) = u$. Please note that the lattice operation on relations correspond to component-wise operations on the matrices and that composition is matrix multiplication based on meet and join (instead of multiplication and addition). For example, if we use the linear lattice $L_3 = \{0, a, 1\}$ with three elements and $A = \{0, 1, 2, 3, 4\}$, then we may define $Q : A \to A$ as the relation $Q(x,y) = 1$ if $y = (x+1) \mod 5$, $Q(x,y) = a$ if $y = (x+2) \mod 5$ and $Q(x,y) = 0$ otherwise. The matrix of Q and $Q;Q$ are given resp. computed as:

$$Q = \begin{array}{c} \\ 0\\1\\2\\3\\4 \end{array}\!\!\begin{array}{c} 0\;1\;2\;3\;4 \\ \begin{pmatrix} 0&1&a&0&0\\0&0&1&a&0\\0&0&0&1&a\\a&0&0&0&1\\1&a&0&0&0 \end{pmatrix}\end{array}, \quad Q;Q = \begin{pmatrix} 0&1&a&0&0\\0&0&1&a&0\\0&0&0&1&a\\a&0&0&0&1\\1&a&0&0&0 \end{pmatrix};\begin{pmatrix} 0&1&a&0&0\\0&0&1&a&0\\0&0&0&1&a\\a&0&0&0&1\\1&a&0&0&0 \end{pmatrix} = \begin{pmatrix} 0&0&1&a&a\\a&0&0&1&a\\a&a&0&0&1\\1&a&a&0&0\\0&1&a&a&0 \end{pmatrix}.$$

We obtain a special case of **Rel**(L) if we choose the Heyting algebra L to be the 2-element Boolean algebra $\mathbb{B}$ of the truth value. The Dedekind category **Rel**($\mathbb{B}$) is actually isomorphic to the category **Rel** of sets and binary relations. We will identify the two categories and see **Rel** as a special case of a Dedekind category of L-fuzzy relations.

Throughout the paper we will use the axioms and some basics facts such as monotonicity of the operations without mentioning. In addition, we will need the notion of the domain of a relation occasionally. The domain of a relation $Q : A \to B$ is the partial identity given by the elements where Q is defined. Relation-algebraically we define

$\mathrm{dom}(Q) = \mathbb{I}_A \sqcap Q; Q^{\smile}$ and $\mathrm{cod}(Q) = \mathrm{dom}(Q^{\smile}) = \mathbb{I}_B \sqcap Q^{\smile}; Q$. A proof of the following lemma can be found in [1,3–5].

Lemma 1. *Suppose $\mathcal{R}$ is a Dedekind category, $Q : A \to B$ and $R : B \to C$. Then we have:*

1. $\mathrm{dom}(Q) = \mathbb{I}_A \sqcap Q; \mathbb{T}_{BA}$.
2. $\mathrm{dom}(Q); Q = Q$.
3. $\mathrm{dom}(Q; R) \sqsubseteq \mathrm{dom}(Q)$.
4. *If* $i \sqsubseteq \mathbb{I}_A$, *then* $\mathrm{dom}(i) = i$.

It is possible to characterize the lattice of membership degrees used by relations between two objects of an arbitrary Dedekind category. In this paper we use the notion scalar relation that was introduced by Furusawa and Kawahara [2].

Definition 2. *Suppose $\mathcal{R}$ is a Dedekind category. A relation $\alpha : A \to A$ is called a scalar on A iff $\alpha \sqsubseteq \mathbb{I}_A$ and $\mathbb{T}_{AA}; \alpha = \alpha; \mathbb{T}_{AA}$.*

In the case of L-fuzzy relations a scalar relation α is a partial identity for which we have $\alpha(a, a) = u$ for some fixed u from L. Therefore, any scalar corresponds to exactly one element from L. Obviously in **Rel** we have exactly two scalars $\perp\!\!\!\perp_{AA}$ and $\mathbb{I}_A$ verifying that **Rel** is based on the Boolean algebra of the truth values.

L-fuzzy relations that only use the smallest and greatest element 0 and 1 as truth values are called crisp. These relations correspond to the relations from **Rel** in an obvious manner. However, crisp relations cannot be characterized within an arbitrary Dedekind category [8]. Therefore the following definition introduces arrow categories. Arrow categories add two operations to Dedekind categories. The relation $R^{\uparrow}$ is the smallest crisp relation that contains R (also called support), and $R^{\downarrow}$ is the greatest crisp relation included in R (also called kernel) [7–9].

Definition 3. *An arrow category $\mathcal{A}$ is a Dedekind category with $\mathbb{T}_{AB} \neq \perp\!\!\!\perp_{AB}$ for all A, B and two operations $\uparrow$ and $\downarrow$ satisfying:*

1. $R^{\uparrow}, R^{\downarrow} : A \to B$ *for all* $R : A \to B$.
2. $(^{\uparrow}, ^{\downarrow})$ *forms a Galois correspondence, i.e.,* $Q^{\uparrow} \sqsubseteq R$ *iff* $Q \sqsubseteq R^{\downarrow}$ *for all* $Q, R : A \to B$.
3. $(R^{\smile}; S^{\downarrow})^{\uparrow} = R^{\uparrow\smile}; S^{\downarrow}$ *for all* $R : B \to A$ *and* $S : B \to C$.
4. $(Q \sqcap R^{\downarrow})^{\uparrow} = Q^{\uparrow} \sqcap R^{\downarrow}$ *for all* $Q, R : A \to B$.
5. *If* $\alpha_A \neq \perp\!\!\!\perp_{AA}$ *is a non-zero scalar then* $\alpha_A^{\uparrow} = \mathbb{I}_A$.

A relation in an arrow category that satisfies $R^{\uparrow} = R$, or equivalently $R^{\downarrow} = R$, is called crisp. Notice that this abstract notion of crispness is equivalent to the notion of crispness defined above for concrete L-fuzzy relations.

Given an arrow category $\mathcal{A}$ it can be shown that the set of scalar relations Scalar(A) for each object A forms a complete Heyting algebra [8]. Furthermore, these Heyting algebras are all isomorphic. Together with the observation that scalars correspond to exactly one element of the underlying lattice in the case of L-fuzzy relations we will say that $\mathcal{A}$ is an arrow category over L if L is isomorphic to Scalar(A) for an (all) object A in $\mathcal{A}$.

It is important to mention that the identity, the smallest, and the greatest relation are crisp, and that crisp relations are closed under all operations of a Dedekind category, i.e., the crisp relations form a sub-Dedekind category.

The following lemma will be needed several times in Theorem 5 showing that any arrow category can be embedded into an arrow category in which all crisp partial equivalence relation split.

Lemma 2. *Suppose $\mathcal{A}$ is an arrow category, $Q : A \to B$ is crisp, and $R : B \to C$. Then we have:*

1. $Q;R^{\downarrow} \sqsubseteq (Q;R)^{\downarrow}$.
2. *If Q is, in addition, univalent, i.e., $Q^{\smile};Q \sqsubseteq \mathbb{I}_B$, then the inclusion in 1. is an equation.*

Proof. 1. First of all, we have

$$\begin{aligned}(Q;R^{\downarrow})^{\uparrow} &= Q^{\uparrow};R^{\downarrow} \\ &= Q;R^{\downarrow} && Q \text{ crisp} \\ &\sqsubseteq Q;R.\end{aligned}$$

This immediately implies the assertion.

2. We only have to show the inclusion $\sqsupseteq$. We compute

$$\begin{aligned}Q^{\smile};(Q;R)^{\downarrow} &\sqsubseteq (Q^{\smile};Q;R)^{\downarrow} && \text{by 1. since} Q^{\smile} \text{ is crisp} \\ &\sqsubseteq R^{\downarrow}. && Q \text{ univalent}\end{aligned}$$

This implies

$$\begin{aligned}(Q;R)^{\downarrow} &= \mathrm{dom}((Q;R)^{\downarrow});(Q;R)^{\downarrow} && \text{Lemma 1(2)} \\ &\sqsubseteq \mathrm{dom}(Q;R);(Q;R)^{\downarrow} \\ &\sqsubseteq \mathrm{dom}(Q);(Q;R)^{\downarrow} && \text{Lemma 1(3)} \\ &\sqsubseteq Q;Q^{\smile};(Q;R)^{\downarrow} \\ &\sqsubseteq Q;R^{\downarrow}, && \text{see above}\end{aligned}$$

i.e., the inclusion $\sqsupseteq$. □

3 Relational Sums

We start our investigation with the so-called relational sum. This construction is an abstract version of the disjoint union of two sets.

Definition 4. *Suppose $\mathcal{R}$ is a Dedekind category. The relational sum of two objects A and B is an object $A+B$ together with two relations $\iota : A \to A+B$ and $\kappa : B \to A+B$ so that the following equations hold*

$$\iota;\iota^{\smile} = \mathbb{I}_A, \quad \kappa;\kappa^{\smile} = \mathbb{I}_B, \quad \iota;\kappa^{\smile} = \perp\!\!\!\perp_{AB}, \quad \iota^{\smile};\iota \sqcup \kappa^{\smile};\kappa = \mathbb{I}_{A+B}.$$

For the purpose of this paper we will call a relational sum in an arrow category $\mathcal{A}$ crisp if the two injections are crisp relations, i.e., if we have $\iota^{\downarrow} = \iota$ and $\kappa^{\downarrow} = \kappa$.

A relational sum is unique up to isomorphism. In fact, a relational sum is a biproduct in $\mathcal{R}$ and a coproduct in the subcategory of maps. In the concrete Dedekind categories **Rel**(L) a crisp relational sum is given by the disjoint union of two sets together with the corresponding crisp injections

$$\iota(a,c) = \begin{cases} 1 \text{ if } c = a \\ 0 \text{ otherwise} \end{cases}, \quad \kappa(b,c) = \begin{cases} 1 \text{ if } c = b \\ 0 \text{ otherwise} \end{cases}$$

with $a \in A, b \in B$, and $c \in A + B$. This indicates that the uniformity of arrow categories does not restrict the existence of (crisp) relational sums. On the other hand, **Rel**(L) may also have a relational sum of A and B where the injections are not crisp. This is possible because the notion of crispness is not preserved by (internal) isomorphisms [8]. When we now consider a suitable substructure of **Rel**(L) the situation can get even worse in the sense that the resulting category has relational sums of A and B but none of them has crisp injections as the following example shows.

Example 1. We will use the Boolean algebra $B_4 = \{0, a, b, 1\}$ with four elements as the lattice L and construct $\mathcal{A}_1$ as a substructure of **Rel**(B_4). Please note that in **Rel**(B_4) the set of relations between two objects is a Boolean algebra, i.e., has complements. The objects of $\mathcal{A}_1$ are the two sets $A = \{*\}$ and $B = \{*_1, *_2\}$. Please note that B can be seen as the disjoint union of A with itself, i.e., B together with the suitable crisp injections is a crisp relational sum of A and A in **Rel**(B_4). Furthermore, any relational sum of A with A must use B as the object due to fact that the relational sum is unique up to isomorphism. Now, consider the relations $\iota, \kappa : A \to B$ defined by

$$\iota = \begin{array}{c} \\ * \end{array}\!\begin{array}{c} *_1\ *_2 \\ \left[\,a\ \ b\,\right] \end{array}, \quad \kappa = \begin{array}{c} \\ * \end{array}\!\begin{array}{c} *_1\ *_2 \\ \left[\,b\ \ a\,\right] \end{array}.$$

It is easy to verify that these relations form a relational sum. For example, using the fact that a is the complement of b in B_4 we obtain

$$\begin{aligned}
\iota;\iota^{\smile} &= \begin{array}{c} \\ * \end{array}\!\begin{array}{c} *_1\ *_2 \\ \left[\,a\ \ b\,\right] \end{array};\begin{array}{c} \\ *_1 \\ *_2 \end{array}\!\begin{array}{c} * \\ \left[\begin{array}{c} a \\ b \end{array}\right] \end{array} \\
&= \begin{array}{c} \\ * \end{array}\!\begin{array}{c} * \\ \left[\,1\,\right] \end{array} \\
&= \mathbb{I}_A, \\
\iota^{\smile};\iota \sqcup \kappa^{\smile};\kappa &= \begin{array}{c} \\ *_1 \\ *_2 \end{array}\!\begin{array}{c} * \\ \left[\begin{array}{c} a \\ b \end{array}\right] \end{array};\begin{array}{c} \\ * \end{array}\!\begin{array}{c} *_1\ *_2 \\ \left[\,a\ \ b\,\right] \end{array} \sqcup \begin{array}{c} \\ *_1 \\ *_2 \end{array}\!\begin{array}{c} * \\ \left[\begin{array}{c} b \\ a \end{array}\right] \end{array};\begin{array}{c} \\ * \end{array}\!\begin{array}{c} *_1\ *_2 \\ \left[\,b\ \ a\,\right] \end{array} \\
&= \begin{array}{c} \\ *_1 \\ *_2 \end{array}\!\begin{array}{c} *_1\ \ *_2 \\ \left[\begin{array}{cc} a & 0 \\ 0 & a \end{array}\right] \end{array} \sqcup \begin{array}{c} \\ *_1 \\ *_2 \end{array}\!\begin{array}{c} *_1\ \ *_2 \\ \left[\begin{array}{cc} b & 0 \\ 0 & b \end{array}\right] \end{array}
\end{aligned}$$

$$= \begin{matrix} & *_1 & *_2 \\ *_1 & 1 & 0 \\ *_2 & 0 & 1 \end{matrix}$$
$$= \mathbb{I}_B.$$

In addition, both relations are not crisp since $\iota^{\downarrow} = \kappa^{\downarrow} = \perp\!\!\perp_{AB}$. Now, we define $\mathcal{A}_1$ to be the smallest structure containing ι and κ that is closed under meet, join, complement, converse, composition, the arrow operations, and contains all constants related to the two objects A and B. This is well-defined since all ingredients we are working with are finite, i.e., we are working with a finite set of relations between finite sets with a finite lattice of truth values. Furthermore, the resulting structure will have relative pseudo-complements and residuals due to the finiteness of the structure. With other words, $\mathcal{A}_1$ is an arrow category. It turns out that $\mathcal{A}_1$ has a total of 26 relations split up as

$$|\mathcal{A}_1[A,A]| = 2, |\mathcal{A}_1[A,B]| = |\mathcal{A}_1[B,A]| = 4, |\mathcal{A}_1[B,B]| = 16.$$

In particular, we have $\mathcal{A}_1[A,B] = \{\perp\!\!\perp_{AB}, \top\!\!\top_{AB}, \iota, \kappa\}$. This shows that B together with ι and κ is the only relational sum in $\mathcal{A}_1$ of A with itself, i.e., $\mathcal{A}_1$ has a relational sum, but no crisp relational sum, of A with itself. □

The example above shows that we cannot always assume that there is a crisp version of a relational sum in a given arrow category. On the other hand, $\mathcal{A}_1$ is embedded in the larger arrow category **Rel**(B_4) which does have a crisp version. We now want to show that this is always the case so that requiring the injections to be crisp is not an actual restriction. We start with the following theorem embedding a Dedekind category $\mathcal{R}$ into a category of matrices over $\mathcal{R}$. A proof of this theorem in various versions can be found in [1,5,6].

Theorem 1. *Suppose $\mathcal{R}$ is a Dedekind category. Then the category of matrices $\mathcal{R}^+$ over $\mathcal{R}$ defined by*

1. *The objects of $\mathcal{R}^+$ are pairs (f, I) where I is a non-empty set and f is a function from I to the objects of $\mathcal{R}$.*
2. *Given two objects (f, I) and (g, J) a morphism in $\mathcal{R}^+$ from (f, I) to (g, J) is a function Q from $I \times J$ to the relations in $\mathcal{R}$ so that $Q(i, j) : f(i) \to g(j)$.*

is a Dedekind category. Furthermore, the functor $E : \mathcal{R} \to \mathcal{R}^+$ defined by $E(A) = (A, \{*\})$ for objects and $E(Q)(*, *) = Q$ is a full embedding of Dedekind categories.

We now extend the previous result to arrow categories.

Theorem 2. *Suppose $\mathcal{A}$ is an arrow category. Then $\mathcal{A}^+$ together with the operations $Q^{\downarrow}(i, j) = Q(i, j)^{\downarrow}$ and $Q^{\uparrow}(i, j) = Q(i, j)^{\uparrow}$ for $Q : (f, I) \to (g, J)$ is an arrow category. Furthermore, E is a full embedding of arrow categories.*

Proof. First of all, we have $\top\!\!\top_{(f,I)(g,J)} \neq \perp\!\!\perp_{(f,I)(g,J)}$ since I and J are not empty and we have $\top\!\!\top_{(f,I)(g,J)}(i, j) = \top\!\!\top_{f(i)g(j)} \neq \perp\!\!\perp_{f(i)g(j)} = \perp\!\!\perp_{(f,I)(g,J)}(i, j)$ for all $i \in I$ and $j \in J$. We now show the properties listed in Definition 3.

1. This is satisfied by definition.
2. We have

$$\begin{aligned} Q^{\uparrow}(i,j) \sqsubseteq R(i,j) &\Leftrightarrow (Q(i,j))^{\uparrow} \sqsubseteq R(i,j) \\ &\Leftrightarrow Q(i,j) \sqsubseteq (R(i,j))^{\downarrow} \\ &\Leftrightarrow Q(i,j) \sqsubseteq R^{\downarrow}(i,j) \end{aligned}$$

for all suitable i and j so that $Q^{\uparrow} \sqsubseteq R \Leftrightarrow Q \sqsubseteq R^{\downarrow}$ follows.

3-4. These properties are shown similarly to 2.

5. Suppose $\alpha : (f, I) \to (f, I)$ is a non-zero scalar. From $\alpha \sqsubseteq \mathbb{I}_{(f,I)}$ we obtain $\alpha(i, j) = \perp\!\!\!\perp_{f(i)f(j)}$ for $i \neq j$ and $\alpha(i, i) \sqsubseteq \mathbb{I}_{f(i)}$. Furthermore, for $i, j \in I$ we have

$$\begin{aligned} \alpha(i,i); \top\!\!\!\top_{f(i)f(j)} &= \alpha(i,i); \top\!\!\!\top_{(f,I)(f,I)}(i,j) \\ &= \bigsqcup_{k\in I} \alpha(i,k); \top\!\!\!\top_{(f,I)(f,I)}(k,j) && \alpha(i,k) = \perp\!\!\!\perp_{f(i)f(k)} \text{ for } i \neq k \\ &= (\alpha; \top\!\!\!\top_{(f,I)(f,I)})(i,j) \\ &= (\top\!\!\!\top_{(f,I)(f,I)}; \alpha)(i,j) && \alpha \text{ is a scalar} \\ &= \bigsqcup_{k\in I} \top\!\!\!\top_{(f,I)(f,I)}(i,k); \alpha(k,j) \\ &= \top\!\!\!\top_{(f,I)(f,I)}(i,j); \alpha(j,j) && \alpha(k,j) = \perp\!\!\!\perp_{f(k)f(j)} \text{ for } j \neq k \\ &= \top\!\!\!\top_{f(i)f(j)}; \alpha(j,j). \end{aligned}$$

For $i = j$ this implies that $\alpha(i, i)$ is a scalar. In addition, we have

$$\begin{aligned} \alpha(i,i) &= \mathrm{dom}(\alpha)(i,i) && \text{Lemma 1(4)} \\ &= (\mathbb{I}_{(f,I)} \sqcap \alpha; \top\!\!\!\top_{(f,I)(f,I)})(i,i) && \text{Lemma 1(1)} \\ &= \mathbb{I}_{(f,I)}(i,i) \sqcap (\alpha; \top\!\!\!\top_{(f,I)(f,I)})(i,i) \\ &= \mathbb{I}_{f(i)} \sqcap \bigsqcup_{j\in I} \alpha(i,j); \top\!\!\!\top_{f(j)f(i)} \\ &= \mathbb{I}_{f(i)} \sqcap \alpha(i,i); \top\!\!\!\top_{f(i)f(i)} && \alpha(i,j) = \perp\!\!\!\perp_{f(i)f(j)} \text{ for } i \neq j \\ &= \mathbb{I}_{f(i)} \sqcap \alpha(i,i); \top\!\!\!\top_{f(i)f(j)}; \top\!\!\!\top_{f(j)f(i)} && \mathcal{A} \text{ uniform} \\ &= \mathbb{I}_{f(i)} \sqcap \top\!\!\!\top_{f(i)f(j)}; \alpha(j,j); \top\!\!\!\top_{f(j)f(i)}. && \text{see above} \end{aligned}$$

If $\alpha(j, j) = \perp\!\!\!\perp_{f(j)f(j)}$ for some $j \in J$, then the computation above implies $\alpha(i, i) = \perp\!\!\!\perp_{f(i)f(i)}$ for every $i \in I$. The latter is impossible since α is non-zero, i.e., there is an $i \in I$ with $\alpha(i, i) \neq \perp\!\!\!\perp_{f(i)f(i)}$. Therefore, $\alpha(i, i)$ is a non-zero scalar for every $i \in I$, i.e., $\alpha(i, i)^{\uparrow} = \mathbb{I}_{f(i)}$. This implies

$$\begin{aligned} \alpha^{\uparrow}(i,j) &= \alpha(i,j)^{\uparrow} \\ &= \begin{cases} \perp\!\!\!\perp^{\uparrow}_{f(i)f(j))} & \text{if } i \neq j \\ \alpha(i,i)^{\uparrow} & \text{otherwise} \end{cases} \\ &= \begin{cases} \perp\!\!\!\perp_{f(i)f(j))} & \text{if } i \neq j \\ \mathbb{I}_{f(i)} & \text{otherwise} \end{cases} \\ &= \mathbb{I}_{(f,I)}(i,j), \end{aligned}$$

i.e., $\alpha^{\uparrow} = \mathbb{I}_{(f,I)}$.

Last but not least, we have to show that E preserves the arrow operations. We have

$$E(Q^{\downarrow})(*,*) = Q^{\downarrow} = E(Q)(*,*)^{\downarrow} = E(Q)^{\downarrow}(*,*),$$

and similar calculation for the support. □

It remains to show that $\mathcal{A}^{+}$ has crisp relational sums.

Theorem 3. *Suppose $\mathcal{A}$ is an arrow category. Then $\mathcal{A}^{+}$ has crisp relational sums for all pairs of objects.*

Proof. Suppose (f, I) and (g, J) are objects in $\mathcal{A}^{+}$. Then the pair $(h, I + J)$ where $I + J$ is the disjoint union of the sets I and J and $h : I + J \to \mathrm{Obj}_{\mathcal{A}}$ is defined by

$$h(x) = \begin{cases} f(x) \text{ if } x \in I \\ g(x) \text{ if } x \in J \end{cases}$$

is also an object of $\mathcal{A}^{+}$. Now, we define $\iota : (f, I) \to (h, I + J)$ and $\kappa : (g, J) \to (h, I + J)$ by

$$\iota(i,x) = \begin{cases} \mathbb{I}_{f(i)} & \text{if } x = i \\ \perp\!\!\!\perp_{f(i)h(x)} & \text{otherwise} \end{cases}, \quad \kappa(j,x) = \begin{cases} \mathbb{I}_{g(j)} & \text{if } x = j \\ \perp\!\!\!\perp_{g(j)h(x)} & \text{otherwise} \end{cases}.$$

The injections are crisp by definition. A proof of the remaining properties of a relational sum can be found in [1,5,6]. □

Please note that if $\mathcal{A}$ had a non-crisp relational sum $(A +_{\mathrm{nc}} B, \iota_{\mathrm{nc}}, \kappa_{\mathrm{nc}})$ of A and B, then $A +_{\mathrm{nc}} B$ (after embedding into $\mathcal{A}^{+}$) is isomorphic to the crisp relational sum $(A +_{\mathrm{c}} B, \iota_{\mathrm{c}}, \kappa_{\mathrm{c}})$ from the theorem above. This isomorphism $i : A +_{\mathrm{c}} B \to A +_{\mathrm{nc}} B$ cannot be crisp, because if i would be crisp, then we would obtain from $\iota_{\mathrm{c}}; i = \iota_{\mathrm{nc}}$ that ι_{nc} is crisp since crisp relations are closed under composition. In Example 1 we embed $\mathcal{A}_1$ into $\mathbf{Rel}(B_4)$ and obtain the isomorphism $i : B \to B$ as

$$i = \begin{array}{c} \\ *_1 \\ *_2 \end{array}\!\!\begin{array}{c} \;\;*_1 \;\; *_2 \\ \begin{bmatrix} a & b \\ b & a \end{bmatrix} \end{array}.$$

4 Splittings

There is already a well developed theory on splitting of idempotents in categories. In allegories it is important to restrict ourself to symmetric idempotent relations, or partial equivalence relations, because the symmetry guarantees the existence of the converse operation in $\mathcal{R}_{\mathcal{E}}$ (see Theorem 4).

A splitting is a combination of a subset and forming equivalence classes. It is based on a given partial equivalence relation, i.e., a relation that is idempotent and symmetric.

Definition 5. *Suppose $\mathcal{R}$ is a Dedekind category, and $Q : A \to A$ is a partial equivalence relation, i.e., we have $Q; Q = Q$ and $Q^{\smile} = Q$. The splitting of Q is an object B together with a relation $R : B \to A$ so that*

$$R; R^{\smile} = \mathbb{I}_B, \quad R^{\smile}; R = Q.$$

In **Rel** every partial equivalence relation splits. The object B is the set of all equivalence classes of elements for which Q is defined. In more details, if $x \in A$ we define the equivalence class of x by $[x] = \{y \in A \mid Q(x, y)\}$. Then $B = \{[x] \mid x \in A \text{ and } [x] \neq \emptyset\}$ and $R : B \to A$ is defined by $R([x], y)$ iff $y \in [x]$ iff $Q(x, y)$.

Since **Rel** without the empty set as an object is isomorphic to the subcategory of crisp relations in **Rel**(L) for any L, the same construction can be used in order to show that every non-empty crisp partial equivalence relation in **Rel**(L) has a splitting. Unfortunately, we will also have partial equivalence relations that do not split if the lattice L is bigger than $\mathbb{B}$. This is a consequence of the following lemma.

Lemma 3. *Suppose $\mathcal{A}$ is an arrow category, and $\alpha : A \to A$ is an ideal relation, i.e., $\mathbb{T}_{AA}; \alpha; \mathbb{T}_{AA} = \alpha$. If a relation $R : B \to A$ splits α, then $\alpha = \mathbb{T}_{AA}$.*

Proof. Assume $R : B \to A$ splits α, i.e., we have $R; R^{\smile} = \mathbb{I}_B$ and $R^{\smile}; R = \alpha$. Then we compute

$$\begin{aligned}
\alpha &= \mathbb{T}_{AA}; \alpha; \mathbb{T}_{AA} \\
&= \mathbb{T}_{AB}; \mathbb{T}_{BA}; \alpha; \mathbb{T}_{AB}; \mathbb{T}_{BA} && \mathcal{A} \text{ uniform} \\
&= \mathbb{T}_{AB}; \mathbb{T}_{BA}; R^{\smile}; R; \mathbb{T}_{AB}; \mathbb{T}_{BA} && R \text{ splits } \alpha \\
&= \mathbb{T}_{AB}; \mathbb{T}_{BB}; \mathbb{T}_{BB}; \mathbb{T}_{BA} && R \text{ total} \\
&= \mathbb{T}_{AB}; \mathbb{T}_{BA} \\
&= \mathbb{T}_{AA}, && \mathcal{A} \text{ uniform}
\end{aligned}$$

verifying that $\alpha = \mathbb{T}_{AA}$. □

First of all, the previous lemma shows that $\perp\!\!\!\perp_{AA}$ does not split in any arrow category. Ultimately, this is a consequence of the axiom $\mathbb{T}_{AB} \neq \perp\!\!\!\perp_{AB}$ for all objects A and B of arrow categories that prevents the empty set to be an object of a concrete arrow category. However, $\perp\!\!\!\perp_{AA}$ is not the only relation that does not split in arrow categories. If the lattice L has an element a with $a \neq 0$ and $a \neq 1$, then the ideal relation $\alpha(x, y) = a$ does not split in **Rel**(L) either due to the lemma above.

We would now like to characterize the class of partial equivalence relations that may have a splitting in an arrow category.

Definition 6. *Let $\mathcal{A}$ be an arrow category. A relation $Q : A \to B$ is called pseudo crisp (p-crisp) iff there are relations $R : A' \to A$ and $S' : B' \to B$ with $R^{\smile}; R = Q; Q^{\smile}$ and $S^{\smile}; S = Q^{\smile}; Q$ so that $R; Q; S^{\smile}$ is crisp.*

With other words, a relation is p-crisp if it can be transformed into a crisp relation with equivalent domain and codomain.

Lemma 4. *Let $\mathcal{A}$ be an arrow category. Then we have:*

1. *Every crisp relation is p-crisp.*
2. *If $Q : A \to A$ is a partial equivalence relation, then Q is p-crisp iff there is a relation $R : A' \to A$ with $R^{\smile}; R = Q$ so that $R; Q; R^{\smile}$ is crisp.*
3. *If $Q : A \to A$ is a partial equivalence relation that has a splitting, then Q is p-crisp.*

Proof. 1. Suppose $Q : A \to B$ is crisp. If we choose $R = Q^{\smile}$ and $S = Q$, then we obtain $R^{\smile};R = Q^{\smile\smile};Q^{\smile} = Q;Q^{\smile}$ and $S^{\smile};S = Q^{\smile};Q$. Furthermore, $R;Q;S = Q^{\smile};Q;Q^{\smile}$ is crisp because Q is and crisp relations are closed under composition and converse.

2. We first show the implication $\Rightarrow$. Since Q is p-crisp we have two relations $R : A' \to A$ and $S' : B' \to B$ with $R^{\smile};R = Q;Q^{\smile}$ and $S^{\smile};S = Q^{\smile};Q$ so that $R;Q;S^{\smile}$ is crisp. First of all we get, $R^{\smile};R = Q;Q^{\smile} = Q;Q = Q$ because Q is a partial equivalence relation. Furthermore, we have

$$\begin{aligned} R;Q;R^{\smile} &= R;Q;Q^{\smile};Q;Q^{\smile};R^{\smile} && Q \text{ is a partial equivalence relation} \\ &= R;Q;S^{\smile};S;Q^{\smile};R^{\smile} && Q \text{ is p-crisp} \\ &= (R;Q;S^{\smile});(R;Q;S^{\smile})^{\smile} \end{aligned}$$

so that $R;Q;R^{\smile}$ is crisp because $R;Q;S^{\smile}$ is and crisp relations are closed under composition and converse. The opposite implication is immediate by choosing $S = R$.

3. Suppose Q has a splitting, i.e., there is a relation $R : B \to A$ so that $R;R^{\smile} = \mathbb{I}_B$ and $R^{\smile};R = Q$. Then we have $R^{\smile};R = Q = Q;Q^{\smile}$ and $R^{\smile};R = Q = Q^{\smile};Q$. Furthermore, we compute $R;Q;R^{\smile} = R;R^{\smile};R;R^{\smile} = \mathbb{I}_B$, i.e., $R;Q;R^{\smile}$ is crisp. □

The previous lemma has shown that a partial equivalence relation has to be p-crisp in order to have a splitting. The next lemma shows that these relations will automatically split if all crisp partial equivalence relations do. With other words, if all crisp equivalence relations split in an arrow category, then all p-crisp partial equivalence relation will split, and no other partial equivalence relation will.

Lemma 5. *Let $\mathcal{A}$ be an arrow category. Then the following are equivalent:*

1. *Every crisp partial equivalence splits in $\mathcal{A}$.*
2. *Every p-crisp partial equivalence splits in $\mathcal{A}$.*

Proof. 1. $\Rightarrow$2.: Assume that every crisp partial equivalence splits in $\mathcal{A}$ and that $Q : A \to A$ is a p-crisp partial equivalence relation. Then Lemma 4(2) shows that there is a relation $R : A' \to A$ $R^{\smile};R = Q$ so that $R;Q;R^{\smile}$ is crisp. Now we want to show that $R;Q;R^{\smile}$ is a partial equivalence relation. This relation is symmetric because Q is. The second property is shown by

$$\begin{aligned} (R;Q;R^{\smile});(R;Q;R^{\smile}) &= R;Q;Q;Q;R^{\smile} \\ &= R;Q;R^{\smile}. && Q \text{ is a partial equivalence relation} \end{aligned}$$

Now suppose that $S : B \to A'$ splits $R;Q;R^{\smile}$, i.e., we have $S;S^{\smile} = \mathbb{I}_B$ and $S^{\smile};S = R;Q;R^{\smile}$. We have

$$\begin{aligned} S;R;R^{\smile} &= S;S^{\smile};S;R;R^{\smile} \\ &= S;R;Q;R^{\smile};R;R^{\smile} \\ &= S;R;Q;Q;R^{\smile} \\ &= S;R;Q;R^{\smile} && Q \text{ is a partial equivalence relation} \\ &= S;S^{\smile};S \\ &= S. \end{aligned}$$

This implies

$$
\begin{aligned}
S;R;(S;R)^{\smile} &= S;R;R^{\smile};S^{\smile} \\
&= S;S^{\smile} && \text{see above} \\
&= \mathbb{I}_B, \\
(S;R)^{\smile};S;R &= R^{\smile};S^{\smile};S;R \\
&= R^{\smile};R;Q;R^{\smile};R \\
&= Q;Q;Q \\
&= Q, && Q \text{ is a partial equivalence relation}
\end{aligned}
$$

showing that $S;R$ is a splitting of Q.

2. $\Rightarrow$ 1.: This is trivial since every crisp relation is p-crisp. □

In the second part of this section we want to concentrate on the splitting of a crisp partial equivalence relation $Q : A \to A$. As outlined above, in **Rel**(L) a splitting of Q is given by the set of equivalence classes of elements for which Q is defined and the relation $R : B \to A$ can be chosen as $R([x], y)$ iff $y \in [x]$ iff $Q(x, y)$, i.e., the relation R is crisp. As for relational sums this needs not to be the case as the following example shows.

Example 2. As in the previous example, we will use the Boolean algebra B_4 with four elements as the lattice L. The objects of $\mathcal{A}_2$ are the two sets $A = \{1, 2, 3, 4\}$ and $B = \{x, y\}$. Now consider the crisp partial equivalence relation $Q : A \to A$ and the relations $R_1, R_2 : B \to A$ defined in matrix form by

$$
Q = \begin{array}{c} \\ 1 \\ 2 \\ 3 \\ 4 \end{array}\!\!\begin{array}{c} \begin{array}{cccc} 1 & 2 & 3 & 4 \end{array} \\ \begin{bmatrix} 1 & 1 & 0 & 0 \\ 1 & 1 & 0 & 0 \\ 0 & 0 & 1 & 0 \\ 0 & 0 & 0 & 0 \end{bmatrix} \end{array}, \quad R_1 = \begin{array}{c} \\ x \\ y \end{array}\!\!\begin{array}{c} \begin{array}{cccc} 1 & 2 & 3 & 4 \end{array} \\ \begin{bmatrix} a & a & b & 0 \\ b & b & a & 0 \end{bmatrix} \end{array}, \quad R_2 = \begin{array}{c} \\ x \\ y \end{array}\!\!\begin{array}{c} \begin{array}{cccc} 1 & 2 & 3 & 4 \end{array} \\ \begin{bmatrix} b & b & a & 0 \\ a & a & b & 0 \end{bmatrix} \end{array}.
$$

Obviously R_1 and R_2 are not crisp but it is easy to verify that both relations are a splitting of Q. For example, we computing

$$
\begin{aligned}
R_1;R_1^{\smile} &= \begin{array}{c} \\ x \\ y \end{array}\!\!\begin{array}{c} \begin{array}{cccc} 1 & 2 & 3 & 4 \end{array} \\ \begin{bmatrix} a & a & b & 0 \\ b & b & a & 0 \end{bmatrix} \end{array};\begin{array}{c} \\ 1 \\ 2 \\ 3 \\ 4 \end{array}\!\!\begin{array}{c} \begin{array}{cc} x & y \end{array} \\ \begin{bmatrix} a & b \\ a & b \\ b & a \\ 0 & 0 \end{bmatrix} \end{array} \\
&= \begin{array}{c} \\ x \\ y \end{array}\!\!\begin{array}{c} \begin{array}{cc} x & y \end{array} \\ \begin{bmatrix} 1 & 0 \\ 0 & 1 \end{bmatrix} \end{array} \\
&= \mathbb{I}_B, \\
R_1^{\smile};R_1 &= \begin{array}{c} \\ 1 \\ 2 \\ 3 \\ 4 \end{array}\!\!\begin{array}{c} \begin{array}{cc} x & y \end{array} \\ \begin{bmatrix} a & b \\ a & b \\ b & a \\ 0 & 0 \end{bmatrix} \end{array};\begin{array}{c} \\ x \\ y \end{array}\!\!\begin{array}{c} \begin{array}{cccc} 1 & 2 & 3 & 4 \end{array} \\ \begin{bmatrix} a & a & b & 0 \\ b & b & a & 0 \end{bmatrix} \end{array}
\end{aligned}
$$

$$= \begin{array}{c} \\ 1 \\ 2 \\ 3 \\ 4 \end{array} \begin{array}{c} 1\;2\;3\;4 \\ \begin{bmatrix} 1&1&0&0 \\ 1&1&0&0 \\ 0&0&1&0 \\ 0&0&0&0 \end{bmatrix} \end{array}$$
$$= Q.$$

Now, we define $\mathcal{A}_2$ to be the smallest structure containing Q, R_1 and R_2 that is closed under meet, join, complement, converse, composition, the arrow operations, and contains all constants related to the two objects A and B. As in the previous example this defines an arrow category $\mathcal{A}_2$. It turns out that $\mathcal{A}_2$ has a total of 1168 relations split up as

$$|\mathcal{A}_2[A,A]| = 1024, |\mathcal{A}_2[A,B]| = |\mathcal{A}_2[B,A]| = 64, |\mathcal{A}_2[B,B]| = 16.$$

Among the 64 relations with source B and target A the relation R_1 and R_2 from above are the only relation that split Q. Finally, we would like to mention that the example above is definitely not the smallest example of an arrow category with only non-crisp splittings of a crisp partial equivalence relation. We have chosen this example because Q is a relation with more than one equivalence class, at least one of those is not a singleton, and there is an element for which Q is not defined. □

In the following we will call a splitting R of a crisp partial equivalence relation Q a crisp splitting iff R is crisp. The example above shows that we cannot always assume that there is a crisp splitting for every a crisp partial equivalence relation. But, as in the case of relational sums, $\mathcal{A}_2$ is embedded in the larger arrow category $\mathbf{Rel}(B_4)$ which does have a crisp splitting. We now want to show that this is always the case so that requiring the splitting of a crisp partial equivalence relation to be crisp is not an actual restriction. We start with the following theorem embedding a Dedekind category $\mathcal{R}$ into a Dedekind category providing all splittings. A proof of this theorem in various versions can be found in [1,5,6].

Theorem 4. *Suppose $\mathcal{R}$ is a Dedekind category and $\mathcal{E}$ a class of partial equivalence relations from $\mathcal{R}$. Then the category $\mathcal{R}_\mathcal{E}$ is defined by*

1. *The objects of $\mathcal{R}_\mathcal{E}$ are the elements of $\mathcal{E}$,*
2. *Given two objects $Q_1 : A \to A$ and $Q_2 : B \to B$ a morphism in $\mathcal{R}_\mathcal{E}$ from Q_1 to Q_2 is a relation $R : A \to B$ so that $Q_1;R;Q_2 = R$,*

is a Dedekind category. Furthermore, if $\mathcal{E}$ contains all identities of $\mathcal{R}$, then the functor $E : \mathcal{R} \to \mathcal{R}_\mathcal{E}$ defined by $E(A) = \mathbb{I}_A$ for objects and $E(R) = R$ is a full embedding of Dedekind categories.

We now extend the previous result to arrow categories. However, we need to require that the partial equivalence relations in $\mathcal{E}$ are crisp.

Theorem 5. *Suppose $\mathcal{A}$ is an arrow category and $\mathcal{E}$ a class of crisp partial equivalence relations from $\mathcal{A}$. Then $\mathcal{A}_\mathcal{E}$ together with the arrow operations inherited from $\mathcal{A}$ is an arrow category. Furthermore, if $\mathcal{E}$ contains all identities of $\mathcal{A}$, then E is a full embedding of arrow categories.*

Proof. In order to establish that $\mathcal{A}_\mathcal{E}$ is an arrow category, we only need to verify that $R^\downarrow$ and $R^\uparrow$ for a morphism R with source $Q_1 : A \to A$ and target $Q_2 : B \to B$ are again morphisms of $\mathcal{A}_\mathcal{E}$, i.e., they satisfy $Q_1; R^\downarrow; Q_2 = R^\downarrow$ resp. $Q_1; R^\uparrow; Q_2 = R^\uparrow$. The second property follows immediately from

$$\begin{aligned} R^\uparrow &= (Q_1; R; Q_2)^\uparrow && R \text{ in } \mathcal{A}_\mathcal{E} \\ &= (Q_1^\downarrow; R; Q_2)^\uparrow && Q_1 \text{ crisp} \\ &= Q_1^\downarrow; (R; Q_2)^\uparrow \\ &= Q_1; (R; Q_2^\downarrow)^\uparrow && Q_1 \text{ and } Q_2 \text{ crisp} \\ &= Q_1; R^\uparrow; Q_2^\downarrow \\ &= Q_1; R^\uparrow; Q_2. && Q_2 \text{ crisp} \end{aligned}$$

For the second equation, we first have

$$\begin{aligned} Q_1; R^\downarrow; Q_2 &\sqsubseteq (Q_1; R)^\downarrow; Q_2 && \text{Lemma 2(1)} \\ &\sqsubseteq (Q_1; R; Q_2)^\downarrow && \text{dual of Lemma 2(1)} \\ &= R^\downarrow. && R \text{ in } \mathcal{A}_\mathcal{E} \end{aligned}$$

The opposite inclusion follows from

$$\begin{aligned} R^\downarrow &= (Q_1; R; Q_2)^\downarrow && R \text{ in } \mathcal{A}_\mathcal{E} \\ &= (\mathrm{dom}(Q_1); Q_1; R; Q_2; \mathrm{cod}(Q_2))^\downarrow && \text{Lemma 1(2)} \\ &= (\mathrm{dom}(Q_1); R; \mathrm{cod}(Q_2))^\downarrow && R \text{ in } \mathcal{A}_\mathcal{E} \\ &= ((\mathbb{I}_A \sqcap Q_1); R; (\mathbb{I}_B \sqcap Q_2))^\downarrow && Q_1 \text{ and } Q_2 \text{ are p. equiv. rel.} \\ &= (\mathbb{I}_A \sqcap Q_1); R^\downarrow; (\mathbb{I}_B \sqcap Q_2) && \text{Lemma 2(2)} \\ &\sqsubseteq Q_1; R^\downarrow; Q_2. \end{aligned}$$

This completes the proof. □

It remains to show that $\mathcal{A}_\mathcal{E}$ has crisp splittings for all elements in $\mathcal{E}$.

Theorem 6. *Suppose $\mathcal{A}$ is an arrow category and $\mathcal{E}$ a class of crisp partial equivalence relations from $\mathcal{A}$. If Q is a crisp partial equivalence relation in $\mathcal{A}_\mathcal{E}$ so that Q is in $\mathcal{E}$, then Q has a crisp splitting in $\mathcal{A}_\mathcal{E}$.*

Proof. Suppose $Q : A \to A$ is a crisp partial equivalence relation on the object $Q_1 : A \to A$ in $\mathcal{A}_\mathcal{E}$, i.e., we have $Q = Q_1; Q; Q_1$. In addition assume that Q is in $\mathcal{E}$. Then Q is an object of $\mathcal{A}_\mathcal{E}$ and we can define a relation R by $R = Q$. Since we have $Q; R; Q_1 = Q; Q; Q_1 = Q; Q_1 = Q_1; Q; Q_1; Q_1 = Q_1; Q; Q_1 = Q$ the relation R is a morphism from Q to Q_1 in $\mathcal{A}_\mathcal{E}$. Furthermore, we have $R; R^\smile = Q; Q^\smile = Q = \mathbb{I}_Q$ and $R^\smile; R = Q^\smile; Q = Q$, i.e., R splits Q. Last but not least, R is crisp since Q is. □

As in the case of relational sums we have the following. If $\mathcal{A}$ had a non-crisp splitting $R : B \to A$ of Q, then B (after embedding into $\mathcal{A}_\mathcal{E}$) is isomorphic to the crisp splitting from the theorem above. This isomorphism is not crisp, of course.

5 Summary and Future Work

In this paper we have investigated when crisp versions of relational sums and splittings exist. In particular, we have shown that the additional requirement of crispness is not restriction since a crisp version of each construction can always be generated in an extension of the arrow category. Furthermore, we have characterized the class of partial equivalence relations that may split in an arrow category.

In future work we will investigate the same question for the relational product. This problem is harder since there cannot be general construction that embeds an arrow category into an arrow category with crisp products. This is due to the fact that the existence of products is closely related to the representation problem for these categories.

It would be interesting to find an algebraic characterization, i.e., an equation or inclusion, of pseudo-crispness. The current definition is based on the existence of two specific relations. This is not very handy in showing properties about p-crisp relations or in verifying that a given relation is p-crisp.

Last but not least, we would like to mention that a questions similar to the ones in this paper do not make much sense for the construction of a relational power. This is due to the fact the L-fuzzy power set, i.e., the set of all L-fuzzy subsets, is different from the regular power set. For details we refer to [10].

References

1. Freyd, P., Scedrov, A.: Categories, Allegories. North-Holland Mathematical Library, vol. 39, North-Holland, Amsterdam (1990)
2. Kawahara, Y., Furusawa, H.: Crispness and Representation Theorems in Dedekind Categories. DOI-TR 143, Kyushu University (1997)
3. Schmidt, G., Ströhlein, T.: Relations and graphs. Discrete mathematics for computer scientists. EATCS Monographs on Theoretical Computer Science. Springer, Heidelberg (1993). https://doi.org/10.1007/978-3-642-77968-8
4. Schmidt G.: Relational Mathematics. Encyplopedia of Mathematics and Its Applications, vol. 132. Cambridge University Press, Cambridge (2011)
5. Winter M.: Strukturtheorie heterogener Relationenalgebren mit Anwendung auf Nichtdetermismus in Programmiersprachen. Dissertationsverlag NG Kopierladen GmbH, München (1998)
6. Winter, M.: A Pseudo representation theorem for various categories of relations. Theory Appl. Categories **7**(2), 23–37 (2000)
7. Winter, M.: A new algebraic approach to L-fuzzy relations convenient to study crispness. INS Inf. Sci. **139**, 233–252 (2001)
8. Winter M.: Goguen Categories - A Categorical Approach to L-Fuzzy Relations. Trends in Logic, vol. 25. Springer, Heidelberg (2007). https://doi.org/10.1007/978-1-4020-6164-6
9. Winter, M.: Arrow categories. Fuzzy Sets Syst. **160**, 2893–2909 (2009)
10. Winter M.: Type-n arrow categories. In: Höfner P., Pous D., Struth G. (eds.): Relational and Algebraic Methods in Computer Science (RAMiCS 2017). LNCS, vol. 11194, pp. 307–322. Springer, Heidelberg (2017)

Change of Base Using Arrow Categories

Michael Winter(✉)

Department of Computer Science, Brock University,
St. Catharines, Ontario L2S 3A1, Canada
mwinter@brocku.ca

Abstract. Arrow categories establish a suitable framework to reason about L-fuzzy relation abstractly. For each arrow category we can identify the Heyting algebra L that is used as the lattice of membership or truth values by the relations of the category. Therefore, arrow categories model the fixed-base approach to L-fuzziness, i.e., all relations of the given arrow category use the same membership values. In this paper we are interested in the process of changing the base, i.e., an operation that allows to switch from an L_1-fuzzy relation to an L_2-fuzzy relation by replacing all membership values from L_1 by values from L_2. We will define and investigate this change of base between two abstract arrow categories for which component-wise reasoning cannot be performed.

1 Introduction

One approach to model uncertainty is given by so-called L-fuzzy relations. An L-fuzzy relation is a relation that uses the values from the Heyting algebra L as truth values instead of the Boolean values *true* and *false*. The smallest element of L is identified with *false*, the greatest element with *true*, and any value in between indicates that two elements are in relationship only up to this given degree. A relation that only uses the greatest and smallest element of L as truth values can be identified with a Boolean valued relation, and is, therfore, called crisp. In applications of L-fuzzy relations one might be interested in changing the Heyting algebra L. For example, type-2 fuzzy controller utilize L-fuzzy relations and $(L \to L)$-fuzzy relations where $L \to L$ is an appropriate set of functions from L to itself. Common to all design principles of those controllers seems to be that the core of the controller is based on $(L \to L)$-fuzzy relations. In order to obtain a crisp output of the controller, the result is first transformed into an L-fuzzy relation by using a so-called type reducer before the result finally converted into a crisp value using a so-called defuzzification [5]. In particular, the type reducer converts an $(L \to L)$-fuzzy relation into an L-fuzzy relation, i.e., this process includes switching from the base $L \to L$ to the base L. In this paper we are interested how a change of base can be defined in the abstract setting of arrow categories and what properties this change might satisfy.

The author gratefully acknowledges support from the Natural Sciences and Engineering Research Council of Canada (283267).

U. Fahrenberg et al. (Eds.): RAMiCS 2021, LNCS 13027, pp. 448–464, 2021.
https://doi.org/10.1007/978-3-030-88701-8_27

Arrow categories establish a suitable framework to reason about L-fuzzy relation abstractly. It is possible to determine the Heyting algebra L that is used as the lattice of membership values for a given arrow category. Therefore, all relation of a given arrow category use the same Heyting algebra L, i.e., arrow categories model the so-called fixed-base approach. Switching the base now needs to be a map between two arrow categories. On the other hand, not every map between two arrow categories should be considered a change of base. Beside the lattice of truth values the two arrow categories might also be different in the purely relational aspect. For example, assume that the first arrow category is the category of infinite sets as objects and L_1-fuzzy relations as morphisms, and the second arrow category is the category of finite sets as objects and L_2-fuzzy relations as morphisms. A map between these category does not only include exchanging L_1 by L_2, it also includes mapping a relation between infinite sets to a relation between finite sets. Such a difference in the relational aspect can be prevented by requiring that the two arrow categories have basically the same crisp relations. Please note that this is not the case in the example above. Here we have the category of binary relations between infinite sets resp. finite sets as the categories of crisp relations. Therefore, we will require that a change of base is between arrow categories so that the subcategories of crisp relation are isomorphic, i.e., basically the same.

2 Mathematical Preliminaries

In this section we want to introduce Dedekind and arrow categories as an abstract framework for L-fuzzy relations [2,10,11]. We use the notation $R : A \to B$ to indicate that a morphism R of a category C has source A and target B. In addition, we write $Q;R$ for the composition of $Q : A \to B$ and $R : B \to C$. Please note that composition has to be read from left to right, i.e., first Q and then R. The identity on A is denoted by $\mathbb{I}_A$.

Definition 1. *A Dedekind category $\mathcal{R}$ is a category satisfying the following:*

1. *For all objects A and B the collection $\mathcal{R}[A, B]$ is a complete Heyting algebra. Meet, join, the implication operation, the induced ordering, the least and the greatest element are denoted by $\sqcap, \sqcup, \to, \sqsubseteq, \mathbb{⊥}_{AB}, \mathbb{T}_{AB}$, respectively.*
2. *There is a monotone operation $\smile$ (called converse) mapping a relation $Q : A \to B$ to $Q^\smile : B \to A$ such that for all relations $Q : A \to B$ and $R : B \to C$ the following holds: $(Q;R)^\smile = R^\smile;Q^\smile$ and $(Q^\smile)^\smile = Q$.*
3. *For all relations $Q : A \to B, R : B \to C$ and $S : A \to C$ the modular law $(Q;R) \sqcap S \sqsubseteq Q;(R \sqcap (Q^\smile;S))$ holds.*
4. *For all relations $R : B \to C$ and $S : A \to C$ there is a relation $S/R : A \to B$ (called the left residual of S and R) such that for all $X : A \to B$ the following holds: $X;R \sqsubseteq S \Leftrightarrow X \sqsubseteq S/R$.*

If L is a complete Heyting algebra, an L-fuzzy relation R between two sets A and B is a function $A \times B \to L$ assigning to each pair of A and B elements a degree of membership in R. The category **Rel**(L) of sets and L-fuzzy relations forms a Dedekind category.

If we define the left residual $Q\backslash R : B \to C$ of two relations $Q : A \to B$ and $R : A \to C$ by $Q\backslash R := (R^{\smile}/Q^{\smile})^{\smile}$ we immediately obtain $X \sqsubseteq Q\backslash R$ iff $Q;X \sqsubseteq R$. Using both residual we define the symmetric quotient as $\mathrm{syQ}(Q,R) = (Q\backslash R) \sqcap (Q^{\smile}/R^{\smile})$. This construction is characterized by $X \sqsubseteq \mathrm{syQ}(Q,R)$ iff $Q;X \sqsubseteq R$ and $R;X^{\smile} \sqsubseteq Q$.

Throughout the paper we will use the axioms and some basics facts such as monotonicity of the operations without mentioning.

The following lemma collects some properties of the residuals and the symmetric quotient that we will need in this paper. A proof can be found in [3,6,7,10].

Lemma 1. *Let be $Q : A \to B$ and $S : A \to C$. Then we have:*

1. $\mathbb{I}_A\backslash S = S$ *and* $\mathbb{T}_{CB}/Q = \mathbb{T}_{CA}$.
2. $\mathrm{syQ}(Q,S)^{\smile} = \mathrm{syQ}(S,Q)$.
3. $Q;\mathrm{syQ}(Q,S) \sqsubseteq S$ *and* $\mathrm{syQ}(Q,S);S^{\smile} \sqsubseteq Q^{\smile}$.

An important class of relations are given by maps (or functions). We call a relation $Q : A \to B$ univalent (or partial function) iff $Q^{\smile};Q \sqsubseteq \mathbb{I}_B$ and total iff $\mathbb{I}_A \sqsubseteq Q;Q^{\smile}$. Q is called a map iff Q is total and univalent.

In order to identify the lattice of membership degrees used by the relations of an arbitrary Dedekind category we use the notion scalar relations that was introduced by Furusawa and Kawahara [4]. A relation $\alpha : A \to A$ is called a scalar on A iff $\alpha \sqsubseteq \mathbb{I}_A$ and $\mathbb{T}_{AA};\alpha = \alpha;\mathbb{T}_{AA}$. In the case of L-fuzzy relations a scalar relation α is a partial identity for which we have $\alpha(a,a) = x$ for all a and some fixed x from L. Therefore, any scalar corresponds to exactly one element from L.

The next definition introduces arrow categories, which add two operations to Dedekind categories. The relation $R^{\uparrow}$ is the smallest crisp relation that contains R (also called support), and $R^{\downarrow}$ is the greatest crisp relation included in R (also called kernel) [9–11].

Definition 2. *An arrow category $\mathcal{A}$ is a Dedekind category with $\mathbb{T}_{AB} \neq \mathbb{L}_{AB}$ for all A, B and two operations $\uparrow$ and $\downarrow$ satisfying:*

1. $R^{\uparrow}, R^{\downarrow} : A \to B$ *for all* $R : A \to B$.
2. $({}^{\uparrow},{}^{\downarrow})$ *forms a Galois correspondence, i.e.,* $Q^{\uparrow} \sqsubseteq R$ *iff* $Q \sqsubseteq R^{\downarrow}$ *for all* $Q,R : A \to B$.
3. $(R^{\smile};S^{\downarrow})^{\uparrow} = R^{\uparrow\smile};S^{\downarrow}$ *for all* $R : B \to A$ *and* $S : B \to C$.
4. $(Q \sqcap R^{\downarrow})^{\uparrow} = Q^{\uparrow} \sqcap R^{\downarrow}$ *for all* $Q,R : A \to B$.
5. *If* $\alpha_A \neq \mathbb{L}_{AA}$ *is a non-zero scalar then* $\alpha_A^{\uparrow} = \mathbb{I}_A$.

First we have the following basic properties in arrow categories. A proof can be found in [8,10,13].

Lemma 2. *Suppose $\mathcal{A}$ is an arrow category, $Q : A \to B$ is crisp, and $R, R' : B \to C$. Then we have:*

1. $R^{\smile\downarrow} = R^{\downarrow\smile}$.
2. $R^{\uparrow\downarrow} = R^{\uparrow}$ *and* $R^{\downarrow\uparrow} = R^{\downarrow}$.
3. $(R \sqcap R')^{\downarrow} = R^{\downarrow} \sqcap R'^{\downarrow}$.
4. $Q; R^{\downarrow} \sqsubseteq (Q; R)^{\downarrow}$.
5. *If* Q *is, in addition, univalent, then the inclusion in 4. is an equation.*

A relation in an arrow category that satisfies $R^{\uparrow} = R$, or equivalently $R^{\downarrow} = R$, is called crisp. It is important to mention that all constants are crisp and that crisp relations are closed under all operations of a Dedekind category, i.e., the crisp relations form a sub-Dedekind category.

Arrow categories are uniform, i.e., they satisfy $\mathbb{T}_{AB}; \mathbb{T}_{BC} = \mathbb{T}_{AC}$ for all objects A, B and C. This actually implies that the sets of scalar relations Scalar(A) for all objects A are isomorphic. Therefore we will say that the $\mathcal{A}$ is an arrow category over L if L is isomorphic to Scalar(A) for any/all objects A in $\mathcal{A}$.

An abstract version of a singleton set is given by a unit. A unit 1 is an object so that $\mathbb{T}_{11} = \mathbb{I}_1$ and $\mathbb{T}_{A1}$ is total for every object A.

The relational product is an abstract version of the Cartesian product of two sets. This construction corresponds to a categorical product in the subcategory of maps.

The relational product of two objects A and B is an object $A \times B$ together with two crisp relations $\pi : A \times B \to A$ and $\rho : A \times B \to B$ so that the following equations hold

$$\pi^{\smile}; \pi \sqsubseteq \mathbb{I}_A, \qquad \rho^{\smile}; \rho \sqsubseteq \mathbb{I}_B, \qquad \pi^{\smile}; \rho = \mathbb{T}_{AB}, \qquad \pi; \pi^{\smile} \sqcap \rho; \rho^{\smile} = \mathbb{I}_{A\times B}.$$

In addition, we define

$$Q \otimes R = Q; \pi^{\smile} \sqcap R; \rho^{\smile}, \qquad S \otimes T = \pi; S \sqcap \rho; T, \qquad U \otimes V = \pi; U; \pi^{\smile} \sqcap \rho; V; \rho^{\smile}.$$

For concrete relations we have $(Q \otimes R)(c, (a, b))$ iff $Q(c, a)$ and $R(c, b)$, i.e., $Q \otimes R$ relates c to a pair (a, b) iff Q relates c to a and R relates c to b. Similar, we have $(S \otimes T)((a, b), d)$ iff $S(a, d)$ and $T(b, d)$ and $(U \otimes V)((a, b), (c, d))$ iff $U(a, c)$ and $V(b, d)$.

A proof of the following lemma can be found in [7,10].

Lemma 3. *Let* $Q : C \to A, R, R' : C \to B, S : A \to D, T : B \to D$ *be relations. Then we have:*

1. $(Q \otimes R); \pi = Q \sqcap R; \mathbb{T}_{BA}$ *and* $(Q \otimes R); \rho = Q; \mathbb{T}_{AB} \sqcap R$.
2. *If* R *is total, then* $(Q \otimes R); \pi = Q$*, and if* Q *is total, then* $(Q \otimes R); \rho = R$.
3. *If* Q *and* R *are univalent (total resp. maps), then so is* $Q \otimes R$.
4. *If* $Q \otimes R$ *is univalent, then* $(Q \otimes R); (S \otimes T) = Q; S \sqcap R; T$.

Notice that (4) of the previous lemma has several immediate consequences such as $(Q \otimes R); (S \otimes T) = Q; S \otimes R; T$ if $Q \otimes R$ is univalent. In the remaining of the paper we will always refer to (4) even if one of these immediate consequences is needed.

The last construction we want to introduce is the relational power. This construction is an abstract version of the set of all L-fuzzy subsets of a set.

Suppose $\mathcal{A}$ be an arrow category over L. An object L^A together with a relation $\varepsilon : A \to L^A$ is called a relational power iff

$$\mathrm{syQ}(\varepsilon,\varepsilon)^{\downarrow} = \mathbb{I}_{L^A} \quad \text{and} \quad \mathrm{syQ}(R,\varepsilon)^{\downarrow} \text{ is total for every } R : A \to B.$$

Our definition of a relational power differs from the definition in [2,7]. For a comparison of the definitions and the reason for using the one presented here in the case of arrow category we refer to [12].

Intuitively, the construction $\mathrm{syQ}(R^{\smile},\varepsilon)^{\downarrow}$ for a relation $R : A \to B$ returns a crisp map relating an element a from A to the L-fuzzy subset M of B so that the degree of any b in M is equal to $R(a,b)$, i.e., the L-fuzzy image operation. We will use the abbreviation $\Lambda(R) = \mathrm{syQ}(R^{\smile},\varepsilon)^{\downarrow}$ and obtain the following lemma. A proof can be found in [12].

Lemma 4. *Let L^B and L^C be a relational power and $R : A \to B$ and $S : L^B \to C$ be a relations. Then we have:*

1. *$\Lambda(R)$ is a map.*
2. *$\Lambda(R);\varepsilon^{\smile} = R$.*
3. *$\Lambda(R);\Lambda(S) = \Lambda(\Lambda(R);S)$.*

The following two relations

$$\mathcal{M}_2 = \Lambda(\varepsilon^{\smile} \otimes \varepsilon^{\smile}) : L^A \times L^A \to L^A, \text{ and } \mathcal{J}_2 = \Lambda(\varepsilon^{\smile};\pi \sqcup \varepsilon^{\smile};\rho) : L^A \times L^A \to L^A$$

define the operations of binary intersection (or meet) and union (or join) for relational powers. In addition, the relation $\mathcal{J} = \Lambda(\varepsilon^{\smile};\varepsilon^{\smile}) : L^{(L^A)} \to L^A$ computes the union of a set of sets.

3 The Object of Truth Values

As mentioned before the set of scalar relations on an object can be seen as the Heyting algebra of membership values used by the relations of the given arrow category. In this section we want to investigate an internal version of this lattice. In order to do so we consider the object L^1 where 1 is the unit of arrow category over L. We will drop the exponent and simply write L instead of L^1.

It is easy to verify that the operations $\alpha \mapsto \Lambda(\alpha)$ and $p \mapsto p;\varepsilon^{\smile}$ from an isomorphism between the scalars on 1 and points of L. The order $(\varepsilon\backslash\varepsilon)^{\downarrow}$ is an internal version of the order on the scalars. We now define the following relations:

$$\mathrm{zero} = \Lambda(\mathbb{L}_{11}), \quad \mathrm{one} = \Lambda(\mathbb{I}_1), \quad \mathrm{down} = \Lambda(\varepsilon^{\smile\downarrow}), \quad \mathrm{up} = \Lambda(\varepsilon^{\smile\uparrow}).$$

The crisp points zero and one correspond to the smallest and greatest element of L, and down and up are internal versions of the arrow operations. First of all, we have

$$\begin{aligned}
\text{one} &= \Lambda(\mathbb{I}_1) \\
&= ((\mathbb{I}_1\backslash\varepsilon) \sqcap (\mathbb{I}_1/\varepsilon^{\smile}))^{\downarrow} \\
&= ((\mathbb{I}_1\backslash\varepsilon) \sqcap (\mathbb{T}_{11}/\varepsilon^{\smile}))^{\downarrow} \\
&= (\varepsilon \sqcap \mathbb{T})^{\downarrow} && \text{Lemma 1(1)} \\
&= \varepsilon^{\downarrow}.
\end{aligned}$$

Furthermore, we will use the following lemma about down and up.

Lemma 5. *For the crisp maps* down, up : $L \to L$ *we have*

1. down *and* up *are idempotent, i.e.,* down; down = down *and* up; up = up,
2. $\text{down}^{\smile}$; down = $\mathbb{I}_L \sqcap$ down *and* $\text{up}^{\smile}$; up = $\mathbb{I}_L \sqcap$ up,
3. $\text{down}^{\smile}$; down = $\text{up}^{\smile}$; up.

Proof. 1. We immediately compute

$$\begin{aligned}
\text{down}; \text{down} &= \Lambda(\text{down}; \varepsilon^{\smile\downarrow}) && \text{Lemma 4(3)} \\
&= \Lambda((\text{down}; \varepsilon^{\smile})^{\downarrow}) && \text{Lemma 2(5)} \\
&= \Lambda(\varepsilon^{\smile\downarrow}) && \text{Lemma 4(2)} \\
&= \text{down}.
\end{aligned}$$

The same property for up is shown analogously.

2. This property is true for any idempotent and univalent relation and was shown in [1].
3. From the two computations

$$\begin{aligned}
\varepsilon^{\downarrow}; \text{up}^{\smile}; \text{up} &= \varepsilon^{\downarrow}; \text{syQ}(\varepsilon, \varepsilon^{\uparrow})^{\downarrow}; \text{up} && \text{Lemma 1(2)} \\
&\sqsubseteq (\varepsilon^{\downarrow}; \text{syQ}(\varepsilon, \varepsilon^{\uparrow}))^{\downarrow}; \text{up} && \text{Lemma 2(4)} \\
&\sqsubseteq (\varepsilon; \text{syQ}(\varepsilon, \varepsilon^{\uparrow}))^{\downarrow}; \text{up} \\
&\sqsubseteq \varepsilon^{\uparrow\downarrow}; \text{up} && \text{Lemma 1(3)} \\
&= \varepsilon^{\uparrow}; \text{up} && \text{Lemma 2(2)} \\
&\sqsubseteq \varepsilon^{\uparrow}; \text{syQ}(\varepsilon^{\uparrow}, \varepsilon) \\
&\sqsubseteq \varepsilon, && \text{Lemma 1(3)}
\end{aligned}$$

$$\begin{aligned}
\text{up}^{\smile}; \text{up}; \varepsilon^{\smile} &= \text{up}^{\smile}; \varepsilon^{\smile\uparrow} && \text{Lemma 4(2)} \\
&= \text{syQ}(\varepsilon, \varepsilon^{\uparrow})^{\downarrow}; \varepsilon^{\smile\uparrow} && \text{Lemma 1(2)} \\
&\sqsubseteq (\text{syQ}(\varepsilon, \varepsilon^{\uparrow}); \varepsilon^{\smile\uparrow})^{\downarrow} && \text{Lemma 2(4)} \\
&\sqsubseteq \varepsilon^{\smile\downarrow} && \text{Lemma 1(3)}
\end{aligned}$$

we obtain $\mathrm{up}^{\smile};\mathrm{up} \sqsubseteq \mathrm{syQ}(\varepsilon^{\downarrow},\varepsilon)$ and, hence, $\mathrm{up}^{\smile};\mathrm{up} = (\mathrm{up}^{\smile};\mathrm{up})^{\downarrow} \sqsubseteq \mathrm{syQ}(\varepsilon^{\downarrow},\varepsilon)^{\downarrow} = \mathrm{down}$. This implies $\mathrm{up}^{\smile};\mathrm{up} \sqsubseteq \mathbb{I}_L \sqcap \mathrm{down}$ since up is univalent. From (2) we conclude $\mathrm{up}^{\smile};\mathrm{up} \sqsubseteq \mathrm{down}^{\smile};\mathrm{down}$. For the converse inclusion consider

$$\begin{aligned}
\varepsilon^{\uparrow};\mathrm{down}^{\smile};\mathrm{down} &= (\varepsilon;\mathrm{down}^{\smile})^{\uparrow};\mathrm{down} && \mathrm{down}^{\smile}\ \text{crisp}\\
&= \varepsilon^{\downarrow\uparrow};\mathrm{down} && \text{Lemma 4(2)}\\
&= \varepsilon^{\downarrow};\mathrm{down} && \text{Lemma 2(2)}\\
&\sqsubseteq \varepsilon^{\downarrow};\mathrm{syQ}(\varepsilon^{\downarrow},\varepsilon) && \\
&\sqsubseteq \varepsilon, && \text{Lemma 1(3)}\\
\mathrm{down}^{\smile};\mathrm{down};\varepsilon^{\smile} &= \mathrm{down}^{\smile};\varepsilon^{\smile\downarrow} && \text{Lemma 4(2)}\\
&\sqsubseteq \mathrm{syQ}(\varepsilon,\varepsilon^{\downarrow});\varepsilon^{\smile\downarrow} && \text{Lemma 1(2)}\\
&\sqsubseteq \varepsilon^{\smile} && \text{Lemma 1(3)}\\
&\sqsubseteq \varepsilon^{\smile\uparrow} &&
\end{aligned}$$

which implies $\mathrm{down}^{\smile};\mathrm{down} \sqsubseteq \mathrm{syQ}(\varepsilon^{\uparrow},\varepsilon)$. Similar to the steps above this finally leads to $\mathrm{down}^{\smile};\mathrm{down} \sqsubseteq \mathbb{I}_L \sqcap \mathrm{up}$ and the second inclusion by (2). □

We are interested in those truth values that are used by crisp relations, i.e., in a vector representing the smallest and greatest element of L. This vector can be defined in two different ways. First of all, we could simply take $\mathrm{zero} \sqcup \mathrm{one}$. On the other hand, these elements are exactly the fixed points of down (or equivalently of up) which motivates the expression $\mathbb{T}_{1L};(\mathbb{I}_L \sqcap \mathrm{down})$ (or equivalently $\mathbb{T}_{1L};(\mathbb{I}_L \sqcap \mathrm{up})$).

Lemma 6. $\mathrm{zero} \sqcup \mathrm{one} \sqsubseteq \mathbb{T}_{1L};(\mathbb{I}_L \sqcap \mathrm{down})$.

Proof. We start by computing

$$\begin{aligned}
\mathrm{zero};\mathrm{down} &= \Lambda(\mathrm{zero};\varepsilon^{\smile\downarrow}) && \text{Lemma 4(3)}\\
&= \Lambda((\mathrm{zero};\varepsilon^{\smile})^{\downarrow}) && \text{Lemma 2(5)}\\
&= \Lambda(\mathbb{L}_{11}) && \text{Lemma 4(2)}\\
&= \mathrm{zero}
\end{aligned}$$

so that

$$\begin{aligned}
\mathrm{zero}^{\smile};\mathrm{zero} &= \mathbb{I}_L \sqcap \mathrm{zero}^{\smile};\mathrm{zero} && \text{zero univalent}\\
&\sqsubseteq \mathbb{I}_L \sqcap \mathrm{zero}^{\smile};\mathrm{zero};\mathrm{down};\mathrm{down}^{\smile} && \text{down total}\\
&\sqsubseteq \mathbb{I}_L \sqcap \mathrm{zero}^{\smile};\mathrm{zero};\mathrm{down}^{\smile} && \text{see above}\\
&\sqsubseteq \mathbb{I}_L \sqcap \mathrm{down}^{\smile} && \text{zero univalent}\\
&= \mathbb{I}_L \sqcap \mathrm{down} && \mathbb{I}_L \sqcap \mathrm{down} \sqsubseteq \mathbb{I}_L
\end{aligned}$$

follows. Similar we obtain $\mathrm{one}^{\smile};\mathrm{one} \sqsubseteq \mathbb{I}_L \sqcap \mathrm{down}$. We conclude

$$\begin{aligned}
\mathrm{zero} \sqcup \mathrm{one} &= \mathbb{T}_{11};\mathrm{zero} \sqcup \mathbb{T}_{11};\mathrm{one} \\
&= \mathbb{T}_{1L};\mathrm{zero}^{\smile};\mathrm{zero} \sqcup \mathbb{T}_{1L};\mathrm{one}^{\smile};\mathrm{one} && \text{zero, one total} \\
&= \mathbb{T}_{1L};(\mathrm{zero}^{\smile};\mathrm{zero} \sqcup \mathrm{one}^{\smile};\mathrm{one}) \\
&= \mathbb{T}_{1L};(\mathbb{I}_L \sqcap \mathrm{down}), && \text{see above}
\end{aligned}$$

i.e., the assertion. □

Unfortunately, so far we were not able to show the opposite inclusion of the property in the previous lemma. Most likely a proof needs to use the last axiom of an arrow category. However, it seems that this is not enough. We suspect that the proof needs the requirement that crisp relations forms a Boolean algebra because of the $\sqcup$ on the bigger side of the inclusion. Furthermore, additional relational products might be needed in order to avoid non-standard models. In the remainder of the paper we will use the relation $\mathbb{I}_L \sqcap \mathrm{down}$ rather than $\mathrm{zero} \sqcup \mathrm{one}$.

4 Covectorization

A relation $v : 1 \to A$ can be seen as a subset of A. Such a relation is called a vector. A covector is the converse of a vector, i.e., a relation $c : A \to 1$. Our goal in this section is to covectorize arbitrary relations, i.e., establishing an isomorphism between relations $R : A \to B$ and covectors $v : A \times B \to 1$. In addition, we will define operations on vectors that perform the usual relational operations applicable to the corresponding relations.

Given a relation $R : A \to B$ and a covector $v : A \times B \to 1$ we define $\mathrm{cov}(R) : A \times B \to 1$ and $\mathrm{rel}(v) : A \to B$ by

$$\begin{aligned}
\mathrm{cov}(R) &:= (R \otimes \mathbb{I}_B);\mathbb{T}_{B1}, \\
\mathrm{rel}(v) &:= \pi^{\smile};(v;\mathbb{T}_{1B} \sqcap \rho).
\end{aligned}$$

The first lemma establishes the isomorphism mentioned above.

Lemma 7. *The operations* $\mathrm{cov} : \mathcal{R}[A, B] \to \mathcal{R}[A \times B, 1]$ *and* $\mathrm{rel} : \mathcal{R}[A \times B, 1] \to \mathcal{R}[A, B]$ *are monotonic and inverse to each other, i.e., we have* $\mathrm{rel}(\mathrm{cov}(R)) = R$ *and* $\mathrm{cov}(\mathrm{rel}(v)) = v$ *for all* $R : A \to B$ *and* $v : A \times B \to 1$.

We now define two relations $\mathrm{swap}_{AB} : A \times B \to B \times A$, and $\mathrm{comp}_{ABC} : (A \times B) \times (B \times C) \to A \times C$ that we will use for internalization of the operations on relations. These relations are defined by

$$\begin{aligned}
\mathrm{swap}_{AB} &:= \rho \otimes \pi, \\
\mathrm{comp}_{ABC} &:= (\mathbb{I}_{(A \times B) \times (B \times C)} \sqcap \pi;\rho;\pi^{\smile};\rho^{\smile});(\pi \otimes \rho).
\end{aligned}$$

The next lemma shows how the usual operations on relations can be performed directly on the corresponding covectors. A proof of this lemma can be found in [14].

Lemma 8. *Let be $Q, R : A \to B$, and $S : B \to C$. Then we have:*

1. $\mathrm{cov}(Q^{\downarrow}) = \mathrm{cov}(Q)^{\downarrow}$,
2. $\mathrm{cov}(Q^{\uparrow}) = \mathrm{cov}(Q)^{\uparrow}$,
3. $\mathrm{cov}(Q \sqcup R) = \mathrm{cov}(Q) \sqcup \mathrm{cov}(R)$,
4. $\mathrm{cov}(Q \sqcap R) = \mathrm{cov}(Q) \sqcap \mathrm{cov}(R)$,
5. $\mathrm{cov}(Q^{\smile}) = \mathrm{swap}; \mathrm{cov}(Q)$,
6. $\mathrm{cov}(R; S) = \mathrm{comp}^{\smile}; (\mathrm{cov}(R) \otimes \mathrm{cov}(S))$.

In addition, we obtain the following.

Corollary 1. *Let be $u, v : A \times B \to 1$, and $w : B \times C \to 1$. Then we have:*

1. $\mathrm{rel}(u \sqcup v) = \mathrm{rel}(u) \sqcup \mathrm{rel}(v)$,
2. $\mathrm{rel}(u \sqcap v) = \mathrm{rel}(u) \sqcap \mathrm{rel}(v)$,
3. $\mathrm{rel}(\mathrm{swap}; u) = \mathrm{rel}(u)^{\smile}$,
4. $\mathrm{rel}(\mathrm{comp}^{\smile}; (u \otimes w)) = \mathrm{rel}(u); \mathrm{rel}(w)$.

The previous corollary is an immediate consequence of Lemma 7 and 8. If we now combine covectorization with the relational power, then we get the following properties. Please note that if $Q : A \to B$, then $\Lambda(\mathrm{cov}(Q)) : A \to B \to L$ so that we can now modify the degree of membership by simply composing this relation with a map starting in L.

Lemma 9. *Let be $Q, R : A \to B$, and $S : B \to C$. Then we have:*

1. $\Lambda(\mathrm{cov}(Q^{\downarrow})) = \Lambda(\mathrm{cov}(Q)); \mathrm{down}$,
2. $\Lambda(\mathrm{cov}(Q^{\uparrow})) = \Lambda(\mathrm{cov}(Q)); \mathrm{up}$,
3. $\Lambda(\mathrm{cov}(Q \sqcup R)) = (\Lambda(\mathrm{cov}(Q)) \otimes \Lambda(\mathrm{cov}(Q))); \mathcal{J}_2$,
4. $\Lambda(\mathrm{cov}(Q \sqcap R)) = (\Lambda(\mathrm{cov}(Q)) \otimes \Lambda(\mathrm{cov}(Q))); \mathcal{M}_2$,
5. $\Lambda(\mathrm{cov}(Q^{\smile})) = \mathrm{swap}; \Lambda(\mathrm{cov}(Q))$,
6. $\Lambda(\mathrm{cov}(R; S)) = \Lambda(\mathrm{comp}^{\smile}; (\Lambda(\mathrm{cov}(R)) \otimes \Lambda(\mathrm{cov}(S))); \mathcal{M}_2); \mathcal{J}$.

Proof. 1. We immediately compute

$$
\begin{aligned}
\Lambda(\mathrm{cov}(Q^{\downarrow})) &= \Lambda(\mathrm{cov}(Q)^{\downarrow}) && \text{Lemma 8(1)}\\
&= \Lambda((\Lambda(\mathrm{cov}(Q)); \varepsilon^{\smile})^{\downarrow}) && \text{Lemma 4(2)}\\
&= \Lambda(\Lambda(\mathrm{cov}(Q)); \varepsilon^{\smile\downarrow}) && \text{Lemma 2(5)}\\
&= \Lambda(\mathrm{cov}(Q)); \Lambda(\varepsilon^{\smile\downarrow}) && \text{Lemma 4(3)}\\
&= \Lambda(\mathrm{cov}(Q)); \mathrm{down}.
\end{aligned}
$$

2. Similar to (1) we obtain

$$\begin{aligned}
\Lambda(\mathrm{cov}(Q^{\uparrow})) &= \Lambda(\mathrm{cov}(Q)^{\uparrow}) && \text{Lemma 8(2)}\\
&= \Lambda((\Lambda(\mathrm{cov}(Q)); \varepsilon^{\smile})^{\uparrow}) && \text{Lemma 4(2)}\\
&= \Lambda(\Lambda(\mathrm{cov}(Q)); \varepsilon^{\smile\uparrow}) && \Lambda(\mathrm{cov}(Q)) \text{ crisp}\\
&= \Lambda(\mathrm{cov}(Q)); \Lambda(\varepsilon^{\smile\uparrow}) && \text{Lemma 4(3)}\\
&= \Lambda(\mathrm{cov}(Q)); \mathrm{up}.
\end{aligned}$$

3. Again we immediately have

$$\begin{aligned}
&\Lambda(\mathrm{cov}(Q \sqcup R))\\
&= \Lambda(\mathrm{cov}(Q) \sqcup \mathrm{cov}(R)) && \text{Lemma 8(3)}\\
&= \Lambda(\Lambda(\mathrm{cov}(Q)); \varepsilon^{\smile} \sqcup \Lambda(\mathrm{cov}(R)); \varepsilon^{\smile}) && \text{Lemma 4(2)}\\
&= \Lambda((\Lambda(\mathrm{cov}(Q)) \sqcup \Lambda(\mathrm{cov}(R))); \varepsilon^{\smile})\\
&= \Lambda(((\Lambda(\mathrm{cov}(Q)) \otimes \Lambda(\mathrm{cov}(Q))); \pi\\
&\qquad \sqcup (\Lambda(\mathrm{cov}(Q)) \otimes \Lambda(\mathrm{cov}(Q))); \rho); \varepsilon^{\smile}) && \text{Lemma 3(2)}\\
&= \Lambda((\Lambda(\mathrm{cov}(Q)) \otimes \Lambda(\mathrm{cov}(Q))); (\pi \sqcup \rho); \varepsilon^{\smile})\\
&= (\Lambda(\mathrm{cov}(Q)) \otimes \Lambda(\mathrm{cov}(Q))); \Lambda((\pi \sqcup \rho); \varepsilon^{\smile}) && \text{Lemma 4(3)}\\
&= (\Lambda(\mathrm{cov}(Q)) \otimes \Lambda(\mathrm{cov}(Q))); \mathcal{J}_2.
\end{aligned}$$

4. Similar to (3) we have

$$\begin{aligned}
&\Lambda(\mathrm{cov}(Q \sqcap R))\\
&= \Lambda(\mathrm{cov}(Q) \sqcap \mathrm{cov}(R)) && \text{Lemma 8(4)}\\
&= \Lambda(\Lambda(\mathrm{cov}(Q)); \varepsilon^{\smile} \sqcap \Lambda(\mathrm{cov}(R)); \varepsilon^{\smile}) && \text{Lemma 4(2)}\\
&= \Lambda((\Lambda(\mathrm{cov}(Q)) \otimes \Lambda(\mathrm{cov}(Q))); (\varepsilon^{\smile} \otimes \varepsilon^{\smile})) && \text{Lemma 3(4)}\\
&= (\Lambda(\mathrm{cov}(Q)) \otimes \Lambda(\mathrm{cov}(Q))); \Lambda(\varepsilon^{\smile} \otimes \varepsilon^{\smile}) && \text{Lemma 4(3)}\\
&= (\Lambda(\mathrm{cov}(Q)) \otimes \Lambda(\mathrm{cov}(Q))); \mathcal{M}_2.
\end{aligned}$$

5. We immediately compute

$$\begin{aligned}
\Lambda(\mathrm{cov}(Q^{\smile})) &= \Lambda(\mathrm{swap}; \mathrm{cov}(Q)) && \text{Lemma 8(5)}\\
&= \mathrm{swap}; \Lambda(\mathrm{cov}(Q)). && \text{Lemma 4(3)}
\end{aligned}$$

6. From the calculation

$$\begin{aligned}
&\Lambda(\mathrm{cov}(R; S))\\
&= \Lambda(\mathrm{comp}^{\smile}; (\mathrm{cov}(R) \otimes \mathrm{cov}(S))) && \text{Lemma 8(6)}\\
&= \Lambda(\mathrm{comp}^{\smile}; \Lambda(\mathrm{cov}(R) \otimes \mathrm{cov}(S)); \varepsilon^{\smile}) && \text{Lemma 4(2)}\\
&= \Lambda(\Lambda(\mathrm{comp}^{\smile}; \Lambda(\mathrm{cov}(R) \otimes \mathrm{cov}(S))); \varepsilon^{\smile}; \varepsilon^{\smile}) && \text{Lemma 4(2)}
\end{aligned}$$

$$= \Lambda(\mathrm{comp}^{\smile}; \Lambda(\mathrm{cov}(R) \otimes \mathrm{cov}(S))); \Lambda(\varepsilon^{\smile}; \varepsilon^{\smile}) \qquad \text{Lemma 4(3)}$$
$$= \Lambda(\mathrm{comp}^{\smile}; \Lambda(\mathrm{cov}(R) \otimes \mathrm{cov}(S))); \mathcal{J}$$
$$= \Lambda(\mathrm{comp}^{\smile}; \Lambda(\Lambda(\mathrm{cov}(R)); \varepsilon^{\smile} \otimes \Lambda(\mathrm{cov}(S)); \varepsilon^{\smile})); \mathcal{J} \qquad \text{Lemma 4(2)}$$
$$= \Lambda(\mathrm{comp}^{\smile}; \Lambda((\Lambda(\mathrm{cov}(R)) \otimes \Lambda(\mathrm{cov}(S))); (\varepsilon^{\smile} \otimes \varepsilon^{\smile}))); \mathcal{J} \qquad \text{Lemma 3(4)}$$
$$= \Lambda(\mathrm{comp}^{\smile}; (\Lambda(\mathrm{cov}(R)) \otimes \Lambda(\mathrm{cov}(S))); \Lambda(\varepsilon^{\smile} \otimes \varepsilon^{\smile})); \mathcal{J} \qquad \text{Lemma 4(3)}$$
$$= \Lambda(\mathrm{comp}^{\smile}; (\Lambda(\mathrm{cov}(R)) \otimes \Lambda(\mathrm{cov}(S))); \mathcal{M}_2); \mathcal{J}$$

we obtain the assertion. □

5 Change of Base

In this section we want to develop a mechanism for changing the base of L-fuzzy relations in the abstract setting of arrow categories. Therefore, we start with two arrow categories $\mathcal{A}_1$ and $\mathcal{A}_2$ each with a unit, relational products, and relational powers over the Heyting algebras L_1 and L_2, respectively. A change of base now is an appropriate operation that maps relations from $\mathcal{A}_1$ to $\mathcal{A}_2$ based on some map between L_1 and L_2. However, since a change of base is supposed to exchange membership values only and keep the structural characteristics (the matrix structure) of a relation the same, we normally would require that $\mathcal{A}_1^{\downarrow}$ and $\mathcal{A}_2^{\downarrow}$ are equal, i.e., are the same Dedekind category. However, we will generalize this slightly by requiring that those two Dedekind categories are isomorphic with the same objects, i.e., we will require that there are functors $H : \mathcal{A}_1 \to \mathcal{A}_2$ and $H^{-1} : \mathcal{A}_2 \to \mathcal{A}_1$ that are the identity on objects, preserve all operations and constants of $\mathcal{A}_1^{\downarrow}$ resp. $\mathcal{A}_2^{\downarrow}$, and are inverse to each other. Please note that this includes the requirement that the unit and all products are the same in both categories, and that H and H^{-1} map projections to projections. On the other hand, these functors will not preserve relational powers since, even though the corresponding objects are available in both categories, the relational power is a not construction in $\mathcal{A}_1^{\downarrow}$ resp. $\mathcal{A}_2^{\downarrow}$ due to the fact that ε is not crisp. For example, for a given object A, $\mathcal{A}_1$ has the relational power L_1^A with the relation $\varepsilon_1 : A \to L_1^A$. The object L_1^A is also an object of $\mathcal{A}_2$ since all four categories $\mathcal{A}_1$, $\mathcal{A}_2$, $\mathcal{A}_1^{\downarrow}$, and $\mathcal{A}_2^{\downarrow}$ have the same objects. But this object is not a relational power in $\mathcal{A}_2$ (unless the two categories are already the same). The relational power of A in $\mathcal{A}_2$ is the object L_2^A together with $\varepsilon_2 : A \to L_2^A$. Furthermore, H is not defined for ε_1 since the relation is not crisp (unless L_1 is trivial). The main reason for the generalization outlined above is that we would like to make explicit in the relational expression of this section where we switch from $\mathcal{A}_1$ to $\mathcal{A}_2$ and, hence, which operation belongs to which category.

We now start with a crisp map $f : L_1 \to L_2$ in $\mathcal{A}_1$. Please note that $H(f)$ is morphism in $\mathcal{A}_2$ so that we could, alternatively, start with a crisp map between L_1 and L_2 in $\mathcal{A}_2$. The we define

$$F(Q) := \mathrm{rel}(H(\Lambda_1(\mathrm{cov}(Q)); f); \varepsilon_2^{\smile})$$

for every $Q : A \to B$ in $\mathcal{A}_1$. Please note that $F(Q) : A \to B$ in $\mathcal{A}_2$, and that we use the indices 1 and 2 indicating in which category the relation is defined if desired. The definition of F is visualized in the Fig. 1 where solid arrows are crisp relations and $R = H(\Lambda_1(\text{cov}(Q)); f); \varepsilon_2^{\smile}$.

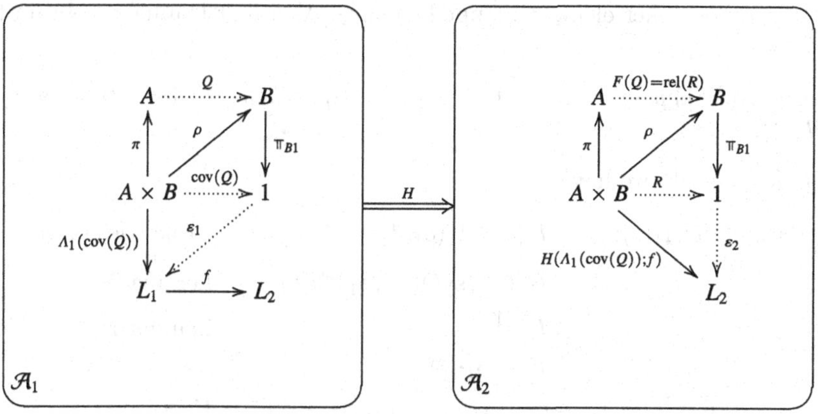

Fig. 1. Visualization of the definition of F.

Without any further assumption on f we immediately obtain that F preserves the converse operation.

Theorem 1. *F preserves converse, i.e., $F(Q^{\smile}) = F(Q)^{\smile}$ for all relations $Q : A \to B$.*

Proof. We have

$$\begin{aligned} F(Q^{\smile}) &= \text{rel}(H(\Lambda_1(\text{cov}(Q^{\smile})); f); \varepsilon_2^{\smile}) \\ &= \text{rel}(H(\text{swap}_1; \Lambda_1(\text{cov}(Q)); f); \varepsilon_2^{\smile}) && \text{Lemma 9(5)} \\ &= \text{rel}(\text{swap}_2; H(\Lambda_1(\text{cov}(Q)); f); \varepsilon_2^{\smile}) && \text{Properties of } H \\ &= \text{rel}(H(\Lambda_1(\text{cov}(Q)); f); \varepsilon_2^{\smile})^{\smile} && \text{Corollary 1(3)} \\ &= F(Q)^{\smile}, \end{aligned}$$

i.e., the assertion. □

In the remainder of this section we would like to investigate under which assumption on f we obtain nice properties of the change of base F. As a first assumption on f we would like to consider that f maps the smallest and greatest element of L_1 to the ones of L_2. We choose the following two conditions postulating that f maps the fixed points of down_1 to the fixed points of down_2 and the greatest element of L_1 uniquely to the greatest element of L_2. Please note that we have to formulate this as a condition involving H and the two categories, of course.

P1: $H((\mathbb{I}_{L_1} \sqcap \mathrm{down}_1); f) = H(f); (\mathbb{I}_{L_2} \sqcap \mathrm{down}_2)$,
P2a: $H(\mathrm{one}_1; f) = \mathrm{one}_2$,
P2b: $\mathrm{one}_2; H(f^{\smile}) = H(\mathrm{one}_1)$.

Condition **P2a** requires that one_1 is mapped to one_2. This condition already implies the inclusion $\sqsupseteq$ of **P2b**. The opposite inclusion adds the requirement that there is no other element mapped to one_2. As a first lemma we obtain the following.

Lemma 10. *Suppose $c : A \to 1$ in $\mathcal{A}_1$ is crisp. Then we have $H(\Lambda_1(c); f) = \Lambda_2(H(c))$.*

Proof. First of all, we have

$$\begin{aligned}
H(c)^{\smile}; H(\Lambda_1(c); f) &= H(c^{\smile}; \Lambda_1(c); f) && \text{Properties of } H\\
&\sqsubseteq H((c^{\smile}; \mathrm{syQ}(c^{\smile}, \varepsilon_1))^{\downarrow}; f) && \text{Lemma 2(4)}\\
&\sqsubseteq H(\varepsilon_1^{\downarrow}; f) && \text{Lemma 1(3)}\\
&= H(\mathrm{one}_1; f) && \text{see Sect. 3}\\
&= \mathrm{one}_2 && \textbf{P2a}\\
&= \varepsilon_2^{\downarrow}, && \text{see Sect. 3}
\end{aligned}$$

$$\begin{aligned}
H(\Lambda_1(c); f); \varepsilon_2^{\smile} &= H(\Lambda_1(c); (\mathbb{I}_{L_1} \sqcap \mathrm{down}_1); f); \varepsilon_2^{\smile}\\
&= H(\Lambda_1(c)); H((\mathbb{I}_{L_1} \sqcap \mathrm{down}_1); f); \varepsilon_2^{\smile} && \text{Properties of } H\\
&= H(\Lambda_1(c)); H(f); (\mathbb{I}_{L_2} \sqcap \mathrm{down}_2); \varepsilon_2^{\smile} && \textbf{P1}\\
&= H(\Lambda_1(c)); H(f); \varepsilon^{\smile\downarrow}_2\\
&= H(\Lambda_1(c)); H(f); \mathrm{one}_2^{\smile} && \text{see Sect. 3}\\
&= H(\Lambda_1(c)); H(\mathrm{one}_1^{\smile}) && \textbf{P2b}\\
&= H(\Lambda_1(c); \mathrm{one}_1^{\smile}) && \text{Properties of } H\\
&= H(\Lambda_1(c); \varepsilon^{\smile\downarrow}_1) && \text{see Sect. 3}\\
&= H((\Lambda_1(c); \varepsilon_1^{\smile})^{\downarrow}) && \text{Lemma 2(5)}\\
&= H(c^{\downarrow}) && \text{Lemma 4(2)}\\
&= H(c) && c \text{ crisp}
\end{aligned}$$

so that $H(\Lambda_1(c); f) \sqsubseteq \mathrm{syQ}(H(c)^{\smile}, \varepsilon_2)$ follows. This immediately implies $H(\Lambda_1(c); f) = H(\Lambda_1(c); f)^{\downarrow} \sqsubseteq \mathrm{syQ}(H(c)^{\smile}, \varepsilon_2)^{\downarrow} = \Lambda_2(H(c))$. This inclusion is an equality because both sides are maps. □

Now we are ready to verify our first property of the change of base.

Theorem 2. *If f satisfies* **P1**, **P2a** *and* **P2b**, *then F extends H, i.e., $F(Q) = H(Q)$ for all crisp relations $Q : A \to B$.*

Proof. First of all, notice that Q is crisp iff $\text{cov}(Q)$ is crisp. Then we compute

$$\begin{aligned} F(Q) &= \text{rel}(H(\Lambda_1(\text{cov}(Q)); f); \varepsilon_2^{\smile}) \\ &= \text{rel}(\Lambda_2(H(\text{cov}(Q))); \varepsilon_2^{\smile}) && \text{Lemma 10} \\ &= \text{rel}(\Lambda_2(\text{cov}(H(Q))); \varepsilon_2^{\smile}) && \text{Properties of } H \\ &= \text{rel}(\text{cov}(H(Q))) && \text{Lemma 4(2)} \\ &= H(Q), && \text{Lemma 7} \end{aligned}$$

showing that F extends H. □

Next we would like to show that F preserves joins if f does this internally with respect to L_1, L_2 and the corresponding join relations. Therefore, we consider the condition

P3: $H(\mathcal{J}_2^1; f) = H(f \otimes f); \mathcal{J}_2^2$.

If H is the identity, then **P3** is just the regular equation requiring preservation of binary joins. Please note again that we use 1 and 2 to indicate in which category the binary join relations are defined.

Theorem 3. *If f satisfies* **P3***, then F preserves binary joins, i.e., $F(Q \sqcup R) = F(Q) \sqcup F(R)$ for all relations $Q, R : A \to B$.*

Proof. We obtain

$$\begin{aligned} &F(Q \sqcup R) \\ &= \text{rel}(H(\Lambda_1(\text{cov}(Q \sqcup R)); f); \varepsilon_2^{\smile}) \\ &= \text{rel}(H((\Lambda_1(\text{cov}(Q)) \otimes \Lambda_1(\text{cov}(R))); \mathcal{J}_2^1; f); \varepsilon_2^{\smile}) && \text{Lemma 9(3)} \\ &= \text{rel}(H((\Lambda_1(\text{cov}(Q)) \otimes \Lambda_1(\text{cov}(R))); (f \otimes f)); \mathcal{J}_2^2; \varepsilon_2^{\smile}) && \textbf{P3} \\ &= \text{rel}(H((\Lambda_1(\text{cov}(Q)); f \otimes \Lambda_1(\text{cov}(R)); f)); \mathcal{J}_2^2; \varepsilon_2^{\smile}) && \text{Lemma 3(4)} \\ &= \text{rel}(H((\Lambda_1(\text{cov}(Q)); f \otimes \Lambda_2(\text{cov}(R)); f)); (\pi \sqcup \rho); \varepsilon_2^{\smile}) && \text{Lemma 4(2)} \\ &= \text{rel}(H((\Lambda_1(\text{cov}(Q)); f) \otimes H(\Lambda_1(\text{cov}(R)); f)); (\pi \sqcup \rho); \varepsilon_2^{\smile}) && \text{Properties of } H \\ &= \text{rel}(H(\Lambda_1(\text{cov}(Q)); f); \varepsilon_2^{\smile} \sqcup H(\Lambda_1(\text{cov}(R)); f); \varepsilon_2^{\smile}) && \text{Lemma 3(2)} \\ &= \text{rel}(H(\Lambda_1(\text{cov}(Q)); f); \varepsilon_2^{\smile}) \sqcup \text{rel}(H(\Lambda_1(\text{cov}(R)); f); \varepsilon_2^{\smile}) && \text{Corollary 1(1)} \\ &= F(Q) \sqcup F(R). \end{aligned}$$

This completes the proof. □

We obtain a similar result if we exchange joins with meets. Due to lack of space we only formulate the results.

P4: $H(\mathcal{M}_2^1; f) = H(f \otimes f); \mathcal{M}_2^2$.

Similar to **P3** the condition **P4** requires that f preserves binary meets.

Theorem 4. *If f satisfies* **P4**, *then F preserves binary meets, i.e., $F(Q \sqcap R) = F(Q) \sqcap F(R)$ for all relations $Q, R : A \to B$.*

Last but not least, we are interested in F preserving composition. Therefore, we consider

P5: $H(\Lambda_1(\varepsilon_1^{\smile\downarrow};\varepsilon_1^{\smile});f) = \Lambda_2(H(\varepsilon_1^{\smile\downarrow};f));\mathcal{J}^2$.

This property needs some explanation. Intuitively, **P5** requires that joins of sets of L_1 elements are preserved. This needs to be formulated in a context where all sets are always L_1 (resp. L_2) fuzzy subsets. The relations on both sides of the equation are crisp and relate an L_1-fuzzy subset M of L_1 with an element of L_2. On the right hand side, the relation $\Lambda_1(\varepsilon_1^{\smile\downarrow};\varepsilon_1^{\smile})$ is similar to $\mathcal{J}^1$ except that for M it computes the join in L_1 of all elements from $M^\downarrow$ (instead of M), i.e., of all elements from L_1 included in M with degree 1. That result is mapped by f. On the left hand side, we map the elements from $M^\downarrow$ by f, and then compute their join in L_2.

Theorem 5. *If f satisfies* **P4** *and* **P5**, *then F preserves composition, i.e., $F(R;S) = F(R);F(S)$ for all relations $R : A \to B$ and $S : B \to C$.*

Proof. First of all, we obtain for a crisp relation $Q : D \to L_1$

$$\begin{aligned}
\Lambda_1(Q);\mathcal{J}^1 &= \Lambda_1(\Lambda_1(Q);\varepsilon_1^{\smile};\varepsilon_1^{\smile}) && \text{Lemma 4(3)}\\
&= \Lambda_1(Q;\varepsilon_1^{\smile}) && \text{Lemma 4(2)}\\
&= \Lambda_1(Q^\downarrow;\varepsilon_1^{\smile}) && \text{crisp}\\
&= \Lambda_1((\Lambda_1(Q);\varepsilon_1^{\smile})^\downarrow;\varepsilon_1^{\smile}) && \text{Lemma 4(2)}\\
&= \Lambda_1(\Lambda_1(Q);\varepsilon_1^{\smile\downarrow};\varepsilon_1^{\smile}) && \text{Lemma 2(5)}\\
&= \Lambda_1(Q);\Lambda_1(\varepsilon_1^{\smile\downarrow};\varepsilon_1^{\smile}). && \text{Lemma 4(3)}
\end{aligned}$$

We use the abbreviations $P = \mathrm{comp}^{\smile};(\Lambda_1(\mathrm{cov}(R)) \otimes \Lambda_1(\mathrm{cov}(S)))$ and $Q = P;\mathcal{M}_2^1$. Notice that P and Q are both crisp. First of all, we have

$$\begin{aligned}
&H(P;(f \otimes f));(\varepsilon_2^{\smile} \oslash \varepsilon_2^{\smile})\\
&= H(\mathrm{comp}^{\smile};(\Lambda_1(\mathrm{cov}(R)) \otimes \Lambda_1(\mathrm{cov}(S)));(f \otimes f));(\varepsilon_2^{\smile} \oslash \varepsilon_2^{\smile})\\
&= H(\mathrm{comp}^{\smile};(\Lambda_1(\mathrm{cov}(R));f \otimes \Lambda_1(\mathrm{cov}(S));f));(\varepsilon_2^{\smile} \oslash \varepsilon_2^{\smile}) && \text{Lemma 3(4)}\\
&= \mathrm{comp}^{\smile};(H(\Lambda_1(\mathrm{cov}(R));f) \otimes H(\Lambda_1(\mathrm{cov}(S));f));(\varepsilon_2^{\smile} \oslash \varepsilon_2^{\smile}) && \text{Properties of } H\\
&= \mathrm{comp}^{\smile};(H(\Lambda_1(\mathrm{cov}(R));f);\varepsilon_2^{\smile} \oslash H(\Lambda_1(\mathrm{cov}(S));f);\varepsilon_2^{\smile}). && \text{Lemma 3(4)}
\end{aligned}$$

Combining the results so far we obtain

$$\begin{aligned}
F(R;S) &= \mathrm{rel}(H(\Lambda_1(\mathrm{cov}(R;S));f);\varepsilon_2^{\smile})\\
&= \mathrm{rel}(H(\Lambda_1(Q);\mathcal{J}^1;f);\varepsilon_2^{\smile}) && \text{Lemma 9(6)}
\end{aligned}$$

$$
\begin{aligned}
&= \mathrm{rel}(H(\Lambda_1(Q);\Lambda_1(\varepsilon_1^{\smile\downarrow};\varepsilon_1^{\smile});f);\varepsilon_2^{\smile}) && \text{see above}\\
&= \mathrm{rel}(H(\Lambda_1(Q));H(\Lambda_1(\varepsilon_1^{\smile\downarrow};f));\varepsilon_2^{\smile}) && \text{Properties of } H\\
&= \mathrm{rel}(H(\Lambda_1(Q));\Lambda_2(H(\varepsilon_1^{\smile\downarrow};f));\mathcal{J}^2;\varepsilon_2^{\smile}) && \textbf{P5}\\
&= \mathrm{rel}(H(\Lambda_1(Q));\Lambda_2(H(\varepsilon_1^{\smile\downarrow};f));\varepsilon_2^{\smile};\varepsilon_2^{\smile}) && \text{Lemma 4(2)}\\
&= \mathrm{rel}(H(\Lambda_1(Q));H(\varepsilon_1^{\smile\downarrow};f);\varepsilon_2^{\smile}) && \text{Lemma 4(2)}\\
&= \mathrm{rel}(H(\Lambda_1(Q);\varepsilon_1^{\smile\downarrow};f);\varepsilon_2^{\smile}) && \text{Properties of } H\\
&= \mathrm{rel}(H((\Lambda_1(Q);\varepsilon_1^{\smile})^{\downarrow};f);\varepsilon_2^{\smile}) && \text{Lemma 2(5)}\\
&= \mathrm{rel}(H(Q^{\downarrow};f);\varepsilon_2^{\smile}) && \text{Lemma 4(2)}\\
&= \mathrm{rel}(H(Q;f);\varepsilon_2^{\smile}) && Q \text{ crisp}\\
&= \mathrm{rel}(H(P;\mathcal{M}_2^1;f);\varepsilon_2^{\smile}) &&\\
&= \mathrm{rel}(H(P);H(\mathcal{M}_2^1;f);\varepsilon_2^{\smile}) && \text{Properties of } H\\
&= \mathrm{rel}(H(P);H(f\otimes f);\mathcal{M}_2^2;\varepsilon_2^{\smile}) && \textbf{P4}\\
&= \mathrm{rel}(H(P;(f\otimes f));\mathcal{M}_2^2;\varepsilon_2^{\smile}) && \text{Properties of } H\\
&= \mathrm{rel}(H(P;(f\otimes f));(\varepsilon_2^{\smile}\otimes\varepsilon_2^{\smile})) && \text{Lemma 4(2)}\\
&= \mathrm{rel}(\mathrm{comp}^{\smile};(H(\Lambda_1(\mathrm{cov}(R));f);\varepsilon_2^{\smile} &&\\
&\qquad \otimes H(\Lambda_1(\mathrm{cov}(S));f);\varepsilon_2^{\smile})) && \text{see above}\\
&= \mathrm{rel}(H(\Lambda_1(\mathrm{cov}(R));f);\varepsilon_2^{\smile});\mathrm{rel}(H(\Lambda_1(\mathrm{cov}(S));f);\varepsilon_2^{\smile}) && \text{Corollary 1(4)}\\
&= F(R);F(S),
\end{aligned}
$$

i.e., the assertion. □

Summarizing the results of this section we obtain the following corollary.

Corollary 2. *If f satisfies* **P1-P5**, *then F extends H and preserves converse, binary joins and meets, and composition.*

Please notice that in the situation of the previous corollary F also preserves all constants because F extends H and all constants are crisp.

References

1. Berghammer, R., Winter, M.: Decomposition of relations and concept lattices. Fundamenta Informaticae **126**, 37–82 (2013)
2. Freyd, P., Scedrov, A.: Categories, Allegories. North-Holland Mathematical Library, vol. 39, North-Holland, Amsterdam (1990)
3. Furusawa, H., Kahl, W.: A study on symmetric quotients. University of the Federal Armed Forces Munich, Bericht Nr. 1998-06 (1998)
4. Kawahara Y., Furusawa, H.: Crispness and representation theorems in dedekind categories. DOI-TR 143, Kyushu University (1997)

5. Mendel, J.M.: Uncertain Rule-Based Fuzzy Logic Systems: Introduction and New Directions. Prentice-Hall, Hoboken (2001)
6. Schmidt, G., Ströhlein, T.: Relations and graphs. Discrete Mathematics for Computer Scientists. EATCS Monographs on Theoretical Computer Science. Springer, Heidelberg (1993). https://doi.org/10.1007/978-3-642-77968-8
7. Schmidt, G.: Relational Mathematics. Encyclopedia of Mathematics and Its Applications, vol. 132. Cambridge University Press, Cambridge (2011)
8. Winter, M.: Strukturtheorie heterogener Relationenalgebren mit Anwendung auf Nichtdetermismus in Programmiersprachen. Dissertationsverlag NG Kopierladen GmbH, München (1998)
9. Winter, M.: A new algebraic approach to L-Fuzzy relations convenient to study crispness. INS Inf. Sci. **139**, 233–252 (2001)
10. Winter, M.: Goguen Categories - A Categorical Approach to L-Fuzzy Relations. Trends in Logic, vol. 25, Springer, Cham (2007). https://doi.org/10.1007/978-1-4020-6164-6
11. Winter, M.: Arrow categories. Fuzzy Sets Syst. **160**, 2893–2909 (2009)
12. Winter, M.: Type-n arrow categories. In: Höfner, P., Pous, D., Struth, G. (eds.) RAMICS 2017. LNCS, vol. 10226, pp. 307–322. Springer, Cham (2017). https://doi.org/10.1007/978-3-319-57418-9_19
13. Winter, M.: Relational sums and splittings in categories of L-Fuzzy relations. In: Fahrenberg, U., Gehrke, M., Santocanale, L., Winter, M. (eds.) RAMiCS 2021. LNCS, vol. 13027, pp. 435–449. Springer, Cham (2021). https://doi.org/10.1007/978-3-030-88701-8_26
14. Winter, M.: Fixed Point Operators in Heyting Categories (in preparation)

Automated Reasoning for Probabilistic Sequential Programs with Theorem Proving

Kangfeng Ye(✉), Simon Foster, and Jim Woodcock

University of York, York, UK
{kangfeng.ye,simon.foster,jim.woodcock}@york.ac.uk

Abstract. Semantics for nondeterministic probabilistic sequential programs has been well studied in the past decades. In a variety of semantic models, how nondeterministic choice interacts with probabilistic choice is the most significant difference. In He, Morgan, and McIver's relational model, probabilistic choice refines nondeterministic choice. This model is general because of its predicative-style semantics in Hoare and He's Unifying Theories of Programming, and suitable for automated reasoning because of its algebraic feature. Previously, we gave probabilistic semantics to the RoboChart notation based on this model, and also formalised the proof that the semantic embedding is a homomorphism, and revealed interesting details. In this paper, we present our mechanisation of the proof in Isabelle/UTP enabling automated reasoning for probabilistic sequential programs including a subset of the RoboChart language. With mechanisation, we even reveal more interesting questions, hidden in the original model. We demonstrate several examples, including an example to illustrate the interaction between nondeterministic choice and probabilistic choice, and a RoboChart model for randomisation based on binary probabilistic choice.

1 Introduction

In our previous work [1], we give a probabilistic semantics to RoboChart [2], a domain-specific language for robotics and distinctive in its support for automated verification, based on He, Morgan and McIver's relational model [3]. The semantics of the model is the theory of designs in Hoare and He's Unifying Theories of Programming (UTP) [4]. The model embeds standard designs in probabilistic designs through the weakest completion solution [3] which is defined on the weakest prespecification [5] and a forgetful function [1,3]. In this paper, we present our mechanisation of the probabilistic semantics in Isabelle/UTP [6], an implementation of UTP in the Isabelle/HOL theorem prover [7].

The main contributions of this work include (1) the formalisation of the proof of the homomorphism for the embedding of sequential composition, which is not addressed in our previous work [1]; and (2) the theory of probabilistic designs in Isabelle/UTP for automated reasoning of probabilistic nondeterministic sequential programs. All definitions and theorems in this paper are mechanised and accompanying icons (🌐) link to corresponding repository artifacts.

U. Fahrenberg et al. (Eds.): RAMiCS 2021, LNCS 13027, pp. 465–482, 2021.
https://doi.org/10.1007/978-3-030-88701-8_28

The remainder of this paper is organised as follows. In Sect. 2, we present an implementation of a randomisation algorithm in RoboChart, based on a binary probabilistic choice. Section 3 briefly describes the syntax of *pGCL* [8] (that is used in our mechanisation) with a program for the algorithm (with recursion), and the relational semantics in UTP. In Sect. 4, we describe how to represent probability distributions, and define weakest prespecification and a forgetful function in Isabelle/UTP. We show the embedding is a homomorphism on the structure of standard programs (Sect. 5), probabilistic choice (Sect. 6), nondeterministic choice (Sect. 7), and sequential composition (Sect. 8). We demonstrate the automated reasoning with three examples in Sect. 9, review related work in Sect. 10, and conclude in Sect. 11.

2 RoboChart

We consider a randomisation algorithm that aims to choose an integer number from a set of integers with equal probability. Let the set be $[0..N)$, that is, integers from 0 to $(N-1)$. We implement the algorithm in an operation ChooseUniform in RoboChart, as shown in Fig. 1. The core of RoboChart is a subset of UML state machines that allows modelling of robotic controllers. We use this model as an example for automated reasoning about probabilistic programs.

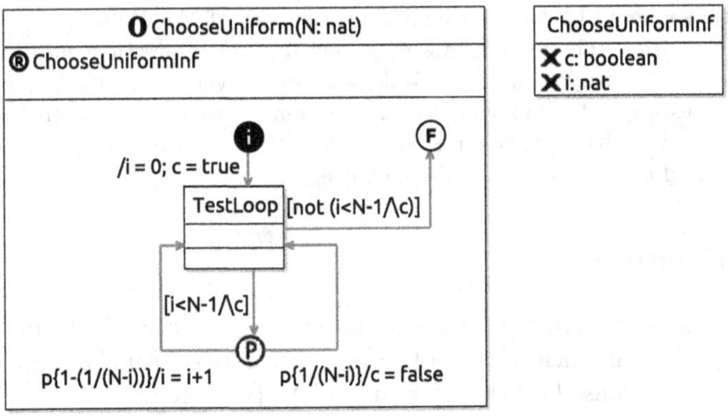

Fig. 1. A RoboChart model: uniform distribution algorithm.

The operation has one parameter N, denoting the size of the set, and has access (write and read) to two variables: c of type boolean and i of type nat, through the requested interface ChooseUniformInf. The operation is defined by a finite state machine having several nodes: one initial junction (❶), a state TestLoop, a probabilistic junction (Ⓟ), and a final state (Ⓕ). Transitions connect nodes and are optionally labelled with a guard (a boolean expression e inside brackets, [e]), a probability value (an expression e inside braces after p, p{e}), and an action (a statement act after /, /act). This state machine, in general, implements the algorithm in three stages: (1) initialisation by the action of the transition from ❶ to TestLoop; (2) iteration from TestLoop to Ⓟ, then back to TestLoop; and (3) termination by the transition from TestLoop to Ⓕ. The iteration is guarded by the condition if i has not reached (N − 1) and c is true. The termination is guarded by the negation of the condition. In each iteration, at Ⓟ, the state machine has probability (1/(N − i)) to update c to false (by the right transition) and so terminate next, and probability (1 − 1/(N-i)) to increase i (by the left transition) and so another iteration or terminate next depending on the condition (i < (N − 1)). An accompanying tool for RoboChart ensures that the probabilities on outgoing transitions of a probabilistic junction add up to 1 through validation of well-formedness conditions of RoboChart models. If they do not add up to 1, then an error is displayed.

3 Probabilistic Programs

Syntax. The abstract syntax of a nondeterministic probabilistic sequential programming language is given below.

$$P ::= \bot \mid I\!I \mid x := e \mid P \lhd b \rhd Q \mid P \sqcap Q \mid P \oplus_r Q \mid P\,;\ Q \mid \mu X \bullet P(X)$$

This probabilistic language introduces in the standard language a probabilistic choice operator $P \oplus_r Q$ which chooses between P and Q with probability r and $(1 - r)$ respectively. The syntax in the standard language includes abort $\bot$, skip $I\!I$, conditional $\lhd \_ \rhd$, and other common constructors.

The uniform distribution algorithm in Fig. 1 is implemented as such a probabilistic program shown in Definition 3.1.

Definition 3.1 (The uniform distribution program).

$$\begin{aligned} &\mathit{ChooseUniform}(N) \\ &\triangleq i := 0;\ c := \mathit{true}\,;\ \mu X \bullet \begin{pmatrix} ((c := \mathit{false}) \oplus_{1/(N-i)} (i := i+1))\,;\ X \\ \lhd (i < (N-1) \land c) \rhd I\!I \end{pmatrix} \end{aligned}$$

Semantics. The semantics of probabilistic programs is given in terms of probabilistic designs [1,3], being lifted from standard designs in UTP [4,9] via weakest completion semantics [3]. UTP employs Hehner's predicative style [10] to treat programs as predicates. It uses the alphabetised relational calculus to encode programs as relations between initial variable observations (x) and subsequent observations (x'). Relations are alphabetised predicates of which each is accompanied by its alphabet (a set of typed variable declarations). The alphabet of a predicate P is divided into the input alphabet ($in\alpha P = \{x, y, ...\}$) and the output alphabet ($out\alpha P = \{x', y', ...\}$).

UTP designs [4] are a subset of alphabetised predicates and denoted as $P \vdash R$: precondition-postcondition pairs. Designs have an additional variable ok in their alphabets to record the termination of programs. We use S to denote the state space of a program, containing only user variables and excluding ok.

An alphabet induces a state space. Probabilistic designs are defined as $p(s) \vdash R(s, prob')$, where $s \in S$ and $prob' \in PROB$ (probabilistic state space). Here, we use p instead of P, to denote that the precondition of a probabilistic design is a condition ($out\alpha\, p = \emptyset$). $PROB$ represents a set of (discrete) probability distributions over S: $PROB \triangleq S \rightarrow [0, 1]$. We are using discrete distributions because the only way to introduce them in the probabilistic programs is through the binary probabilistic choice operator. For each distribution $prob$ in $PROB$, its probabilities sum to 1: $\sum_{s \in S} prob(s) = 1$.

Both standard and probabilistic designs, which are used to give semantics to probabilistic programs, have their preconditions being conditions. In Isabelle/UTP, such designs are named normal designs [11] and denoted by $p \vdash_n R$.

Alphabetised predicates are presented in Isabelle/UTP through alphabetised expressions $[V, S]$*uexpr*, which are parametric over the value type V and the observation space S, and defined as total functions $S \rightarrow V$. Alphabetised predicates are boolean expressions: $[S]$*upred* $\triangleq$ [*bool*, S]*uexpr*. Relations are predicates over a product space: $[S_1, S_2]$*urel* $\triangleq$ $[S_1 \times S_2]$*upred*, where S_1 and S_2 are the initial and final observation space. Designs are relations with an additional ok variable: $[S_1, S_2]$*rel_des* $\triangleq$ $[[S_1]$*des*, $[S_2]$*des*]*urel*, where *des* introduces the ok variable into alphabets. If S_1 is the same as S_2, we use $[S]$*hrel_des* for $[S, S]$*rel_des*.

4 Probabilistic Designs in Isabelle/UTP

We use the weakest completion solution [3] to embed standard designs D into probabilistic designs through an operator $\mathcal{K}$, where $\mathcal{K}(D) \triangleq D/\rho$, the weakest prespecification of ρ through D. The non-homogeneous design ρ (with alphabet $\{ok, prob, ok', s'\}$) is a forgetful function to retract states from probabilistic states and defined as $\rho \triangleq (\textbf{true} \vdash prob(s') > 0)$: the probability of arriving at state s' is $prob(s')$; this is replaced by the possibility ($prob(s') > 0$) of arriving at that state. So this function forgets the probability $prob$ in its initial observation space and retracts state s' in its final observation space. The embedding ($\mathcal{K}(D)$) is the weakest probabilistic design to make $\mathcal{K}(D)$; ρ a refinement of D.

We present the representation of probabilistic state spaces in Isabelle/UTP in Sect. 4.1. Then we describe our implementation of weakest prespecification and ρ in Isabelle/UTP in Sect. 4.2.

4.1 Representation of Probabilistic State Spaces

Isabelle/UTP provides a semantic framework for verification based on UTP, which is implemented in Isabelle/HOL, a generic proof system.

We use probability mass functions (PMFs) [12] in Isabelle/HOL to represent discrete probabilistic distributions. The type of PMFs ($[\alpha]pmf$, parametric over α) is a set of probability measure spaces. A measure space is a tuple $(\Omega, \mathcal{A}, \mu)$, where Ω is a set, $\mathcal{A}$ is a σ-algebra on Ω, and μ is a measure function from $\mathcal{A}$ to positive real numbers. A probability measure space is a measure space with its measure being 1 ($\mu(\Omega) = 1 < \infty$), and so finite.

We declare an observational variable $prob : [\alpha]pmf$ to represent distributions in probabilistic designs and use an **alphabet** command [6] to construct a probabilistic state space, namely $[\alpha]\mathsf{prss}$, a record type parametric over α: **alphabet** $[\alpha]\mathsf{prss} = prob :: [\alpha]pmf$. In the definition, *prob* is a program variable whose type is $[\alpha]pmf$. Our probabilistic designs, therefore, are (non-homogeneous) designs: $[S_1, S_2]\mathsf{rel\_pdes} \triangleq [S_1, [S_2]\mathsf{prss}]\mathsf{rel\_des}$, with initial observation space (state space S_1) to final observation space (probabilistic state space $[S_2]\mathsf{prss}$ over state space S_2). They are non-homogeneous because the initial observation and final observation are over different spaces: state space versus probabilistic state space. For probabilistic programs defined in this paper, S_1 and S_2 are the same, and so probabilistic designs are actually homogeneous: $[S]\mathsf{hrel\_pdes} \triangleq [S, S]\mathsf{rel\_pdes}$. Here, "homogeneous" means the final probabilistic observation space is over the same state space as the initial observation space. We define $[S]\mathsf{hrel\_hpdes} \triangleq [[S]\mathsf{prss}, [S]\mathsf{prss}]\mathsf{rel\_des}$ for homogeneous designs whose initial and final observation spaces are over probabilistic distributions.

4.2 Weakest Prespecification and Forgetful Function

Weakest prespecification is the generalisation of weakest precondition from a condition to a relation.

Definition 4.1. *The weakest prespecification of K through Y is defined as:* $Y/K \triangleq \neg\,((\neg\, Y);\ K^{-})$ *where* $^{-}$ *is a relational converse operator.*

We note different notations are used for weakest prespecification in literature: $K \backslash Y$ in [5,13] and Y/K in [3,4]. Our mechanisation of weakest prespecification is based on [5] and so uses $\backslash$, but the mechanisation of probabilistic designs is based on [3]. We, therefore, use an abbreviation to relate $/$ to $\backslash$. The weakest prespecification satisfies two theorems below.

Theorem 4.2. $Y \sqsubseteq (P;\ K) \Leftrightarrow (Y/K) \sqsubseteq P$

This theorem shows that a program P is a refinement of the weakest prespecification of K through Y, if and only if a specification Y is implemented by sequential composition of P and K. The refinement relation $S \sqsubseteq P$ in UTP is defined as P implies S universally (for all alphabets of S and P): $S \sqsubseteq P \triangleq [P \Rightarrow S]$. For probabilistic designs, we rename predicates into $D \sqsubseteq (P;\ \rho) \Leftrightarrow (D/\rho) \sqsubseteq P$, which is interpreted as: a probabilistic design P implements the embedding of a standard design D into probabilistic designs through ρ if and only if the retraction of P through ρ to a standard design implements D. Another theorem is related to normal designs which our semantics relies on.

Theorem 4.3. $(p \vdash_n Q)/(\textbf{\textit{true}} \vdash_n R) = (p \vdash_n Q/R)$

The weakest prespecification operator of two normal designs, when the precondition of its first design is ***true***, can be moved into their postconditions.

The forgetful function ρ (**fp** in our mechanisation) is a non-homogeneous design, which is reflected in our definition below.

Definition 4.4. *The forgetful function* **fp** : $[[S]\mathsf{prss}, S]\mathsf{rel\_des}$ *is a normal design:* $\mathbf{fp} \triangleq (\textbf{\textit{true}} \vdash_n prob(\mathbf{v}') > 0)$ *where* $\mathbf{v}$ *denotes all variables in the input alphabet of a program except the ok variable, and* $\mathbf{v}'$ *are similar but in the output alphabet. We also say* $\mathbf{v}$ *or* $\mathbf{v}'$ *represent a state in state space.*

From this definition, we know **fp** is a relation between the initial probabilistic observation space and the final standard observation space.

5 Embedding Standard Programs

A design D is embedded into probabilistic designs through $\mathcal{K}$ defined below.

Definition 5.1 (Embedding). $\mathcal{K}(D) \triangleq D/\,\mathbf{fp}$

Here, embedding is the weakest prespecification of **fp** through D. So $\mathcal{K}(D)$ is the weakest probabilistic design related to D, and can be undone by retraction:

Theorem 5.2. *Let* D *be a normal design.* $\mathcal{K}(D);\ \mathbf{fp} = D$

In other words, embedding a standard design into a probabilistic design, and then retracting returns the original design. Embedding is also monotonic:

Theorem 5.3. $P \sqsubseteq Q \Rightarrow \mathcal{K}(P) \sqsubseteq \mathcal{K}(Q)$

We show that embedded standard designs are probabilistic designs:

Theorem 5.4. *We fix the predicate* $R : [S_1, S_2]\mathsf{urel}$*, then*

$$\mathcal{K}\,(p \vdash_n R) = (p \vdash_n (\Sigma_a i \in S_2 \mid (R\ \textbf{\textsf{wp}}\,(\mathbf{v} = i)) \bullet pmf\,(prob', i)) = 1)$$

Hence, embedding a standard normal design (LHS) is simplified to a normal design (RHS) with this theorem. Here, the form $(\Sigma_a x \in X \bullet exp(x))$ is a summation of the expression exp for all elements in set X. The symbol Σ_a denotes the summation over a possible infinite set. In the predicate $(R\ \textbf{\textsf{wp}}\,(\mathbf{v} = i))$, $\textbf{\textsf{wp}}$ is the weakest precondition operator [14]. The predicate characterises the weakest precondition for R to be guaranteed to achieve $(\mathbf{v} = i)$. We recall that $\mathbf{v}$ denotes all variables in state space, and so in other words, this predicate is simply a condition characterising when a given state i is a possible final state for R, which is equal to $\exists\, s \in S_1 \bullet R(s, i)$.

In the expression part of the summation, $pmf\,(prob', i)$, the function pmf returns the probability measure of the single state i in the distribution $prob'$. We use coercion in Isabelle/HOL to simplify its syntax further to $prob'(i)$. We also use $prob'(X)$ to denote $(\Sigma_a x \in X \bullet prob'(i))$, the probability measure of a set of states. We use the simpler syntax in the rest of the paper.

From the theorem, we know that the precondition p is unchanged after embedding. The postcondition is a condition such that the probabilities of all the final states of R sum to 1 and so $prob'$ is a distribution.

The embedding of abort, skip, assignment, and conditional is given in the theorem below.

Theorem 5.5.

$$\mathcal{K}(\bot_D) = \bot_D \qquad (\bot_D \triangleq \textbf{\textit{false}} \vdash \textbf{\textit{false}}) \quad (1)$$
$$\mathcal{K}(x :=_D e) = (\textbf{\textit{true}} \vdash_n prob'(\mathbf{v}[e/x]) = 1) \quad (x :=_D e \triangleq \textbf{\textit{true}} \vdash \mathbf{v}' = \mathbf{v}[e/x]) \quad (2)$$
$$\mathcal{K}(I\!I_D) = (\textbf{\textit{true}} \vdash_n prob'(\mathbf{v}) = 1) \qquad (I\!I_D \triangleq \textbf{\textit{true}} \vdash \mathbf{v}' = \mathbf{v}) \quad (3)$$
$$\mathcal{K}(P \lhd b \rhd Q) = \mathcal{K}(P) \lhd b \rhd \mathcal{K}(Q) \quad (4)$$

Embedding the design abort (1, defined on the right) is still itself. Embedding an assignment $x :=_D e$ (2, defined on the right) is a probabilistic design with precondition ***true*** and postcondition establishing that the probability of the state with e substituted for x (the values of other variables are unchanged) is equal to 1, that is, embedding an assignment results in a point distribution: from each initial state $\mathbf{v}$, $prob'$ in the final state has probability 1 for the state $\mathbf{v}[e/x]$. Based on the fact that $prob'$ is a distribution, this also implies the probabilities for other states in $prob'$ are 0. A design skip is a special form of assignment and so is its embedding (3). $\mathcal{K}$ distributes through conditional, shown in (4).

6 Distributions Combination and Probabilistic Choice

In this section, we introduce an operator to combine probability distributions, and then use this operator to construct probabilistic choice.

6.1 Distribution Combinations

Definition 6.1 (Distributions combination). *We fix $P : [S]pmf$, $Q : [S]pmf$, and $r : \mathbb{R}$, and define a distribution plus operator $+_r$ to merge two distributions P and Q based on the weight r:*

$$P +_r Q \triangleq join\_pmf\ (pmf\_of\_list\ [(P, r), (Q, 1 - r)])$$

This combination is essentially a join from a distribution of type $[[S]pmf]pmf$ constructed from a list with two elements using *pmf_of_list*: the first element is a pair from P to weight r and the second one is a pair from Q to weight $1 - r$. This join flattens two distributions based on their measure functions. The distribution combination satisfies the theorem below.

Theorem 6.2. *We fix $i : S$ and assume $r \in [0..1]$, then*

$$(P +_r Q)\, i = P(i) * r + Q(i) * (1 - r)$$

The probability of a particular state i in the combined distribution is a weighted sum of P and Q based on their weights r and $1 - r$.

This combination operator also satisfies several theorems below.

Theorem 6.3.

$$P +_r Q = Q +_{(1-r)} P \quad \text{(quasi-commutative)} \qquad P +_0 Q = Q \quad \text{(zero)}$$
$$P +_r P = P \quad \text{(idempotent)} \qquad P +_1 Q = P \quad \text{(unit)}$$

Because $+_r$ is idempotent, we know that the set of discrete distributions represented by $[S]pmf$ is convex-closed [15]: the combination of a distribution with itself with any weight is still in the set. The operator is also quasi-associative:

Theorem 6.4 (Quasi-associativity). *We fix* $r_1, r_2, w_1, w_2 : \mathbb{R}$, *and assume* $w_1 \in [0..1]$, $w_2 \in [0..1]$, $(1 - w_1) * (1 - w_2) = (1 - r_2)$, *and* $w_1 = r_1 * r_2$. *Then*

$$P +_{w_1} (Q +_{w_2} R) = (P +_{r_1} Q) +_{r_2} R$$

This will be used to prove associativity of probabilistic choice in Theorem 6.8.

6.2 Probabilistic Choice

As shown in [1], we use UTP's parallel-by-merge scheme [4, Chap. 7], $P \parallel_M Q$, to define probabilistic choice. The two parallel programs P and Q share the same initial observation space $\mathbf{v}$, then establish their own final observation spaces individually as $0.\mathbf{v}'$ and $1.\mathbf{v}'$. A merge predicate M then describes how $\mathbf{v}$, $0.\mathbf{v}'$ and $1.\mathbf{v}'$ are merged. A conjunction of three separate programs is sequentially composed with M. The three programs include the two parallel programs P and Q, and one program to copy the initial observation space. The final observation spaces from P, Q, and the copy program, therefore, are referred to as $0.\mathbf{v}$, $1.\mathbf{v}$, and $\mathbf{v}$ in M. We start with the definition of a distribution merge operator:

Definition 6.5. $PM(r) \triangleq (prob' = 0.prob +_r 1.prob)$

The merge predicate establishes that the final distribution is the combination of the distribution ($0.prob$) from the first program and the distribution ($1.prob$) from the second program with weight r. Probabilistic choice is defined below.

Definition 6.6 (Probabilistic choice). *We fix* $P, Q : [S]$*hrel_pdes*, $r : \mathbb{R}$.

$$P \oplus_r Q \triangleq \left(P \parallel^D_{\mathbf{PM}(r)} Q\right) \lhd r \in (0..1) \rhd (Q \lhd r = 0 \rhd (P \lhd r = 1 \rhd \top_D))$$

The probabilistic choice is defined as a conditional:

- if r is not in the open interval (0..1), the choice is defined as follows:
  - if r is equal to 0, the choice is Q;
  - if r is equal to 1, the choice is P;
  - if r is not 0 and 1, the choice is the design miracle $\top_D$: a miraculous or infeasible specification, defined as (***true*** $\vdash$ ***false***).
- if r is in the open interval (0..1), the choice is between P and Q by parallel-by-merge: a design parallel composition ($\parallel^D_{\mathbf{PM}(r)} \triangleq \parallel_{\mathbf{DM}(\mathbf{PM}(r))}$) using the merge predicate $PM(r)$ to merge $prob$ and another design merge predicate $\mathbf{DM}$ to merge ok.

We note that the definition of probabilistic choice is not simply a parallel composition between P and Q, but a conditional to characterise two special cases:

$r = 0$ and $r = 1$. This is because of the definition of parallel composition. Let $P = (p \vdash_n \mathcal{P})$ and $Q = (q \vdash_n \mathcal{Q})$, then

$$P \parallel^D_{\mathbf{PM}(r)} Q = \begin{pmatrix} (p \vee (q \wedge \neg \mathit{Pre}(\mathcal{Q}))) \wedge \\ (q \vee (p \wedge \neg \mathit{Pre}(\mathcal{P}))) \end{pmatrix} \vdash_n \left(\mathcal{P} \parallel_{\mathbf{PM}(r)} \mathcal{Q}\right)$$

Here $\mathit{Pre}(\mathcal{P})$ is the predicate characterising the domain of a UTP relation $\mathcal{P}$. We note that r is not in the precondition. The parallel composition, therefore, cannot be simplified to P or Q for the special cases as its precondition has to take the other operand Q or P into account. Zero and unit, however, are important algebraic properties for probabilistic choice, and so we define it as conditional. The probabilistic choice satisfies the theorems below.

Theorem 6.7 (Quasi-commutative, zero, unit). *We assume* $r \in [0..1]$.

$$P \oplus_r Q = Q \oplus_{(1-r)} P \qquad P \oplus_0 Q = Q \qquad P \oplus_1 Q = P$$

Theorem 6.8 (Quasi-associativity). *We fix* $r_1, r_2, w_1, w_2 : \mathbb{R}$, *and assume* $w_1, w_2 \in [0..1]$, $(1 - w_1) * (1 - w_2) = (1 - r_2)$, $w_1 = r_1 * r_2$, *and that* P, Q, *and* R *are probabilistic designs. Then* $P \oplus_{w_1} (Q \oplus_{w_2} R) = (P \oplus_{r_1} Q) \oplus_{r_2} R$

Probabilistic choice, therefore, is quasi-associative, with the weights adjusted as specified. This is basically the extension of quasi-associativity (Theorem 6.4) of the distribution combination $+_r$.

Probabilistic choice is also left-distributive and right-distributive over nondeterministic choice and conditional.

Theorem 6.9. *We assume* $r \in [0..1]$, P, Q, *and* R *are normal designs.*

$$P \oplus_r (Q \sqcap R) = (P \oplus_r Q) \sqcap (P \oplus_r R) \qquad \text{(distl-nondeterminism)}$$
$$(Q \sqcap R) \oplus_r P = (Q \oplus_r P) \sqcap (R \oplus_r P) \qquad \text{(distr-nondeterminism)}$$
$$P \oplus_r (Q \lhd b \rhd R) = (P \oplus_r Q) \lhd b \rhd (P \oplus_r R) \qquad \text{(distl-conditional)}$$
$$(Q \lhd b \rhd R) \oplus_r P = (Q \oplus_r P) \lhd b \rhd (R \oplus_r P) \qquad \text{(distr-conditional)}$$

Even though $+_r$ is idempotent (Theorem 6.3), $\oplus_r$ is not idempotent in general. This is due to the parallel-by-merge scheme used in its definition. $\parallel_M$ is not idempotent in general. Consider, for example, $P \oplus_r P$ for $r \in (0..1)$. The probabilistic choice is just $P \parallel^D_{\mathbf{PM}(r)} P$, according to its definition. Based on the definitions of the parallel-by-merge scheme [4, Chap. 7] and the merge predicate, $P(s, 0.\mathit{prob}')$, $P(s, 1.\mathit{prob}')$, and $\mathit{prob}' = 0.\mathit{prob}' +_r 1.\mathit{prob}'$ are established. The parallel composition $P(s, 0.\mathit{prob}') \parallel^D_{\mathbf{PM}(r)} P(s, 1.\mathit{prob}')$ is equal to $P(s, \mathit{prob}')$ (and so idempotent) only if the probability distributions in the final observation space of P are convex-closed. If the final observation space of P is a single distribution (so P is deterministic), then $P \oplus_r P = P$ because a singleton set is convex-closed with respect to $+_r$. If, for example, the final observation space of P contains two distributions (so P is nondeterministic), then for any $0 < r < 1$, $P \oplus_r P \neq P$. We note that the set of distributions in the embedding of nondeterministic choice, as illustrated in Theorem 7.1, is convex-closed.

6.3 Merge Witness Distributions

In [1], we define two projections $\mathcal{F}$ and $\mathcal{G}$ for decomposition of a probabilistic program into the probabilistic choice of two subprograms. The functions $\mathcal{F}$ and $\mathcal{G}$ map a distribution *prob* over a set of states S into a distribution $0.prob$ over a subset A of S and another distribution $1.prob$ over a subset B of S separately with $A \cup B = S$. They, therefore, are projections.

This decomposition is useful when implementing a probabilistic program with multiway probabilistic choice, such as in the Reactive Modules formalism [16], in this *pGCL* (with only binary probabilistic choice). This has been illustrated in [1, Sect. 7]. This decomposition is also useful to provide witnesses for the merge predicate, and the witnesses are indeed required later to prove Theorem 7.1, an important distribution theorem for nondeterministic choice, as demonstrated in its proof in [1, Sect. 8].

We define the projection $\mathcal{F}$ below. $\mathcal{G}$ shares the same definition (but with different arguments in its applications).

Definition 6.10. *We fix* $A, B : [S]set$*, and* $p : [S]pmf$*.*

$$\begin{aligned}
&\mathcal{F}(A, B, p) \triangleq measure_of \\
&\quad \left(space(p), sets(p), \lambda\, C \bullet p\,(C \cap (A - B)) * ratio(A, B, p) + p\,(C \cap A \cap B) \right) \\
&ratio(A, B, p) \triangleq (p\,(B - A) + p\,(A - B))\,/\,p\,(A - B)
\end{aligned}$$

The result of $\mathcal{F}$ is a measure space constructed by *measure_of* from the space of the probability measure space p (by *space*), the σ-algebra of p (by *sets*), and a measure function (a curried function). In this function, we use $A - B$ to denote set difference between A and B. Indeed $\mathcal{F}$ defines a probability distribution:

Theorem 6.11. *We fix* $P : [S]pmf$ *and* $A, B : [S]set$*, and assume* $P(A \cup B) = 1$, $P(A - B) > 0$*, and* $P(B - A) > 0$*, then* $prob\_space\,(\mathcal{F}\,(A, B, P))$.

The constructed measure space by $\mathcal{F}$ is a probability space (*prob_space*), that is, its measure sums to 1.

7 Nondeterministic Choice

In predicative programming, including UTP, nondeterministic choice is defined simply as $P \sqcap Q \triangleq P \vee Q$, and, therefore, $P \vee Q \sqsubseteq P$. This is reflected in the semantics of the embedding of nondeterministic choice. We show $\mathcal{K}$ distributes through nondeterministic choice in the theorem below.

Theorem 7.1. *We assume* P *and* Q *are normal designs.*

$$\mathcal{K}\,(P \sqcap Q) = \left(\bigsqcap r \in [0..1] \bullet (\mathcal{K}\,(P) \oplus_r \mathcal{K}\,(Q)) \right)$$

This theorem demonstrates that an embedding of the nondeterministic choice of two standard designs P and Q is just the nondeterministic choice of the probabilistic choices of the embeddings of P and Q in all possible ways (all possible weights from 0 to 1 inclusive). For the two special cases 0 and 1 for r, according to Theorem 6.7, they are just $\mathcal{K}(Q)$ and $\mathcal{K}(P)$ respectively.

In relational semantics for probabilistic programs here, nondeterministic choice is refined by probabilistic choice with any particular weight in $[0..1]$:

Theorem 7.2. *We assume $r \in [0..1]$, P and Q are normal designs.*

$$\mathcal{K}(P \sqcap Q) \sqsubseteq \mathcal{K}(P) \oplus_r \mathcal{K}(Q)$$

So probabilistic choice refines nondeterministic choice, which is the most significant difference of this relational model from others.

8 Sequential Composition

In our previous work [1], we did not complete formalisation of the proof presented in [3] that $\mathcal{K}$ is a homomorphism for sequential composition. In particular, a Kleisli lifting operator $\uparrow$ [3, Def. 3.11] is defined to lift a probabilistic design to a design taking $(ok, prob)$ to $(ok', prob')$. A probabilistic design, therefore, is able to be sequentially composed with this lifted design because sequential composition requires the output alphabet of the first operand to be equal to the input alphabet of the second operand (and so two probabilistic designs are not allowed to be sequentially composed). The lifting operator $\uparrow$ is defined below.

Definition 8.1. $\uparrow P \triangleq \mathit{kleisli\_lift2}\left(\lfloor pre_D(P) \rfloor_<, pre_D(P) \wedge post_D(P)\right)$

The operator $\uparrow$ has one parameter: a probabilistic design $P : [S]\mathsf{hrel\_pdes}$, and is defined using another auxiliary function *kleisli_lift2*. The first argument $(\lfloor pre_D(P) \rfloor_<)$ to that function is the design precondition $(pre_D(P))$ of P with its output alphabet dropped (by $\lfloor \_ \rfloor_<$), and so the argument is a condition. The second argument is the postcondition of P.

The *kleisli_lift2*, defined below, has two parameters: q of type $[S]\mathsf{upred}$, and R of type $[S]\mathsf{hrel\_pdes}$, and characterises a homogeneous design of type $[S]\mathsf{hrel\_hpdes}$, whose initial and final observation spaces are both over probabilistic distributions.

Definition 8.2.

$$\begin{array}{l}
\mathit{kleisli_lift2}\,(q, R) \\
\triangleq \left(\begin{array}{l}
\left(\mathit{prob}\left(\llbracket q \rrbracket_p\right) = 1\right) \vdash \\
\exists\, Q \bullet \left(\begin{array}{l}
(\forall\, ss \bullet \mathit{prob}'(ss) = \Sigma_a t \bullet \mathit{prob}(t) * (Q(t))(ss))) \wedge \\
\left(\forall\, s \bullet \left(\begin{array}{l}
\neg\ (\mathit{prob}(\mathbf{v}') > \mathbf{0} \wedge \mathbf{v}' = s)\,; \\
(\neg\, R;\ (\forall\, t \bullet \mathit{prob}(t) = (Q(s))(t)))
\end{array} \right) \right)
\end{array} \right)
\end{array} \right)
\end{array}$$

Generally, this definition establishes that if the program starts in every possible state satisfying q, then it terminates with a distribution from that state which makes R hold for that state and that distribution.

The precondition of the definition of *kleisli_lift*2 means that the initial observation space *prob* is a distribution over all states satisfying the predicate q, where $[\![q]\!]_p$ denotes extraction of the characteristic set of q.

The postcondition of the definition is an existential quantification over Q, a function of type $S \rightarrow [S]pmf$. The predicate part of $\exists\, Q$ is a conjunction of two predicates. The first predicate establishes that for any state ss, its probability in the final observation space *prob*$'$ is equal to the summation of the products of the probability of each state t in the initial observation space *prob* and the probability of ss in the distribution $Q(t)$ corresponding to t. This predicate characterises Q. The second predicate establishes that for any state s, if its probability in the initial distribution *prob* is larger than 0, then R must be satisfied with its initial observation state s (the initial observation of R is $\mathbf{v}$, which is the same as $\mathbf{v}'$ in the precedent predicate where $\mathbf{v}'$ is just s because of $\mathbf{v}' = s$) and its final observation distribution *prob*$'$ being a distribution $Q(s)$. This predicate characterises R based on Q.

We now define sequential composition of probabilistic designs.

Definition 8.3 (Sequential composition). $P \,;_p\, Q \triangleq P \,;\, \uparrow Q$.

$\mathcal{K}$ distributes through sequential composition in the theorem below.

Theorem 8.4. *We fix* $P, Q : [S]$*hrel_des and assume* P *and* Q *are normal designs, and* S *is finite. Then* $\mathcal{K}\,(P;\; Q) = \mathcal{K}\,(P) \;;_p \mathcal{K}\,(Q)$.

We note that the assumption, S is finite, is necessary to prove this theorem. This assumption is hidden in the original proof [3, Theorem 3.12] when giving the witness function $f(u, v)$, where a cardinality $\#$ is used.

The $\uparrow$ satisfies the theorems below.

Theorem 8.5. *We assume* P *and* Q *are probabilistic designs.*

$$\uparrow (\mathcal{K}\,(\mathit{I\!I}_D)) = (\textbf{true} \vdash_n prob' = prob) \qquad \text{(skip)}$$

$$P \,;_p \mathcal{K}\,(\mathit{I\!I}_D) = P = \mathcal{K}\,(\mathit{I\!I}_D) \;;_p P \qquad \text{(left/right unit)}$$

$$P \sqsubseteq Q \Rightarrow \uparrow P \sqsubseteq \uparrow Q \qquad \text{(monotonic)}$$

$$(\uparrow P) \textit{ is a normal design} \qquad \text{(normal design)}$$

The lifted probabilistic skip is simply a skip, that is, its initial and final observation spaces are the same. The probabilistic skip is both a left unit and a right unit (left/right unit). The operator $\uparrow$ is also monotonic. We note the definition of $\uparrow$ is not a normal design (see Definition 8.2), and just a general design $\vdash$. We use it to ease type constraints in the definition. The operator $\uparrow$, however, is proved to be a normal design ( ).

9 Examples

The first example illustrates the proof of a probabilistic program with sequential composition, conditional, and probabilistic choice from Hehner's work [17, Sect. 6]. We define $P1$, $P2$, and $P3$ used in Theorem 9.1.

$$P1 \triangleq (\mathcal{K}\,(x :=_{D} 0) \oplus_{1/3} \mathcal{K}\,(x :=_{D} 1))$$
$$P2 \triangleq (\mathcal{K}\,(x :=_{D} x + 2) \oplus_{1/2} \mathcal{K}\,(x :=_{D} x + 3))$$
$$P3 \triangleq (\mathcal{K}\,(x :=_{D} x + 4) \oplus_{1/4} \mathcal{K}\,(x :=_{D} x + 5))$$

In $P1$, x is assigned 0 or 1 with probability 1/3 or 2/3. $P2$ has the equal probability to increase x by 2 or 3. $P3$ increases x by 4 or 5 with probability 1/4 or 3/4. The semantics of the composition of $P1$, $P2$, and $P3$ is shown below.

Theorem 9.1.

$$P1 \;;_p\; (P2 \lhd x = 0 \rhd P3) =$$
$$\left(\textbf{\textit{true}} \vdash_n \begin{pmatrix} \mathit{prob}'[2/x] = 1/6 \wedge \mathit{prob}'[3/x] = 1/6 \wedge \\ \mathit{prob}'[5/x] = 1/6 \wedge \mathit{prob}'[6/x] = 1/2 \end{pmatrix}\right)$$

The probabilistic program is equal to a normal design whose precondition is ***true*** and postcondition establishes that in the final observation space, the value of x is 2, 3, and 5 each with probability 1/6, and 6 with probability 1/2. The result is the same as that of [17].

The second example originates in [15] to illustrate how probabilistic choice interacts with nondeterministic choice in the relational semantics. It is also discussed in [17, Sect. 10] about nondeterminism. We define below P, a nondeterministic choice between x assigned to 0 and 1, and Q, a probabilistic choice between y assigned to 0 and 1.

$$P \triangleq (\mathcal{K}\,(x :=_{D} 0) \sqcap \mathcal{K}\,(x :=_{D} 1)) \qquad Q \triangleq (\mathcal{K}\,(y :=_{D} 0) \oplus_{1/2} \mathcal{K}\,(y :=_{D} 1))$$

We now consider sequential composition of P and Q in either order with the aim of establishing $(x = y)$. If Q is after P, then

Theorem 9.2.

$$P \;;_p\; Q = \left(\textbf{\textit{true}} \vdash_n \begin{pmatrix} (\mathit{prob}'[0, 0/x, y] = 1/2 \wedge \mathit{prob}'[0, 1/x, y] = 1/2) \vee \\ (\mathit{prob}'[1, 0/x, y] = 1/2 \wedge \mathit{prob}'[1, 1/x, y] = 1/2) \end{pmatrix}\right)$$

The theorem shows that the probability of establishing $(x = y)$ is 1/2: both alternatives of the disjunction in the postcondition have the same probability to establish $(x = y)$. Informally, the nondeterministic choice in P cannot take advantage of the value of x (because P is executed first), and, therefore, the probability of establishing $(x = y)$ in $(P;\ Q)$ is determined by Q, no matter which value of x is chosen. If P is after Q, then

Theorem 9.3.

$$Q \,;_p P = \left(\textbf{true} \vdash_n \begin{pmatrix} (prob'[0, 0/x, y] = 1/2 \wedge prob'[0, 1/x, y] = 1/2) \vee \\ (prob'[1, 0/x, y] = 1/2 \wedge prob'[0, 1/x, y] = 1/2) \vee \\ (prob'[0, 0/x, y] = 1/2 \wedge prob'[1, 1/x, y] = 1/2) \vee \\ (prob'[1, 0/x, y] = 1/2 \wedge prob'[1, 1/x, y] = 1/2) \end{pmatrix} \right)$$

The theorem demonstrates that the probability of establishing $(x = y)$ can be 0 (the second alternative of the disjunctions in the postcondition), 1/2 (the first and fourth alternative), and 1 (the third alternative). Informally, the nondeterministic choice in P can now take advantage of the value of x because y has been probabilistically determined in Q (so the probability of y being 0 or 1 in each alternative is 1/2). P, therefore, can choose (1) (2) x opposite to the value of y, and so the probability of establishing $(x = y)$ is 0; (3) x the same as the value of y, and so the probability of establishing $(x = y)$ is 1; (4) x always 0, and so the the probability of establishing $(x = y)$ is 1/2; and (5) x always 1, and so the the probability of establishing $(x = y)$ is 1/2. The choice between the four cases is nondeterministic and represented as disjunctions in the theorem.

Hehner [17] describes four varieties of nondeterminism: angelic, demonic, oblivious, and prescient. For $Q \,;_p P$, if P is an angelic choice, the result corresponds to the third alternative; if P is a demonic choice, the result corresponds to the second alternative; if P is a oblivious choice, the results corresponds to the first and the fourth alternative. $Q \,;_p P$ in Theorem 9.3, therefore, is more abstract and can be refined into angelic, demonic, and oblivious choice. P defined above is not prescient, and so it does not know the future value of Q in $(P \,;_p Q)$. The program, therefore, has probability 1/2 to achieve $x = y$, shown in Theorem 9.2.

The third example is the algorithm in Fig. 1 and its probabilistic program in Definition 3.1. First, we find and define an invariant for the recursion.

Definition 9.4 (Invariant).

$$\begin{aligned} & \mathit{ChooseUniform\_inv}(N) \\ & \triangleq \left(\textbf{true} \vdash_n \left(\begin{pmatrix} c \wedge i < (N-1) \Rightarrow \\ \begin{pmatrix} \begin{pmatrix} \forall j < (N-i-1) \bullet \\ prob'(\mathbf{v}[j+i, \mathit{false}/i, c]) = 1/(N-i) \end{pmatrix} \wedge \\ prob'(\mathbf{v}[N-1, \mathit{true}/i, c] = 1/(N-i))) \end{pmatrix} \end{pmatrix} \wedge \\ (\neg\, (c \wedge i < (N-1)) \Rightarrow prob'(\mathbf{v}) = 1) \right) \right) \end{aligned}$$

The postcondition of the definition is a conjunction: (1) the first conjunct establishes that if c is true and i is less than $(N - 1)$, the value of i in the final state space $\mathbf{v}$ is a uniform distribution in close interval $[i, N - 1]$ (each with probability $1/(N - i)$); (2) the second conjunct corresponds to the termination of the program, the probability of the final state being the same as the initial state $\mathbf{v}$ is 1. $\mathit{ChooseUniform\_inv}(N)$ indeed is an invariant for the recursion.

Theorem 9.5 (Invariant). *We assume N is larger or equal to 1.*

$$ChooseUniform\_inv(N) \sqsubseteq (\mu X \bullet ChooseUniformBody(N, X))$$

Here *ChooseUniformBody* is an abbreviation for the body of the recursion in Definition 3.1. Finally, *ChooseUniform*(N) is proved to be a uniform distribution.

Theorem 9.6. *We assume N is larger or equal to 1.*

$$\left(\textbf{true} \vdash_n \begin{pmatrix} (\forall j \bullet j < (N-1) \Rightarrow (prob'(\mathbf{v}[j, false/i, c] = 1/N))) \wedge \\ prob'(\mathbf{v}[(N-1), true/i, c]) = 1/N \end{pmatrix}\right) \\ \sqsubseteq ChooseUniform(N)$$

The program *ChooseUniform*(N) satisfies a contract, the left-hand side of $\sqsubseteq$, that (1) the probability of finally arriving a state, of which i is less than $(N-1)$ and c is *false*, is $1/N$; and (2) the probability of finally arriving a state, of which i is $(N-1)$ and c is *true*, is also $1/N$. We, therefore, conclude the program implements a uniform distribution given N.

10 Related Work

Hurd [18] developed a formal framework in High-Order Logic (HOL) [19], a predecessor of Isabelle/HOL [7], for modelling and verification of probabilistic algorithms using theorem proving. The work uses mathematical measure theory to represent probability space. Hurd et al. [20] also mechanised *pGCL* in HOL based on the quantitative logic [21], enabling verification of partial correctness of probabilistic programs. Our mechanisation is also based on measure and probability theory in HOL of Isabelle and uses the same notation *pGCL*. We, however, mechanise the relational semantics of *pGCL* in the theory of designs in UTP, enabling reasoning about total correctness.

Audebaud et al. [22] use the monadic interpretation of randomised programs for probabilistic distributions (instead of measure theory) and mechanise their work in the Coq theorem prover [23]. They consider only probabilistic choice (without nondeterminism) in a functional language with recursion, not in a nondeterministic probabilistic imperative program setting like us.

Cock [24] presents a shallow embedding of *pGCL* with Isabelle/HOL for proof automation. The work is based on McIver and Morgan's interpretation of a *pGCL* program as an expectation transformer from post-expectations to pre-expectations [25]. Its mechanisation uses real numbers ($\mathbb{R}$) in Isabelle/HOL as a type for probabilities, which improves automation. By contrast, we use measure theory and PMFs in Isabelle/HOL to encode probability distributions. Additionally, we base our formalisation on Isabelle/UTP (instead of the shallow embedding of Cock's work) which enables modelling at high-level abstraction and unification of semantics with other paradigms such as time and reactive systems (this is important in order to capture the semantics of RoboChart).

11 Conclusions

Previously, we gave the probabilistic semantics to RoboChart based on He, Morgan and McIver's relational model for *pGCL*, and formalised its proof. In this paper, we present a mechanisation of the proof in Isabelle/UTP to enable automated reasoning for probabilistic programs.

We use measure theory and probability mass functions in Isabelle/HOL to represent probability distributions. Our mechanisation shows that (1) PMFs are convex closed, (2) the probabilistic choice is not idempotent in general, and (3) embedding distributes through sequential composition for finite state space.

Based on the mechanisation, we use several examples to illustrate the automated reasoning, including the randomisation algorithm in RoboChart. We note this notation is general enough to capture other distributions, and not restricted to uniform distributions illustrated here.

As illustrated by the probabilistic nondeterministic programs in Theorems 9.2 and 9.3, computations of probabilistic programs are related to those of imperative programs. Probability information has become predicates over the *prob* variable and nondeterminism becomes disjunctions of these predicates, which, therefore, enables us to reason about probabilistic programs using general designs or relational facilities in UTP, such as contract-based reasoning [26]. This is the way in the relational model to tackle reasoning complexity introduced in probabilistic programs.

Our immediate future work is to lift probabilistic designs into UTP's reactive theory to unify the semantics of reactive, time, and probability in RoboChart.

Acknowledgements. This work is funded by the EPSRC projects RoboCalc (Grant EP/M025756/1), RoboTest (Grant EP/R025479/1), and CyPhyAssure (CyPhyAssure Project: https://www.cs.york.ac.uk/circus/CyPhyAssure/) (Grant EP/S001190/1). The icons used in RoboChart have been made by Sarfraz Shoukat, Freepik, Google, Icomoon and Madebyoliver from www.flaticon.com, and are licensed under CC 3.0 BY.

References

1. Woodcock, J., Cavalcanti, A., Foster, S., Mota, A., Ye, K.: Probabilistic semantics for RoboChart. In: Ribeiro, P., Sampaio, A. (eds.) UTP 2019. LNCS, vol. 11885, pp. 80–105. Springer, Cham (2019). https://doi.org/10.1007/978-3-030-31038-7_5
2. Miyazawa, A., Ribeiro, P., Li, W., Cavalcanti, A., Timmis, J., Woodcock, J.: RoboChart: modelling and verification of the functional behaviour of robotic applications. Softw. Syst. Model. **18**(5), 3097–3149 (2019)
3. Jifeng, H., Morgan, C., McIver, A.: Deriving probabilistic semantics via the 'weakest completion'. In: Davies, J., Schulte, W., Barnett, M. (eds.) ICFEM 2004. LNCS, vol. 3308, pp. 131–145. Springer, Heidelberg (2004). https://doi.org/10.1007/978-3-540-30482-1_17
4. Hoare, C.A.R., He, J.: Unifying Theories of Programming. Prentice-Hall, Hoboken (1998)

5. Hoare, C.A.R., He, J.: The weakest prespecification. Inf. Process. Lett. **24**(2), 127–132 (1987)
6. Foster, S., Baxter, J., Cavalcanti, A., Woodcock, J., Zeyda, F.: Unifying semantic foundations for automated verification tools in Isabelle/UTP. Sci. Comput. Program. **197**, 102510 (2020)
7. Nipkow, T., Wenzel, M., Paulson, L.C.: Isabelle/HOL: A Proof Assistant for Higher-order Logic. Springer, Cham (2002). https://doi.org/10.1007/3-540-45949-9
8. McIver, A., Morgan, C.: Introduction to pGCL: Its logic and Its Model. In: Abstraction, Refinement and Proof for Probabilistic Systems. Springer, New York, January 2005. https://doi.org/10.1007/0-387-27006-X_1
9. Woodcock, J., Cavalcanti, A.: A tutorial introduction to designs in unifying theories of programming. In: Boiten, E.A., Derrick, J., Smith, G. (eds.) IFM 2004. LNCS, vol. 2999, pp. 40–66. Springer, Heidelberg (2004). https://doi.org/10.1007/978-3-540-24756-2_4
10. Hehner, E.C.R.: A Practical Theory of Programming. Texts and Monographs in Computer Science. Springer, New York (1993). https://doi.org/10.1007/978-1-4419-8596-5
11. Guttmann, W., Möller, B.: Normal design algebra. J. Log. Algebraic Meth. Program. **79**(2), 144–173 (2010)
12. Hölzl, J., Lochbihler, A.: Probability Mass Function. Technical Report https://isabelle.in.tum.de/library/HOL/HOL-Probability/Probability_Mass_Function.html
13. Hoare, C.A.R., He, J.: The weakest prespecification. Technical Report PRG44, OUCL, June 1985
14. Dijkstra, E.W.: Guarded commands, nondeterminacy and formal derivation of programs. Commun. ACM **18**(8), 453–457 (1975)
15. Jifeng, H., Seidel, K., McIver, A.: Probabilistic models for the guarded command language. Sci. Comput. Program. **28**(2–3), 171–192 (1997)
16. Alur, R., Henzinger, T.A.: Reactive modules. Formal Meth. Syst. Des. **15**(1), 7–48 (1999). https://doi.org/10.1023/A:1008739929481
17. Hehner, E.C.R.: Probabilistic predicative programming. In: Kozen, D. (ed.) MPC 2004. LNCS, vol. 3125, pp. 169–185. Springer, Heidelberg (2004). https://doi.org/10.1007/978-3-540-27764-4_10
18. Hurd, J.: Formal verification of probabilistic algorithms. Technical report, University of Cambridge, Computer Laboratory (2003)
19. Gordon, M.J.C., Melham, T.F. (eds.): Introduction to HOL: A Theorem Proving Environment for Higher Order Logic. Cambridge University Press, Cambridge (1993)
20. Hurd, J., McIver, A., Morgan, C.: Probabilistic guarded commands mechanized in HOL. Theor. Comput. Sci. **346**(1), 96–112 (2005)
21. Morgan, C., McIver, A., Seidel, K.: Probabilistic predicate transformers. ACM Trans. Program. Lang. Syst. (TOPLAS) **18**(3), 325–353 (1996)
22. Audebaud, P., Paulin-Mohring, C.: Proofs of randomized algorithms in Coq. Sci. Comput. Program. **74**(8), 568–589 (2009)
23. The Coq development team: The Coq Proof Assistant. https://coq.inria.fr. Accessed 20 May 2021
24. Cock, D.: Verifying Probabilistic Correctness in Isabelle with pGCL. In Cassez, F., Huuck, R., Klein, G., Schlich, B., (eds.): Proceedings Seventh Conference on Systems Software Verification, SSV 2012, Sydney, Australia, 28–30 November 2012, volume 102 of EPTCS, pp. 167–178 (2012)

25. McIver, A., Morgan, C.: Abstraction, Refinement and Proof for Probabilistic Systems. Monographs in Computer Science. Springer, New York (2005). https://doi.org/10.1007/b138392
26. Ye, K., Foster, S., Woodcock, J.: Compositional assume-guarantee reasoning of control law diagrams using UTP. In: Adamatzky, A., Kendon, V. (eds.) From Astrophysics to Unconventional Computation. ECC, vol. 35, pp. 215–254. Springer, Cham (2020). https://doi.org/10.1007/978-3-030-15792-0_10

Domain Range Semigroups and Finite Representations

Jaš Šemrl(✉)

University College London, Gower Street, London WC1E 6BT, UK
j.semrl@cs.ucl.ac.uk

Abstract. Relational semigroups with domain and range are a useful tool for modelling nondeterministic programs. We prove that the representation class of domain-range semigroups with demonic composition is not finitely axiomatisable. We extend the result for ordered domain algebras and show that any relation algebra reduct signature containing domain, range, converse, and composition, but no negation, meet, nor join has the finite representation property. That is any finite representable structure of such a signature is representable over a finite base. We survey the results in the area of the finite representation property.

Keywords: Domain-range semigroups · Demonic composition · Finite representation property

1 Introduction

Formal reasoning about programs and their correctness is an important, yet a demonstrably difficult task and many well known approaches have been proposed. Algebraically speaking, a deterministic program is a partial function mapping from the state space to itself. Generalising this, to account for nondeterminism, we can say that a program (deterministic or nondeterministic) is a binary relation over the state space. This ability to naturally express such concepts motivates the endeavour of formalising the logic of binary relations.

A formalisation of this sort is found in Relation Algebra, obtained by extending the language of Boolean Algebra with operations specific to binary relations. This enables us to reason about the behaviour of binary relations in an abstract manner. However, these algebras are also very badly behaved, with an abundance of undecidability results, see [6, Part V]. A possible way of combating this is by dropping some operations from the language, sacrificing the ability to encapsulate the behaviour of relational calculus in exchange for decidability of certain decision problems. We will formally define some of these and how to prove positive properties later in this section.

Here we examine some of these favourable properties, or lack thereof, for languages containing domain and range, and put them in the bigger context of

J. Šemrl—The author thanks Professor Robin Hirsch for supervision and insightful conversations about the work presented.

U. Fahrenberg et al. (Eds.): RAMiCS 2021, LNCS 13027, pp. 483–498, 2021.
https://doi.org/10.1007/978-3-030-88701-8_29

relation algebra reduct languages. We chose this subset of languages as they were found useful in algebraically reasoning about correctness of nondeterministic programs, see Sect. 2 for more details.

But first, some definitions. Let X be a base set. *Domain* (D) and *range* (R) are operations, defined for some relation $R \subseteq X \times X$ as

$$\mathrm{D}(R) = \{(x, x) \mid \exists y : (x, y) \in R\} \quad \mathrm{R}(R) = \{(y, y) \mid \exists x : (x, y) \in R\}$$

and together with composition, they form the signature of domain-range semigroups. However, relational composition is not always interpreted in the same way. Two examples of interpretations include the *angelic* or ordinary composition (denoted ;) and *demonic* composition (denoted $*$), defined below for $R, S \subseteq X \times X$

$$R; S = \{(x, z) \mid \exists y : (x, y) \in R \wedge (y, z) \in S\}$$
$$R * S = \{(x, y) \in R; S \mid \forall z : (x, z) \in R \Rightarrow (z, z) \in \mathrm{D}(S)\}$$

Whilst the first definition seems pretty intuitive, the second one may appear a bit odd, even arbitrary, so let us have a closer look. The operation is motivated in the behaviour of a nondeterministic machine when the demon is in control of nondeterminism. Imagine the relations R, S were programs over the state space X. The pair $(x, y) \in R; S$ is included in $R * S$ if and only if there is no run from R to some z from which S aborts or loops forever, i.e. $(z, z) \notin \mathrm{D}(S)$. Should such a run exist, the demon will take the opportunity and abort the computation. For more details on this refer to [9].

Any $\{D, R, ;\}$- or $\{D, R, *\}$-structure $\mathcal{S}$ with an underlying set $S \subseteq \wp(X \times X)$ for some base X and operations interpreted relationally (as defined above) is *proper*. Let τ be a signature of operations that are well defined for binary relations. The *representation class* for τ, denoted $R(\tau)$, is the class of all proper τ-structures, closed under isomorphic copies. An isomorphism θ that maps a representable structure to a proper structure is called a *representation*.

A representation is finite if the base set X of the proper image is finite. If all finite members of $R(\tau)$ have finite representations, we say that the signature has the *finite representation property* (FRP).

The two properties described above are of special interest to us. This is because they both guarantee the decidability of determining membership in $R(\tau)$ for finite structures, also known as the *representability decision problem.* Although the properties both ensure decidability of the said decision problem, they in no way follow from each other. This provides us with two non trivial questions for each Relation Algebra reduct language that, given either is answered affirmatively, provide us with a decidability guarantee.

Here, we answer [12, Question 4.9] and show that $R(D, R, *)$ is not finitely axiomatisable. We do so by defining a two-player game that corresponds to a recursively enumerable axiomatisation of the representation class. Then we show that for each finite subset of this axiomatisation has a non-representable model. By compactness of first order logic, we are able to reach a contradiction under the assumption of finite axiomatisability.

Then we show that any relation algebra reduct signature containing domain, range, converse and composition, but no negation, meet, nor join has the finite representation property. This is an extension of a previous finite representation property result for ordered domain algebras [5]. We conclude by putting the result in a larger context of finite representation property for all reduct signatures of relation algebra. We survey the existing results and raise some open questions in the area.

2 Motivation and Context

In this section we take a closer look at the related work and motivate the problems. We have seen that structures of relations provide us with a natural way of formally reasoning about nondeterministic programs [4]. In [3], a good intuition on how to use structures with domain and range to model program control flow using semigroups with domain and range – functional for deterministic, and relational for nondeterministic programs. This allows us to express partial correctness equationally.

However, to extend this to total correctness, we have to turn to the demon. Demonic calculus was introduced to model the behaviour of programs, should the demon be in control of making nondeterministic decisions. Recently, it has been shown we may take this to our advantage and introduce equations to model total correctness. One such approach expresses total correctness using the domain and demonic composition [8] and another using ordinary composition and the bottom element of the demonic lattice [9].

These applications motivate our search for computational guarantees. As we have discussed, this includes looking for finite axiomatisability of the representation class and the finite representation property. A major negative result is shown with $R(\mathrm{D}, \mathrm{R}, ;)$ and $R(\mathrm{D}, ;)$ having no finite axiomatisation [7].

Both of these two signatures have the finite representation property open. However, one may add the partial ordering, converse, the identity, and the empty relation to obtain the signature of ordered domain algebras. Surprisingly, this signature has both the finite representation property, as well as a finitely axiomatisable representation class [5]. Another interesting result is the axiomatisation of $R(\mathrm{D}, *)$ is not only finite, but also the same as that of representable domain semigroups of partial functions [11]. Furthermore, the equational theories of both $R(\mathrm{D}, \mathrm{R}, ;)$ and $R(\mathrm{D}, \mathrm{R}, *)$ are finitely axiomatisable [12].

Finally, it is important to note that although the finite axiomatisability of the representation class and the finite representation property both guarantee the decidability of the representation decision problem, neither is stronger or weaker than the other. We have seen an example of a signature with both properties in the ordered domain algebras, as well as the full signature of relation algebras with neither property. However, you can find signatures with finitely axiomatisable representation class but no FRP, like meet-lattice semigroups [2,14], and semigroups with demonic refinement [10] with FRP, but non finitely axiomatisable representation class.

3 Networks and Representation by Games

In this section we outline a representation game that will help us prove the non finite axiomatisability result of $R(\mathrm{D}, \mathrm{R}, *)$. This argument is based on [6], but defined for this specific signature. The proofs presented are outlines, however, they are more detailed in parts where it is necessary to show the argument can be feasibly used to show results for demonic composition. For full details of proofs, see [6, Chapter 7].

On an intuitive level, this approach entails defining a game where a player is challenged to build a representation, on a step by step basis over a predetermined number of moves. The design of our game must be such that the player challenged will have a winning strategy if and only if they can survive the game of any length.

We then, for every natural number, define a formula that corresponds to a winning strategy for a game of that length. This means that we have defined a recursively enumerable theory that axiomatises the representation class.

In later sections we define, for each length of the game, an unrepresentable structure where the player challenged has got the winning strategy. This will enable us to use the compactness of first order logic to reach a contradiction under the assumption of finite axiomatisability.

Now, we will define these concepts more formally. A network $\mathcal{N} = (N, \bot, \top)$ where $\bot, \top : N \times N \to \wp(\mathcal{S})$ and $\mathcal{S}$ is some $\{\mathrm{D}, \mathrm{R}, *\}$-structure. We say it is *consistent* if and only if

$$\forall x, y \in N : \top(x, y) \cap \bot(x, y) = \emptyset$$

$$\forall x, y \in N, \forall s, t \in \mathcal{S} : \Big(s \in \top(x, y) \wedge \big(s = \mathrm{D}(t) \vee s = \mathrm{R}(t)\big)\Big) \Rightarrow x = y$$

Now, let us define for any $a, b \in \mathcal{S}$ the two networks $\mathcal{N}_{ref}[a, b]$ and $\mathcal{N}_{nref}[a, b]$ as follows

$$\mathcal{N}_{ref}[a, b] = (\{x\}, \{(x, x) \mapsto \{b\}\}, \{(x, x) \mapsto \{a\}\})$$
$$\mathcal{N}_{nref}[a, b] = (\{x, y\}, \{(x, y) \mapsto \{b\}\}, \{(x, y) \mapsto \{a\}\})$$

And all other pairs map to $\emptyset$ for $\top, \bot$.

We also define two operations $+_\top[\mathcal{N}, x, y, a], +_\bot[\mathcal{N}, x, y, a]$ which take a network $\mathcal{N} = (N, \bot, \top)$, some $x, y \in N \dot{\cup} \{x_+\}$ and some $a \in \mathcal{S}$ and return

$$+_\top[\mathcal{N}, x, y, a] = (N \cup \{x, y\}, \bot, \top^+)$$
$$+_\bot[\mathcal{N}, x, y, a] = (N \cup \{x, y\}, \bot^+, \top)$$

where $\top^+(v, w)$ is the same as $\top(v, w)$, or $\emptyset$ (if $\top(v, w)$ is undefined), for all v, w, except for x, y where a is also added to $\top^+(x, y)$. Similarly for $\bot^+$.

A network $\mathcal{N}' = (N', \top', \bot')$ is said to *extend* $\mathcal{N} = (N, \top, \bot)$, denoted $\mathcal{N} \subseteq \mathcal{N}'$ if and only $N \subseteq N'$ and for all $x, y \in N$ we have $\top(x, y) \subseteq \top'(x, y), \bot(x, y) \subseteq \bot'(x, y)$. Clearly, both $+_\top, +_\bot$ for $\mathcal{N}$ with any operands are extensions of $\mathcal{N}$. Furthermore, observe how inconsistency is inherited under extensions.

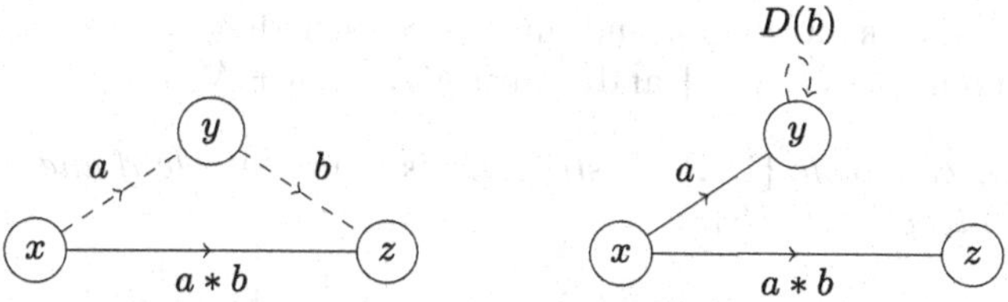

Fig. 1. Witness move (left) and composition-domain move (right)

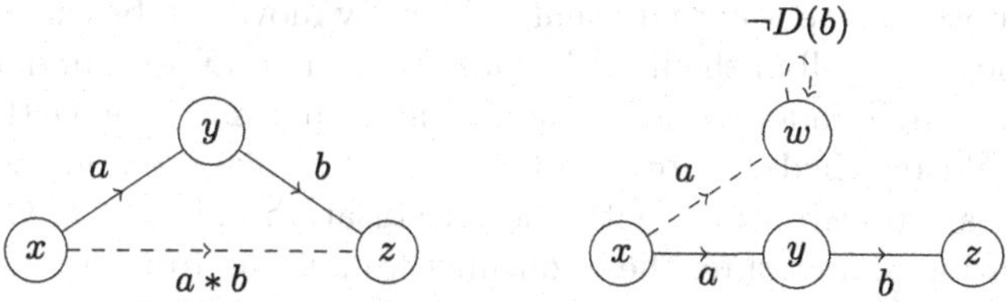

Fig. 2. Composition move

We can now define a game for a $\{D, R, *\}$-structure $\mathcal{S}$. It is played by two players $\forall, \exists$, we will call them Abelard and Eloise. The game, denoted $\Gamma_n(\mathcal{S})$, starts with the initialisation (zeroth) move and then continues for n moves where $0 < n \leq \omega$. Let $k \leq n$. At kth move $\forall$ challenges $\exists$ to return a $\mathcal{N}_k$ such that $\mathcal{N}_0 \subseteq \mathcal{N}_1 \subseteq ... \subseteq \mathcal{N}_n$. $\forall$ wins the game if and only if $\exists$ introduces an inconsistent network.

Initialisation. $\forall$ picks a pair $a \neq b \in \mathcal{S}$ and $\exists$ returns $\mathcal{N}_0$ that is an extension of $\mathcal{N}_{ref}[a,b]$, $\mathcal{N}_{nref}[a,b]$, $\mathcal{N}_{ref}[b,a]$ or $\mathcal{N}_{nref}[b,a]$.

Witness Move. $\forall$ picks a pair of nodes x, z in the network $\mathcal{N}_k$ and a pair of elements $a, b \in \mathcal{S}$ such that $a * b \in \top(x,z)$. $\exists$ picks a $y \in N \dot{\cup} \{x_+\}$ and returns $\mathcal{N}_{k+1} \supseteq +_\top[+_\top[\mathcal{N}, x, y, a], y, z, b]$, see Fig. 1 left.

Composition-Domain Move. $\forall$ picks, some x, y, z with $a \in \top(x,y)$ and $a * b \in \top(x,z)$ and $\exists$ must return $\mathcal{N}_{k+1} \supseteq +_\top[\mathcal{N}, y, y, \mathrm{D}(b)]$, see Fig. 1 right.

Composition Move. $\forall$ picks some $x, y, z \in \mathcal{N}_k$ along with a, b such that $a \in \top(x,y)$ and $b \in \top(y,z)$. $\exists$ has a choice between returning $\mathcal{N}_{k+1} \supseteq +_\top[\mathcal{N}, x, z, a * b]$ (Fig. 2 left) and $\mathcal{N}_{k+1} \supseteq +_\bot[+_\top[\mathcal{N}, x, w, a], w, w, \mathrm{D}(b)]$ where she picks a $w \in N \dot{\cup} \{x_+\}$ (Fig. 2 right).

Domain-Range Move. $\forall$ picks $x, y \in N_n, a \in \mathcal{S}$ such that $a \in \top(x,y)$ and $\exists$ must return $\mathcal{N}_{k+1} \supseteq +_\top[+_\top[\mathcal{N}, x, x, \mathrm{D}(a)], y, y, \mathrm{R}(a)]$.

Domain Move. $\forall$ picks a node x and an $a \in \mathcal{S}$ such that $\mathrm{D}(a) \in \top(x,x)$ and $\exists$ must pick a node $y \in N \dot{\cup} \{x_+\}$ and return $\mathcal{N}_{k+1} \supseteq +_\top[\mathcal{N}, x, y, a]$

Range Move. $\forall$ picks a node y and an $a \in \mathcal{S}$ such that $\mathrm{R}(a) \in \top(y, y)$ and $\exists$ must pick a node $x \in N\dot{\cup}\{x_+\}$ and return $\mathcal{N}_{k+1} \supseteq +_\top[\mathcal{N}, x, y, a]$

Lemma 1. *A countable* $\{\mathrm{D}, \mathrm{R}, *\}$*-structure is representable if and only if* $\exists$ *has a winning strategy for* $\Gamma_\omega(S)$.

Proof. If the structure is representable, $\exists$ can play the game by mapping the responses from the representation. Conversely, if $\exists$ has a winning strategy for $\Gamma_\omega(\mathcal{S})$, she must also have the winning strategy for any length of the game where $\forall$ schedules moves in the way that eventually every move will be called and the $\top$ label of the network will in the limit be closed under composition, domain-range moves and saturated under witness, domain and range moves. Since the structure is countable, $\forall$ can schedule moves in this manner. Take the limit network, call it $\mathcal{N}_\omega[a \neq b]$, after such a play with the initialisation pair $a \neq b$. Observe how due to saturation and closure, the $\top$ outlines a mapping from $\mathcal{S}$ to $N \times N$ that represents $\mathrm{D}, \mathrm{R}, *$ correctly and ensures that a, b map to different relations. Thus a disjoint union $\dot{\bigcup}_{a\neq b}\mathcal{N}_\omega[a \neq b]$ is a representation of $\mathcal{S}$. □

Lemma 2. *For every* $n < \omega$*, there exists a first order formula* σ_n *such that* $\exists$ *has a winning strategy for* $\Gamma_n(\mathcal{S})$ *if and only if* $\mathcal{S} \models \sigma_n$*. Furthermore, the first order theory* $\Sigma = \{\sigma_i \mid i < \omega\}$ *axiomatises* $R(\mathrm{D}, \mathrm{R}, *)$.

Proof. Let us define a *variable network* in a slightly different manner with the mappings $\top, \bot : N \times N \to \wp(\mathrm{Vars})$. A valuation $v : \mathrm{Vars} \to \mathcal{S}$ defines a conventional network $v(\mathcal{N})$. This allows us to define a formula $\phi_n(\mathcal{N})$ in a way that, together with a valuation $v : \mathcal{S} \to \mathrm{Vars}$, $\exists$ can survive the *conservative* play of the game for n more moves, starting from $v(\mathcal{N})$. By conservative, we mean that $\exists$ plays the network requested without proper extensions.

In the base case, observe how $v(\mathcal{N})$ only needs to be consistent and thus

$$\phi_0(\mathcal{N}) = \bigwedge_{\substack{x \neq y \in N \\ s \in \top(x,y)}} \neg\exists t : s = \mathrm{D}(t) \vee s = \mathrm{R}(t) \;\wedge \bigwedge_{\substack{x, y \in N \\ s \in \top(x,y) \\ t \in \bot(x,y)}} s \neq t$$

In the induction case, if $\phi_n[\mathcal{N}]$ signifies that $\exists$ can survive for n more moves, simply define ϕ_{n+1} as

$$\phi_{n+1}(\mathcal{N}) = \bigwedge_{\substack{x, z \in N \\ s \in \top(x,z)}} \forall t, u : s = t * u \Rightarrow \bigvee_{y \in N\dot{\cup}\{x_+\}} \phi_n(+_\top[+_\top[\mathcal{N}, x, y, a], y, z, b])$$

$$\wedge \bigwedge_{\substack{x, y, z \in N \\ t \in \top(x,y),\, u \in \top(y,z)}} \forall s : s = t * u \to \Bigg(\phi_n(+_\top[\mathcal{N}, x, z, s]$$

$$\vee\ \forall v : v = \mathrm{D}(t) \Rightarrow \bigvee_{w \in N\dot{\cup}\{x_+\}} \phi_n(+_\bot[+_\top[\mathcal{N}, x, w, t], w, w, v]) \Bigg)$$

$$\wedge \bigwedge_{\substack{x,y,z \in N \\ t \in \mathsf{T}(x,y) \\ s \in \mathsf{T}(x,z)}} \forall u, v : (s = t * u \wedge v = \mathrm{D}(u)) \Rightarrow \phi_n(+_{\mathsf{T}}[\mathcal{N}, y, y, v])$$

$$\wedge \bigwedge_{\substack{x,y \in N \\ s \in \mathsf{T}(x,y)}} \forall t, u : \Big(t = \mathrm{D}(s) \wedge u = \mathrm{R}(s)) \Rightarrow$$

$$\phi_n(+_{\mathsf{T}}[+_{\mathsf{T}}[\mathcal{N}, x, x, t], y, y, u]\Big)$$

$$\wedge \bigwedge_{\substack{x \in N \\ s \in \mathsf{T}(x,x)}} \forall t : \mathrm{D}(t) = s \Rightarrow \bigvee_{y \in N \dot\cup \{x_+\}} \phi_n(+_{\mathsf{T}}[\mathcal{N}, x, y, t])$$

$$\wedge \bigwedge_{\substack{y \in N \\ s \in \mathsf{T}(y,y)}} \forall t : \mathrm{R}(t) = s \Rightarrow \bigvee_{x \in N \dot\cup \{x_+\}} \phi_n(+_{\mathsf{T}}[\mathcal{N}, x, y, t])$$

Thus $\exists$ can win a conservative game $\Gamma_n(\mathcal{S})$ if and only if $\mathcal{S} \models \sigma_n$ where

$$\sigma_n = \forall s, t : s \neq t \Rightarrow \Big(\phi_n(\mathcal{N}_{ref}[a, b]) \vee \phi_n(\mathcal{N}_{nref}[a, b])$$

$$\vee \phi_n(\mathcal{N}_{ref}[b, a]) \vee \phi_n(\mathcal{N}_{nref}[b, a])\Big)$$

Since inconsistencies in networks are inherited in extensions, it is true that for countable structures if $\exists$ has a winning strategy for conservative plays of $\Gamma_n(\mathcal{S})$, she will also have a winning strategy for any play of $\Gamma_n(\mathcal{S})$. Furthermore, as inconsistency is inherited in extensions, if $\mathcal{S} \models \Sigma$, $\exists$ has a winning strategy for $\Gamma_\omega(\mathcal{S})$. Thus for all countable $\mathcal{S}$, $\mathcal{S} \in R(\mathrm{D}, \mathrm{R}, *)$ if and only if $\mathcal{S} \models \Sigma$. As the representation class is pseudoelementary, it is closed under elementary equivalence, and by Löwenheim-Skolem Theorem, we conclude $\mathcal{S} \models \Sigma$ is both sufficient and necessary for membership, even for uncountable structures. □

4 Demonic Refinement

Before we move on to defining structures used to prove non finite axiomatisability, we will quickly have a look at the demonic lattice. We discuss in Sect. 2 that the demonic lattice has found use in algebraically modelling total correctness. However, in this section, it will help us show that the structures we will use in the argument are in fact non-representable.

We do so by defining demonic refinement, the partial ordering predicate arising from the demonic lattice. Furthermore, we observe that even though the predicate is not in the signature, some pairs of elements of a representable $\{\mathrm{D}, \mathrm{R}, *\}$-structure will always be represented as demonic refinement pairs.

Now assume that a $\{\mathrm{D}, \mathrm{R}, *\}$-structure has a cycle of elements where each element is a demonic refinement of its successor. As the predicate is a partial order, it means by antisymmetry and transitivity that these distinct elements will map the same binary relation in any representation and thus the structure is not representable.

Now let us define demonic refinement for $R, S \subseteq X \times X$ as

$$R \sqsubseteq S \iff (\mathrm{D}(S) \subseteq \mathrm{D}(R) \wedge \mathrm{D}(S); R \subseteq S)$$

This is motivated, again, with the demon in control of nondeterminism. Imagine R, S were programs over the state space X. If the demon is given the choice to run R or S, he will always run S. This is because when we are outside the domain of S, running S rather than R will result abort and when in the domain of S it will maximise the odds of reaching an erroneous state.

Now we recursively define a predicate $\preceq$ using infinitary $\{\mathrm{D}, \mathrm{R}, *\}$-formula such that for every structure $\mathcal{S}$ with a representation θ we will have $\forall s, t \in \mathcal{S} : s \preceq t \Rightarrow s^\theta \sqsubseteq t^\theta$. We take advantage of the fact that sometimes non domain elements may compose to a domain element, and define $\preceq_1$. Then we inductively close the predicate under monotonicity and transitivity. More formally, we say that

$$s \preceq_1 t \iff \exists u, v : \mathrm{D}(u * v) = u * v \wedge s = \mathrm{R}(u * \mathrm{D}(v)) \wedge t = s * v * u$$

$$s \preceq_{n+1} t \iff \begin{pmatrix} \exists s', t', u, v : s' \preceq_n t' \wedge s = u * s' * v \wedge t = u * t' * v \\ \vee \exists v : s \preceq_n v \wedge v \preceq_n t \end{pmatrix}$$

and $\preceq = \bigcup_{n<\omega} \preceq_n$

Lemma 3. *For any $s, t \in \mathcal{S}$, if $s \preceq t$, it is true that for any representation θ we have $s^\theta \sqsubseteq t^\theta$.*

Proof. We show this by induction over n.

In the base case, we see that there exists a u, v such that $u * v = \mathrm{D}(u * v)$ and $s = \mathrm{R}(u * \mathrm{D}(v))$ and $t = s * v * u$. First see how if $(x, x) \in t^\theta$, there must exist a witness for $s * v * u$ and since s is a range element, it must hold that $(x, x) \in s^\theta$. Since $\mathrm{D}(s^\theta) = s^\theta$, we have $\mathrm{D}(t^\theta) \subseteq \mathrm{D}(s^\theta)$. Furthermore, assume that $(x, x) \in \mathrm{D}(t^\theta)$ and $(x, x) \in s^\theta$. See how there must exist a y such that $(y, x) \in (u * \mathrm{D}(v))^\theta$. There must also exist a z such that $(x, z) \in v^\theta$. Since $(y, y) \in \mathrm{D}(u * \mathrm{D}(v))^\theta$, we can see that $(y, z) \in (u * v)^\theta$ and since $u * v$ is a domain element, $y = z$. And because $(x, x) \in \mathrm{D}(t)^\theta$ and because $(x, z) \in (s * v)^\theta$ and $(z, x) \in u^\theta$, we conclude $(x, x) \in (s * v * u)^\theta = t^\theta$.

The induction case follows from the fact that $\sqsubseteq$ is transitive as well as left and right monotone for $*$ as discussed in [10]. □

The use of refinement cycles may seem similar to [7] where the predicate $\triangleleft$ is defined as the monotone, transitive closure of $\mathrm{D}(s); \mathrm{D}(t) \triangleleft \mathrm{D}(t)$ to signify ordinary inclusion ($\leq$) for the angelic signature. However, for the demonic signature, $\triangleleft$ can be simply described as $\mathrm{D}(s) * t \triangleleft t$ as the following axiom is sound

$$\forall s, t : \mathrm{D}(s * \mathrm{D}(t)) * s = s * \mathrm{D}(t)$$

Thus, $\triangleleft$ does not show useful when trying to show $R(\mathrm{D}, \mathrm{R}, *)$ is not finitely axiomatisable, as avoiding cycles of $\triangleleft$ can be described in a single axiom.

5 $R(\mathrm{D}, \mathrm{R}, *)$ is Not Finitely Axiomatisable

We can now define the non representable structures for every $n < \omega$ for which $\exists$ will have a winning strategy in Γ_n. First we use the demonic refinement predicate, defined in Sect. 4, to show these are not representable as they include a refinement cycle. Then we show by induction that $\exists$ will have a winning strategy for n moves in the representation game. Using the compactness trick, we show that the representation class is not finitely axiomatisable.

For every $n < \omega$, let $N = 2n + 1$. Define a $\{\mathrm{D}, \mathrm{R}, *\}$-structure $\mathcal{S}_n$, with the following underlying set

$$\{0, d, r\} \cup \{m_i, \varepsilon_i, a_i, b_i, c_i, d_i, ac_i, acd_i, cdb_i, db_i, ab_i \mid 0 \le i < N\}$$

$0, d, r, m_i, \varepsilon_i$ are the domain-range elements, idempotent with respect to composition, and disjoint, i.e. composition of two distinct domain-range elements evaluates to 0. We now examine domain-range elements, see visualisation in Fig. 3. For all $i < N$, we have

$$\begin{aligned} d &= \mathrm{D}(a_i) = \mathrm{D}(ac_i) = \mathrm{D}(acd_i) = \mathrm{D}(ab_i) \\ m_i &= \mathrm{D}(c_i) = \mathrm{D}(b_i) = \mathrm{D}(cdb_i) = \mathrm{R}(a_i) = \mathrm{R}(d_i) = \mathrm{R}(cd_i) \\ \varepsilon_i &= \mathrm{D}(d_i) = \mathrm{D}(db_i) = \mathrm{R}(c_i) = \mathrm{R}(ac_i) \\ r &= \mathrm{R}(ab_i) = \mathrm{R}(cdb_i) = \mathrm{R}(db_i) = \mathrm{R}(b_i) \end{aligned}$$

The reader may find it helpful to pay close attention to Fig. 3 while we define the compositions. First, we say that

$$d_i * c_i = \varepsilon_i \qquad c_i * d_i = cd_i \qquad cd_i * cd_i = cd_i$$

for every $i < N$. Furthermore, some elements will result in a composition with an index increasing by one, namely

$$a_i * cdb_i = ab_{i+1} \qquad acd_i * cdb_i = ab_{i+1} \qquad ac_i * db_i = ab_{i+1} \qquad acd_i * b_i = ab_{i+1}$$

for $i < N$ where $+$ denotes addition modulo N. Composition results below are defined more naturally

$$\begin{array}{lllll} cd_i * c_i = c_i & d_i * cd_i = d_i & a_i * b_i = ab_i & a_i * c_i = ac_i & a_i * cd_i = acd_i \\ c_i * db_i = cdb_i & d_i * b_i = db_i & ac_i * d_i = acd_i & acd_i * c_i = ac_i & cd_i * b_i = cdb_i \end{array}$$

All other compositions are either the mandatory domain-range compositions or they evaluate to 0.

The following two Lemmas will now show that although $\mathcal{S}_n$ is not representable, $\exists$ will be able to maintain consistency in the network for n moves.

Lemma 4. *$\mathcal{S}_n$ is not $\{\mathrm{D}, \mathrm{R}, *\}$-representable.*

Proof. Observe how $m_i \preceq c_i * d_i$ and thus $ab_i = a_i * m_i * b_i \preceq a_i * c_i * d_i * b_i = ab_{i+1}$ for all $i < N$. This means by transitivity of $\preceq$ that for all $i, j < N$ we have $ab_i \preceq ab_j$. Now assume that there existed a representation θ. We would have $ab_i^\theta \sqsubseteq ab_j^\theta, ab_j^\theta \sqsubseteq ab_i^\theta$, even where $i \neq j$. Since $\sqsubseteq$ is antisymmetric, we would have $ab_i^\theta = ab_j^\theta$ for $i \neq j$. Therefore, no such θ can exist. □

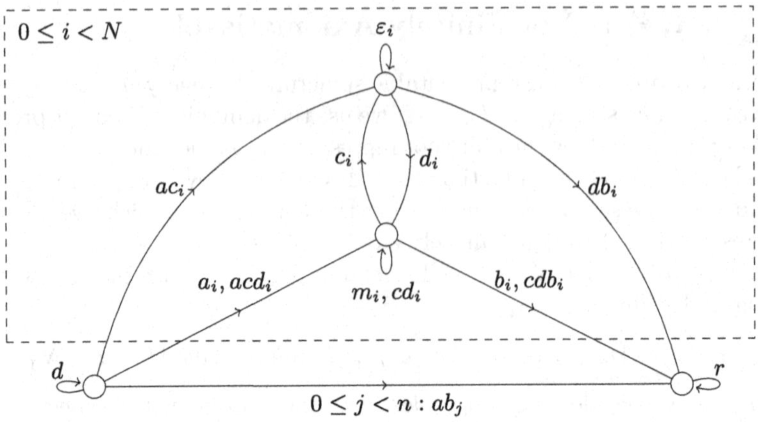

Fig. 3. Visualisation of $\mathcal{S}_n$

Lemma 5. *For all* $n < \omega$, $\exists$ *has a winning strategy for* $\Gamma_n(\mathcal{S}_n)$

Proof. First see how $\exists$ may play in a way that she returns a network that is closed under composition, composition domain and domain-range moves. For composition moves, she always chooses to add $a * b$ to the label, rather than adding a node with $D(b)$ in its $\bot$ label. Furthermore, she may set the $\top$ label in a way that for all (x, y)

$$\top_{k+1} \subseteq \top_k(x, y) \cup \{ab_{i+1} \mid ab_i \in \top_k(x, y)\}$$

where $+$ is modulo N and

$$a_i \in \top_k(x,y) \Rightarrow acd_i \in \top_k(x,y) \quad m_i \in \top_k(x,y) \Rightarrow cd_i \in \top_k(x,y)$$
$$b_i \in \top_k(x,y) \Rightarrow cdb_i \in \top_k(x,y) \qquad 0 \notin \top(x,y)$$

as well as ensure that domain-range elements are only added to reflexive edge $\top$ labels. If $m_i \in \top(x, x)$, there exists at most one y such that $c_i \in \top(x, y) \vee d_i \in \top(y, x)$ and if $cd_i \in \top(x, x')$, the y must be the same for x, x'. to prevent compositional closure from adding m_i to a $\top(z, w), z \neq w$.

In the base case, observe how for every $s \neq t$, it is possible to find either s or t to put in $\top(x, y)$ of the initialisation network. Without loss, if $s = 0$ or $s = a_i, t = acd_i$ or $s = m_i, t = cd_i$ or $s = b_i, t = cdb_i$ she has to play t. Otherwise, she is free to play either s or t, making sure that she plays the reflexive network if and only if she opts to play a domain-range element.

In the induction case, as the network is closed under domain-range, composition and composition-domain moves, $\forall$'s only non-redundant move options are composition, domain, and range.

For the domain move, $\exists$ may add a new node unless c_i is requested on some x. In that case, she must pick to add c_i to $\top(x, y)$ to the designated y if such y exists, otherwise create such a y and close it under all the necessary moves to maintain

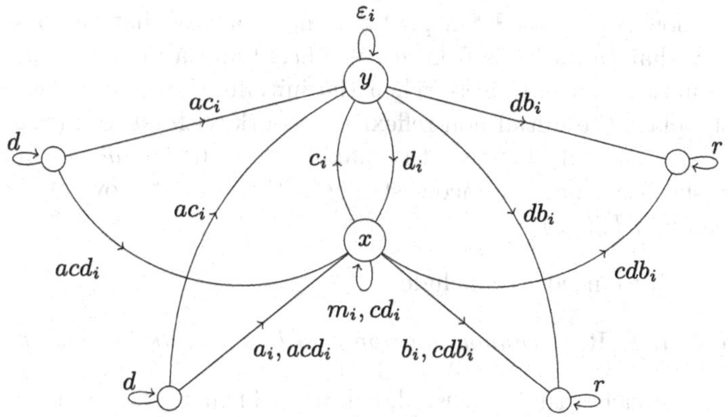

Fig. 4. Compositions with cd_i

the induction hypothesis. All the compositions resulting in cd_i, acd_i, cdb_i are included in the appropriate labels due to her strategy, see Fig. 4. Otherwise the move can be satisfied with a new node, satisfying all the properties in ∃'s strategy. Similarly, the argument can be constructed for range moves including d_i or otherwise.

In case a witness move is called and the left operand is c_i or the right operand is d_i (or both), the witness node returned must be the designated y and the induction hypothesis is maintained (again, see Fig. 4). If the witness move has cd_i as an operand, she makes sure to designate the appropriate y, again preserving the induction hypothesis. All other non-redundant operations result in ab_i. If the index of the operands is $i-1$, ∃ may ensure she does not include $a_{i-1} * b_{i-1}$ witness (see Fig. 5 left). Finally, for operands with index i she adds a witness node with m_i, cd_i in the reflexive label, a_i, acd_i on the left and b_i, cdb_i on the right (see Fig. 5 right). This covers all the possible non-redundant witness moves, but results in ab_{i+1} being added to the label. In any case, the induction hypothesis is maintained.

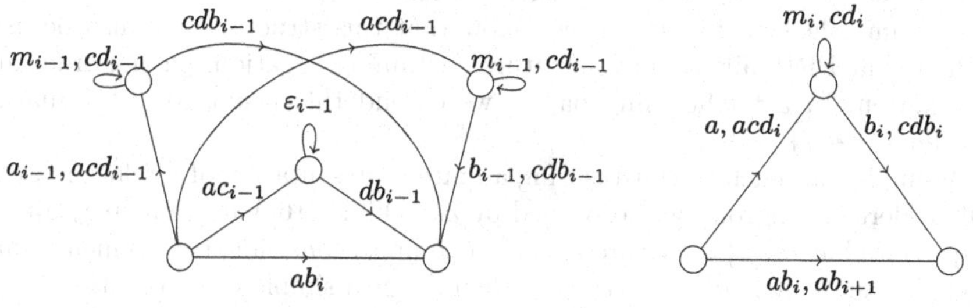

Fig. 5. Witness moves for ab_i with $i-1$ left and i on the right

We have now seen that $\exists$ can play a game in a way that the only possible inconsistency that can arise is from $ab_i \in \top(x,y)$ and also in $\bot(x,y)$. Without loss, this situation can only arise when the initialisation pair is ab_0, ab_{n+1}. In this case she plays the initial non-reflexive network with $ab_0 \in \top(x,y), ab_{n+1} \in \bot(x,y)$. As she can only increase the maximal i such that $ab_i \in \top(x,y)$ by 1 each move, she introduces an inconsistency at the $n+1$st move at the earliest. Thus she can win $\Gamma_n(\mathcal{S}_n)$. □

This gives us all we need to conclude

Theorem 1. *$R(\mathrm{D},\mathrm{R},*)$ cannot be axiomatised by a finite first order theory.*

Proof. Suppose such a theory existed, call it Ψ. Then $R(\mathrm{D},\mathrm{R},*)$ is axiomatised by a single axiom $\psi = \bigwedge_{\psi' \in \Psi} \psi'$. Thus $\Sigma \cup \{\neg\psi\}$ is not consistent as, by Lemma 2, Σ ensures that any model of it is representable and $\neg\psi$ ensures it is not. Now look at any finite subtheory $\Omega \subseteq \Sigma \cup \{\neg\psi\}$. Observe how, since it is finite, there exists $n < \omega$ such that for all $m > n$ we have $\sigma_m \notin \Omega$. Thus $\mathcal{S}_n \models \Omega$ as by Lemmas 5, 2 we have $\mathcal{S}_n \models \sigma_i, i \leq n$, and by Lemmas 1, 4 we have $\mathcal{S}_n \models \neg\psi$. By compactness of first order logic, we conclude the Theory $\Sigma \cup \{\neg\psi\}$ is consistent and we have reached a contradiction. □

6 Finite Representation Property

We have now seen that both the angelic and demonic representable domain-range semigroups cannot be axiomatised finitely. However, it remains unknown if all finite members of $R(\mathrm{D},\mathrm{R},;)$ and $R(\mathrm{D},\mathrm{R},*)$ have the finite representation property. Although the finite axiomatisability (or lack thereof) is known for a number of representation classes [13], FRP remains largely unknown for signatures with composition. In this section we discuss some existing results and extend FRP result for ordered domain algebras [5].

The known results regarding FRP are summarised in Table 1. The signatures $\{;\}, \{1',;\}, \{\mathrm{D},*\}$ are well known examples where Cayley representation for groups can be used to represent the structure over a finite base. Neuzerling shows that any signature containing meet and composition fails to have FRP using Point Algebra [14]. In [10] we show that this structure can also be used to show that FRP fails for any signature containing negation, partial order and composition. In a forthcoming paper, we extend this result to any signature containing $\{-,;\}$.

A simple approach to constructing a finite representation of a relational partially ordered semigroup was proposed by Zareckiĭ in [16] where one may amend a representable $\{\leq,;\}$-structure $\mathcal{S}$ with a compositional identity element e and only add the mandatory (e,e) to $\leq$ to then define a simple representation θ over the base $\mathcal{S}$ with

$$(s,t) \in a^\theta \iff t \leq s;a$$

The inclusion of e ensures faithfulness as for $a \not\leq b$ $(e,a) \in a^\theta \setminus b^\theta$ and the associativity and monotonicity ensure that $\leq,;$ are correctly represented.

Table 1. Signatures with composition where FRP is known

FRP	No FRP
$\{;\}, \{1', ;\}, \{\mathrm{D}, *\}$	$\{\cdot, ;\} \subseteq \tau$ [14]
$\{\leq, ;\}, \{\sqsubseteq, *\}$ [16]	$\{-, ;\} \subseteq \tau$
$\{0, 1, \mathrm{D}, \mathrm{R}, \leq, 1', \smile, ;\}$ [5]	
$\{\sqsubseteq, ;\}$ [10]	
$\{\leq, \backslash, /, ;\}$ [15]	

Egrot and Hirsch [5] amend the idea to represent the ordered domain algebras, the signature $\{0, \mathrm{D}, \mathrm{R}, \leq, 1', \smile, ;\}$ where 0 is the empty relation (bottom element of the Boolean lattice), $1'$ is the relational identity and $\smile$ is the relational converse. They represent the structures in $R(0, \mathrm{D}, \mathrm{R}, \leq, 1', \smile, ;)$ over the base of subsets of the structure, rather than its elements.

However, their result can be adapted for a wider range of signatures. Below we present an outline of the proof for the following theorem.

Proposition 1. *For any signature* $\{\mathrm{D}, \mathrm{R}, \smile, ;\} \subseteq \tau \subseteq \{0, 1, \mathrm{D}, \mathrm{R}, \leq, 1', \smile, ;\}$, $R(\tau)$ *has the finite representation property.*

Proof. We can, for any representable τ-structure $\mathcal{S}$, define a partial ordering $\leq$ (even if $\leq \notin \tau$) as the set of all pairs where $s \leq t$ if and only if for all representations θ, $s^\theta \leq t^\theta$. Similarly, one can define at most one element 0 (again even if $0 \notin \tau$) that will always be represented as an empty relation.

This means that we can define the set of closed sets $\mathcal{G}$ as the set of all $\emptyset \subsetneq S \subseteq \mathcal{S} \setminus \{0\}$ such that for $\mathrm{D}(S) = \prod_{s \in S} \mathrm{D}(s)$ and similarly $\mathrm{R}(S)$, we have $(\mathrm{D}(S); S; \mathrm{R}(S))^\uparrow = S$ where $\uparrow$ is upward closure with respect to $\leq$. Then define a mapping $\rho : \mathcal{S} \to \wp(\mathcal{G} \times \mathcal{G})$ such that $(S, T) \in a^\rho$ if and only if $S; a \subseteq T$ and $T; \breve{a} \subseteq S$.

The mapping is faithful as for $a \not\leq b$, $(\mathrm{D}(a), a) \in a^\rho$ as $a; \breve{a} \geq \mathrm{D}(a)$, but not in b as that would mean $a \leq \mathrm{D}(a); b \leq b$. It represents $\leq$ correctly by monotonicity of ; over $\leq$ and $0, 1$ correctly as 1 is the top element with respect to ordering and $a; 0 = 0; a = 0$, for all a. Domain and range are correctly represented as if there is an outgoing/incoming edge from S with $a/\breve{a}$, then $S; a; \breve{a} \subseteq S$ and since $\mathrm{R}(\breve{a}) = \mathrm{R}(a; \breve{a}) = \mathrm{D}(a)$, $S; \mathrm{D}(a) \subseteq S$ and thus $\mathrm{D}(a) = \mathrm{R}(\breve{a})$ is included in (S, S). Furthermore if $\mathrm{R}(a) = \mathrm{D}(\breve{a})$ is included in (S, S) then $(S; \breve{a})^\uparrow$ ensures that there is an incoming edge with a and an outgoing edge with $\breve{a}$. Finally, domain elements are only on reflexive nodes as if $(S; \mathrm{D}(a))^\uparrow = S$ so if $(S, T) \in \mathrm{D}(a)$ then $S \subseteq T \subseteq S$ and similarly $(S, T) \in (1')^\rho$ if and only $S = T$. Converse is correctly represented as $\breve{\breve{a}} = a$. Finally $a^\rho; b^\rho \leq (a; b)^\rho$ by monotonicity and $(a; b)^\smile = \breve{a}; \breve{b}$ and $(a; b)^\rho \leq a^\rho; b^\rho$ as if $(S, T) \in (a; b)^\rho$, $\left(S; a; \mathrm{D}(\breve{a}; \breve{T}) \cup T; a; \mathrm{R}(S; a)\right)^\uparrow$ is an appropriate witness for the composition. □

Note that the second part of the proof where we show that ρ is indeed a representation is an outline. This is because the argument closely follows that in [5, Section 6], refer to it for more detail.

Finally, Rogozin shows that one can embed residuated semigroups into relational quantales in [15] and we show in [10] that a Zareckiĭ representation can be modified in a way to represent semigroups with demonic refinement. The latter was the first example of a signature with composition without a finitely axiomatisable representation class, but with FRP.

7 Problems

In this section we look at some open problems and outline the difficulties with showing the finite representation property.

We begin with the observation that e in the Zareckiĭ representation, as defined in Sect. 6, is not represented as the true relational identity element, i.e. $1' = \{(x, x) \mid x \in X\}$, as for some $a \lneq a'$ we will have $(a', a) \in e^\theta$. Thus this good behaviour does not extend to the signature of $\{1', \leq, ;\}$, with $R(\leq, 1', ;)$ non-finitely axiomatisable [7] and FRP unknown.

$R(\leq, 1', ;)$ suffers from the same problem as $R(\mathrm{D}, \mathrm{R}, *)$ and $R(\mathrm{D}, \mathrm{R}, ;)$. That is, some elements are always represented as *partial functions*, that is, for any representation θ over X, if $(x, y) \in f^\theta, (x, z) \in f^\theta$ then $y = z$. Simple examples of that include the domain-range elements, as well as those $f \leq 1'$. However, composition makes for some more interesting examples, like c_i in $\mathcal{S}_n$ in Sect. 5 or in $R(\mathrm{D}, \mathrm{R}, ;)$, $\mathrm{R}(a); b$ will always be represented as a partial function if $\mathrm{D}(a; b) = a; b$. This is illustrated in Fig. 6, from left to right, observe how for any representation θ if $(x, y) \in (R(a); b)^\theta$ then $(x, x) \in R(a)^\theta$, so there must exist a z such that $(z, x) \in a^\theta$. As $a; b = \mathrm{D}(a; b)$ and by composition, z must be the same as y. Similarly, for any outgoing z with $(x, z) \in (R(a); b)^\theta$, it has to be the case that $y = z$.

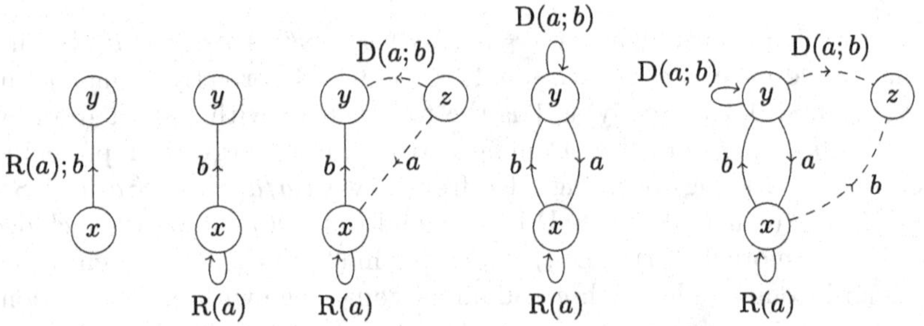

Fig. 6. Partial-functional nature of $R(a); b$ when $a; b = \mathrm{D}(a; b)$

Every function in the signature of domain-range algebras comes with a converse. More specifically, if $D(a; b) = a; b$ then not only is $R(a); b$ a function, but $a; D(b)$ is its well defined converse. Unfortunately, this does not enable us to use represent structures over a finite base in the same way as the structures in Proposition 1.

It is true that partial functions, their converses and arbitrary compositions of those have their converse well defined. But take an a with its converse defined and say $a = b; c$ and $\mathrm{R}(b) = \mathrm{D}(c)$. Observe that converses of b, c not defined. Both b and c have a *partial converse*. That is, for every representation θ, $(b^\theta)^\smile \leq (c; \breve{a})^\theta$ and $(c^\theta)^\smile \leq (\breve{a}; b)^\theta$, but the $\geq$ inclusions do not necessarily hold, see Fig. 7.

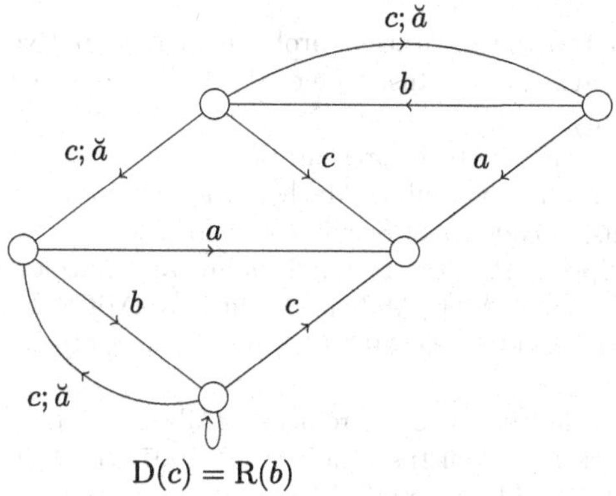

Fig. 7. Partial converse of b, i.e. $b \leq c; \breve{a}$, but $\breve{c}; a \not\leq b$, where a, b, c are elements of a domain range semigroup

This enables us to define the partial converse of $s \in \mathcal{S}$ to be the set $\mathrm{C}(s) \subseteq \mathcal{S}$ where $\mathrm{C}(s)$ is the set of all $s' \in \mathcal{S}$ such that $(s^\theta)^\smile \leq (s')^\theta$, for any representation θ. However, as we have seen there is no guarantee that $\mathrm{C}(\mathrm{C}(s)) = s^\uparrow$. Furthermore, $\mathrm{C}(t); \mathrm{C}(s) \subseteq \mathrm{C}(s; t)$ but not necessarily $\mathrm{C}(t); \mathrm{C}(s) \supseteq \mathrm{C}(s; t)$. As the proof of FRP for ordered domain algebras heavily relies on both $\breve{\breve{a}} = a$ and $(a; b)^\smile = \breve{b}; \breve{a}$, the same representation cannot be used for converse-free signatures.

Adding join ($+$) to the signature adds additional difficulty. The class of representable join-lattice semigroups $R(+, ;)$ was shown non-finitely axiomatisable in [1], with the finite representation property remaining open. Similar to the case where $1'$ is added to the signature of $\{\leq, ;\}$, this slight modification completely breaks the Zareckiĭ representation. That is because $+$ is not necessarily *distributive*, i.e. if $a \leq b + c$ there exists some $b' \leq b$ and $c' \leq c$ such that $a = b' + c'$.

For distributive lattices, one can define the Zareckiĭ representation over the set of minimal non-0 elements and preserve all operations in a faithful manner. However, no signature including $\{+, ;\}$ has been shown to have the finite representation property for its representation class thus far.

The problems raised in this section can be summarised below

Problem 1. Do converse-free (ordered) domain-range semigroups have the finite representation property? How about their demonic counterparts?

Problem 2. Do signatures containing the join-semilattice and composition have the finite representation property?

Problem 3. Does $R(\leq, 1', ;)$ have FRP? How about $R(\leq, 1', \smile, ;)$ or $R(\leq, \smile, ;)$?

References

1. Andréka, H.: On the representation problem of distributive semilattice-ordered semigroups. In: Mathematical Institute of the Hungarian Academy of Sciences, p. 174 (1988, preprint)
2. Bredihin, D.A., Schein, B.M.: Representations of ordered semigroups and lattices by binary relations. In: Colloquium Mathematicum, vol. 39, pp. 1–12. Institute of Mathematics Polish Academy of Sciences (1978)
3. Desharnais, J., Jipsen, P., Struth, G.: Domain and antidomain semigroups. In: Berghammer, R., Jaoua, A.M., Möller, B. (eds.) RelMiCS 2009. LNCS, vol. 5827, pp. 73–87. Springer, Heidelberg (2009). https://doi.org/10.1007/978-3-642-04639-1_6
4. Dijkstra, E.W., Scholten, C.S.: Predicate Calculus and Program Semantics. Springer, New York (2012). https://doi.org/10.1007/978-1-4612-3228-5
5. Hirsch, R., Egrot, R.: Meet-completions and representations of ordered domain algebras. J. Symb. Log. (2013)
6. Hirsch, R., Hodkinson, I.: Relation Algebras by Games. Elsevier, Amsterdam (2002)
7. Hirsch, R., Mikulás, S.: Axiomatizability of representable domain algebras. J. Logic Algebraic Programm. **80**(2), 75–91 (2011)
8. Hirsch, R., Mikulás, S., Stokes, T.: The algebra of non-deterministic programs: demonic operators, orders and axioms. arXiv preprint arXiv:2009.12081 (2020)
9. Hirsch, R., Šemrl, J.: Demonic lattices and semilattices in relational semigroups with ordinary composition. In: Proceedings of the 36th Annual Symposium on Logic in Computer Science, LICS, Rome, Italy. IEEE (2021)
10. Hirsch, R., Šemrl, J.: Finite representability of semigroups with demonic refinement. Algebra Univers. **82**(2), 1–14 (2021). https://doi.org/10.1007/s00012-021-00718-5
11. Hirsch, R., Stokes, T.: Axioms for signatures with domain and demonic composition. Algebra Univers. **82**(2), 1–19 (2021). https://doi.org/10.1007/s00012-021-00719-4
12. Jackson, M., Mikulás, S.: Domain and range for angelic and demonic compositions. J. Log. Algebraic Meth. Program. **103**, 62–78 (2019)
13. Mikulás, S.: Axiomatizability of algebras of binary relations. In: Classical and New Paradigms of Computation and Their Complexity Hierarchies, pp. 187–205. Springer, Cham (2004). https://doi.org/10.1007/978-1-4020-2776-5_11
14. Neuzerling, M.: Undecidability of representability for lattice-ordered semigroups and ordered complemented semigroups. Algebra Univers. **76**(4), 431–443 (2016). https://doi.org/10.1007/s00012-016-0409-9
15. Rogozin, D.: The finite representation property for representable residuated semigroups. arXiv preprint arXiv:2007.13079 (2020)
16. Zareckiĭ, K.A.: The representation of ordered semigroups by binary relations. Izvestiya Vysšhikh. Uchebnykh. Zavedeniı. Matematika **6**(13), 48–50 (1959)

Author Index

Aguzzoli, Stefano 1
Alpay, Natanael 19

Bannister, Callum 37
Berghammer, Rudolf 54, 72
Bianchi, Matteo 1

Calk, Cameron 90, 108
Conradie, Willem 126

Doumane, Amina 144

Fahrenberg, Uli 90
Fernandez, Alexandre 159
Foster, Simon 465
Fussner, Wesley 176

Glück, Roland 192
Goranko, Valentin 126
Goubault, Eric 108
Guatto, Adrien 309
Guttmann, Walter 209, 225

Hoare, Tony 325
Höfner, Peter 37

Jipsen, Peter 19, 126
Johansen, Christian 90

Kulczynski, Mitja 72
Kurucz, Agi 241
Kuznetsov, Stepan L. 258

Lucas, Christophe 275

Maignan, Luidnel 159
Malbos, Philippe 108
Marquès, Jérémie 292
Metcalfe, George 309
Mio, Matteo 275
Möller, Bernhard 325

Naaf, Matthias 344
Nester, Chad 362

O'Hearn, Peter 325

Pinzón, Carlos 413
Pous, Damien 378

Quintero, Santiago 413

Ramírez, Sergio 413
Robinson-O'Brien, Nicolas 225
Rot, Jurriaan 378
Ryzhikov, Vladislav 241

Santocanale, Luigi 396
Santschi, Simon 309
Savateev, Yury 241
Šemrl, Jaš 483
Spicher, Antoine 159
Struth, Georg 37, 90
Sugimoto, Melissa 19

Valencia, Frank 413
van Gool, Sam 309

Wagemaker, Jana 378
Winter, Michael 433, 448
Woodcock, Jim 465

Ye, Kangfeng 465

Zakharyaschev, Michael 241
Ziemiański, Krzysztof 90
Zuluaga Botero, William 176

Printed in the United States
by Baker & Taylor Publisher Services